Compendium of Plant Genomes

Series Editor

Chittaranjan Kole, Raja Ramanna Fellow, Government of India, ICAR-National Research Center on Plant Biotechnology, Pusa, New Delhi, India

Whole-genome sequencing is at the cutting edge of life sciences in the new millennium. Since the first genome sequencing of the model plant Arabidopsis thaliana in 2000, whole genomes of about 100 plant species have been sequenced and genome sequences of several other plants are in the pipeline. Research publications on these genome initiatives are scattered on dedicated web sites and in journals with all too brief descriptions. The individual volumes elucidate the background history of the national and international genome initiatives; public and private partners involved; strategies and genomic resources and tools utilized; enumeration on the sequences and their assembly; repetitive sequences; gene annotation and genome duplication. In addition, synteny with other sequences, comparison of gene families and most importantly potential of the genome sequence information for gene pool characterization and genetic improvement of crop plants are described.

Interested in editing a volume on a crop or model plant?

Please contact Prof. C. Kole, Series Editor, at ckoleorg@gmail.com

More information about this series at http://www.springer.com/series/11805

Dario Cantu · M. Andrew Walker
Editors

The Grape Genome

Springer

Editors
Dario Cantu
Department of Viticulture and Enology
University of California, Davis
Davis, CA, USA

M. Andrew Walker
Department of Viticulture and Enology
University of California, Davis
Davis, CA, USA

ISSN 2199-4781 ISSN 2199-479X (electronic)
Compendium of Plant Genomes
ISBN 978-3-030-18600-5 ISBN 978-3-030-18601-2 (eBook)
https://doi.org/10.1007/978-3-030-18601-2

This Springer imprint is published by the registered company Springer Nature Switzerland AG
The registered company address is: Gewerbestrasse 11, 6330 Cham, Switzerland

This book series is dedicated to my wife Phullara, and our children Sourav, and Devleena
Chittaranjan Kole

Harold Olmo (left) and Al Koyama (center), his grape breeding assistant of many years, and Andy Walker (right) under the Winkler Vine in the UC Davis vineyards in 2003 (Picture by Daniel Ng)

This book is dedicated to the memory of Harold P. Olmo. He was the leading figure in grape genetics and breeding for 40 years and had a remarkable influence on viticulture across the globe. His extensive travels (by car, train, foot, and horse) through Afghanistan and Iran collecting grapes, Prunus *and other horticultural crops while avoiding disasters, gunshots, angry tribal disputes, earned him the moniker "The Indiana Jones of Viticulture". He released wine grapes, table grapes, raisin grapes and rootstocks, and was an excellent ampelographer. May his inspirational viticultural spirit live on.*

Preface to the Series

Genome sequencing has emerged as the leading discipline in the plant sciences coinciding with the start of the new century. For much of the twentieth century, plant geneticists were only successful in delineating putative chromosomal location, function, and changes in genes indirectly through the use of a number of "markers" physically linked to them. These included visible or morphological, cytological, protein, and molecular or DNA markers. Among them, the first DNA marker, the RFLPs, introduced a revolutionary change in plant genetics and breeding in the mid-1980s, mainly because of their infinite number and thus potential to cover maximum chromosomal regions, phenotypic neutrality, absence of epistasis, and codominant nature. An array of other hybridization-based markers, PCR-based markers, and markers based on both facilitated construction of genetic linkage maps, mapping of genes controlling simply inherited traits, and even gene clusters (QTLs) controlling polygenic traits in a large number of model and crop plants. During this period, a number of new mapping populations beyond F2 were utilized, and a number of computer programs were developed for map construction, mapping of genes, and for mapping of polygenic clusters or QTLs. Molecular markers were also used in the studies of evolution and phylogenetic relationship, genetic diversity, DNA fingerprinting, and map-based cloning. Markers tightly linked to the genes were used in crop improvement employing the so-called marker-assisted selection. These strategies of molecular genetic mapping and molecular breeding made a spectacular impact during the last one and a half decades of the twentieth century. But still, they remained "indirect" approaches for elucidation and utilization of plant genomes since much of the chromosomes remained unknown and the complete chemical depiction of them was yet to be unraveled.

Physical mapping of genomes was the obvious consequence that facilitated the development of the "genomic resources" including BAC and YAC libraries to develop physical maps in some plant genomes. Subsequently, integrated genetic–physical maps were also developed in many plants. This led to the concept of structural genomics. Later on, the emphasis was laid on EST and transcriptome analysis to decipher the function of the active gene sequences leading to another concept defined as functional genomics. The advent of techniques of bacteriophage gene and DNA sequencing in the 1970s was extended to facilitate sequencing of these genomic resources in the last decade of the twentieth century.

As expected, the sequencing of chromosomal regions would have led to too much data to store, characterize, and utilize with the-then available computer software could handle. But the development of information technology made the life of biologists easier by leading to a swift and sweet marriage of biology and informatics, and a new subject was born—bioinformatics.

Thus, the evolution of the concepts, strategies, and tools of sequencing and bioinformatics reinforced the subject of genomics—structural and functional. Today, genome sequencing has traveled much beyond biology and involves biophysics, biochemistry, and bioinformatics!

Thanks to the efforts of both public and private agencies, genome sequencing strategies are evolving very fast, leading to cheaper, quicker, and automated techniques right from clone-by-clone and whole-genome shotgun approaches to a succession of second-generation sequencing methods. The development of software of different generations facilitated this genome sequencing. At the same time, newer concepts and strategies were emerging to handle sequencing of the complex genomes, particularly the polyploids.

It became a reality to chemically—and so directly—define plant genomes, popularly called whole-genome sequencing or simply genome sequencing.

The history of plant genome sequencing will always cite the sequencing of the genome of the model plant Arabidopsis thaliana in 2000 that was followed by sequencing the genome of the crop and model plant rice in 2002. Since then, the number of sequenced genomes of higher plants has been increasing exponentially, mainly due to the development of cheaper and quicker genomic techniques and, most importantly, the development of collaborative platforms such as national and international consortia involving partners from public and/or private agencies.

As I write this preface for the first volume of the new series "Compendium of Plant Genomes," a net search tells me that complete or nearly complete whole-genome sequencing of 45 crop plants, eight crop and model plants, eight model plants, 15 crop progenitors and relatives, and three basal plants is accomplished, the majority of which are in the public domain. This means that we nowadays know many of our model and crop plants chemically, i.e., directly, and we may depict them and utilize them precisely better than ever. Genome sequencing has covered all groups of crop plants. Hence, information on the precise depiction of plant genomes and the scope of their utilization are growing rapidly every day. However, the information is scattered in research articles and review papers in journals and dedicated Web pages of the consortia and databases. There is no compilation of plant genomes and the opportunity of using the information in sequence-assisted breeding or further genomic studies. This is the underlying rationale for starting this book series, with each volume dedicated to a particular plant.

Plant genome science has emerged as an important subject in academia, and the present compendium of plant genomes will be highly useful both to students and teaching faculties. Most importantly, research scientists involved in genomics research will have access to systematic deliberations on the plant genomes of their interest. Elucidation of plant genomes is of interest not only for the geneticists and breeders, but also for practitioners of an array of plant science disciplines, such as taxonomy, evolution, cytology, physiology, pathology, entomology, nematology, crop production, biochemistry, and obviously bioinformatics. It must be mentioned that information regarding each plant genome is ever-growing. The contents of the volumes of this compendium are, therefore, focusing on the basic aspects of the genomes and their utility. They include information on the academic and/or economic importance of the plants, description of their genomes from a molecular genetic and cytogenetic point of view, and the genomic resources developed. Detailed deliberations focus on the background history of the national and international genome initiatives, public and private partners involved, strategies and genomic resources and tools utilized, enumeration on the sequences and their assembly, repetitive sequences, gene annotation, and genome duplication. In addition, synteny with other sequences, comparison of gene families, and, most importantly, the potential of the genome sequence information for gene pool characterization through genotyping by sequencing (GBS) and genetic improvement of crop plants have been described. As expected, there is a lot of variation of these topics in the volumes based on the information available on the crop, model, or reference plants.

I must confess that as the series editor, it has been a daunting task for me to work on such a huge and broad knowledge base that spans so many diverse plant species. However, pioneering scientists with a lifetime experience and expertise on the particular crops did excellent jobs editing the respective volumes. I myself have been a small science worker on plant genomes since the mid-1980s and that provided me the opportunity to personally know several stalwarts of plant genomics from all over the globe. Most, if not all, of the volume editors are my longtime friends and colleagues. It has been highly comfortable and enriching for me to work with them on this book series. To be honest, while working on this series I have been and will remain a student first, a science worker second, and a series editor last. And I must express my gratitude to the volume editors and the chapter authors for providing me the opportunity to work with them on this compendium.

I also wish to mention here my thanks and gratitude to Springer staff particularly, Dr. Christina Eckey and Dr. Jutta Lindenborn, for the earlier set of volumes and presently Ing. Zuzana Bernhart for all their timely help and support.

I always had to set aside additional hours to edit books beside my professional and personal commitments—hours I could and should have given to my wife, Phullara, and our kids, Sourav, and Devleena. I must mention that they not only allowed me the freedom to take away those hours from them but also offered their support in the editing job itself. I am really not sure whether my dedication of this compendium to them will suffice to do justice to their sacrifices for the interest of science and the science community.

New Delhi, India Chittaranjan Kole

Preface

Grapevines (*Vitis vinifera*) have been a source of food and wine since their domestication nearly 8000 years ago. Grape is one of the most important horticultural crops in the world, with over 7 million hectares planted worldwide. In addition to its economic value, grapevine is a model organism for the study of perennial fruit crops and non-climacteric fruit ripening. Its economic and scientific importance made *V. vinifera* an obvious early candidate for genomic sequencing. The two draft genome references released in 2007 were the second publicly available genomes of a woody species and the fourth of a flowering plant. The genome assembly of the experimental inbred line released by "The French–Italian Public Consortium for Grapevine Genome Characterization," PN40024, has served as reference for thousands of genetic and transcriptomic studies. Now over a decade since its release, the PN40024 genome is still a valuable resource to the grapevine community thanks to the continuous effort of the Consortium to improve its structure and annotation.

However, it was understood that a single reference genome was inadequate for studying the function of non-reference cultivar genomes. Seminal work in Tannat and other wine grape cultivars showed substantial unshared gene content between grape cultivars. Recent advancements in sequencing technologies and bioinformatics have made it feasible to generate genome references for other cultivars of equivalent or greater quality than that of PN40024. The genome assemblies of Cabernet Sauvignon, Chardonnay, Carménère, and Zinfandel were released in the last two years. A *V. riparia* genome assembly was released when this book was in the final stages of production; we expect many more genome references for *Vitis* species to be publicly available in the next few years, including those of North American and Asian accessions that are being produced in our laboratories as part of National Science Foundation (1741627) and USDA National Institute of Food and Agriculture (2017-51181-26829) projects. Our research groups have been contributing to the recent advancements in *V. vinifera* genomics. This has been possible because of support from E. & J. Gallo Winery, J. Lohr Vineyards and Wines, Dolce Winery, the Louis P. Martini Endowment in Viticulture, Viña San Pedro, Concha y Toro, UC Davis Chile Life Sciences Innovation Center, and the Chilean Economic Development Agency, and the collaboration between our groups and the scientists at Pacific Biosciences, specifically Paul Peluso, Jason Chin, David Rank, Kristin Mars, and Emily Hatas.

Today, grape cultivation, sustainability, and security rely heavily on North American *Vitis* species as sources of resistance to abiotic and biotic stresses. This reliance originated in the 1860s when the European wine industry was saved by the use of North American species as rootstocks. Currently, more than a dozen North American and Central Asian varieties are used in breeding programs as sources of resistance to abiotic and biotic stresses, either for rootstocks or hybridized with *V. vinifera* for the scion. We expect that genetic diversity, breeding, and biotechnology will play a critical role for sustaining viticulture when faced with a changing climate and other challenges as they arise.

The sixteen chapters of this volume provide a comprehensive review of early and ongoing efforts to discern the genetics, genomics, and breeding of the grapevine. We are grateful to all the authors for their contributions. We would like to thank Prof. Chittaranajan Kole, Editor-in-Chief of the Genome Compendium Series, for inviting us to contribute this volume as well as Naresh Kumar Mani, Manopriya Saravana, and the staff at Springer for their help. We would also like to thank Jadran Francisco Garcia Navarrete, Mélanie Massonnet, Rosa Figueroa-Balderas, Amanda Vondras, and Summaira Riaz for helping review and edit the chapters. Dario would also like to thank his wife, Annegret, and daughters, Amanda and Adele, for their infinite patience and support during the two-year journey that turned an idea into a table of contents and finally into a book.

Davis, USA

Dario Cantu
M. Andrew Walker

Contents

Abbreviations

(s)PLS	(Sparse) Partial least square regression
1-MCP	1-Methylcyclopropene
2,4-D	2,4-Dichlorophenoxyacetic acid
2-CEPA	2-Chloroethylphosphonic acid
2-DE	Two-dimensional electrophoresis
4CL	4-Coumarate-CoA ligase
5mC	5 Methylcytosine
AB	Advanced backcross
ABA	Abscisic acid
ABC	ATP-binding cassette
ABF	Abscisic acid response element-binding factor
AB-QTL	Advanced backcross QTL
AC	After Christ
ACC	1-Aminocyclopropane-1-carboxylic acid
ACO	1-Aminocyclopropane-1-carboxylic acid oxidase
ACR	AC-rich
ACS	ACC synthase
AD	Anno Domini
aDNA	Ancient DNA
ADP	Adenosine diphosphate
AFLP	Amplified fragment length polymorphism
AGAP	Amélioration génétique et adaptation des plantes méditerranéennes et tropicales
AI	Acidic invertases
AIL	AINTEGUMENTA-like
AM	Association mapping
amiRNAs	Artificial miRNAs
ANR	Anthocyanidin reductase
ANT	AINTEGUMENTA
AOC	Appellation d'Origine Contrôlée
AOS	Allene oxide synthase
AP2/ERF	APETALA 2/ethylene-responsive element-binding factor
APHIS	Animal and Plant Health Inspection Service
APT	Adenine phosphoribosyl transferase
AQUILO	AcQUIred tolerance to LOw temperatures

AraNet	Probabilistic functional gene network of Arabidopsis thaliana
ARF	Auxin response factor
ARS	Agricultural Research Service
ATAC-seq	Assay for transposase accessible chromatin sequencing
ATP	Adenosine triphosphate
ATRX	Arabidopsis trithorax-related protein
AUDPC	Area under the disease progress curve
AuxRE	Auxin response elements
AVA	American viticultural areas
AVG	Aminoethoxyvinylglycine
B	Billions
BAC	Bacterial artificial chromosome
BAH	Bromo-adjacent homology
BAP	6-Benzylaminopurine
BC	Before Christ
BCE	Before Common Era
BeYDV	*Bean yellow dwarf virus*
Bgh	*Blumeria graminis* f. sp. *hordeii*
BGs	β-Glucosidases
BLAST	Basic local alignment search tool
BOINC	Berkeley Open Infrastructure for Network Computing
BR	Brassinosteroid
BSA	Bulked segregant analysis
BUSCO	Benchmarking universal single-copy orthologs
bZIP	Basic leucine zipper domain
C4H	Cinnamate-4-hydroxylase
CAT	Chloramphenicol acetyltransferase
CC	Coiled coil
CCA	Canonical correlation analysis
CCoAOMT	Caffeoyl-CoA 3-O-methyltransferase
cDNA	Complementary DNA
cDNA-AFLP	cDNA-amplified fragment length polymorphism
CDS	Coding sequence
CEC	Cation exchange capacity
CG(s)	Candidate gene(s)
ChIP-seq	Chromatin immunoprecipitation sequencing
CHK	Cytokinin histidine kinase
Chr	Chromosome
CHS	Chalcone synthase
CIMIS	California Irrigation Management Information System
CIRAD	Centre de coopération internationale en recherche agronomique pour le développement
CKX	Cytokinin oxidase/dehydrogenase

CLF	Curly leaf
cM	CentiMorgans
CMT(s)	Chromomethylase(s)
CNR	Colorless non-ripening
CO_2	Carbon dioxide
COI	Coronatine insensitive
COLOMBOS	COLlection Of Microarrays for Bacterial OrganismS
COMT	Caffeic acid 3-O-methyltransferase
CORFO	Chilean Economic Development Agency
COST	European Cooperation in Science and Technology
CP	Coat protein
CRE(s)	Cis-regulatory element(s)
CRISPR/Cas9	Clustered regularly interspaced short palindromic repeats/Cas9-associated protein
CT	Computed tomography
CTAB	Cetyltrimethylammonium bromide
DAP-seq	DNA-affinity-purified sequencing
DART-MS	Direct analysis in real-time-mass spectrometry
DB(s)	Database(s)
DCL	Dicer-like ribonuclease III
DEF	Deficiens
DFR	Dihydroflavonol reductase
DGE	Differentially expressed gene
DM	Downy mildew
DME	Demeter
DML	Demeter-like protein
DMR(s)	Differentially methylated region(s)
DNA GL(s)	DNA glycosylase lyase(s)
DNA	Deoxyribonucleic acid
DNase	Deoxyribonuclease
DPA	Diphenylamine
dpi	Days post-inoculation
DR	Decamers
DRM(s)	Domain rearranged methyltransferase(s)
dsRNA(s)	Double-strand RNA(s)
EBI	European Bioinformatics Institute
ELISA	Enzyme-linked immunosorbent assays
ELIXIR-EXCELERATE	European life sciences infrastructure for biological information
EMBL	European Molecular Biology Laboratory
EMPHASIS	The European Infrastructure for Multi-Scale Plant Phenomics and Simulation
ENCODE	Encyclopedia of DNA elements
eQTL	Expression QTL
ERF(s)	Ethylene response factor(s)
ESC	Extra sex comb

ESFRI	European Strategy Forum on Research Infrastructures
ESI	Electrospray ionization
EST(s)	Expressed sequence tag(s)
ETI	Effector-triggered immunity
eTM	Endogenous target mimics
ETR	EThylene receptor
EZ	Enhancer of zeste
F	Female
F3'5'Hs	Flavonoid-3',5'-hydroxylases
F3H	Flavanone 3-hydroxylase
F3'Hs	Flavonoid-3'-hydroxylases
FAIR	Findability, accessibility, interoperability, and reusability
FAO	Food and Agricultural Organization of the United Nations
FAOSTAT	Food and Agriculture Organization Corporate Statistical Database
FAS	Fatty acid synthase
FD	Flavescence dorée
FDp	FD-phytoplasma
FEELnc	Flexible extraction of long non-coding RNA software
FEM	Fondazione Edmund Mach
FIE	Fertilization-independent endosperm
flb	Fleshless berry mutation
FLC	Flowering Locus C
Flp	Flippase
FLS	Flavonol synthases
FLS(s)	Flavonol synthase(s)
FRT	Flippase recognition target
fruitENCODE	Fruit encyclopedia of DNA elements
FT	Flowering Locus T
FUL	Fruitfull
FUM	Fumarase
GA(s)	Gibberellin(s)
GABA	Gamma-aminobutyric acid
GAox	GA-oxidases
GbM	Gene body methylation
Gbp	Gygabase pairs
GBS	Genotyping by sequencing
GC	Gas chromatography
GC-MS	Gas chromatography–mass spectrometry
GCN(s)	Gene co-expression network(s)
gDNA	Genomic DNA
GENCODE	Genomic encyclopedia of DNA elements
GEO	Gene expression omnibus

GFF	General feature format
GFLV	Grapevine fanleaf virus
GFP	Green fluorescent protein
GL	Glycosylase lyases
GLRaV	Grapevine leafroll-associated virus
g_m	Mesophyll conductance
GMO	Genetically modified organism
GPS	Global Positioning System
GrapeIS	Grape Information System
GRBaV	Grapevine red blotch-associated virus
gRNA	Guide RNA
GRSPaV	Grapevine rupestris stem pitting-associated virus
GS	Genomic selection
g_s	Stomatal conductance
GS/MS	Gas chromatography–mass spectrometry
GST	Glutathione-S-transferase
GTD(s)	Grapevine trunk disease(s)
GUS	Beta-glucuronidase
GWAS	Genome-wide association scans/studies
GxE	Genotype by environment interaction
H	Hermaphrodite
H_2O_2	Hydrogen peroxide
HAT(s)	Histone acetyltransferase(s)
HB	HD-Zip homeobox
HDAC(s)	Histone deacetylase(s)
HDMT(s)	Histone demethylase(s)
HDP1	Harbinger transposon-derived protein 1
HGAP	Hierarchical genome assembly process
Hi-C	Genome-wide chromatin conformation capture protocol
HMT(s)	Histone methyl transferase(s)
HMW	High molecular weight
HPLC	High-performance liquid chromatography
HPLC-MS	High-performance liquid chromatography-mass spectrometry
HPTM(s)	Histone post-translational modification(s)
HR	Hypersensitive response
HRM	High-resolution melting analysis
HS	Headspace
HT	High-throughput
HTML	Hypertext Markup Language
HTs	Hexose transporters
HTS	High-throughput sequencing
HY5	Elongated hypocotyl 5
HYH	HY5 homologue
IAA	Indole-3-acetic acid
IBMP	3-Isobutyl-2-methoxypyrazineare

ICP-MS	Inductively coupled plasma mass spectrometry
IDM1	Increase in DNA methylation 1
IGGP	International Grape Genome Program
indels	Single-base insertions or deletions
INTEGRAPE	Data integration to maximize the power of omics for grapevine improvement
IPCC	Intergovernmental Panel on Climate Change
IPMP	3-Isopropyl-2-methoxypyrazine
IPT	Isopentenyltransferase
IR	Infrared light
Iso-Seq	Isoform sequencing
IT	Information technologies
iTRAQ	Isobaric tags for relative and absolute quantitation
ITS	Internal transcribed spacer region
JA	Jasmonic acid
JA-Ile	Jasmonic acid—isoleucine
JAZ	Jasmonate ZIM domain
JGI	Joint Genomics Institute
JMJ12	Jumonji domain-containing protein 12
JMT	S-adenosyl-l-methionine:jasmonic acid carboxyl methyltransferase
K Ha	Thousand hectares
K	Potassium
kb	Kilobases
K_{leaf}	Leaf hydraulic conductance
KT	Thousand tons
KYP	Kryptonite
LAR	Leucoanthocyanidin reductases
LB	Left T-DNA border
LBD	Lateral Organ Boundaries Domain
LC	Liquid chromatography
LC-DAD	LC-Diode array detector
LD	Linkage disequilibrium
LDOX	Leucoanthocyanidin dioxygenase
LGN(s)	Local gene network(s)
LiCl	Lithium chloride
L-IdnDH	L-Idonate dehydrogenase
lncRNA(s)	Long non-coding RNA(s)
LOB	Lateral Organ Boundaries Domain family
LOD	Logarithm of the odds
LOG	Phosphoribohydrolase 'Lonely Guy'
LOX	Lipoxygenase
Lp_r	Root hydraulic conductivity
LRR	Leucine-rich repeat domain
LTQ	Linear trap quadrupole
M	Male
MAPK(s)	Mitogen-activated protein kinase(s)

MAS	Marker-assisted selection
Mb	Megabases
MBD7	Methyl CpG-binding protein 7
MDH	Malate dehydrogenase
MDS	Multidimensional scaling
ME	Malic enzyme
MEA	Medea
MeDIP-seq	Methyl DNA immunoprecipitation sequencing
MeJA	Methyl jasmonate
MEMS	Methylation monitoring sequence
MEP	2C-Methyl-D-erythritol-4-phosphate
MET1	Methyltransferase 1
MFA	Multiple factor analysis
Mg	Magnesium
MH	Million hectares
microCT	Micro-computed tomography
MIP	Major intrinsic protein family
miRNA	MicroRNA
ML	Maximum likelihood
MLO	Mildew resistance Locus O
MML	Modified maximum likelihood
MNase-seq	Micrococcal nuclease sequencing
MP	Movement protein
Mpa	Megapascal
MPs	Methoxypyrazines
MRG	Morf-related gene
mRNA	Messenger RNA
MS	Mass spectrometry
MS/MS	Tandem mass spectrometry
MSAP(s)	Methylation-sensitive amplification polymorphism(s)
MSI1	Multicopy suppressor of IRA 1 protein
MT	Million tons
mtDNA	Mitochondrial DNA
MVA	Cytosolic mevalonate
My	Million years
Mya	Million years ago
NaCl	Sodium chloride
NADPH	Nicotinamide adenine dinucleotide phosphate
NaOAc	Sodium acetate
NB	Nucleotide-binding site
NBT(s)	New breeding technique(s)
NCBI	National Center for Biotechnology Information
NCED	9-cis-epoxycarotenoid dioxygenase
ncRNA	Non-coding RNA
N_e	Effective population size

NES^2RA	Network expansion by stratified variable subsetting and ranking aggregation
ng	Nanograms
NGO	Non-governmental organization
NGS	Next-generation sequencing
NI	Neutral invertases
NMR	Nuclear magnetic resonance
NO	Nitric oxide
NOA	Naphthoxyacetic acid
NOR	Non-ripening
NSF PGRP	National Science Foundation: Plant Genome Research Program
nuDNA	Nuclear DNA
O_2^-	Oxygen
OIV	International Organization of Vine and Wine
OPR	12-Oxophytodienoate reductase
ORCAE	Online Resource for Community Annotation of Eukaryotes
P5CS	1-Pyrroline-5-carboxylate synthetase
PA	Polyamide
PAL	Phenylalanine ammonia lyase
PAM	Protospacer adjacent motif
PBA	Pedigree-based analysis
PCA	Principal component analysis
PCD	Programmed cell death
PcG	Polycom group proteins
PCR	Polymerase chain reaction
PD	Pierce's disease
PDH	Proline dehydrogenase
PDO	Protected designations of origin
PdR	PD resistance locus
PDS	Phytoene desaturase
PEP	Phosphoenolpyruvate
PEPC	Phosphoenolpyruvate carboxylase
PFGE	Pulse field gel electrophoresis
PGDBj	Plant Genome DataBase Japan
PGIP	Polygalacturonase-inhibiting protein
PIP	Plasma membrane Intrinsic Protein
PlantGDB	Plant Genome Database
PLEXdb	Plant Expression Database
PM	Powdery mildew
PODC	The Plant Omics Data Center
PP2C	2C protein phosphatases
PR	Pathogenesis-related proteins
PRC2	Polycomb repressive complex 2
pre-miRNA	Precursor miRNA
pri-miRNA	Primary microRNA

PSII	Photosystem II
PTI	Pattern-triggered immunity
PTM(s)	Post-translational modification(s)
PVP	Polyvinylpyrrolidone
qPCR	Quantitative PCR
QqQ	Triple quadrupole
QTL(s)	Quantitative trait locus/loci
RAD	Restriction-site associated DNA
RAPD	Random amplification of polymorphic DNA
RB	Right T-DNA border
Rcg	Resistance to crown gall
Rda	Resistance to diaporthe ampelina
RdDM	RNA-directed DNA methylation
rDNA	Ribosomal DNA
Refseq	Reference sequence database
Ren	Resistance to erysiphe necator
RFLP	Restriction fragment length polymorphism
RFO(s)	Raffinose family of oligosaccharide(s)
RGAs	Resistance genes analogous
R-genes	Resistance genes
rin	Ripening inhibitor
RIN	RNA integrity number
RISC	RNA interference silencing complex
RK	Receptor kinase
RNA	Ribonucleic acid
RNAi	RNA interference
RNase A	Ribonuclease A
RNA-seq	RNA sequencing
ROS	Reactive oxygen species
ROS1	Repressor of Silencing 1
Rpv	Resistance to plasmopara viticola
RR	Response regulators
RT-qPCR	Reverse transcription quantitative PCR
RUBISCO	Ribulose-1,5-bisphosphate carboxylase/oxygenase
RuBP	Ribulose 1,5-bisphosphate
Run	Resistance to uncinula necator
SAR	Systemic acquired resistance
SAS	Statistical analysis Software
SBP	SQUAMOSA promoter-binding protein
SBP-box/SPL	SQUAMOSA promoter-binding protein-like transcription factor
SCAR	Sequence-characterized amplified region
SE	Somatic embryogenesis
siRNA	Small interfering RNA
SLAM	Simultaneous localization and mapping
SMC	Sequential Markovian coalescent

SMRT	Single-molecule real-time sequencing
sNCGGa	Super-Nomenclature Committee for Grape Gene Annotation
SNP(s)	Single-nucleotide polymorphism(s)
SnRK(s)	Serine/Threonine-protein kinase(s)
SPE	Solid-phase extractions
SPME	Solid-phase microextraction
SRA	SET- and RING-associated
sRNA(s)	Small RNA(s)
sRNA-Seq	Small RNA sequencing
S-SAP	Sequence-specific amplification polymorphism
SSCP	Single-strand conformation polymorphism
SSE	Sum of square errors
SSR(s)	Simple sequence repeat(s)
STSs	Stilbene synthase genes
Su(z)12	Suppressor of zeste 12
SUTs	Sucrose transporters
SUVH	Suppressor of variegation homologue
SWEET	Sugars will eventually be exported transporter
SWN	Swinger
T/Ha	Tons per hectare
TAA1/TAR	TRYPTOPHAN AMINOTRANSFERASE OF ARABIDOPSIS1/TRYPTOPHAN AMINOTRANSFERASE RELATED
TAGL	Tomato agamous-like
TALE	Transcription activator-like effector
TALEN	Transcription activator-like effector nuclease
T-DNA	Transfer DNA
TDZ	Thidiazuron
TE(s)	Transposable element(s)
TF(s)	Transcription factor(s)
TFBS(s)	Transcription factor-binding site(s)
TG	Translucent green
Ti	Tumor inducing
TILLING	Targeting-induced local lesions in genomes
TIP	Tonoplasm intrinsic protein
TIR	Toll/Interleukin-1 receptor
TM	Thompson Seedless
TMT	Tandem mass tagging
TPSs	Terpene synthases
TQD	Triple quadrupole
transPLANT	Trans-national infrastructure for plant genomic science
TS	Thompson Seedless
TSS	Transcriptional start site
TT8	Theban Tomb 8

TTB	Alcohol and Tobacco Tax and Trade Bureau (US Department of the Treasury)
UAS	Unmanned aerial systems
UFGT	UDP-glucose flavonoid-3-O-glucosyltransferase
UGTs	UDP-glucosyltransferases
UHPLC	Ultra-high-performance liquid chromatography
UNIPROT	Universal Protein Resource
UPD	Uridine diphosphate
URGI	Unité de Recherche Génomique Info
USDA	United States Department of Agriculture
USDA/FAS	United States Department of Agriculture: Foreign Agricultural Service
USDA/NASS	United States Department of Agriculture National Agricultural Statistics Service Information
UTR	Untranslated region
UV	Ultraviolet light
VESPUCCI	Vitis Expression Studies Platform Using COLOMBOS Compendia Instances
VIB	Vlaams Instituut voor Biotechnologie
VIGS	Virus-induced gene silencing
VIM	Variant in methylation
VOC(s)	Volatile organic compound(s)
VPD	Vapor pressure deficit
VR	*Vinifera* x *Rotundifolia*
VTCdb	ViTis Co-expression DataBase
WET	Wine Equalization Tax
WGBS	Whole-genome bisulfite sequencing
WT	Wild type
WUE	Water-use efficiency
ya	Years ago
Y_{leaf}	Leaf water potential
YUC	YUCCA genes
ZEP	Zeaxanthin epoxidase
ZF	Zinc finger
ZFN	Zinc finger nuclease

1 Grapes in the World Economy

Julian M. Alston and Olena Sambucci

Abstract

With a farm gate value in 2016 of US$68 billion, grapes are the world's third most valuable horticultural crop (after potatoes and tomatoes). Cultivation of grapes for fruit and wine began at least 7000 years ago in the Near East, and over the millennia, thousands of cultivars have been developed and selected for particular purposes. Nowadays, grapes are grown all around the world, but mainly in places having a temperate, Mediterranean-style climate, and they are used to produce diverse consumer products including wine, table grapes, raisins, grape juice concentrate and distillate for various industrial uses as well as making fortified wine and brandy. The cultivars of grapes used to make these diverse products are likewise diverse, but a relatively small number account for the vast majority of production in each major category. Predominantly, European *V. vinifera* scions are grown on rootstock from phylloxera-resistant Native American species. Particular cultivars are valuable to farmers in particular applications for their agronomic traits and fruit-quality traits, which together determine the value of the crop and the cost of producing it. These values can be conditioned by consumer preferences for attributes of the production process and by government policies including trade taxes, alcohol excise taxes, and regulations over production practices or limiting yields. Evolving demands for traits create demands for work by viticulturists and other scientists to understand the grape genome and work with it.

1.1 Grapes in the World Economy

Archeological evidence suggests stone-age people were making wine from grapes in Georgia and Armenia 8000 years ago, and grapes have been cultivated for winemaking for at least 7000 years (McGovern 2003)—well before the time of the "Epic of Gilgamesh," set in Mesopotamia around 2100 BCE, which is the first written account of grapes and wine. Over the millennia, and especially during the past 500 years, *Vitis vinifera* grapevines originating from the Near East have spread to all four corners of the world. Thousands of cultivars have

J. M. Alston (✉) · O. Sambucci
Agricultural and Resource Economics, University of California Davis, One Shields Ave, Davis, CA 95616, USA
e-mail: julian@primal.ucdavis.edu

O. Sambucci
e-mail: sambucci@primal.ucdavis.edu

D. Cantu and M. A. Walker (eds.), *The Grape Genome*,
Compendium of Plant Genomes, https://doi.org/10.1007/978-3-030-18601-2_1

been generated and selected for particular purposes; and thousands more are known, including many wild varieties.[1]

Grapes are grown for diverse end uses, beyond wine production. *V. vinifera* grapes, along with non-*vinifera* varieties or hybrids, are eaten as fresh table grapes, dried to make raisins, or crushed either to produce grape juice concentrate, or to be fermented and distilled for industrial use as well as for use in making alcoholic beverages; and they are used as ornamental plants. These diverse end uses call for different varietal traits, and thus many diverse varieties, but a relatively small number account for the vast majority of production in each major category. Predominantly, European *V. vinifera* scions are grown on rootstock from phylloxera-resistant American species such as *Vitis aestivalis*, *rupestris*, and *riparia*. Although the genus includes a total of 79 "accepted" species (The Plant List: Vitis 2018), predominantly from North America and the Near East, the vast majority of today's cultivated grapes are varieties of *V. vinifera*, and only a few varieties from other species and some hybrids are of commercial significance.

Grapes are significant in the global economy. In 2016, the world produced 77.4 million tonnes (MT) of grapes (worth some $68.3 billion at the farm) from 7.1 million hectares (MH) of vineyard—a 50 percent increase over the 52.0 MT produced from 9.5 MH in 1966. These grapes are used to produce food and wine at retail worth several times the farm value of the grapes themselves. Over the 50 years, 1966–2016, global average yields almost doubled, from 5.5 to 10.9 tonnes per hectare (T/Ha), and the farm value of grape production grew from $29.6 billion to $44.3 billion in real (2004–2006 international dollar) terms, even though the total vineyard area shrank by one-quarter.[2] Changes in grape cultivars contributed directly to the growth in yield, production, and economic value, and while many other aspects of grape production also changed—including where in the world grapes are grown, how, and for which end uses—these aspects are all chosen jointly with varieties.

Looking to the future, the demand for genetic innovation in grape production will depend in part on the patterns of growth in demand for grape products. Growth in population and per capita incomes would be expected to cause an increase in demand for all grape products, with a relative increase in the demand for more income-elastic fresh versus dried grapes and premium versus more basic wine. Where that growth is to take place around the world will matter, too. In the context of a market driven by broad shifts in final consumer demand, growers will continue to demand cultivars of scions (and rootstocks) that produce fruit with desired quality attributes and have desired agronomic attributes: higher yielding, resistant to pests and diseases, and tolerant of environmental stresses.

This chapter provides an introductory overview of the economic geography (and, where relevant, economic history) of the cultivation of grapes around the world with an eye to how these aspects relate to the grape genome, which is the broader subject of the volume. We discuss the patterns of production of grapes for each of the main end uses, and how they have been changing, and the roles of genetic traits of cultivars as contributors to those patterns. We consider the value of particular traits to producers in specific settings and how these values are influenced by evolving market demand for product and process attributes of food and beverage products, government policy as a conditioning factor, and the changing natural environment, including the ever-present and evolving pests and diseases and, more recently, climate. The chapter begins with an overview of grape production around the world in terms of where grapes are grown, and recent trends in production and utilization.

[1] In the preface to their book describing 1368 varieties of wine grapes, Robinson, Harding, and Vouillamoz (2012, p. viii) suggest the "total number of different vine varieties is about 10,000."

[2] Statistics reported in this section are based primarily on FAOSTAT (2018); Table 1.1 includes more detailed data for 2016.

Table 1.1 Area, volume, yield, and value of grape production in 2016, by regions and countries

Region and country	Total area (K Ha)	Volume (KT)	Yield (T/Ha)	Value ($ M)	Average unit value ($/T)
Africa	**349.6**	**4882.5**	**14.0**	**3463.7**	**709**
Egypt	74.9	1716.8	22.9	567.9	331
South Africa	120.5	2008.8	16.7	1780.1	886
Americas	**1001.4**	**13,659.4**	**13.6**	**12,747.5**	**933**
Argentina	223.9	1758.4	7.9	358.7	204
Brazil	77.0	984.5	12.8	596.6	606
Chile	203.1	2473.6	12.2	4455.0	1801
Peru	27.9	690.0	24.7	490.9	711
North America	**421.9**	**7188.6**	**17.0**	**5236.8**	**728**
USA	409.9	7097.7	17.3	5130.3	723
Asia	**2122.6**	**28,918.4**	**13.6**	**22,249.9**	**769**
Uzbekistan	135.1	1642.3	12.2	489.4	298
China and HK	843.4	14,842.7	17.6	14,007.2	944
Afghanistan	82.5	874.5	10.6	*392.7*	*449*
India	122.0	2590.0	21.2	1837.1	709
Iran	207.3	2450.0	11.8	801.8	327
Turkey	435.2	4000.0	9.2	1967.3	492
Europe	**3446.9**	**27,797.1**	**8.1**	**28,325.3**	**1019**
Romania	175.1	736.9	4.2	523.9	711
Greece	112.3	990.3	8.8	771.3	779
Italy	668.1	8201.9	12.3	3311.9	404
Portugal	175.0	773.9	4.4	1463.6	1891
Spain	920.1	5934.2	6.4	4487.9	756
France	757.2	6247.0	8.2	14,496.1	2320
Germany	100.0	1225.6	12.3	1298.3	1059
Oceania	**176.4**	**2181.4**	**12.4**	**1506.4**	**691**
Australia	136.3	1772.9	13.0	991.1	559
World total	**7096.7**	**77,438.9**	**10.9**	**68,292.9**	**882**

Notes Value and average unit value for Afghanistan (in italics) calculated as weighted averages for the region
Sources Created by the authors using data from FAOSTAT (2018) and USDA/FAS (2018a)

1.1.1 Grape Production and Utilization

Table 1.1 and Fig. 1.1 provide statistics on the production of grapes around the world in terms of area of vineyard, average yield, production, total value of production, and average unit value, drawing on data from FAOSTAT (2018).[3]

In 2016, the world had a total of 7.1 MH planted to grapes. Five countries (Spain, China, France, Italy, and Turkey) accounted for 3.6 MH, just over half the total area, and just 15 countries accounted for 5.5 MH, more than three-quarters.

[3]We draw on various sources for data, including the International Organization of Vine and Wine (OIV), the Food and Agricultural Organization of the United Nations (FAO), the United States Department of Agriculture Foreign Agriculture Service (USDA/FAS), Anderson and Aryal (2013), and Anderson and Pinilla (2018). The Appendix provides more detailed data tables and some discussion of the different data sources.

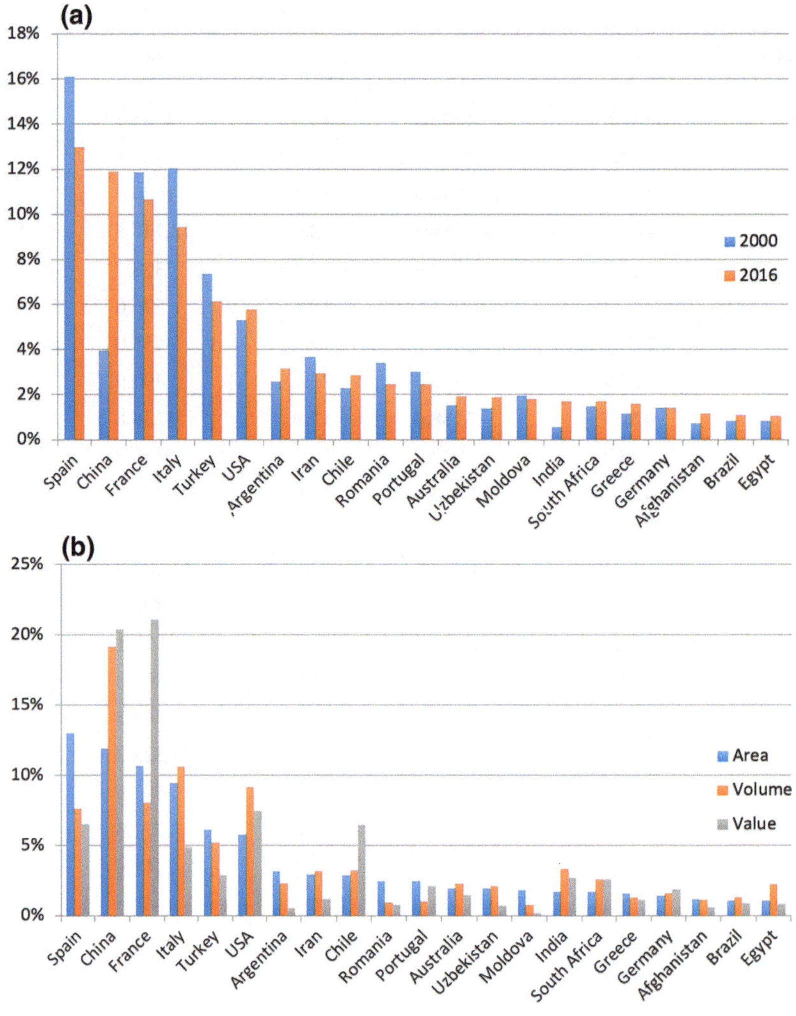

Fig. 1.1 Global distribution of grape area in 2000 and 2016, and area, production volume and value in 2016—top 20 countries by area in 2016. *Source* Created by the authors using data from FAOSTAT (2018). **a** National shares of global grape area, 2000 and 2016, %. **b** National shares of global grape area, production volume, and value, 2016, %

Total production, also, is concentrated among a few countries, but the ranking is slightly different reflecting differences in end uses and average yields. The top five countries in terms of quantity produced (now China, Italy, the USA, Spain, and France) accounted for 42.2 MT, more than half of the total of 77.4 MT, and just 15 countries accounted for 63.8 MT, more than four-fifths of the total. Country rankings change again when we look at value of production, reflecting differences in average unit values among countries, especially for wine grapes. In terms of value of production, the top five countries are France, China, the USA, Spain, and Chile.

These country rankings reflect both the historical origins of grape production in the Old World and the development of grape production in the New World, especially in recent decades. Whether in the New World or the Old World, grapes are grown in mid-latitude regions where temperatures during the growing season average 13–21 °C (Jones 2006), predominantly

Table 1.2 Production from top 20 grape-producing countries and world, 2000 and 2016

Country	2000		2016			Growth in production 2000–2016
	Production	Share of world total	Production	Share of world total	Cumulative share	
	KT	%	KT	%	%	%
China	3281.7	5.2	14,763.0	19.1	19.1	349.9
Italy	8869.5	14.0	8201.9	10.6	29.7	−7.5
USA	6973.8	11.0	7097.7	9.2	38.8	1.8
France	7762.6	12.2	6247.0	8.1	46.9	−19.5
Spain	6539.8	10.3	5934.2	7.7	54.6	−9.3
Turkey	3600.0	5.7	4000.0	5.2	59.7	11.1
India	1130.0	1.8	2590.0	3.3	63.1	129.2
Chile	1899.9	3.0	2473.6	3.2	66.3	30.2
Iran	2097.2	3.3	2450.0	3.2	69.4	16.8
South Africa	1454.7	2.3	2008.8	2.6	72.0	38.1
Australia	1311.4	2.1	1772.9	2.3	74.3	35.2
Argentina	2459.9	3.9	1758.4	2.3	76.6	−28.5
Egypt	1075.1	1.7	1716.8	2.2	78.8	59.7
Uzbekistan	624.2	1.0	1642.3	2.1	80.9	163.1
Germany	1361.0	2.1	1225.6	1.6	82.5	−10.0
Greece	667.6	1.1	990.3	1.3	83.8	48.3
Brazil	1024.5	1.6	984.5	1.3	85.0	−3.9
Afghanistan	330.0	0.5	874.5	1.1	86.2	165.0
Portugal	913.6	1.4	773.9	1.0	87.2	−15.3
Romania	1295.3	2.0	736.9	1.0	88.1	−43.1
Other	8881.0	14.0	9196.4	11.9	100.0	3.6
World	**63,552.7**	**100.0**	**77,438.9**	**100.0**		**21.8**

Source Created by the authors using data from FAOSTAT (2018)

in river valleys near the coast, often with a Mediterranean-type climate. Since growing season duration and temperatures have a major influence on grape ripening and fruit quality, within this broad landscape particular cultivars have been developed to be grown for particular end uses and in specific regions and sub-regions (see, e.g., Jones 2018).

The economic geography of grape production has been shifting over time, reflecting changes in both supply and demand for grape products among diverse countries. On the supply side, along with changes in technology of production and in the availability of labor and other inputs, changes in climate have begun to influence where particular cultivars can profitably be grown for particular end uses. On the demand side, along with changes in other sociodemographic factors, changes in income have implications for the mixture of grape products demanded given relatively high income elasticities of demand for premium wine versus basic wine, and for fresh versus dried grapes (see, e.g., Fuller and Alston 2012).

Between 2000 and 2016, total production of grapes worldwide grew by 22 percent, from 63.5 MT to 77.4 MT (Table 1.2).[4] However, that growth was not shared evenly among countries. Among the world's top producers, production by the three predominant Old World producers (Italy, Spain, and especially France) shrank, while it grew among the New World countries, and especially in Asia. China increased its production more than threefold and rose from the seventh-ranked to become the world's largest producer of grapes during this period. China now accounts for one-fifth of the world's total production of grapes, almost twice as much as the next-ranked country. Since the increase in China was predominantly in quantities of table grapes, whereas the declines in Europe were predominantly in quantities of wine grapes, the relative importance of table grapes has grown in the world. These changes in where grapes are produced and for what purposes have contributed to the increases in global average yields and changes in other aspects of the global grape industry.

Detailed data are not available on a consistent basis describing the patterns of grape production by end use of grapes, partly because some grape varieties can be used for diverse end uses, including drying for raisins, packing as table grapes for fresh consumption, and crushing for making grape juice concentrate, distillate, or wine. Some multipurpose grape varieties—such as Thompson Seedless—have been grown and used in significant quantities for any and all of these end uses, but complete data typically are not available on the utilization of these varieties. In some places, data are available only on production by varieties, classified according to their predominant use, and some of the available estimates might be better described as "guesstimates," so we must exercise caution in interpreting data on the allocation of grape acreage and volume of production among end uses. Nevertheless, the broad picture today is as shown in Figs. 1.2 and 1.3.

China accounted for the lion's share of growth in table grape production over the past 20 years, and now dominates global production of table grapes, with its share approaching half of the world total ("Appendix 1" Table 1.6). According to USDA/FAS data, global production of table grapes increased from 13.0 MT (3.7 MT from China) in 2001/02 to 24.3 MT (11.2 MT from China) in 2017/18. India ranks second (3.0 MT in 2016) and also has experienced rapid recent growth. The top five "countries" (here, counting the European Union as one country) accounted for almost 80% of the total volume of table grape production in 2016, and the top ten accounted for almost 94%. Data are available on raisin production in tonnes dried weight from USDA/FAS, which we converted to an estimate of fresh weight equivalent using a factor of 4:1. In 2017/18, according to these data, global production of raisins was 1.2 MT dried weight (4.9 MT fresh weight), up about 22% over the quantity produced in 2001/02. Turkey has replaced the USA as the world's largest raisin producer, China has risen from fifth to replace Iran as the third largest, and Uzbekistan has risen from last to fifth among the top twelve listed in "Appendix 1" Tables 1.7 and 1.8). Some of these patterns reflect a more general drift in demand toward fresh fruit and away from dried (and canned) fruit, associated with rising per capita incomes. In the USA, at least, over the 50 years 1976–2016, per capita consumption of table grapes trended up, along with fresh fruit in total, while per capita consumption of raisins trended down or stayed flat, along with dried fruit in total.

Of the total grape production in 2016 (77.4 MT in Table 1.1), an estimated 24.3 MT (31.3%) were table grapes ("Appendix 1" Table 1.6) and 4.9 MT (6.3%) were used to produce raisins ("Appendix 1" Table 1.8), leaving 48.2 MT (62% of the total) to be crushed—mainly for making wine. The total grape crush can include significant quantities used for grape juice

[4] In this part, we consider data since 2000 from FAOSTAT (2018) for making detailed comparisons. While data are available for earlier years, they are less complete in terms of country coverage and less accurate for some countries, and more so the farther back we go.

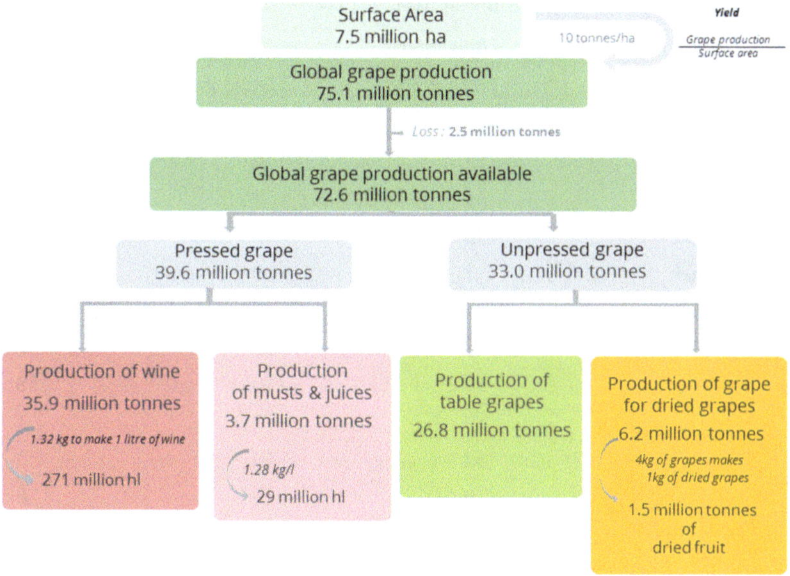

Fig. 1.2 Schematic representation of the global vitiviniculture situation 2014. *Source* Provided courtesy of OIV (pers. comm. Nicolas Goldschmidt, July 2018)

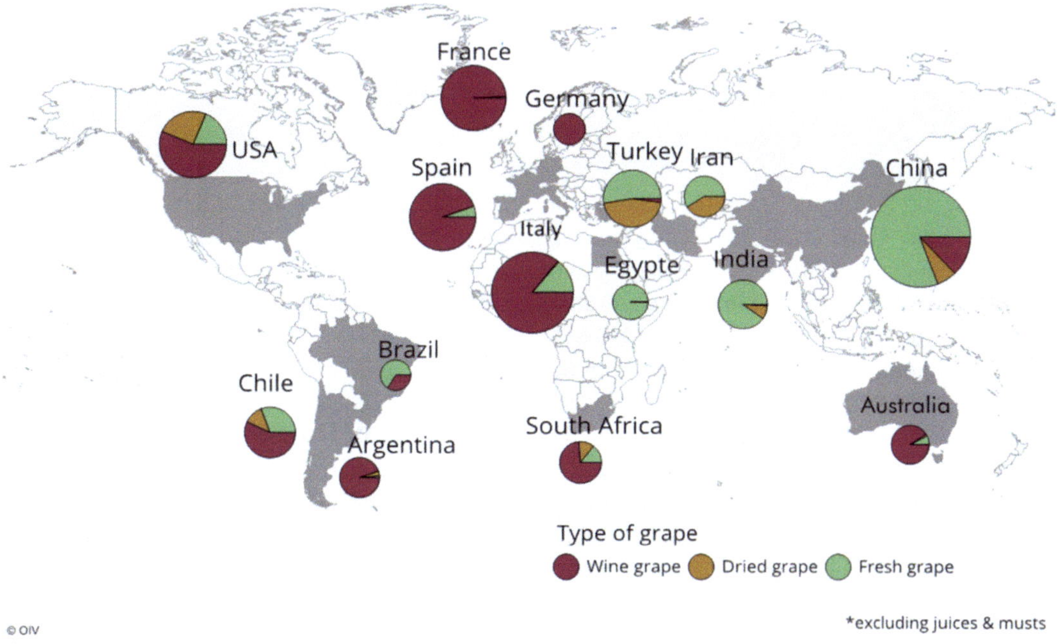

Fig. 1.3 Vine-growing areas and utilization of grape production in 2015. *Source* Provided courtesy of OIV (pers. comm. Nicolas Goldschmidt, July 2018)

Table 1.3 Grape production: grapes intended for all uses, 2015

Country	Production	Utilization		
		Fresh grapes	Dried grapes	Wine grapes
	MT	%		
China	13.7	83	6	12
Italy	8.2	15	0	85
USA	7.3	40	18	42
France	6.4	1	0	99
Spain	6.0	5	0	87
Turkey	4.0	48	50	2
India	2.6	32	10	58
Iran	2.2	89	10	1
Chile	2.2	32	10	52
South Africa	1.9	9	13	78
Australia	1.8	9	13	78
Argentina	1.8	2	55	77
Egypt	1.6	100	0	0
Uzbekistan	1.3	81	15	3
Germany	1.2	0	0	100
Brazil	1.1	67	0	33
World	**77.3**	**36**	**8**	**47**

Source OIV (2017a). http://www.oiv.int/public/medias/5479/oiv-en-bilan-2017.pdf

concentrate or distillation—around 30% of the total California grape crush in recent years.[5] If wine production globally took the same share (70%) of the total crush volume as in California, the quantity used for wine would be 33.7 MT, 43% of total production of grapes. This is in the range of estimates from other sources, but smaller. In its 2017 Statistical Report, OIV (2017a, p. 8) reported shares of grapes utilized in three categories—fresh, dried, and wine for the top 16 grape-producing countries, and the world as a whole in 2015 (Table 1.3). Of the global total of 77.3 MT, almost half (47%, or 36.3 MT) was for wine. Anderson and Pinilla (2018, p. 179, Table 131) estimate 52% of global grape production was used for wine over the period 2010–2016. The shares among end uses vary considerably among countries, some of which are heavily specialized in fresh grapes or wine grapes, while others produce a mixture (Table 1.3).

1.1.2 Many Diverse Varieties

Combining the variation in mixture of end uses with other sources of variation, the total number of varieties grown is large and the varietal mix varies considerably from one country to another —even when they are close neighbors. Recently, the OIV (2017b) published provisional estimates of total area planted to the main varieties of grapes in 2015 (Table 1.4). They reported that thirteen varieties accounted for more than one-third of the world's vineyard area, and thirty-three varieties accounted for one-half of the total. The top three varieties in this ranking

[5]For example, Alston et al. (2018b) deduced that, of the total California crush volume, on average for the years 2000 to 2016, 14.5% was used for grape juice concentrate, 15.8% was fermented to make distillate, and 69.6% was used to make wine.

Table 1.4 Top 35 grape varieties, total area planted in 2015

Variety	Planted area	End use
	K Ha	
Kyoho	365	Table grapes
Cabernet Sauvignon	341	Red wine
Sultanina (Sultana, Thompson Seedless)	273	Table, drying, and wine
Merlot	266	Red wine
Tempranillo	231	Red wine
Arien	218	White wine, brandy
Chardonnay	210	White wine
Syrah (Shiraz)	190	Red wine
Red Globe	159	Table grapes
Grenache Noir (Garnacha Tinta)	163	Red wine
Sauvignon Blanc	123	White wine
Pinot Noir (Blauer Burgunder)	112	Red wine
Trebbiano Toscano (Ugni Blanc)	111	White wine, brandy
Rkatsiteli	75	White wine
Riesling	64	White wine
Bobal	63	Red wine
Sangiovese	60	Red wine
Mourvèdre	56	Red wine
Malbec (Cot)	55	Red wine
Pinot Gris	54	White wine
Cabernet Franc	53	Red wine
Carignan Noir	51	Red wine
Viura	48	White wine
Concord	37	Juice, table, and wine
Alicante Bouschet	35	Red wine
Zinfandel (Primitivo)	35	Red wine
Aligote	35	White wine
Muscat of Alexandria	34	Table, drying, and wine
Chenin Blanc	33	White wine
Colombard	32	White wine
Muscat Blanc à Petits Grains	32	White wine
Cereza	29	White wine
Montepulciano	28	Red wine
Gamay Noir	27	Red wine
Glera	27	White wine
Total	**3740**	

Source OIV (2017b) http://www.oiv.int/en/oiv-life/the-distribution-of-the-worlds-grapevine-varieties-new-oiv-study-available are Kyoho, a table grape variety grown in China, Cabernet Sauvignon, a red wine grape variety, and Sultanina (aka Sultana or Thompson Seedless) a truly multipurpose grape, predominant among varieties used for dried grapes. Together these three varieties accounted for almost 1 million hectares, about one-seventh of the total. The next ten varieties are all wine grape varieties, except for Red Globe, a table grape variety, and Trebbiano Toscano (aka Ugni Blanc), used for both wine and brandy. The OIV report also indicates that the mix of varieties grown varies considerably among countries. To illustrate (in Table 1.5), we present data on the top ten varieties for each of the top five grape producers in 2015, taken from OIV (2017a, b).

Anderson and Aryal (2013) complied a "Database of Regional, National and Global Winegrape Bearing Areas by Variety, 2000 and 2010," which also includes details on bearing areas for multipurpose grape varieties used predominantly for other purposes, and some specialist table grape varieties. The dataset covers some 2000 varieties (of which almost 1300 are "primes" and the rest are their synonyms) and spans over 600 regions in 44 countries that together account for 99 percent of global wine production (Anderson 2014, p. 251). Along with the data, Anderson and Aryal (2013) present summaries—both variety-by-variety (showing areas planted in 2000 and 2010 for the main countries growing each variety) and country-by-country (showing the varietal mix for 2000 and 2010 for each important variety).

Drawing on these data, Anderson (2014) (see also Anderson 2010a, b, 2013) presents some analysis of the evolving varietal mix around the world. This analysis highlights the great diversity among countries and sub-regions within countries, in terms of the mix of grape varieties grown, and the considerable persistence of those differences in spite of the effects of globalization in making it easier to move plant materials around the world to better match genetics to the production environment. Particular varieties tend to be associated with particular places, and places tend to be specialized in particular varieties to a greater extent than can easily be justified by agronomic considerations alone.[6]

Nevertheless, Anderson (2014) documents several ways in which the distribution of wine grape varieties has been shifting. First, the varietal mix has become more concentrated (less diverse) for the world as a whole and in both the New World and the Old World. In particular, between 1990 and 2010 the global wine grape area devoted to varieties of French origin increased from 26% to 36% (in the New World, from 53% to 67%); varieties of Spanish and Italian origin account for a further, largely unchanged, 40%. Second, the rankings of individual varieties changed markedly—for instance, Cabernet Sauvignon and Merlot jumped to first and second from eighth and seventh—such that the list of the world's top 35 varieties in 2010 shows a quite different mix and ranking compared with 1990. Third, the global share of red varieties grew from 49% to 56% between 2000 and 2010. Anderson (2010a, b, 2014) also provides some more detailed analysis of the roles of particular varieties in the global picture, the roles of particular countries and regions, and the extent to which particular countries and regions are becoming more or less specialized in specific varieties, and more or less similar or dissimilar.[7]

In the case of wine grapes, although their relative importance may be changing, the varieties in use are predominantly traditional European varieties, typically hundreds of years old. The picture with table grapes is very different, partly because table grape producers are less committed to traditional *V. vinifera* varieties and more likely to adopt non-*vinifera* varieties and hybrids, leaving much greater scope for innovation. Here, varietal innovation is proceeding apace, including private varieties developed and

[6]Among other things, this outcome reflects efforts by producers to develop a reputation for the production of high-quality wines, sometimes through the development of collective "brands" associated with regions and varieties, as discussed later in this chapter.

[7]More recently, Anderson (2016) provides a detailed analysis of changes in Australia's grape varietal mix relative to the world as a whole, and Alston et al. (2015) do likewise for the USA.

Table 1.5 Top ten grape varieties in each of the top five producing countries, total area in 2015

Spain		China		France		Italy		USA	
Variety	Area	Variety	Area	Variety	Area	Variety	Area	Variety	Area
	K Ha		K Ha		K Ha		K Ha		K Ha
Arien	217	Kyoho	365	Merlot	112	Sangiovese	54	Thompson S.	60
Tempranillo	203	Red Globe	146	Trebbiano	82	Montepulciano	27	Chardonnay	43
Bobal	62	Cab. Sauv.	60	Grenache	81	Glera	27	Cab. Sauv.	41
Grenache	62	Carmenère	8	Syrah	64	Pinot Gris	25	Concord	34
Viura	46	Merlot	7	Chardonnay	51	Merlot	24	Pinot Noir	25
Mouverdre	43	Cab. Franc	3	Cab. Sauv.	33	Italia	22	Merlot	21
Alicante B.	26	Chardonnay	3	Cab. Franc	33	Cattarratto B.C.	21	Zinfandel	19
Pardina	25	Riesling	2	Carignan N.	33	Trebbiano	21	Syrah	9
Cab. Sauv.	20	Syrah	1	Pinot Noir	32	Chardonnay	20	Pinot Gris	8
Syrah	20	Pinot Noir	1	Sauv. Blanc	30	Barbera	18	Colombard	8
Others	250	Others	233	Others	240	Others	379	Others	174
Total	**974**		**830**		**806**		**682**		**443**

Notes Cab. Sauv. Is Cabernet Sauvignon; Alicante B. is Alicante Bouschet; Cab. Franc is Cabernet Franc; Carignan N. is Carignan Noir; Sauv. Blanc is Sauvignon Blanc; Cattarratto B.C. is Cattarrati Bianco Comune; Thompson S. is Thompson Seedless
Source OIV (2017b) http://www.oiv.int/en/oiv-life/the-distribution-of-the-worlds-grapevine-varieties-new-oiv-study-available

owned by individual producers as well as public varieties developed by grape breeders supported by government or a mixture of government and industry funding. Raisin grape varieties are changing too, but much less quickly and the varietal mix is much less diverse.

California illustrates the global phenomenon. The California Grape Acreage Report (USDA/NASS 2018a) lists area planted in California in 2017 for each of more than seventy table grape varieties, of which fourteen had at least 1000 acres planted and together accounted for the lion's share (71%) of the total.[8] As one indicator of the rapid rate of varietal change, all of the bearing and non-bearing acreage for many varieties was planted at least ten years previously, while for many others, all of the current acreage was planted within the past five years. Varieties that had the largest share of bearing acreage in 2016 (Flame Seedless, 18.5%; Crimson Seedless, 11.2%; Red Globe, 9.2%) had much smaller shares of non-bearing acreage (a combined total of 11.1%) compared with some up-and-coming varieties (Scarlet Royal, 12.2%, Autumn King, 10.4%; Allison, 9.2%). The California Grape Acreage Report (USDA/NASS 2018a) lists area planted in California in 2017 in total and individually for just six specific raisin varieties—Thompson Seedless, Selma Pete, Fiesta, DOVine, Sultana, and Black Corinth. Three of these varieties together accounted for 98% of the total planted area: Thompson Seedless (86%), Fiesta (8%), and Selma Pete (4%).

1.1.3 The Value of Diverse Varieties

Genetics by Environment by Management (G x E x M) interactions determine the value of particular wine grape varieties in particular locations, as can be illustrated by detailed US data on wine grapes. Within the USA, in 2014 five varieties (Chardonnay, Cabernet Sauvignon, Merlot, Pinot noir, and Zinfandel) accounted for 52.3% of the total volume and 63.2% of the total value of wine grape production from the four states (California, Washington, Oregon, and New York) that dominate national production. As discussed in detail by Alston et al. (2015), these five varieties predominate in several of the main production regions, but the emphasis varies among the premium price regions and some regions are quite different. In particular, California's hot Southern San Joaquin Valley (dominated by French Colombard and Rubired used to produce grape juice concentrate as well as bulk wine) and New York (dominated by non-*vinifera* American varieties, Concord and Niagara) are quite unlike the other regions climatically and in terms of their grape varietal mix. In terms of total bearing area, Chardonnay is the most important wine grape variety nationally and is highly ranked throughout the premium regions, but the North Coast region is especially known for its Cabernet Sauvignon, which is its most important variety and increasingly so, and likewise in Washington. The cooler coastal regions are relatively specialized in Chardonnay and Pinot Noir and other cool climate varieties. Zinfandel is more significant in terms of bearing area and value of production in the Northern San Joaquin Valley and other mid-price regions.

Prices vary systematically among regions—the North Coast region has generally higher prices than other regions for all varieties and the Southern San Joaquin Valley has generally lower prices.[9] In addition, prices vary systematically among varieties—among the higher-quality (higher-priced) varieties grown in significant quantity—Cabernet Sauvignon generally is ranked higher than Chardonnay, and Zinfandel generally is ranked lower. But the sizes of the premia, and even the rankings of varieties, vary among regions. For

[8]The California Table Grape Commission (2018) refers to a total of 85 varieties currently in production and provides details on the top 17. http://www.grapesfromcalifornia.com/docs/2016-variety-chart-and-merchandising-guide.pdf.

[9]In 2016 in Napa County, the average yield was 7.9 tonnes/ha and the average crush price was $5155/tonne, almost ten times the average crush price in the Southern San Joaquin Valley where the average yield was 40.5 tonnes/ha. The other regions were distributed between these extremes with higher yields being generally associated with lower prices per tonne, as described by Alston et al. (2018a, b).

example, Pinot Noir ranks above Cabernet Sauvignon almost everywhere, but not in Oregon where Pinot Noir is by far the dominant variety, nor in the Napa–Sonoma region; Chardonnay is ranked above Cabernet Sauvignon in the Central Coast region. These regional averages mask important variation within regions; prices for the same variety in the same crush district (of which California has 17) can vary considerably, even within a season. For example, in the California Grape Crush Report (2017) (USDA/NASS 2018b) the statewide average price for wine grapes purchased for crushing was $880/tonne in 2017, and the statewide average price for Cabernet Sauvignon was $1553/tonne. In that same year for crush district 4 (Napa), the average price was $5748/tonne for all grapes purchased for crushing and $8260 for Cabernet Sauvignon (a weighted average across some 35,000 tonnes). But that average price for Cabernet Sauvignon in district 4 reflected prices that ranged from less than $2000/tonne, for 80 tonnes in four lots, up to more than $40,000/tonne for 40 tonnes in five lots.

Prices of grapes fundamentally determine the value of land used to grow them. In the prime parts of the Napa Valley, in 2017, land suitable for commercial vineyards was valued at $500,000/ha and more, and, when planted with vines, $750,000/ha and more (see, e.g., California Chapter of the American Association of Farm Managers & Real Estate Appraisers 2017). Much of this value is attributable to potential to grow premium wine grapes; otherwise, similar farmland nearby sells for very much less. The same kinds of price variation for grapes and land to grow them can be seen among and within regions around the world, especially the premium wine-producing regions—such as in France, which has the highest priced vineyards in the world. In the Champagne region, for example, vineyard prices *average* well more than one million $/Ha (see, e.g., Gaeta and Corsinovi 2014); likewise, in premium locations in Bordeaux or Burgundy vineyards can command prices exceeding two million $/Ha, but within each of these regions prices range enormously, in multiples of up to 100 times the lower-end prices, as discussed by Franson (2013).

1.1.4 The Demand for Varieties

Particular varieties are valuable to farmers in particular applications for their agronomic traits (such as timing of harvest, yield, disease resistance, or cold tolerance) and fruit quality traits (such as seedlessness for table grapes, flavor profile for wine grapes, or sugar content for juice grapes), which together determine the value of the crop and the cost of producing it. These values for the inherent attributes of the fruit and products it is used to make can be conditioned by consumer preferences for attributes of the production process (e.g., organic or GMO-free; particular varietal names; geographic location of production) and government policies including trade taxes, alcohol excise taxes, and regulations over production practices or yields such as those associated with European Protected Designations of Origin for wine. These diverse determinants of value are to some extent intertwined with one another, owing to events going back 500 years, and more.

The "Columbian Exchange" was a mixed blessing for the world of wine. Sailing in 1524 at the behest of the King of France—some 32 years after Columbus landed at Hispaniola in the Caribbean—the Florentine navigator, Giovanni da Verrazzano, was the first European to explore the East Coast of what is now the USA. Da Verrazzano and the other early explorers of the North American East Coast would have seen grapes growing in profusion and must have imagined great possibilities for producing wine in the New World. They probably did not realize that the Native American grapes were not well-suited for producing high quality table wine. Nor could they know that the American grapes had co-evolved with numerous pests and diseases—including phylloxera, Pierce's disease (and its vectors), powdery mildew, downy mildew, and black rot, among others—which would present great obstacles to the establishment of an industry based on what would prove to be highly susceptible European *V. vinifera* varieties. Indeed, it would take several centuries, and many failed attempts to establish a wine industry in Colonial America, and subsequently the USA, before these barriers to the development of an American wine industry based on *V. vinifera* could be understood and overcome.[10] On the other side of this exchange was the movement of American vine stock and American pests and diseases to Europe and the rest of the world—eventually with devastating effects as *V. vinifera* grapevines became exposed to new pests and diseases against which they had little natural defense. Perhaps the best-known example is phylloxera, the cause of the "great wine blight" epidemic that devastated most of the world's vineyards in the late nineteenth and early twentieth centuries, with lasting effects on viticulture around the world.

Nowadays, phylloxera is managed at reasonably low cost by grafting scions of susceptible cultivars onto resistant rootstocks, and by employing preventive measures to avoid introducing it in places that have never had it (such as Chile and South Australia). In contrast, the fungal diseases, downy mildew and powdery mildew, which are also American natives, continue to impose massive costs on grape producers around the world every year. Meanwhile, some other American natives—like Pierce's disease, vectored by native and introduced sharpshooters—impose costs and restrict the scope of production in America, but have not yet spread to the rest of the world.[11] Other fungal diseases, such as Botrytis or trunk diseases such as Esca and Eutypa dieback, which are also important in America and affect vineyards worldwide, might have spread with *V. vinifera* grapes from the Old World, and new invasive pest and disease species are a perennial concern for grape growers everywhere.

Pest- and disease- management problems are economically significant in the grape industries worldwide. For example, Sambucci et al. (2019) estimated that, in California, the statewide cost of powdery mildew management in 2015 was about $239 million, including the costs of pesticide materials and application. These "pecuniary" costs represent about 5% of total revenue for growers on average, but may be more like 20% of revenue for growers of the most susceptible varieties (e.g., Chardonnay) in the cooler locations where disease prevalence and pressure is higher (e.g., California's Central Coast). In addition, Sambucci et al. (2019) reported that powdery mildew management accounts for 89% (by weight) of restricted material (pesticide, mostly sulfur) applications by grape growers, and eliminating powdery mildew would significantly reduce the environmental burden from disease management in grapes. These environmental externalities and the other "nonpecuniary" costs to growers from having to use chemical pesticides are hard to quantify but are no doubt significant. Similar patterns can be seen in the grape industries in other countries: pests and diseases are a major concern, as are the pesticides that represent a significant share of costs of production, and alternatives are being actively sought.

All of these problems invite genetic solutions. Grape breeders in several places have recently developed hybrid varieties that are resistant to some of the currently most concerning diseases, including powdery mildew and Pierce's disease. Further work is well underway to develop a greater scientific understanding of the issues and seeking to develop the means to extend the number of resistant varieties and introduce resistance genes to a wider range of grapes in ways that will be commercially attractive to growers (e.g., the *Vitis*Gen2 project: https://www.vitisgen2.org/). Until that happens, and even afterward, at least some growers will remain heavily reliant on the use of pesticides as damage-mitigation technologies. In particular, some growers may be reluctant to adopt disease-resistant varieties, or other novel

[10] Lapsley et al. (2018) review the American history drawing heavily on Pinney (1989, 2005). Other chapters in Anderson and Pinilla (2018) discuss the parallel history in other countries.

[11] Tumber et al. (2014) estimated that the cost of Pierce's disease in California was approximately $104.4 million per year, of which $56.1 million was the cost of lost production and vine replacement borne by grape growers, and $48.3 million was spent to fund Pierce's disease activities undertaken by various government agencies, the nursery and citrus industries, and the University of California system. Alston et al. (2013) found that the cost to producers and consumers would be much higher in the absence of the Pierce's Disease Control Program.

varieties, for fear of market resistance from consumers or market intermediaries who value traditional *V. vinifera* varietal names or object to the methods used to create new varieties.[12]

Evidence from stated preference surveys, market experiments, and consumer purchasing behavior indicates that, everything else equal, consumers prefer food and beverage products to be produced without using pesticides that entail risk to the environment, farmworkers, or food safety and health (e.g., see Lusk and Briggeman 2009; Loureiro et al. 2005; Baker 1999). Reflecting these concerns, governments around the world are imposing regulations that restrict or disallow the use of certain pesticides, and no doubt the list of restricted chemicals will continue to grow. In addition, pesticides that have been useful may become less useful as pests develop resistance to them.

These forces reinforce the demand for alternative pest and disease control technologies to supplement or replace the existing pesticides, including resistant varieties. In addition to demanding products made with varieties that require less pesticide, consumers (and market intermediaries) demand various fruit quality traits (of which there are many that can be changed through genetics), a lower product price for a given quality of product (e.g., from higher-yielding varieties that enable lower-cost production), and extended seasonal availability for fresh fruit. And growers demand varieties that produce fruit with quality attributes that consumers and intermediate buyers will value and also have desirable agronomic attributes such as high yield and low cost of production, tolerance of abiotic stresses such as high and low temperatures and drought, and resistance to pests and diseases.

However, a particular challenge with genetic innovations in grape production (whether for pest- and disease-resistance traits or for other reasons) is that a new variety produced by conventional cross-breeding cannot use a traditional *V. vinifera* name. This can be a substantial disadvantage in wine production where varietal names play a unique role in defining product designations and can attract a large premium. In many situations, growers will not find it profitable to forego the premium for, say, Chardonnay and grow an otherwise identical grape variety that cannot be called Chardonnay but has some other desired trait such as powdery mildew resistance. This problem arises in the wine industry regardless of whether a new variety is a hybrid or the result of crossing *vinifera* varieties, but not to the same extent in other parts of the grape industry. Indeed, many of the new and popular table grape varieties are hybrids.

Methods of modern biotechnology such as genetic engineering or gene editing might be used to enable certain traits in existing varieties, but it remains to be seen what these novel versions of existing varieties could be called, lawfully, or how they would be received in the marketplace.[13] It would be reasonable to anticipate some political action by the NGOs that have actively opposed other genetically engineered products to discourage farmers from growing and market intermediaries from selling genetically engineered grapes and products made with them if such varieties become available. Some wine-producing jurisdictions (e.g., South Australia, several counties in Northern California, and much of Europe) have already regulated to disallow production of genetically engineered crops. These same jurisdictions tend also to be ones where people appear to be actively

[12]While we have focused on pest- and disease- resistance traits in this section, the same issues arise in the development of new varieties that are more tolerant of environmental stresses such as heat, cold, or drought. We are also conscious of the fact that we have paid scant attention to the distinctions between traits that can be introduced through genetic innovations in rootstocks versus scions.

[13]In the European Union, at least, the current indications appear unfavorable. On July 25, 2018, the EU Court of Justice ruled that plants created with new gene-editing techniques should be regulated as genetically modified plants. While the market worldwide has accepted the use of non-*vinifera* rootstocks with *V. vinifera* scions, it remains to be seen which parts of the market—if any—will accept genetically modified rootstocks.

concerned over the environmental and human health consequences of pesticides, which leaves producers in those regions (and consumers of their products) facing a dilemma that may well get worse if existing heavily used pesticides become less effective or less acceptable in the market or both.

1.1.5 Government Intervention

Governments intervene heavily in agriculture worldwide, in a host of ways, but the production of grapes for making wine attracts more regulation than most of agriculture for two reasons. First, the market for wine grapes is influenced indirectly because they are used to produce alcohol, the most heavily regulated and taxed part of the food and beverage sector, whether as sin taxes or as a source of revenue (see, e.g., Anderson 2010b). These indirect effects can be quite substantial, since the taxes and regulations entail significant impositions. Second, the wine industry itself has sought specific rules and regulations governing the production and marketing of wines and the varieties of grapes used to produce them in particular places, and governments have legislated accordingly. Both kinds of government intervention have had substantial implications for the demand for varieties and for varieties with particular traits at times.

In history, trade tariffs and excise taxes on alcohol have been important as a source of government revenue for financing government and as a political issue. For example, prior to 1913 the USA did not have any permanent income tax, and between about 1865 and 1915, about 70% of internal revenue (and about 40% of total government revenue) was raised as excise taxes, mainly on alcohol (in particular, whiskey); the rest was mainly from tariffs. In 1913, the 16th Amendment to the Constitution was ratified, permanently legalizing an income tax—a necessary precursor for Prohibition (1920–1933), which was to eliminate the main alternative internal revenue source for the US government (see, e.g., Okrent 2010). Both the excise taxes and the Prohibition that made them irrelevant for 13 years have had implications for the production and consumption of other forms of alcohol, as well as wine and the grapes used to produce it. As discussed by Alston et al. (2018a, b) and Lapsley (1996), Prohibition banned the sale of alcohol but not the sale of grapes to be used for home winemaking, which encouraged an increase in production in California of grapes that would be suitable for transportation to the major East Coast markets and use in home winemaking. It took some time after Repeal to replace these varieties with others, better suited to making high quality table wine. The same authors discuss various other US tax policies that have had consequences for the structure of the US wine-producing industry and implicitly for the pattern of wine grape production and the demand for grape varieties and varietal traits.

Other countries offer different examples of the role of government policies in shaping the markets for wine and the grapes used to produce them. Writing in the eighteenth and nineteenth centuries, the Classical economists, Adam Smith and David Ricardo, developed important economic ideas in the context of British trade tariffs—including the concept of comparative advantage. As discussed by Nye (2007) during the seventeenth and eighteenth centuries, the extended conflict with France caused Britain to turn away from French wine toward wine from Spain and Portugal, and away from wine to beer and spirits (especially, gin). At least partly as a source of war finance, Britain imposed tariffs on imported wine. The fact that these were specific (per unit) tariffs rather than percentage (or *ad valorem*) tariffs meant that they represented a higher percentage tax on cheaper French wine, to the advantage of the British brewers and distillers and reinforcing the establishment of Britain as a beer- and gin-drinking nation, especially among the working class, but with relatively little consequence for the wealthy British consumers of fine claret from Bordeaux. Britain's entry to the European Common Market in the 1970s eliminated remaining trade barriers between Britain and Europe, facilitating the more recent growth in the UK wine market, a pattern that may be disrupted by Brexit, possibly to the advantage of non-European suppliers; likewise, the recent introduction of tariffs in China on wine from the USA.

While per unit taxes distort consumption in one way, *ad valorem* tariffs distort them in another. In Australia, the "wine equalization tax" (WET) is 29% of the wholesale value of wine (Anderson 2010b), which amounts to a considerable sum per bottle on fine wine compared with the lowest priced wines. Economists have argued that this tax is inefficient and distortionary if the purpose is as a "sin tax" to discourage excessive alcohol consumption (see, e.g., Freebairn 2010). James and Alston (2002) compare the consequences for the balance of production and consumption—across market segments from premium to bulk—between ad valorem and per unit taxes in the context of the Australian wine market. Another example of this phenomenon is the encouragement to produce bulk wine created by the US duty drawback policy (see, e.g., Sumner et al. 2012). Such policy-induced changes in the balance of types of wine produced have indirect implications for the demand for wine grape varieties and traits.

Producers and consumers of wine are not numbered among the enthusiastic supporters of wine taxes, let alone Prohibition. However, some other forms of government intervention have been introduced at the behest of producers, and possibly to the benefit of consumers, and these policies sometimes have direct connection to grape varieties. Specifically, here, we are referring to Protected Designations of Origin (PDOs), such as the French *appellation d'origine contrôlée* (AOC), which was the first European PDO system.[14] The AOC was conceived as a geographic indication certified by the government: "Products covered by AOC labels are controlled by the state to ensure both their territorial origin and their conformity to precise rules for production and processing that guarantee their 'typicity,' or distinctive character" (Barham 2003, p. 128). Currently under the AOC system over 300 different PDOs exist for French wines, including 57 in Bordeaux alone.

Livat et al. (2018) discuss various perspectives on the economic rationale for PDOs for wine, all related to the economics of imperfect information. Wine is an "experience" good (since quality is difficult for the consumer to assess prior to purchase) with a wide range of product quality, wine markets exhibit imperfect information, and it can be costly to acquire information about quality. In such a setting, it can be in the interests of a group of producers to create a collective "brand" and to provide some assurance to consumers that the branded product will meet certain quality standards; consumers, too, stand to benefit. PDO systems like the AOC apply this concept where the "brand" applies to products (in this case, wine) from a particular defined geographic origin. This has a particular logic, in the case of wine, given the association of quality with terroir.[15] Producers want to differentiate their products from those of their competitors in the eye of consumers and earn a premium from doing so, but they also want to claim credit for particular attributes and to enjoy the benefits from collective reputation associated with their region of production. Wine PDOs capture these attributes that wine producers aim to use to differentiate their products. In today's wine market

[14]Laws passed in 1919, 1927, and 1935 allowed the creation of the current system; the first French law on viticultural designations of origin dates to 1905 (Chevet et al. 2018, p. 69–73). Meloni and Swinnen (2013) discuss the political and policy context in which quality regulations were introduced, with their essential purpose at the time being to create a barrier to entry and restrict competition from surging imports, especially Algerian wine. This situation arose in the aftermath of the "Great French Wine Blight" from phylloxera, which led to the development of the Algerian wine industry to replace the great loss of production capacity in France during the period of the 1850s–1870s.

[15]In his provocatively titled book, *Terroir and Other Myths of Winegrowing*, Matthews (2015) challenges some of the conventional wisdom in this context. Hedonic studies by economists have produced a mixture of results on the value of terroir (see, e.g., the extensive listing of studies and discussion by Haeck et al. 2018). Nevertheless, there appears to be a clear general association of quality and price with the place of production for wine, and producers perceive returns to creating a collective reputation associated with a PDO. See Frick and Simmoins (2013) and studies they cite regarding the economics of collective reputation for wine.

as many as 1239 different wine PDOs exist, and information about PDOs is included with other information on wine labels (International Organization of Vine and Wine).[16]

In the case of wine, in addition to being produced in a defined geographic area, qualifying for an AOC may also require wine to conform to technological restrictions, such as the grape varieties used to produce it, the maximum yields per hectare, planting density of the vineyard, the (minimal) alcohol percentage, or particular vinicultural practices used (see, e.g., Coates 2001). Thus, for example, to qualify for the Pomerol AOC (which is found within the right-bank region of Bordeaux) the only permitted grape varieties are Merlot, Cabernet Franc, Cabernet Sauvignon, and Malbec—i.e., strictly red wine varieties. Yields are restricted to a maximum of 42 hectoliters/hectare, and the finished wine must contain at least 10.5% alcohol by volume. Other regulations apply to the planting density and the spacing between the rows, and the wine may be subject to quality tests.

The total planted area in the AOC is fixed, and this, combined with the maximum yield for the PDO, restricts the total supply from the PDO. Even if the yield restriction and the limitation on total quantity do not result in a price premium compared with other wine, the quality assurance should command a premium, if the AOC system works as intended. The work by Livat et al. (2018) finds that this does not appear to be so and conclude that the 57 different PDOs for the Bordeaux region may be too many for the system to provide useful information to consumers.

Nevertheless, the system in France has been emulated in the main wine-producing regions throughout Europe, and, in the premium producing regions in France, Spain, and Italy, to qualify for the PDO growers must produce according to the relevant rules and regulations; in particular, this means producing the designated varieties. That aspect of the PDO system imposes severe strictures on the opportunities and incentives for growers in those regions to stray from the varietal mix that is typical for their AOC, let alone adopt new varieties that would not qualify, a potentially serious problem in years to come as the world warms. Moreover, the imposition of yield limitations is a disincentive to develop and adopt higher-yielding varieties from among those that would qualify for the PDO.

Other countries have adopted PDO systems that do not impose the same kinds of technological restrictions, aiming to capitalize on the economics of collective reputation. For example, in the USA, in 1983 the Federal Government responded to industry desire to place more precise vineyard locations on wine labels by creating "American Viticultural Areas" (AVAs—see US Treasury/TTB 2013). AVAs are defined geographic areas that may be quite large and cross state or county lines, or may be quite small and lie within a county or, in some cases, another AVA. The Napa Valley AVA is, for instance, a large AVA located within Napa County, and the Oakville AVA is a much smaller AVA that is located within the Napa Valley AVA. In 2018, the USA had a total of 242 AVAs (TTB 2018). Today, wineries may identify the grapes used in a wine as coming from an AVA if 85 percent of the grapes were grown in the AVA. There is no restriction on the grape varieties that may be used, nor on allowable yields for the resulting wines to qualify for an AVA, but varieties tend to be associated with AVAs (such as Cabernet Sauvignon with the Napa Valley AVA and its sub-appellations; or Pinot Noir with the Willamette Valley AVA and its sub-appellations), and some wineries do market their wines as having been produced from low-yielding vines.

The direct linkage, by regulation, of specific grape varieties to particular geographic locations through PDOs is an Old World phenomenon; the PDO implicitly indicates which varieties (from a relatively short list) could have been used to make the wine and even some ideas about the likely emphasis in the blend. In the New World, where such regulations do not exist, many wine labels specify the main grape varieties used to produce the wine, directly connecting the grape varieties used to the product in ways that convey

[16]See http://www.oiv.int/en/databases-and-statistics/database.

a sense of value associated with particular varieties. Wine is marketed with varietal content often used as a primary dimension for organizing the retail display, and the value of particular varieties may be associated with specific places of origin (see, e.g., Kwon et al. 2008 for an illustration using data on 8800 California wines). If the label does not refer to a geographically narrow PDO, the product may be seen as implicitly lower-value, generic wine.

The marketing of varietal wines as such is a comparatively recent phenomenon, largely having developed over the past 50 years among the New World producers, several of which have a "signature" variety associated with them such as Australia and Shiraz, South Africa and Pinotage, Uruguay and Tannat, Argentina and Malbec, or New Zealand and Sauvignon Blanc. Some Old World producers also have begun to provide information about grape varieties on the labels, and perhaps, we would have seen more of this if information about varieties were not conveyed implicitly in information about the PDO already on the label for much of the wine. In any event, implicitly or explicitly, varieties per se have value both in general (e.g., Anderson 2014 identifies "premium" varieties in terms of the prices they command) and in conjunction with particular places and sometimes particular producers. This fact has implications for the potential for making varietal innovations for wine grapes in those places where it is the existing varieties with their indelible names that attract the premia. The same is not true for other end uses of grapes, however.

1.2 Conclusion

Growth in population and per capita income leads to increases in demand for grapes and all the products they are used to produce. The evolving patterns of consumer demand also reflect trends and cycles in which types of alcoholic beverages and which types of wine within that category are more or less popular, some of which is driven by demographic change. We also observe a rising demand for "process attributes" of grapes and products made with them, expressed in demand for products carrying eco-labels such as "organic" or "biodynamic" or "sustainable" or "fair trade" or "non-GMO."

Evolving consumer demand is one set of forces driving the demand for different types of grapes with different bundles of traits, including agronomic traits that will facilitate the use of production processes that qualify for eco-labels. Another set of forces is the public policy processes that are applying increasingly stringent restrictions on the use of pesticides and other agricultural chemicals in vineyards, increasing the demand for alternatives, including resistant varieties. In addition, changes in supply of other agricultural inputs such as irrigation water and farm labor—in terms of reliability of availability as well as normal availability and price—and similarly, natural inputs such as rainfall and solar energy, give rise to demand for new varieties: varieties that are more tolerant of environmental stresses and more suitable for production in a mechanized system, or more suitable than traditional varieties given changes in climate. Finally, even if nothing changes, growers are looking for varieties that are more profitable to grow compared with their existing varieties under existing conditions—varieties that are higher yielding, more resistant to pests and diseases, more resilient to environmental stresses, with greater amounts of more-desired fruit-quality attributes, and so on.

In short, the demand for grapevines that have particular combinations of attributes, including various agronomic and fruit quality traits, is a derived demand—derived from the final demand for the final consumer products made with the fruit, the costs of making those products with the fruit, and the costs of growing the fruit with those vines. In the case of table grapes and raisin grapes, the post-farm value chain is relatively short and simple, but for some wines it is a complex and expensive process over many years. The challenge for grape breeders is to find ways to effectively introduce new desired traits without foregoing too much in terms of existing traits that producers also value, such that it is profitable for producers to adopt the new varieties.

In turn, the demand for varietal innovations—and for investments in science to create those new varieties—is also a derived demand. It depends on the demands for attributes, the supply of attributes from the existing stock of varieties, and the costs of innovation. As discussed, one of the important attributes of existing varieties used for winemaking is the varietal name, each of which comes with a bundle of attributes that cannot be changed without changing the name at the same time. This is an important constraint on varietal innovation when the value of existing varieties, entailed in their names, is large relative to the value of other traits that might also be desired such as higher yield, resistance to pests and diseases, or fruit quality attributes. This constraining effect of demand for existing names for wine grapes can help account for the fact that varietal innovation has been more rapid in the table grape industry, which has been growing faster (and partly because of those same varietal innovations), compared with the wine grape industry.

Acknowledgements The work for this project was partly supported by a Specialty Crop Research Initiative Competitive Grant, Award No. 2017-51181-26829 (the *Vitis*Gen 2 project) of the USDA National Institute of Food and Agriculture. The authors are grateful for this support and for helpful comments and advice provided by Kym Anderson, Jim Lapsley, and Brad Rickard. Views expressed are the authors' alone.

Appendix 1: Data Resources

Agricultural data are available from a variety of public sources for individual countries and for global aggregates. All these sources depend to some extent on national data agencies, which are

Table 1.6 Production of table grapes by country, selected marketing years, 2001/02–2017/18

Country	2001/02	2004/05	2007/08	2010/11	2013/14	2016/17	2017/18
Thousand tonnes (KT)							
China	3679	4025	4647	6200	8085	10,800	11,200
India	1184	1565	1735	1235	2585	2784	3000
Turkey	1568	1663	1920	2150	2200	2350	2120
Uzbekistan	516	642	900	1206	1579	1580	1580
European Union	1839	1706	1977	2090	1816	1666	1450
Brazil	1300	1233	1421	1495	1454	980	980
USA	784	801	835	865	1013	943	935
Chile	754	855	1185	1215	1055	916	900
Peru	136	170	223	297	500	605	638
Mexico	176	233	266	215	260	256	290
South Africa	224	238	259	245	252	334	280
Korea, South	422	381	334	269	269	269	269
Ukraine	200	230	300	320	320	260	260
Australia	100	100	99	93	72	179	200
Others	148	150	250	228	204	199	201
World Total	**13,030**	**13,990**	**16,350**	**18,122**	**21,663**	**24,120**	**24,302**

Notes The USA and Mexico are on a May–April marketing year. All other northern hemisphere countries are on a June–May marketing year. Southern hemisphere producer countries of Argentina, Chile, Peru, and South Africa are on an October–September marketing year, and Australia and Brazil are on a calendar year indicated as the second year of the split year. Some countries may include raisin-type and/or table-type grapes. Countries are ordered according to total production in 2017/18
Sources Created by the authors using online data from USDA/FAS (2018a), available at https://apps.fas.usda.gov/psdonline/ and described by USDA/FAS (2018b) available at https://apps.fas.usda.gov/psdonline/circulars/fruit.pdf

Table 1.7 Production of raisins by country, selected marketing years, 2001/02–2017/18

Country	2001/02	2004/05	2007/08	2010/11	2013/14	2016/17	2017/18
Thousand tonnes dried weight (KT)							
Turkey	220.0	300.0	250.0	250.0	242.6	310.0	295.0
USA	378.4	251.6	326.6	358.2	368.4	297.7	275.0
China	85.0	95.0	150.0	135.0	165.0	185.0	190.0
Iran	115.0	154.0	166.0	147.0	160.0	170.0	160.0
Uzbekistan	12.0	28.0	37.0	26.0	18.0	73.0	75.0
Chile	45.0	55.9	67.4	72.5	69.2	59.0	60.0
South Africa	40.5	30.4	40.2	23.5	46.0	55.0	55.0
Argentina	21.0	27.0	28.0	34.0	20.5	31.0	40.0
Afghanistan	16.0	18.0	24.5	31.0	31.0	26.0	30.0
Australia	34.0	28.5	11.0	7.4	10.0	18.0	20.0
European Union	28.0	30.0	10.0	11.0	10.0	10.0	10.0
Mexico	13.1	7.5	8.5	8.3	10.0	9.0	10.0
World total	**1008.0**	**1025.9**	**1119.1**	**1103.8**	**1150.7**	**1243.7**	**1220.0**

Notes The marketing year begins in August of the first year for northern hemisphere countries and January of the second year for southern hemisphere countries. Countries are ordered according to total production in 2017/18
Sources Created by the authors using online data from USDA/FAS (2018a), available at https://apps.fas.usda.gov/psdonline/ and described by USDA/FAS (2018c) available at: https://downloads.usda.library.cornell.edu/usda-esmis/files/8p58pc92q/cz30pt133/2j62s532b/raiswm-09-22-2017.pdf

not all equally reliable. Grape production is for the most part concentrated among higher-income countries that have comparatively reliable data resources, but even so, inconsistencies can arise (e.g., Alston et al. 2018a, b find substantial differences between alternative US sources of data on grape production in California).

It is not always possible to resolve such inconsistencies in terms of differences in definitions of variables, or assumptions, or to decide which source is more reliable. The fact that grapevines are long-lived perennials means issues arise about how to count non-bearing acreage and knowing if it is included in the data accurately. The fact that the product (e.g., wine) is often made within vertically integrated businesses, so the farm product is not traded on markets as such, adds to data gathering issues, including the challenge of determining whether grapes were used for fresh consumption, dried, or crushed, and if crushed whether destined for wine or other uses.

In this chapter, we make use of data from various sources, including (1) the International Organization of Vine and Wine (OIV) website: http://www.oiv.int/en/databases-and-statistics, (2) the Food and Agricultural Organization of the United Nations, FAO), FAOSTAT website: http://www.fao.org/faostat/en/#data, (3) the United States Department of Agriculture Foreign Agriculture Service (USDA/FAS) website: https://www.fas.usda.gov/data, and (4) data on global wine markets compiled by Anderson and Pinilla (2018), available at the website: https://www.adelaide.edu.au/press/titles/global-wine-markets/. We are conscious of discrepancies among these sources and do our best to make use of the best source for each purpose in ways that make for consistent comparisons within the chapter.

Table 1.8 Production of raisins by country, selected marketing years, 2001/02–2017/18

Country	2001/02	2004/05	2007/08	2010/11	2013/14	2016/17	2017/18
Thousand tonnes fresh weight (KT)							
Turkey	880.0	1200.0	1000.0	1000.0	970.5	1240.0	1180.0
USA	1513.5	1006.2	1306.3	1432.6	1473.6	1191.0	1100.0
China	340.0	380.0	600.0	540.0	660.0	740.0	760.0
Iran	460.0	616.0	664.0	588.0	640.0	680.0	640.0
Uzbekistan	48.0	112.0	148.0	104.0	72.0	292.0	300.0
Chile	180.0	223.6	269.4	290.0	276.8	236.0	240.0
South Africa	162.1	121.6	160.8	93.9	184.0	220.0	220.0
Argentina	84.0	108.0	112.0	136.0	82.0	124.0	160.0
Afghanistan	64.0	72.0	98.0	124.0	124.0	104.0	120.0
Australia	136.0	114.0	44.0	29.6	40.0	72.0	80.0
European Union	112.0	120.0	40.0	44.0	40.0	40.0	40.0
Mexico	52.4	30.0	34.0	33.2	40.0	36.0	40.0
World total	**4032.0**	**4103.4**	**4476.5**	**4415.3**	**4603.0**	**4975.0**	**4880.0**

Notes Fresh weight quantities are based on dried weight quantities in Table A-1.2, multiplied by a factor of 4, assuming an average of 4 tonnes fresh weight per tonne dried weight. The marketing year begins in August of the first year for northern hemisphere countries and January of the second year for southern hemisphere countries. Countries are ordered according to total production in 2017/18

Sources Created by the authors using online data from USDA/FAS (2018a), available at https://apps.fas.usda.gov/psdonline/ and described by USDA/FAS (2018c) available at: https://downloads.usda.library.cornell.edu/usda-esmis/files/8p58pc92q/cz30pt133/2j62s532b/raiswm-09-22-2017.pdf

References

Alston JM, Fuller K, Kaplan J, Tumber K (2013) The economic consequences of pierce's disease and related policy in the California winegrape industry. J Agric Resour Econ 38(2):269–297

Alston JM, Anderson K, Sambucci O (2015) Drifting towards bordeaux? The evolving varietal emphasis of U.S. wine regions. J Wine Econ 10(3):349–378

Alston JM, Lapsley JT, Sambucci O, Sumner DA (2018a) United States. In: Anderson K, Pinilla V (eds) Wine's evolving globalization: comparative histories of the old and new world. Cambridge University Press, Cambridge, pp 410–440

Alston JM, Lapsley JT, Sambucci O (2018b) Grape and wine production in California. In: Goodhue R, Martin P, Wright B (eds) California agriculture: dimensions and issues (in process). Giannini Foundation of Agricultural Economics, Berkeley

Anderson K (2010a) Varietal intensities and similarities of the World's Wine Regions. J Wine Econ 5(2):270–309

Anderson K (2010b) Excise and import taxes on wine versus beer and spirits: an international comparison. Econ Pap 29(2):215–228

Anderson K (2013) Which winegrape varieties are grown where? A global empirical picture. University of Adelaide Press, Adelaide

Anderson K (2014) Changing varietal distinctiveness of the World's Wine Regions: evidence from a new global database. J Wine Econ 9(3):249–272

Anderson K (2016) Evolving varietal and quality distinctiveness of Australia's Wine Regions. J Wine Res 27 (3):173–192

Anderson K, Aryal NR (2013) Database of regional, national and global winegrape bearing areas by variety, 2000 and 2010. Wine Economics Research Centre, University of Adelaide, Adelaide

Anderson K, Pinilla V (eds) (2018) Wine's evolving globalization: comparative histories of the old and new world. Cambridge University Press, Cambridge

Baker G (1999) Consumer preferences for food safety attributes in fresh apples: market segments, consumer characteristics, and marketing opportunities. J Agric Resour Econ 24(1):80–97

Barham E (2003) Translating terroir: the global challenge of French AOC labeling. J Rural Stud 19(1):127–138

California Chapter of the American Association of Farm Managers & Real Estate Appraisers. (2017). 2017 Trends in agricultural land and lease values.

Ebook. http://www.calasfmra.com/db_trends/2017 Trends_ebook.pdf
Chevet J-M, Fernandéz E, Giraud-Héraud E, Pinilla V (2018) France. In: Anderson K, Pinilla V (eds) Wine's evolving globalization: comparative histories of the old and new world. Cambridge University Press, Cambridge, pp 5–91
Coates C (2001) An encyclopedia of the wines and domaines of France. University of California Press, Oakland
Food and Agriculture Organization of the United Nations Statistics (FAOSTAT) (2018) Data retrieved from: http://www.fao.org/faostat/en/#data
Franson P (2013) The price of prestige: prime French vineyard properties soar while ordinary land languishes. Wines & Vines, September 2013. https://www.winesandvines.com/sections/printout_article.cfm?article=feature&content=121185#
Freebairn JW (2010) Special taxation of alcoholic beverages to correct market failures. Econ Pap 29(2): 200–214
Frick B, Simmoins R (2013) The impact of individual and collective reputation on wine prices: empirical evidence from the mosel valley. J Bus Econ 2:119–201
Fuller KB, Alston JM (2012) The demand for California winegrapes. J Wine Econ 7(2):192–212
Gaeta D, Corsinovi P (2014) Economics, governance, and politics in the wine market: European developments. Palgrave-Macmillan, New York
Haeck C, Meloni G, Swinnen J (2018, July) The value of terroir. a historical analysis of Bordeaux and Champagne, the World's first geographical indications. LICOS Center for Institutions and Economic Performance & Department of Economics, University of Leuven (KU Leuven)
International Organization of Vine and Wine (OIV) (2017a). 2017 World vitivinicultural situation: statistical report on world viticviniculture. http://www.oiv.int/public/medias/5479/oiv-en-bilan-2017.pdf. Accessed 10 Jan 2018
International Organization of Vine and Wine (OIV) (2017b) Distribution of the World's grapevine varieties. Focus OIV 2017. International Organization of Vine and Wine (OIV), 2017. http://www.oiv.int/en/oiv-life/the-distribution-of-the-worlds-grapevine-varieties-new-oiv-study-available Accessed July 8 2018
James JS, Alston JM (2002) Taxes and quality: a market-level analysis. Aust J Agric Resour Econ 46 (3):417–445
Jones GV (2006) Climate and terroir: impacts of climate variability and change on wine. In: Macqueen RW, Meinert LD (eds) Fine wine and terroir: the geoscience perspective. Geoscience Canada reprint series number 9. Geological Association of Canada, St John's, pp 203–216
Jones GV (2018) The climate component of terroir. Elements (Special Issue on Terroir: Science Related to Grape and Wine Quality) 14(3):167–172
Kwon OS, Hyunok Lee H, Sumner DA (2008) Appellation, variety, and the price of California wines. ARE Update 11(4):15–19
Lapsley JT (1996) Bottled Poetry: Napa winemaking from prohibition to the modern era. University of California Press, Berkeley
Lapsley JT, Alston JM, Sambucci O (2018) Structural features of the U.S. wine industry. In: Alonso Ugaglia A, Albisu LM, Cardebat J-M, Corsi A, Gil C, Mazzarino S (eds) The wine industry worldwide. Palgrave-Macmillan, New York
Livat F, Alston JM, Cardebat J-M (2018) Do denominations of origin provide useful quality signals? The case of Bordeaux wines. Econ Model. https://doi.org/10.1016/j.econmod.2018.06.003
Loureiro ML, McCluskey JJ, Mittelhammer RC (2005) Will consumers pay a premium for eco-labeled apples? J Consum Affairs 36(2):203–219
Lusk JL, Briggeman BC (2009) Food values. Am J Agric Econ 91(1):184–196
Matthews MA (2015) Terroir and other myths of winegrowing. University of California Press, Oakland
McGovern P (2003) Ancient wine: the search for the origins of viniculture. Princeton University Press, Princeton
Meloni G, Swinnen J (2013) The political economy of European wine regulations. J Wine Econ 8(3):244–284
Nye JVC (2007) War, wine, and taxes: the political economy of Anglo-French Trade, 1689–1900. Princeton University Press, Princeton
Okrent D (2010) Last call: the rise and fall of prohibition. Scribner, New York
Pinney T (1989) A history of Wine in America: from beginning to prohibition, vol 1. University of California Press, Berkeley
Pinney T (2005) A history of wine in America: from prohibition to the present, vol 2. University of California Press, Berkeley
Robinson J, Harding J, Vouillamoz J (2012) Wine grapes: a complete guide to 1,368 vine varieties including their origins and flavours. AllenLane, London
Sambucci O, Alston JM, Fuller KB, Lusk J (2019) The pecuniary and non-pecuniary costs of powdery mildew, and the potential value of resistant varieties in California grapes. Am J Enol Vitic 70(2):177–187
Sumner DA, Lapsley JT, Rosen-Molina JT (2012) Economics of wine import duty and excise tax drawbacks. Agric Resour Econ Update 15(4):1–4
The Plant List: Vitis (2018) Version 1.1 published on the internet. http://www.theplantlist.org/1.1/browse/A/Vitaceae/Vitis/. Accessed 1 July 2018
Tumber KP, Alston JM, Fuller KB (2014) Pierce's disease costs California $104 Million per year. Calif Agric 68(1):20–29
U.S. Department of The Treasury/Alcohol and Tobacco Tax and Trade Bureau (U.S. Treasury/TTB) (2013). American viticultural area (AVA). https://www.ttb.gov/wine/ava.shtml. Accessed 12 July 2016
U.S. Department of The Treasury/Alcohol and Tobacco Tax and Trade Bureau (U.S. Treasury/TTB) (2018)

Established American viticultural areas. https://www.ttb.gov/wine/us_by_ava.shtml. Accessed 14 Aug 2018

U.S. Department of Agriculture/National Agricultural Statistics Service (USDA/NASS) (2018a) California Grape Acreage Report, 2017. https://www.nass.usda.gov/Statistics_by_State/California/Publications/Specialty_and_Other_Releases/Grapes/Acreage/2019/201904gabtb00.pdf. Accessed August 14 2019

U.S. of Agriculture/National Agricultural Statistics Service (USDA/NASS) (2018b) Grape Crush Report, Final, 2017. https://www.nass.usda.gov/Statistics_by_State/California/Publications/Specialty_and_Other_Releases/Grapes/Crush/Final/2017/201703gcbtb00.pdf. Accessed August 14 2019

U.S. Department of Agriculture Foreign Agricultural Service (USDA/FAS) (2018a) Production, supply, and distribution online database. https://apps.fas.usda.gov/psdonline/. Accessed July 1 2018

U.S. Department of Agriculture Foreign Agricultural Service (USDA/FAS) (2018b) Fresh apples, grapes, and pears: world markets and trade. https://apps.fas.usda.gov/psdonline/circulars/fruit.pdf. Accessed July 1 2018

U.S. Department of Agriculture Foreign Agricultural Service (USDA/FAS) (2018c) Raisins: world markets and trade. September 2017. https://downloads.usda.library.cornell.edu/usda-esmis/files/8p58pc92q/cz30pt133/2j62s532b/raiswm-09-22-2017.pdf. Accessed August 17 2019

2 Grape Taxonomy and Germplasm

M. Andrew Walker, Claire Heinitz, Summaira Riaz and Jacob Uretsky

Abstract

This chapter provides a grape breeder's perspective on the *Vitis* germplasm and taxonomic relationships among the species. It reviews current taxonomic perspectives and how the species are organized. It also discusses the evolution of the grape species and the most widely cultivated *V. vinifera*, as it was moved and selected by birds and people. The introduction of *V. vinifera* into the New World had an impact on the North American *Vitis* species and encouraged breeding to combine the fruit quality traits of *V. vinifera* with the disease and pest resistance of the North American grape species. The introduction of pests and diseases from North America to Europe, and from there around the world, had a very large influence on grape breeding both for rootstock and scion cultivars. The chapter focuses on the North American Vitis and their past, present, and future use in grape breeding.

M. A. Walker (✉) · C. Heinitz · S. Riaz · J. Uretsky
Department of Viticulture and Enology, University of California Davis, One Shields Ave, Davis, CA 95616, USA
e-mail: awalker@ucdavis.edu

C. Heinitz
e-mail: Claire.Heinitz@ARS.USDA.GOV

S. Riaz
e-mail: snriaz@ucdavis.edu

J. Uretsky
e-mail: juretsky@ucdavis.edu

2.1 Introduction

Grapes are one of the most widely cultivated and highest value horticultural crops. They are grown throughout the temperate regions of the planet ranging from hot dry desert environments, to tropical climates, to very cold areas where the vines must be buried during dormancy. The fruit is used for wine, table grape, and raisin production. The vast majority of the cultivated grapes are cultivars of *Vitis vinifera* L., which are considered to have the highest fruit quality. Although these cultivars have desirable fruit, a wide range of pests and diseases impacts their cultivation. Fortunately, resistance to most of these pests and diseases exists within the *Vitis* species.

Vitis vinifera has thousands of cultivars, and many are specifically adapted to the wide range of climates these cultivars are grown in. When humans migrated out of Africa into Central Asia and Central and Western Europe, they encountered wild forms of *V. vinifera*. The European forms of wild *V. vinifera* are within the subspecies *sylvestris* and the Asiatic forms are within the subspecies *caucasia*. These wild forms of *V. vinifera* are now rare and were killed by imported pests like grape phylloxera (*Daktulosphaira vitifoliae*) or diseases like powdery (*Erysiphe necator*) and downy mildew (*Plasmopara viticola*), and severe grazing pressure from goats and sheep. One of the key defining characteristics of wild *V. vinifera* is that they, like

D. Cantu and M. A. Walker (eds.), *The Grape Genome*,
Compendium of Plant Genomes, https://doi.org/10.1007/978-3-030-18601-2_2

all other *Vitis* species, are dioecious with male or female flowered vines. Hermaphrodism was likely one of the first traits selected for in ssp. *sylvestris* and ssp. *caucasia* populations (see Chap. 3). This is a rare trait that would have been self-limiting given the highly heterozygous nature of grape and the exposure of deleterious recessive alleles when the hermaphroditic seed was germinated. However, the advantages of hermaphrodism—excellent fruit set, larger clusters and berries—and the fact that the preferred *vinifera* selections were easily rooted and maintained clonally made hermaphrodism a desired trait.

In wild populations of *V. vinifera* and the vast majority of *Vitis* species, the individuals are composed of heterozygous male (M) female (F) or homozygous FF at the sex locus resulting in populations of progeny that are half male and half female. Pollination is by bees or wind. If you sow seeds from a *Vitis* spp. cross in a greenhouse or nursery row, the progeny will be 50/50 MF. However, in the wild male (MF) individuals far outnumber female (FF). This imbalance is most likely due to the physiological sink created by the developing fruit and seed in female vines, which prevents adequate starch storage in the trunk and roots, and preparation for untimely abiotic stress such as drought or cold. This imbalance is further encouraged by grazing animals attracted by the fruit on female vines in summer and fall.

The dioecious nature of wild grape species helps maintain heterozygosity and genetic diversity. It also increases the chances of hybridization and development of hybrid forms where sympatric species exist. The grape berry is also a mechanism for migration. The berries ripen during the fall in time for the southbound migration of birds. This migration favors the small-berried species as they are the most attractive to birds, which only peck at the larger-berried *Vitis* spp. The seeds that these birds carry and deposit are agents of migration and can introduce unique individuals and their alleles to populations they fly by or visit.

Vitis vinifera cultivars were introduced to North America by the first Europeans who sailed across the Atlantic. Few of these cultivars survived introduction into their new environment. They were killed by grape phylloxera, mildew diseases, and cold winter temperatures. To the south, Pierce's disease was an additional and common killer of newly introduced *V. vinifera* cultivars. There are examples of chance hybrids between female flowered wild vines and *V. vinifera* pollen. "Concord" was such a chance hybrid between *V. labrusca* and *V. vinifera* and became an important commercial success. These chance and intentional hybrids have been recognized with their own cultigen name *V.* x *labruscana*. There were also apparently chance hybrids between *V. aestivalis* and *V. vinifera* such as "Norton" ("Cynthiana") and "Jacquez" ("Black Spanish" or "Lenoir") in the southeastern USA.

The first *V. vinifera* cultivar to survive in North America for an extended time was "Mission" now known to be "Listan Prieto" (Milla Tapia et al. 2007), a grape from the Canary Islands where westward-bound explorers and settlers stopped for water, food, and citrus before heading to the New World. This cultivar was imported into central Mexico, likely with Muscat of Alexandria and from there it moved west and south with missionaries from the Catholic Church. "Mission" was the first *V. vinifera* cultivar in California and is still grown to a limited extent (408 acres in 2017). It is also grown in Peru, Chile, and Argentina, where it is known as "Criolla" or "Pais" and exists in several forms. One of the most interesting is a hybrid of Criolla x Muscat of Alexandria that is widely grown as "Torrontes" or "Torrontes Riojano" (Agüero et al. 2003).

Almost all European *V. vinifera* cultivars are homozygous or heterozygous for hermaphrodism (HH or HF). Olmo states that the Central Asian cultivars have a higher percentage of female (FF) individuals (Olmo 1995). In warm dry climates and close plantings to encourage cross-pollination, these female cultivars are effectively pollinated. However, female vines are usually selected against due to poor set.

The *V. vinifera* cultivars were selected and bred regionally. Negrul (1938) was the first to sort them into three mostly geographic groupings. These groups also have morphological (ampelographic) and genetic connections. Negrul called them proles—occidentalis for western European forms,—orientalis for the Middle Eastern Central Asian forms, and—pontica for the eastern European and intermediate forms.

There have been many germplasm studies examining the genetic relationships among the *V. vinifera* cultivars. These studies confirm many of the ecogeographic groupings previously based on leaf and fruit morphology. One of the most interesting aspects of these studies is proof that many of what were considered to be ancient cultivars that evolved in a given region are in fact crosses between other often well-known cultivars (e.g., Bowers et al. 1999). And the vast majority are crosses of cultivars with self-fertile hermaphroditic flowers. Thus, someone intentionally made these crosses and had to emasculate the maternal parent otherwise the progeny would have been selfings.

The *Vitis* species are thought to have three centers of origin: the region between the Black Sea and Caspian Sea—the "Fertile Crescent," North America, and Asia. There is one species in the Fertile Crescent *V. vinifera* ssp. *sylvestris* and ssp. *caucasia* and the many feral and introgressive forms of these dioecious species in crosses with the cultivated and feral hermaphroditic forms. Olmo collected *V. vinifera* ssp. *sylvestris* across Afghanistan, Iran, and Iraq in the 1950s and many of these male accessions produce straggly clusters with a few berries and viable seeds. They are clearly not cultivars, but they highlight the problem of defining specimens as true forms of *V. vinifera* ssp. *sylvestris* as opposed to feral forms of cultivated *V. vinifera* or hybrids between these taxa. *Vitis jacquemontii* (*V. lanata*) is also found on the western edge of Middle East and along the southwestern foothills of the Himalayan Mountains. There are accessions of this species that also appear to be feral forms of *V. vinifera*, and one (DVIT 1815) collected by Olmo that appears to be correctly identified.

This confusion between cultivated and wild forms of *V. vinifera* also occurs in North America where viticulture encroaches on the native *Vitis* species. This is particularly true with *V. californica* in California, where there appear to be very few pure wild forms and instead most wild grapevines appear to be *V. californica* x *V. vinifera* hybrids (Dangl et al. 2015). The ornamental grapevine "Rogers Red" is a good example (Dangl et al. 2010). This hybridity also occurs in southern California where most of the *V. girdiana* are hybrids with *V. vinifera* (Wada 2008). It also occurs in old mining camps across the southwestern USA. These camps are often associated with springs or permanent sources of water. The southwestern US *Vitis* species found in such areas as are often hybrids with *V. vinifera* cultivars that were brought and grown there by miners and settlers. These hybrids may be prone to diseases that their progenitor wild species resisted, and they also bring the hermaphroditic allele into the wild dioecious forms. These hermaphrodites produce more fruit and seeds and over time reduce, or eliminate, the members of pure native species populations.

The Journal of Systematics and Evolution recently published a special issue on the systematics of Vitaceae. The family was organized into 16 genera and about 950 species (Lu et al. 2018). The introductory paper of a recent conference on Vitaceae ends with this statement "Most genera of Vitaceae need to be taxonomically revised with new bioinformatic tools (Wen et al. 2018) and further integrative systematic studies are especially needed for *Ampelocissus* Planch., *Cayratia* Juss., *Cissus* L., *Cyphostemma*, *Tetrastigma*, and *Vitis*" (Lu et al. 2018).

Vitis is often divided into two subgenera *Vitis* and *Muscadinia*, but others prefer to keep the two taxa as separate genera. Examinations of genomic differences between *Vitis* and *Muscadinia* clearly separate the two taxa but not to the extent to which other genera in Vitaceae diverge (for example, *Ampelopsis*, *Parthenocissus*, or *Cyphostemma* are genetically and morphologically distinct from *Vitis* or *Muscadinia*. *Vitis* and *Muscadinia* species look similar, but *Vitis* spp. have striate bark that shreds in strips, shoots with discontinuous pith (diaphragms at the nodes), branched tendrils, and $2n = 38$

chromosomes. *Muscadinia* species have stellate (non-shredding) bark with lenticels, continuous pith (lack diaphragms), simple unbranched tendrils, and $2n = 40$ chromosomes. This last feature is of great importance. Crosses within *Vitis* species are interfertile, as are crosses within *Muscadinia* species. Crosses between the two taxa are difficult and more successful with *Vitis* as the maternal parent, as the pollen tubes of *Vitis* grow poorly if at all in *Muscadina* styles (Lu and Lamikanra 1996). However, these offspring are almost all sterile with 39 chromosomes.

Overcoming this sterility has been a target of breeder efforts for over 100 years as these hybrids would have the potential to possess the outstanding resistance that *M. rotundifolia* has to pests and diseases (phylloxera, root-knot and dagger nematodes, fanleaf virus, powdery and downy mildew, and Pierce's disease). Fortunately, breeders in North Carolina (Dearing 1917; Detjen 1919), Olmo (1971, 1986) at the University of California Davis, and Bouquet (1980) kept at efforts to produce fertile *Vinifera* x *Rotundifolia* (VR) hybrids. These fertile hybrids have been used to introgress *M. rotundifolia* genes for powdery and downy mildew resistance into *V. vinifera* backgrounds, although they have not been as successful in breeding Pierce's disease (PD) resistance across multiple generations of backcrosses.

Vitis has about 70 species in the Northern Hemisphere with the two major centers of origin in North America (about 30 species) and the other eastern Asia. The East Asian species have recently been detailed in the Flora of China (Chen et al. 2007), which lists 37 species (30 as endemic) and many varieties. These species are also discussed in the special issue of the Journal of Systematics and Evolution 56 (2018). The review presented here will focus on the North American *Vitis* because of their widespread use in breeding for pest and disease resistance (at the rootstock and scion level), and for abiotic stress adaptation (primarily lime, drought, and salt tolerance). They will continue to have great importance in breeding.

The North American *Vitis* were divided into nine series by Munson (1909) in which he included 26 species. Bailey (1934) divided *Vitis* into six series and included 30 species. He also discusses the history of *Vitis* taxonomy, nomenclature, hybridity, and variation. He recognized the latter issues as potential problems to defining North American *Vitis* and stated "The North American *Vitis* are difficult to confine in a key!"

Galet (1988) divided *Vitis* into 11 series and included 32 North American species, and two series of Chinese species (Flexuosae and Spinosae) in which he included 23 species and a telling "etc." after the listing within Flexuosae. More recently, the Chinese Vitaceae have been described in the Flora of China (Chen et al. 2007), which details 37 species and many varieties. The Chinese *Vitis* species are beyond the scope of this review, but a few will be dealt with when appropriate for their specific disease resistances. Moore and Wen in the Flora of North America (http://www.efloras.org/florataxon.aspx?flora_id=1&taxon_id=134649) recognized 19 species including *V. rotundifolia* and *V. vinifera*, but they did not include Mexican species that have been described by Comeaux (1987a, b, 1991), and Comeaux and Lu (2000).

Comeaux et al. (1987) arranged the North American *Vitis* into series following Munson, although Comeaux combined Labruscae and Coriaceae. He listed 6 Series, added *Muscadinia* as a sub-genus, and listed 25 species, 5 stable hybrid forms, 7 species from Mexico and subdivided *V. aestivalis* into 5 varieties and *V. cinerea* into 4 varieties. His grouping makes good functional sense, accommodates the wide variation seen in *Vitis* species, and recognizes the Mexican species (although they need additional attention).

Many of the *Vitis* species are sympatric with one or more species and all are fully interfertile with the exception of *Muscadinia*, which form sterile hybrids with *Vitis* species. When collecting grape in the wild, it can be difficult to identify a specimen given natural variation and the high potential for hybrids where the ranges of two or more species overlap. One key habitat restriction for US *Vitis* is water. It has to be readily available through high levels of rainfall, rivers, streams perennial creeks, springs, and catchment basins.

Many of these watershed areas are under human-related habitat destruction. The impact on water and the resulting habitat restriction is even more clear west of the 98th Meridian that runs north south near Dallas TX. To the west of this line, rainfall is less abundant and drought common. To the east, rainfall amounts are greater and more consistent. Water in the western USA is also critical for birds and they are the primary dispersal agent for *Vitis* seed.

Grape habitat is also under attack by highway departments all over the south and Midwest where the abundant vegetation is mowed or sprayed to convert the roadsides into grass swards for easier maintenance via broadleaf herbicides and mowers. Most grape species can tolerate some mowing but broadleaf herbicides are very damaging. Many of the grapes I have collected from roadsides in the past 20 years have had herbicide damage on them. Urbanization of the southwest and south is also negatively impacting grapevines. Cities have greatly expanded and it is hard to find the grapevines that Grapevine Texas was named for. *Vitis shuttleworthii* was relatively common in central Florida 25 years ago and now development in and around Tampa and Orlando have greatly reduced its habitat. Large ranches existed across Texas and more and more are being subdivided, fenced and sold as smaller ranchettes, and occupied by people who remove grapes from fences and trees.

The *Vitis* species are kept separate by differences in their habitat preferences, geographical barriers, and phenological differences in flowering dates. The desert regions of the southwestern USA create the best geographical barriers, but even in these desert environments, where permanent water sources exist wild grapevines will usually be found. The southwestern deserts are interspersed with tall, widely scattered mountains. These mountains (called sky islands) act as barriers for rain clouds and are far more mesic environments than the surrounding deserts would indicate. *Vitis* species are found on the north- and east-facing slopes of sky islands or wherever permanent springs or streams exist in the desert regions.

2.2 North American *Vitis*

The next section of this chapter describes the North American *Vitis*. They will be grouped into six series following Comeaux (1984).

2.2.1 Series Labruscae

This group contains large-berried species with thick leathery leaves and dense wooly tomentum. Their berries are not swallowed by small birds so the seeds have limited potential for long-distance dispersal compared to that of smaller-berried grape species.

***Vitis labrusca* L.** (the fox grape, swamp grape, and northern muscadine) grows across the eastern USA in moist, sandy alluvial soils. It is the only *Vitis* species with continuous tendrils (at every node rather than the tendril tendril skip seen in most species) it also has small prickles on first year shoots. The fruit has a strong distinctive smell of methyl anthranilate and can be very astringent until well ripened. This species has a very long history of use and many natural and selected hybrids of *V. labrusca* x *V. vinifera* occurred and gave rise to the eastern US grape industry (Hedrick et al. 1908). Such hybrids created or selected by human activity are called cultigens and the cultigen created by *V. labrusca* x *V. vinifera* crossings is designated as *V.* x *labruscana*. There are many examples of these in hybrids such as "Concord" and "Catawba" (Huber et al. 2016). These chance hybrids would have been relatively common as *V. vinifera* pollen from vines dying of cold weather or diseases, brought about by overly humid summers, was blown or carried by bees to the pistillate flowers of wild *V. labrusca*. *Vitis* x *labruscana* cultivars are relatively easy to root from hardwood cuttings, likely due to their *V. vinifera* parentage. Pure *V. labrusca* like most North American *Vitis* can be difficult to root from hardwood cuttings. *Vitis labrusca* can also be found hybridized with *V. riparia* to create *V.* x *novae-angliae* Fewrnald, which has less astringent fruit and is more easily rooted from dormant cuttings than *V. labrusca*. This variety "Clinton" is an early variety of this cultigen used in attempts to combat the phylloxera crisis in France but it was soon replaced with better quality hybrid direct producers.

***Vitis candicans* Engelm.** is known as the mustang grape in Texas and is a very common vine. It is very strong liana and is one of the few *Vitis* species able to outcompete the tree it grows upon. It is abundant on fences and trees across central to eastern Texas and is found from the southern to the northern borders of the state. It is not found in west Texas where it is likely too dry. This species is hard to root from dormant cuttings, but makes a vigorous and long-lived vine and can grow to heights of 100 feet. *Vitis candicans* is also known for its huge crops of large berries that are used for home/garage winemaking. However, the berries are very astringent and have calcium oxalate raphides in the juice which can irritate the back of the throat until the berries are fully ripe. The mustang grape is prone to attack by leaf-folder moths (*Desmia funeralis*), which can defoliate wild vines. *Vitis candicans* is a source for strong nematode resistance as can be seen in hybrids produced from it. However, its excessive vigor and poor rooting ability have limited its use in breeding.

***Vitis shuttleworthii* House** (*V. coriacea*) might be best considered as an eastern form of *V. candicans*. This grape was once common across Florida and was known as the Florida, leatherleaf, or calloossa grape; the last common name referring to the Native American tribe of southwest Florida. This species is now rare, a victim of urban development. The fruit is less astringent than *V. candicans*. It has been used to breed PD resistant wine (Mortenson) and table (Fennell) grapes. The leaves are very distinctive with dark green upper surface and wooly white tomentum below, making it very easy to spot along the backroads! *Vitis candicans* has very tough skins and the berries remain bitter and under ripe until late in the season. The berries often drop to the ground in large numbers where they seem to ripen and attract small animals. Munson (1909) reports that *V. shuttleworthii* berries are less astringent than *V. candicans* and that birds are fond of the fruit "scarcely allowing it to ripen."

***Vitis* x *doaniana* Munson** (Doan's grape) is a hybrid of *V. candicans* x *V. acerifolia*. This grape grows along the Red River border of Texas and Oklahoma and is common south of Lawton OK and north of Wichita Falls, TX. *Vitis* x *doaniana* has leaves and stems with the dense tomentum of *V. candicans* and thick leathery leaves with distinct, sharp teeth on the margins. The berries are small and the teeth on the leaf margin are more like *V. acerifolia* than *V. candicans*. Selections of this species have excellent nematode resistance (*Meloidogyne* spp. and *Xiphinema index*) and excellent chloride tolerance. It is a very vigorous vine and propagates more easily than *V. candicans*.

Vitis candicans also produces natural hybrids with *V. rupestris* creating *V.* x *champinii*. ***Vitis* x *champinii* Planch**. (Champin's grape) is the parent species of the grape rootstock "Ramsey" (incorrectly called "Salt Creek," which is a cultivar of *V. doaniana* (Loomis and Lider 1971) and DogRidge, are both selections of Munson's (1909). *Vitis* x *champinii* is found across Texas but is no longer common. I have collected it from Del Rio, TX, north of San Antonio, west of Austin and around Fort Hood east to Temple. Many of the forms in the Hill Country of central Texas appear to be *V. candicans* hybrids with *V. monticola* or *V. berlandieri*. The increasing scarcity of *V.* x *champinii* may be related to the lack of *V. rupestris*, which is now very rare in Texas and only common in southern Missouri. *Vitis rupestris* is more of a small shrub than a vine and is prone to severe grazing damage from cattle. It appears that new accessions of *V.* x *champinii* may not be forming because of the greatly limited range of *V. rupestris*.

2.2.2 Series Aestivales

***Vitis aestivalis* Michx.** (summer grape or pigeon Grape) is a variable species that ranges across the eastern USA from northern Texas east of the 98th meridian to Florida and north to Michigan and east to New York. This species has relatively large, fruity berries with a more pleasant flavor than *V. labrusca*. The leaves have short rounded teeth, are often 3 (occasionally 5) lobed, and usually have glaucous abaxial surfaces, and the stems are round in outline. There have been many named varieties of *V. aestivalis* to account for the morphological differences within this species. Munson recognized *V. lincecumii* Buckley, *V. bicolor* Leconte. (*V. argentifolia*), *V. aestivalis,* and *V. simpsonii* Munson. The Flora of North America merges all of these taxa under *V. aestivalis*. Other important forms include *V. rufotomentosa* Small, which is distinguished by its high resistance to the dagger nematode, *Xiphinema index* (Kunde et al. 1968).

Three varieties of *V. aestivalis* were recognized by Moore and Wen (2016): *V. aestivalis* var. *aestivalis, V. aestivalis* var. *lincecumii*, which is relatively common in northeast Texas north to Oklahoma, and Arkansas, and east to Louisiana; *V. aestivalis* var. *bicolor*, which is scattered across the central and northern states; and var. *aestivalis* which is found across the central and eastern states. This species, particularly var. *lincecumii* (the Post Oak Grape) has been extensively used in table and wine grape breeding and was the foundation of Munson's breeding efforts. It also gained fame in its hybrid form with *V. vinifera*—*V.* x *bourquina* Munson ex Viala (also known as *V. bourquiniana*). "Lenoir" (aka "Jacquez," "Black Spanish") is a cultivar of *bourquina* that has performed well against Pierce's disease in the southern USA, although new selections from hybrids of *V. vinifera* and *V. arizonica* have far better quality and PD resistance and are nearing release (Riaz et al. 2018). Jaeger crossed *V. lincecumii* with *V. rupestris* to create "Munson" (also known as Jaeger 70), which can be found as the source of disease resistance and dark color in many of the hybrid direct producers produced in France to combat Phylloxera in the late 1800 s (Viala and Ravaz 1903).

Vitis aestivalis var. *argentifolia* Munson (also known as var. *bicolor*) is found across the northeastern USA in wetter areas. This grape has smaller berries than the other taxa of this series and has not been used as much in breeding. In the southeastern USA, the tomentum on *V. aestivalis* becomes more rufous and the vines seem to intergrade with *V. cinerea*. These red-brown forms of *V. aestivalis* include *V. simpsonii*, *V. smalliana*, and *V. rufotomentosa*.

A new Mexican species *V. nesbittiana* (Comeaux 1987a, b) was found in the cloud forests near Xalapa, Veracruz, Mexico. It has bicolor leaves with glaucous lower leaf surfaces. This species seems to have a relatively restricted range, but it may be more widely distributed.

The *V. aestivalis* forms are not common over much of their range and are usually in soils that drain well or in sunny areas on ridges. They are not as common as the ubiquitous *V. riparia*, *V. vulpina,* and *V.* cinerea.

2.2.3 Series Cinerascentes

Vitis cinerea* var. *cinerea Engelm ex Millardet (the ashy-leaved grape) is a common species across the southern USA. The leaves are usually cordate, with short teeth and white/gray tomentum on the abaxial leaf surfaces (hence the common name the ashy-leaf grape) and felty white young shoots. The berries are black and relatively small, and the clusters are very loose and long conical in shape. The vines are very vigorous and climb high in trees. This species group tends to be very difficult to root and the fruit ripens very late and remains very acidic until ripe. It is very common from eastern Texas to Florida and on the banks and bottomlands of the Missouri and Mississippi Rivers. It thrives in moist soils.

There are three varieties within *V. cinerea*: *V. cinerea* var. *helleri* (L.H. Bailey) M.O. Moore, var. *baileyana* (Munson) Comeaux, and var. *floridiana* Munson. They are spatially separated from *V. cinerea* var. *cinerea*. ***Vitis cinerea* var. *helleri*** or ***V. cinerea* var. *berlandieri* (Planch.) Comeaux** is more commonly, but less correctly known as ***V. berlandieri***—one of the three *Vitis* species that the French used to create the rootstocks that solved the phylloxera crisis. This crisis resulted from the importation of grape phylloxera (*Daktulosphaira vitifoliae*) from the USA into

France in the 1850 s. This grape root aphid rapidly spread and killed the own-rooted *V. vinifera* cultivars—all of which are very susceptible to the root feeding of this insect. French scientists reasoned that grafting *vinifera* cultivars on American *Vitis* species that had evolved with this insect would solve the problem. They had many American *Vitis* species to work with, but only two root easily from dormant cuttings: *V. riparia* and *V. rupestris*. These two species were used as rootstocks onto which the European *V. vinifera* cultivars were grafted and the vineyards were replanted. However, in a few years, many vineyards began to decline because the *V. riparia* and *V. rupestris* rootstocks were unable to supply the iron needs of the *vinifera* scions on the common limestone-based soils. To solve this problem, the French contacted Munson who suggested that *V. berlandieri* might help with lime tolerance. This species is found on the limestone soils of central Texas. Unfortunately, *V. berlandieri* roots poorly from hardwood cuttings so French scientists and nursery owners began hybridizing *V. berlandieri* with *V. riparia* and with *V. rupestris* to create the phylloxera resistant, lime-tolerant rootstocks that are still used.

The current range of *V. berlandieri* is mostly confined to the Edwards Plateau in central Texas. The Brazos River serves as the delineation between *V. cinerea* and *V. berlandieri*. It is also reported to be found in northern Mexico, although these populations are diminishing from over-grazing, and rerouting of water sources and rights along the Rio Grande. It can be found from Fort Davis in west Texas to the west of the Brazos River in eastern Texas, but is most abundant in the Hill Country west of Austin. *Vitis berlandieri* in Texas is being threatened by the expansion of Austin and its suburbs and by the subdivision of what were large ranches on 1000 s of acres to small "ranchettes" of 1 to 5 acres. Overly zealous roadside brush removal by owners of these ranchettes, and by highway departments, is making *V. berlandieri* much less common than it was in the recent past. This species has great breeding value. It is a source of drought, chloride, and lime tolerance and has strong resistance to phylloxera and Pierce's disease.

***Vitis cinerea* var. *baileyana* (Munson) Comeaux** is an eastern form of *V. cinerea*. It grows at moderate elevations along mountain streams. Munson (1909) suggests that it is the "connecting link between" *V. vulpina* and *V. cinerea* and it does have a mixture of both species' appearances. This species roots poorly and its fruit is very acidic until ripe. It has not been used in breeding.

***Vitis cinerea* var. *floridiana* Munson** is a form of *V. cinerea* from the southeastern USA. Several taxa have been included within this variety, *V. simpsonii, V. rufotomentosa*, and *V. sola*, their leaf shape and degree of tomentum on the stems and leaves are variable. Accessions of *V. rufotomentosa* have proven to be good sources of dagger nematode resistance and have been used in the UC Davis grape rootstock breeding program.

The next set of *V. cinerea* taxa are from Mexico. These species are not well known, although there are ethnobotany studies on the fruit (Franco-Mora and Cruz-Castillo 2012; Tobar-Reyes et al. 2009). ***Vitis bourgaeana* Planch.** is from the central east states of Mexico and south into Central America. Its leaves are deeply 5 to 7 lobed. It has not been used in breeding. ***Vitis biformis* Rose** is scattered across Mexico and its leaves are also lobed, and it was been described as similar to *V. berlandieri* with shorter teeth. *Vitis biformis* was originally collected northwest of Mexico City in Guanajuato. It is not recognized as a valid taxon in the "Checklist of the native vascular plants of Mexico" (Villaseñor 2016). ***Vitis tiliifolia* Humb. & Bonpl. ex Schult.** (*V. caribaea*) is found from central Mexico, south to northern South America, and throughout the Caribbean. It is the most widespread of the Mexican *Vitis* according to Villaseñor (2016) who lists it in 24 states. Some of these incidences may have been the result of similarities between other members of this series. This species has been used in breeding for PD resistance, but stronger forms of PD resistance exist in other Mexican and US *Vitis*. ***Vitis peninsularis* Jones** is found scattered across Baja California Sur. ***Vitis blancoii* Munson** has been re-evaluated by Comeaux and Lu (2000) and is found in southern Mexico along streams at high elevation (3500 to 7000 feet). Forms of this species in northern Mexico and the USA are now considered to be *V. cinerea* var. *tomentosa* (Planchon) Comeaux. This taxon also intergrades with V. *berlandieri* (*V. cinerea* var. *helleri*) producing intermediate forms.

2.2.4 Series Vulpinae

There are two species in the Vulpinae series: *V. vulpina* (*cordifolia*) and *V. palmata* (*rubra*). ***Vitis vulpina* L.** (the frost grape) has a very wide distribution from Texas to Florida and north to Pennsylvania and Missouri. This species appears to be very similar to *V. riparia* and they have been confused and misnamed since being originally described. The shoot tips of *V. vulpina* are open while those of *V. riparia* are enclosed in the young leaves. The stems of *V. vulpina* are smooth and waxy while those of *V. riparia* often have short bristles, the fruit of *V. vulpina* ripens very late, and the vines of *V. vulpina* are much larger climbing high into trees. In addition, although their ranges overlap, *V. vulpina* is most abundant in the south and *V. riparia* is more northerly and widespread in the northern states and Canada. Comeaux, Munson, and Bailey grouped ***V. palmata* Vahl** (*V. rubra*) with *V. vulpina*. *Vitis palmata* (the catbird grape) is uncommon and grows along the Mississippi and Missouri rivers in or near side streams and swampy areas south to east Texas where I have collected it near Canton City. It is a very attractive vine with red stems and new shoots and could be regarded as an ornamental. It appears to be a weak *V. riparia* with 3- to 5-lobed leaves and long sharp teeth. It has not been used in breeding; although it roots well from dormant cuttings, it is a relatively weak vine.

2.2.5 Series Precoces

The Precoces Series houses *V. riparia*, *V. rupestris*, and *V. acerifolia*. All of these species root well from dormant cuttings, and *V. riparia* and *V. rupestris* root almost as well as *V. vinifera*. As mentioned above, ***V. riparia* Michx.** (the riverside grape) is very widely spread and is a major part of the flora of the northern states. It can be found along the eastern slope of the Rocky Mountains south to northern New Mexico, where it hybridizes with *V. arizonica*, and east to the Atlantic coast. It is very common in the northeastern USA.

***Vitis rupestris* Scheele** (the sand or rock grape) was once common in rocky creek beds and sand bars and was found from Pennsylvania to the Rio Grande in Texas. It is now almost entirely restricted to southern Missouri and eastern Kansas and a few rare sites in Oklahoma and Texas. This species usually grows as a shrub or small vine. This growth habit led to its demise from much of its range as cattle grazed it to extinction while they moved west. It is now found in isolated areas protected from cattle and in only a small part of its original range. *Vitis rupestris* is also unusual because of its deep, penetrating root system. These roots allow *V. rupestris* to survive in highly erosive streams where intense stream flows wash away silts, sand and plants leaving coarse gravel. *Vitis rupestris*' shrubby habit and very deep roots prevent it from being swept or eroded away. It roots very well from dormant cuttings, but produces short canes with short internodes, and frequent lateral shoots. As a rootstock, it promotes vigorous growth and is drought adapted due to its ability to extract water more deeply in the soil profile. *Vitis rupestris* also has strong foliar disease resistance and very dark black-red juice, leading to its use as a parent in the breeding of the hybrid direct producers in France. It has also been used to breed red-juiced teinturier grape varieties for blending and concentrate production (e.g., "Rubired" and "Scarlet"). This important species is at risk of extinction as its range continues to shrink.

***Vitis acerifolia* Rafinesque** (*longii*) is found from west and north Texas, across Oklahoma to Kansas and westward to southeast Colorado and northeast New Mexico. This species' range and its appearance suggest that it is a hybrid form between *V. riparia* and *V. candicans* with additional introgression from *V. arizonica*. As mentioned above, *V.* x. *doaniana* is the result of hybridization of *V. acerifolia* x *V. candicans*. The common name of *V. acerifolia* is the canyon grape and it often found in dry heavily eroded sandy creek beds, where its deep plunging roots mine water and hold it in place. This species roots well from dormant cuttings and has good resistance to calcareous soils, and we have found it has good drought tolerance and excellent chloride tolerance, but it has only moderate resistance to phylloxera. This last trait has limited its use as a grape rootstock, but with careful selection, it will be possible to utilize its beneficial traits. *Vitis acerifolia* is another species that is threatened by state highway departments. These departments are making conscious effort to convert the roadsides and shoulders to grass swards and away from their current mixture of woody plants, herbs, and grasses, which often contains grape species. On a recent collection trip pursuing *V. acerifolia* from Amarillo to Kansas, almost every plant we collected from had obvious broadleaf phenoxy herbicide damage.

2.2.6 Series Occidentales

The next series of grape species contains some of the least well understood and studied North American taxa. There are two California grape species in this group: *V. californica* and *V. girdiana*. ***Vitis californica*** **Benth.** ranges from the Tehachapi Mountains in the southern San Joaquin Valley to southern Oregon, and I have found a large vine near a creek south of Eugene, OR. Within California, this species grows from about 1200 m elevation in the Sierra Nevada mountain range to the Coastal Range. It is not found within about 15 km of the ocean, presumably because it is too cold in the spring to set seed. This species is almost extinct in California due to the frequent hybridization with *V. vinifera* and their hermaphroditic progeny's ability to out compete the native vines. The first generation of these crosses occurs in the same way mentioned above in regard to *V.* x. *labruscana* (*V. labrusca* x *V. vinifera*) and *V.* x *bourguina* (*V. aestivalis* x *V. vinifera*). These unintentional crosses would often involve female flowered wild vines and pollen from hermaphroditic cultivated grapes. The hybrid progeny will be either 50 or 100% hermaphroditic in the next generation and capable of producing far more seed. This genetic erosion has been studied with *V. californica* collections made across its range—DNA analysis revealed that the vast majority were hybrids with *V. vinifera* (Dangl et al. 2010; Wada 2008).

The hybrid types have lobed leaves, less tomentum on leaves and stems, perfect flowers, large berries, and well-filled clusters. This extinction of *V. californica* likely first began with the Spanish missionaries who came north from Mexico, establishing 21 Missions and planting the Mission (*V. vinifera* cv. Listan Prieto) grape. They had already moved this grape across Mexico, damaging the wild species there too. Munson (1909) mentions the presence of these *V. californica* x *V. vinifera* hybrids. Interestingly, the lack of phylloxera resistance of *V. californica* is often emphasized, and however, when we tested accessions that looked like pure *V. californica* they were resistant, while the hybrids were susceptible (Grzegorczyk and Walker 1998). Accessions of *V. californica* may have resistance to diseases like Armillaria root rot as they coevolved in wooded areas. However, many of the now hybrid *V. californica* will be compromised by the pest and disease susceptibility of *V. vinifera*.

The southern California wild grape, ***V. girdiana*** **Munson** ranges along the Pacific Coast from Santa Barbara (including the Channel Islands to Baja California Sur and east through southern Nevada and north to southwestern Utah. It was named after H.H. Gird who sent Munson specimens he collected from near Fallbrook, CA (Munson 1909). There are still many *V. girdiana* plants scattered across Fallbrook where Gird Ranch still exists. Across the coastal portion of its range, this species is threatened by habitat destruction. Most recently, the renovation of Hwy76 has brought more people and disrupted the wild grape habitat along the San Luis Rey River.

Vitis girdiana has been considered a variant of *V. californica*, but it is unique (Wada and Walker 2009). These two species are kept separate by the

Tehachapi Mountain Range that separates the northern and southern portion of the state, although we found a population of *V. girdiana* x *V. californica* hybrids in the southeastern Sierra Nevada east of Isabella Lake (Wada 2008). The leaves and stems of *V. girdiana* are more tomentose, the berries less glaucous more black, and the leaves are more three-lobed. *Vitis californica* has less tomentose leaves and stems, larger more glaucous berries, and the leaves are more rounded. As is the case with *V. californica*, hybrids with *V. girdiana* and *V. vinifera* are common and occur across its range. There are reports of very large grapevines planted on overhead arbors at many of the Catholic missions along the southern California coast. From the descriptions, some of these vines are the Mission grape variety mentioned above. However, some of these famous Viña Madre vines had long loose clusters of small black berries, were more tomentose, and were likely *V. girdiana* x "Mission" hybrids. Contrary to Munson's belief that "Little or nothing, probably, of value, can be gained in any way from this species," we have found that desert forms can have high levels of chloride tolerance and strong resistance to Pierce's disease.

***Vitis arizonica* Engelm.** is a quite variable and appears to intergrade with *V. girdiana* in its western range, with *V. riparia* in its northeastern range, and *V. acerifolia*, *V. candicans*, and *V. cinerea* in its eastern range. Pure forms of *V. arizonica* have small tomentose gray-green cordate to partially 3-lobed leaves. Individuals and populations with mostly glabrous leaves exist (*V. arizonica* forma *galvinii*, *V. arizonica* var. *glabra*, and *V. treleasei*) and can be hard to distinguish from each other. As a group, the vines are usually brushy with short shoots. Although it grows in the arid southwestern USA and northern and northwestern Mexico, it is usually found in relatively mesic areas on the eastern and northern flanks of "sky island" mountains and near springs, streams, and catchments. Although Munson stated that "…nor does there seem much of value in this species," we have found it to be a very valuable source of resistance to Pierce's disease (Riaz et al. 2006, 2007, 2018; Krivanek et al. 2005), and the dagger nematode (*Xiphinema index*) vector of fanleaf disease, and of chloride tolerance. We are currently examining a large collection (over 700 accessions) of southwestern *Vitis* I have collected for resistance to PD and chloride tolerance. The genetic diversity of this collection has been studied with SSR markers (Heinitz 2016) and are now part of a whole genome resequencing project to optimize grape breeding and characterize these resistance genes (NSF PGRP grant #1741627).

***Vitis monticola* Buckley** (the sweet mountain grape) was also put in the Cordifoliae (Vulpinae) series by Galet (1988) Moore (1991) and Munson (1909), but in the Occidentales series by Comeaux (1984).

This species is found on Cretaceous limestone hills in the counties west of Austin TX and east of Killeen TX. It grows on rocky ridges without much soil usually nearby *V. candicans* and *V. berlandieri*, both of which are confined to deeper soils. *Vitis monticola* can be found growing on mesquite and juniper in very dry areas—no other grape seems to be as drought tolerant as its habitat suggests. However, it is very hard to propagate from woody cuttings and it appears to tolerate drought by limiting growth, a trait that persists even when it is planted on deep fertile soils here at UC Davis. This species is very distinct and appears a bit like a small form of *V. riparia* or a less acutely toothed *V. palmata*. Hybrid forms with *V. monticola* and *V. candicans* and *V. berlandieri* are relatively common in the Texas Hill Country. *Vitis monticola* is not a common species and is at risk due to the rapid suburbanization of this region of Texas. Munson (1909) noted "It truly is a remarkable and distinct species. It is well worthy of cultivation as an ornamental vine." as is *V. palmata*.

Two new Mexican species are also included in the Occidentales series by Comeaux (1991)—***V. bloodworthiana* Comeaux** and ***V. jaegeriana* Comeaux**. The former was found at higher elevations of the Sierra Madre Occidental (western range) in Sinaloa and Durango, while the latter was found at higher elevations of the Sierra Madre Oriental (eastern range) in San Luis Potosi. These species have been compared to other members of this series and although they appear quite different, they were grouped in this series based on the lenticels found on their fruit. The leaves of these new species are usually narrow cordate, *V. bloodworthiana* leaves tend toward 3 lobed, and *V. jaegeriana* are usually entire. *Vitis bloodworthiana* has dark red pigmentation of the shoot tips, and young leaves and stems, and would make a nice ornamental vine. *Vitis jaegeriana* has a brown to red-brown pubescent shoot tips. No species has been used for grape breeding in the USA, although based on their habitat they should be tested for PD resistance. Both the southwestern US *Vitis* species and the Mexican *Vitis* are in need of thorough taxonomic and genetic analysis.

2.3 Subgenus *Muscadinia*

Muscadinia has been considered to be a subgenus of *Vitis* by all current taxonomists, although some grape breeders felt that its genetic, anatomic, and morphological differences, the strength and breadth of its pest and disease resistance, and the sterility of *Vitis* x *Muscadinia* hybrids supported a generic rank for *Munsoniana*. Harold Olmo (1995) felt *Muscadinia* deserved this status as did Alain Bouquet (Mullins et al. 1992), two of grape breeding's most notable practitioners.

There are three taxa within *Muscadinia*: *M. rotundifolia, M. munsoniana,* and *M. popenoi*. ***Muscadinia rotundifolia*** **Michx**. (the Muscadine grape, white bronze forms are called scuppernong) grows across the southeastern USA from eastern Texas to northern Arkansas, east to Virginia and Florida, and has been reported in the state of Veracruz (Comisión Nacional de Fruticultura 1973). This grape species was cultivated in the USA well before the arrival of European settlers and is the foundation of a fruit industry in the southeastern USA. Modern muscadine cultivars are large berried, fruit more uniformly and are self-fertile (Olien 1990). The species is very resistant to most pests and diseases that affect *V. vinifera* cultivars, but because it has 2n = 40 chromosomes it makes sterile hybrids with all *Vitis* species (2n = 38). There have been a few fertile *V. vinifera* x *V. rotundifolia* hybrids produced (VR hybrids—produced by Olmo (1971, 1986), Bouquet (1980), and Bloodworth et al. (1980), which have been used to introgress resistance to powdery and downy mildew into *V. vinifera* cultivars and breeding lines. A genetic map was created within *M. rotundifolia* from a cross of "Fry" x "Trayshed" (Riaz et al. 2012), and several *M. rotundifolia*-based maps have been created to develop resistance markers for breeding powdery mildew-resistant winegrapes.

Muscadinia **var.** ***munsoniana*** **S. ex M) Comeaux** is largely restricted to the southern half of Florida. It was regarded as a valid species by Munson, but Comeaux and the Flora North America regard it as a variety of *M. rotundifolia*. This form of *rotundifolia* has smaller leaves, sharper teeth, and smaller berries. It is almost ever-blooming in southern Florida. Munson mentioned that *V. munsoniana*'s ever-blooming character allowed it to make hybrids with *Vitis shuttleworthii* and other species due to overlapping blooming periods. He felt it would make an excellent resistant parent in crosses with *V. vinifera*. It may be that Munson did not appreciate the sterility barrier, or perhaps *M. munsoniana* should be examined closely for its potential to form more and better *Vitis* x *Muscadinia* hybrids.

Muscadinia popenoi Fennell is listed in the state of Puebla (Comisión Nacional de Fruticultura 1973) and noted by Galet (1988) in the state of Oaxaca at the Isthmus of Tehuantepec. It is regarded as a separate species and has longer more cordate leaves with shorter teeth. It has not been used in breeding and is poorly known.

References

Agüero CB, Rodriguez JG, Martinez LE, Dangl GS, Meredith CP (2003) The parentage of Torrontés cultivars in Argentina. Am J Enol Vitic 54:318–321

Bailey LH (1934) Vites peculiares ad Americum Borealem. Genet Herb 3:149–244

Bloodworth PJ, Nesbitt WB, Barker KR (1980) Resistance to root-knot nematodes in Euvitis x Muscadinia hybrids. In: Third international symposium on grape breeding, University of California, Davis

Bouquet A (1980) *Vitis* x *Muscadinia* hybridization: a new way in grape breeding for disease resistance in France. In: Proceedings of the 3rd international symposium on grape breeding. University of California, Davis, USA, June 1980, pp 42–61
Bowers JE, Boursiquot JM, This P, Chu K, Johansson H, Meredith CP (1999) Historical genetics: the parentage of Chardonnay, Gamay and other wine grapes of north eastern France. Science 289:1562–1565
Chen ZD, Ren H, Wen J (2007) Vitaceae. In: Wu Z-y, Hong D-Y, Raven PH (eds) Flora of China, vol 12. Science Press, Beijing, pp 173–222
Comeaux BL (1984) Taxonomic studies on certain native grapes of eastern North Carolina. Ph.D. dissertation, North Carolina State University, Raleigh 178p
Comeaux BL (1987a) Studies on *Vitis champinii*. In: Proceedings of the Texas grape growers association 11th annual conference. San Antonio, TX, pp 158–162
Comeaux BL (1987b) A new *Vitis* (Vitaceae) from Veracruz, Mexico. SIDA 12:273–287
Comeaux BL (1991) Two new *Vitis* (Vitaceae) from mountainous Mexico. SIDA 14:459–466
Comeaux BL, Lu J (2000) Distinction between *Vitis blancoii* and *Vitis cinerea* var. *tomentosa* (Vitaceae). SIDA 19:123–131
Comeaux BL, Nesbitt WB, Fantz PR (1987) Taxonomy of the native grapes of North Carolina. Castanea 52:197–215
Comisión Nacional de Fruticultura (1973) Vides nativas de Mexico. Comisión Nacional de Fruticultura, Mexico
Dangl GS, Lou Mendum M, Yang J et al (2015) Hybridization of cultivated *Vitis vinifera* with wild *V. californica* and *V. girdiana* in California. Ecology and Evolution 5:5671–5684
Dangl GS, Raiche R, Sim S, Yang J, Golino DA (2010) Genetic composition of the ornamental grape Roger's Red. Am J Enol Viticult 61:266–271
Dearing C (1917) Muscadine grape breeding: the native grape of the southeastern United States has been hybridized successfully with the European grape—valuable self-fertile varieties produced. A new possibility for the cut-over pine lands of the south. J Hered 8:409–424
Detjen LR (1919) Some F1 hybrids of *V. rotundifolia* with related species and genera. N C Agric Exp Stat Technol Bull 18:1–50
Franco-Mora O, Cruz-Castillo JG (2012) La vid silvestre en México. Estado de México, Universidad Autónoma del Estado de México, Toluca
Galet P (1988) Cépages et vignobles de France Tome 1. Les Vignes Américaines, 2nd edn. C. Déhan, Montpellier [France]
Grzegorczyk W, Walker MA (1998) Evaluating resistance to grape phylloxera in Vitis species with an in vitro dual culture system. Am J Enol Viticult 49:17–22
Hedrick UP, Booth NO, Dorsey MJ et al (1908) The Grapes of New York. JB Lyon Company, Albany
Heinitz C (2016) Characterization of *Vitis* species from the southwest United States and Mexico for breeding and conservation. Ph.D. dissertation University of California, Davis
Huber F, Röckel F, Schwander F et al (2016) A view into American grapevine history: Vitis vinifera cv. 'Semillon' is an ancestor of 'Catawba' and 'Concord'. Vitis J Grapevine Res 55:53–56. https://doi.org/10.5073/vitis.2016.55.53-56
Krivanek AF, Famula TR, Tenscher A, Walker MA (2005) Inheritance of resistance to *Xylella fastidiosa* within a *Vitis rupestris* x *Vitis arizonica* hybrid population. Theor Appl Genet 111:110–119
Kunde RM, Lider LA, Schmitt RV (1968) A test of *Vitis* resistance to *Xiphinema index*. Am J Enol Viticult 19:30–36
Loomis NH, Lider LA (1971) Nomenclature of the 'Salt Creek' grape. J Fruit Var Hort Digest 25:41–43
Lu J, Lamikanra O (1996) Barriers to interspecific crosses between *Muscadinia* and *Euvitis*. Hort Sci 31:269–271
Lu L-M, Ickert-Bond S, Wen J (2018) Recent advances in systematics and evolution of grape family Vitaceae. J Syst Evol 56:259–261
Milla Tapia A, Cabezas JA, Cabellp F, Lacombe T, Martinez-Zapater JM, Hinrichsen P, Cervera MT (2007) Determining the Spanish origin of representative ancient American grapevine varieties. Am J Enol Vitic 58:242–251
Moore MO (1991) Classification and Systematics of Eastern North American *Vitis* L. Vitaceae North of Mexico. Sida 14:339–367
Moore MO, Wen J (2016) Vitaceae. In: Flora of North America Editorial Committee (ed) Flora of North America North of Mexico, vol 12. Oxford University Press, New York
Mullins MG, Bouquet A, Williams LE (1992) Biology of the grapevine. Cambridge University Press, Cambridge
Munson TV (1909) Foundations of American grape culture. Orange Judd Company, Evanston
Negrul AM (1938) Evolution of cultivated forms of grapes. C R Acad Sci USSR 18:585–588
Olien WC (1990) The muscadine grape: botany, viticulture, history, and current industry. HortSci 25:732–739
Olmo HP (1971) Vinifera rotundifolia hybrids as wine grapes. Am J Enol Viticult 22:87–91
Olmo HP (1986) The potential role of vinifera x rotundifolia hybrids in grape variety improvement. Experientia 42:921–926
Olmo HP (1995) Grapes. In: Smartt J, Simmonds NW (eds) Evolution of crop plants. Longman Scientific and Technical, Harlow
Riaz S, Hu R, Walker MA (2012) A framework genetic map of Muscadinia rotundifolia. Theor Appl Genet 125:1195–1210
Riaz S, Huerta-Acosta K, Tenscher C, Walker MA (2018) Genetic characterization of *Vitis* germplasm collected from the southwestern United States and Mexico to

expedite Pierce's disease resistance breeding. Theor Appl Genet 131:1589–1602
Riaz S, Krivanek AF, Xu K, Walker MA (2006) Refined mapping of the Pierce's disease resistance locus, *PdR1*, and *Sex* on an extended genetic linkage map of *Vitis rupestris* x *V. arizonica*. Theor Appl Genet 113:1317–1329
Riaz S, Vezzulli S, Harbertson ES, Walker MA (2007) Use of molecular markers to correct grape breeding errors and determine the identity of novel sources of resistance to *Xiphinema index* and Pierce's disease. Am J Enol Viticult 58:494–498
Tobar-Reyes JR, Franco-Mora O, Morales-Rosales EJ, Cruz-Castillo JG (2009) Contenido de resveratrol en hojas de vides silvestres (*Vitis* spp.) mexicanas. Rev la Fac Ciencias Agrar 41:127–137
Viala P, Ravaz L (1903) American Vines (resistant stock): their adaptation, culture, grafting and propagation. Freygang-Leary Company, San Francisco
Villaseñor JL (2016) Checklist of the native vascular plants of Mexico. Rev Mex Biodiver 87:559–902
Wada EB (2008) Systematics and evolution of *Vitis*. University of California, Davis
Wada EB, Walker MA (2009) Vitaceae. The Jepson Manual 2nd edn. On line at http://ucjeps.berkeley.edu/tjm2/review/treatments/vitaceae.html
Wen J, Lu L, Nie Z-L, Liu X-Q, Zhang N, Ickert-Bond S, Gerrath J, Manchester SR, Boggan J, Chen Z-D (2018) A new phylogenetic tribal classification of the grape family Vitaceae. J Syst Evol 56:262–272

Evolutionary Genomics and the Domestication of Grapes

3

Yongfeng Zhou, Aline Muyle and Brandon S. Gaut

Abstract

We summarize aspects of the domestication of grapevines (*Vitis vinifera* ssp. *sativa*) from its wild ancestor (*Vitis vinifera* ssp. *sylvestris*) by focusing on the first three stages of the domestication process. The first stage is the management of the wild plant by humans, prior to purposeful cultivation. Both archeological and genetic evidence suggest that man interacted with grapes prior to the onset of agriculture. These interactions may have extended to 20,000 year ago (ya) in the Transcaucasus region, the primary center of grapevine domestication. The second stage of domestication is purposeful cultivation. For most annual crops, this stage is defined by a strong bottleneck that winnowed and limited genetic diversity. There is, however, little evidence for the history of a strong bottleneck in grapevines and some other perennial crops. Another feature of the second stage is a positive selection for traits associated with cultivation and harvesting. In theory, the genes underlying these traits can be identified using population genomic approaches. Although these approaches have been applied sparingly to grapevines thus far, they have identified numerous genes as targets of selection that likely contribute to agronomic traits. The third stage of domestication is the geographic dispersal of a nascent crop to new locations, where the crop must adapt to new, local environments. This local adaptation is often facilitated by introgression between the crop and locally adapted wild populations. There is ample evidence to indicate that introgression has been an important process in the evolution of the grapevine germplasm. Unfortunately, however, these introgression events tend to complicate questions about the number of origins of the crop; was there one primary origin of the crop or many? Based on the data available to date, we take the view that there was a single domestication event, but this and many other questions about evolution, domestication, and genomics of grapevine require further investigation.

3.1 Introduction

Among domesticated crops, few are as historically important as the grapevine (*Vitis vinifera* spp. *sativa*; hereafter "*sativa*"). Grapes have been

Y. Zhou · A. Muyle · B. S. Gaut (✉)
Department of Ecology and Evolutionary Biology, University of California Irvine, 321 Steinhaus Hall, Irvine, CA 92617, USA
e-mail: bgaut@uci.edu

Y. Zhou
e-mail: yongfez1@uci.edu

A. Muyle
e-mail: amuyle@uci.edu

D. Cantu and M. A. Walker (eds.), *The Grape Genome*,
Compendium of Plant Genomes, https://doi.org/10.1007/978-3-030-18601-2_3

used as a source of food and wine for centuries, and they have particular significance in rituals and religion. The Old Testament, for example, mentions grapes in its first book (Genesis 9:20), detailing Noah's planting of a vineyard after the flood. Of course, the Greeks had a god, Dionysus, who was responsible for creating wine and spreading the art of viticulture. Our reverence for grapes has not waned over the intervening centuries. Grapes are arguably the most important horticultural crop in the world, with 7.1 million hectares producing 77.4 million tons of fruit globally in 2016 (OIV 2015; Migicovsky et al. 2017) (see Chap. 1). The products of grape cultivation—which include table grapes, raisins, juice, wine, and oil—have a value of $68.3B at the farm gate and contribute an estimated $162B annually to the American economy alone (National Grape and Wine Initiative 2007).

We may take grapes for granted, but their history nonetheless remains somewhat enigmatic. Like all crops, there are myriad unanswered questions about their origin and domestication. Some questions pertain to crop history: Where were grapes domesticated? Was the process of domestication rapid or protracted? Did domestication occur only once, or twice or perhaps independently on several occasions? Other questions pertain to the genetic and phenotypic effects of domestication: What genetic changes occurred and how do these changes affect phenotype? A final series of questions pertain to the domesticators: Who were they? How and when did they disperse their new crop to additional locations? How might the history of the crop parallel their culture?

Of course, we cannot answer all of these intriguing questions in a single chapter. We will instead focus on the use of genetic and genomic data to address some of these questions, and we will occasionally rely on examples from other crops while doing so. To lay a foundation, the chapter will begin with a general overview of the process of domestication, which we consider to have occurred in four stages. From there, we will focus on the specific stages to summarize what is—and is not—known about grapevine domestication and also about the evolutionary dynamics that have shaped genetic diversity of the crop.

3.2 Four Stages of Domestication

Thousands of plant species have been modified morphologically for human use (Meyer et al. 2012)—i.e., they have been domesticated. Domesticates are often diverged phenotypically from their wild ancestors by a series of morphological changes that are collectively known as the "domestication syndrome." This syndrome includes phenotypes like enhanced robustness, the production of fewer but larger fruits, changes in photoperiod sensitivity, and altered seed dormancy and dispersal (Hammer 1984; Gepts 2004; Miller and Gross 2011). For *sativa*, the domestication syndrome includes higher sugar content in the berry, increased berry and bunch size, changes in seed morphology, and a shift from a dioecious to the hermaphroditic mating system (This et al. 2006).

The phenotypic changes associated with domestication are the result of a protracted process that modifies patterns of genetic diversity relative to the wild ancestor. The process can be considered to consist of four stages (Gaut et al. 2018) (Fig. 3.1). The first is "management," which reflects human stewardship and harvesting of wild plant populations, presumably by hunter-gatherers, prior to their purposeful cultivation. The recognition of this stage is relatively nascent and somewhat speculative, but it conforms to a growing body of evidence that humans greatly affected flora, as well as fauna, prior to the onset of agriculture. Stage 2 is purposeful cultivation, a process that undoubtedly included selection for desirable traits, like fruit size and taste, but also unintended selection for traits associated with growth conditions and responses to stresses. Because it is likely that only a subset of wild populations was cultivated, this stage often features a dramatic domestication bottleneck that reduces and repatterns genetic diversity relative to the wild ancestor. Stage 3 is

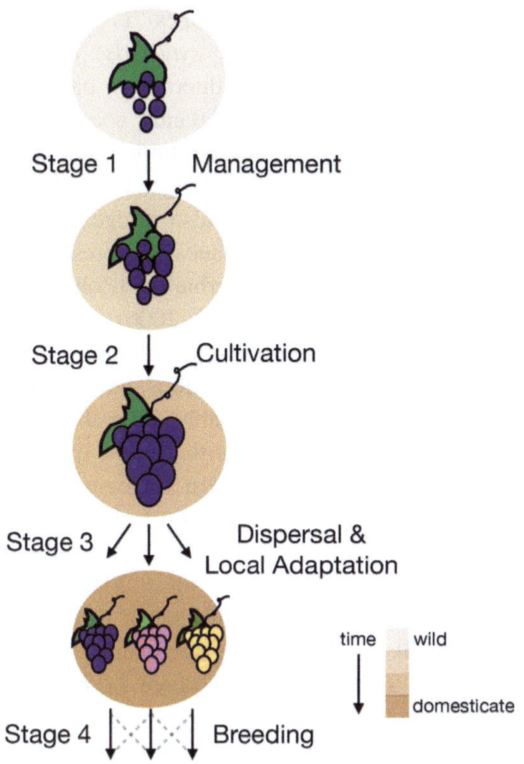

Fig. 3.1 Schematic of the domestication process, illustrating a hypothesized process of domestication that includes the four stages. The grape cluster at the top represents *sylvestris*

geographic expansion, which occurs when the incipient crop is dispersed to new locations. These new locations present novel biotic and abiotic stresses that further drive crop adaptation. Interestingly, some of this adaption may be due to hybridization with—and introgression from—locally adapted wild plants. Finally, stage 4 is modern, deliberate breeding, a process that has occurred only over the last few hundred years for most crops (Meyer and Purugganan 2013). We will ignore the last stage within this chapter, but it is of course fundamentally important for understanding modern crop germplasm.

It is worth emphasizing that artificial selection acts continually throughout the four stages of domestication, and this selection can be either conscious or unconscious. Conscious selection, as defined by Darwin, refers to an attempt "…to modify a breed according to some predetermined standard" (Darwin 1868). In other words, it is breeding to a type or a concept—for example, larger and sweeter fruits. In contrast, unconscious selection is a consequence of humans changing the conditions under which a species is grown, without emphasis on a particular trait or a predetermined goal (Ross-Ibarra et al. 2007). In Darwin's view—and that of other students of domestication (Vavilov 1992; Zeven 1973)—unconscious selection is analogous to natural selection, even though it is caused by humans. Indeed, authors have argued that some artificial selection is no more potent than selection in the wild (Purugganan and Fuller 2009). We included this information to make the point that domestication affects both obvious phenotypes (e.g., sugar content and bunch architecture in grapes) but also less obvious morphological traits through unconscious selection.

3.2.1 Stage 1: Management Prior to Cultivation

The shift from hunter-gatherer to agricultural societies altered the course of human history. This shift is typically considered to have begun ~ 10,000 years ago (Purugganan and Fuller 2009) and occurred roughly contemporaneously across diverse regions of the globe, such as the Fertile Crescent and Mexico. The earliest domesticated crops include some of the cereals—such as maize, barley and the progenitors of wheat—that continue to be staples in the human diet. Curiously, perennial crops typically were domesticated later than annual crops; most perennial crops were domesticated ~ 2000 years ago (ya) vs 4000–5000 ya for annual crops (Meyer et al. 2012). Given this time frame, grapes are likely to have been among the first cultivated perennial crops because the earliest evidence of wine production has been dated to the 8000–7800 ya from sites in the Kvemo province of modern-day Georgia (McGovern et al. 2017). These new dates precede earlier evidence for wine production in northwestern Iran ~ 7400 ya (McGovern et al. 1996a, b), where the volume of wine containers strongly suggest that grapevines had been domesticated already (McGovern et al. 2017).

The question is whether this ~ 8000 ya date represents the first interactions of man with grapes or, instead, whether there had been a lengthy human history with grapes prior to purposeful cultivation. If so, how long was this history? This is of course an interesting academic question, but it also has practical implications for understanding extant genetic diversity. If, for example, humans had an extended history of gathering wild grapes prior to purposeful cultivation, then they likely exerted unconscious selection on wild populations, thereby affecting genetic diversity in those wild populations prior to purposeful cultivation.

This idea of a long association between humans and a crop lineage—i.e., stage 1 of the protracted domestication model (Fig. 3.1)—is relatively new and quite speculative. The idea originates, in part, from the analysis of fossil data, particularly the fossilized remains of the rachis of grain crops. The rachis is important because it can indicate whether a plant had non-shattering seeds, which is a key indication of harvesting and therefore cultivation. The benefit of fossil data is that a series of fossils can illustrate the progression of phenotypes over time. For example, Allaby and co-authors have used this approach to study a progression of the rice rachis, some of which were dated to 10,000 ya. Using this data, they have argued that selection for a non-shattering rachis in rice began in the Pleistocene (Allaby et al. 2017), which potentially predates accepted dates for agricultural settlement. The important point is that their data suggest that humans may have been altering plants within wild populations (stage 1, Fig. 3.1) prior to Neolithic revolution (stage 2, Fig. 3.1). This idea complements the ongoing recognition that humans altered plant species and ecosystems long before the onset of agriculture (Boivin et al. 2016; Allaby et al. 2017; Roberts et al. 2017).

Is it possible, then, that humans and grapes have a commensal history that extends beyond 8000 years and that this prolonged history impacted the extant crop? The answer is "yes," based on two pieces of evidence. The first piece is *prima facie*: Wild grapes (*V. vinifera* subsp. *sylvestris;* hereafter *sylvestris*) are distributed throughout the Mediterranean basin and across Eurasia, from the Atlantic coast to the western Himalayas. Their distribution includes a foray into central Europe along the Rhine and Danube waterways, and it also bridges Europe and Asia via the Transcaucasus (present-day Georgia, Armenia, and Azerbaijan) (Zohary and Spiegel-Roy 1975; This et al. 2006). Assuming that the current geographic distribution of *sylvestris* reflects earlier distributions, humans have roamed these regions probably since the initial replacement of the Neanderthals. For example, some regions of the Southern Caucasus mountains, near the sites of the earliest fossil evidence for wine (McGovern et al. 2017), contain evidence of human habitation for > 20,000 years (Adler and Tushabramishvili 2004). It therefore seems unlikely that hunter-gatherers did not take advantage of this obvious source of nutrition. Supporting this contention, carbonized grape seeds (pips) have been found in prehistoric sites throughout Europe, likely reflecting material collected from the wild (Zohary and Spiegel-Roy 1975; Zohary 1996).

The second line of evidence is based on population genetic analyses of whole-genome resequencing data, using an approach called the sequential Markovian coalescent (SMC). SMC analyses employ genomic data to estimate the history of a taxon's effective population size (N_e) through time, from the present until to far into the past. Zhou et al. (2017) applied this approach with population genomic data from a sample of 14 grape cultivars and nine putatively wild *sylvestris* accessions. The results were intriguing, for at least three reasons. First, the wild and cultivated accessions could be discriminated based on genetic evidence, indicating that they are indeed different, as expected. Second, the two samples were estimated to have diverged ~ 22,000 years ago. This lengthy time frame probably reflects that the wild sample in the study did not represent the exact *sylvestris* population(s) that were ultimately cultivated. Most importantly, the inferred population histories of the two samples differed markedly. The wild sample exhibited fluctuating population sizes over time, and these fluctuations corresponded roughly with histories of glacial maxima and minima. In contrast, the *sativa* sample had an apparent history of a long population decline that started ~ 20,000 ya and led eventually to a mild domestication bottleneck ~ 8000 ya. This protracted N_e decline of ~ 12,000 years is potential evidence that humans gradually favored a subset of *sylvestris* from the proto-*sativa* lineage, which led to more narrow genetic base over time. This inference comes with important caveats, both because SMC methods have shortcomings that can prove misleading [see (Gaut et al. 2018)] and because a more recent study made conclusions at odds with some but not all of these conclusions (Liang et al. 2019). Nonetheless, population genomic data tentatively support a long population decline in the history of domesticated grapevine that corresponds roughly with the time of some habitation in the Transcaucasus (Adler and Tushabramishvili 2004).

What does it matter if there has been a prolonged history of human management (stage 1; Fig. 3.1) prior to purposeful cultivation (stage 2; Fig. 3.1)? First, it informs human history because it suggests the common-sense idea that interactions between humans and plants extend in time beyond the agricultural revolution of the Neolithic and likely impacted the genetic diversity available to those purposeful cultivators. Second, a lengthier history provides an extended time frame for unconscious selection on traits. This challenges the current paradigm because typically the traits that differ between crops and their wild relatives are assumed to have evolved rapidly during the onset of agriculture (stage 2).

3.2.2 Stage 2: Purposeful Cultivation

No matter the duration of the history between man and grapes, one thing is certain: Grapes were eventually cultivated purposefully by agricultural settlements. As mentioned above, a reasonable hypothesis is that viticulture began ~ 8000 years ago somewhere in the Transcaucasus (McGovern et al. 2017), but many questions about this transition remain. Did this stage of domestication occur once, somewhere near or in the Caucasus Mountains, or did it occur multiple independent times in different regions? Did this purposeful cultivation result in a strong genetic bottleneck, as appears to be common for many species? Finally, what traits mark this transition and what are the underlying genetic causes? In this section, we address these questions, focusing again on insights gleaned from, and the limitations of, genetic data.

Where and how many times were grapes domesticated? One compelling hypothesis about grape cultivation is that it occurred ~ 8000 ya somewhere in the Transcaucasus (i.e., modern-day Georgia and/or neighboring regions). This area has been described as the "world center" of the Eurasian grape, where *sylvestris* had its greatest diversity (Vavilov 1992). Genetic data support this view that *sylvestris* genetic diversity is elevated in the Transcaucasus (Ekhvaia and Akhalkatsi 2010; Imazio et al. 2013; Ekhvaia et al. 2014). Also, as noted by McGovern et al. (2017), the hypothesized origin of *vinifera* in the Transcaucasus is further supported by observations that: *i*) Some Western European cultivars are more closely related to *sylvestris* accessions from this region than to *sylvestris* from Western Europe, and *ii*) cultivars from Georgia also have a close relationship to those from Western Europe (Vouillamoz et al. 2006). Finally, a recent exhaustive study of microsatellite (SSR) diversity also strongly implicates samples of *sylvestris* from Georgia in the major cultivation event (Riaz et al. 2018).

But was there only one center of grapevine domestication, or were there two or maybe even several? There is some evidence to suggest that grapevine was domesticated independently more than once. The argument for multiple domestications came originally from the morphological differentiation between cultivars from the Near East and the Western Mediterranean (Negrul 1938). Further genetic support of the multiple domestication hypotheses has come from genetic studies. For example, Arroyo-García et al. (2006)

genotyped 1201 wild and cultivated samples with chloroplast microsatellite (SSR) markers (Arroyo-García et al. 2006). All of their genetic analyses grouped cultivars into two clusters; one cluster grouped with *sylvestris* from the Western Mediterranean and the other grouped with *sylvestris* from the Near East. Based on this evidence, the authors concluded that their data support genetic contributions of both eastern and western *sylvestris* population to grapevine cultivars. However, they also reasonably noted that they could not conclude whether these genetic contributions constituted two distinct, independent domestication events because their observed genetic patterns could also be caused by introgression between *sativa* and wild *sylvestris* in distinct locations (see stage 3; below).

Another recent study has genotyped ~ 1400 accessions with 20 nuclear SSR markers; importantly this study included *sylvestris* samples from the Transcaucasus region (Riaz et al. 2018). Overall, patterns of genetic relatedness for their study confirm a main domestication event in the Transcaucasus, but there are also clear signals that *sylvestris* from other regions have contributed genetically to cultivars. Based on these genetic contributions, Riaz et al. (2018) suggest there were "…at least two separate domestication events that gave raise to the cultivated grape; one derived from the Transcaucasia wild grape, and another from the wild grapes of Western Europe" (Riaz et al. 2018). Multiple grapevine domestication events have also been suggested by patterns of genetic diversity in the sex-determining region of chromosome 2 (Picq et al. 2014).

In practice, however, it is remarkably difficult to differentiate between independent domestication events (stages 1 and 2) from local introgression events (stage 3; see below). Two other crops—olives and Asian rice—illustrate this difficulty. Olive (*Olea europaea* ssp. *europaea*) is like grapevine in that it is a perennial crop, and it too has been studied primarily with chloroplast markers and SSRs. Based on thorough genetic sampling of cultivars and wild accessions, the data have been interpreted to indicate as many as nine domestication events (Breton et al. 2009), although the hypothesis has more recently been modified to include a single primary domestication event (Besnard et al. 2013). However, some data suggest the possibility of a second, minor domestication event near the Grecian peninsula or, alternatively, hybridization of the crop with local wild populations that left a genetic footprint in the cultivars of that region (Diez et al. 2015). At this point, the "jury is still out" on a definitive model of olive domestication (Besnard and Rubio de Casas 2016; Díez and Gaut 2016); despite extensive study, and it is not clear that any definitive answers are forthcoming.

The domestication of Asian rice (*Oryza sativa*) has probably been studied more thoroughly, with more genomic data, than any other crop. Initial studies were based on SSRs or SNPs in nuclear genes, and they strongly suggested that Asian rice was domesticated twice, in India and in China (Londo et al. 2006; Caicedo et al. 2007). Later analyses with more data pointed to a single domestication event (Molina et al. 2011). Now, however, the field is reaching a consensus on a relatively complex domestication scenario that involves a single, initial domestication event in China. This scenario posits that the incipient crop was dispersed from China to the Himalayas, where it hybridized to distinct wild populations and created *Indica* rice (Huang et al. 2012; Choi et al. 2017). The important point is that the origin of rice in India was not wholly independent from the first domestication event in China because it relied upon alleles that had originated in China (Choi et al. 2017). These alleles encoded traits crucial for cultivation.

These examples illustrate that it can be quite difficult to infer the number of domestication events, and this inference is often conflated with introgression events that occurred after the geographic expansion of the crop. Based on the available genetic data from the grapevine, it seems reasonable to continue to hypothesize a primary event in the Transcaucasus or a nearby region, followed by the distribution of the incipient crop throughout the Mediterranean, where it hybridized with local *sylvestris* populations. There is strong support for local introgression events between *sativa* and *sylvestris*, based on the observations that some cultivars tend to group genetically with local wild accessions (Myles et al. 2011; Riaz et al. 2018) and that evolutionary models require substantial levels of gene flow to explain extant patterns of diversity (Riaz et al. 2018). However, given the nuances and complexities of cases like Asian rice, we stop short of concluding that there have been multiple independent domestication events in grapevine. In our opinion, the question merits further study using approaches that utilize whole-genome resequencing data.

Did cultivation lead to a domestication bottleneck? One of the hallmarks of the second stage of domestication is a strong genetic bottleneck. This stems from the fact that early farmers probably based the incipient crop on a limited sample of wild individuals from a particular region or population, and then, they only propagated the best individuals for ensuing generations (Doebley et al. 2006). The resulting genetic bottleneck reduced genetic diversity in a crop compared to its wild relative, perhaps leaving useful genetic variants behind. Domestication bottlenecks have been studied in great depth using population genetic approaches. For example, the maize domestication bottleneck has been studied for more than three decades, based first on isozyme data (Doebley 1989), then single-nucleotide polymorphism data (Eyre-Walker et al. 1998; Wright et al. 2005) and eventually whole-genome data (Hufford et al. 2012; Beissinger et al. 2016). Taken together, this work has suggested that the bottleneck in maize was fairly severe, such that less than 10% of the progenitor population was retained during domestication (Wright et al. 2005; Beissinger et al. 2016).

Annual crops like maize typically undergo severe domestication bottlenecks, but they tend to be less pronounced for perennial crops. On average, annual crops retain 60% of the diversity of their wild relatives (Miller and Gross 2011), but perennial fruit crops retain ~ 95% of the genetic diversity within their wild progenitor. In this respect, *sativa* is similar to apples (Cornille et al. 2012) and cherries (Mariette et al. 2010) in exhibiting little to no loss of genetic diversity compared to their wild progenitors. Based on whole-genome sequences, for example, the total amount of genetic diversity within *vinifera* is ~ 94% that of *sylvestris* (Zhou et al. 2017) (Fig. 3.2). Although the value of 94% is likely to vary somewhat with the samples under comparison, it strongly suggests that the domestication bottleneck for grapes was mild (Myles et al. 2011; Zhou et al. 2017). Indeed, population modeling suggests that 33–50% of the progenitor population was retained during grape domestication, which is of sufficient size to sample most

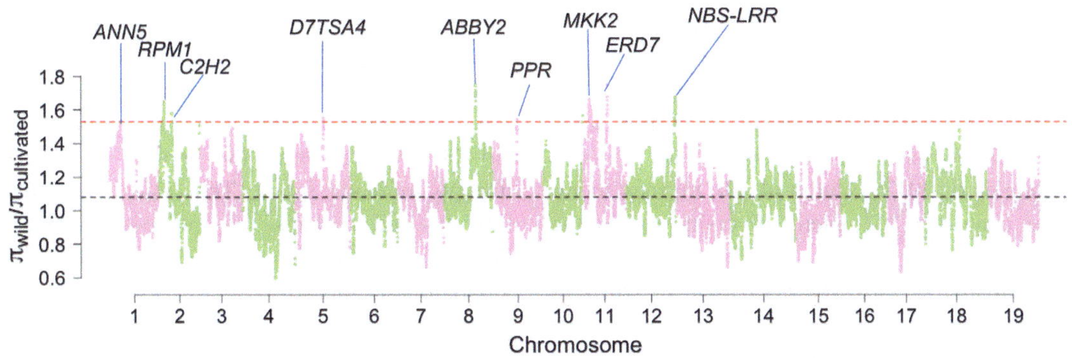

Fig. 3.2 Plot of genetic diversity for samples of *sativa* cultivars and *sylvestris* accessions from (Zhou et al. 2017). The *y*-axis represents genetic diversity in the wild sample compared to the cultivated sample; the dashed line at 1.06 represents the average across all 19 chromosomes. The colored points represent diversity within 20 kb sliding windows, and different colors represent different chromosomes. The peaks represent putative regions of selective sweeps, where the diversity in the *sativa* sample is substantially lower than that of the *sylvestris* samples, and we gave labeled a few genes with strong evidence for a selective sweep (Please note that we have inverted the *y*-axis relative to the discussion in the text to more easily highlight selective sweep regions)

of the extent genetic diversity from the progenitor. We note, however, that all of these conclusions depend critically on the wild sample that is contrasted to the sample of domesticates.

Overall, the lack of dramatic bottleneck effects may help explain why perennial fruit crops like grape tend to exhibit fewer phenotypic shifts during domestication than do annual crops (Meyer et al. 2012). But why do perennials in general, and grapes specifically, tend to have less dramatic bottlenecks? There are at least five reasons (Gaut et al. 2015). One is that perennials tend to cross-pollinate and have high inbreeding depression so that particularly strong bottlenecks are untenable (McClure et al. 2014). That is, very small populations promote inbreeding, and severe inbreeding depression could cause small populations to crash. A second is that perennials tend to have been domesticated more recently—both in terms of years and in generations—providing less time for large losses of diversity to accrue. A third is that perennials have overlapping generations, a feature that is also likely to reduce the severity of bottlenecks. Fourth, hybridization between different species has often played a central role in the origin and diversification of perennials. While hybridization between species may not have played a large role in the early domestication of grapevine, repeated introgression of the crop with wild populations has reintroduced genetic diversity from *sylvestris* (and other *Vitis* species) into *sativa* (see stage 3, below). Finally, many perennials are like grapevines in that they are propagated clonally. Somatic mutations can accumulate during clonal propagation, particularly when they are recessive (Zhou et al. 2017; Gaut et al. 2018); they therefore contribute to genetic diversity within the crop.

Genomic regions that contribute to morphological differences between sativa *and sylvestris:* The study of bottlenecks is important because they are one of the two major forces that shape genetic diversity within an incipient crop. Since grapes did not undergo a severe bottleneck, we now turn to the second force, unconscious and conscious selection. Selection helps the incipient crop adapt to new growing conditions, and it also drives morphological divergence between the crop and its wild progenitor. In this section, we consider what is known about the genes and genomic regions that appear to have been under selection in *sativa*. The identification of these regions is the basis for a "bottom-up" approach to identify the genetic variants that contribute to morphological divergence (Ross-Ibarra et al. 2007).

The tools of evolutionary genomics can identify some of the genomic regions that contribute to this divergence. The basic approach starts with a comparison of genetic diversity between wild and cultivated populations, preferably across the entire genome. From these diversity comparisons, one can detect regions of aberrantly low diversity in the cultivated crop relative to the wild crop (Fig. 3.2). In theory, these regions have been subjected to a "selective sweep," whereby either unconscious or conscious selection has removed (or swept away) genetic diversity by favoring a particular beneficial allele. In practice, the identification of these swept regions—which is also called "selective sweep mapping"—has a number of limitations that must always be kept in mind. One is that differences in genetic diversity between the crop and the wild relative can be caused by other mechanisms, such as genetic drift. Another is that it is difficult to identify a swept region in the domesticate if that same region has lower than average diversity in the wild sample. Finally, the results are always dependent on the samples being used. One typically assumes that the samples from the wild progenitor populations and the crop represent the breadth of genetic diversity of each taxon. That said, it is often difficult—and perhaps impossible—to identify wild samples that represent the exact progenitor populations of the crop.

Genome-wide searches for selective sweeps have been applied to numerous crops, but only a handful of studies have taken this approach in grapes. For example, Myles et al. (2011) genotyped 950 *vinifera* accessions and 59 *sylvestris* samples with the Vitis9KSNP chip and then scanned the genome for potential regions of selective sweeps. They found at least one candidate sweep region, a 5 Mb region on chromosome 17 that contains at least 20 genes. Another study used the same data to contrast wine and table grapes, hoping that they could identify the loci that differentiate the two germplasm sets. They found evidence for selective sweeps near the flower sex locus, for berry skin color, berry size, and muscat aroma (Migicovsky et al. 2017). Similarly, Marrano et al. (2018) sought to identify signals of selection between *sylvestris* and *sativa* samples using genotypes based on a 20 K SNP array. They found regions of significant differentiation between their *sylvestris* and *sativa* samples that encompass as many as ~ 2000 genes, a number that sounds very high but is only slightly higher than the number of genes estimated to have experienced selection during maize domestication (Wright et al. 2005; Hufford et al. 2012). Marrano et al. (2018) investigated the function of a subset of their set of 2000 candidate genes and found they were enriched for functions in metabolism and responses to environmental stimuli.

It is important to note that limited-representation data—such as SNP array or GBS data—are expected to be helpful for many applications, such as genome-wide association (Laucou et al. 2018) and phylogenetic (Klein et al. 2018) analyses. However, as mentioned by Myles et al. (2011), SNP array data are severely underpowered for selective sweep mapping. Moreover, SNP arrays usually contain ascertainment biases that can mislead population genetic analyses if the biases are not properly corrected. Such biases may contribute to the observation that *sylvestris* samples sometimes have lower genetic diversity than *sativa* samples (Marrano et al. 2018), which is typically not expected for wild vs. crop contrasts.

Whole-genome resequencing data are superior to both array and GBS data because of the high marker density and the lack of ascertainment bias. Several studies have reported resequencing data from *vinifera* cultivars (e.g., (Di Genova et al. 2014; Xu et al. 2016)), but to our knowledge only two studies have focused on identifying sweep regions (Fig. 3.1) (Zhou et al. 2017; Liang et al. 2019). For example, Zhou et al (2017) compared genomic data from *sativa* and *sylvestris* and identified hundreds of candidate genes that may have been targets of selection. The candidates included genes implicated in berry development and/or quality, including the sugar transporter *SWEET1* gene; a leucoanthocyanidin dioxygenase (*LDOX*) gene that may be involved in proanthocyanidin accumulation; genes potentially involved in berry softening and flowering-time genes, including a *Phytochrome C* homolog. Interestingly, separate identification of selected genes in the *sylvestris* sample identified fewer selected genes, and they were implicated in distinct functions from the selected genes in the *sativa* sample (Zhou et al. 2017).

Overall, the search for the genomic regions affected by selection in *sativa* is just beginning. The approach nonetheless has the potential to yield valuable insights into the types of genes and biochemical networks that have been key determinants of agronomic phenotypes.

The curious case of sex: One of the major phenotypic shifts that occurred during grapevine domestication was a transition in the mating system. At some point during the domestication process, grapes transitioned from dioecy (i.e., separate male and female individuals) in *sylvestris* to hermaphroditic individuals in *sativa*. This shift is particularly dramatic given that all extant wild *Vitis* species are dioecious. Hence, dioecy has been maintained since the origin of the genus, which is estimated to have occurred from ~ 18Mya (Wan et al. 2013) to ~ 39Mya (Liu et al. 2016).

A switch to hermaphroditism provides immediate advantages for cultivation. In hermaphrodites, all individuals can contribute to fruiting and to pollination. In contrast, only half of the population bears fruit in a dioecious species. In agricultural settings, dioecy means that most males must be removed from the fields as soon as they can be identified. For that reason, dioecy is particularly disadvantageous for agricultural productivity in perennial crops (e.g., date palms, persimmons, and kiwifruit), where first flowering takes many years and sex can be identified only after first flowering. To circumvent this problem, substantial efforts have been focused on identifying molecular markers that allow for earlier gender identification for these crops (Cherif et al. 2013; Akagi et al. 2014; Zhang et al. 2015). Even so, males are still needed to fertilize females; in kiwifruit, for example, orchards are commonly planted with 13% males (McNeilage and Steinhagen 1998).

The timing of the switch to hermaphroditism in grapes is unknown, but we assume it occurred early and rapidly during stage 2 of domestication. Our conjecture about rapidity has precedence in strawberry (Liston et al. 2014). The modern cultivated strawberry *Fragaria × ananassa* is a self-compatible hermaphrodite octoploid species that originated through the hybridization of two American species, *F. virginiana* and *F. chiloensis* (Liston et al. 2014). The two species are dioecious and subdioecious (where individuals can be either male, female, or hermaphrodite), respectively. Early *F. × ananassa* cultivars had separate sexes, but it has been documented that unconscious selection rapidly selected for hermaphroditism in the nineteenth century (Darrow 1966). [As a historical aside, the dioecious strawberry species *F. moschata* had been previously cultivated in Europe, but gardeners would remove males that lacked fruit because they thought them "sterile." These same gardeners unwittingly caused female production to be sporadic, due to the lack of pollen. In 1766, Duchesne uncovered the existence of separate sexes and thereby improved strawberry productivity (Duchesne 1766).]

In grapevine, Oberle (1938) proposed a model by which sex is determined by two tightly linked genes: one for female sterility and another for male sterility (Oberle 1938). In this model, males arise from a dominant female sterility *So* allele, while females result from a recessive male sterility mutation, the *sp* allele (Fig. 3.3). This model is identical to the "two-gene model" of sex chromosome evolution in plants proposed by Charlesworth and Charlesworth (1978). In this model, dioecious plants are formed from hermaphroditic plants by a two-step process. The first step is likely the evolution of a recessive male sterility mutation (in Oberle's nomenclature, *Sp*- > *sp*; Fig. 3.3), which would lead to a gynodioecious population consisting of females and hermaphrodites. The second step is the formation of a dominant female sterility mutation (*so*- > *So* in Fig. 3.3). If the two male and female loci are tightly linked, alleles at the two loci represent proto-sex chromosomes, where males are heterozygous for the M and F haplotypes and females are homozygous for the F haplotype (Fig. 3.3).

Interestingly, the two-locus model may not be universal because a single gene can determine the sex of individuals in persimmons and artificially dioecious cucurbits (Akagi et al. 2014; Boualem et al. 2015; Renner 2016). In this context, it is worth noting that Carbonneau (1983) proposed another model for grapevine sex determination (Carbonneau 1983). The model involves a single locus with a M (male)

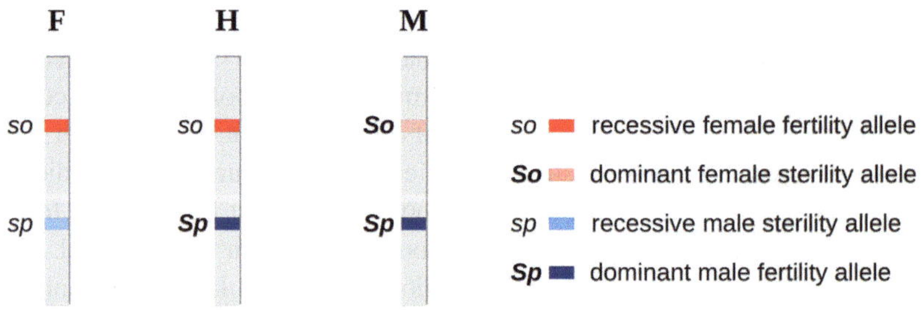

Fig. 3.3 Hypothesized three sex locus haplotypes in grapes according to the two-locus model for sex determination. The female, hermaphroditic, and male haplotype are denoted by F, H, and M, respectively. The recessive *so* allele on the F and H haplotypes confers female fertility when homozygous, but the dominant *So* allele suppresses female function. The recessive *sp* allele on the F haplotype causes male sterility when homozygous, but the dominant *Sp* allele provides male fertility on both the H and M haplotypes

haplotype that is dominant over an H (hermaphrodite) haplotype, which is in turn dominant over an F (female) haplotype (i.e., M > H > F). However, this single-locus model cannot explain some observations from crosses —i.e., the rare appearance of males in crosses between hermaphrodites or between hermaphrodites and females. Neither can it explain the deficit of females in crosses between hermaphrodites or between hermaphrodites and females. For this reason, Carbonneau (1983) hypothesized that a second locus interacted epistatically with the first H/M/F locus and also that the epistatically acting locus linked (to within 0.3 centimorgans) of the sex locus. It is worth noting that the Carbonneau (1983) model is compatible with the two-gene model if its sex locus includes the two hypothetical sex-determining genes without any recombination between them. The model does, however, predict that another locus interacts epistatically with the H/M/F locus.

The classic two-locus model predicts that recombination between the *sp* and *so* genes will produce hermaphrodites and neutered individuals that lack both male and female fertility. That is, recombination between the two loci would lead to males, females, hermaphrodites and neutered plants that have *So/sp* haplotypes, as observed in *Fragaria* (Spigler et al. 2008). Hermaphrodite grapevines are rarely observed in wild populations, and those that have been observed are likely escapees from domestication (Arnold et al. 1998). A few non-flowering vines have also been observed in the wild, but it is difficult to ascertain whether they are neuters or have grown in flowering-limiting conditions. If the two-locus model in Fig. 3.3 is correct, these observations suggest that recombination between the two loci is exceedingly rare and therefore that the three M, H and F haplotypes are divergent enough to prevent recombination.

Assuming a two-locus system of dioecy in *Vitis*, domestication reverted a stable, dioecious mating system that had existed for at least $\sim$ 18 My to the ancestral hermaphroditic state. This could occur, presumably, through the knockout of the dominant female sterility (*So*) mutation of the M haplotype, creating a hermaphroditic (H) haplotype (Fig. 3.3). This is consistent with the fact that the H haplotype has been found to be closer to the M than to the F haplotype (Picq et al. 2014). Under this three haplotype model (Antcliff 1980), females are homozygous for the F haplotype, hermaphrodites can be either HF or HH, and males can be MM, MF or MH.

To date, two sex-determining genes have been found in *Asparagus* (Harkess et al. 2017) and *Fragaria* (Tennessen et al. 2018), but neither the *So* knockout mutation nor the *so* and *sp* genes have been identified in grapevine. However, it is known that the sex-determining region maps to a 152 kb region of chromosome 2, between chromosomal position 4.90 Mb and 5.04 Mb in the Pinot Noir PN40024 12X (v. 2.1) reference genome (Fechter et al. 2012). Unfortunately, the reference is heterozygous for the F and H haplotypes, causing the assembly of this locus into two separate scaffolds. It has been assumed that the F haplotype is represented on chromosome 2 of the reference and that the H haplotype is located on unassigned scaffold_233 (Picq et al. 2014). However, the chromosome 2 assembly lacks the well-studied candidate sex-determining gene adenine phosphoribosyl transferase (*APT3*) (Fechter et al. 2012), and the position of others genes is still approximate (Picq et al. 2014). Further work will be required to better assemble the three *Vitis* sex haplotypes.

Once the M, H, and F haplotypes are assembled, it will be useful to the estimate the time of split among them. Given the age of dioecy in *Vitis*, we expect the F and M haplotypes to be highly diverged (reflecting the $\geq$ 18Mya conservation of dioecy in the genus), but we also expect the H and M haplotypes to have diverged more recently, i.e., approximating the $\sim$ 8000 year time frame of cultivation. There is interesting precedence for this approach because the sequencing of the sex chromosomes in papaya led to an estimated divergence time of $\sim$ 4000 years between the male Y chromosome and the hermaphroditic Y^h chromosome. This time frame coincides with the rise of the Mayan civilization and the origin of papaya cultivation (VanBuren et al. 2015).

Interestingly, population genomic analyses of chromosome 2 have uncovered two peaks of divergence between *sativa* and *sylvestris* accessions (Zhou et al. 2017). The first peak corresponds to the sex-determining region identified by Picq et al. (2014) and contains ~ 13 genes (Picq et al. 2014). The second peak is close to the first, at positions 5.20 Mb to 5.33 Mb. It is not clear why there are two peaks. Each could, for example, contain one of the *so* and *Sp* sex-determining loci. Alternatively, if the sex-determining genes both lie within the first peak, then the second peak could house the epistatic locus hypothesized by (Carbonneau 1983) or perhaps even a group of genes with antagonistic sex effects, as observed in the pseudoautosomal region of *Silene latifolia* (Qiu et al. 2013).

Altogether, the two peaks contain ~ 45 genes, some of which exhibit sex-biased expression (Zhou et al. 2017) (Table 3.1). For example, the first peak contains six genes overexpressed in F flowers, including *VviFSEX*, which may abort stamen development and thus be the *sp* male sterility locus (Coito et al. 2017). The second peak has four genes that exhibit biased sex expression: one gene has a higher F expression, two have higher H expression, and one has higher M expression (Zhou et al. 2017). All are reasonable candidates for sex determination, but the search continues.

3.2.3 Stage 3: Geographic Expansion with Introgression as a Means of Local Adaptation

Our discussion about the complexities of differentiating single vs. multiple domestication effects is intricately tied to the third stage of domestication: the geographic expansion of the crop to new locations. Geographic expansion requires crops to adapt to new environments. In theory, adaptation could occur either through selection within the crop (i.e., on standing genetic variation or on new genetic mutations) or through adaptive introgression with local, cross-fertile wild plants. Adaptive introgression is the introgression of genomic regions that have positive

Table 3.1 Gene expression analysis of genes within the sex determination region that have significantly different expression between sexes (M = male, F = female, H = hermaphrodite)

Gene ID[a]	Functional_annotation[b]	Peak[c]	M vs. F	M vs H	F vs. H
VIT_02s0154g00130	Exostosin (Xyloglucan galactosyltransferase KATAMARI 1)	1	F	–	F
VIT_02s0154g00140	3-oxoacyl-[acyl-carrier-protein] synthase 3 A, cpl precursor	1	F	–	F
VIT_02s0154g00160	FMO family protein	1	F	–	F
VIT_02s0154g00170	Flavin-containing monooxygenase 3	1	F	–	F
VIT_02s0154g00190	Flavin-containing monooxygenase 3	1	F	H	F
VIT_02s0154g00200	*VviFSEX*(Unknown protein)	1	F	–	F
VIT_02s0154g00380	Unknown	2	F	–	F
VIT_02s0154g00310	Protease inhibitor/seed storage/lipid transfer protein (LTP)	2	–	H	H
VIT_02s0154g00480	Heat shock protein MTSHP	2	–	H	H
VIT_02s0154g00370	YbaK/prolyl-tRNA synthetase associated region	2	M	M	–

The sex with higher expression is indicated. The results are from (Zhou et al. 2017)
[a]Gene IDs are taken from the *V. vinifera* genome annotation in Ensembl Plants
[b]Functional annotation is based on VitisNet functional annotations (Grimplet et al. 2009)
[c]As described in the text, peak number 1 spans from approximately the 4.90 Mb position on chromosome 2 to the position 5.04 Mb. Peak number 1 is located nearby on the same chromosome, spanning 5.20 Mb to 5.33 Mb

fitness consequences (Suarez-Gonzalez et al. 2018).

Most crops undergo geographic expansion from their center of origin, but it is not clear how often crops introgress adaptively with local populations after expansion. An obvious requirement is that the crop must be able to hybridize with its wild progenitor (or a close wild relative), and those wild relatives must be distributed in areas where the crop is dispersed. These conditions certainly hold for grapevines, both because *sativa* x *sylvestris* crosses are fertile and because *sylvestris* was distributed throughout the Mediterranean, where grapes may have been initially dispersed. In fact, all *Vitis* taxa are interfertile, which provides numerous opportunities for introgression events.

When hybridization occurs between wild and cultivated accessions, it introduces large genomic regions from the wild into the cultivated background. In most cases, introgressed regions will be purged rapidly from the cultivated germplasm because they do not confer an adaptive benefit. Alternatively, the introgressed region may bring locally adapted alleles into the genetic background of the cultivar, thereby assisting crop establishment in the local environments. Interestingly, the introgressed region may not need to have better adapted alleles per se because local alleles could be beneficial to the crop by two other mechanisms. First, it could drive synergistic epistatic interactions within the cultivated genetic background or, second, it could increase fitness by reducing the genetic load. The latter is true because wild populations often harbor higher genetic diversity and maintain larger effective population sizes (N_e) than crop populations (Gaut et al. 2015). Since mutation load is expected to be correlated with N_e, regions introgressed from the wild are expected to reduce the mutation load (Moyers et al. 2018; Gaut et al. 2018). If an introgressed region remains in the crop germplasm, it is expected to eventually decrease in size (Janzen et al. 2018), due to the recurrent backcrossing of the hybridized individual with other cultivated germplasm. Interestingly, the clonal propagation of grapevines is likely to slow this process; hence, we hypothesize that the size of introgressed regions is larger in grapevine than in most other sexually propagated crop species.

There are now several methods to infer introgressed regions from genetic and genomic diversity data. Some of these methods were pioneered in the analysis of human data, where it has become apparent that small (< 100 kb) vestiges of ancestral hybridization events with Neanderthals remain with modern human genomes (Sankararaman et al. 2014). In principle, introgression events can be detected using population genetic tools such as TreeMix (Pickrell and Pritchard 2012) and the ABBA–BABA test (Green et al. 2010), and then, they can be localized using the genome-wide sliding windows of the f_d statistic (Martin et al. 2015) or a tool for inferring local ancestry (Maples et al. 2013). To date, the most compelling crop example comes from maize, where introgression was detected from the wild relative *Zea mays* ssp. *mexicana* to maize in the Mexican highlands (Hufford et al. 2013). Subsequent work found that at least one introgression is ~ 15 Mb in length on chromosome 3 (Wang et al. 2017). The case for adaptive introgression is particularly compelling when an introgressed region overlaps with obvious candidate genes for local adaptation. In the maize chromosome 3 example, the putatively introgressed region contains an inversion that is closely associated with flowering-time variation among maize landraces (Romero Navarro et al. 2017).

As we have mentioned, there is compelling genetic evidence for introgression from *sylvestris* into *sativa* based on a number of studies and a variety of molecular markers (Arroyo-García et al. 2006; Myles et al. 2011; Riaz et al. 2018). For example, Myles et al. (2011) have shown that *sativa* cultivars from Western Europe tend to cluster more closely to *sylvestris* accessions from the same region, strongly suggesting some history of genetic exchange. Given the distribution of *sylvestris,* it seems reasonable to assume that most early introgression events with *sativa* involved populations of *sylvestris*. However, as *sativa* cultivars were distributed more widely, so was the opportunity for introgression with other wild *Vitis* species. For example, the Koshu Cultivar appears to owe ~ 70% of its genetic identity to *sativa* and the remaining ~ 30% to wild Chinese *Vitis* species. Similarly, the wild Amur grape (*Vitis amurensis* Rupr.) from Northeast Asia is the apparent source of downy mildew resistance for some *sativa* cultivars (Venuti et al. 2013). Based on these examples, it is apparent that introgression from wild *Vitis* to *sativa* has played a prominent role in shaping *sativa* germplasm and has been a crucial aspect of the *sativa* domestication process.

We conclude with two final points. The first is that we have focused on introgression into *sativa*, but it is also clear that genetic introgression can go the opposite direction—i.e., from the crop into wild populations and species. For example, a low level of pollen-mediated gene flow has been detected from *sativa* to *sylvestris* using chloroplast markers (Di Vecchi-Staraz et al. 2009). This pollen-flow has the potential to contaminate *sylvestris* gene pools, thereby polluting an important genetic resource for grapevine breeding. Another study has detected introgression from *Vitis* species used as rootstocks into *sylvestris* (Schröder et al. 2015). The ongoing phenomenon of introgression from cultivated germplasm into *sylvestris* needs to be considered in the context of the conservation of wild European grape populations. Second, we believe that our understanding of historical introgression among *Vitis* species—from *sylvestris* and other wild *Vitis* into *sativa*, from *sativa* into wild populations, and among wild *Vitis* species—is in its infancy because the existing studies have relied primarily on non-genomic approaches. More widespread application of genomic approaches will help elucidate the dynamics of adaptive introgression in grapevine and may yield clues into its agronomic effects.

Acknowledgements This work was supported by an NSF Plant Genome Research Program grant to BSG (1741627). YZ has been supported in part by an International Postdoctoral Exchange Fellowship Program and also by NSF grant to BSG (1542703). AM is supported by postdoctoral fellowships from EMBO and from the Human Frontiers in Science program.

References

Adler DS, Tushabramishvili N (2004) Middle palaeolithic patterns of settlement and subsistence in the southern caucasus. In: Conard (ed) Settlement dynamics of the middle paleolithic and middle stone age, vol II. Kerns Verlag, Tübingen, pp 91–132

Akagi T, Henry IM, Tao R, Comai L (2014) Plant genetics. A Y-chromosome–encoded small RNA acts as a sex determinant in persimmons. Science 346:646–650

Allaby RG, Stevens C, Lucas L, Maeda O, Fuller DQ (2017) Geographic mosaics and changing rates of cereal domestication. Philos Trans R Soc Lond B Biol Sci 372:20160429

Antcliff AJ (1980) Inheritance of sex in Vitis. Annales de l'Amelioration des Plantes 30:113–122

Arnold C, Gillet F, Gobat JM (1998) Occurrence of the wild vine *Vitis vinifera* ssp. silvestris in Europe. Vitis 37:159–170

Arroyo-García R, Ruiz-García L, Bolling L, Ocete R, López MA, Arnold C, Ergul A, Söylemezoğlu G, Uzun HI et al (2006) Multiple origins of cultivated grapevine (*Vitis vinifera* L. ssp. sativa) based on chloroplast DNA polymorphisms. Mol Ecol 15:3707–3714

Beissinger TM, Wang L, Crosby K, Durvasula A, Hufford MB, Ross-Ibarra J (2016) Recent demography drives changes in linked selection across the maize genome. Nat Plants 2:16084

Besnard G, Khadari B, Navascues M, Fernandez-Mazuecos M, El Bakkali A, Arrigo N, Baali-Cherif D, Brunini-Bronzini de Caraffa V, Santoni S et al (2013) The complex history of the olive tree: from Late Quaternary diversification of Mediterranean lineages to primary domestication in the northern Levant. Proc Biol Sci 280:20122833

Besnard G, Rubio de Casas R (2016) Single vs multiple independent olive domestications: the jury is (still) out. New Phytol 209:466–470

Boivin NL, Zeder MA, Fuller DQ, Crowther A, Larson G, Erlandson JM, Denham T, Petraglia MD (2016) Ecological consequences of human niche construction: Examining long-term anthropogenic shaping of global species distributions. Proc Natl Acad Sci USA 113:6388–6396

Boualem A, Troadec C, Camps C, Lemhemdi A, Morin H, Sari M-A, Fraenkel-Zagouri R, Kovalski I, Dogimont C et al (2015) A cucurbit androecy gene reveals how unisexual flowers develop and dioecy emerges. Science 350:688–691

Breton C, Terral JF, Pinatel C, Medail F, Bonhomme F, Berville A (2009) The origins of the domestication of the olive tree. C R Biol 332:1059–1064

Caicedo AL, Williamson SH, Hernandez RD, Boyko A, Fledel-Alon A, York TL, Polato NR, Olsen KM, Nielsen R et al (2007) Genome-wide patterns of nucleotide polymorphism in domesticated rice. PLoS Genet 3:1745–1756

Carbonneau A (1983) Stérilités mâle et femelle dans le genre Vitis. I. Modélisation de leur hérédité Agronomie 3:635–644

Charlesworth B, Charlesworth D (1978) A model for the evolution of dioecy and gynodioecy. Am Nat 988:975–997

Cherif E, Zehdi S, Castillo K, Chabrillange N, Abdoulkader S, Pintaud JC, Santoni S, Salhi-Hannachi A, Glémin S et al (2013) Male-specific DNA markers provide genetic evidence of an XY chromosome system, a recombination arrest and allow the tracing of paternal lineages in date palm. New Phytol 197:409–415

Choi JY, Platts AE, Fuller DQ, Hsing YI, Wing RA, Purugganan MD (2017) The rice paradox: multiple origins but single domestication in Asian rice. Mol Biol Evol 34:969–979

Coito JL, Ramos MJ, Cunha J, Silva HG, Amâncio S, Costa MM, Rocheta M (2017) VviAPRT3 and VviFSEX: two genes involved in sex specification able to distinguish different flower types in *Vitis*. Front Plant Sci 8:98

Cornille A, Gladieux P, Smulders MJ, Roldan-Ruiz I, Laurens F, Le Cam B, Nersesyan A, Clavel J, Olonova M et al (2012) New insight into the history of domesticated apple: secondary contribution of the European wild apple to the genome of cultivated varieties. PLoS Genet 8:e1002703

Darrow GM (1966) The strawberry: History, breeding and physiology. Holt, Rinehart and Winston , New York, USA

Darwin C (1868) The variation of animals and plants under domestication. John Murray, London, UK

Di Genova A, Almeida AM, Muñoz-Espinoza C, Vizoso P, Travisany D, Moraga C, Pinto M, Hinrichsen P, Orellana A et al (2014) Whole genome comparison between table and wine grapes reveals a comprehensive catalog of structural variants. BMC Plant Biol 14:7

Di Vecchi-Staraz M, Laucou V, Bruno G, Lacombe T, Gerber S, Bourse T, Boselli M, This P (2009) Low level of pollen-mediated gene flow from cultivated to wild grapevine: consequences for the evolution of the endangered subspecies *Vitis vinifera* L. subsp. silvestris. J Hered 100:66–75

Díez CM, Gaut BS (2016) The jury may be out, but it is important that it deliberates: a response to Besnard and Rubio de Casas about olive domestication. New Phytol 209:471–473

Diez CM, Trujillo I, Martinez-Urdiroz N, Barranco D, Rallo L, Marfil P, Gaut BS (2015) Olive domestication and diversification in the Mediterranean Basin. New Phytol 206:436–447

Doebley J (1989) Isozymic evidence and the evolution of crop plants. In: Soltis DE, Soltis PS (eds) Isozymes in plant biology. Chapman and Hall, London, pp 165–191

Doebley JF, Gaut BS, Smith BD (2006) The molecular genetics of crop domestication. Cell 127:1309–1321

Duchesne AN (1766) Histoire naturelle des fraisiers. Muséum national d'Histoire naturelle, Paris

Ekhvaia J, Akhalkatsi M (2010) Morphological variation and relationships of Georgian populations of Vitis vinifera L. subsp. sylvestris (CC Gmel.) Hegi. Flora-Morphol Distrib Funct Ecol Plants 205:608–617

Ekhvaia J, Gurushidze M, Blattner FR, Akhalkatsi M (2014) Genetic diversity of *Vitis vinifera* in Georgia: relationships between local cultivars and wild grapevine, *V. vinifera* L. subsp. *sylvestris*. Genet Resour Crop Evol 61:1507–1521

Eyre-Walker A, Gaut RL, Hilton H, Feldman DL, Gaut BS (1998) Investigation of the bottleneck leading to the domestication of maize. Proc Natl Acad Sci USA 95:4441–4446

Fechter I, Hausmann L, Daum M, Sörensen TR, Viehöver P, Weisshaar B, Töpfer R (2012) Candidate genes within a 143 kb region of the flower sex locus in *Vitis*. Mol Genet Genomics 287:247–259

Gaut BS, Díez CM, Morrell PL (2015) Genomics and the contrasting dynamics of annual and perennial domestication. Trends Genet 31:709–719

Gaut BS, Seymour DK, Liu Q, Zhou Y (2018) Demography and its effects on genomic variation in crop domestication. Nat Plants 4:512

Gepts P (2004) Crop domestication as a long-term selection experiment. Plant Breed Rev 24:1–44

Green RE, Krause J, Briggs AW, Maricic T, Stenzel U, Kircher M, Patterson N, Li H, Zhai W et al (2010) A draft sequence of the Neandertal genome. Science 328:710–722

Grimplet J, Cramer GR, Dickerson JA, Mathiason K, Van Hemert J, Fennell AY (2009) VitisNet: "Omics" integration through grapevine molecular networks. PLoS ONE 4:e8365

Hammer K (1984) Das Domestikationssyndrom. Kulturpflanze 32:11–34

Harkess A, Zhou J, Xu C, Bowers JE, Van der Hulst R, Ayyampalayam S, Mercati F, Riccardi P, McKain MR et al (2017) The asparagus genome sheds light on the origin and evolution of a young Y chromosome. Nat Commun 8:1279

Huang X, Kurata N, Wei X, Wang ZX, Wang A, Zhao Q, Zhao Y, Liu K, Lu H et al (2012) A map of rice genome variation reveals the origin of cultivated rice. Nature 490:497–501

Hufford MB, Lubinksy P, Pyhajarvi T, Devengenzo MT, Ellstrand NC, Ross-Ibarra J (2013) The genomic signature of crop-wild introgression in maize. PLoS Genet 9:e1003477

Hufford MB, Xu X, van Heerwaarden J, Pyhajarvi T, Chia JM, Cartwright RA, Elshire RJ, Glaubitz JC, Guill KE et al (2012) Comparative population genomics of maize domestication and improvement. Nat Genet 44:808–811

Imazio S, Maghradze D, De Lorenzis G, Bacilieri R, Laucou V, This P, Scienza A, Failla O (2013) From the cradle of grapevine domestication: molecular overview and description of Georgian grapevine (*Vitis vinifera* L.) germplasm. Tree Genet Genomes 9:641–658

Janzen T, Nolte AW, Traulsen A (2018) The breakdown of genomic ancestry blocks in hybrid lineages given a finite number of recombination sites. Evolution 72:735–750

Klein LL, Miller AJ, Ciotir C, Hyma K, Uribe-Convers S, Londo J (2018) High-throughput sequencing data clarify evolutionary relationships among North American Vitis species and improve identification in USDA Vitis germplasm collections. Am J Bot 105:215–226

Laucou V, Launay A, Bacilieri R, Lacombe T, Adam-Blondon AF, Bérard A, Chauveau A, de Andrés MT, Hausmann L et al (2018) Extended diversity analysis of cultivated grapevine *Vitis vinifera* with 10 K genome-wide SNPs. PLoS ONE 13:e0192540

Liang Z, Duan S, Sheng J et al (2019) Whole-genome resequencing of 472 Vitis accessions for grapevine diversity and demographic history analyses. Nat Commun 10:1190

Liston A, Cronn R, Ashman TL (2014) Fragaria: a genus with deep historical roots and ripe for evolutionary and ecological insights. Am J Bot 101:1686–1699

Liu XQ, Ickert-Bond SM, Nie ZL, Zhou Z, Chen LQ, Wen J (2016) Phylogeny of the *Ampelocissus–Vitis* clade in Vitaceae supports the New World origin of the grape genus. Mol Phylogenet Evol 95:217–228

Londo JP, Chiang YC, Hung KH, Chiang TY, Schaal BA (2006) Phylogeography of Asian wild rice, *Oryza rufipogon*, reveals multiple independent domestications of cultivated rice, *Oryza sativa*. Proc Natl Acad Sci USA 103:9578–9583

Maples BK, Gravel S, Kenny EE, Bustamante CD (2013) RFMix: a discriminative modeling approach for rapid and robust local-ancestry inference. Am J Hum Genet 93:278–288

Mariette S, Tavaud M, Arunyawat U, Capdeville G, Millan M, Salin F (2010) Population structure and genetic bottleneck in sweet cherry estimated with SSRs and the gametophytic self-incompatibility locus. BMC Genet 11:77

Marrano A, Micheletti D, Lorenzi S, Neale D, Grando MS (2018) Genomic signatures of different adaptations to environmental stimuli between wild and cultivated. Hortic Res 5:34

Martin SH, Davey JW, Jiggins CD (2015) Evaluating the use of ABBA-BABA statistics to locate introgressed loci. Mol Biol Evol 32:244–257

McClure KA, Sawler J, Gardner KM, Money D, Myles S (2014) Genomics: a potential panacea for the perennial problem. Am J Bot 101:1780–1790

McGovern P, Jalabadze M, Batiuk S, Callahan MP, Smith KE, Hall GR, Kvavadze E, Maghradze D, Rusishvili N et al (2017) Early Neolithic wine of Georgia in the South Caucasus. Proc Natl Acad Sci U S A 114:E10309–E10318

McGovern PE, Glusker DL, Exner LJ, Voigt MM (1996a) Neolithic resinated wine. Nature 381:481

McGovern PE, Fleming SJ, Katz SH (1996b) The origins and ancient history of wine. Gordon and Breach, New York

McNeilage MA, Steinhagen S (1998) Flower and fruit characters in a kiwifruit hermaphrodite. Euphytica 101:69–72

Meyer RS, DuVal AE, Jensen HR (2012) Patterns and processes in crop domestication: an historical review and quantitative analysis of 203 global food crops. New Phytol 196:29–48

Meyer RS, Purugganan MD (2013) Evolution of crop species: genetics of domestication and diversification. Nat Rev Genet 14:840–852

Migicovsky Z, Sawler J, Gardner KM, Aradhya MK, Prins BH, Schwaninger HR, Bustamante CD, Buckler ES, Zhong GY et al (2017) Patterns of genomic and phenomic diversity in wine and table grapes. Hortic Res 4:17035

Miller AJ, Gross BL (2011) From forest to field: perennial fruit crop domestication. Am J Bot 98:1389–1414

Molina J, Sikora M, Garud N, Flowers JM, Rubinstein S, Reynolds A, Huang P, Jackson S, Schaal BA et al (2011) Molecular evidence for a single evolutionary origin of domesticated rice. Proc Natl Acad Sci USA 108:8351–8356

Moyers BT, Morrell PL, McKay JK (2018) Genetic costs of domestication and improvement. J Hered 109:103–116

Myles S, Boyko AR, Owens CL, Brown PJ, Grassi F, Aradhya MK, Prins B, Reynolds A, Chia JM et al (2011) Genetic structure and domestication history of the grape. Proc Natl Acad Sci USA 108:3530–3535

National Grape and Wine Initiative (2007) The impact of wine, grapes and grape products on the american economy: family businesses building value. https://www.ngwiorg/economic-impact-study_226html

Negrul AM (1938) Evolucija kuljturnyx form vinograda. Dokl Akad Nauk SSSR 8:585

Oberle GD (1938) A genetic study of floral morphology and function in cultivated forms of vitis N.Y. State Agric. Exp. Sta., Technical Bulletin, Geneva

OIV (2015) OIV report on the world vitivinicultual situation. http://www.oiv.int/en/oiv-life/2015-oiv-report-on-the-world-vitivinicultural-situationnbsp

Pickrell JK, Pritchard JK (2012) Inference of population splits and mixtures from genome-wide allele frequency data. PLoS Genet 8:e1002967

Picq S, Santoni S, Lacombe T, Latreille M, Weber A, Ardisson M, Ivorra S, Maghradze D, Arroyo-Garcia R et al (2014) A small XY chromosomal region explains sex determination in wild dioecious *V. vinifera* and the reversal to hermaphroditism in domesticated grapevines. BMC Plant Biol 14:229

Purugganan MD, Fuller DQ (2009) The nature of selection during plant domestication. Nature 457:843–848

Qiu S, Bergero R, Charlesworth D (2013) Testing for the footprint of sexually antagonistic polymorphisms in the pseudoautosomal region of a plant sex chromosome pair. Genetics 194:663–672

Renner SS (2016) Pathways for making unisexual flowers and unisexual plants: moving beyond the "two mutations linked on one chromosome" model. Am J Bot 103:587–589

Riaz S, De Lorenzis G, Velasco D, Koehmstedt A, Maghradze D, Bobokashvili Z, Musayev M, Zdunic G, Laucou V et al (2018) Genetic diversity analysis of cultivated and wild grapevine (*Vitis vinifera* L.) accessions around the Mediterranean basin and Central Asia. BMC Plant Biol 18:137

Roberts P, Hunt C, Arroyo-Kalin M, Evans D, Boivin N (2017) The deep human prehistory of global tropical forests and its relevance for modern conservation. Nat Plants 3:17093

Romero Navarro JA, Willcox M, Burgueño J, Romay C, Swarts K, Trachsel S, Preciado E, Terron A, Delgado HV et al (2017) A study of allelic diversity underlying flowering-time adaptation in maize landraces. Nat Genet 49:476–480

Ross-Ibarra J, Morrell PL, Gaut BS (2007) Plant domestication, a unique opportunity to identify the genetic basis of adaptation. Proc Natl Acad Sci USA 104 (Suppl 1):8641–8648

Sankararaman S, Mallick S, Dannemann M, Prüfer K, Kelso J, Pääbo S, Patterson N, Reich D (2014) The genomic landscape of Neanderthal ancestry in present-day humans. Nature 507:354–357

Schröder S, Kortekamp A, Heene E, Daumann J, Valea I, Nick P (2015) Crop wild relatives as genetic resources–the case of the European wild grape. Can J Plant Sci 95:905–912

Spigler RB, Lewers KS, Main DS, Ashman TL (2008) Genetic mapping of sex determination in a wild strawberry, Fragaria virginiana, reveals earliest form of sex chromosome. Heredity (Edinb) 101:507–517

Suarez-Gonzalez A, Lexer C, Cronk QCB (2018) Adaptive introgression: a plant perspective. Biol Lett 14:20170688

Tennessen JA, Wei N, Straub S, Govindarajulu R, Liston A, Ashman T-L (2018) Repeated translocation of a gene cassette drives sex chromosome turnover in strawberries. PLoS Biol 16:e2006062

This P, Lacombe T, Thomas MR (2006) Historical origins and genetic diversity of wine grapes. Trends Genet 22:511–519

VanBuren R, Zeng F, Chen C, Zhang J, Wai CM, Han J, Aryal R, Gschwend AR, Wang J et al (2015) Origin and domestication of papaya Yh chromosome. Genome Res 25:524–533

Vavilov N (1992) Origin and geography of cultivated plants. Cambridge University Press, Cambridge

Venuti S, Copetti D, Foria S, Falginella L, Hoffmann S, Bellin D, Cindrić P, Kozma P, Scalabrin S et al (2013) Historical introgression of the downy mildew resistance gene Rpv12 from the Asian species Vitis amurensis into grapevine varieties. PLoS ONE 8:e61228

Vouillamoz JF, McGovern PE, Ergul A, Söylemezoğlu G, Tevzadze G, Meredith CP, Grando MS (2006) Genetic characterization and relationships of traditional grape cultivars from Transcaucasia and Anatolia *Plant Genetic*. Resources 4:144–158

Wan Y, Schwaninger HR, Baldo AM, Labate JA, Zhong GY, Simon CJ (2013) A phylogenetic analysis of the grape genus (*Vitis L.*) reveals broad reticulation and concurrent diversification during neogene and quaternary climate change. BMC Evol Biol 13:141

Wang L, Beissinger TM, Lorant A, Ross-Ibarra C, Ross-Ibarra J, Hufford MB (2017) The interplay of demography and selection during maize domestication and expansion. Genome Biol 18:215

Wright SI, Bi IV, Schroeder SG, Yamasaki M, Doebley JF, McMullen MD, Gaut BS (2005) The effects of artificial selection on the maize genome. Science 308:1310–1314

Xu Y, Gao Z, Tao J, Jiang W, Zhang S, Wang Q, Qu S (2016) Genome-wide detection of SNP and SV variations to reveal early ripening-related genes in grape. PLoS ONE 11:e0147749

Zeven AC (1973) Dr. Th. H. Englebrecht's views on the origin of cultivated plants. Euphytica 22:279–286

Zhang Q, Liu C, Liu Y, VanBuren R, Yao X, Zhong C, Huang H (2015) High-density interspecific genetic maps of kiwifruit and the identification of sex-specific markers. DNA Res 22:367–375

Zhou Y, Massonnet M, Sanjak JS, Cantu D, Gaut BS (2017) Evolutionary genomics of grape (*Vitis vinifera* ssp. vinifera) domestication. Proc Natl Acad Sci USA 114:11715–11720

Zohary D, Spiegel-Roy P (1975) Beginnings of fruit growing in the old world. Science 187:319–327

Zohary D (1996) The mode of domestication of the founder crops of Southwest Asian agriculture. In: The origins and spread of agriculture and pastoralism in Eurasia, pp 142–158

4 Grape Archaeology and Ancient DNA Sequencing

Maria Rosa Guasch-Jané

Abstract

The cultivation and domestication of the grape appear to have occurred between 7000 and 4000 BC. The archaeological and historical evidences suggest that the domestication of the grapevine took place in the Near East. Nevertheless, whether a single origin or secondary independent grapevine domestications occurred and where they happened remains so far unanswered. Wine has had an important role in religious rituals since antiquity. In mythology and theology, wine was symbolic of the power to revitalize and rebirth. In ancient Egypt, wine was daily served to the gods by the Pharaoh and the priests in ritual ceremonies in the Egyptian temples. In daily life, wine was an enjoyable drink consumed by the elite in festivals, banquets and funerals. Further, the grape was one of the most important fruits in the classical Mediterranean civilizations and grapevines and the wine were widely spread through trade sea routes. This chapter presents an overview of the archaeological evidence for wine culture in the ancient Near East, Egypt and the Mediterranean region. It also presents a discussion of the chemical and morphological research methods and paleogenomic analyses that have been applied to ancient grape and plant material.

4.1 Grape Archaeology

The cultivation and domestication of the grapevine appear to have occurred between 7000 and 4000 BC. The place and period of the original domestication and biogeographical history of *Vitis vinifera* L. (domesticated grapevine) remain largely unknown, and it is likely that secondary independent domestications took place in a complex, long-term and multi-locale process (Zohary 1996; Grassi et al. 2003; This et al. 2006; Arroyo-García et al. 2006; Terral et al. 2010; Bouby et al. 2013; Pagnoux et al. 2015; see Chap. 3).

Despite the important corpus of bioarchaeological, morphological, historical and genetic data available, the identity of former cultivars, history biogeography and mechanisms of grapevine domestication remain obscure (Terral et al. 2010). The archaeological and historical evidences suggest that primo-domestication of the vine occurred in the Near East, before spreading to adjacent regions such as Egypt and Lower Mesopotamia (c.3500–3000 BC), and then further dispersal around the Mediterranean (This et al. 2006). However, there is evidence for secondary domestications in Sicily

M. R. Guasch-Jané (✉)
Sorbonne University, 1, Rue Victor Cousin, 75005 Paris, France
e-mail: mariarosaguasch@hotmail.com

D. Cantu and M. A. Walker (eds.), *The Grape Genome*,
Compendium of Plant Genomes, https://doi.org/10.1007/978-3-030-18601-2_4

(Grassi et al. 2003) and in the Western Mediterranean (Arroyo-García et al. 2006). Crucial unanswered questions regarding whether the process was rapid or slow and the related geographical area had single or multiple-origins remain (Bouby et al. 2013; Zhou et al. 2017; see Chap. 3).

Two forms of the grapevine coexist in Eurasia and north of Africa: cultivated (*Vitis vinifera* subssp. *vinifera*) and wild (*Vitis vinifera* subssp. *silvestris*) vines (Renfrew 1996; This et al. 2006; Arroyo-García et al. 2006). Critical was the shift from sexual reproduction—in the wild—to vegetative propagation—under domestication—and the change from dioecious to a hermaphroditic plant, able to pollinate itself (Zohary 1996; This et al. 2006; Zohary et al. 2012; Zhou et al. 2017; see Chap. 3).

The historical separation into subspecies was based on morphological differences, and the wild form is believed to be the ancestor of present cultivars (Zohary 1996; Zohary et al. 2012). Nevertheless, the ancestral cultivars and the process by which they diversified through time are not well known. Resolving this issue would be important for understanding the origin of current grape cultivars and the processes involved in the domestication of woody plant species.

Wine was the earliest fermented beverage since, contrary to beer production, grapes only need to have their skins broken open to release the juice (Singleton 1996). Although exactly where wine was first made is still uncertain, early evidence of wine production is suggested by several research studies (McGovern et al. 2017; Valamoti et al. 2007; Garnier and Valamoti 2016).

The archaeological evidences in regard to winemaking mainly include iconography (paintings, reliefs and mosaics), texts (ostraca and papyrus), artefacts (wine jars, cups and strainers), wine presses and cellars, as well as organic material (grape pips or berries, vine wood and wine remains). These materials had been recorded from a diversity of archaeological contexts such as houses, burials, winemaking installations, storage rooms and even ancient Mediterranean shipwrecks. Wine jars in shipwrecks, mostly dated to the end of the Late Bronze Age and discovered through underwater archaeology, have revealed ancient sea trade networks and the transport of wine along the Mediterranean. Indeed, amphorae studies are important for world economic history.

Wine has also had an important role in religious rituals since antiquity, and grape was one of the most important fruits in the classical Mediterranean world. Wine was a drink of the gods in ancient Egypt (Osiris), Greece (Dionysus) and Rome (Bacchus), and its flavour and alcoholic content was highly appreciated. In daily life, wine was an enjoyable drink which was first consumed by the elite in festivals, banquets and funerals and later widely extended to the Mediterranean region mainly by Phoenicians, Greeks and Romans.

4.1.1 Archaeological Evidences of the Wine Culture in the Near East and Mediterranean Region

The earliest evidence of winemaking has been recently discovered (McGovern et al. 2017) in two Neolithic villages of Georgia on pottery fragments from Shulaveris Gora and Gadachrili Gora (c. 6000 BC). The first evidence for wine was reported from pottery jars found at Hajji Firuz Tepe in the Zagros Mountains in northern Iran (c. 5400 BC; McGovern et al. 1996). In a house at the Greek Neolithic village site of Dikili Tash (c. 4300 BC), a charred grapevine (*Vitis vinifera*) and jars containing grape pips with skins attached were found (Valamoti et al. 2007; Garnier and Valamoti 2016). The oldest winery was located in the Areni-1 cave in southern Armenia (c. 4000 BC) containing wine jars, grape pips and skins (Barnard et al. 2011). Prehistoric grapes from the Palaeolithic/Mesolithic (c. 11000 BC), early Neolithic (c. BC) and late Neolithic (c. BC) onwards have been recovered from Greece (Renfrew 1966, 1996).

These finds do not provide definitive evidence for grapevine cultivation (Zohary and Hopf

1993; Zohary 1996). However, an early exploitation of wild grapes is suggested, as well as the selection of different types for different purposes, such as raisins, dessert grapes and different types of wine and vinegar (Renfrew 1996). In ancient Greek shipwrecks, DNA evidence from grape, oil and different herbs in amphoras has been suggested (Hanson and Foley 2008; Foley et al. 2012). In Egypt, inscriptions on wine amphoras—two-handled jars—included relevant information about the harvest and wine production (see 4.1.4 below). In Greco-Roman stamped amphoras, the origin of a wine from Rhodian, Thasian, Cypriot, Cnidian or Egyptian sources has been documented (Panagou 2016). The explanation would be that those wine amphoras, for example, from the Thasos island, were produced for foreign markets (Tzochev 2016). Sometimes only the name of a person related to the amphora production, that could possibly be the pottery maker, is indicated (Tzochev 2016).

Phoenician and Greek trade networks distributed the wines and spread the cultivated grapevine across the Mediterranean. Amphoras discovered inside shipwrecks have allowed the study of sea routes, ancient trade and gift exchange, for instance, between the Aegean and Cyprus (Pulak 2001; Demesticha 2011). Furthermore, great economic activity is represented by Greek and Roman coins depicting grapes and god Dionysus (BMCollection 1 2019; BMCollection 2 2019). Roman presses have been reconstructed by experimental archaeologists to understand the winemaking procedures.

The wine, bread and oil, the so-called Mediterranean triad, were important agricultural crops for the economy and food sustainability and become the basic food of ancient Greece and the Roman Empire. They continue to have a fundamental role throughout the Mediterranean region today. The extensive Bronze Age cultivation of olives and grapes is documented by the appearance of numerous presses and by the remains of storage facilities for olive oil and wine (Zohary et al. 2012).

The role of wine in ancient Greece was described by Plato and Xenophon of Athens in the 'symposia': wine parties where wealthy men enjoyed drinking wine in special seats while discussing philosophy. Athenaeus of Naucratis—a Greek city in the Nile Delta—who lived during the second century–third century BC, described in his book *Deipnosophistae* (Sophists at dinner) that the wine from Thasos island in the Aegean was the most expensive. In Greek and Roman times, wine was usually diluted before consumption. Varieties of special vessels were used in Greece for mixing wine with water (*krater*), to cool before consumption (*psykter*) and serving wine (*olpe*).

Wine-drinking scenes were represented in mosaics such as at the House of Dionysus, dated to c. 200 BC, in Paphos, Cyprus. The Roman writer Columella (c. 4–70 AC) in *De Re Rustica* (On Rural Affairs) described how to plant and prune the grapevine, and how to produce, give flavour and preserve the wine (Columella 2012). Roman naturalist Pliny the Elder (c. 23–79 AC) in his book *Historia naturalis* (Natural history) reported on viticulture, varieties of vines and on the Italian and foreign wines. Early reports of archaic Egyptian wines (McGovern et al. 2009) note that they were flavoured with aromatic herbs or spices. In ancient Greece and Rome, resins could be added for preservation. The modern Greek retsina wine with resin from Aleppo pine is currently a protected designation of origin in Europe (EU Reg 2010).

In antiquity, wine was used for the re-establishment of good health and doctors considered it as a remedy (Jouanna 2012). The oldest references describing the medicinal role of wine are Mesopotamian medical texts written on Sumerian cuneiform tablets (c. 2000 BC) and Egyptian medical papyri (c. 1800 BC); the latter believed to have been copied from earlier texts possibly dating back to c. 3000 BC. In addition, the Greek physician Hippocrates of Kos (c. 460–370 BC) recommended wine as part of a healthy diet. Ancient Greek medicine knowledge could have originated in the Greek Alexandrian medical school (established c. 300 BC) whose teachings and writings were later spread across the Mediterranean region.

4.1.2 Wine Culture in Ancient Egypt

The ancient Egyptian wine culture is one of the world's most ancient and possesses the most extensive records. In Egypt, wine was a luxury drink that was mostly served by the royal family and nobles, and it was enjoyed in banquets with guests and music for entertainment. Meanwhile, the common people had only access to wine in annual religious celebrations. The symbolism of wine and its relationship with the funerary world was first documented in the Early Dynastic Period (c. 3000 BC) with large quantities of jars found in royal tombs at Abydos and Saqqara (Meyer Ch 1986; Murray 2000). These reports provide evidence that wine production in Egypt was highly sophisticated by the beginning of this period. Wine was mainly produced in the delta region from the beginning of the Early Dynastic, and it was later expanded to the Nile Valley and the Western oases. During the New Kingdom Period (1539–1075 BC), the inscriptions on jars indicate the wine came mainly from vineyards in the eastern and western Nile delta.

In Egyptian mythology, wine symbolized the blood of Osiris, the god of the underworld, the dead and the resurrection. God Osiris—who died in a violent death—was 'foremost of the westerners', the first to undergo resurrection (Griffiths 1982; Hornung 1996), and 'Lord of Wine' in the late Old Kingdom (2575–2150 BC) Pyramid Texts (Allen 2005). A relation was established between Osiris and wine because of the timing of grape harvest with the annual flood of the Nile river, which turned to a wine red colour caused by ferruginous sediments from the Ethiopian highlands (Poo 1986). Grapes and wine were considered a symbol of resurrection as represented in the Book of the Dead of the royal scribe Nakht (Fig. 4.1a), and this idea persists in Christian Coptic iconography today.

4.1.3 Grape and Wine Iconography from Ancient Egypt

The Egyptian iconographic records include vineyards, bunches of grapes, viticulture and wine production scenes, divinities, wine-offering rituals and funerary banquets. Among the representations, wall paintings, reliefs and stone reliefs or wooden panels are found. Offering tables with food and wine for the deceased have been found in Early Dynastic tombs, whether carved in stone walls or wood reliefs. The earliest is the wooden panel from Hesyre's tomb (Spencer 1993) dated to the Third Dynasty (c. 2650 BC) at Saqqara.

In the earliest funerary offering liturgies, the Pyramid Texts—carved on the inner walls—at the burial chamber of King Unis (c. 2325 BC) of the Old Kingdom Period (2575–2150 BC) at Saqqara, a list of five wines is recorded: Delta wine, wine in *abesh* jar, Buto wine, Mariut wine and Pelusium wine (Allen 2005). These wines were presented to the deceased King during the food-offering ritual to help him ascend to heaven for rebirth and became standard features in the decoration of royal tombs until the Roman Period (c. 395 AD). In the Egyptian temples, a daily offering of two bowls of wine to the Gods was made by the Pharaoh and the priests in religious rituals (Fig. 4.1b). Wine was also offered during great occasions such as the foundation and coronation ceremonies, and the *Heb Sed*—royal jubilee—and Valley festivals (Poo 1995). Moreover, in the walls of New Kingdom Theban royal tombs, such as Horemheb's tomb [KV57] and Nefertari's tomb [QV66], two bowls of wine are offered to the Gods, once again documenting the importance that wine had for the ancient Egyptian civilization.

Viticulture and winemaking scenes were depicted on the walls of the Egyptian private tombs from the Old Kingdom period (2575–2150 BC) through the Greco-Roman Period (332 BC–395 AD). The 'Study of the viticulture and oenology scenes in Egyptian tombs', a 2011 to 2014 scientific project, has recorded and studied these scenes, together with the texts associated with the images, and a mission to Egypt was permitted to photograph them (Guasch-Jané 2016; EGYWINE 2019). The main steps of the harvest and wine production are represented, and they are unique (Fig. 4.2a, b). The scenes-detail database with 92 records of scenes in tombs

(a)

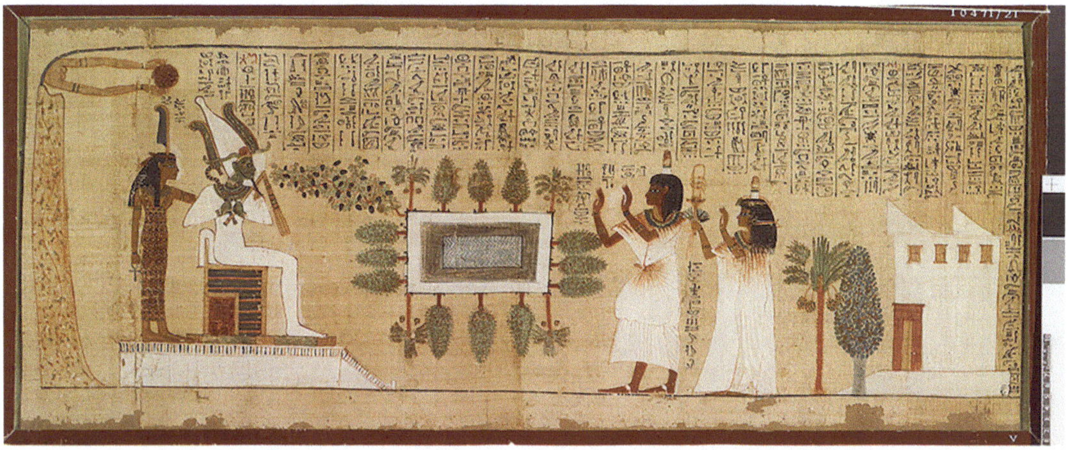

(b)

Fig. 4.1 **a** Royal scribe Nakht and his wife in the garden, in front of their house, adoring Osiris; the vine leads to the nose of Osiris, the resurrection god, symbolizing rebirth. Book of the Dead of Nakht, sheet 21 [EA 10471,21] at the British Museum in London, UK. ©Maria Rosa Guasch-Jané, with permission of the British Museum. **b** King Tutmosis III (1479–1425 BC) is offering two bowls of wine to the god Horus. Temple of Queen Hatshepsut at Deir el-Bahari, Western Thebes, Egypt. Dynasty 18. ©Maria Rosa Guasch-Jané

(Guasch-Jané 2016) is included in a georeferenced archaeological map of Egypt (Fig. 4.3a, b) that is presented in the project's dedicated website 'Wine of Ancient Egypt' (Fig. 4.3c, EGY-WINE 2019). The viticulture scenes include the steps represented in the tombs' scenes are such as taking care of the vine, grape harvest and counting the baskets. In the winemaking scenes, the steps represented are transporting grapes to the press, pressing grapes, heating and filtering

(a)

(b)

Fig. 4.2 **a** Viticulture and winemaking scene in two registers. From left to right, first register: workers pick up grapes, press them in a vat, sack-press to extract more juice, and finally counting the baskets of grapes; second register: counting wine jars, filling the jars, sealing and, finally, goats cleaning the vine. Tomb of Amenemhat [BH2] at Beni Hasan, Egypt. Dynasty 12. ©EGYWINE Project 2019. **b** Viticulture and winemaking scene. To the right, workers pick up dark grapes and put them in a basket. To the left, workers are pressing the grapes with their feet; besides, the red must is coming out from the press to a deposit and sealed wine jars are on top. Tomb of Nakht [TT52] at Sheikh Abd El-Gurna, Western Thebes, Egypt. Dynasty 18. [Shedid and Seidel (1996) The Tomb of Nakht, Mainz, p 57]

(elaboration of *shedeh*), pressing the remains in a sack-press, filling wine jars, fermentation, offerings to the goddess Renenutet, grape and wine tasting, sealing and labelling wine jars, counting the jars, transporting wine jars to a cellar, refrigeration during fermentation and storage of jars in the cellar. The sack-press consisted of a bag made of linen through the looped ends of which two poles were placed; the poles were twisted by two workers on each side the fifth man in the middle trying to keep the two sticks separated to allow squeezing of the grape berry remains, skins, pips and stalks in the bag (EGYWINE 2019). The sack-press might have permitted wines of different quality to be distinguished. This type of press evolved to fix a pole on one side so that fewer workers were needed—e.g. for the tomb of Baqet II at Beni Hasan, dynasty 12, Middle Kingdom Period (EGYWINE 2019).

In many cases, only the essential parts are represented in the scenes to achieve the supposed magical effect of making wine available for the deceased in the afterlife. The scenes of viticulture

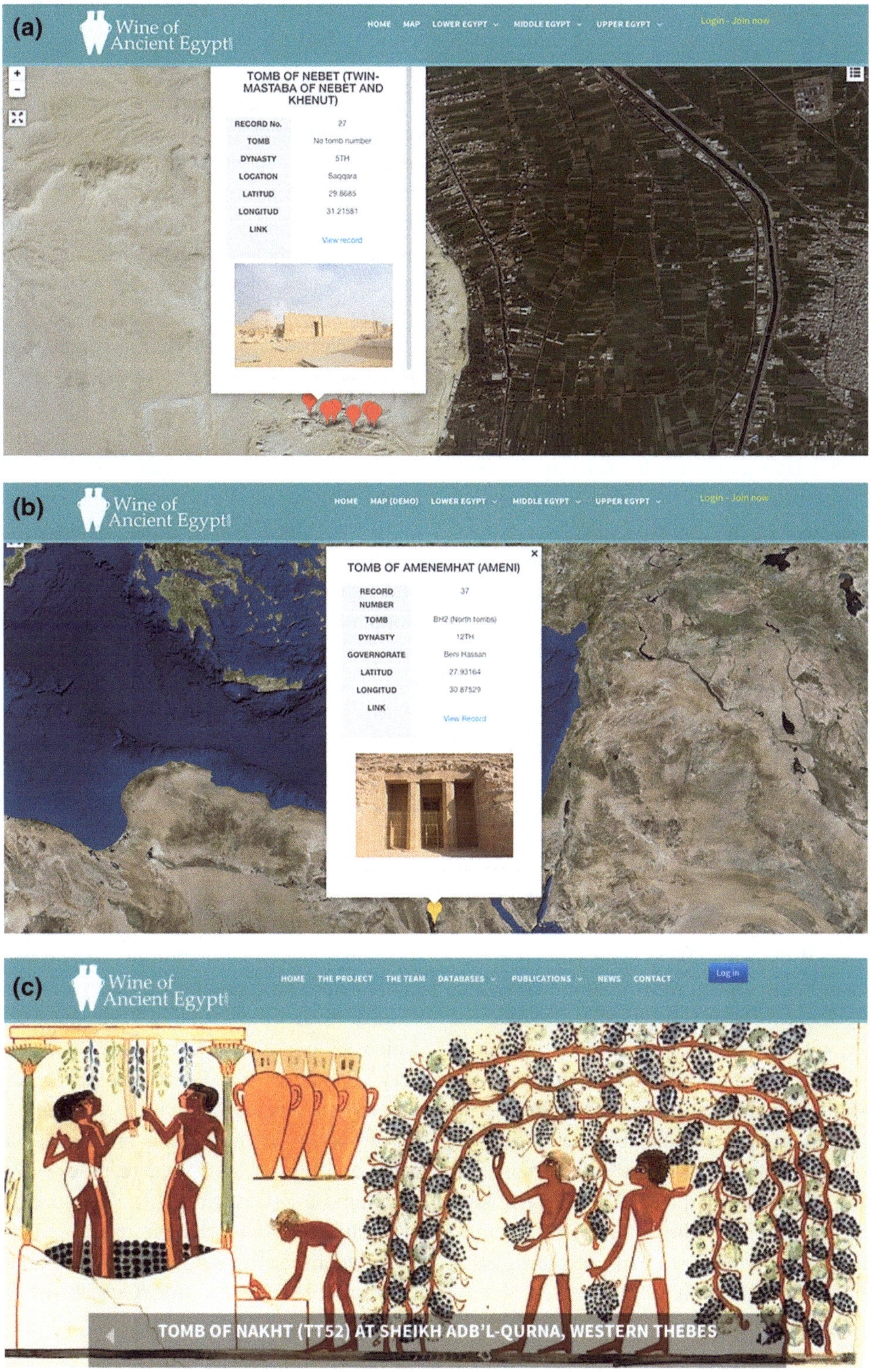

Fig. 4.3 **a** Archaeological map of Egypt with the location of the viticulture and winemaking scenes in the Egyptian private tombs, as shown on the website 'Wine of Ancient Egypt' with the entry record for the tomb of Nebet at Saqqara (Upper Egypt). ©EGYWINE Project 2019. **b** Archaeological map of Egypt showing the entry record for the wine scene in the tomb of Amenemhat at Beni Hasan (Middle Egypt). ©EGYWINE Project 2019. **c** The website 'Wine of Ancient Egypt' dedicated to the multidisciplinary study of the ancient Egyptian wine culture. ©EGYWINE Project 2019

and winemaking from the Egyptian private tombs are an extraordinary source of information for investigations into the evolution of winemaking in Egypt during three thousand years. Their record and study become an important tool for the future documentation and preservation of the archaeological heritage of Egypt.

An exceptional scene in the tomb of Huya at el-Amarna shows the Amarna royal family—Pharaoh Akhenaten, Queen Nefertiti and Queen Mother Tiye—being served wine at dinner (Davies 1905). Moreover, scenes of wine drinking by guests at parties with servants and musicians were popular paintings in private Theban funerary tombs of the Eighteenth Dynasty (1539–1292 BC) as shown in Nebamun's tomb chapel (Parkinson 2008).

Grapes and grapevine leaves are represented on ceilings and ornamental friezes of tombs from the New Kingdom Period (1539–1075 BC) at the Theban necropolis (Cherpion 1999). Sennefer has a painted vine symbolizing the rebirth of the deceased (Desroches-Noblecourt 1985). Wine jars surrounded with grapes and vine leaves for decoration and refreshment were also represented (Parkinson 2008).

With respect to the divinities related to wine, Osiris (Fig. 4.1a) and Renenutet are often found. The snake goddess Renenutet (Thermuthis in Greek) was a goddess of the harvest who was honoured in shrines erected in harvest fields and vineyards (Wilkinson 2003). The cult to Renenutet ensured the supply of wine to the deceased, allowing the enjoyment in the afterlife, and her presence with offerings, libations and hymns to guarantee good wine production. Renenutet was a protective goddess in the viticulture and winemaking scenes of the private tombs at Western Thebes from the Eighteenth Dynasty. In the tomb of Mentiwy [TT172] at El-Khokha, Renenutet is controlling wine production. Shezmu was the wine-press deity responsible for the wine production (Poo 1995).

4.1.4 Wine Archaeology from Ancient Egypt

The Egyptian archaeological artefacts and texts relevant to wine include wine jars and wine inscriptions, ostraca (inscribed pottery shards), cups and strainers, statues of divinities and papyrus, and grape and wine remains have been found. Nevertheless, in an archaeological context there is a lack of wine presses or wine deposits for study.

The earliest grapes found in Egypt come from the Predynastic Period (4000–3100 BC) from archaeological sites in Tell Ibrahim Awad and Tell el-Farain in the Nile Delta and from the Tomb U-j at Umm el-Qa'ab at Abydos (Murray 2000). During the First and Second Dynasties (c. 2950–2650 BC), pottery wine jars were placed in royal tombs at Abydos and Saqqara as funerary offerings for the deceased. Some of the jars and stoppers were inscribed with the name of the vineyard where the wine was produced and the king's name (Petrie 1901; Emery 1958; Spencer 1993). These jars are large, about 1 m in high, and have a mud sealing on top, for instance, the jar (Fig. 4.4a) from the tomb of Hemaka at Saqqara, which was dated to the reign of Den of the First Dynasty (c. 2950–2775 BC).

During the New Kingdom Period (1539–1075 BC), two-handled wine jars' amphoras (Fig. 4.4b, c) were written in hieratic—cursive—script by hand in black ink to indicate details about the harvest and wine production: the year of reign—vintage year—the name of the product—*irep*, which is wine, or *shedeh*—the quality—good, very good or excellent—and sweetness, the provenance—delta, western oases, Menfis, etc.—the property—royal, temple or private—and the name and title of the winemaker—ex. chief vintner Ramose—(Guasch 2010). The specific order of writing this information might indicate that well-established rules regarding presentation and labelling of wine existed (Guasch 2010). The

Fig. 4.4 **a** Wine jar with mud stopper [*Journal d'Entrée* number 69772 of the Egyptian Museum in Cairo] from Hemaka's tomb at Saqqara, Egypt. Dynasty 1. ©Maria Rosa Guasch-Jané, with permission of the Egyptian Museum in Cairo. **b** Wine amphora [JE 62303] with hieratic writing: 'Year 4, wine from the Estate-of-Aten in the Western River, chief vintner Nen' from Tutankhamen's tomb [KV62] at Western Thebes, Egypt. Dynasty 18. ©Maria Rosa Guasch-Jané, with permission of the Egyptian Museum in Cairo. **c** Wine inscription in Tutankhamun's amphora [JE 62305] bearing the inscription: 'Year 4, *shedeh* of very good quality of the Estate-of-Aten of the Western River, chief vintner Kha^{c}y'. Dynasty 18. ©Maria Rosa Guasch-Jané, with permission of the Egyptian Museum in Cairo

same information is found in a modern bottle of wine, and the consumer will consider the same parameters when choosing a bottle of wine (Guasch 2010).

Furthermore, Egyptian wine jars were protected with a clay capsule—or lid—that covered the mouth of the jar to prevent contamination of the contents and a mud sealing—or stopper—on top, which closed the jar completely, and enclosed the whole of the neck of the amphora (Guasch 2010). A seal in hieroglyphic script was stamped on the mud stopper indicating the product, the quality and the origin—or estate—on the stopper, which was a summary of the data on the hieratic inscription (Guasch 2010). This information on the jar and/or the stopper recorded the economic circuit of the wine: origin, production and destination (Tallet 1998).

In the tomb of Tutankhamun (1332–1322 BC), discovered intact by Howard Carter in 1922 in the Valley of Kings [KV62] at Western Thebes, 26 wine amphorae—two-handled jars—and seven one-handled jars of attenuated form were found (Holthoer 1993; Guasch 2010). All of them were found in the annexe chamber except for three amphorae that had been placed inside the burial chamber lying on the ground, between the outermost shrine and the walls, surrounding the mummified body of the King to the east, west and south, respectively (Guasch-Jané 2011). Chemical analysis revealed that those three amphoras contained three different types of wine: red (Guasch-Jané et al. 2004), white (Guasch-Jané et al. 2006b) and *shedeh*, a red grape wine with a different preparation (Guasch-Jané et al. 2006a). A ritual use of the three wines in the royal burial chamber to rebirth was suggested, while the rest of wines found in the annexe might had been offerings of the usual kind for the sustenance of the King in his afterlife (Guasch-Jané 2011).

No textual references to white wine—or to red wine—from the Dynastic Period (3100–343 BC) have yet been found. The first mention of the presence of white wine in Egypt is from the ancient Greek writer Athenaeus of Naucratis (third century AD) in *Deipnosophistae* (Sophists at dinner) described the Mareotis wine in the area of Lake Mariout near Alexandria, north-west coast of Egypt, as 'excellent, white and enjoyable, aromatic, easy to assimilate, fine and does not go to one's head apart from also being diuretic…' (Athenaeus 1961). Earlier, the Latin poet Virgil, who lived during the first century BC, in his book *Georgicon* (Georgics) detailed a list of vineyards and highlighted the vines from Thasos and the white grapes from Mariut (Virgil 1586). Food including grapes and different kinds of fruits, and wine jars and bronze strainers for serving wine, were found at the tomb of Kha and Merit [TT8] and dated c.1350 BC, discovered by Ernesto Schiapparelli in 1906 at Deir el-Medina (Vassilika 2010). A few thousands of wine inscriptions have been found at the Theban necropolis, mainly in the archaeological sites of Deir el-Medina—the city of the tomb builders—the Ramesseum temple of Ramses II (1279–1213 BC), and the royal palace of Malkata of Amenhotep III (1390–1353 BC). Moreover, Late Roman stamped amphoras have been found, for instance, in Naukratis (BM Collection3 2019c). The Egyptian temples were associated with large numbers of vineyards in the Nile Delta region. For example, in the Papyrus Harris I a total of 433 vineyards were overseen by the Theban temples during the reign of Ramesses III (1187–1156 BC). In the medical papyri, such as the Ebers papyrus (c. 1600 BC), grapes and wine were included in the pharmaceutical formulations.

The 'Ancient Egypt's wine rebirth' research project (EGYWINE 2016–2018) studies how Egyptian wines were made to understand winemaking history, and advance the conservation of this heritage (EGYWINE DB 2019). The EGYWINE project collects and documents archaeological evidence, mainly pottery and organic material, of the entire process of grape cultivation and wine production in Egypt utilizing various scientific disciplines: archaeology, paleogenomics, history, oenology and semantics to reveal the Egyptian footprint on the history of wine culture. Following the previous studies, EGYWINE analyses the following five aspects: (a) the viticulture and winemaking scenes depicted on the walls of the private tombs from the Old Kingdom (2680–2160 BC) to the Greco-Roman Period (332 BC–395 AD); (b) the ancient Egyptian wine jars typology and material to know how the jars were made; (c) the wine inscriptions to reveal ancient winemaking procedures; (d) the ancient Egyptian bacteria and yeasts involved in the fermentation process and preservation of wine; and (e) the study of amphora wines. Furthermore, EGYWINE is recording and studying the main concentration of wine jars and wine inscriptions from the Predynastic Period to the New Kingdom Period (3800–1069 BC), and a database for the wine jars and wine inscriptions, which will be accessible through a dedicated website (EGYWINE 2019).

4.2 Chemical and Morphometric Analyses of Grapes and Wine Samples

Methodological, technical and analytical advances have provided new insights into research on archaeological grapes and wines. To study the colour type of ancient Egyptian wines, a method (Guasch-Jané et al. 2004) to detect archaeological residues of wine was developed using liquid chromatography and mass spectrometry in tandem (LC/MS/MS). Two biomarkers were identified in archaeological wine samples: tartaric acid, a distinctive grape marker, and syringic acid derived from malvidin-3Glu, which is primarily responsible for the red colour of grapes and young wines (Guasch-Jané et al. 2004; Guasch-Jané 2008). Malvidin-3-glucoside is the predominant anthocyanin of *Vitis vinifera* (Eurasian) grapes, and whether polymerized or not, it is partially converted upon alkaline fusion to syringic acid (Singleton 1996).

The results of analysing samples from Tutankhamun's wine amphoras confirmed that in Egypt, during the New Kingdom Period (1539–1075 BC), three different grape-derived products were made: red wine (Guasch-Jané et al. 2004), white wine (Guasch-Jané et al. 2006b) and *shedeh*, being a red grape wine with a different preparation (Guasch-Jané et al. 2006a). The study also revealed that both red and white wines were given the name *irp* and added new information to the inscriptions on these amphoras—the type of wine stored (Guasch-Jané 2008). The origin of the *shedeh*—an Egyptian word with no translation—was a mystery over the last century, with both pomegranates and grapes been proposed as the raw material (Tallet 1995; Guasch-Jané 2008). The elaboration of *shedeh* wines is mentioned in the Papyrus Salt 825 [British Museum EA10051] of the Late Period (715–332 BC). It was filtered and heated, but its botanical source remained unknown due to damaged papyrus (Guasch-Jané 2008). The analysis (Guasch-Jané et al. 2006a) confirmed that *shedeh* was a red grape wine and settled the discussion about its botanical source that lasted over a hundred years.

Geometric morphometric studies (Terral et al. 2010; Milanesi et al. 2011, 2014; Pagnoux et al. 2015) of archaeological grape seeds using elliptic Fourier transform method combined with multivariate statistical methods have been developed in recent years. Morphometric studies are non-invasive and considered ideal for rare and valuable archaeobotanical remains (Milanesi et al. 2011). Shape comparison between current forms and archaeological material may elucidate the timing of domestication events, origins of cultivars, exchange and cultural interactions (Terral et al. 2010). However, according to Zohary et al. (2012) pip morphology cannot be regarded as a completely safe diagnostic trait for distinguishing between wild and domesticated *Vitis* remains in archaeological excavations. The external profile of archaeological and modern grape seeds is a good phenotypic descriptor to investigate the origin and diffusion of *Vitis vinifera* L. (Milanesi et al. 2014). Whereas size, shape and colour of berries are phenotypic traits, which might have been traditionally selected by humans, seed shape was probably not a target of selective pressures (Terral et al. 2010). The characterization of the seed shape and size of modern and archaeological material has allowed investigation of grapevine diversity (Pagnoux et al. 2015), established hypotheses of relationships, discriminated among different groups of grape varieties, and discriminated between domesticated and wild subspecies of the grapevine (Terral et al. 2010; Bouby et al. 2013). Seeds of wild and domesticated grapevine display dissimilarities which allow the discrimination between both subspecies: *V. vinifera* grapevines have small and roundish seeds with short stalks, while pips from cultivars are more elongated, with longer stalks (Pagnoux et al. 2015).

Nevertheless, morphometric analysis is not enough to establish the descent of modern material from palaeobotanical specimens and genetic analysis of ancient grape seeds would enable new comparisons of ancient profiles and contemporary cultivars (Milanesi et al. 2011, 2014). Phylogenetic comparison of palaeobotanic and modern materials will be sustained by improved methods, larger databases and interdisciplinary studies (Milanesi et al. 2014). However, degradation must be considered as most of the pips recovered from Greece (Renfrew 1996) are preserved by carbonization, particularly the wild samples that are very small, while the cultivated ones are both carbonized and mineralized. However, not all ancient pips are carbonized (Renfrew 1996). Shape characterization combined with genetic data should allow a better understanding of the changes that have occurred during domestication (Terral et al. 2010). Additionally, new application of 3D scanning technology to pottery, wine jars for instance, is an invaluable tool for archaeologists to investigate typologies and differences among production technologies (Karasik et al. 2018). The interdisciplinary project Viniculture (2017–2020) investigates grapes and wines from the Neolithic to the Middle Ages in France using methodological advances in morphogeometry and genomics to describe the grapevine's diversity and analyse their spatial and chronological dynamics. Plant remains and archaeological containers (pottery and wood jars) collected according to strict sampling procedures are analysed, and this interdisciplinary approach combines archaeobotany, geometric morphometrics, archaeogenetics, biochemistry and experimental archaeology (Bouby 2017).

4.3 Ancient DNA Sequencing

The invention of polymerase chain reaction, or PCR, technology allows to obtain millions of copies of the few remaining ancient DNA [aDNA] molecules, and greatly increased experiments using aDNA. The invention of PCR and subsequent modifications to next-generation or high-throughput sequencing (NGS or HTS) technologies, as for whole-genome sequencing, have accomplished the identification of mitochondrial genomes from human (Green et al. 2010; Krause et al. 2010) and animal (Palkopoulou et al. 2015) fossils. aDNA research is rapidly developing particularly within vertebrates, especially humans. Recent advances in sequencing technologies have permitted plant aDNA analyses from fossil samples that enable the molecular reconstruction of palaeofloras (Parducci et al. 2017). Nevertheless, difficulties must be considered, and some wrong results resulting from sample contamination or false positives have appeared, especially with regard to the study of aDNA from human remains (Gilbert et al. 2005), probably because such aDNA is the most investigated. NGS/HTS technologies and sequence data analysis have increased the single-gene and whole-genome sequences of plant genomes, although they are difficult to assemble because of their large size and complex, high ploidy, high heterozygosity and the presence of a large number of repeat sequences (Basantani et al. 2017).

The most relevant aspects to be considered when undertaking aDNA studies are selection and sampling of archaeological material, authentication of ancient origin, contamination and false positives, DNA preservation, new technological approaches and improved methods for aDNA extraction and analysis, as well as the existence of large databases for phylogenetic comparisons and advanced statistical methods for data analysis.

A major concern is contamination of laboratories and equipment. An aDNA laboratory work needs extensive multi-strategy measures to minimize sample contamination: an isolated and exclusively dedicated ancient DNA facility, rigorously separated from work involving modern DNA; a protocol to maintain a sterile environment including all personnel wearing protective full-body suits, hood, gloves, mask and shoe covers at all times; treatment of the laboratory equipment and materials with bleach; decontamination with nightly UV irradiation and isolation from post-PCR laboratories (Hofreiter et al. 2001; Champlot et al. 2010; Fulton 2012; Bennett et al. 2014).

How easy is to generate erroneous data through contamination? When categorizing risk in aDNA studies, it is widely accepted that human aDNA data have the most problems and are categorized as high risk, while plant aDNA is medium risk (Gilbert et al. 2005). The authenticity of samples in aDNA studies can be proved by patterns of damage identification: the major factor is cytosine deamination, C-to-T transitions, causing nucleotide misincorporations in ancient DNA, and being of shorter sequence length (Briggs et al. 2007; Sawyer et al. 2012). A significant list of criteria of authenticity to avoid or prevent contamination from exogenous DNA has been established (Pääbo et al. 2004; Gilbert et al. 2005).

4.3.1 DNA Preservation of Archaeological Material

Sampling and selection of archaeological material are crucial steps for achieving useful results. Before proceeding to the selection of archaeological material for study, some questions should be considered: What type of substrates exist for a botanical species to study? Are the samples in good preservation? To whom do the samples belong? Are museums' authorizations for sampling possible? Sources of damage or even degradation causing high DNA fragmentation of plant material from museums have been identified, and an exponential relation between length of the fragment and year of the collection has been confirmed (Weiss et al. 2016). According to Milanesi et al. (2014), the use of seeds from archaeological museums is difficult because DNA studies are destructive of specimens. To avoid problems of sampling permissions, a non-destructive method was proposed for teeth, bones and skin samples up to 146 years old; although no damage to the specimens was detected, the amount of extractable DNA decreases with increasing numbers of successive extractions (Rohland et al. 2004; Rohland and Hofreiter 2007). Highly degraded DNA can be due to conditions of the substrate provenance, storage or preservation, and degradation may influence analytical success (Lindahl 1993). Ancient DNA might have different types of damage, such as strand breaks, DNA crosslinks and oxidative or hydrolytic lesions, and the knowledge of the effects on DNA is still limited; however, aDNA is invariably of shorter length (Pääbo et al. 2004).

DNA preservation in human, animal and plant remains depends on the types of substrates. For instance, bones and teeth preserve DNA quite well and are abundant in the fossil record, much better than soft tissue. In the recent years, the sources of aDNA have been extended to include archaeological artefacts and archaeobotanical remains (Green and Speller 2017), dental calculus (Weyrich et al. 2017), palaeofeces and coprolites (Wood et al. 2012; Bennett et al. 2016), hair (Rasmussen et al. 2011) or even ostrich eggshells (Demarchi et al. 2016). Recently, good preservation of high endogenous DNA is revealed on the petrous bone, cochlea (Pinhasi et al. 2015; Margaryan et al. 2018). Studies on coprolites, often found in caves in dry areas, are useful for the diet and ecology of extinct animals (Wood et al. 2012).

Plant material can vary from small seeds—as in case of grape pips—to large stone fruits or complete cereals. Furthermore, the reproducibility within the same individual is impossible to fulfil in the case of small seeds (Schlumbaum et al. 2008). The DNA from seeds tends to be preserved in archaeological sites only when they are charred, desiccated, frozen or deposited in anoxic conditions (Green and Speller 2017) and are among the most highly prized aDNA sources

(Di Donato et al. 2018). DNA preservation in plant archaeology depends foremost on the provenance and storage through time. Cold, dry and/or low oxygen environments are beneficial for DNA survival (Schlumbaum et al. 2008). Exposure to high temperatures, such as in the case of charring, can heavily fragment the DNA, while higher temperatures and longer exposure cause a greater destruction apart from spontaneous DNA decay (Lindahl 1993). Nevertheless, DNA was identified in charred archaeological wheat seeds (Allaby et al. 1997).

4.3.2 Ancient DNA Studies in Plant Archaeology

In the recent years, ancient DNA analyses of wheat (Bilgic et al. 2016), barley (Mascher et al. 2016), grapevines (Wales et al. 2016), pollen and other plant fossils from lake sediments (Parducci et al. 2017) and from historic plant collections from herbarium archives such as the olive family (Zedane et al. 2016) have been reported. The analysis performed on aDNA can shed light on phylogenetic questions concerning evolution, domestication and improvement of plant species as well as to help resolve problems related to the origin of the material and external contamination (Di Donato et al. 2018).

Ancient genomes from desiccated archaeobotanical remains provide information regarding the origin, early domestication and subsequent migration of crop species (Mascher et al. 2016). For instance, ancient charred wheat is reported to be similar to contemporary hexaploid wheat species, suggesting an early transitory state of hexaploid wheat agriculture from the Fertile Crescent towards Europe crossing present-day Turkey (Bilgic et al. 2016). The study of domestication and early crop evolution has largely been limited to the identification of key phenotypic, morphological and genetic changes between extant crops and their wild relatives (Da Fonseca et al. 2015). Documenting ancient diffusion routes of domesticated crops and how they were modified when introduced into new regions has long been a challenge (Da Fonseca et al. 2015). The use of nuclear DNA population genetic analysis of maize enabled the differentiation of selective forces during domestication and its adaptation to the climatic and cultural environment of the southwest USA (Da Fonseca et al. 2015).

Ancient DNA analyses can add new perspectives for the study of ancient plant populations and will provide higher taxonomic resolution and more precise estimation of abundance and relationships; however, key questions and challenges remain for plant aDNA studies (Parducci et al. 2017). One key question is the suitability of the chloroplast genome (plastome) to address archaeological and evolutionary investigations (Wales et al. 2016). In plant aDNA research, ribosomal DNA [rDNA] genes are of interest for ancient DNA research (Zedane et al. 2016), whereas plant mitochondrial [mtDNA] studies are rarer (Di Donato et al. 2018). Advanced molecular technologies for investigating ancient nuclear DNA [nuDNA] have been able to reveal a much greater potential since nuDNA carries several important loci (Wales et al. 2016; Di Donato et al. 2018). However, nuDNA is more susceptible to degradation and some polynucleotides are more damaged than others (Weiss et al. 2016).

4.4 Future Perspectives

Advances in DNA extraction methodology and sequencing technology have allowed for the study of archaeological plant remains. Ancient genome studies of grapevine might shed light on significant questions concerning the origin, evolution and domestication of grape, on the history of viticulture and how aDNA degrades or persists.

The type of substrate used for aDNA extraction is essential and crucial for the success of these projects. In the case of grapevine, the substrates would mainly be wood, seeds and wine samples from pottery jars. The majority of models predicting DNA degradation and fragmentation have been based on ancient bone, and understanding the methods by which DNA may bind to non-organic substrates like pottery is of particular importance (Green and Speller 2017). According to Nistelberger et al. (2016), although HTS of charred archaeobotanical specimens remains relatively unexplored, charred plant material appears to be largely incompatible with these technologies and false positives might occur. An open question about the analysis is whether researchers should extract seeds individually or in bulk due to the limited sizes of most archaeobotanical remains; however, no consistent differences in the quality of data resulted from archaeological seeds (Wales et al. 2016). Although the extraction of a single seed is preferable because only one genetic signature is present, in practice, if DNA yields are very low, then insufficient endogenous DNA may be available for library preparation or genetic characterization; these multi-seed samples may consequently have a mixed signal from multiple individuals (Wales et al. 2016). Without doubt, sample preservation is critical.

The application of new paleogenomic approaches to well-documented temporal sequences of archaeological assemblages opens a new chapter in the study of domestication. It is now possible to move beyond a simple distinction of 'wild' versus 'domesticated' and track sequence changes in a wide range of genes over the course of thousands of years (Da Fonseca et al. 2015).

Library construction through a double-stranded method (Bennett et al. 2014) generated high-resolution genomes from ancient DNA samples and appears to recover a greater fraction of endogenous ancient material. However, a direct comparison of results from different library construction methods on a diversity of ancient DNA samples was lacking (Bennett et al. 2014) and whether they are more suitable for ancient plant material is still under discussion. Recently, a more detailed and comprehensive comparison of library preparation methods for highly degraded DNA has been developed (Gansauge et al. 2017). An in-depth exploration of the suitability of splinted DNA ligation for single-stranded DNA library preparation (Gansauge et al. 2017) shows that it can be utilized for more robust and less costly single-stranded library preparation while increasing the proportion of mapped sequences in ancient DNA libraries. Mitochondrial genomes have played a key role in many ancient DNA research projects focused on extinct hominids (Hofreiter et al. 2001; Rogaev et al. 2006) and prehistoric humans (Green et al. 2009; Krause et al. 2010). However, it is unclear how useful plastomes may be at elucidating questions related to plant evolution, crop domestication and the prehistoric movement of botanical products through trade and migration (Wales et al. 2016).

The grape plastome provides limited intraspecific phylogenetic resolution for aDNA research (Wales et al. 2016). The plastome network generated from modern samples has a relatively limited amount of genetic diversity, suggesting phenotypically and genotypically divergent lineages of grapes are not differentiated at the plastome level, ultimately diminishing the value of the grape plastome as a suitable locus for intraspecific phylogenetic analyses (Wales et al. 2016).

In contrast, the grape nuclear genome shows great promise for archaeological samples and preliminary analyses demonstrate that individual grape specimens can be compared to modern varieties, showing their genetic affiliations (Wales et al. 2016). The recovery of five microsatellite loci from ancient grape seed samples demonstrated good nuclear DNA preservation (Cappellini et al. 2010). Hydrolytic damage is reported in the seed storage proteins as well as the basis for the development of a protein approach for species or sub-species attribution of archaeological seeds to integrate DNA-based methods (Cappellini et al. 2010). According to Wales et al. (2016), analysis of nuclear genomic DNA recovered from archaeological samples reveals a much greater potential for understanding ancient viticulture, including domestication events, genetic introgressions from local wild populations and the origins and histories of cultivars.

Increasing evidence for epigenetic variation within and among natural plant populations has led to much speculation about its role in the evolution of plant phenotypes; however, we still have a very limited understanding of the evolutionary potential of epigenetic variation, in particular in comparison to DNA sequence-based variation (Henderson and Jacobsen 2007; Zhang et al. 2018). Epigenetic changes in plants and animals might have accompanied their extinction or their domestication (Orlando and Willersleb 2014). Epigenetic inheritance can be important for adaptation to new environments, especially in cases where available genetic variation is limited (Lind and Spagopoulou 2018). Epigenetic variation has the potential to create phenotypic variation that is stable and substantial and thus of evolutionary significance (Zhang et al. 2018).

As bioinformatic methods improve, more genomic and metagenomic information from unconventional substrates will be recovered (Green and Speller 2017). The applicability of combined use of morphogeometric and archaeological DNA analyses and comparing different molecular markers to reveal DNA variation, namely simple sequence repeats (SSRs) and single nucleotide polymophisms (SNPs), is promising for deciphering the intricate history of grapevine domestication (Bacilieri et al. 2017). Targeted enrichment and shotgun sequencing of 10,000 SNP loci have been performed by Ramos-Madrigal et al. (2019) to genotype 28 archaeological grape seeds dating to the Iron Age, Roman era and medieval period. Multidimensional scaling (MDS) was used to investigate whether archaeological samples were more closely related to wild accessions or domesticated varieties. The results show that most archaeological seeds were related to wine grapes from Western Europe, and wild ancestries are primarily associated with central and Western European vines (Ramos-Madrigal et al. 2019).

Furthermore, an innovative method (Karasik et al. 2018) for the morphological discrimination between grape varieties using high-resolution 3D scanning has been developed and verified using genetic methods. The 3D seed morphological tool enables separation, with high statistical certainty, between different *Vitis vinifera* varieties. It can detect morphological differences between previously considered 'synonym' couples, thus allowing investigation of new questions that were not accessible before (Karasik et al. 2018).

Acknowledgements This work has been developed thanks to the 'Ancient Egypt's Wine Rebirth' EGYWINE Project (2016–2018) which received funding from the European Union's Horizon 2020 research and innovation program under the Marie Sklodowska-Curie Grant Agreement No 699858.

References

Allaby RG, O'Donoghue K, Sallarès R et al (1997) Evidence for the survival of ancient DNA in charred wheat seeds from European archaeological sites. Anc Biomol 1(2):119–129

Allen (2005) The ancient Egyptian pyramid texts. In: Der Manuelian P (ed) Writings from the ancient world 23. Leiden, pp 26, 107

Athenaeus (1961) Athenaeus I:33 d–f. The Deipnosophists, with an English translation by CB Gulik. In: Loeb classical library, 1. London, pp 146–147

Arroyo-García R, Ruiz-García L, Bolling L et al (2006) Multiple origins of cultivated grape vine (*Vitis vinifera* L. *sativa*) based on chloroplast DNA polymorphism. Mol Ecol 15:3707–3714

Bacilieri R et al (2017) Potential of combining morphometry and ancient DNA information to investigate grapevine domestication. Veg Hist Archaeobot 26:345–356

Barnard H, Dooley AN, Areshian G et al (2011) Chemical evidence for wine production around 4000BCE in the Late Chalcolitic Near Eastern highlands. JAS 38:977–984

Basantani MK, Gupta D, Mehrota R et al (2017) An update on bioinformatics resources for plant genomics research. Curr Plant Biol 11–12:33–40

Bennett EA, Massilani D, Lizzo G et al (2014) Library construction for ancient genomics: Single strand or double strand? Biotechniques 56:289–300

Bennett EA, Gorgé O, Grange T et al (2016) Coprolites, paleogenomics and bone content analysis. In: Fernández-Jalvo Y, King T, Yepiskoposyan L et al (eds) Azokh cave and the transcaucasian corridor. Vertebrate paleobiology and paleoanthropology. Springer, Cham, pp 271–286

Bilgic H, Hakki EE, Pandey A et al (2016) Ancient DNA from 8400 year-old Çatalhöyük wheat: implications from the origin of Neolithic agriculture. PLoS ONE 11(3):e0151974

Briggs AW, Stenzel U, Johnson PLF et al (2007) Patterns of damage in genomic DNA sequences from a Neanderthal. PNAS 104(37):14616–14621

Bouby L, Figueiral I, Bouchette A et al (2013) Bioarchaeological insights into the process of domestication of grapevine (*Vitis vinifera* L.) during Roman times in Southern France. PLoS ONE 8(5):e63195

Bouby L (2017) Diversité de la vigne et des vins archéologiques: le programme Viniculture. ArchéOrient-Le Blog, 10 février 2017. https://archeorient.hypotheses.org/7096. Accessed 25 June 2019

BMCollection 1 (2019) Greek silver coin with a bunch of grapes represented in the obverse (c. 600–501 BC). British Museum number 1928,0120.71. http://www.britishmuseum.org/research/collection_online/collection_object_details.aspx?objectId=1281067&partId=1&searchText=grapes+greek&page=1. Accessed 27 June 2019

BMCollection 2 (2019) Roman copper alloy coin with Dionysus with bunch of grapes in right hand represented in the reverse (c. 198–217 AC). British Museum number 2004,0405.179. http://www.britishmuseum.org/research/collection_online/collection_object_details.aspx?objectId=3286994&partId=1&searchText=coin+dionysus&page=2. Accessed 27 June 2019

BMCollection 3 (2019) White plaster wine-amphora stopper with impression from Naukratis (4th–7th C). British Museum number 1888,0712.43. http://www.britishmuseum.org/research/collection_online/collection_object_details.aspx?objectId=60938&partId=1&searchText=Naukratis+wine+amphorae&page=1. Accessed 27 June 2019

Cappellini E, Thomas M, Gilbert P et al (2010) A multidisciplinary study of archaeological grape seeds. Naturwissenschaften 97:205–217

Champlot S, Berthelot C, Pruvost M et al (2010) An efficient multistrategy DNA decontamination procedure of PCR reagents for hypersensitive PCR applications. PLoS ONE 5:e13042

Cherpion N (1999) Deux Tombes de la XVIIIe Dynastie à Deir el-Medina, Nos 340 (Amenemhat) et 354 (anonyme). MIFAO 144. Cairo, pp 24–27 **(fig 6–13)**

Columella (2012) De Re Rustica 1878. In: Martí Escayol MA (ed) De Re Rustica. Vilafranca del Penedès, pp 79–85

Davies NG (1905) The Rock Tombs of El Amarna, Part III: The Tombs of Huya and Ahmes. In: Griffith FLl (ed) ASE fifteenth memoir (Reprinted 2004). London, p 7 Pl VI

Demarchi B, Hall S, Roncal-Herrero T et al (2016) Protein sequences bound to mineral surfaces persist into deep time. eLife 5:e17092

Demesticha S (2011) The 4th-century-BC Mazotos Shipwreck, Cyprus: a preliminary report. IJNA 40(1): 39–59

Di Donato A, Filippone E, Ercolano MR et al (2018) Genome sequencing of ancient plant remains: findings, uses and potential applications for the study and improvement of modern crops. Front Plant Sci 9:441

Da Fonseca RR, Smith BD, Wales N et al (2015) The origin and evolution of maize in the Southwestern United States. Nat Plants 1:14003

Desroches-Noblecourt C (1985) Reconstitution du Caveau de Sennefer dit 'Tombe aux Vignes': Thèbes-Ouest. Paris, p 9

EGYWINE DB (2019) Database of the scenes of viticulture and winemaking from the Ancient Egyptian private tombs. EGYWINE, Wine of Ancient Egypt, research project. http://www.wineofancientegypt.com. Accessed 26 June 2019

Emery WB (1958) Great tombs of the first dynasty III. London, p 32 Pl 37(2,7), 66 Pl. 78(21,24)

EU Reg (2010) European Union Regulations, Commission Regulation (EU) No 401/2010 of 7 May 2010. Amending and correcting Regulation (EC) No 607/2009 laying down certain detailed rules for the implementation of Council Regulation (EC) No 479/2008 as regards protected designations of origin and geographical indications, traditional terms, labelling and presentation of certain wine sector products. http://www.wipo.int/wipolex/en/text.jsp?file_id=214930#LinkTarget_517. Accessed 27 June 2019

Foley BP, Hansson MC, Kourkoumelis DP et al (2012) Aspects of ancient Greek trade re-evaluated with amphora DNA evidence. JAS 39:389–398

Fulton TL (2012) Setting up an ancient DNA laboratory. In: Shapiro B, Hofreiter M (eds) Ancient DNA: methods and protocols. Methods in molecular biology. Humana Press, pp 1–11

Gansauge MT, Gerber T, Glocke I et al (2017) Single-stranded DNA library preparation from highly degraded DNA using T4 DNA ligase. Nucleic Acids Res 45(10):e79

Garnier N, Valamoti SM (2016) Prehistoric wine-making at Dikili Tash (Northern Greece): integrating residue analysis and archaeobotany. JAS 74:195–206

Gilbert MTP, Bandelt HJ, Hofreiter M et al (2005) Assessing ancient DNA studies. Trends Ecol Evol 20(10):541–544

Grassi F, Labra M, Imazio S et al (2003) Evidence of a secondary grapevine domestication centre detected by SSR analysis. Theor Appl Genet 107(7):1315–1320

Green RE, Briggs AW, Krause J, Prüfer K, Burbano HA, Siebauer M, Pääbo S (2009) The Neandertal genome and ancient DNA authenticity. EMBO J 28(17): 2494–2502

Green RE, Krause J, Briggs AW et al (2010) A draft sequence of the Neandertal genome. Science 328 (5979):710–722

Green EJ, Speller CF (2017) Novel substrates as sources of ancient DNA: prospects and hurdles. Genes 8(180): 1–26

Griffiths JG (1982) Osiris. In: Helck, W, Otto, E, Westendorf W (eds) Lexikon der Ägyptologie IV. Wiesbaden, pp 623–633

Guasch-Jané MR, Ibern-Gómez M, Andrés-Lacueva C et al (2004) Liquid chromatography with mass spectrometry in tandem mode applied for the identification of wine markers in residues from ancient Egyptian vessels. Anal Chem 76:1672–1677

Guasch-Jané MR, Andrés-Lacueva C, Jáuregui O et al (2006a) First evidence of white wine in ancient Egypt from Tutankhamun's tomb. JAS 33:1075–1080

Guasch-Jané MR, Andrés-Lacueva C, Jáuregui O et al (2006b) The origin of the ancient Egyptian drink *shedeh* revealed using LC/MS/MS. JAS 33:98–101

Guasch-Jané MR (2008) Wine in ancient Egypt: a cultural and analytical study. BAR S1851, Oxford, pp 29–31, 39–41, 64

Guasch MR (2010) On Egyptian wine marketing. In: Hudecz A, Petrik M (eds) Commerce and economy in ancient Egypt, proceedings of the 3rd international congress for young egyptologists 25–27 September 2009, Budapest. BAR S2131, Oxford, pp 63–69

Guasch-Jané MR (2011) The meaning of wine in Egyptian tombs: the three amphorae from Tutankhamun's Burial chamber. Antiquity 85(329):851–858

Guasch-Jané MR (2016) An interdisciplinary study on the ancient Egyptian wines: the EGYWINE project. In: Ioannides M, Fink E, Moropoulou A, et al (eds) Proceedings of the 6th international conference EURO-MED 2016, Nicosia (Cyprus) October 31–November 5 2016, digital heritage. Progress in cultural heritage: documentation, preservation and protection, EuroMed 2016, Part I, LNCS vol 10058, pp 737–748

Hanson MC, Foley BP (2008) Ancient DNA fragments inside classical Greek amphoras reveal cargo of 2400-year-old shipwreck. JAS 35:1169–1176

Henderson IR, Jacobsen SE (2007) Epigenetic inheritance in plants. Nature 447(7147):418–424

Hofreiter M, Serre D, Poinar HN et al (2001) Ancient DNA. Nat Rev Genet 2:353–359

Holthoer R (1993) The pottery. In: Baines J (ed) Stone vessels, pottery and sealings from the Tomb of Tutankhamun. Oxford, pp 43–56, 64–67

Hornung E (1996) Conceptions of god in ancient Egypt: the one and the many. New York, pp 152–153

Jouanna J (2012) Wine and medicine in ancient Greece. In: Van der Eijk Ph (ed) Greek medicine from Hippocrates to Galen: selected papers. Studies in ancient medicine 40. Leiden–Boston, pp 173–194

Karasik A, Rahimi O, David M et al (2018) Development of a 3D seed morphological tool for grapevine variety identification, and its comparison with SSR analysis. Sci Rep 8:6545

Krause J, Fu Q, Good JM et al (2010) The complete mitocondrial DNA genome of an unknown hominin from southern Siberia. Nature 464:894–897

Lind MI, Spagopoulou F (2018) Evolutionary consequences of epigenetic inheritance. Heredity 121:205–209

Lindahl T (1993) Instability and decay of the primary structure of DNA. Nature 362:709–715

Margaryan A, Hansen HB, Rasmussen S et al (2018) Ancient pathogen DNA in human teeth and petrous bones. Ecol Evol 8:3534–3542

Mascher M, Schuenemann V, Davidovich U et al (2016) Genomic analysis of 6,000-year-old cultivated grain illuminates the domestication history of barley. Nat Genet 48(9):1089–1093

Meyer Ch (1986) Wein. In: Helck W, Otto E, Westendorf W (eds) Lexikon der Ägyptologie VI. Wiesbaden, pp 1169–1182

Milanesi C, Antonucci F, Menesatti P et al (2011) Morphology and molecular analysis of ancient grape seeds. IANSA II(2):95–100

Milanesi C, Constantini L, Firmati M et al (2014) Geometric morphometry and archaeobotany: characterization of grape seeds (*Vitis vinifera* L.) by analysis of form. Open Access Library Journal 1:e634

McGovern PE, Jalabatze M, Batiuk S et al (2017) Early Neolithic wine of Georgia in the South Caucasus. Proc Natl Acad Sci USA 114(48):E10309–E10318

McGovern PE, Mirzoian A, Hall GR (2009) Ancient Egyptian herbal wines. PNAS 106(18):7361–7366

McGovern PE, Glusker DL, Exner LJ et al (1996) Neolithic resinated wine. Nature 381(6):480–481

Murray MA (2000) Viticulture and wine production. In: Nicholson PT, Shaw I (eds) Ancient Egyptian materials and technology, vol 23. Cambridge University Press, Cambridge, pp 577–608

Nistelberger HM, Smith O, Wales N et al (2016) The efficacy of high-throughput sequencing and target enrichment on charred archaeobotanical remains. Sci Rep 6:37347

Orlando L, Willersleb E (2014) An epigenetic window into the past? Science 345(6196):511–512

Pääbo S, Poinar H, Serre D et al (2004) Genetic analyses from ancient DNA. Annu Rev Genet 38:645–679

Pagnoux C, Bouby L, Ivorra S et al (2015) Inferring the agrobiodiversity of *Vitis vinifera* L. (grapevine) in ancient Greece by comparative shape analysis of archaeological and modern seeds. Veget Hist Archaeobot 24(1):75–84

Palkopoulou E, Mallich S, Skoglund P et al (2015) Complete genomes reveal signatures of demographic and genetic declines in the wooly mammoth. Curr Biol 25(10):1395–1400

Parducci L, Bennet KD, Ficetola GF et al (2017) Ancient plant DNA in lake sediments. New Physiol 214:924–942

Panagou T (2016) Patterns of amphora stamp distribution: tracking down export tendencies. In: Harris EM, Lewis DM, Woolmer M (eds) The ancient Greek economy: markets, households and city states. Cambridge University Press, Cambridge, pp 207–229

Parkinson R (2008) The painted tomb-chapel of Nebamun. Cairo, pp 84–87

Petrie WMF (1901) The royal tombs of the earliest dynasties. Memoir of the Egypt exploration fund 21. London, p 54 Pl. 20(163), Pl. 23(193,196)

Pinhasi R, Fernandes D, Sirak K et al (2015) Optimal ancient DNA yields from the inner ear part of the human petrous bone. PLoS ONE 10(6):e0129102

Poo MCh (1995) Wine and wine offering in the religion of ancient Egypt. In: Martin GT (ed) Studies in egyptology. London-New York, pp 5–7, 9–12, 149–153

Poo MCh (1986) Weinopfer. In: Helck W, Otto E, Westendorf W (eds) Lexikon der Ägyptologie VI. Wiesbaden, pp 1186–1190

Pulak C (2001) The cargo of the Uluburun ship and evidence for trade with the Aegean and beyond. In: Bonfante L, Karageorghis V (eds) Italy and Cyprus in antiquity 1500–450 BC. Nicosia, pp 13–60

Ramos-Madrigal J, Wiborg Runge AK, Bouby L et al (2019) Palaeogenomic insights into the origins of French grapevine diversity. Nat Plants 5:595–603

Rasmussen M, Guo X, Wang Y et al (2011) An aboriginal Australian genome reveals separate human dispersals into Asia. Science 334(6052):94–98

Renfrew JM (1966) A report on recent finds of carbonized cereal grains and seeds from Prehistoric Thessaly. Thessalika 5:21–36

Renfrew JM (1996) Palaeoethnobotanical finds of *Vitis* from Greece. In: McGovern PE, Fleming SJ, Katz SH (eds) The origins and ancient history of wine. Gordon and Breach Science Publishers, Amsterdam, pp 255–267

Rogaev EI, Moliaka YK, Malyarchuk BA et al (2006) Complete mitochondrial genome and phylogeny of pleistocene mammoth Mammuthus primigenius. PLoS Biol 4:e73

Rohland N, Siedel H, Hofreiter M et al (2004) Non-destructive DNA extraction method for mitochondrial DNA analyses of museum specimens. Biotechniques 36(5):1–6

Rohland N, Hofreiter M (2007) Comparison and optimization of ancient DNA extraction. Biotechniques 42 (3):343–352

Sawyer S, Krause J, Guschanski K et al (2012) Temporal patterns of nucleotide misincorporations and DNA fragmentation in ancient DNA. PLoS ONE 7(3):e34131

Schlumbaum A, Tensen M, Jaenicke-Després V (2008) Ancient plant DNA in archaeobotany. Veg Hist Archaeobot 17:233–244

Singleton VL (1996) An enologist's commentary on ancient wines. In: McGovern PE, Fleming SJ, Katz SH (eds) The origins and ancient history of wine. Gordon and Breach Science Publishers, Amsterdam, pp 67–77

Spencer AJ (1993) Early Egypt: the rise of civilisation in the Nile Valley. London, p 90 (fig 69), 107 (fig 81)

Terral JF, Tabard E, Bouby L et al (2010) Evolution and history of grapevine (*Vitis vinifera*) under domestication: new morphometric perspectives to understand seed domestication syndrome and reveal origins of ancient European cultivars. Ann Bot 105:443–455

Tallet P (1995) Le *shedeh*, etude d'un procédé de vinification en Égypte ancienne. BIFAO 95:459–492

Tallet P (1998) Quelques aspects de l'économie du vin en Égypte ancienne, au nouvel empire. In: Grimal N, Menu B (eds) Le Commerce en Égypte Ancienne, BdE 121. Cairo, pp 241–267

This P, Lacombe T, Thomas MR (2006) Historical origins and genetic diversity of wine grapes. Trends Genet 22(9):511–519

Tzochev Ch (2016) Markets, amphora trade and wine industry: the case of Thasos. In: Harris EM, Lewis DM, Woolmer M (eds) The ancient Greek economy: markets, households and city states. Cambridge University Press, pp 230–253

Valamoti SM, Mangafa M, Koukouli-Chrysanthaki C et al (2007) Grape-pressings from northern Greece: The earliest wine in the Aegean? Antiquity 81:54–61

Vassilika E (2010) The Tomb of Kha, the Architect. Florence, p 12, 86

Virgil (1586) Georgicon II. In: P Virgilii Maronis. Bucolica, Georgica et Aeneis. Venice, p 91

Wales N, Ramos Madrigal J, Cappellini E et al (2016) The limits and potential of paleogenomic techniques for reconstructing grapevine domestication. JAS 72:57–70

Weiss CL, Schuenemann VJ, Devos J et al (2016) Temporal patterns of damage and decay kinetics of DNA retrieved from plant herbarium specimens. R Soc Open Sci 3:160239

Weyrich LS, Duchene S, Soubrier J et al (2017) Neanderthal behaviour, diet and disease inferred from ancient DNA in dental calculus. Nature 544:357–361

Wood JR, Wilmshurst JM, Wagstaff SJ et al (2012) High-resolution coproecology: using coprolites to reconstruct the habits and habitats of New Zealand's extinct upland Moa (*Megalapteryx didinus*). PLoS ONE 7(6):e40025

Wilkinson RH (2003) The complete gods and goddesses of ancient Egypt. London, pp 128–129, 225–226

Zedane L, Hong-Wa C, Murienne J et al (2016) Museomics illuminate the history of an extinct, paleoendemic plant lineage (*Hesperelaea, Oleaceae*) known from an 1875 collection from Guadalupe Island, Mexico. Biol J Lin Soc 117:44–57

Zhang YY, Latzel V, Fischer M et al (2018) Understanding the evolutionary potential of epigenetic variation: a comparison of heritable phenotypic variation in epiRILs, RILs, and natural ecotypes of Arabidopsis thaliana. Heredity 121:258–265

Zhou Y, Massonnet M, Sanjak JS, Cantu D, Gaut BS (2017) Evolutionary genomics of grape (Vitis viniferassp. vinifera) domestication. Proceedings of the national academy of sciences, 201709257. https://doi.org/10.1073/pnas.1709257114

Zohary D (1996) The domestication of the grapevine *Vitis vinifera* L. in the Near East. In: McGovern PE, Fleming SJ, Katz SH (eds) The origins and ancient history of wine. Gordon and Breach Science Publishers, Amsterdam, pp 23–43

Zohary D, Hopf M (1993) Domestication of plants in the Old World: the origin and spread of cultivated plants in West Asia, Europe and the Nile Valley. Oxford University Press, Oxford, p 145

Zohary D, Hopf M, Weiss E (2012) Domestication of plants in the Old World: The origin and spread of cultivated plants in Southwest Asia, Europe and the Mediterranean Basin. Oxford University Press, Oxford, pp 121–125

Strategies for Sequencing and Assembling Grapevine Genomes

5

Rosa Figueroa-Balderas, Andrea Minio, Abraham Morales-Cruz, Amanda M. Vondras and Dario Cantu

Abstract

Though grape transcriptomics has expanded dramatically over the last ten years, few additional novel genomic resources were developed since the release of the PN40024 reference genome in 2007. This is partly because of the difficulty associated with assembling grape genomes. Despite a relatively small genome size of ~500 Mb and modest repeat content, high sequence and structural heterozygosity makes assembling grape genomes particularly challenging. Without assemblies representative of the genetic diversity within the cultivated germplasm, identifying cultivar-specific functions not represented in the PN40024 genome has remained elusive. Now, third-generation sequencing technologies and long-range scaffolding methods have made it possible to relatively inexpensively and rapidly generate highly contiguous and complete grape genomes. This chapter will describe the challenges associated with the isolation of high-quality nucleic acids suitable for long-read sequencing and provide an overview of the sequencing and assembling approaches that can be used to successfully reconstruct grape genomes.

R. Figueroa-Balderas · A. Minio · A. Morales-Cruz · A. M. Vondras · D. Cantu (✉)
Department of Viticulture and Enology, University of California Davis, One Shields Ave, Davis, CA 95616, USA
e-mail: dacantu@ucdavis.edu

R. Figueroa-Balderas
e-mail: rfiguero@ucdavis.edu

A. Minio
e-mail: aminio@ucdavis.edu

A. Morales-Cruz
e-mail: abmora@ucdavis.edu

A. M. Vondras
e-mail: amvondras@ucdavis.edu

5.1 Introduction

The French–Italian Public Consortium for Grapevine Genome Characterization released the first grapevine genome assembly in 2007 (Jaillon et al. 2007). This was the second genome assembly of a woody species (Tuskan et al. 2006) and the fourth assembly of a flowering plant genome (The Arabidopsis Genome Initiative 2000; Goff et al. 2002; Tuskan et al. 2006). Despite its limitations, this first attempt to reconstruct the grape genome remains a valuable resource to the grapevine community and was the basis of investigating molecular markers, studying species, cultivar and clonal diversity, researching evolution and domestication events (see Chap. 3) and was a reference for hundreds of transcriptomic studies (see Chap. 13).

The advent of the next-generation sequencing (NGS) dramatically improved the efficiency of

D. Cantu and M. A. Walker (eds.), *The Grape Genome*, Compendium of Plant Genomes, https://doi.org/10.1007/978-3-030-18601-2_5

sequencing and reduced the costs of the sequencing process, driving forward genomic and transcriptomic studies of grapevine. Though 454 pyrosequencing data were the first produced with a high-throughput technology, Solexa/Illumina technology was used more extensively for genome assembly (Di Genova et al. 2014), genotyping by sequencing (Myles et al. 2010; Hyma et al. 2015; Yang et al. 2016; Cardone et al. 2016; Zhou et al. 2017), transcriptome reconstruction (Da Silva et al. 2013; Venturini et al. 2013; Jiao et al. 2015; Gambino et al. 2017), and expression profiling (see Chap. 13). With the exception of its transcriptomic applications, however, short reads are inadequate for grape genome reconstruction because of their inability to resolve repetitive regions. This was the case for the Thompson Seedless cultivar: Despite high sequencing coverage, short reads could not resolve repetitive regions of the genome, leading to a highly fragmented assembly (Di Genova et al. 2014). In addition, resequencing fails to resolve complex structural variants (Da Silva et al. 2013; Gambino et al. 2017), hindering the characterization of cultivar-specific gene space and genome structures.

Newer (third-generation) long-read sequencing technologies make it possible to generate phased de novo genome assemblies with greater contiguity and completeness than before. This technology was applied to sequence the Cabernet Sauvignon, Chardonnay, and Carmenere genomes that resulted in phasing $\sim$74% of the diploid structure of the genome and produced reference assemblies that were $\sim$30 times less fragmented in comparison with the PN40024 assembly (Chin et al. 2016; Roach et al. 2018; Minio et al. 2019b; Zhou et al. 2018). The transcriptome of Cabernet Sauvignon was also sequenced with long-read methods (Minio et al. 2019a). The combination of expressed gene isoform data generated in long reads and short Illumina reads allowed to generate a tissue-specific reference transcriptome without a reference genome (Minio et al. 2019a).

These advances in technology at decreased cost are responsible for a greater abundance and quality of genomic resources available for grapevine. They also helped to overcome the grape-specific impediments to genome sequencing and assembly. Although the haploid genome size of the grapevine is relatively small (about 500 Mb, Jaillon et al. 2007), its highly heterozygous nature of cultivated *V. vinifera* genotypes make them challenge for genome assemblies (Aradhya et al. 2003; Laucou et al. 2011). Vegetative propagation of grapevines in part contributes to maintain the heterozygosity over generations. Moreover, it is also associated with the accumulation of recessive deleterious mutations and structural variants that encompass $\sim$15% of genes (Zhou et al. 2017, 2018). Assemblies of highly heterozygous genomes are often more fragmented than other genomes of comparable similar size and complexity (Yu et al. 2005; Argout et al. 2011). Prior to sequencing and assembly, there are grape-specific challenges associated with nucleic acid isolation. Grapes accumulate high levels of complex secondary metabolites, especially in older tissues (Murray and Thompson 1980; Couch and Fritz 1990; Fang et al. 1992). Persistent polysaccharides and polyphenols in nucleic acid extracts can severely compromise downstream applications, reduce PCR and enzymatic reaction efficiency, and impair sequencing library preparation (Ausubel et al. 1994; Healy et al. 2014). High-quality nucleic acid extracts are essential starting materials for long read and isoform sequencing, optical maps, and chromatin interaction studies. This chapter will discuss challenges associated with isolating high-quality nucleic acids from grapevines, sequencing and assembling grapevine genomes, and strategies to overcome these issues in part or entirely.

5.2 Major Factors Influencing Nucleic Acids Isolation from Grapevine

Next-generation sequencing (NGS) technologies require high-quality nucleic acids as starting material (Endrullat et al. 2016), which help ensure even coverage with minimal sequencing bias (Healy et al. 2014). Nucleic acid quality (yield, purity, and integrity) is measured quantitatively and qualitatively. NGS protocols typically require at least 50 ng to begin with; using less than the amount recommended risks reducing final library yield and sequencing performance (Aigrain et al. 2016). Metabolites co-extracted during isolation, like polysaccharides and polyphenols, can detrimentally affect or inhibit downstream enzymatic reactions, interfering with library preparation and sequencing (Ausubel et al. 1994; Healy et al. 2014). Similarly, oxidative contaminants in nucleic acid extracts increase the risk of degradation during shearing steps (Costello et al. 2013). Successful NGS requires high-integrity nucleic acids, else degradation products will interfere with polymerase activity during sequencing, especially when sequencing long inserts, and affect sequencing coverage (Mayjonade et al. 2016).

We developed an optimized protocol for the extraction of high-molecular-weight (HMW) genomic DNA (gDNA) from grape leaves (Chin et al. 2016). The protocol is based on methods originally reported by Japelaghi et al. (2011) and Healy et al. (2014). It was modified to increase yield and improve the removal of interfering metabolites while preserving DNA integrity. Overviews of these DNA/RNA isolation procedures are shown in Figs. 5.1 and 5.2. This section will discuss methods to prepare high-quality nucleic acid extracts from grapevine for NGS.

5.2.1 Tissue Collection

Choosing the right starting plant material is critical for efficient nucleic acid isolation. Young leaf tissue is the best source of gDNA because it has more cells per unit area and typically contains lower concentrations of secondary metabolites and phenolic compounds (Murray and Thompson 1980; Doyle and Doyle 1987; Peterson et al. 1997; Iandolino et al. 2004). For high-molecular-weight genomic DNA, young leaves no more than one and a half-inch long are recommended to minimize secondary metabolites, waxy leaf coating and to maximize yield (Doyle and Doyle 1987; Lutz et al. 2011). Proper tissue storage is also important. Freshly collected tissue, if not immediately used for nucleic acids isolation, should be immediately frozen in liquid nitrogen and stored at − 80 °C to avoid degradation by endonucleases (Ribeiro and Lovato 2007; Varma et al. 2007; Knebelsberger and Stöger 2012).

5.2.2 Tissue Disruption

Initially, leaf tissue is ground to a fine powder using a pre-chilled mortar and pestle and liquid nitrogen, which minimizes gDNA degradation (Murray and Thompson 1980). Hard woody tissues and berries are challenging to grind with a mortar and pestle, so using a mechanical mill to crush frozen tissues is recommended instead (Lodhi et al. 1994; Bhattacharjee et al. 2004). This is particularly important for RNA extractions, for which finely ground tissue help to maximize yield (Yockteng et al. 2013). Because gDNA has a tendency to get easily sheared, delicacy is required in its preparation for downstream applications. Genomic DNA is sensitive to repetitive pipetting (Murray and Thompson 1980; Doyle and Doyle 1987), vortexing, and violent shaking. Using wide-bore pipet tips can help minimize mechanical damage to DNA (Fig. 5.1; Sahu et al. 2012; Healy et al. 2014). In contrast, vortexing or violent shaking does not reduce RNA integrity (Fig. 5.2; Iandolino et al. 2004; Xiao et al. 2015).

5.2.3 Removal of Contaminants

Polysaccharides and polyphenols are the most problematic metabolites that persist in nucleic acid extracts from grapevine leaves (Lodhi et al. 1994; Hanania et al. 2004; Marsal et al. 2013). These reduce yield and can impair the activities of ligases, endonucleases, and polymerases (Fang et al. 1992; Kim et al. 1997; Sharma et al. 2002; Iandolino et al. 2004). During cell lysis, polyphenols are readily oxidized and can form

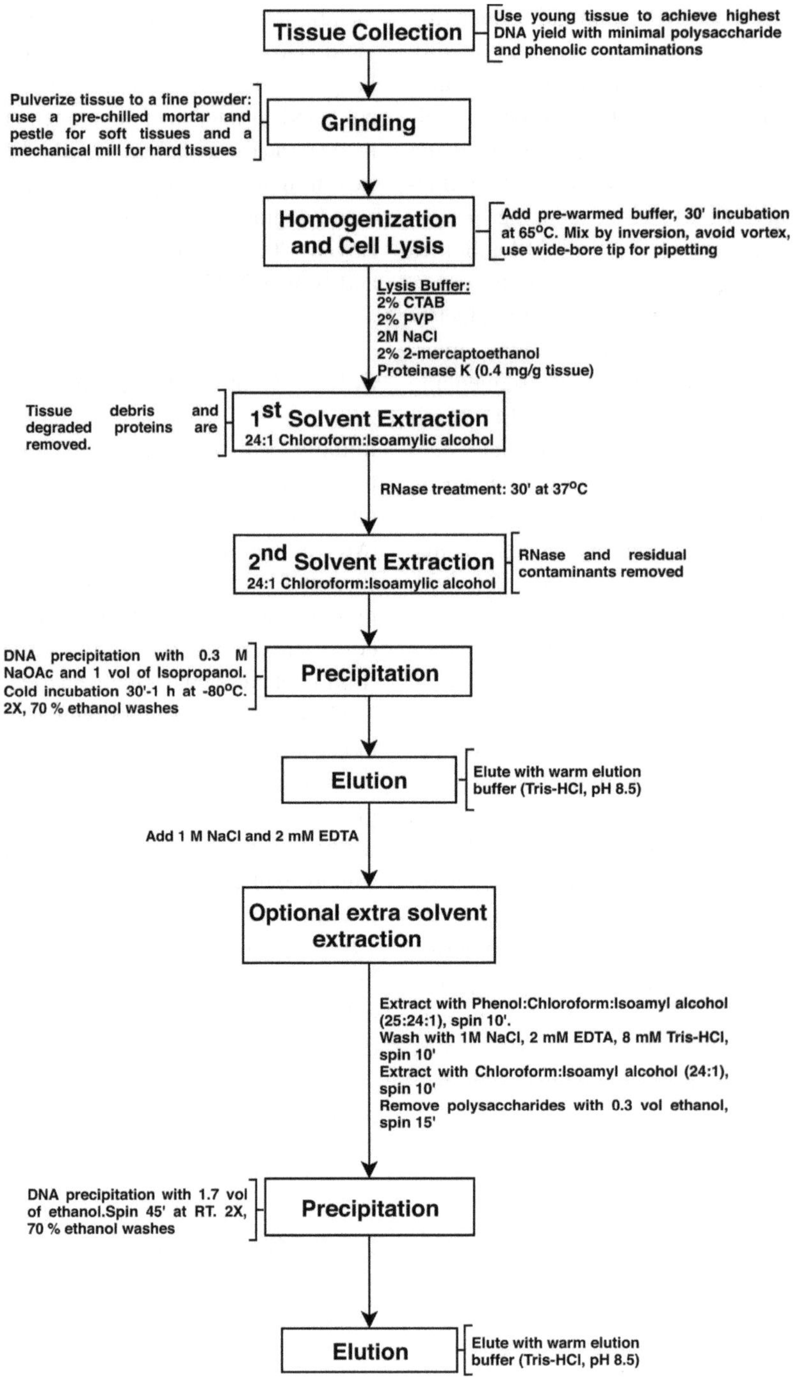

Fig. 5.1 Workflow of DNA isolation from grapevine

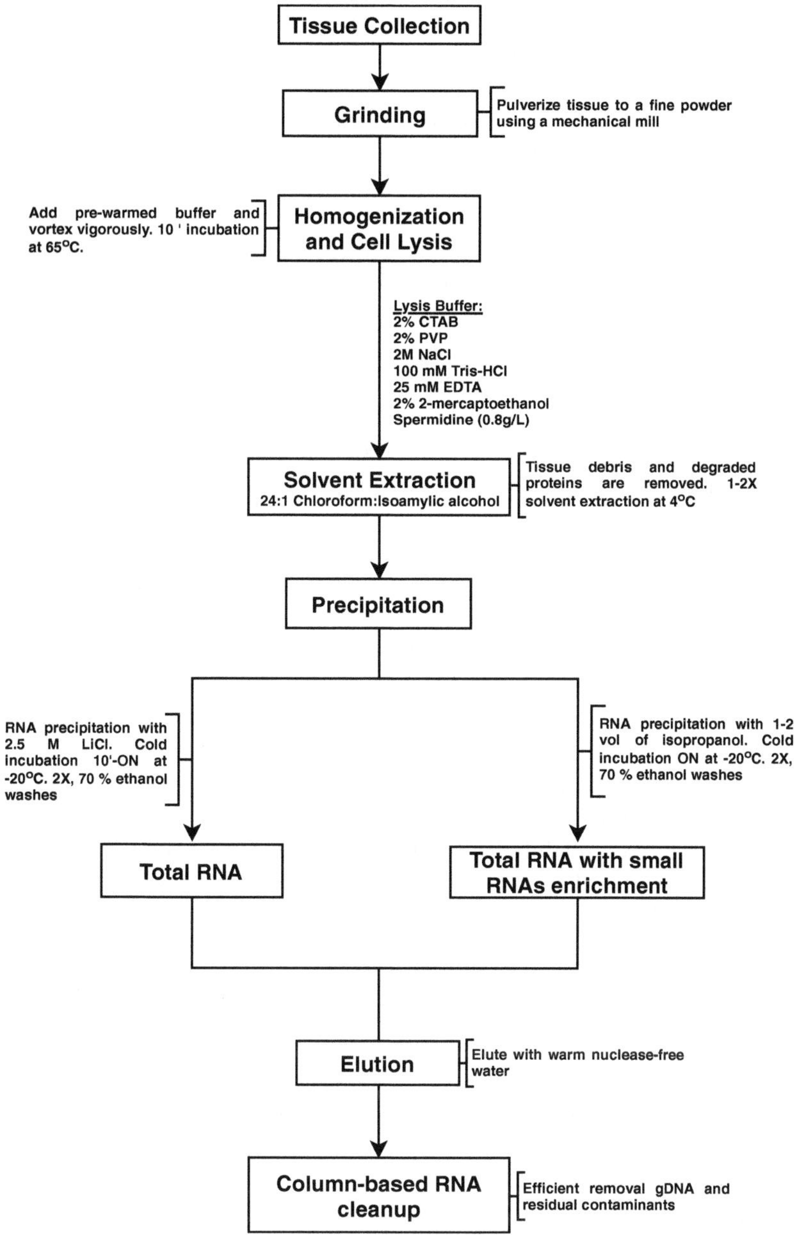

Fig. 5.2 Workflow of RNA isolation from grapevine

complexes that can co-precipitate with nucleic acids and causes preparations to brown (Newbury and Possingham 1977; Kim et al. 1997; Peterson et al. 1997; Varma et al. 2007). Polysaccharides can also co-precipitate with nucleic acids and impart a sticky and highly viscous consistency that is challenging to remove (Newbury and Possingham 1977; Maliyakal 1992; Lodhi et al. 1994; Iandolino et al. 2004; Varma et al. 2007; Sahu et al. 2012). The abundance of these metabolites varies with developmental stage and exposure to biotic/abiotic stresses (Braidot et al. 2008). Other than polysaccharides and polyphenols, proteins can also co-precipitate with nucleic acids during isolation that may inhibit restriction endonucleases and contribute to DNA shearing and degradation (Hanania et al. 2004; Varma et al. 2007). In addition to high levels of polysaccharides and polyphenols and degradative nucleases, berries are highly acidic and, when ripe, have a relatively lower number of expressed genes (Iandolino et al. 2004; Reid et al. 2006; Massonnet et al. 2017; Minio et al. 2019a). Together, these variables can reduce DNA/RNA integrity, purity, and yield if left unaddressed (Iandolino et al. 2004; Reid et al. 2006; Romieu 2010; Yang et al. 2011).

These problems can be circumvented with good technique and optimized extraction solutions. Warm extraction buffers containing high concentrations (at least 2%) of cetyltrimethylammonium bromide (CTAB) and polyvinylpyrrolidone (PVP) will minimize the carryover of phenolic and polysaccharides, prevent polyphenol oxidation, and help precipitate nucleic acids (Figs. 5.1 and 5.2; Iandolino et al. 2004; Varma et al. 2007; Japelaghi et al. 2011). High-salt concentrations (1.5–2 M NaCl) in combination with CTAB are often used to increase the solubilization of polysaccharides in ethanol for their subsequent removal (Fang et al. 1992; Peterson et al. 1997; Varma et al. 2007; Japelaghi et al. 2011; Healy et al. 2014). Adding Proteinase K (0.4 mg/g tissue), NaCl (2–2.5 M), and 2-mercaptoethanol (2%) to the extraction buffer will help remove proteins and inactivate nucleases released during tissue disruption and homogenization (Iandolino et al. 2004; Varma et al. 2007). Furthermore, an additional high-salt/phenol/chloroform wash can remove persistent polysaccharides, though can cause up to 50% loss of the sample (Mayjonade et al. 2016; VanBuren et al. 2015).

Complete removal of residual contaminant RNA/DNA from nucleic acids preparations is crucial for NGS technologies. The contaminant RNA could be eliminated efficiently by addition of ribonuclease A (RNase A). Up to 20 mg of RNase A per gram of tissue between chloroform: isoamyl alcohol extractions efficiently removes residual RNA and minimize extra manipulations that could contribute to degradation and loss (Fig. 5.1; Chin et al. 2016). Similarly, and extra cleaning step with deoxyribonuclease (DNase) is recommended that removes lingering gDNA from RNA extracts (Fig. 5.2; Iandolino et al. 2004; Blanco-Ulate et al. 2017).

5.2.4 Nucleic Acid Precipitation

The selective precipitation of nucleic acids differs for DNA and RNA and requires further consideration (Marsal et al. 2013; Rezadoost et al. 2016; Xiao et al. 2015). For gDNA, 0.3 M of sodium acetate (NaOAC) is added to neutralize the negative charges on DNA, making them more stable and less water-soluble (Fig. 5.1; Tan and Chen 2005). Adding one volume of alcohol, such as ethanol or isopropanol, will cause DNA to precipitate because it is not soluble in alcohol (Fig. 5.1; Green and Sambrook 2016, 2017). Cold incubation at − 20 °C or − 80 °C will increase DNA recovery. This step should last no longer than one hour; longer incubations will also increase the precipitate of CTAB and NaCl from the solution (Healy et al. 2014). High-molecular-weight RNA will precipitate in cold lithium chloride (LiCl, up to 2.5 M), a step which also helps remove polysaccharides. If small RNAs are sought after, an isopropanol precipitation step should be used instead (Fig. 5.2; Iandolino et al. 2004).

5.2.5 Nucleic Acids Quantity and Quality Evaluation

The quantity, quality, and integrity of nucleic acids should be assessed prior to NGS. Quantity is typically assessed by measuring the fluorescence of a dye that binds specifically to double-stranded DNA or RNA (Nakayama et al. 2016). UV absorbance can be used as a preliminary estimation of nucleic acids quantity and quality, but it is not recommended to evaluate quantity because of its relatively poor specificity and general overestimation of sample concentration (Varma et al. 2007). UV absorbance ratios ($A_{260/280}$ and $A_{260/230}$ ratios) provide information

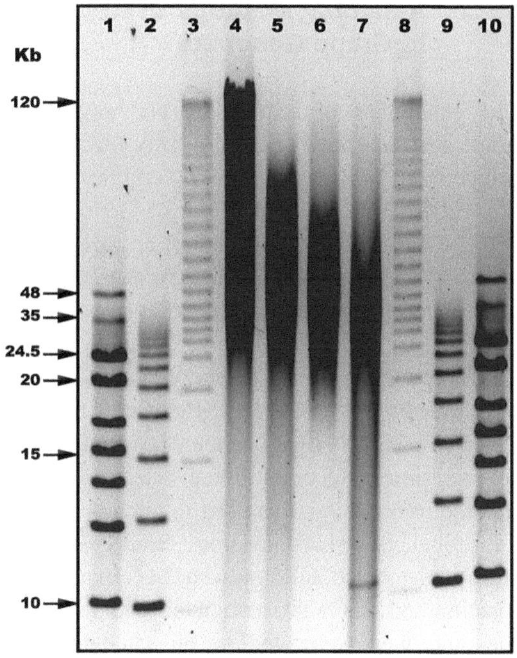

Fig. 5.3 Evaluation of HMW gDNA quality and SMRTbell template libraries using pulse field gel electrophoresis (PFGE). Lane 1, 10: High Range DNA ladder (Thermo Scientific); Lane 2, 9: 2.5 Kb Ladder (BioRad); Lane 3, 8: 5 Kb Ladder (BioRad); Lane 4: HMW gDNA suitable for SMRTbell template libraries; Lane 5: HMW gDNA sheared with a 26 gauge needles (10× shears); Lane 6: SMRTbell template final library (>20 Kb SMRTbell template library); and Lane 7: Example of fragmented or degraded gDNA, not appropriate for large insert SMRTbell template libraries (small fragments at 10 Kb)

about the purity of the sample. Optimal $A_{260/280}$ will differ slightly for DNA (>1.8) and RNA (>2.0), and optimal $A_{260/230}$ is greater than 2.0 (Cheng et al. 2000; Heptinstall and Rapley 2000). Nucleic acid integrity can be verified by electrophoresis. Conventional electrophoresis does not adequately resolve high-molecular-weight fragments. Pulse field gel electrophoresis (PFGE) is preferred to evaluate DNA integrity prior to long-insert sequencing (Fig. 5.3; Guzmán and Ecker 1988). RNA integrity can be estimated with a non-denaturing bleached gel made with between 1% and 2% agarose (Aranda et al. 2012). When high-quality RNA is needed for sequencing libraries, microcapillary-based electrophoresis instruments that generate an RNA integrity number (RIN) are appropriate to assess RNA quality. RIN values greater than 8.0 are usually indicative of high-quality RNA adequate for isoforms sequencing (An et al. 2018).

5.3 Sequencing and Assembly of Grape Genomes

The sequencing and comparative analyses of crop genomes provide critical information about their origins, domestication events, and the basis of valuable traits (Edwards and Batley 2010; Feuillet et al. 2011; Morrell et al. 2012; Michael and Jackson 2013; Thottathil et al. 2016). Diverse technologies were developed to read the succession of nucleotides that form the polymeric DNA molecule. The advent of newer sequencing platforms was associated with increased read length, quality, and an exponential decline in sequencing cost per base pair (https://www.genome.gov/sequencingcostsdata/). However, all of the technologies available to date share the same limitation: They cannot sequence complete chromosomes. Given this limitation, a reconstruction of the sequenced fragments is required to create a genome assembly. High-quality assemblies have relatively little fragmentation, with reads assembled into few long sequences. Measurements like N50 and NG50 values are widely used to evaluate assembly quality and indicate an assembly's contiguity (Earl et al. 2011; Bradnam et al. 2013; Ekblom and Wolf 2014). Another important criterion that describes assembly quality is gene space completeness, estimated as the number of highly conserved, single-copy orthologous genes detected in the assembly (e.g., BUSCO, Simão et al. 2015).

Plant genome assemblies are challenging to construct because of their large size and high repetitive content (Morrell et al. 2012). In addition, the grape genome is high heterozygous with significant differences in sequence between parental genotypes that make the assembly of the genome challenging (Aradhya et al. 2003; Jaillon et al. 2007; Velasco et al. 2007; Minio et al. 2017; Zhou et al. 2017). The divergence of haplotype sequences causes ambiguity during the assembly procedure and increases assembly fragmentation as a consequence. This section will discuss methods that mitigate the challenges associated with sequencing and assembling grape genomes.

5.3.1 Sequencing Methods Used in Grape Genomics

Reads from the Sanger sequencer and the 454 pyrosequencer were used to create the first grape genome assemblies (Jaillon et al. 2007; Velasco et al. 2007). The birth of second-generation sequencing technologies and their short reads was accompanied by decreased costs and increasing throughput and sequencing quality. However, short reads are not suitable for assembling the highly repetitive genome of grape. Moreover, short reads deliver highly fragmented assemblies that underrepresent repetitive content (Di Genova et al. 2014). High heterozygosity further makes assemblies fragmented because assemblers have difficulty in generating consensus at heterozygous loci (Claros et al. 2012; Kajitani et al. 2014; Safonova et al. 2015; Pryszcz and Gabaldón 2016). The advent of long-read sequencing technologies like Pacific Bioscience single-molecule real-time sequencing (PacBio SMRT) and Oxford Nanopore technologies greatly improved grape genome assemblies. Though error rates are higher than the short-read technologies, the longer reads resolve the ambiguity of repetitive regions, delivering more contiguous assemblies. By applying diploid-aware assemblers like FALCON-Unzip, genome heterozygosity can also be represented in the assembly (Chin et al. 2016). So far, PacBio SMRT has been successfully used to reconstruct three grape cultivars: Cabernet Sauvignon (Chin et al. 2016), Chardonnay (Roach et al. 2018; Zhou et al. 2018), and Carmenere (Minio et al. 2019b).

5.3.2 Assembly Methods Used in Grape Genomics

The challenges of sequencing and assembling grape genomes have resulted in very few genome projects and publicly available assemblies (Jaillon et al. 2007; Velasco et al. 2007; Di Genova et al. 2014; Chin et al. 2016; Roach et al. 2018; Zhou et al. 2018). Greater than 12% of the cultivated grape genome sequence is heterozygous (Jaillon et al. 2007; Velasco et al. 2007). Jaillon et al. (2007) reduced genome heterozygosity to ~7 % by inbreeding *V. vinifera* cv. Pinot noir to create PN40024. The PN40024 genome has been revised several times since its initial release. Genetic linkage information was used to anchor some of the sequences to chromosomes. Some sequences were connected to a chromosome but not a specific location and others not associated with any chromosome were relegated to an "undetermined chromosome" (Jaillon et al. 2007). The quality of the anchoring process was greatly influenced by the limited contig length and the ambiguity of assembly. The high fragmentation observed in PN40024 was caused in part by short-read sequences unable to distinguish identical regions in the genome longer than the length of the reads. A consequence of the high fragmentation, the assembly has gaps in sequence (16 Mb, over 3%) that intersect protein-coding genes (Venturini et al. 2013; Da Silva et al. 2013). Despite its limitations, this reference was and continues to be a valuable resource for grapevine studies.

Almost a decade after the publication of the PN40024 genome, genome assemblies of Cabernet Sauvignon (Chin et al. 2016), Chardonnay (Roach et al. 2018; Zhou et al. 2018), and Carmenere (Minio et al. 2019a, b) were released taking full advantage of single-molecule real-time (SMRT) technology. The availability of long-read sequencing methods allows to assemble genomes with high contiguity, thanks to the development of dedicate methods, like HGAP (Chin et al. 2013), Canu (Walenz et al. 2017), and wtdbg2 (Ruan and Li 2019) among the others that are able to handle such an information regardless of the high sequencing noise. Furthermore, thanks to the employment of diploid-aware software for the genome reconstruction, the Cabernet Sauvignon, Chardonnay, and Carmenere assemblies are able to represent their haplotype diversity. The FALCON-Unzip diploid-aware software first creates a highly continuous "primary" haploid assembly and alternative paths (associated contigs). Next, reads are mapped on the primary assembly, structural variants distinguishing the two haplotypes are phased, and divergent "haplotigs" are generated by the software relative to the primary assembly. This strategy produced genome assemblies that are more contiguous and complete than PN40024 and include haplotype-specific gene sequences that are endemic to the highly heterozygous species. For example, with an N50 of 2.1 Mb, the Cabernet Sauvignon contigs are 20 times less fragmented than the original PN40024 assembly, despite PN40024 being significantly less heterozygous. The 368 Mb of haplotigs phased the diploid structure of ~74% of the whole Cabernet Sauvignon genome. Although FALCON-Unzip produces primary assemblies with outstanding contiguity, the phasing process does have limitations. Over a given region, extremely divergent haplotypes can be difficult to correctly identify as such; both of these sequences can be incorrectly assigned to the primary assembly. This inflates the estimated haploid genome size (123% of the expected genome size for Cabernet Sauvignon primary contigs, 121% for Chardonnay).

5.4 Conclusions

Staggering improvements in the genome assembly quality of grape have been achieved. These advances are the result of the combined improvement in nucleic acid isolation methods, library preparation tools, sequencing technologies, and available assembling algorithms. Technical challenges remain, including accurately representing both complete haplotypes that compose the diploid and highly heterozygous grape genome. Integrating combinations of sequencing technologies could be a key to further improving genome assemblies. Proximity ligation sequencing data obtained through the capture of genome-wide chromosome conformation (Hi–C) can be used to assist the scaffolding of the genome assembly, like in the case of a hybrid approach adopted for Chardonnay in Zhou et al. 2018. Non-sequencing-based technologies, like optical and next-generation mapping, can also serve to increase contiguity and better evaluate the structural differences between haplotypes (Pendleton et al. 2015). These valuable tools provide unprecedented opportunities to understand and improve grapevine.

Acknowledgements This work was supported by J. Lohr Vineyards and Wines, E. & J. Gallo Winery, Dolce Winery, the Louis P. Martini Endowment in Viticulture, and the NSF Grant #1741627. Part of this work was carried out in collaboration with UC Davis Chile and funded by the Chilean Economic Development Agency (CORFO).

References

Aigrain L, Gu Y, Quail MA (2016) Quantitation of next generation sequencing library preparation protocol efficiencies using droplet digital PCR assays a systematic comparison of DNA library preparation kits for Illumina sequencing. BMC Genom 17:458

An D, Cao HX, Li C, Humbeck K, Wang W (2018) Isoform sequencing and state-of-art applications for unravelling complexity of plant transcriptomes. Genes 9(1):43. https://doi.org/10.3390/genes9010043

Aradhya MK, Dang GS, Prins BH et al (2003) Genetic structure and differentiation in cultivated grape *Vitis vinifera* L. Genet Res 81:179–192. https://doi.org/10.1017/s0016672303006177

Aranda PS, LaJoie DM, Jorcyk CL (2012) Bleach gel: a simple agarose gel for analyzing RNA quality. Electrophoresis 33(2):366–369

Argout X, Salse J, Aury JM et al (2011) The genome of *Theobroma cacao*. Nat Genet 43:101–108. https://doi.org/10.1038/ng.736

Ausubel FM, Brent R, Kingston et al (1994) Current protocols in molecular biology. Wiley, New York, pp 2.0.1–2.14.8

Bhattacharjee R, Kolesnikova-Allen M, Aikpokpodion P, Taiwo S, Ingelbrecht I (2004) A semi-automated rapid method of extracting genomic DNA for molecular marker analysis in cocoa, *Theobroma cacao* L. Plant Mol Biol Report 22:435a–435h458

Blanco-Ulate B, Hopfer H, Figueroa-Balderas R et al (2017) Red blotch disease alters grape berry development and metabolism by interfering with the transcriptional and hormonal regulation of ripening. J Exp Bot 68(5):1225–1238

Bradnam KR, Fass JN, Alexandrov A et al (2013) Assemblathon 2: evaluating de novo methods of genome assembly in three vertebrate species. Gigascience 2:10

Braidot E, Zancani M, Petrussa E et al (2008) Transport and accumulation of flavonoids in grapevine (Vitis vinifera L.). Plant Signal Behav 3(9):626–632

Cardone MF, D'Addabbo P, Alkan C et al (2016) Inter-varietal structural variation in grapevine genomes. Plant J 88:648–661

Cheng SH, Moore BD, Seemann JR (2000) Purification of uncontaminated, intact plant RNA. In: Rapley R (ed) The nucleic acid protocols handbook. Humana Press, Totowa

Chin CS, Peluso P, Sedlazeck FJ et al (2016) Phased diploid genome assembly with single-molecule real-time sequencing. Nat Methods 13:1050–1054. https://doi.org/10.1038/nmeth.4035

Chin CS, Alexander DH, Marks P et al (2013) Nonhybrid, finished microbial genome assemblies from long-read SMRT sequencing data. Nat Methods 10(6):563–569. https://doi.org/10.1038/nmeth.2474

Claros MG, Bautista R, Guerrero-Fernández D et al (2012) Why assembling plant genome sequences is so challenging. Biology (Basel) 1:439–459. https://doi.org/10.3390/biology1020439

Costello M, Pugh TJ, Fennell TJ et al (2013) Discovery and characterization of artifactual mutations in deep coverage targeted capture sequencing data due to oxidative DNA damage during sample preparation. Nucleic Acids Res 41(6):e67

Couch JA, Fritz PJ (1990) Isolation of DNA from plants high in polyphenolics. Plant Mol Biol Rep 8:8–12

Da Silva C, Zamperin G, Ferrarini A et al (2013) The high polyphenol content of grapevine cultivar tannat berries is conferred primarily by genes that are not shared with the reference genome. Plant Cell 25:4777–4788. https://doi.org/10.1105/tpc.113.118810

Di Genova A, Almeida AM, Muñoz-Espinoza C et al (2014) Whole genome comparison between table and wine grapes reveals a comprehensive catalog of structural variants. BMC Plant Biol 14:7. https://doi.org/10.1186/1471-2229-14-7

Doyle JJ, Doyle JL (1987) A rapid DNA isolation procedure for small quantities of fresh leaf tissue. Phytochem Bull 19:11–15

Earl D, Bradnam K, St. John JN et al (2011) Assemblathon 1: a competitive assessment of de novo short read assembly methods. Genome Res 21:2224–2241. https://doi.org/10.1101/gr.126599.111

Edwards D, Batley J (2010) Plant genome sequencing: applications for crop improvement. Plant Biotechnol J 8:2–9. https://doi.org/10.1111/j.1467-7652.2009.00459.x

Ekblom R, Wolf JBW (2014) A field guide to whole-genome sequencing, assembly and annotation. Evol Appl. https://doi.org/10.1111/eva.12178

Endrullat C, Glökler J, Franke P, Frohme M (2016) Standardization and quality management in next-generation sequencing. Appl Transl Genom 10:2–9. https://doi.org/10.1016/j.atg.2016.06.001

Fang G, Hammar S, Grumet R (1992) A quick and inexpensive method for removing polysaccharides from plant genomic DNA. Biotechniques 13(1):52–54

Feuillet C, Leach JE, Rogers J et al (2011) Crop genome sequencing: lessons and rationales. Trends Plant Sci 16:77–88. https://doi.org/10.1016/j.tplants.2010.10.005

Gambino G, Dal Molin A, Boccacci P et al (2017) Whole-genome sequencing and SNV genotyping of 'Nebbiolo' (*Vitis vinifera* L.) clones. Sci Rep 7:17294

Goff SA, Ricke D, Lan T-H et al (2002) A draft sequence of the rice genome (*Oryza sativa* L. ssp. japonica). Science 296:92–100. https://doi.org/10.1126/science.1068275

Green MR, Sambrook J (2016) Precipitation of DNA with ethanol. Cold Spring Harb Protoc. https://doi.org/10.1101/pdb.prot093377

Green MR, Sambrook J (2017) Precipitation of DNA with isopropanol. Cold Spring Harb Protoc. https://doi.org/10.1101/pdb.prot093385

Guzmán P, Ecker JR (1988) Development of large DNA methods for plants: molecular cloning of large segments of Arabidopsis and carrot DNA into yeast. Nucleic Acids Res 16(23):11091–11105

Hanania U, Velcheva M, Sahar N et al (2004) An improved method for isolating high-quality DNA vitis nuclei. Plant Mol Biol Rep 22:173–177

Healy A, Furtado A, Cooper T, Henry R (2014) Protocol: a simple method for extracting next-generation sequencing quality genomic DNA from recalcitrant plant species. Plant Methods 10(1):1–8

Heptinstall J, Rapley R (2000) Spectrophotometric analysis of nucleic acids. In: Rapley R (ed) The Nucleic acid protocols handbook. Humana Press, Totowa

Hyma KE, Barba P, Wang M et al (2015) Heterozygous mapping strategy (HetMappS) for high resolution genotyping-by-sequencing markers: a case study in grapevine (N. A. Tinker, Ed.). PLoS ONE 10: e0134880

Iandolino AB, Goes da Silva F, Lim H, Choi H, Williams LE, Cook DR (2004) High-quality Rna, cDNA, and derived EST libraries from grapevine (*Vitis vinifera* L.). Plant Mol Biol Rep 22:269–278

Jaillon O, Aury J-M, Noel B et al (2007) The grapevine genome sequence suggests ancestral hexaploidization in major angiosperm phyla. Nature 449:463–467. https://doi.org/10.1038/nature06148

Japelaghi RH, Haddad R, Garoosi GA (2011) Rapid and efficient isolation of high quality nucleic acids from

plant tissues rich in polyphenols and polysaccharides. Mol Biotechnol 49:129–137. https://doi.org/10.1007/s12033-011-9384-8
Jiao C, Gao M, Wang X, Fei Z (2015) Transcriptome characterization of three wild Chinese Vitis uncovers a large number of distinct disease related genes. BMC Genom 16:223
Kajitani R, Toshimoto K, Noguchi H et al (2014) Efficient de novo assembly of highly heterozygous genomes from whole-genome shotgun short reads. Genome Res 24:1384–1395. https://doi.org/10.1101/gr.170720.113
Kim CS, Lee CH, Shin JS, Chung YS, Hyung NI (1997) A simple and rapid method for isolation of high quality genomic DNA from fruit trees and conifers using PVP. Nucleic Acids Res 25:1085–1086
Knebelsberger T, Stöger I (2012) DNA extraction, preservation, and amplification. In: Kress W, Erickson D (eds) DNA Barcodes. Methods in molecular biology (methods and protocols). Humana Press, Totowa
Laucou V, Lacombe T, Dechesne F et al (2011) High throughput analysis of grape genetic diversity as a tool for germplasm collection management. Theor Appl Genet 122(6):1233–1245
Lodhi MA, Ye G-N, Weeden NF, Reisch BI (1994) A simple and efficient method for DNA extraction from grapevine cultivars and *Vitis* species. Plant Mol Biol Rep 12:6–13
Lutz KA, Wang W, Zdepski A, Michael TP (2011) Isolation and analysis of high quality nuclear DNA with reduced organellar DNA for plant genome sequencing and resequencing. BMC Biotechnol 11:54. https://doi.org/10.1186/1472-6750-11-54
Maliyakal JE (1992) An efficient method for isolation of RNA and DNA from plants containing polyphenolics. Nucleic Acids Res 20:2381
Marsal G, Baiges I, Canals JM, Zamora F, Fort F (2013) Comparison of the efficiency of some of the most usual DNA extraction methods for woody plants in different tissues of *Vitis vinifera* L. J Int Sci Vigne Vin 47:227–237
Massonnet M, Fasoli M, Tornielli GB et al (2017) Ripening transcriptomic program in red and white grapevine varieties correlates with berry skin anthocyanin accumulation. Plant Physiol 174(4):2376–2396
Mayjonade B, Gouzy J, Donnadieu C et al (2016) Extraction of high-molecular-weight genomic DNA for long-read sequencing of single molecules. Biotechniques 61:203–205. https://doi.org/10.2144/000114460
Michael TP, Jackson S (2013) The first 50 plant genomes. Plant Genome 6:1–7
Minio A, Lin J, Gaut BS, Cantu D (2017) How single molecule real-time sequencing and haplotype phasing have enabled reference-grade diploid genome assembly of wine grapes. Front Plant Sci 8:1–6. https://doi.org/10.3389/fpls.2017.00826
Minio A, Massonnet M, Figueroa-Balderas R et al (2019a) Iso-seq allows genome-independent transcriptome profiling of grape berry development. G3 Genes Genomes Genet 9:755–767
Minio A, Massonnet M, Figueroa-Balderas R et al (2019b) Diploid genome assembly of the wine grape carmenere. G3 Genes Genomes Genet 9:1331–1337
Morrell PL, Buckler ES, Ross-Ibarra J (2012) Crop genomics: advances and applications. Nat Rev Genet 13(2):85–96. https://doi.org/10.1038/nrg3097
Murray MG, Thompson WF (1980) Rapid isolation of high molecular weight DNA. Nucleic Acids Res 8:4321–4325
Myles S, Chia J-M, Hurwitz B et al (2010) Rapid genomic characterization of the genus Vitis. PLoS ONE 5:e8219. https://doi.org/10.1371/journal.pone.0008219
Nakayama Y, Yamaguchi H, Einaga N, Esumi M (2016) Pitfalls of DNA quantification using DNA-binding fluorescent dyes and suggested solutions. PLoS ONE 11(3):e0150528. https://doi.org/10.1371/journal.pone.0150528
Newbury HJ, Possingham JV (1977) Factors affecting the extraction of intact ribonucleic acid from plant tissues containing interfering phenolic compounds. Plant Physiol 60:543–547
Pendleton M, Sebra R, Pang AWC et al (2015) Assembly and diploid architecture of an individual human genome via single-molecule technologies. Nat Methods 12:780–786
Peterson DG, Boehm KS, Stack SM (1997) Isolation of milligram quantities of nuclear DNA from tomato (*Lycopersicon esculentum*), a plant containing high levels of polyphenolic compounds. Plant Mol Biol Rep 15(2):148–153
Pryszcz LP, Gabaldón T (2016) Redundans: An assembly pipeline for highly heterozygous genomes. Nucleic Acids Res 44:e113. https://doi.org/10.1093/nar/gkw294
Reid KE, Olsson N, Schlosser J, Peng F, Lund ST (2006) An optimized grapevine RNA isolation procedure and statistical determination of reference genes for real-time RT-PCR during berry development. BMC Plant Biol 6:27. https://doi.org/10.1186/1471-2229-6-27
Rezadoost MH, Kordrostami M, Kumleh HH (2016) An efficient protocol for isolation of inhibitor-free nucleic acids even from recalcitrant plants. 3 Biotech 6(1):61
Ribeiro RA, Lovato MB (2007) Comparative analysis of different DNA extraction protocols in fresh and herbarium specimens of the genus *Dalbergia*. Gen Mol Res 6:173–187
Roach MJ, Johnson DL, Bohlmann J et al (2018) Population sequencing reveals clonal diversity and ancestral inbreeding in the grapevine cultivar Chardonnay. PLoS Genet 14:e1007807. https://doi.org/10.1371/journal.pgen.1007807
Romieu C (2010) RNA extraction from young, acidic berries and other organs from *Vitis vinifera* L. In: Delrot S, Medrano H, Or E, Bavaresco L, Grando S

(eds) Methodologies and results in grapevine research. Springer, Dordrecht
Ruan J, Li H (2019) Fast and accurate long-read assembly with wtdbg2. *BioRxiv*, 530972. https://doi.org/10.1101/530972
Safonova Y, Bankevich A, Pevzner PA (2015) dipSPAdes: assembler for highly polymorphic diploid genomes. J Comput Biol A J Comput Mol Cell Biol 22:528–545. https://doi.org/10.1089/cmb.2014.0153
Sahu SK, Thangaraj M, Kathiresan K (2012) DNA extraction protocol for plants with high levels of secondary metabolites and polysaccharides without using liquid nitrogen and phenol. ISRN Mol Biol 2012:205049. https://doi.org/10.5402/2012/205049
Sharma AD, Gill PK, Singh P (2002) DNA isolation from dry and fresh samples of polysaccharide-rich plants. Plant Mol Biol Rep 20:415. https://doi.org/10.1007/BF02772129
Simão FA, Waterhouse RM, Ioannidis P et al (2015) BUSCO: assessing genome assembly and annotation completeness with single-copy orthologs. Bioinformatics. https://doi.org/10.1093/bioinformatics/btv351
Tan ZJ, Chen SJ (2005) Nucleic acid helix stability: effects of salt concentration, cation valence and size, and chain length. Biophys J 90(4):1175–1190
The Arabidopsis Genome Initiative (2000) Analysis of the genome sequence of the flowering plant *Arabidopsis thaliana*. Nature 408:796–815
Thottathil GP, Jayasekaran K, Othman AS (2016) Sequencing crop genomes: a gateway to improve tropical agriculture. Trop Life Sci Res. https://doi.org/10.1111/evo.12191.1
Tuskan GA, Difazio S, Jansson S et al (2006) The genome of black cottonwood, *Populus trichocarpa* (Torr. & Gray). Science 313(5793):1596–1604
VanBuren R, Bryant D, Edger PP et al (2015) Single-molecule sequencing of the desiccation-tolerant grass *Oropetium thomaeum*. Nature 527:508–511
Varma A, Padh H, Shrivastava N (2007) Plant genomic DNA isolation: an art or a science. Biotechnol J 2:386–392
Velasco R, Zharkikh A, Troggio M et al (2007) A high quality draft consensus sequence of the genome of a heterozygous grapevine variety. PLoS ONE 2:e1326. https://doi.org/10.1371/journal.pone.0001326
Venturini L, Ferrarini A, Zenoni S et al (2013) De novo transcriptome characterization of *Vitis vinifera* cv. Corvina unveils varietal diversity. BMC Genom 14:41. https://doi.org/10.1186/1471-2164-14-41
Walenz BP, Koren S, Bergman et al (2017) Canu: scalable and accurate long-read assembly via adaptive k-mer weighting and repeat separation. Genome Res 27(5):722–736. https://doi.org/10.1101/gr.215087.116
Xiao H, Kim WS, Meng B (2015) A highly effective and versatile technology for the isolation of RNAs from grapevines and other woody perennials for use in virus diagnostics. Virol J 12:171. https://doi.org/10.1186/s12985-015-0376-3
Yang C, Li F, Ji X, Wang J (2011) An effective method for RNA extraction from grapevine berry skins. Afr J Biotechnol 10(45):9032–9035
Yang S, Fresnedo-Ramírez J, Wang M et al (2016) A next-generation marker genotyping platform (AmpSeq) in heterozygous crops: a case study for marker-assisted selection in grapevine. Hortic Res 3:16002. https://doi.org/10.1038/hortres.2016.2
Yockteng R, Almeida AM, Yee S, Andre T, Hill C, Specht CD (2013) A method for extracting high-quality RNA from diverse plants for next-generation sequencing and gene expression analyses. Appl Plant Sci. https://doi.org/10.3732/apps.1300070
Yu J, Wang J, Lin W et al (2005) The genomes of *Oryza sativa*: a history of duplications. PLoS Biol 3:e38. https://doi.org/10.1371/journal.pbio.0030038
Zhou Y, Massonnet M, Sanjak JS et al (2017) Evolutionary genomics of grape (*Vitis vinifera* ssp. vinifera) domestication. Proc Natl Acad Sci USA 114:11715–11720. https://doi.org/10.1073/pnas.1709257114
Zhou Y, Minio A, Massonnet M, et al (2018) Structural variants, clonal propagation, and genome evolution in grapevine (*Vitis vinifera*). *bioRxiv* 508119; doi:https://doi.org/10.1101/508119

The Grapevine Genome Annotation

6

Jérôme Grimplet and Grant R. Cramer

Abstract

The release of the grapevine genome sequence has allowed the generation of invaluable data on gene function, providing tools for a better understanding of the plant biology. To capitalize on this information, the annotation of the genome has been an ongoing effort performed by the research community on that species. Annotation initiatives can take the form of automatic annotation with gene prediction performed in silico based on the knowledge of other species and transcriptomic data as well as manual curation and integration of results from the literature. The International Grape Genome Program created recently a committee to harmonize the annotation process. The primary aims of the committee are to provide a unified high quality and highly accessible annotation of grapevine genes. To reach that objective, standard nomenclature for locus identifiers and conventions for a gene naming system were set up. Genome annotation is a work in progress because of new improved annotation technologies and new discoveries of structural components and functions within the genome. As technology and knowledge on genome functioning improves, it is expected that new challenges and perspectives will arise in the field of genome annotation such as the integration of the role of non-coding areas of the genome or the integration of polymorphic diversity within cultivars.

J. Grimplet (✉)
Unidad de Hortofruticultura, Centro de Investigación y Tecnología Agroalimentaria de Aragón, Instituto Agroalimentario de Aragón-IA2 (CITA-Universidad de Zaragoza), Zaragoza 50059, Spain
e-mail: jgrimplet@cita-aragon.es

G. R. Cramer
Department of Biochemistry and Molecular Biology, University of Nevada, Reno 89557, USA
e-mail: cramer@unr.edu

6.1 Introduction

Knowledge of the structure and sequence of the genome has become an invaluable tool in grapevine biological studies. Recent progress in genomic techniques has enabled whole-genome sequencing for many species, producing great quantities of data to mine for significant discoveries (see Chap. 5). However, complete and accurate annotation of both the structure and the function of genome features are necessary to reduce false-negative (from missing annotation) and false-positive (from incorrect annotation) errors in genetic and genomic studies (Steward et al. 2017). An annotated set of genes (or a gene sequence that is part of a set) is a tool used on a daily basis by all researchers in molecular biology, who are not necessarily fully aware of the level of accuracy (structural and functional) and

D. Cantu and M. A. Walker (eds.), *The Grape Genome*,
Compendium of Plant Genomes, https://doi.org/10.1007/978-3-030-18601-2_6

exhaustiveness of the annotation. To date, the total number of genes in the whole grapevine genome is far from being fully known, an annotation for many of them is computer-predicted automatically, which presents serious limitations. Therefore, the actual level of quality of the annotation of the grapevine genome should be put in perspective with the most manually curated genomes such as the human genome. Initially, the first publication reported that there were approximately 30,000 to 40,000 protein-coding genes (Lander et al. 2001; Venter et al. 2001). Through additional curation performed by the GENCODE consortium and the development of knowledge of genome functioning, this initial assertion was reassessed. The number of protein-coding genes dropped to 19,901 in version 28 of the human genome giving a total of 58,381 genes including long and short non-coding genes and pseudogenes (https://www.gencodegenes.org/stats/archive.html). Even so, the human genome is still not considered to be fully annotated (Southan 2017). The workforce in grapevine is much smaller than that for the human genome or for other model plants, such as Arabidopsis or rice; nevertheless, we can benefit from previous experience and knowledge for these species to provide a better genome curation.

6.2 Automatic Annotation in Grapevine

The development of the grapevine genome annotation is strongly linked to genome sequencing and the curation of the genome structure of the reference genome, PN40024 (Table 6.1). The reference genome has been a central tool for the research performed on grapevine for more than a decade (Jaillon et al. 2007). The sequence of the PN40024 genome is a near homozygous line related to the Pinot Noir cultivar.

The grape gene annotation is continuously evolving both for structural and functional annotations as additional analyses are performed. There have been two major updates, one in 2010 upgrading the structural sequence from an 8X with a 12X coverage (genome 12Xv0, Adam-Blondon et al. 2011) and one in 2017 with reassembly of the 12X sequence [genome 12Xv2 (Canaguier et al. 2017)]. Paralleling the release of these updates, sequence annotations and gene predictions were developed, each utilizing different algorithms for prediction leading to significantly different sets of genes. The gene prediction released with the original 8X sequence was performed in the frame of the same initiative using the GAZE software (Howe et al. 2002) for gene prediction; the initial set contained 30,434 genes. For the 12Xv0, several sets of predicted genes have been developed and used. The V0 gene prediction was developed following the same modality as the 8X prediction and was immediately merged into the V1 annotation with a prediction performed with the JIGSAW software (Allen and Salzberg 2005) by CRIBI (http://genomes.cribi.unipd.it/grape/). This annotation (CRIBIv1, 29,971 genes) was the basis for designing the first (and sole) microarray platform that included potentially the whole set of genes encountered in the genome. It was, for example,

Table 6.1 History of the grapevine reference genome assembly and annotation

Reference genome version	Annotation version	Responsible institution
8x	Genoscope 8x	Genoscope, France
12Xv1	Genoscope 12X	Genoscope
	CRIBI v1	CRIBI, Italy
	CRIBI v2	CRIBI
	RefSeq annotation	NCBI, USA
12Xv2	VCost	VIB, Belgium, COST action FA1106, EU
	VCost v3	IGGP, International

used for the grapevine gene expression atlas (Fasoli et al. 2012) and numerous other studies (Ghan et al. 2015; Nicolas et al. 2014). Gene expression studies in following years have evolved with the RNAseq method, which has allowed the use of any set of genes, because transcriptomes can be easily built de novo for expression data analysis by the RNAseq method. Several versions were used in the different publications using RNAseq (e.g., Venturini et al. 2013; Jiao et al. 2015; Gambino et al. 2017; see Chap. 13).

The CRIBI team, focusing on the identification of alternative variants, updated the v1 version into a v2 (Vitulo et al. 2014) adding 2258 coding genes and 3336 putative long non-coding RNAs. This CRIBI v2 gene prediction version is not to be confused with the genome 12Xv2 sequence version; the CRIBI v2 gene prediction was performed on the genome 12Xv1, which has now been reassembled to the 12Xv2. The RefSeq prediction produced by the Gnomon–NCBI eukaryotic gene prediction tool (Souvorov et al. 2010) at NCBI identified 27,043 putative genes. The 12Xv2 genome prediction was performed by the VIB in the frame of a European COST program using the Eugene software (Foissac et al. 2008) was accessible through the ORCAE annotation platform (Sterck et al. 2012). It is common to see recent grape publications using outdated versions of the annotation such as the V0 because of the lack of visibility or accessibility of newer annotations. A central Web site with the latest annotations is needed to reduce confusion and facilitate comparisons.

Each of these versions was built independently of each other, each capable of identifying different genes. In an effort to standardize the annotation and provide interoperability between different versions, it was later chosen to merge all of the previous annotations into a unified version (COSTv3), which was released with the publication of the 12Xv2 genome update (Canaguier et al. 2017). This version at the time of the publication included 42,414 gene models. However, this version is regularly updated with manually curated data. The latest iteration is available at the URGI Web site (https://urgi.versailles.inra.fr/Species/Vitis/Annotations). This latest version of the reference genome annotation was performed under the auspices of the Super-Nomenclature Committee for Grape Gene Annotation (sNCGGa) (Grimplet et al. 2014), an emanation of the International Grape Genome Program. The committee regulates the incorporation of curated data by the community of researchers. Allowing the use of a dynamic version of the annotation should be favored over a static version because genome annotation needs to be constantly improved. In addition, in transcriptomic studies, RNAseq allows greater flexibility in the set of studied genes since it is not constrained to a predefined design of probes as for a microarray. The GFF files used for read counts can be constantly and flexibly updated. Old data can be reanalyzed in light of newer gene predictions. The latest version of the annotation also includes in the GFF file all the previous IDs used for a given gene in older annotations, which was done in order to facilitate cross-platform data comparison. The nomenclature scheme was also designed to facilitate the easy incorporation of new genes, as discussed later. Table 6.2 shows the example of the *MYBA1* gene, involved in berry color. It is a well-described gene in grapevine since the earliest eras of sequencing. Since its discovery in 2002, 10 different IDs have been used to identify this gene.

The merging of the datasets was performed by comparing the sequences produced with the different algorithms and highlighted the discrepancies between results that were generated. The decision-making process for choosing the best sequence within the three annotations used indicators of quality related to the frequency of nearly perfect overlapping sequences in public repositories. For example, *MYBA1* coding sequence was correctly annotated in three annotations (Fig. 6.1). The RefSeq sequence was kept in the final set because it included the longest UTRs.

Many predicted genes (16,444 genes) were only detected in only one of the three algorithms

Table 6.2 Nomenclature history of gene MYBA1

ID for MYBA1	Genome release name	Annotation name/source	Position for MYBA1 on chr02	Date
Before whole-genome sequencing				
AB073013	GenBank ID	Initial publication (Kobayashi)		2002
1620959_s_at 1615798_at	GeneChip probeset	EST from various cultivars. DFCI gene index version 4		2003
VVTU17547_at	Grapegen probeset	EST from various cultivars. DFCI gene index v5 + GrapeGen project EST		2007
Whole-genome sequencing				
GSVIVP00038762001 and GSVIVP00038763001	8X	Genoscope 8X	12448352–12449010 12449064–12449458	2007
GSVIVT01022659001	12Xv1	Genoscope 12X	14239792–14240808	2010
VIT_02s0033g00410	12Xv1	CRIBI 12Xv1	14239792–14240808	2010
Vitvi02g01019	12Xv2	ORCAE annotation	14351795–14352913	2013
VIT_202s0033g00410	12Xv1	CRIBI 12Xv2	14239584–14240983	2014
LOC100233098	12Xv1	RefSeq	14239789–14240887	2014
Vitvi02g01019	12Xv2	12Xv3	14351795–14352913	2017

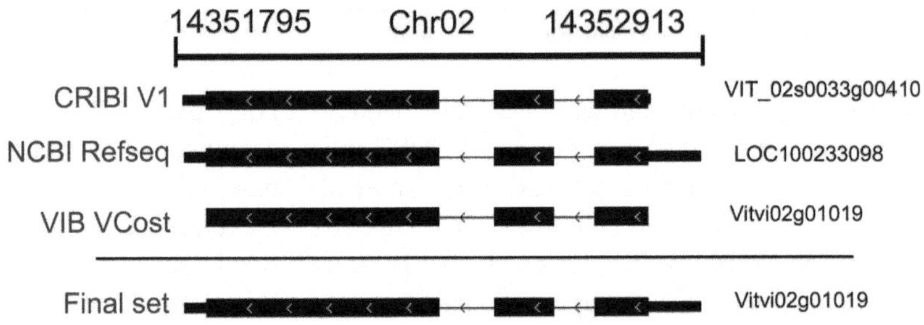

Fig. 6.1 Alignment between the gene models from the three annotations for MYBA1

(JIGSAW/GAZE, GNOMON or Eugene), and many genes (15,288) were detected in all the three annotations (Fig. 6.2). The differences were related to many different parameters. Some algorithms seem to perform better to detect small exons; some are more adapted for loci that contain long introns. The variation is also related to the set of transcript sequences used to correct the gene models; this type of analysis might include bias toward genes expressed in tissues used to produce the data. In addition, many structural discrepancies were also encountered for genes identified in several annotations, which has implicated that manual validation is necessary for many of them. For example, based on the gene positions in each annotation, 5761 loci contained more than one gene in one annotation and only one gene in another annotation.

6.3 New Gene Discovery Through Manual Curation

Automatic annotation is an essential tool for gene prediction but cannot guarantee exhaustive and completely accurate prediction for all the genes models; in fact, the automatic annotation has many inaccuracies and only should be considered a crude estimate until the gene prediction has

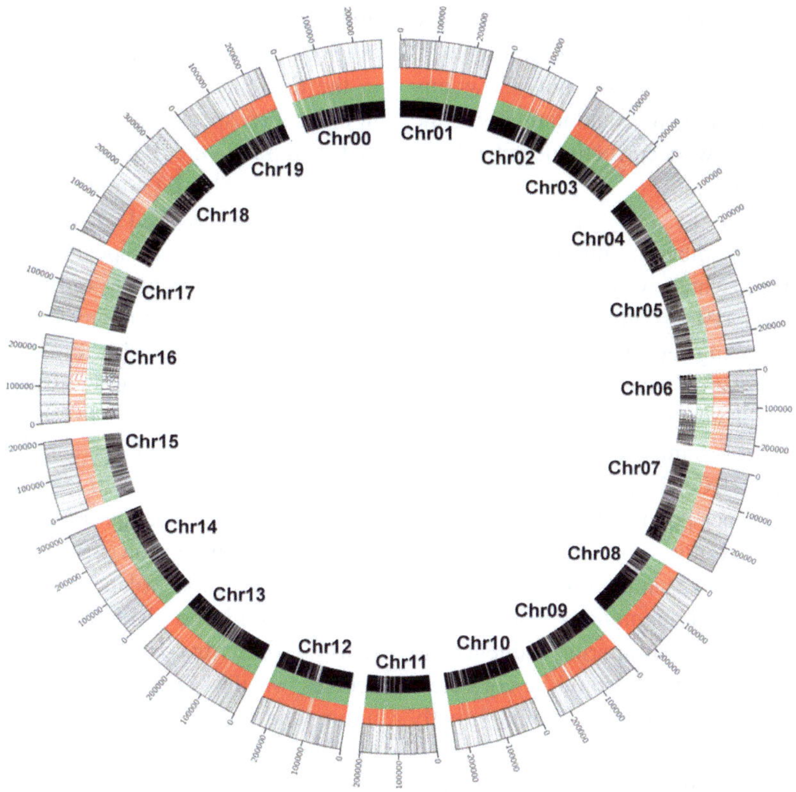

Fig. 6.2 Coverage over the genome 12xv2 of the different annotations. The outer circle represents the gene position on the unified v3. Inner circles: Red, V1. Green, VCost. Black, RefSeq

been manually inspected with other data such as full-length transcripts or ESTs. Several studies based on manual curation were incorporated into the current set of predicted genes, highlighting the benefit of manual annotation.

In a previous study (Grimplet et al. 2016b), a methodology based heavily on manual curation for the annotation of MADS-box transcription factor family was used; however, it is nearly impossible to streamline this approach at high throughput. Nevertheless, the manual inspection must be done, bit by bit. This type of approach also allows the discovery of new genes for specific conditions. The strategy was to target all the MADS-box motifs along the whole-genome sequence to identify all loci of this family of transcription factors. The surrounding genome region was examined and compared to other species to determine if they would correspond to the putative genes. This strategy proved fruitful for a subfamily of Type-1 MADS-box genes poorly described in any plant species; these grapevine gene sequences were fairly distant from genes from other species, besides the MADS-box motif itself. It may be only present in few embryo cells as in Arabidopsis (Bemer et al. 2010) for which no tissue-specific expression or RNAseq data have been incorporated in any grapevine prediction models. Furthermore, single-cell RNA sequencing has never been published in grapevine, explaining why their expression was never detected. As a consequence, automatic annotation software failed to detect genes of this subfamily because they share little homology to other species and there was not enough expression data to validate the models. In contrast, the same strategy was used on the GRAS family transcription factor (Grimplet et al. 2016a), but this approach did not show any benefits in improving the annotation of the family, all the genes were correctly detected by at least one algorithm.

Other additions to the latest annotation include the curated stilbene synthase and terpene synthase families, which were curated in a previous publication (Parage et al. 2012) and the MYB family (Wong et al. 2016). Structural curation and annotation were also performed (G. R. Cramer, unpublished results) for the AP2/ERF transcription factor family using the bioinformatics tools at ORCAE (Sterck et al. 2012). There were 130 genes previously identified in the v1 annotation, and this was increased to 152 genes in the current v3 annotation. In addition, many of the structural annotations were discovered to be incorrect. By comparing sequences to existing ESTs, the structures could be corrected. Full-length transcript sequences would greatly facilitate improvements in gene model predictions in the future, although this may get more complicated with alternative splicing data.

6.4 Gene Nomenclature for Improving Data Description and Interoperability

In order to normalize the nomenclature of the annotation of grapevine genes so that everyone could be working on the same page, the IGGP commissioned the sNCGGa (Grimplet et al. 2014) to define a set of rules for the naming and annotation of the genes. The guidelines focused on the definitions on the one hand of the nomenclature of unique alphanumeric loci identifiers, and on the other hand, the nomenclature of genes names with a short identifier and a longer identifier that should describe the function.

For the unique alphanumeric identifier, the rules were defined in the context of previous attempts of nomenclature in other species and in grapevine to construct a future-proof model taking into account probable new gene discoveries in the reference genome and production of data in other *Vitis* species and cultivars. Each element that composes the ID should contain relevant information. The ID for the 8X version of the annotation and the RefSeq annotation was generated with no species-specific information, thus providing codes with the only requirement that they are not redundant in any species. The CRIBI v1 and v2 incorporated a reference to the taxonomic family and the chromosome number in the locus ID in addition to the other information from the original scaffold name and a number sequentially assigned relative to its chromosome position. These grapevine models for identifiers were largely based on models used for Arabidopsis and the Solanaceae for its general structure. With the vision of providing an annotation system that could be applied to all the species of the *Vitis* genus, interoperable and nonredundant, the sNCGGa recommends the use of a unique prefix for each species of the family corresponding to the five-letter prefixes published in the UniProt database controlled vocabulary of species (https://www.uniprot.org/docs/speclist). For the reference genome, belonging to the *Vitis vinifera* species, that identifier is "Vitvi." For *Vitis riparia,* it is Vitri or *Vitis labrusca* is Vitla. The prefix should be followed by the chromosome number and a single letter for the molecule (gene: g; transcript: t; protein: p). Finally, a five-letter code uniquely attributed within each chromosome, with no reference to the position, in order to easily incorporate new genes on the not-yet allocated numbers. For Solanaceae and Arabidopsis, genes were allocated to every decimal position on the genome (at1g00010, at1g00020, etc.) to allow incorporation of newly discovered genes in between. As the number of potentially new genes between two genes is not known it was decided by the sNCGGa to not use gaps between numbers. Furthermore, structural rearrangements in future structural annotations would also mess up a logically ordered numbering system. It was decided that newly discovered genes should simply be allocated to the next available number for a chromosome. It also has the advantage to simplify the nomenclature during the annotation curation process after the merging of loci compared to sequential numbering, the two loci can be replaced and classified as "synonyms" of the new locus.

The main recommendations for the rules for the functional identifiers (short and long versions) were primarily to simplify gene names and not have multiple names with other genes described in the literature. It is therefore less important that authors or curators are fully aware of previous works and nomenclature of the genes related to their studied models. Other elements of the guidelines favored the use of nomenclature in previous studies whenever possible. However, using new names is acceptable as long as it is not redundant. This may occur for several reasons. If a single gene was detected as having been published under different names including redundant names with another gene it would be preferable to provide an unequivocal new name. It may specifically happen in an article annotating entire genes families disregarding the previous annotation. This issue could also typically occur in whole gene family annotation using a common prefix describing the family followed by the chromosome and chromosome position to attribute sequential number for each gene. Such an approach is not recommended because it is not sustainable when new genes are discovered. It also does not provide any potentially useful information on orthology in other species or subfamilies. A certain degree of liberty should be left to the authors for the gene nomenclature, which also depends on the studied genes families; universal rules cannot be defined for all circumstances.

One proposition is to use the names echoing Arabidopsis orthologs. This approach has the advantage to narrow down potential functional roles for the grapevine genes, but in practice encountering clear orthologs using phylogenetics tools is not always possible, specifically in the large gene family. In these families, some genes have orthologs, some do not, which force the curator to use a hybrid system, with genes named after their orthologs, and others newly named. For large gene families, many include a subfamily hierarchy in previously annotated species, which could be used for naming. This approach was used to annotate the MADS, GRAS, and LOB subfamilies.

The major intrinsic protein (MIP) family was annotated in three studies from different genome versions. An example of good practice in the context of the previously annotated family, Wong et al. recently annotated the major intrinsic protein (MIP) gene family in grapevine (Wong et al. 2018) following the guidelines of the sNCGGa. The grapevine MIPs were previously annotated (Shelden et al. 2009; Fouquet et al. 2008), and both previous nomenclatures were considered for the reannotation. Shelden et al. also took into consideration the previous nomenclature, for the already described genes, they either used previous nomenclature (Fouquet et al. 2008) or corrected genes they considered mis-annotated; thus, the affiliation between names is clear. All three studies followed the classical model for this family to categorize the members according to subfamilies describes by their localization, e.g., P(lasma membrane)IP, T (onoplastic)IP. Wong et al. identified new genes in the v3 annotation as compared to the previous study. These genes did not have clear orthologs within Arabidopsis; in these cases, they assigned these gene names with new numbers within the subfamilies. One of the subfamilies (XIP) do not exist in Arabidopsis. The authors performed some changes in the nomenclature for the name uniformity and reflection of phylogenetic similarities (Table 6.3). For example, PIP1 (Vitvi12g01740), PIP1-5 (Vitvi15g01109), and a new gene were renamed (Vitvi18g02210), respectively, PIP1-2c, PIP1-2a, and PIP1-2b. Others were renamed to reflect orthology with Arabidopsis.

There is, however, an occurrence highlighting the risks of renaming a gene to fit the orthology of Arabidopsis (Italic in Table 6.3). Vitvi13g00255 was originally annotated as TIP1-1

Table 6.3 Genes labeled differently within the three annotations of the MIP family

Wong et al. (2018)	Shelden et al. (2009)	Fouquet et al. (2008)	Locus ID
VviNIP1-2	VvNIP1;1		Vitvi10g00639
VviNIP4-1	VvNIP3;1		Vitvi14g00966
VviNIP8-1	VvNIP2;1	VvNIP4;1	Vitvi14g01952
Not identified	**VvNIP4;1**	**VvNIP8;1**	Vitvi11g01601
VviPIP1-2a	VvPIP1;5		Vitvi15g01109
VviPIP1-2b	Not identified		Vitvi18g02210
VviPIP1-2c	VvPIP1		Vitvi12g01740
VviPIP1-4	VvPIP1;4	VvPIP1;2	Vitvi15g01110
VviPIP2-7	VvPIP2;2		Vitvi03g00155
VviTIP1-1	***VvTIP1;3***		Vitvi06g01346
VviTIP1-3	***VvTIP1;1***		Vitvi13g00255

Bold-labeled genes indicates conflicting names between annotation versions

(Shelden et al. 2009). Wong et al. renamed it as vviTIP1-3 since it is an ortholog to Arabidopsis TIP1-3. The name TIP1-3 name was previously assigned to Vitvi06g01346. This created redundancy since TIP1-3 corresponds to two different genes in the function of the publication and can cause confusion. Vitvi11g01601 is another confusing gene (Bold in Table 6.3); it was also not detected as a MIP in Wong et al., but it was described with two different names in the two older studies (NIP4-1 and NIP8-1) that were assigned to two other genes in Wong et al. Thus, it would be useful to have a committee that could review such reannotations and sort through possible or unforeseen confusion by the changes. Ideally, the authors of all of the publications concerned would be involved in sorting things out so that the community can know that agreement has been reached and use correct annotations in the future.

6.5 Proteogenomics-Based Annotation

Annotation of the grapevine genome using proteogenomics was performed by (Chapman and Bellgard 2017). This study resulted in the identification of 54 proteins different from the 12Xv2 annotation, incorporating 106 novel peptides when compared to this version. We compared these 54 proteins to the latest release of the grapevine genome annotation, 15 were identical to the putative proteins from this version, the annotation of 23 of them was not improved, and it improved the annotation for 14 of them. The relatively high number of genes not improved was not related to the efficiency of the proteogenomics technique by itself. The authors used a different gene prediction algorithm, Augustus (Stanke et al. 2006) than the ones already included in the reference annotation (JIGSAW-GAZE, Eugene and Gnomon). Augustus is a tool particularly well adapted for the inclusion of constraint (such as a known peptide), which is suitable for the proteogenomic analysis. For some loci, Augustus was the algorithm providing the best prediction but not for other loci. Augustus also tends to deliver shorter and a greater number of predicted proteins than the other algorithms. It predicted 84,948 genes (Chapman and Bellgard 2017), twice as many as the current v3 annotation. Proteogenomics is a valuable tool for genome annotation with an important potential in the future.

6.6 Annotation of Non-coding Transcriptome

So far, little work has been performed on the annotation of grapevine non-coding RNA (ncRNA); some non-coding data from the RefSeq annotation were integrated into the latest annotation (around 2000). According to what has been observed in other species a much higher number of ncRNA is expected in grapevine. There are currently 28,468 ncRNA in GENCODE v28 for human and 17,855 for mouse (https://www.gencodegenes.org/stats/current.html). The role of ncRNA has be highlighted in plant responses to biotic and abiotic stresses (Wang et al. 2017; Nejat and Mantri 2018), but plant ncRNA are poorly identified; Arabidopsis is the only plant species present in the NONCODE database with 3763 transcripts (http://www.noncode.org/analysis.php). New tools have been developed in recent years for annotating ncRNA such as FEELnc (Wucher et al. 2017) or CPAT (Wang et al. 2013) and significant improvements in the knowledge of the grapevine non-coding transcriptome is expected.

6.7 Future Perspective for Improving the Annotation

The proposed set of rules for the gene denomination were recommended with the unique requirement of not providing a redundant name. The most recommended method in the guideline is the construction of a phylogenetic tree with grapevine and Arabidopsis genes for a family. However, in many cases, high levels of duplication of genes were observed in grapevine after the divergence with Arabidopsis resulting in little value of reporting the Arabidopsis annotation of a single gene in a whole subfamily. Moreover, in transcription factors families such as the LOB or GRAS, subfamilies with specific roles, or a conserved motif were well documented in other species. Performing phylogenetic analysis on a multi-species level and attributing genes names related to the subfamilies provides a very informative way of annotation.

More and more publications involving grapevine annotation comply with the nomenclature rules. However, there are still many issues regarding the integration of these data into the reference genome annotation. Several factors are slowing the integration process:

1. many recent annotation analyses are still performed on an outdated version of the gene, even when following the nomenclature,
2. there are tools available for transposition of the annotation from older to the latest version of the genome, but it is adding an extra step in the integration process, and
3. relatively, few people actually take the time to input their annotation data into the annotation platform. An option to simplify the procedure for the authors would be to send the information relative to a gene annotation as gff format to the committee. Regularly, the committee would validate the data and integrate them into the updated reference genome annotation.

Many genome cultivars have been sequenced in recent years, and many more are expected to come (Chin et al. 2016; Roach et al. 2018; Minio et al. 2019a, b; see Chap. 5). The rules drafted by the sNCGGa were designed to consider this and provide guidelines to annotate several alleles for a single gene but so far, the annotation platform is only dedicated to the reference genome.

6.8 Conclusions

Genome sequencing and annotations are works in progress. There are constant improvements in technologies and a need for uniformity of the nomenclature. Therefore, annotations must be flexible and researchers need to have access to the latest versions. We propose that the current International Grape Genome Program (vitaceae.org) host a Web site that gives researchers access (or links) to the latest annotations. We also propose that there is a grape community responsibility to voluntarily help in the correction and updates of this information. Such improvements can be performed at the *Vitis* site in ORCAE (http://bioinformatics.psb.ugent.be/orcae/). This has to be an ongoing effort as it is expected to continue for many years to come due to the massive efforts for manual curation and the expected sequencing of many more grape genomes.

References

Adam-Blondon AF, Jaillon O, Vezzulli S, Zharkikh A, Troggio M, Velasco R (2011) Genome sequence initiatives. In: Adam-Blondon A-F, Martinez-Zapater JM, Kole C (eds) Genetics, genomics, and breeding of grapes. Science Publishers, New York, pp 211–234. https://doi.org/10.1201/b10948-10

Allen JE, Salzberg SL (2005) JIGSAW: integration of multiple sources of evidence for gene prediction. Bioinformatics 21(18):3596–3603. https://doi.org/10.1093/bioinformatics/bti609

Bemer M, Heijmans K, Airoldi C, Davies B, Angenent GC (2010) An atlas of type I MADS box gene expression during female gametophyte and seed development in Arabidopsis. Plant Physiol 154(1):287–300. https://doi.org/10.1104/pp.110.160770

Canaguier A, Grimplet J, Di Gaspero G, Scalabrin S, Duchene E, Choisne N, Mohellibi N, Guichard C, Rombauts S, Le Clainche I, Berard A, Chauveau A, Bounon R, Rustenholz C, Morgante M, Le Paslier MC, Brunel D, Adam-Blondon AF (2017) A new version of the grapevine reference genome assembly (12X.v2) and of its annotation (VCost.v3). Genom Data 14(Supplement C):56–62. https://doi.org/10.1016/j.gdata.2017.09.002

Chapman B, Bellgard M (2017) Plant proteogenomics: improvements to the grapevine genome annotation. Proteomics 17(21):1700197. https://doi.org/10.1002/pmic.201700197

Chin CS, Peluso P, Sedlazeck FJ, Nattestad M, Concepcion GT, Clum A, Dunn C, O'Malley R, Figueroa-Balderas R, Morales-Cruz A, Cramer GR, Delledonne M, Luo C, Ecker JR, Cantu D, Rank DR, Schatz MC (2016) Phased diploid genome assembly with single-molecule real-time sequencing. Nat Methods 13(12):1050–1054. https://doi.org/10.1038/nmeth.4035

Fasoli M, Dal Santo S, Zenoni S, Tornielli GB, Farina L, Zamboni A, Porceddu A, Venturini L, Bicego M, Murino V, Ferrarini A, Delledonne M, Pezzotti M (2012) The grapevine expression atlas reveals a deep transcriptome shift driving the entire plant into a maturation program. Plant Cell 24(9):3489–3505. https://doi.org/10.1105/tpc.112.100230

Foissac S, Gouzy J, Rombauts S, Mathe C, Amselem J, Sterck L, de Peer YV, Rouze P, Schiex T (2008) Genome annotation in plants and fungi: EuGene as a model platform. Curr Bioinform 3(2):87–97

Fouquet R, Léon C, Ollat N, Barrieu F (2008) Identification of grapevine aquaporins and expression analysis in developing berries. Plant Cell Rep 27(9):1541–1550. https://doi.org/10.1007/s00299-008-0566-1

Gambino G, Dal Molin A, Boccacci P, Minio A, Chitarra W, Avanzato CG, Tononi P, Perrone I, Raimondi S, Schneider A, Pezzotti M, Mannini F, Gribaudo I, Delledonne M (2017) Whole-genome sequencing and SNV genotyping of 'Nebbiolo' (*Vitis vinifera* L.) clones. Sci Rep 7(1):17294. https://doi.org/10.1038/s41598-017-17405-y

Ghan R, Van Sluyter SC, Hochberg U, Degu A, Hopper DW, Tillet RL, Schlauch KA, Haynes PA, Fait A, Cramer GR (2015) Five omic technologies are concordant in differentiating the biochemical characteristics of the berries of five grapevine (*Vitis vinifera* L.) cultivars. BMC Genom 16(1):946. https://doi.org/10.1186/s12864-015-2115-y

Grimplet J, Adam-Blondon A-F, Bert P-F, Bitz O, Cantu D, Davies C, Delrot S, Pezzotti M, Rombauts S, Cramer G (2014) The grapevine gene nomenclature system. BMC Genom 15(1):1077

Grimplet J, Agudelo Romero P, Teixeira R, Martinez Zapater JM, Fortes AM (2016a) Structural and functional analysis of the GRAS gene family in grapevine indicates a role of GRAS proteins in the control of development and stress responses. Front Plant Sci 7:353. https://doi.org/10.3389/fpls.2016.00353

Grimplet J, Martinez-Zapater JM, Carmona MJ (2016b) Structural and functional annotation of the MADS-box transcription factor family in grapevine. BMC Genom 17(1):80. https://doi.org/10.1186/s12864-016-2398-7

Howe KL, Chothia T, Durbin R (2002) GAZE: a generic framework for the integration of gene-prediction data by dynamic programming. Genome Res 12(9):1418–1427. https://doi.org/10.1101/gr.149502

Jaillon O, Aury JM, Noel B, Policriti A, Clepet C, Casagrande A, Choisne N, Aubourg S, Vitulo N, Jubin C, Vezzi A, Legeai F, Hugueney P, Dasilva C, Horner D, Mica E, Jublot D, Poulain J, Bruyere C, Billault A, Segurens B, Gouyvenoux M, Ugarte E, Cattonaro F, Anthouard V, Vico V, Del Fabbro C, Alaux M, Di Gaspero G, Dumas V, Felice N,

Paillard S, Juman I, Moroldo M, Scalabrin S, Canaguier A, Le Clainche I, Malacrida G, Durand E, Pesole G, Laucou V, Chatelet P, Merdinoglu D, Delledonne M, Pezzotti M, Lecharny A, Scarpelli C, Artiguenave F, Pe ME, Valle G, Morgante M, Caboche M, Adam-Blondon AF, Weissenbach J, Quetier F, Wincker P, French-Italian Public Consortium for Grapevine Genome C (2007) The grapevine genome sequence suggests ancestral hexaploidization in major angiosperm phyla. Nature 449(7161):463–467. https://doi.org/10.1038/nature06148

Jiao C, Gao M, Wang X, Fei Z (2015) Transcriptome characterization of three wild Chinese Vitis uncovers a large number of distinct disease related genes. BMC Genom 16:223. https://doi.org/10.1186/s12864-015-1442-3

Lander ES, Linton LM, Birren B, Nusbaum C, Zody MC, Baldwin J, Devon K, Dewar K, Doyle M, FitzHugh W, Funke R, Gage D, Harris K, Heaford A, Howland J, Kann L, Lehoczky J, LeVine R, McEwan P, McKernan K, Meldrim J, Mesirov JP, Miranda C, Morris W, Naylor J, Raymond C, Rosetti M, Santos R, Sheridan A, Sougnez C, Stange-Thomann Y, Stojanovic N, Subramanian A, Wyman D, Rogers J, Sulston J, Ainscough R, Beck S, Bentley D, Burton J, Clee C, Carter N, Coulson A, Deadman R, Deloukas P, Dunham A, Dunham I, Durbin R, French L, Grafham D, Gregory S, Hubbard T, Humphray S, Hunt A, Jones M, Lloyd C, McMurray A, Matthews L, Mercer S, Milne S, Mullikin JC, Mungall A, Plumb R, Ross M, Shownkeen R, Sims S, Waterston RH, Wilson RK, Hillier LW, McPherson JD, Marra MA, Mardis ER, Fulton LA, Chinwalla AT, Pepin KH, Gish WR, Chissoe SL, Wendl MC, Delehaunty KD, Miner TL, Delehaunty A, Kramer JB, Cook LL, Fulton RS, Johnson DL, Minx PJ, Clifton SW, Hawkins T, Branscomb E, Predki P, Richardson P, Wenning S, Slezak T, Doggett N, Cheng JF, Olsen A, Lucas S, Elkin C, Uberbacher E, Frazier M, Gibbs RA, Muzny DM, Scherer SE, Bouck JB, Sodergren EJ, Worley KC, Rives CM, Gorrell JH, Metzker ML, Naylor SL, Kucherlapati RS, Nelson DL, Weinstock GM, Sakaki Y, Fujiyama A, Hattori M, Yada T, Toyoda A, Itoh T, Kawagoe C, Watanabe H, Totoki Y, Taylor T, Weissenbach J, Heilig R, Saurin W, Artiguenave F, Brottier P, Bruls T, Pelletier E, Robert C, Wincker P, Smith DR, Doucette-Stamm L, Rubenfield M, Weinstock K, Lee HM, Dubois J, Rosenthal A, Platzer M, Nyakatura G, Taudien S, Rump A, Yang H, Yu J, Wang J, Huang G, Gu J, Hood L, Rowen L, Madan A, Qin S, Davis RW, Federspiel NA, Abola AP, Proctor MJ, Myers RM, Schmutz J, Dickson M, Grimwood J, Cox DR, Olson MV, Kaul R, Raymond C, Shimizu N, Kawasaki K, Minoshima S, Evans GA, Athanasiou M, Schultz R, Roe BA, Chen F, Pan H, Ramser J, Lehrach H, Reinhardt R, McCombie WR, de la Bastide M, Dedhia N, Blocker H, Hornischer K, Nordsiek G, Agarwala R, Aravind L, Bailey JA, Bateman A, Batzoglou S, Birney E, Bork P, Brown DG, Burge CB, Cerutti L, Chen HC, Church D, Clamp M, Copley RR, Doerks T, Eddy SR, Eichler EE, Furey TS, Galagan J, Gilbert JG, Harmon C, Hayashizaki Y, Haussler D, Hermjakob H, Hokamp K, Jang W, Johnson LS, Jones TA, Kasif S, Kaspryzk A, Kennedy S, Kent WJ, Kitts P, Koonin EV, Korf I, Kulp D, Lancet D, Lowe TM, McLysaght A, Mikkelsen T, Moran JV, Mulder N, Pollara VJ, Ponting CP, Schuler G, Schultz J, Slater G, Smit AF, Stupka E, Szustakowki J, Thierry-Mieg D, Thierry-Mieg J, Wagner L, Wallis J, Wheeler R, Williams A, Wolf YI, Wolfe KH, Yang SP, Yeh RF, Collins F, Guyer MS, Peterson J, Felsenfeld A, Wetterstrand KA, Patrinos A, Morgan MJ, de Jong P, Catanese JJ, Osoegawa K, Shizuya H, Choi S, Chen YJ, Szustakowki J (2001) Initial sequencing and analysis of the human genome. Nature 409(6822):860–921. https://doi.org/10.1038/35057062

Minio A, Massonnet M, Figueroa-Balderas R, Vondras AM, Blanco-Ulate B, Cantu D (2019a) Iso-Seq allows genome-independent transcriptome profiling of grape berry development. G3 Genes Genomes Genet 9 (3):755–767. https://doi.org/10.1534/g3.118.201008

Minio A, Massonnet M, Figueroa-Balderas R et al (2019b) Diploid genome assembly of the wine grape carmenere. G3 Genes Genomes Genet 9:1331–1337

Nejat N, Mantri N (2018) Emerging roles of long non-coding RNAs in plant response to biotic and abiotic stresses. Crit Rev Biotechnol 38(1):93–105. https://doi.org/10.1080/07388551.2017.1312270

Nicolas P, Lecourieux D, Kappel C, Cluzet S, Cramer G, Delrot S, Lecourieux F (2014) The basic leucine zipper franscription factor ABSCISIC ACID RESPONSE ELEMENT-BINDING FACTOR2 is an important transcriptional regulator of abscisic acid-dependent grape berry ripening processes. Plant Physiol 164(1):365–383. https://doi.org/10.1104/pp.113.231977

Parage C, Tavares R, Rety S, Baltenweck-Guyot R, Poutaraud A, Renault L, Heintz D, Lugan R, Marais GA, Aubourg S, Hugueney P (2012) Structural, functional, and evolutionary analysis of the unusually large stilbene synthase gene family in grapevine. Plant Physiol 160(3):1407–1419. https://doi.org/10.1104/pp.112.202705

Roach MJ, Johnson DL, Bohlmann J, van Vuuren HJJ, Jones SJM, Pretorius IS, Schmidt SA, Borneman AR (2018) Population sequencing reveals clonal diversity and ancestral inbreeding in the grapevine cultivar Chardonnay. PLoS Genet 14(11):e1007807. https://doi.org/10.1371/journal.pgen.1007807

Shelden MC, Howitt SM, Kaiser BN, Tyerman SD (2009) Identification and functional characterisation of aquaporins in the grapevine, *Vitis vinifera*. Funct Plant Biol 36(12):1065–1078. https://doi.org/10.1071/FP09117

Southan C (2017) Last rolls of the yoyo: Assessing the human canonical protein count [version 1; referees: 1 approved, 2 approved with reservations], vol 6. vol 448

Souvorov A, Kapustin Y, Kiryutin B, Chetvernin V, Tatusova T, Lipman D (2010) Gnomon–NCBI eukaryotic gene prediction tool

Stanke M, Keller O, Gunduz I, Hayes A, Waack S, Morgenstern B (2006) AUGUSTUS: ab initio prediction of alternative transcripts. Nucleic Acids Res 34 (Web Server issue):W435–W439. https://doi.org/10.1093/nar/gkl200

Sterck L, Billiau K, Abeel T, Rouze P, Van de Peer Y (2012) ORCAE: online resource for community annotation of eukaryotes. Nat Methods 9(11):1041. https://doi.org/10.1038/nmeth.2242

Steward CA, Parker APJ, Minassian BA, Sisodiya SM, Frankish A, Harrow J (2017) Genome annotation for clinical genomic diagnostics: strengths and weaknesses. Genome Med 9(1):49. https://doi.org/10.1186/s13073-017-0441-1

Venter JC, Adams MD, Myers EW, Li PW, Mural RJ, Sutton GG, Smith HO, Yandell M, Evans CA, Holt RA, Gocayne JD, Amanatides P, Ballew RM, Huson DH, Wortman JR, Zhang Q, Kodira CD, Zheng XH, Chen L, Skupski M, Subramanian G, Thomas PD, Zhang J, Gabor Miklos GL, Nelson C, Broder S, Clark AG, Nadeau J, McKusick VA, Zinder N, Levine AJ, Roberts RJ, Simon M, Slayman C, Hunkapiller M, Bolanos R, Delcher A, Dew I, Fasulo D, Flanigan M, Florea L, Halpern A, Hannenhalli S, Kravitz S, Levy S, Mobarry C, Reinert K, Remington K, Abu-Threideh J, Beasley E, Biddick K, Bonazzi V, Brandon R, Cargill M, Chandramouliswaran I, Charlab R, Chaturvedi K, Deng Z, Di Francesco V, Dunn P, Eilbeck K, Evangelista C, Gabrielian AE, Gan W, Ge W, Gong F, Gu Z, Guan P, Heiman TJ, Higgins ME, Ji RR, Ke Z, Ketchum KA, Lai Z, Lei Y, Li Z, Li J, Liang Y, Lin X, Lu F, Merkulov GV, Milshina N, Moore HM, Naik AK, Narayan VA, Neelam B, Nusskern D, Rusch DB, Salzberg S, Shao W, Shue B, Sun J, Wang Z, Wang A, Wang X, Wang J, Wei M, Wides R, Xiao C, Yan C, Yao A, Ye J, Zhan M, Zhang W, Zhang H, Zhao Q, Zheng L, Zhong F, Zhong W, Zhu S, Zhao S, Gilbert D, Baumhueter S, Spier G, Carter C, Cravchik A, Woodage T, Ali F, An H, Awe A, Baldwin D, Baden H, Barnstead M, Barrow I, Beeson K, Busam D, Carver A, Center A, Cheng ML, Curry L, Danaher S, Davenport L, Desilets R, Dietz S, Dodson K, Doup L, Ferriera S, Garg N, Gluecksmann A, Hart B, Haynes J, Haynes C, Heiner C, Hladun S, Hostin D, Houck J, Howland T, Ibegwam C, Johnson J, Kalush F, Kline L, Koduru S, Love A, Mann F, May D, McCawley S, McIntosh T, McMullen I, Moy M, Moy L, Murphy B, Nelson K, Pfannkoch C, Pratts E, Puri V, Qureshi H, Reardon M, Rodriguez R, Rogers YH, Romblad D, Ruhfel B, Scott R, Sitter C, Smallwood M, Stewart E, Strong R, Suh E, Thomas R, Tint NN, Tse S, Vech C, Wang G, Wetter J, Williams S, Williams M, Windsor S, Winn-Deen E, Wolfe K, Zaveri J, Zaveri K, Abril JF, Guigo R, Campbell MJ, Sjolander KV, Karlak B, Kejariwal A, Mi H, Lazareva B, Hatton T, Narechania A, Diemer K, Muruganujan A, Guo N, Sato S, Bafna V, Istrail S, Lippert R, Schwartz R, Walenz B, Yooseph S, Allen D, Basu A, Baxendale J, Blick L, Caminha M, Carnes-Stine J, Caulk P, Chiang YH, Coyne M, Dahlke C, Mays A, Dombroski M, Donnelly M, Ely D, Esparham S, Fosler C, Gire H, Glanowski S, Glasser K, Glodek A, Gorokhov M, Graham K, Gropman B, Harris M, Heil J, Henderson S, Hoover J, Jennings D, Jordan C, Jordan J, Kasha J, Kagan L, Kraft C, Levitsky A, Lewis M, Liu X, Lopez J, Ma D, Majoros W, McDaniel J, Murphy S, Newman M, Nguyen T, Nguyen N, Nodell M, Pan S, Peck J, Peterson M, Rowe W, Sanders R, Scott J, Simpson M, Smith T, Sprague A, Stockwell T, Turner R, Venter E, Wang M, Wen M, Wu D, Wu M, Xia A, Zandieh A, Zhu X (2001) The sequence of the human genome. Science 291(5507):1304–1351. https://doi.org/10.1126/science.1058040

Venturini L, Ferrarini A, Zenoni S, Tornielli GB, Fasoli M, Dal Santo S, Minio A, Buson G, Tononi P, Zago ED, Zamperin G, Bellin D, Pezzotti M, Delledonne M (2013) De novo transcriptome characterization of *Vitis vinifera* cv. Corvina unveils varietal diversity. BMC Genom 14:41. https://doi.org/10.1186/1471-2164-14-41

Vitulo N, Forcato C, Carpinelli EC, Telatin A, Campagna D, D'Angelo M, Zimbello R, Corso M, Vannozzi A, Bonghi C, Lucchin M, Valle G (2014) A deep survey of alternative splicing in grape reveals changes in the splicing machinery related to tissue, stress condition and genotype. BMC Plant Biol 14:99. https://doi.org/10.1186/1471-2229-14-99

Wang J, Meng X, Dobrovolskaya OB, Orlov YL, Chen M (2017) Non-coding RNAs and their roles in stress response in plants. Genomics Proteom Bioinform 15(5):301–312. https://doi.org/10.1016/j.gpb.2017.01.007

Wang L, Park HJ, Dasari S, Wang S, Kocher J-P, Li W (2013) CPAT: Coding-Potential Assessment Tool using an alignment-free logistic regression model. Nucleic Acids Res 41(6):e74–e74. https://doi.org/10.1093/nar/gkt006

Wong DCJ, Schlechter R, Vannozzi A, Höll J, Hmmam I, Bogs J, Tornielli GB, Castellarin SD, Matus JT (2016) A systems-oriented analysis of the grapevine R2R3-MYB transcription factor family uncovers new insights into the regulation of stilbene accumulation. DNA Res Int J Rapid Publ Rep Genes Genomes 23 (5):451–466. https://doi.org/10.1093/dnares/dsw028

Wong DCJ, Zhang L, Merlin I, Castellarin SD, Gambetta GA (2018) Structure and transcriptional regulation of the major intrinsic protein gene family in grapevine. BMC Genom 19(1):248. https://doi.org/10.1186/s12864-018-4638-5

Wucher V, Legeai F, Hédan B, Rizk G, Lagoutte L, Leeb T, Jagannathan V, Cadieu E, David A, Lohi H, Cirera S, Fredholm M, Botherel N, Leegwater PAJ, Le Béguec C, Fieten H, Johnson J, Alföldi J, André C, Lindblad-Toh K, Hitte C, Derrien T (2017) FEELnc: a tool for long non-coding RNA annotation and its application to the dog transcriptome. Nucleic Acids Res 45(8):e57–e57. https://doi.org/10.1093/nar/gkw1306

7 Molecular Mapping of Grapevine Genes

Silvia Vezzulli, Agnès Doligez and Diana Bellin

Abstract

In this chapter, we review the history of grapevine genetics and gene mapping. Genetic markers are introduced considering both sequence-based and sequence-independent approaches used for variant discovery. We provide a survey of genotyping tools, from low- to high-throughput platforms. We describe general principles of map building and implementation, highlighting specificities for outbred species such as the grapevine. Then, we review the different approaches applied for QTL identification according to the genetic material, from bi-parental progenies, pedigree-supported segregating populations, to germplasm collection. In particular, our emphasis is on the relevance of such studies for the dissection of a complex trait. We describe the difficult process of identifying genes responsible for QTLs and the few cases of QTL cloning. Many years have passed from the first grapevine marker isolation, the development of genetic and physical maps, until the deciphering of the genome sequence. With such a wealth of detailed information on wild and cultivated grapevines, we discuss how data sharing and multidisciplinary data integration are the current challenges that the scientific community faces to effectively translate knowledge into practice.

S. Vezzulli (✉)
Research and Innovation Centre, Fondazione Edmund Mach, Via E. Mach 1, 38010 San Michele all'Adige, Italy
e-mail: silvia.vezzulli@fmach.it

A. Doligez
UMR AGAP, University of Montpellier-CIRAD-INRA-Montpellier SupAgro, Montpellier, France
e-mail: agnes.doligez@inra.fr

A. Doligez
UMT Geno-Vigne®, IFV-INRA-Montpellier SupAgro, Montpellier, France

D. Bellin
Dipartimento di Biotecnologie, Università di Verona, Strada Le Grazie 15, 37134 Verona, Italy
e-mail: diana.bellin@univr.it

7.1 Introduction

Several milestones have been achieved in the molecular mapping of genes and quantitative trait loci (QTLs) in grapevine (*Vitis* spp.). At the beginning of the 1990s, Thomas and Scott (1993) isolated the first microsatellite marker, and Lodhi et al. (1995) built the first genetic map of grapevines. Linkage mapping allowed to detect the first QTLs for berry-related traits (Doligez et al. 2002). The first physical map came more than 10 years later (Moroldo et al. 2008), immediately followed by the release of the first grapevine genome assemblies based on Sanger sequencing (Jaillon et al. 2007) and a combination of Sanger sequencing and pyrosequencing (Velasco et al. 2007). The implementation of

D. Cantu and M. A. Walker (eds.), *The Grape Genome*,
Compendium of Plant Genomes, https://doi.org/10.1007/978-3-030-18601-2_7

single-molecule real-time sequencing has more recently allowed to assemble the diploid genomes of "Cabernet Sauvignon" (Chin et al. 2016; Minio et al. 2017), "Chardonnay" (Roach et al. 2018), and "Carménère" (Minio et al. 2019).

Many grape harvests have passed, and a page of the molecular history of the grapevine has been written. We now need to face the challenge of dissection of complex traits, such as resistance to biotic and abiotic stress, or oenological characteristics. To help translate genetic knowledge into grapevine improvement, we argue that funding should now focus on multidisciplinary approaches that bridge genetics, physiology, biochemistry, phytopathology, and agronomy, as well as the private and public sectors. For instance, to dissect complex traits, it is crucial to begin combining high-throughput genotyping approaches with high-resolution trait phenotyping methods (see Chap. 10).

7.2 The Variety of Genetic Markers: Development and Screening

Genetic markers are biological features that are determined by allelic forms and are used to tag and track genetic variation. Genetic polymorphisms among individuals linked to phenotypic traits can be used to expedite studies of inheritance and diversity as well as breeding activities (Xu 2010). The oldest genetic markers are the morphological (or classical) markers, which themselves are phenotypic traits. Unlike in herbaceous and staple crops, in grapevine very few morphological markers have been described. These are mainly related to flower sex (Dalbó et al. 2000; Costantini et al. 2008; Salmaso et al. 2008; Marguerit et al. 2009), pigmentation differences (e.g. berry colour, Fischer et al. 2004), seedlessness (Mejía et al. 2007; Costantini et al. 2008), and presence/absence of specific tissues or organs (e.g. berry flesh, Fernandez et al. 2006). Biochemical markers are a different type of markers and include allelic variants of enzymes, which are called isozymes. Differences in enzymes detected by electrophoresis and staining were exploited mainly to characterize grapevine germplasm (e.g. Subden et al. 1987; Ortiz et al. 2004), including somatic mutants (de Oliveira Collet et al. 2005). The major disadvantages of morphological and biochemical markers are that they may be limited in number and that are influenced by environmental factors or developmental stage of the plant (Winter and Kahl 1995). Nowadays, most grapevine genetic markers are DNA (or molecular) markers. Unlike phenotypic and biochemical markers, molecular markers are more abundant, stable, and independent from environmental factors and conserved throughout all cells of an organism (Xu 2010). Table 7.1 lists the main DNA marker types that are currently available for *Vitis* spp. and their features.

7.2.1 Genetic Variant Discovery

The literature provides different classifications of DNA markers. Here, we describe the four types of markers, which have been most widely employed in grapevine genetics. We also provide information on how markers were developed and whether they were designed based on prior sequence information or not. The development and adoption of sequence-based markers started in the pre-genomic era thanks to Sanger sequencing of ad hoc regions but was boosted by the release of the first grapevine genomes (Jaillon et al. 2007; Velasco et al. 2007).

Random Amplified Polymorphic DNA (RAPD; Williams et al. 1990) refers to the utilization of a single and random-sequence oligonucleotide primer for the simultaneous low-stringency amplification of several discrete DNA fragments. This type of dominant marker, i.e. that does not allow the discrimination between the homozygous and heterozygous forms, does not depend on the sequence information of the target. In grapevine, RAPD markers were initially applied to DNA fingerprinting (e.g. Xu et al. 1995) and genetic diversity studies (e.g. Qu et al. 1996) but were later abandoned because of low reproducibility (Jones et al. 1997). To overcome such limitation, there was a remarkable effort to convert RAPD to the more useful sequence characterized amplified region

Table 7.1 Features of the most employed DNA markers in grapevine

	RAPD	AFLP	SSR	SNP
Template	gDNA	gDNA/cDNA	gDNA	gDNA
Amount of DNA required	1–100 ng	1–100 ng	1–50 ng	10–50 ng
Quality of DNA required	Low	High	Medium–High	High
Type of polymorphism	Single base changes/indels	Single base changes/indels	Changes in the length of repeats	Single base changes/indels
Type of probes/primers	10 bp random nucleotides	Specific sequences	Specific sequences	Allele-specific PCR primers or probes
Prior sequence information	No	No	Yes	No/Yes
Polymorphic information content	High	High	High	Medium
Loci multiplex power	Medium	High	Medium–High	Medium–High
Inheritance	Dominant	Dominant/Co-dominant	Co-dominant	Co-dominant
Detection system	Gel staining	Radioactive/Fluorescence	Gel staining/Fluorescence	Gel staining/Fluorescence
Automation	Medium	High	High	High
Assay throughput	Low–Medium	Medium	Medium–High	High
Reproducibility	Low	Medium	High	High
Suitability	Diversity	Diversity and mapping	All purposes (including MAS)	All purposes (including MAS)

(SCAR) markers (This et al. 1997). Besides a limited application to genetic mapping (Lodhi et al. 1995; Dalbó et al. 2000; Fischer et al. 2004), RAPD markers were used in phylogenetic analyses (e.g. Vidal et al. 1998; Benjak et al. 2005) and were combined with co-dominant markers to characterize genetic backgrounds (e.g. Pollefeys and Bousquet 2003).

Amplified fragment length polymorphism (AFLP; Zabeau and Vos 1993; Vos et al. 1995) markers are based on the selective PCR amplification of restriction fragments from a total double-digest of DNA. In grapevine, most AFLP analyses are applied on gDNA, few studies used cDNAs (Polesani et al. 2008); both gDNA and cDNA types do not rely on prior sequence information. AFLP markers were initially developed and employed as dominantly inherited (e.g. Sensi et al. 1996), subsequently scored as co-dominant markers thanks to a high-resolution ^{33}P detection (Troggio et al. 2007). An AFLP-derived method based on the selective amplification of transposable elements was developed based on transposable element sequence information (S-SAP, sequence-specific amplification polymorphism; e.g. Labra et al. 2004). For grapevine, AFLP markers were developed to increase the saturation of genetic maps both along the whole genome (e.g. Doligez et al. 2002) or locally (e.g. Pauquet et al. 2001) before whole genome assemblies were available. In addition, somaclonal variation was analysed with these molecular markers, independently (Baránek et al. 2009) or in combination with a methylation-sensitive amplification polymorphism (MSAP) approach (Schellenbaum et al. 2008). AFLP markers were widely used for fingerprinting of genetic resources (e.g. Cervera et al. 1998) and intra-varietal characterization (e.g. Blaich et al. 2007; Anhalt et al. 2011), also together with microsatellite markers (e.g. Fossati et al. 2001; Cretazzo et al. 2010).

Microsatellites, also known as simple sequence repeats (SSRs, Tautz and Renz 1984), are tandemly repeated units of short (1–6 bp long) nucleotide motifs. Di-, tri-, and tetra-nucleotide repeats are widely distributed through the genome of plants. In addition to being co-dominant, an important feature of SSR markers is their high level of allelic variation, making them highly informative as genetic markers. Microsatellites were first discovered in grapevine by constructing and screening enriched small-insert clone libraries (e.g. Thomas and Scott 1993; Bowers et al. 1996, 1999; Di Gaspero et al. 2005). With the advent of the genome sequencing projects, numerous SSRs have been directly identified on contig sequences (e.g. Cipriani et al. 2008). Over 400 grapevine SSRs are currently publicly available. SSR markers are widely established markers for the identification of grapevine cultivars. Through an international effort, the grapevine research community defined a reference set of microsatellite markers and analysis protocols for cultivar identification (This et al. 2004; Maul et al. 2012). Many SSR-based germplasm characterizations and diversity studies are reported at all taxon levels, from *Vitis* species (e.g. Fernández et al. 2008) to *V. vinifera* (e.g. Cipriani et al. 2010), including the discrimination of somatic mutations (Migliaro et al. 2017). SSR markers also enable a wide range of applications, comprising the analysis of ancient DNA (e.g. Gismondi et al. 2016), domestication (e.g. Imazio et al. 2013), and population structure (e.g. Bacilieri et al. 2013). Studies on genetic relatedness and pedigree reconstruction were performed using nuclear SSR markers both at large scale (e.g. Lacombe et al. 2013) and at key cultivar level (e.g. Bowers and Meredith 1997). Chloroplast microsatellite polymorphisms were also developed and used to demonstrate the maternal inheritance of chloroplast (e.g. Arroyo-García et al. 2002, 2006). Finally, because of their multiallelic nature, reproducibility and transferability (due to highly conserved flanking sequences) across diverse genetic backgrounds, SSRs have been extensively used in mapping studies.

Single nucleotide polymorphisms (SNPs) are differences in individual nucleotide bases between DNA sequences (Ganal et al. 2009). Single base insertions or deletions (indels) in the genome are also considered as SNPs. Co-dominantly inherited as microsatellites, SNPs differentiate from SSR markers for their greater "informativeness". Marker informativeness can be evaluated by using two main criteria: (1) the number of alleles (i.e. markers with a larger number of alleles are more likely to be polymorphic within any given germplasm set); (2) the minor allele frequency, a measure used to assess informativeness of SNP loci and related to expected heterozygosity when the number of alleles is two (biallelic marker), as it is usually the case for SNPs (Jones et al. 2007).

A variety of approaches have been adopted for discovering SNPs in grapevine, falling into three categories: (1) in vitro, when new sequence data are generated; (2) in silico, when relying on the analysis of available sequence data; (3) indirect, when the base sequence of the polymorphism remains unknown (Edwards et al. 2007). In the last decade, computational approaches have dominated SNP discovery methods due to the advent of next-generation sequencing (NGS, Varshney et al. 2009) and consequent ever-increasing grapevine sequence information in public databases. In vitro approaches include, for example, the first SNP identification based on Sanger sequencing of expressed and BAC-end regions (e.g. Salmaso et al. 2008; Vezzulli et al. 2008b) and the more recent restriction site associated DNA (RAD) sequencing (Marrano et al. 2017). Web-based tools, such as SNiPlay (Dereeper et al. 2011), have been developed for in silico SNP identification and analysis. Indirect SNP discovery strategies that do not depend on prior sequence knowledge have been also used. These methods rely on the detection of base change through differences in the pattern under denaturing conditions (e.g. SSCP, Troggio et al. 2008) and in melting temperature (e.g. HRM, Emanuelli et al. 2014).

Following the first report (Owens 2003), SNPs have been widely deployed in grapevine research. Besides few cases reporting the use of SNPs as diagnostic markers for cultivar identity analysis (e.g. Nicolè et al. 2013), this type of markers has proved useful to define haplotype diversity (Riahi et al. 2013) and to perform linkage disequilibrium (LD) and parentage analysis (e.g. This et al. 2007; Ghaffari et al. 2014). Coupled with SSRs, SNP markers have been used for assessing population genetic structure (Myles et al. 2011; Emanuelli et al. 2013; Laucou et al. 2018) and for high-resolution mapping (e.g. Troggio et al. 2007; Teh et al. 2017). Unlike SSRs, contradictory results have been reported about SNP transferability between cultivated and wild grapevines. The transferability of SNPs discovered in "Pinot noir" to 37 non-*vinifera Vitis* accessions was only 2.3% (Vezzulli et al. 2008a). Conversely, SNPs identified by comparison of ten *V. vinifera* cultivars, six wild *Vitis* accessions, and the near-homozygous line PN40024 were shared by 24.3% (Myles et al. 2010). The reported discrepancy in SNP occurrence among studies could be due to different experimental designs and genome distributions of the studied SNPs; indeed, it is known that SNP frequency varies along the genome and is higher along intergenic than intragenic regions (Salmaso et al. 2004).

7.2.2 Molecular Marker Localization

All DNA markers occupy specific genomic positions within chromosomes called "loci" (singular form "locus"). According to their localization, which is crucial for further applications, they can be classified into random, gene-targeted, or functional markers (Andersen and Lübberstedt 2003). Among all described grapevine molecular markers, there are examples for all the categories. Most markers are "random markers" (namely anonymous or neutral) with no effect on the expression of the target trait. This is the case of classical AFLP markers and SSR markers, which are more abundant in intergenic regions, also belong to this category. "Gene-targeted" SSR and SNP markers were also reported. These include markers developed from information on gene sequence (e.g. Mejía et al. 2011) or expressed sequence tag (EST) (e.g. Decroocq et al. 2003; Kayesh et al. 2013). In addition to markers identified within gene regions, gene tags located in close proximity to genes were also identified. For example, the sex of grapevine flowers is currently targeted through a marker tightly linked to the sex locus (Battilana et al. 2013).

The discovery of "functional markers" that are causal of phenotype variation has been reported only for few cases. Emanuelli et al. (2014) have identified functional markers for the *VvDXS* gene, which is responsible for the muscat flavour. Kobayashi et al. (2004) detected the insertion of the *Gret1* retrotransposon into the *VvMybA1* promoter and associated it with the loss of anthocyanin synthesis function in white-berried varieties. To date, the presence/absence of *Gret1*, along with its homozygous/heterozygous state, is universally adopted as a functional marker for the characterization, discrimination, and prediction of berry skin colour (e.g. Walker et al. 2006; Migliaro et al. 2014).

7.2.3 Genotyping Tools

Molecular markers can be grouped based on the genotyping tools used to detect them. The first group includes markers that require a PCR-based genotyping (RAPD, AFLP, SSR), while the second group includes markers based on hybridization (e.g. array) or sequencing (e.g. minisequencing, genotyping by sequencing). The latter group comprises mostly SNPs. Genotyping based on SNP markers evolved quickly from low-throughput (minisequencing or SNaPshot™, e.g. Troggio et al. 2008; Battilana et al. 2013) to mid-throughput (SNPlex™, e.g. Pindo et al. 2008; Cabezas et al. 2011) methods. In the last decade, NGS enabled the generation of high-throughput genotyping systems. Important progress has been achieved thanks to the introduction of array-based technologies, allowing the screening of several thousands of SNPs per assay.

Myles et al. (2010) designed the first SNP array (Illumina Vitis9KSNP chip) by using a panel of 17 genomic DNA samples from *V. vinifera* cultivars and wild *Vitis* species. Myles et al. (2015) validated the use of this array-based genotyping approach to identify large-effect QTLs. Miller et al. (2013) analysed SNP genotype and hybridization data to measure the effects of ascertainment bias and to reconstruct evolutionary relationships among *Vitis* species. The second high-throughput SNP array (Illumina Vitis18KSNP chip) was produced as part of the GrapeReSeq Consortium (Le Paslier et al. 2013) and then deployed to deeply characterize genetic resources (De Lorenzis et al. 2015; Sunseri et al. 2018), to assess genetic variability among cultivars and biotypes of the same cultivar (Mercati et al. 2016), and to perform and refine parentage analyses (Laucou et al. 2018). Overall, from these studies, it was clear that the application of array-based technologies to population genetics may underestimate the real genetic diversity of the investigated populations, especially when the discovery panel is evolutionarily divergent from the studied accessions. Genotyping by sequencing (GBS) was also applied to grapevine genetics; GBS was first used to discover SNPs in an F1 population; SNPs were located along the reference genome and successfully tested for trait association (Barba et al. 2014). GBS provided opportunities to generate high-resolution genetic maps at a low cost; however, for a highly heterozygous species like grapevine, missing data and heterozygote under-calling complicated the creation of linkage maps. To overcome these limitations of GBS-based genotyping, Hyma et al. (2015) developed HetMappS, which corrects for genotyping errors associated with heterozygosity, independently of parental genotypes. SNP markers generated with a GBS approach and linked to a resistance locus have been validated by Sequenom MassARRAY (Smith et al. 2018a).

The evolution of genotyping tools has practical implication, impacting breeding activities. Marker-assisted selection (MAS) is often employed in breeding programs for wine grapes, table grapes, and rootstocks, to accelerate and enhance cultivar development, via parental selection prior to crossing and progeny selection during the juvenile phase (Töpfer et al. 2011). SSRs currently represent the marker system of choice; whereas, SNP markers, being amenable to high-throughput detection formats and platforms, hold the potential to become the preferred marker system in the future (Mammadov et al. 2012). To bridge the gap between marker development and MAS implementation, a novel practical strategy with a semi-automated pipeline that incorporates trait-associated SNP discovery, low-cost genotyping through amplicon sequencing (AmpSeq) and decision-making, has been recently developed (Yang et al. 2016b).

7.3 Parental, Consensus and Integrated Genetic Maps

Besides diversity and pedigree studies, the most extensive use of DNA markers is for building genetic (or linkage) maps. In grapevine, numerous parental and consensus genetic maps have been developed, mainly with the aim of QTL detection. Genetic maps served as reference for marker and gene localization, before the first release of whole genome sequences (Doligez et al. 2006a; Vezzulli et al. 2008b), and helped improve physical anchoring (Troggio et al. 2007) and measure recombination rates across *Vitis* species (Lowe et al. 2009; Delame et al. 2018).

7.3.1 General Map Building Principles and Implementation

Two main steps are required to build a genetic map based on the segregation of DNA markers in a single bi-parental population. First, markers have to be grouped into linkage groups based on two-point recombination rates. Then, within each linkage group, markers have to be ordered and distances estimated, based on multipoint recombination rates. Since it is computationally intractable to compare all possible orders of markers within a group, several different algorithms are used to efficiently explore the space of all possible orders. For grapevine, the most common methods are (1) the regression mapping procedure implemented in the software JoinMap (Stam 1993), which adds loci one by one, finding the best position with a goodness of fit measure based on the minimum sum of square errors (SSE); (2) the multipoint maximum likelihood (ML)-based algorithms implemented in MapMaker (Lander et al. 1987), JoinMap (Jansen et al. 2001), and Carthagene (de Givry et al. 2005); (3) the modified maximum likelihood (MML)-based algorithm implemented in TMAP (Cartwright et al. 2007), which incorporates possible genotype errors; (4) the algorithm computing the minimum spanning tree of the graph associated with the genotype data implemented in MSTMap (Wu et al. 2008).

Once markers have been ordered with one of these algorithms, the map can be locally refined by several methods (e.g. ripple to test all possible permutations in a sliding window, implemented in most software). Finally, when the best marker order has been selected, a mapping function is applied to convert recombination rates into genetic distances, the most widely used being the Kosambi function (Kosambi 1944). It is also possible to build an integrated map from multiple populations. Two different strategies can be used. Genotypic data sets from all populations can be analysed jointly using mapping algorithms analogous to those for single populations implemented in JoinMap (Van Ooijen 2006), Carthagene (de Givry et al. 2005), or MultiPoint (Ronin et al. 2012). Alternatively, individual genetic maps can be merged using graph theory or a more recent algorithm based on linear programming and implemented in LPmerge (Endelman and Plomion 2014), with large gains in computational efficiency and no loss in map accuracy.

7.3.2 Specificities for Outbred Species

In grapevine, as in other heterozygous species, populations used for genetic mapping mainly result from a cross between two different parents. Recombinations thus occur independently in each parent, and these populations are called pseudo-F1 populations. It is then possible to build each parental map separately, by using the marker segregation information for each parent while ignoring segregation in the other parent. This is the pseudo-testcross strategy, first proposed by Grattapaglia and Sederoff (1994). According to this, markers segregating only in one parent can be used together with the markers segregating in both parents, which are re-coded to keep the segregation information for each parent. The genotypic classes for which parental origin cannot be determined are set to missing data. Since linkage phases in parents are unknown, all genotypic data have to be re-coded as "mirror" before linkage analysis, by exchanging alleles (i.e. genotypes "A" re-coded "H" and vice versa). This re-coded data set is then analysed together with the original one, as for a classical backcross, to determine linkage groups. Twice the expected number of linkage groups is obtained, with homologous groups containing the same markers but in the opposite phase. In a pseudo-F1 population, it is also possible to build a consensus map by estimating recombinations between all markers whatever their segregation type (Ritter et al. 1990), which yields more precise recombination estimates (Ritter and Salamini 1996). Among above-cited software, only JoinMap, Carthagene, and TMAP can build such consensus maps. Being essentially biallelic, SNPs are not informative enough for consensus mapping. It is therefore recommended to derive multiallelic markers from haplo-blocks of individual SNPs whenever possible.

7.3.3 An Overview of Published *Vitis* Genetic Maps

More than 160 maps (including parental and consensus ones) have been published since the first one presented by Lodhi et al. (1995). The number of markers per map has drastically increased with the advent of NGS-derived markers (Fig. 7.1). Map density has been continuously increasing over the years, reaching

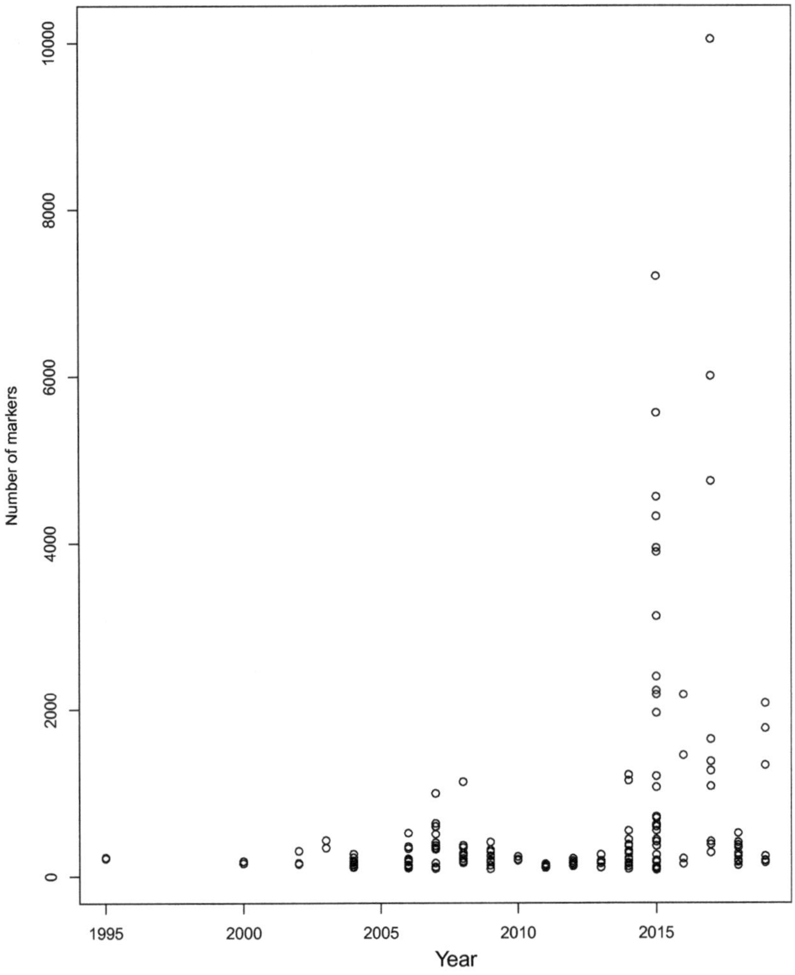

Fig. 7.1 Trend of the number of markers per map

values of mean distance between markers as low as 0.1 cM (Fig. 7.2). Most maps have a total length between 1000 and 1500 cM (Fig. 7.3), which could therefore be considered as the "reference" range for *Vitis* map length, even though several factors of genetic or environmental origin can affect it. Very short maps probably correspond to unsaturated maps; whereas, most very long ones (over 1800 cM), which show very high marker densities (less than 2 cM), probably result from genotyping errors and/or difficulty in ordering high numbers of markers in small populations with low recombination information. Indeed, most of these long maps were obtained from populations including less than 190 offspring individuals.

7.4 Quantitative Trait Loci (QTLs) Mapping Studies

Genetic maps have been widely used in grapevine to assist the detection of QTLs associated to traits of agronomic interest. Indeed, as in many other important crops, phenotypic traits mainly show complex quantitative inheritance in grapevine, being under polygenic control, with small additive or dominant effects of each individual gene on the variation of the trait. By applying QTL analysis, taking advantage of available populations and genetic maps, segments of the genome most probably carrying polymorphisms involved in the traits of interest, and thus with

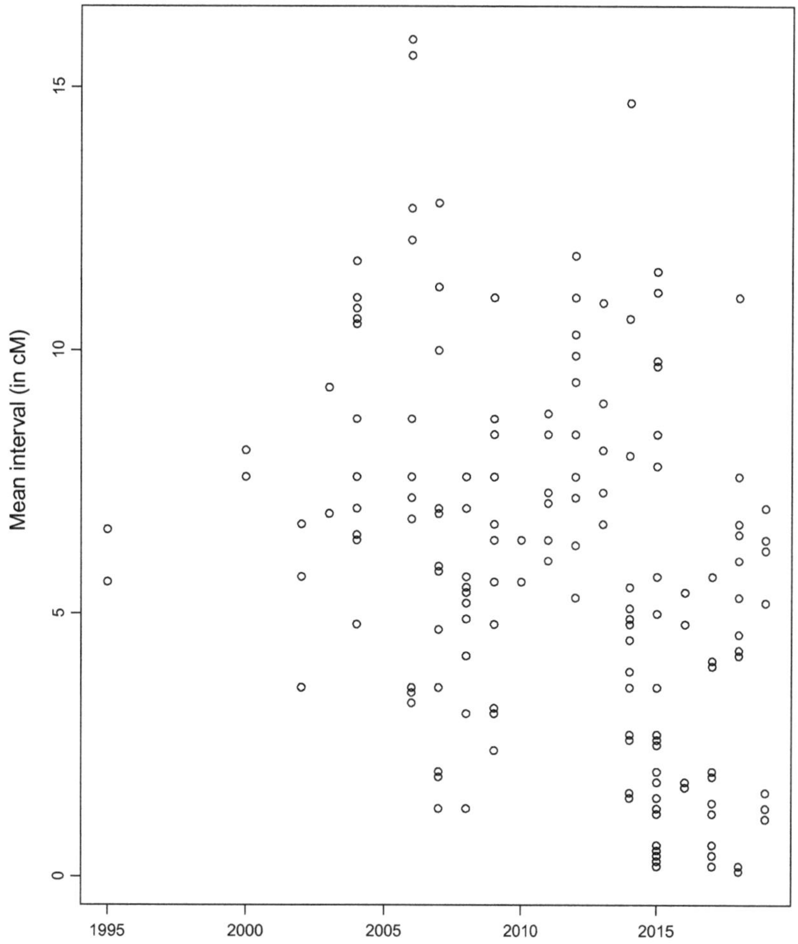

Fig. 7.2 Distribution of mean distance between markers

potential for breeding applications, can be identified. This approach provides valuable information about the specific architecture of the genetic control of each studied phenotypic trait. Even though numerous QTL studies are reported in the grapevine literature, only in few cases have such analyses led to the identification of the causative polymorphisms. Most successful QTL cloning strategies seem to rely on combined approaches including association studies in germplasm collections, in addition to local marker saturation and fine mapping in larger populations.

7.4.1 QTL Detection Approaches

Bi-parental segregating populations. Seventy-six literature records presenting grapevine QTL studies in bi-parental segregating populations have been published so far (Table 7.2). These references include 90 QTL studies, relying on 50 cross populations constituted on average by 166 offspring individuals (ranging from 40 to 424 in the different populations). Mainly F1 cross populations are used in these studies, with a few exceptions. Four populations obtained by selfing were used for mapping QTLs associated to pathogen resistance, berry terpenol content, vegetative, and oenological traits (Duchêne et al. 2009; Garris et al. 2009; Blasi et al. 2011; Blanc et al. 2012; Yang et al. 2016a). A cross population derived from microvine was also characterized in the frame of a QTL study (Houel et al. 2015). Interestingly, more than half of the segregating populations used for QTL mapping were obtained crossing parents coming from different

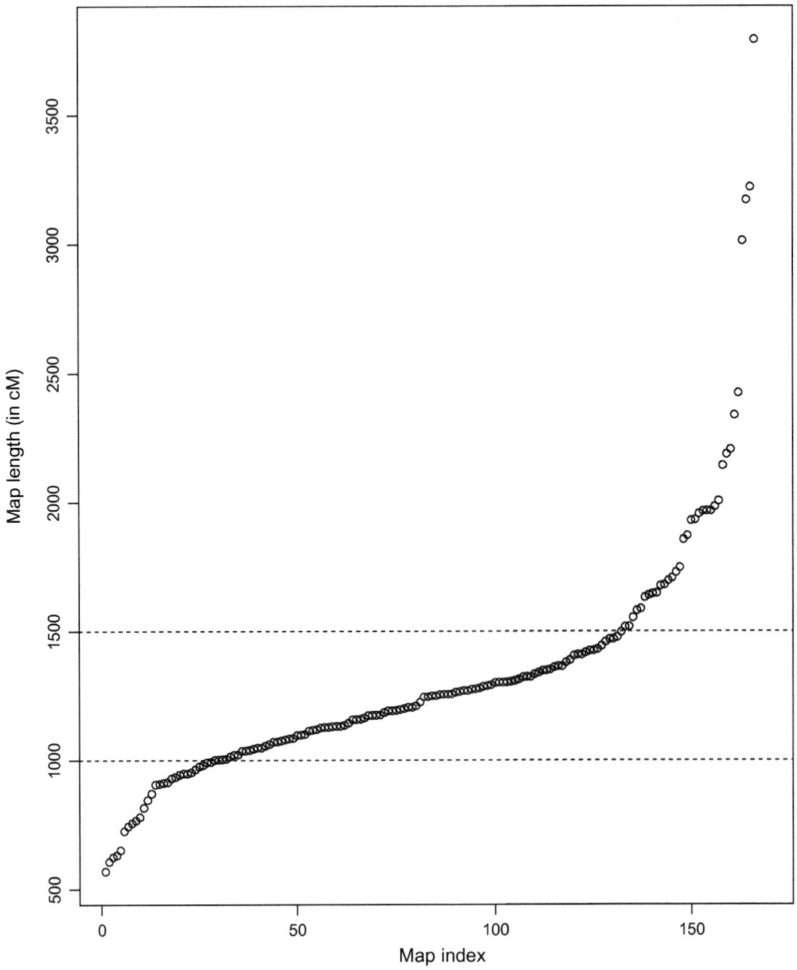

Fig. 7.3 Distribution of map length

Vitis species, since genetic resistance to pathogens, a largely studied trait, is mainly introgressed from non-*vinifera* species (Eibach et al. 2007).

Both genotypic and phenotypic data are required for QTL detection. Recent progress on genotyping tools has already been described, as well as the use of these data for linkage map building. Lately, efforts to develop high-throughput semi-automated or fully automated strategies for phenotyping grapevines have started (Bigard et al. 2018; Tello et al. 2018; Kicherer et al. 2015, 2017a, b; Oerke et al. 2016; Rose et al. 2016; Coupel-Ledru et al. 2016). The most widely used software for QTL mapping in grapevine has been MapQTL, which can perform QTL detection in bi-parental populations of heterozygous diploid species (Van Ooijen 2006). MapQTL remains popular even though a more recent software running under R environment, the qtl package, is being introduced for QTL analysis in grapevine too (Arends et al. 2010), allowing to detect also QTL×QTL interactions. MapQTL requires as input files the genotypic and phenotypic data for all offspring individuals, as well as the genetic linkage maps. Interval mapping, composite interval mapping, as well as nonparametric methods can be selected for computation. Several QTLs were found with consensus maps only and not with parental maps, emphasizing the need to perform detection using both parental and consensus maps. Most of these indeed showed dominant allelic effects on the consensus map. However, the study of parental

Table 7.2 Main results of QTL detection studies in grapevine

Authors	Year	Trait category	Cross	Pop size	Nb years	Meth	Soft	Main QTLs per trait: trait LGs (max % var expl)
Doligez	2002	BMo, S	Vv MTP2223-27 × Vv MTP2121-30	139	3	C	Q	BW 18 (38%); SN 8, 18 (51%); SW 18 (49%); SDM 18 (40%)
Fisher	2004	R, P, BMo, V	Regent × Vv Lemberger	153	3–4	I	M	PM 15 (65%); DM 5, 18 (70%); V 7, 8, 16; BW 5, 13; AxS 13
Fanizza	2005	C, BMo	Vv Italia × Vv Big Perlon	184	3	C	M	CN 8 (10%), 19; CW 5 (7%), 12, 16, 17; BN 2, 5, 7 (9%), 8, 12, 17; BW 4, 5 (19%), 13, 16, 20
Cabezas	2006	BMo, S	Vv Dominga × Vv Autumn Seedless	118	3	C	M	BW 1, 9, 15, 18 (44%); SW 1, 10, 18 (63%); SN 11 (67%)
Doligez	2006	BMe	Vv MTP2687-85 × Vv Muscat of Hamburg	174	3	C	M, Q	Terp 2, 5 (55%), 13, 16
Krivanek	2006	R	D8909-15 × F8909-17	137	1?	C	M	P 14 (72%)
Mandl	2006	V	Vv Welschriesling × Sirius	92	2	C	P	LMag 11 (56%)
Mejia	2007	P, BMo, S	Vv Ruby Seedless × Vv Sultanina	144	2	I	M	Rip 18 (32%); BW 18 (67%); BD 18 (58%); SW 18 (85%); SN 4 (96%), 16, 18; SDM 18 (64%)
Welter	2007	R, V	Regent × Vv Lemberger	144	1–5	C	M	DM 4, 18 (18%); PM 15 (65%); Lmorph 1 (71%), 2, 5, 6, 7, 8, 10, 11, 12, 13, 15, 16
Costantini	2008	P, BMo, S	Vv Italia × Vv Big Perlon	163	3	C	M	F 1, 2, 6 (21%); V 2, 6, 16 (45%); Rip 6 (17%); F-V 2, 6, 16 (37%); F-R 6 (15%); V-R 2 (22%), 12; BW 1, 12, 18 (43%); SN 2 (23%); SDM 18 (91%); SW 2, 6, 10, 13, 15, 18 (75%)
Xu	2008	R	D8909-15 × F8909-17	188	1	C	M	Xi 17, 19 (60%)
Battilana	2009	BMe	Vv Italia × Vv Big Perlon	163	3	C	M	Terp 5 (84%), 10
Battilana	2009	BMe	Vv Moscato Bianco × Vri Wr 63	174	2	C	M	Terp 2, 5 (92%)
Bellin	2009	R	Vv Chardonnay × Bianca	116	2	C	M	DM 5, 7, 18 (81%)
Duchêne	2009	BMe	Vv Muscat Ottonel × Vv Muscat Ottonel	121	2	C	Q	Terp 1, 5 (87%), 10, 13, 15
Fournier-Level	2009	BMe	Vv Syrah × Vv Grenache (and reverse)	191	2	C	M, Q	Ant 2 (62%)
Garris	2009	V	Vri PI588289 × Seyval	119	2–3	C	Q	GC 1, 2, 11, 12, 13 (97%), 15, 17
Marguerit	2009	R, P, C, F	Vv CS × Vri RGM1995-1	138	1	C	M	DM 9 (34%), 12; Infl Morph 1, 2 (38%), 3, 7, 10, 13, 14, 17, 18; Flow Morph 2 (64%), 6, 7, 12, 19; Fert 2 (15%); F 2 (29%), 7, 14
Zhang	2009	R	Vv V3125 × Börner	188	2	C	M	Phyl 13 (67%)
Doligez	2010	C	Vv MTP2223-27 × Vv MTP2121-30	139	2	C	M, Q	Fert 5 (19%)

(continued)

Table 7.2 (continued)

Authors	Year	Trait category	Cross	Pop size	Nb years	Meth	Soft	Main QTLs per trait: trait LGs (max % var expl)
Doligez	2010	C	Vv MTP2687-85 × Vv Muscat of Hamburg	174	2	C	M, Q	Fert 5 (13%), 14
Blasi	2011	R	Va Ruprecht × Va Ruprecht	232	3	I	M	DM 14 (86%)
Fournier-Level	2011	BMe	Vv Syrah × Vv Grenache (and reverse)	191	1	C	M, Q	AntM 1, 2 (27%)
Moreira	2011	R	Vv Moscato Bianco × Vri Wr 63	174	3	I	M	DM 12 (21%)
Moreira	2011	R	VRH3082 1-42 × SK77 5/3	94	2	I	M	DM 1 (77%)
Blanc	2012	R	Mr Regale × Mr Regale	191	2	C	M	DM 18 (25%); PM 5, 14 (24%)
Duchêne	2012	P	Vv Riesling × Vv Gewürztraminer	188	4	M	Rq	BB 4, 6, 7, 10, 14, 19 (19%); BB-F 2, 6, 7, 14 (39%), 15, 16; F-V 7, 14, 16 (21%), 18
Huang	2012	BMe	Vv Syrah × Vv Grenache (and reverse)	191	2	M	Rq	SkPA 1, 2, 3, 5, 6, 8, 10, 13, 14, 17 (56%), 18; SePA 1, 2, 3, 4, 5, 6, 7, 8, 9, 10, 11, 12, 13, 14, 15, 16, 17 (39%), 18, 19
Marguerit	2012	V, A	Vv Cabernet Sauvignon × Vri Gloire	138	3	M	Mu	Tr 1, 6 (11%), 11, 17; TTSW 3 (21%), 5, 11
Schwander	2012	R	Gf.GA-47-42 × Solaris	265	4*	C	M	DM 5, 9 (50%), 18
Bert	2013	V, A	Vv Cabernet Sauvignon × Vri Gloire	138	2–4	C	M, Q	Chl 1, 2, 4, 5, 6, 8, 9, 10, 11, 12, 13 (45%), 14, 17, 18, 19; SG 1, 2, 4, 5, 6, 7, 9, 10, 11, 12, 13 (45%), 15, 18, 19
Doligez	2013	BMo, S	Vv Syrah × Vv Grenache (and reverse)	191	3	C	M, Q	BW 1, 4, 7, 8, 12, 13, 17 (31%), 18; SN 2 (48%), 4, 13; SW 1, 2 (45%), 4, 13, 19
Doligez	2013	BMo, S	Vv MTP2223-27 × Vv MTP2121-30	139	5	C	M, Q	BW 1, 4, 11, 14, 17, 18 (61%); SN 5, 18 (59%); SW 8, 12, 14, 18 (87%); SDM 17, 18 (84%)
Doligez	2013	BMo, S	Vv MTP2687-85 × Vv Muscat of Hamburg	174	3	C	M, Q	BW 5, 7, 11, 19 (25%); SN 8, 14 (28%); SW 8, 11, 14, 16 (41%); SDM 14 (51%)
Grzeskowiak	2013	P, C	Vv Syrah × Vv Pinot N	170	3–5	C	M	V 2 (44%), 15, 17; Fert 3 (20%)
Guillaumie	2013	V	Vv Cabernet Sauvignon × Vri Gloire	138	2	C	M	IBMP 3, 5, 12 (11%)
Huang	2013	BMe	Vv Syrah × Vv Grenache (and reverse)	191	1	M	Rq	VvUFGT 2 (∼ 35%), 16
Venuti	2013	R	99-1-48 × Vv Pinot N	?	2	I	M	DM 14 (79%)
Venuti	2013	R	Vv Cabernet Sauvignon × 20/3	?	2	I	M	DM 14 (75%), 18
Viana	2013	V, C, BMo, S, BMe, P	D8909-15 × Vv 90-116	111	1	I	Rq?	SS 3 (9%); BW 11 (8%); Ant 2 (12%)
Ban	2014	BMe	626-84 × Iku82	98	4	I	M	Ant 2 (40%), 14
Barba	2014	R	Vru B38 × Vv Chardonnay	71	3	I	Rq	PM 9

(continued)

Table 7.2 (continued)

Authors	Year	Trait category	Cross	Pop size	Nb years	Meth	Soft	Main QTLs per trait: trait LGs (max % var expl)
Correa	2014	C	Vv Ruby Seedless × Vv Sultanina	137	3	C	M	CA 5, 8, 9 (24%), 14, 17, 18
Coupel-Ledru	2014	V, A	Vv Syrah × Vv Grenache (and reverse)	191	2	C	M	Tr 1, 2, 4, 10, 17 (13%), 18; K 1, 2 (12%), 7, 11, 13, 17, 18; Psi 1 (16%), 10, 18; LA 3, 7, 17, 18 (20%)
Fechter	2014	P	Vv V3125 × Börner	202	2–3	C	M	F 1, 11, 16 (29%); V 1 (20%), 11
Fechter	2014	P	GF.GA-47-42 × Villard blanc	151	5	C	M	F 1, 4, 8 (31%), 14
Huang	2014	BMe	Vv Syrah × Vv Grenache (and reverse)	191	2	M	Rq	DFR 1, 3, 6, 8, 9, 13, 18 (32%), 19; LDOX 1, 9, 18 (12%); LAR1 1 (56%), 16, 17; LAR2 3, 16, 17 (70%); ANR 6, 8, 10 (14%), 18
Rex	2014	R	Vv V3125 × Börner	202	6*	C	M	BR 3, 4, 10, 14 (22%), 16
van Heerden	2014	R	Regent × Vv RedGlobe	206	2-3	C	M	DM 18 (62%); PM 15 (44%)
Azuma	2015	BMe	626-84 × Iku82	98	1	C	M	Ant 1 (89%), 2, 6, 7, 13, 14, 18
Carreño	2015	BMo	Vv Ruby Seedless × Vv Moscatuel	78	4	C	M	BF 5, 13 (17%)
Carreño	2015	BMo	Vv Muscat Hamburg × Vv Sugraone	153	2	C	M	BF 1, 4, 9, 10, 18 (20%)
Chen	2015	P, BMe	Beihong × E.S.7-11-49	249	3	C	Q	Fru 4, 11 (10%), 14, 17; Glu 14 (8%); SS 1, 14 (11%), 18; Glu/Fru 2, 3, 7 (11%), 9, 17; Ma 6 (17%), 18; TA 6 (17%), 13, 18; Tar/Ma 18 (16%)
Correa	2015	BMo, S	Vv Ruby Seedless × Vv Sultanina	137	3	C	M	BD 2, 18 (37%); BW 2, 18 (40%); BV 2, 18 (45%); SDM 18 (70%)
Costantini	2015	BMe	Vv Syrah × Vv Pinot N	170	4	C	M	Ant 1, 2 (89%), 4, 6, 7, 8, 9, 10, 12, 17, 18, 19
Guo	2015	BMe	87-1 × 9-22	149	2	I	M	BSA 1, 2, 3, 4, 7, 9, 10 (64%), 12, 13, 14, 16, 19
Herzog	2015	R	GF.GA-47-42 × Villard blanc	151	1?	C	M	BCI 17? (20%)
Houel	2015	P, V, S, BMo, BMe, C	Vv 00C001V0008 × Ugni Blanc flb	129	2–8	M	Rq	IL 5 (18%), 10; PHY 1, 3, 4, 5, 6, 7, 8, 10 (89%), 13, 16, 18, 19; LA 3, 4 (17%), 10, 17, 19; F-V 16 (14%); IF 1 (12%); BW 1, 7 (44%), 8, 10; BN 1,3, 4, 7, 8, 12, 14, 16, 19 (43%); SN 7 (76%); CN 2, 7 (25%); acids 2, 4, 5, 6, 7, 8, 9, 12, 13 (70%), 14, 17, 19; sugars 2, 7, 8, 16 (18%), 17, 19; K 1, 7, 8, 14 (40%); osmotica 10, 19 (14%)
Malacarne	2015	BMe	Vv Syrah × Vv Pinot N	170	4	C	M	Fl 1, 2 (73%), 5, 6, 7, 11, 14, 16, 17, 18
Zhao	2015	BMo, P	87-1 × 9-22	149	1?	I?	M	BW 5 (13%), 6; SS 3 (56%)

(continued)

Table 7.2 (continued)

Authors	Year	Trait category	Cross	Pop size	Nb years	Meth	Soft	Main QTLs per trait: trait LGs (max % var expl)
Ban	2016	BMo, P, S	626-84 × Iku82	98	3–4	C	M	BW 11 (40%); BC 11, 13 (21%); BF 3, 10 (31%); SM 11 (21%); SS 2 (24%); TA 13 (29%)
Cadle-Davidson	2016	R	C81-227 × Y315-43-04	205	2–3?	C	Rq	PM 13 (74%)
Cadle-Davidson	2016	R	Horizon × Vc B9	156	2–4?	C	Rq	PM 2, 3 (24%), 4, 14
Correa	2016	BMo	Vv Ruby Seedless × Vv Sultanina	137	3	C	M	BF 8 (20%), 18
Coupel-Ledru	2016	V, A	Vv Syrah × Vv Grenache (and reverse)	191	2	C	M	NT 1, 4, 8, 13, 17 (24%); DT 1, 2 (13%), 10, 17; Psi 13, 15, 17 (14%); TE 4 (19%), 8, 10, 13, 17, 18; LA 3, 7, 17, 18 (20%); GR 4 (16%), 10, 15, 17, 18
Ochssner	2016	R	Vv V3125 × Börner	202	2–4	C	M	DM 1, 5 (17%), 7
Pap	2016	R	Vv F2-35 × Vp DVIT202	277	2?	C	M	PM 9 (62%), 19
Yang	2016	BMe, P	Vri PI588259 × Seyval	424	1	C, m	Rq, G	SS 1, 6 (19%), Ma 1, 6 (26%), YAN 7 (23%)
Zhao	2016	P	87-1 × 9-22	149	1?	I?	M	SSM 5, 6 (78%), 11, 14, 16, 18
Zyprian	2016	R, P	GF.GA-47-42 × Villard blanc	151	1–6	C	M	DM 1, 9, 11, 12, 14, 15, 17, 18 (58%); PM 8, 9, 14, 15 (19%), 16, 18; V 1, 5, 14, 16 (57%); F-V 5, 9, 16 (64%)
Teh	2017	R	MN1264 × MN1214	147	2	C	Rq	PM 2, 15 (16%)
Teh	2017	R	MN1264 × MN1246	125	2	C	Rq	PM 15 (29%)
Barba	2018	R	Horizon × Illinois 547-1	366	3	I	Rq	PCa 1, 2, 7, 15 (46%)
Barba	2018	R	Horizon × Vc B9	162	2–3	I	Rq	PCa 7, 15 (56%); PCl (23%)
Barba	2018	R	Vv Chardonnay × Vc B9	148	2–3	I	Rq	PCa 15 (80%); PCl 15 (73%)
Clark	2018	R	MN1264 × MN1246	125	1–5?	?	Rq	Phyl 5, 14 (61%)
Divilov	2018	R	Vru B38 × Horizon	215	2	M, m	Rq, Rb	DM 8, 11 (17%), 14, 16, 18
Divilov	2018	R, V	Horizon × Vc B9	162	2	M, m	Rq, Rb	DM 5 (15%), 6, 7, 8; LT 7, 8, 15
Henderson	2018	V	K51-40 × 140 Ruggeri	40	1?	C	M	NaExcl 11 (72%)
Kono	2018	R, V	Vv Muscat of Alexandria × Campbell Early	94	3–1?	I	M	DM 5 (76%), 7, 18; LHD 5 (79%), 7
Kono	2018	R, V	626-84 × Iku82	95	1?	I	M	DM 5 (54%); LHD 5 (88%)
Richter	2018	Bmo, C	GF.GA-47-42 ("Bacchus" × "Seyval") × "Villard blanc"	151	2–4	I	M	PL 1 (28%), 14; BN 10 (17%), 17, 18; BW 10 (17%); CW 2 (17%), 10, 18; BV 12, 17 (20%); PED 1, 11 (24%), 18; RL 2 (12%), 3; RW 18 (14%); SL 3 (11%); TBV 10 (14%); Wing 14 (13%); OIV204 1, 2 (19%), 15, 17

(continued)

Table 7.2 (continued)

Authors	Year	Trait category	Cross	Pop size	Nb years	Meth	Soft	Main QTLs per trait: trait LGs (max % var expl)
Royo	2018	S	Vv Red Globe × Vv Crimson Seedless	292	3	C	M	SW 2, 5, 14, 18 (83%)
Smith	2018a	R	Vc C2-50 × Vv Riesling	90	1?	I	Rq	RKN 18
Smith	2018b	R	Vc C2-50 × Vv Riesling	90	1?	I	Rq	Phyl 14 (?)
Tandonnet	2018	V	Vv CS × Vri RGM1995-1	138	1	C	M	AB 3 (12%); RB 1 (10%), 2, 5; RS 1, 5 (19%); RN-T 9 (21%); RN-S 9 (18%); RN-M 2 (12%); RN-L 1, 5 (20%); A/R 6, 9, 18 (15%)
Bayo-Canha	2019	Bme	Vv monastrell × Vv Syrah	229	6	M	M	TA 1, 2 (18%); SS/TA 1, 2 (20%), 4; Tar 18, 19 (16%); Ma 4, 5, 8, 9, 15 (29%), 17, 18; Tar/Ma 5, 8 (21%), 11
Lin	2019	R	Vv Red Globe × Va Shuangyou	149	5	I	M	DM 15 (64%)
Saptoka	2019	R	Norton × Vv Cabernet Sauvignon	182	2	M	M	DM 18 (34%)
Vezzulli	2019	R, V	Merzling × Vv Teroldego	126	1	I	M	DM 18 (23%); Poly 15 (15%), 17

Authors: only first author is given. Year: year published. Trait category: A: abiotic stress response; *Bme* berry metabolites, *Bmo* berry morphology, *C* cluster-related traits, *F* flower morphology, *P* phenology, *R* pathogen resistance, *S* seeds-related traits, *V* vegetative traits. Cross: *Vv Vitis vinifera*, *Vc Vitis cinerea*, *Vri Vitis riparia*, *Vru Vitis rupestris*, *Mr Muscadinia rotundifolia*, *Va Vitis amurensis*, *Gloire* Gloire de Montpellier, 00C001V0008: Picovine 00C001V0008. Pop size: number of offsprings in genetic maps. Nb years number of years of phenotyping, *: number of experiments, with several experiments in the same year. Meth: QTL detection method, I: simple interval mapping, C: composite interval mapping, M: multiple interval mapping, m: multitrait. Soft: QTL detection software, M: Mapqtl, Q: QTLCartographer, Rq: R/qtl, P: Plabqtl, Mu: MultiQTL, G: Genstat, Rb: R/bnlearn. Main QTLs per trait: trait LGs (max % var expl): all QTLs passing the 5% genome-wide LOD threshold (if not given, we considered a classical LOD threshold of 4 for consensus maps and 2.5 for parental maps), max % var expl: for each trait, the maximum variance observed over years, maps, and LGs is given in parentheses after the corresponding LG, *AB* aerial biomass, *Ant* anthocyanins, *AntM* anthocyanin methylation, *A/R* aerial/root ratio, *AxS* axillary shoot, *BB* budburst, *BB-F* budburst-flowering, *BC* berry cracking, *BCI* berry cuticle impedance, *BD* berry diameter, *BF* berry firmness, *BN* berry number, *BR* black rot, *BSA* berry skin anthocyanidin, *BV* berry volume, *BW* berry weight, *CA* cluster architecture, *Chl* chlorosis, *CL* cluster length, *CN* cluster number, *CW* cluster weight, *DM* downy mildew, *Fl* flavonols, *F* flowering, *Fert* fertility, *F-R* flowering-ripening, *F-V* flowering-veraison, *Flow Morph* flower morphology, *Fru* fructose, *GA* gibberellic acid, *GC* growth cessation, *Glu* glucose, *GR* growth rate, *Glu/Fru* glucose to fructose ratio, *IF* inflorescence appearance-flowering, *IL* internode length, *Infl Morph* inflorescence morphology, *K* hydraulic conductance, *LA* leaf area, *LHD* leaf hair density, *Lmag* leaf magnesium, *LMorph* leaf morphology, *LT* leaf trichomes, *Ma* malic acid, *NaExcl* Na exclusion, *NT* night transpiration, *OIV204* compactness, *P* Pierce, *PA* proanthocyanidin, *PCa* phomopsis on canes, *PCl* phomopsis on clusters, *PED* pedicel length, *PHY* phyllochron, *Phyl* phylloxera, *PL* peduncle length, *PM* powdery mildew, *Poly* polyphenol leaf content, *Psi* water potential, *RB* root biomass, *Rip* ripening, *RL* rachis length, *RN-L* large root number, *RN-M* medium root number, *RN-S* small root number, *RN-T* total root number, *RKN* root knot nematode, *RS* root section, *RW* rachis weight, *SDM* seed dry matter, *SG* shoot growth, *Sless* seedlessness, *SM* sensory maturity, *SN* seed number, *SePA* seed per berry, *SkPA* skin per berry, *SL* shoulder length, *SPC* seed phenolic content, *SS* soluble solids, *SSM* seeds maturity, *SS/TA* ratio total soluble solids to total acidity, *Su* sugar content, *SW* seed weight, *TA* titratable acidity, *Tar* tartaric acid, *Tar/Ma* ratio tartaric acid to malic acid, *TBV* total berry volume, *Terp* terpenols, *Tr* transpiration, *Tar/Ma* tartaric to malic acid ratio, *TTSW* total transpirable soil water, *V* veraison, *V-R* veraison-ripening, *Wing* shoulder presence, *WP* predawn leaf water potential, *Xi* Xiphinema index, *YAN* yeast assimilable nitrogen

maps proved to remain necessary by revealing QTLs otherwise undetected or unstable on the consensus map. This could result from a higher power of additive QTL detection in parental maps, where the sample size of each genotypic class is twice as large as in the consensus map (Doligez et al. 2013). The introduction of high-density genotyping was a breakthrough in the history of QTL mapping. Although increasing map density does not improve the detection power, high-density genotyping can provide more precise localization of QTLs (Stange et al. 2013). In grapevine, numerous examples of dense mapping to saturate specific intervals have been reported (e.g. Mejía et al. 2011; Rex et al. 2014). High-density genotyping coupled with the increase of the progeny size is used for fine mapping; this approach led to the characterization of the powdery mildew resistance locus *Run1* (Pauquet et al. 2001; Barker et al. 2005).

A possible shortcut for QTL detection, and consequent gene tagging, is bulked segregant analysis (BSA), which can be used to identify markers linked to major QTLs for a given trait of interest. Briefly, two pools or "bulks" of DNA samples are combined from 10 to 20 individual plants each, with the most contrasted values for the target trait, from a segregating population; markers found polymorphic between these bulks are likely linked to a major QTL (Michelmore et al. 1991). BSA was applied to study the powdery mildew resistance locus *Ren1* (Hoffmann et al. 2008) and the fleshless berry mutation (Fernandez et al. 2006). A further alternative to expedite QTL detection is the "limited mapping" strategy which relies on (1) generating local genetic maps for new populations, with molecular markers from genomic regions that are already reported to be associated with the trait of interest in previous studies and (2) associating these genomic regions with phenotypic data from these new populations (e.g. Duchêne et al. 2009; Doligez et al. 2010, 2013; Riaz et al. 2011, 2018).

Pedigree-supported segregating populations. Some QTL detections assisted by pedigree information have been performed to validate, by tracing back, resistant haplotypes against grapevine downy mildew (*Rpv10*, Schwander et al. 2012; *Rpv12*, Venuti et al. 2013). Recently, borrowed from animal genetics, an actual pedigree-based analysis (PBA) has been adopted for the dissection of resistance traits with oligogenic basis (Peressotti et al. 2015). PBA is a statistical framework implemented in FlexQTL™ (Bink 2005), which was designed to identify, validate, and use QTL information from pedigree-linked individuals to inform breeding decision-making. With prior pedigree validation and identity by descent analysis, the PBA-based QTL analysis was performed on the basis of the genotypic data of several segregating populations, their pedigree-supported parental genotypes and their validated ancestors, along with the phenotypic data (downy mildew resistance parameters) recorded for the progenies and their parents (the ancestral phenotypes were not necessary). This analysis resulted in the identification of three major QTLs on the overall downy mildew resistance sources, with associated markers. These markers were most often identified in one single cross. Consequently, only one or two favourable alleles of the related QTL were identified and are exploitable for marker-assisted breeding, whereas, a breeding program should include several alleles. Selection for these alleles only means that many favourable genotypes may be ignored, which decreases efficiency and leads to genetic erosion.

Germplasm collections. Association mapping (AM) has also been used for locating QTLs. AM overcomes some limits of bi-parental populations that exploit only the variability present in the parental genotypes and show large LD extent. In AM studies, a set of diverse genotypes derived from germplasm collections and/or breeding programs is used to constitute an association panel for mapping QTLs for target traits. Therefore, multiple alleles are available at each locus, in contrast to at most four alleles of bi-parental populations. However, spurious marker-trait association is often detected because genome-wide LD between unlinked loci may be due to population stratification and multiple levels of relatedness among individuals rather than to tight linkage of markers with QTLs of interest. Diversity panels used in association mapping often have substantial sub-population genetic structure, since they are mixtures of geographically distinct genotypes with varying levels of pedigree relationships (Myles et al. 2009). As a result, subgroups within the diversity panel can differ for mean trait values and also for allele frequencies at many loci. This population substructure can lead to the identification of false-positive marker-trait associations. Although advancement in statistical methods helps to remove the confounding effects of population structure on association tests and to increase QTL detection power in most cases (Yu et al. 2006; Zhang et al. 2010; Korte et al. 2012; Segura et al. 2012; Li et al. 2014), population structure still strongly reduces the power of marker-trait association tests (Camus-Kulandaivelu et al. 2005; Holland 2015).

Genome-wide association scans/studies (GWAS) or association studies at candidate genes (CGs) are both based on germplasm collections and are used for genetic dissection of complex traits. In addition to the higher diversity, they also take advantage of the long history of recombination events in natural populations or during breeding history to identify small haplotype blocks associated with phenotypes of interest across species-scale diversity. Accumulation of recombination events across generations reduces the extent of LD and thus ensures a finer exploration of the genome, provided that marker density is sufficient. As a consequence, the resolution of the QTLs found through a GWAS can directly highlight a few CGs. Because intra-specific and inter-specific LD can vary dramatically, LD assessment in the association panel used is a necessary preliminary step before GWAS itself. The resolution of AM can vary dramatically, from the level of individual genes to several hundred kilobases, depending on the LD in the association panel and other parameters (including population structure). GWAS performance depends on the rate of LD decay; in grapevine, LD has been shown to decay fast while was more extended in the wild *V. vinifera* sub-species (Lijavetzky et al. 2007; Nicolas et al. 2016; Barnaud et al. 2006, 2010; Myles et al. 2011; Marrano et al. 2017, 2018). Only few GWASs have been reported so far, namely about leaf shape and venation patterning (Chitwood et al. 2014), seedlessness (Zhang et al. 2017), acidity (Laucou et al. 2018), berry-related traits (Marrano et al. 2018; Razi et al. 2019; Guo et al. 2019) or domestication traits (Myles et al. 2011; Migicovsky et al. 2017; Marrano et al. 2018). CG approach is used when genes controlling a trait under study are known in related or model crop species. It can be used separately and also in parallel with GW approach. In grapevine, an example of CG prioritization approach is reported about proanthocyanidin synthesis (Carrier et al. 2013). Other CG-based association studies were performed to dissect anthocyanin composition, aroma, and cluster characteristics (e.g. Fournier-Level et al. 2009; Emanuelli et al. 2010; Fernandez et al. 2014; Tello et al. 2015a, b).

7.4.2 Trait Architecture

QTL studies also allow to define the genetic control of phenotypic traits, through dissecting the phenotypic variation and determining the contribution of each QTL. QTLs explaining less than 20% of the total phenotypic variation are considered minor QTLs, while QTLs explaining more than 20% are major QTLs (Davey et al. 2006). QTL studies in grapevine have addressed several phenotypic traits, which can be arbitrarily grouped into nine main categories (Table 7.2, Fig. 7.4).

Disease resistances. The largest number of grape QTL studies aimed to dissect the genetic basis of resistance to pathogens, with downy and powdery mildew resistance being the most studied. These studies revealed mainly oligogenic architecture for resistance to downy mildew. Among 21 QTL studies performed so far, twelve and five studies consistently revealed major contributions from genomic regions located in chromosomes 18 and 14, respectively, explaining from 25% to 86% of the total phenotypic variance (Fischer et al. 2004; Welter et al. 2007; Bellin et al. 2009; Blasi et al. 2011; Blanc et al. 2012; Schwander et al. 2012; Venuti et al. 2013; van Heerden et al. 2014; Zyprian et al. 2016; Divilov et al. 2018; Kono et al. 2018; Sapkota et al. 2019; Vezzulli et al. 2019). Besides these major contributions, several minor loci were also detected in different studies, which, however, were less reproducible across studies (Ochssner et al. 2016; Moreira et al. 2011; Marguerit et al. 2009; Lin et al. 2019). These findings fit with the expected biological basis of plant resistance. Major QTLs are often found to co-locate with genomic regions enriched in resistance genes analogous (RGAs; Donald et al. 2002; Di Gaspero et al. 2007), which are known to mediate gene-for-gene pathogen recognition that leads to effector triggered immunity (ETI). The persistence of few

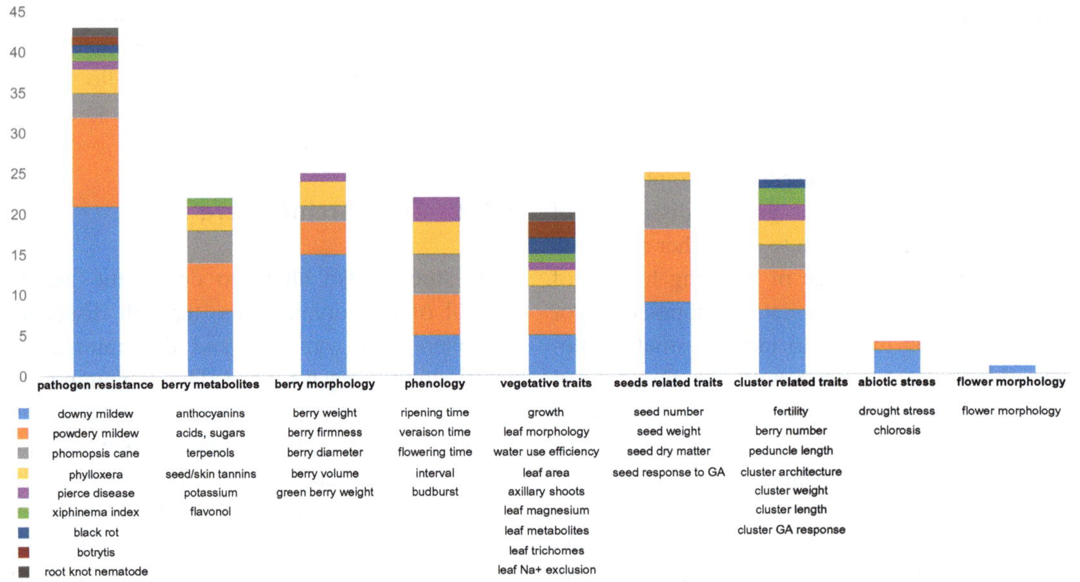

Fig. 7.4 Number of QTL studies available for each grape phenotypic trait. Phenotypic traits have been grouped according to nine different categories. Available QTL studies addressing each of the traits have been counted and are shown separately for each category. Studies addressing more than one trait have been considered for each trait, to provide an overview of most studied grape traits

Rpv3 haplotypes on chromosome 18 across many resistant varieties generated by breeding for downy mildew resistance has already been described by Di Gaspero et al. (2012). Besides the resistance-associated *Rpv3-1* haplotype coming from the "Seibel 4614" lineage (Welter et al. 2007; Bellin et al. 2009; van Heerden et al. 2014), other two *Rpv3* haplotypes have recently been validated in segregating populations, namely *Rpv3-2* derived from "Munson" (Zyprian et al. 2016) and *Rpv3-3* tracing back to "Noah" (Vezzulli et al. 2019). Lately, Foria et al. (2018) demonstrated that the genetic background influences the intensity of genetic resistance in the presence of the same resistance haplotype.

Selection of proper combinations of major QTLs with the appropriate genetic background (best-suited minor QTLs modulating major QTL effects) is critical to obtain high levels of durable field resistance. This is in agreement with recent findings in model species hinting to a quantitative inheritance also for ETI (Iakovidis et al. 2016). A similar scenario compatible with oligogenic trait architecture was also found for other resistance traits. Few major loci for resistance to powdery mildew, Pierce's disease, *Xiphinema index*, phylloxera, and phomopsis cane and leaf spot were detected, some of which emerging consistently from different studies (e.g. chromosome 15 for powdery mildew or phomopsis), together with minor QTLs (Fischer et al. 2004; Krivanek et al. 2006; Welter et al. 2007; Xu et al. 2008; Zhang et al. 2009; Blanc et al. 2012; Barba et al. 2014; Rex et al. 2014; van Heerden et al. 2014; Herzog et al. 2015; Cadle-Davidson et al. 2016; Pap et al. 2016; Zyprian et al. 2016; Teh et al. 2017; Barba et al. 2018; Clark et al. 2018; Smith et al. 2018a, b). Alternative approaches to traditional QTL analysis like BSA or limited mapping strategy have allowed to confirm such loci as well as to identify further sources of resistance at these loci (e.g. Merdinoglu et al. 2003; Hoffmann et al. 2008; Riaz et al. 2011).

Other relevant traits. Many QTL studies focused on other traits of agronomical/economical relevance, mainly related to berry quality and plant phenology. For table grape, reduction of fruit seed content without altering fruit size is an appreciated berry quality trait and therefore a desired breeding goal, as for many other fruit crops (Varoquaux et al. 2000). Several breeding programs have focused on the generation of table grape cultivars, combining seedlessness with other berry quality traits such as large size, muscat flavours, and crispness. The Thompson seedless (TS) cultivar is the main donor of the stenospermocarpic grape seedlessness and most of the commercial table grape varieties descend from this cultivar (Di Genova et al. 2014). QTL analyses have dissected the genetic basis of TS-derived stenopermocarpic seedlessness (Bouquet and Danglot 1996; Doligez et al. 2002; Cabezas et al. 2006; Mejía et al. 2007, 2011; Costantini et al. 2008; Doligez et al. 2013; Correa et al. 2015; Royo et al. 2018). QTL studies consistently revealed the contribution of a major QTL in crosses from seedless varieties, also located on chromosome 18, providing evidence of oligogenic trait architecture. Interestingly, the characterization of the same seed-related traits in seeded varieties revealed the contribution of other major and minor QTLs, which could also potentially be exploited for breeding (Viana et al. 2013; Doligez et al. 2013; Houel et al. 2015; Ban et al. 2016).

Among berry-related traits, berry morphology traits like size and weight are major yield components. Large berries are desirable for table grape. For winemaking, smaller berries are preferred to increase skin-to-flesh ratio and improve the final concentrations of anthocyanins, tannins, and aroma. A positive correlation between berry final weight/size and seed traits has been observed frequently within populations segregating for seedlessness. QTL studies in such populations clearly revealed the same major QTL on chromosome 18, co-localized with those for seed content previously described (Doligez et al. 2002; Fanizza et al. 2005; Cabezas et al. 2006; Mejía et al. 2007; Costantini et al. 2008; Doligez et al. 2013; Carreño et al. 2015; Correa et al. 2015, 2016). A pleiotropic effect on berry size is likely in these cases (Mejía et al. 2011).

The analyses in seeded cross populations and with a specific statistical strategy designed to map residual berry size/weight contributions detected additional QTLs for berry size that were not co-located with QTLs for seeds; a promising discovery that could allow to uncouple seedlessness and berry size (Doligez et al. 2013). Recently, Royo et al. (2018) reported that the origin of seedless grapes was associated with a missense mutation in the MADS-box gene *VviAGL11* (Royo et al. 2018).

The accumulation in grape berries of metabolites like anthocyanins (Fournier-Level et al. 2009, 2011; Viana et al. 2013; Huang et al. 2013, 2014; Ban et al. 2014; Azuma et al. 2015; Guo et al. 2015; Costantini et al. 2015), terpenols (Doligez et al. 2006b; Battilana et al. 2009; Duchêne et al. 2009), tannins (Huang et al. 2012), flavonols (Malacarne et al. 2015), sugars, and acids (Chen et al. 2015; Houel et al. 2015; Yang et al. 2016a; Bayo-Canha et al. 2019) have also been analysed in segregating populations, often by coupling genetics with metabolomics. These quality traits are mainly controlled by minor QTLs with few exceptions. Major QTLs were found for anthocyanin and terpenols: the major QTL for anthocyanin content co-located with the berry colour locus on chromosome 2; the QTL for terpenol content was mapped to chromosome 5. Both QTLs were also analysed by applying approaches exploiting the variation at these loci or at CGs therein in germplasm collections, which led to the identification of phenotype-associated SNP/variants (some of which already demonstrated to be causative for the berry colour by Kobayashi et al. 2004 and Walker et al. 2007) and cloning of the responsible gene (Fournier-Level et al. 2009; Emanuelli et al. 2010).

A number of QTL studies have also addressed plant phenology. Understanding the genetic control of phenological developmental stages (i.e. flowering, veraison, ripening, etc.) is critical for creating cultivars adapted to local climate. In particular, the delay of veraison and ripening is a desirable breeding target, since ripening occurring under very hot summers negatively affects and uncouples berry quality traits. According to these QTL studies, a complex inheritance seems to control these phenology traits, with low contributions scarcely reproducible among studies, even though a few reproducible contributions were found (Table 7.2) (Fischer et al. 2004; Mejía et al. 2007; Costantini et al. 2008; Marguerit et al. 2009; Duchêne et al. 2012; Grzeskowiak et al. 2013; Viana et al. 2013; Fechter et al. 2014; Chen et al. 2015; Houel et al. 2015; Zhao et al. 2015, 2016; Ban et al. 2016; Zhang et al. 2016; Zyprian et al. 2016). We cannot exclude that studies in other genetic backgrounds can reveal also major contributions, since only few studies have been performed for each individual trait. Moreover, it is possible that individual developmental stages are collectively controlled by pleiotropic genes. Therefore, co-location of QTLs for different phenological stages could eventually also be considered in searching for consistent QTLs.

QTL studies also addressed vegetative traits and abiotic stress response (Table 7.2). Even though in rare cases, major genetic controls have emerged, too few studies were performed so far to allow a comprehensive view on the genetic architecture of these traits. Interestingly, a comprehensive QTL study has recently addressed the genetic determinism of cluster architecture, revealing eight genomic regions which collectively can explain 87% of the genetic variance for this trait (Richter et al. 2019).

7.5 Genetic Maps and QTLs: Information Sharing

Recent advance in markers technology has given a strong impulse to plant genotyping and linkage mapping. Consequently, the number of plant studies reporting QTLs has been growing at an impressive pace. In grapevine, the first QTL study relying on high-throughput SNP genotyping, in particular on the NGS-based GBS technology, appeared in 2014; since then more than half of all grapevine available QTL studies have been performed.

7.5.1 Data Integration

Cataloguing, summarizing, and making the plethora of increasing QTL information readily accessible are the next challenge. The large amount of detected QTLs calls for the need to deposit in public databases the raw data (genotypes, phenotypes, and environmental information) of published experiments to avoid losing precious information and guarantee its effective exploitation (Zamir 2013). Assembling and integrating diverse QTL data with other information in a "QTL browser" would (1) enhance our understanding of the genetic regulation of different phenotypes, (2) assist the QTL cloning process and facilitate the application of QTL-derived information in biological research, and (3) reveal QTLs consistent across studies, which are particularly valuable for breeding.

Collection of results from different QTL studies has been implemented both in model plants (Nijveen et al. 2017; Zeng et al. 2007) and in crops. The Gramene database (http://archive.gramene.org/, Ware et al. 2002) was originally developed as a comparative genome mapping and functional genomics database for grasses and rice (*Oryza sativa*). This database has later been extended to other species, now including fourteen ones (Tello-Ruiz et al. 2016). The development of a specific QTL tool was also implemented, which contains the largest online collection of rice QTL-related data in the world. QTLs are aligned to the genomic sequence and can thus be searched as standard genomic features to facilitate comparison of QTL genomic localization and the mining of positional candidate genes according to ontology terms. This tool also integrates information derived either from functional characterizations or studies on association mapping panels, thus assisting and boosting the QTL fine mapping process and validation of genotype–phenotype associations (Ni et al. 2009).

The Plant Genome DataBase Japan (PGDBj, http://pgdbj.jp/index.html?ln=en) is another portal website aiming to integrate plant genome-related information from databases and the literature. PGDBj includes three component

DBs. Among these, the DNA marker DB provides manually and automatically curated information on QTLs and related linkage maps and includes a QTL list for grapevine with chromosomal positions and LOD scores. Unfortunately, the listed genomic regions are not currently up-to-date (Asamizu et al. 2014). Taxa-specific databases have also been produced to collect and manage the growing amount of data and sequence information in rice, wheat, cotton, and in Solanaceae (Ni et al. 2009; Kim et al. 2014; Said et al. 2015a; Tecle et al. 2010). Moreover, specific tools to manage and mine QTL data have also been implemented in these databases (Thongjuea et al. 2009; Smita et al. 2011).

Concerning grapevine, the International Grape Genome Program (IGGP) has promoted and coordinated efforts for the release of genomic resources for the *Vitis* genus, starting from the establishing of a French-Italian public consortium for the reference grapevine genome sequencing (Jaillon et al. 2007; Adam-Blondon et al. 2016). A database hosted by the French National Repository for Plant Genomic Data (URGI, https://urgi.versailles.inra.fr/Species/Vitis) provides access to the whole genome sequencing results from this consortium and to the different versions of genome assembly as well as annotations, including tools for genome browsing. In this grapevine dedicated database, some genotyping data can be also retrieved. Relevant information about SSR markers, like links to external repositories or genomic location through genome browsing, can be easily accessed. A data set of 10,207 SNPs derived by resequencing 783 cultivars, with refined genomic locations, which has been the basis for implementing the development of the 18KSNP array for high-throughput genotyping, has recently been made available through this Internet site (Laucou et al. 2018). Finally, this database also hosts information about public genetic maps, which can be accessed through the suite GnpMap. However, QTL information is still lacking. First efforts aiming to collect and mine literature about grapevine phenotypic, "omics" or QTL data are just starting. New initiatives aiming to coordinate *Vitis* genomic data integrations are being funded (see COST CA17111 INTEGRAPE: Data integration to maximize the power of omics for grapevine improvement, http://www.cost.eu/COST_Actions/ca/CA17111 as an example).

In this context, an additional relevant aspect is the coordination of ongoing phenotyping approaches. Grapevine phenotyping is rapidly improving thanks to advancement in technology (see Chap. 10), which includes the implementation of high-throughput semi-automated and automated methods, besides new statistical and interpretative models, also adapted from other plant species (Kicherer et al. 2015, 2017a, b; Oerke et al. 2016; Rose et al. 2016; Coupel-Ledru et al. 2016; Bigard et al. 2018; Tello et al. 2018). This is expected to largely promote our ability to measure agronomically relevant phenotypes in many individuals at unprecedented accuracy, speed, and costs, both in controlled and field conditions (Houle et al. 2010; Granier and Vile 2014). Some effort to standardize phenotyping protocols across studies and facilitate data/QTLs integration is required; standard phenotyping rules for grapevine have been defined under the direction of international plant phenotyping networks for phenomics (EMPHASIS, ESFRI infrastructure for the synergistic development and long-term operation of phenotyping infrastructure in Europe, https://emphasis.plant-phenotyping.eu/).

An interesting instrument to rationalize and interpret the plethora of QTL information, especially with the goal of providing relevant trait candidates, is QTL meta-analysis (Goffinet and Gerber 2000; Veyrieras et al. 2007). QTL meta-analysis is a statistical framework to project QTLs on a consensus map which allows to identify and mine co-localizing QTLs among independent experiments. Indeed, QTLs detected independently and located in a given region of a chromosome could possibly represent several estimations of the position of one single QTL. This hypothesis can be tested by appropriate statistical tools, which indicate the most likely number of "real" QTLs underlying a pool of QTLs from independent experiments, providing alongside consensus positions for these, narrowing down the QTL confidence intervals. The resulting meta-QTLs are expected to better define the boundaries of the causative genomic intervals by integrating information from different studies. QTL meta-analyses have become popular and they are used both to summarize QTL information about one trait as well as to locally verify the co-location of QTLs between different populations as the first step towards QTL validation and/or prioritization of candidates. Chardon et al. (2004) first applied this approach to study flowering time in maize by synthesizing several QTLs from different mapping populations into meta-QTLs. Subsequent positional cloning and association mapping analysis found in meta-QTL intervals two genes effectively involved in modulating flowering time (Salvi et al. 2007, 2011; Hung et al. 2012). These successful examples confirmed that meta-analysis is a useful method for predicting candidate genes and for developing molecular markers for breeding. Meta-analysis has been successfully used in studying QTLs in other crop species like rice (Khowaja et al. 2009), cotton (Said et al. 2015b), and potato (Danan et al. 2011). In grapevine, this approach has been only recently applied to identify candidate genes for genetic regulation of plant veraison time (Delfino et al. 2018).

Whole repertoire information describing all experimentally supported QTLs for a trait in one species has recently been also condensed into the "trait QTLome" definition (Salvi and Tuberosa 2015; Martinez et al. 2016). A trait QTLome reports the map position, allele identity, and genetic effect in terms of magnitude and type (additive vs. dominant) incorporating all detected QTLs relevant for a specific trait. This information is of pivotal importance for breeders; in fact, it provides essential knowledge driving the selection of the best markers and alleles to be selected. Martinez et al. (2016) assembled a yield QTLome database for maize based on published studies, which summarizes results from several independent mapping experiments, thereby providing information on the high genetic complexity for the inheritance of yield. The QTLome concept has recently been extended to grapevine to describe the overall knowledge on the genetic basis of downy mildew resistance (Buonassisi et al. 2017). QTLome information integrated with high-density chromosome resolution is expected to enable the identification of the most valuable and effective SNP-based haplotypes to guide the selection of the best parental genotypes in breeding programs and the recurrent selection of the best performing individuals.

7.5.2 From Research to Breeding

Although there have been numerous QTL mapping studies for a wide range of traits, relatively few markers have actually been implemented in grapevine breeding programs and routinely employed for MAS. The main reason for this lack of adoption is that genetic markers have not been always reliable in predicting the desired phenotype. Many factors influence the detection of QTLs segregating in a population, namely QTL properties, environmental effects, population size, and experimental error. Generally, the steps required for the development of markers for use in MAS include fine mapping, validation of markers, and, possibly, marker conversion (Collard et al. 2005).

First, more tightly linked markers can be identified with larger population sizes and a greater number of markers. High-resolution (or fine) mapping of QTLs may be used to develop reliable markers for MAS (at least < 5 cM but ideally < 1 cM away from the gene) (Michelmore 1995). Markers should be then validated in independent populations constructed from the same parental genotypes or closely related genotypes to those used in the primary QTL mapping study. Some studies have warned of the danger of assuming that marker-QTL linkages will remain in different genetic backgrounds or in different testing environments (Reyna and Sneller 2001). Remarkably, Pap et al. (2016) validated the SSR markers linked to two novel loci associated to grapevine powdery mildew resistance in hundreds of F1 additional individuals compared to the primary segregating population and almost 1000 seedlings from four pseudo-backcross populations. Under this perspective of QTL stability, it has recently been recovered the concept of advanced backcross-QTL (AB-QTL), which combines QTL analysis and cultivar development by designing a mapping/breeding scheme for the simultaneous identification and introgression of wild haplotypes. AB-QTL relies on segregating populations in which most of the wild parent genome that donates the trait of interest has been purged in early segregating generations by phenotypic selection (Tanksley and Nelson 1996). This is relevant to guarantee QTL stability once the associated markers are screened in derived breeding materials. In fact, favourable QTL alleles identified in early generations often disappear in later backcross generations, once the donor genes that have epistatic interactions with the beneficial QTL alleles are removed from highly *V. vinifera* genetic backgrounds (Di Gaspero and Foria 2015; Foria et al. 2018). Finally, in order to be implemented in breeding programs, markers should be reliable, efficient, and cost-effective. Stable and co-dominant markers are required for MAS. Among these, SNP markers are favoured over SSRs, because they are amenable to high-throughput genotyping platform (see Sect. 7.2). However, to date, there are only few cases of SNP implementation in marker-assisted selection programs of grapevines (Barba et al. 2014; Zyprian et al. 2015).

Unfortunately, only few results from QTL mapping studies were converted into practical genetic improvement in grapevine breeding programs. In this regard, it is relevant to consider differences about trait characteristics and genetic basis. While oligogenic traits, such as disease resistance, are suitable for MAS, QTL stability should be evaluated for fruit quality and phenology before the linked markers are proposed to breeders. Since these complex traits are controlled by several QTLs, it is not always straightforward to determine which QTLs should be selected during breeding. When minor QTLs are chosen for MAS, they should be validated for stability across environments.

For polygenic traits, innovative selection approaches, such as genomic selection (GS), are needed. GS simultaneously estimates the effect of each marker across the entire genome to predict the breeding value of individuals, theoretically capturing more genetic variation for small effects underneath complex traits. Contrary to MAS, the contribution of all genome-wide DNA polymorphisms to the breeding value is accounted for in the diagnostic model during calibration (Jonas and De Koning 2013). In grapevine, GS approach can be advantageous to quickly test in the field candidates for complex traits such as bud break and berry weight. This approach has recently been tested for grapevine in the specific case of bi-parental populations, in order to speed up the selection of genotypes (Flutre et al. 2018).

Acknowledgements The authors apologize to the scientists that are not cited because of space limitation. They gratefully thank Paola Bettinelli (FEM) for literature management.

Bibliography

Adam-Blondon AF, Alaux M, Pommier C, Cantu D, Cheng ZM, Cramer GR, Davies C, Delrot S, Deluc L, Di Gaspero G, Grimplet J, Fennell A, Londo JP, Kersey P, Mattivi F, Naithani S, Neveu P, Nikolski M, Pezzotti M, Reisch BI, Töpfer R, Vivier MA, Ware D, Quesneville H (2016) Towards an open grapevine information system. Hortic Res 3:16056

Andersen JR, Lübberstedt T (2003) Functional markers in plants. Trends Plant Sci 8:554–560

Anhalt UCM, Martínez SC, Rühl E, Forneck A (2011) Dynamic grapevine clones-an AFLP-marker study of the *Vitis vinifera* cultivar Riesling comprising 86 clones. Tree Genet Genomes 7:739–746

Arends D, Prins P, Jansen RC, Broman KW (2010) R/qtl: high-throughput multiple QTL mapping. Bioinformatics 26:2990–2992

Arroyo-García R, Lefort F, De Andrés MT, Ibáñez J, Borrego J, Jouve N, Cabello F, Martínez-Zapater JM (2002) Chloroplast microsatellite polymorphisms in *Vitis* species. Genome 45:1142–1149

Arroyo-García R, Ruiz-García L, Bolling L, Ocete R, López MA, Arnold C, Ergul A, Söylemezoğlu G, Uzun HI, Cabello F, Ibáñez J, Aradhya MK, Atanassov A, Atanassov I, Balint S, Cenis JL, Costantini L, Gorislavets S, Grando MS, Klein BY, McGovern PE, Merdinoglu D, Pejic I, Pelsy F,

Primikirios N, Risovannaya V, Roubelakis-Angelakis KA, Snoussi H, Sotiri P, Tamhankar S, This P, Troshin L, Malpica JM, Lefort F, Martinez-Zapater JM (2006) Multiple origins of cultivated grapevine (*Vitis vinifera* L. ssp. sativa) based on chloroplast DNA polymorphisms. Mol Ecol 15:3707–3714

Asamizu E, Ichihara H, Nakaya A, Nakamura Y, Hirakawa H, Ishii T, Tamura T, Fukami-Kobayashi K, Nakajima Y, Tabata S (2014) Plant genome database Japan (PGDBj): a portal website for the integration of plant genome-related databases. Plant Cell Physiol 55:1–7

Azuma A, Ban Y, Sato A, Kono A, Shiraishi M, Yakushiji H, Kobayashi S (2015) MYB diplotypes at the color locus affect the ratios of tri/di-hydroxylated and methylated/non-methylated anthocyanins in grape berry skin. Tree Genet Genomes 11(2):31

Bacilieri R, Lacombe T, Le Cunff L, Di Vecchi-Staraz M, Laucou V, Genna B, Péros JP, This P, Boursiquot JM (2013) Genetic structure in cultivated grapevines is linked to geography and human selection. BMC Plant Biol 13:25

Ban Y, Mitani N, Hayashi T, Sato A, Azuma A, Kono A, Kobayashi S (2014) Exploring quantitative trait loci for anthocyanin content in interspecific hybrid grape (*Vitis labruscana* × *Vitis vinifera*). Euphytica 198:101–114

Ban Y, Mitani N, Sato A, Kono A, Hayashi T (2016) Genetic dissection of quantitative trait loci for berry traits in interspecific hybrid grape (*Vitis labruscana*× *Vitis vinifera*). Euphytica 211:295–310

Baránek M, Raddová J, Krizan B, Pidra M (2009) Genetic changes in grapevine genomes after stress induced by in vitro cultivation, thermotherapy and virus infection, as revealed by AFLP. Genet Mol Biol 32:834–839

Barba P, Cadle-Davidson L, Harriman J, Glaubitz JC, Brooks S, Hyma K, Reisch B (2014) Grapevine powdery mildew resistance and susceptibility loci identified on a high-resolution SNP map. Theor Appl Genet 127:73–84

Barba P, Lillis J, Stephen Luce R, Travadon R, Osier M, Baumgartner K, Wilcox WF, Reisch BI, Cadle-Davidson L (2018) Two dominant loci determine resistance to Phomopsis cane lesions in F1 families of hybrid grapevines. Theor Appl Genet 131:1173–1189

Barker CL, Donald T, Pauquet J, Ratnaparkhe MB, Bouquet A, Adam-Blondon AF, Thomas MR, Dry I (2005) Genetic and physical mapping of the grapevine powdery mildew resistance gene, *Run1*, using a bacterial artificial chromosome library. Theor Appl Genet 111:370–377

Barnaud A, Lacombe T, Doligez A (2006) Linkage disequilibrium in cultivated grapevine, *Vitis vinifera* L. Theor Appl Genet 112:708–716

Barnaud A, Laucou V, This P, Lacombe T, Doligez A (2010) Linkage disequilibrium in wild French grapevine, *Vitis vinifera* L. subsp. silvestris. Heredity (Edinb) 104:431–437

Battilana J, Costantini L, Emanuelli F, Sevini F, Segala C, Moser S, Velasco R, Versini G, Grando MS (2009) The 1-deoxy-d-xylulose 5-phosphate synthase gene co-localizes with a major QTL affecting monoterpene content in grapevine. Theor Appl Genet 118:653–669

Battilana J, Lorenzi S, Moreira FM, Moreno-Sanz P, Failla O, Emanuelli F, Grando MS (2013) Linkage mapping and molecular diversity at the flower sex locus in wild and cultivated grapevine reveal a prominent ssr haplotype in hermaphrodite plants. Mol Biotechnol 54:1031–1037

Bayo-Canha A, Costantini L, Fernández-Fernández JI, Martínez-Cutillas A, Ruiz-García L (2019) QTLs related to berry acidity identified in a wine grapevine population grown in warm weather. Plant Mol Biol Rep 37:157–169

Bellin D, Peressotti E, Merdinoglu D, Wiedemann-Merdinoglu S, Adam-Blondon A-F, Cipriani G, Morgante M, Testolin R, Di Gaspero G (2009) Resistance to *Plasmopara viticola* in grapevine 'Bianca' is controlled by a major dominant gene causing localised necrosis at the infection site. Theor Appl Genet 120:163–176

Benjak A, Ercisli S, Vokurka A, Maletić E, Pejić I (2005) Genetic relationships among grapevine cultivars native to Croatia, Greece and Turkey. Vitis J Grapevine Res 44:73–77

Bigard A, Berhe DT, Maoddi E, Sire Y, Boursiquot J-M, Ojeda H, Péros J-P, Doligez A, Romieu C, Torregrosa L (2018) *Vitis vinifera* L. fruit diversity to breed varieties anticipating climate changes. Front Plant Sci 9:455

Bink MCAM (2005) FlexQTL software: efficient estimation of identity by descent probabilities and QTL mapping in pedigreed populations. In: Abstract book plant and animal genomes XII conference, 15–19 January, San Diego, USA

Blaich R, Konradi J, Rühl E, Forneck A (2007) Assessing genetic variation among Pinot noir (*Vitis vinifera* L.) clones with AFLP markers. Am J Enol Vitic 4:526–529

Blanc S, Wiedemann-Merdinoglu S, Dumas V, Mestre P, Merdinoglu D (2012) A reference genetic map of *Muscadinia rotundifolia* and identification of *Ren5*, a new major locus for resistance to grapevine powdery mildew. Theor Appl Genet 125:1663–1675

Blasi P, Blanc S, Wiedemann-Merdinoglu S, Prado E, Rühl EH, Mestre P, Merdinoglu D (2011) Construction of a reference linkage map of *Vitis amurensis* and genetic mapping of *Rpv8*, a locus conferring resistance to grapevine downy mildew. Theor Appl Genet 123:43–53

Bouquet A, Danglot Y (1996) Inheritance of seedlessness in grapevine (*Vitis vinifera* L.). Vitis J Grapevine Res 35:35–42

Bowers JE, Dangl GS, Meredith CP (1999) Development and characterization of additional microsatellite DNA markers for grape. Am J Enol Vitic 50:243–246

Bowers JE, Dangl GS, Vignani R, Meredith CP (1996) Isolation and characterization of new polymorphic simple sequence repeat loci in grape (*Vitis vinifera* L.). Genome 39:628–633

Bowers JE, Meredith CP (1997) The parentage of a classic wine grape, Cabernet Sauvignon. Nat Genet 16:84–87

Buonassisi D, Colombo M, Migliaro D, Dolzani C, Peressotti E, Mizzotti C, Velasco R, Masiero S, Perazzolli M, Vezzulli S (2017) Breeding for grapevine downy mildew resistance: a review of "omics" approaches. Euphytica 213:103

Cabezas JA, Cervera MT, Ruiz-García L, Carreño J, Martínez-Zapater JM (2006) A genetic analysis of seed and berry weight in grapevine. Genome 49:1572–1585

Cabezas JA, Ibáñez J, Lijavetzky D, Vélez D, Bravo G, Rodríguez V, Carreño I, Jermakow AM, Carreño J, Ruiz-García L, Thomas MR, Martinez-Zapater JM (2011) A 48 SNP set for grapevine cultivar identification. BMC Plant Biol 11:153

Cadle-Davidson L, Gadoury D, Fresnedo-Ramírez J, Yang S, Barba P, Sun Q, Demmings EM, Seem R, Schaub M, Nowogrodzki A, Kasinathan H, Ledbetter C, Reisch BI (2016) Lessons from a phenotyping center revealed by the genome-guided mapping of powdery mildew resistance loci. Phytopathology 106:1159–1169

Camus-Kulandaivelu L, Veyrieras J-B, Madur D, Combes V, Fourmann M, Barraud S, Dubreuil P, Gouesnard B, Manicacci D, Charcosset A (2005) Maize adaptation to temperate climate: relationship between population structure and polymorphism in the Dwarf8 gene. Genetics 172:2449–2463

Carreño I, Cabezas JA, Martínez-Mora C, Arroyo-García R, Cenis JL, Martínez-Zapater JM, Carreño J, Ruiz-García L (2015) Quantitative genetic analysis of berry firmness in table grape (*Vitis vinifera* L.). Tree Genet Genomes 11:818

Carrier G, Huang YF, Le Cunff L, Fournier-Level A, Vialet S, Souquet JM, Cheynier V, Terrier N, This P (2013) Selection of candidate genes for grape proanthocyanidin pathway by an integrative approach. Plant Physiol Biochem 72:87–95

Cartwright DA, Troggio M, Velasco R, Gutin A (2007) Genetic mapping in the presence of genotyping errors. Genetics 176:2521–2527

Cervera MT, Cabezas JA, Sancha JC, Martínez De Toda F, Martínez-Zapater JM (1998) Application of AFLPS to the characterization of grapevine *Vitis vinifera* L. genetic resources. A case study with accessions from Rioja (Spain). Theor Appl Genet 97:51–59

Chardon F, Virlon B, Moreau L, Falque M, Joets J, Decousset L, Murigneux A, Charcosset A (2004) Genetic architecture of flowering time in maize as inferred from quantitative trait loci meta-analysis and synteny conservation with the rice genome. Genetics 168:2169–2185

Chen J, Wang N, Fang LC, Liang ZC, Li SH, Wu BH (2015) Construction of a high-density genetic map and QTLs mapping for sugars and acids in grape berries. BMC Plant Biol 15:1–14

Chin C-S, Peluso P, Sedlazeck FJ, Nattestad M, Concepcion GT, Clum A, Dunn C, O'Malley R, Figueroa-Balderas R, Morales-Cruz A, Cramer GR, Delledonne M, Luo C, Ecker JR, Cantu D, Rank DR, Schatz MC (2016) Phased diploid genome assembly with single molecule real-time sequencing. Nat Methods 13:1050

Chitwood DH, Ranjan A, Martinez CC, Headland LR, Thiem T, Kumar R, Covington MF, Hatcher T, Naylor DT, Zimmerman S, Downs N, Raymundo N, Buckler ES, Maloof JN, Aradhya M, Prins B, Li L, Myles S, Sinha NR (2014) A modern ampelography: a genetic basis for leaf shape and venation patterning in grape. Plant Physiol 164:259–272

Cipriani G, Marrazzo MT, Di Gaspero G, Pfeiffer A, Morgante M, Testolin R (2008) A set of microsatellite markers with long core repeat optimized for grape (*Vitis* spp.) genotyping. BMC Plant Biol 8:1–13

Cipriani G, Spadotto A, Jurman I, Di Gaspero G, Crespan M, Meneghetti S, Frare E, Vignani R, Cresti M, Morgante M, Pezzotti M, Pe E, Policriti A, Testolin R (2010) The SSR-based molecular profile of 1005 grapevine (*Vitis vinifera* L.) accessions uncovers new synonymy and parentages, and reveals a large admixture amongst varieties of different geographic origin. Theor Appl Genet 121:1569–1585

Clark MD, Teh SL, Burkness E, Moreira L, Watson G, Yin L, Hutchison WD, Luby JJ (2018) Quantitative trait loci identified for foliar phylloxera resistance in a hybrid grape population. Aust J Grape Wine Res 24:292–300

Collard BCY, Jahufer MZZ, Brouwer JB, Pang ECK (2005) An introduction to markers, quantitative trait loci (QTL) mapping and marker-assisted selection for crop improvement: the basic concepts. Euphytica 142:169–196

Correa J, Mamani M, Muñoz-Espinoza C, González-Agüero M, Defilippi BG, Campos-Vargas R, Pinto M, Hinrichsen P (2016) New stable QTLs for berry firmness in table grapes. Am J Enol Vitic 67:212–217

Correa J, Ravest G, Laborie D, Mamani M, Torres E, Muñoz C, Pinto M, Hinrichsen P (2015) Quantitative trait loci for the response to gibberellic acid of berry size and seed mass in table grape (*Vitis vinifera* L.). Aust J Grape Wine Res 21:496–507

Costantini L, Battilana J, Lamaj F, Fanizza G, Grando MS (2008) Berry and phenology-related traits in grapevine (*Vitis vinifera* L.): From Quantitative Trait Loci to underlying genes. BMC Plant Biol 8:38

Costantini L, Malacarne G, Lorenzi S, Troggio M, Mattivi F, Moser C, Grando MS (2015) New candidate genes for the fine regulation of the colour of grapes. J Exp Bot 66:4427–4440

Coupel-Ledru A, Lebon E, Christophe A, Gallo A, Gago P, Pantin F, Doligez A, Simonneau T (2016) Reduced nighttime transpiration is a relevant breeding target for high water-use efficiency in grapevine. Proc Natl Acad Sci 113:8963–8968

Cretazzo E, Meneghetti S, De Andrés MT, Gaforio L, Frare E, Cifre J (2010) Clone differentiation and

varietal identification by means of SSR, AFLP, SAMPL and M-AFLP in order to assess the clonal selection of grapevine: The case study of Manto Negro, Callet and Moll, autochthonous cultivars of Majorca. Ann Appl Biol 157:213–227
Dalbó MA, Ye GN, Weeden NG, Steinkellner H, Sefc KM, Reisch BI (2000) A gene controlling sex in grapevines placed on a molecular-based genetic map. Genome 43:333–340
Danan S, Veyrieras J-B, Lefebvre V (2011) Construction of a potato consensus map and QTL meta-analysis offer new insights into the genetic architecture of late blight resistance and plant maturity traits. BMC Plant Biol 11:16
Davey MW, Kenis K, Keulemans J (2006) Genetic control of fruit vitamin C contents. Plant Physiol 142:343–351
de Givry S, Bouchez M, Chabrier P, Milan D, Schiex T (2005) CARTHAGENE: Multipopulation integrated genetic and radiation hybrid mapping. Bioinformatics 21:1703–1704
De Lorenzis G, Chipashvili R, Failla O, Maghradze D (2015) Study of genetic variability in *Vitis vinifera* L. germplasm by high-throughput *Vitis*18kSNP array: the case of Georgian genetic resources. BMC Plant Biol 15:1–14
de Oliveira Collet SA, Collet MA, de Machado M, Maria de Fátima PS (2005) Differential gene expression for isozymes in somatic mutants of *Vitis vinifera* L. (Vitaceae). Biochem Syst Ecol 33:691–703
Decroocq V, Favé MG, Hagen L, Bordenave L, Decroocq S (2003) Development and transferability of apricot and grape EST microsatellite markers across taxa. Theor Appl Genet 106:912–922
Delame M, Prado E, Blanc S, Guillaume R, Schneider C, Mestre P, Rustenholz C, Merdinoglu D (2018) Variation of recombination rate along the genome in *Vitis vinifera* × *Vitis rotundifolia* interspecific hybrids. In: Abstract book GBG 2018—Bordeaux, France 15–20 July, p 64
Delfino P, Zenoni S, Tornielli G, Crespan M, Gardiman M, Mirella G, Bellin D (2018) An integrated meta-QTL and transcriptomic data mining approach to select candidates controlling veraison time in grapevine. In: Abstract book GBG 2018—Bordeaux, France 15–20 July, p 53
Dereeper A, Nicolas S, Le Cunff L, Bacilieri R, Doligez A, Peros JP, Ruiz M, This P (2011) SNiPlay: A web-based tool for detection, management and analysis of SNPs. Application to grapevine diversity projects. BMC Bioinform 12:1–14
Divilov K, Barba P, Cadle L, Bruce D (2018) Single and multiple phenotype QTL analyses of downy mildew resistance in interspecific grapevines. Theor Appl Genet 131:1133–1143
Di Gaspero G, Cipriani G, Adam-Blondon AF, Testolin R (2007) Linkage maps of grapevine displaying the chromosomal locations of 420 microsatellite markers and 82 markers for R-gene candidates. Theor Appl Genet 114:1249–1263
Di Gaspero G, Cipriani G, Marrazzo MT, Andreetta D, Prado Castro MJ, Peterlunger E, Testolin R (2005) Isolation of (AC)n-microsatellites in *Vitis vinifera* L. and analysis of genetic background in grapevines under marker assisted selection. Mol Breed 15:11–20
Di Gaspero G, Copetti D, Coleman C, Castellarin SD, Eibach R, Kozma P, Lacombe T, Gambetta G, Zvyagin A, Cindrić P, Kovács L, Morgante M, Testolin R (2012) Selective sweep at the *Rpv3* locus during grapevine breeding for downy mildew resistance. Theor Appl Genet 124:277–286
Di Gaspero G, Foria S (2015) Molecular grapevine breeding techniques. In: Reynolds A (ed) Grapevine breeding programs for the wine industry. Woodhead Publishing, p 23–37
Di Genova A, Almeida A, Muñoz-Espinoza C, Vizoso P, Travisany D, Moraga C, Pinto M, Hinrichsen P, Orellana A, Maass A (2014) Whole genome comparison between table and wine grapes reveals a comprehensive catalog of structural variants. BMC Plant Biol 14:7
Doligez A, Adam-Blondon AF, Cipriani G, Di Gaspero G, Laucou V, Merdinoglu D, Meredith CP, Riaz S, Roux C, This P (2006a) An integrated SSR map of grapevine based on five mapping populations. Theor Appl Genet 113:369–382
Doligez A, Audiot E, Baumes R, This P (2006b) QTLs for muscat flavor and monoterpenic odorant content in grapevine (*Vitis vinifera* L.). Mol Breed 18:109–125
Doligez A, Bertrand Y, Dias S, Grolier M, Ballester JF, Bouquet A, This P (2010) QTLs for fertility in table grape (*Vitis vinifera* L.). Tree Genet Genomes 6:413–422
Doligez A, Bertrand Y, Farnos M, Grolier M, Romieu C, Esnault F, Dias S, Berger G, François P, Pons T, Ortigosa P, Roux C, Houel C, Laucou V, Bacilieri R, Péros JP, This P (2013) New stable QTLs for berry weight do not colocalize with QTLs for seed traits in cultivated grapevine (*Vitis vinifera* L.). BMC Plant Biol 13:217
Doligez A, Bouquet A, Danglot Y, Lahogue F, Riaz S, Meredith CP, Edwards KJ, This P (2002) Genetic mapping of grapevine (*Vitis vinifera* L.) applied to the detection of QTLs for seedlessness and berry weight. Theor Appl Genet 105:780–795
Donald TM, Pellerone F, Adam-Blondon A-F, Bouquet A, Thomas MR, Dry IB (2002) Identification of resistance gene analogs linked to a powdery mildew resistance locus in grapevine. TAG Theor Appl Genet 104:610–618
Duchêne E, Butterlin G, Claudel P, Dumas V, Jaegli N, Merdinoglu D (2009) A grapevine (*Vitis vinifera* L.) deoxy-d-xylulose synthase gene colocates with a major quantitative trait loci for terpenol content. Theor Appl Genet 118:541–552
Duchêne E, Butterlin G, Dumas V, Merdinoglu D (2012) Towards the adaptation of grapevine varieties to climate change: QTLs and candidate genes for developmental stages. Theor Appl Genet 124:623–635
Edwards D, Forster JW, Cogan NOI, Batley J, Chagné D (2007) Single Nucleotide Polymorphism Discovery.

In: New York NY (ed) Association mapping in plants. Springer, New York, pp 53–76

Eibach R, Zyprian E, Welter L, Töpfer R (2007) The use of molecular markers for pyramiding resistance genes in grapevine breeding. Vitis 46:120–124

Emanuelli F, Battilana J, Costantini L, Le Cunff L, Boursiquot JM, This P, Grando MS (2010) A candidate gene association study on muscat flavor in grapevine (*Vitis vinifera* L.). BMC Plant Biol 10:241

Emanuelli F, Lorenzi S, Grzeskowiak L, Catalano V, Stefanini M, Troggio M, Myles S, Martinez-Zapater JM, Zyprian E, Moreira FM, Grando MS (2013) Genetic diversity and population structure assessed by SSR and SNP markers in a large germplasm collection of grape. BMC Plant Biol 13:1–17

Emanuelli F, Sordo M, Lorenzi S, Battilana J, Grando MS (2014) Development of user-friendly functional molecular markers for VvDXS gene conferring muscat flavor in grapevine. Mol Breed 33:235–241

Endelman JB, Plomion C (2014) LPmerge: an R package for merging genetic maps by linear programming. Bioinformatics 30:1623–1624

Fanizza G, Lamaj F, Costantini L, Chaabane R, Grando MS (2005) QTL analysis for fruit yield components in table grapes (*Vitis vinifera*). Theor Appl Genet 111:658–664

Fechter I, Hausmann L, Zyprian E, Daum M, Holtgräwe D, Weisshaar B, Töpfer R (2014) QTL analysis of flowering time and ripening traits suggests an impact of a genomic region on linkage group 1 in *Vitis*. Theor Appl Genet 127:1857–1872

Fernandez L, Le Cunff L, Tello J, Lacombe T, Boursiquot JM, Fournier-Level A, Bravo G, Lalet S, Torregrosa L, This P, Martinez-Zapater JM (2014) Haplotype diversity of VvTFL1A gene and association with cluster traits in grapevine (*V. vinifera*). BMC Plant Biol 14:209

Fernandez L, Doligez A, Lopez G, Thomas MR, Bouquet A, Torregrosa L (2006) Somatic chimerism, genetic inheritance, and mapping of the fleshless berry (flb) mutation in grapevine (*Vitis vinifera* L.). Genome 49:721–728

Fernández MP, Núñez Y, Ponz F, Hernáiz S, Gallego FJ, Ibáñez J (2008) Characterization of sequence polymorphisms from microsatellite flanking regions in *Vitis* spp. Mol Breed 22:455–465

Fischer BM, Salakhutdinov I, Akkurt M, Eibach R, Edwards KJ, Töpfer R, Zyprian EM (2004) Quantitative trait locus analysis of fungal disease resistance factors on a molecular map of grapevine. Theor Appl Genet 108:501–515

Flutre T, Bacilieri R, Berger G, Bertrand Y, Boursiquot J-M, Fodor A, Lacombe T, Laucou V, Launay A, Le Cunff L, Romieu C, This P, Péros J-P, Doligez A (2018) VITIRAMA: a program to characterise disease susceptibility in French ampelographic collections. Abstract book GBG 2018—Bordeaux, 15–20 July, p 55

Foria S, Magris G, Morgante M, Di Gaspero G (2018) The genetic background modulates the intensity of *Rpv3*-dependent downy mildew resistance in grapevine. Plant Breed 137:220–228

Fossati T, Labra M, Castiglione S, Failla O, Scienza A, Sala F (2001) The use of AFLP and SSR molecular markers to decipher homonyms and synonyms in grapevine cultivars: The case of the varietal group known as "Schiave". Theor Appl Genet 102:200–205

Fournier-Level A, Le Cunff L, Gomez C, Doligez A, Ageorges A, Roux C, Bertrand Y, Souquet JM, Cheynier V, This P (2009) Quantitative genetic bases of anthocyanin variation in grape (*Vitis vinifera* L. ssp. sativa) berry: a quantitative trait locus to quantitative trait nucleotide integrated study. Genetics 183:1127–1139

Fournier-Level A, Hugueney P, Verries C, This P, Ageorges A (2011) Genetic mechanisms underlying the methylation level of anthocyanins in grape (*Vitis vinifera* L.). BMC Plant Biol 11:179

Ganal MW, Altmann T, Röder MS (2009) SNP identification in crop plants. Curr Opin Plant Biol 12:211–217

Garris A, Clark L, Owens C, McKay S, Luby J, Mathiason K, Fennell A (2009) Mapping of photoperiod-induced growth cessation in the wild grape *Vitis riparia*. J Am Soc Hortic Sci 134:261–272

Ghaffari S, Hasnaoui N, Zinelabidine LH, Ferchichi A, Martínez-Zapater JM, Ibáñez J (2014) Genetic diversity and parentage of Tunisian wild and cultivated grapevines (*Vitis vinifera* L.) as revealed by single nucleotide polymorphism (SNP) markers. Tree Genet Genomes 10:1103–1112

Gismondi A, Di G, Martini F, Sarti L, Crespan M, Martínez-labarga C, Rickards O, Canini A (2016) Grapevine carpological remains revealed the existence of a Neolithic domesticated *Vitis vinifera* L. specimen containing ancient DNA partially preserved in modern ecotypes. J Archaeol Sci 69:75–84

Goffinet B, Gerber S (2000) Quantitative trait loci: a meta-analysis. Genetics 155:463–473

Granier C, Vile D (2014) Phenotyping and beyond: modelling the relationships between traits. Curr Opin Plant Biol 18:96–102

Grattapaglia D, Sederoff R (1994) Genetic linkage maps of Eucalyptus grandis and Eucalyptus urophlla using a pseudo-testcross: mapping strategy and RAPD markers. Genetics 137:1121–1137

Grzeskowiak L, Costantini L, Lorenzi S, Grando MS (2013) Candidate loci for phenology and fruitfulness contributing to the phenotypic variability observed in grapevine. Theor Appl Genet 126:2763–2776

Guo Y, Xue R, Lin H, Su K, Zhao Y, Zhendong L, Ma H, Shi G, Niu Z, Li K, Guo X (2015) Genetic analysis and QTL mapping for fruit skin anthocyanidin in grape (*Vitis vinifera*). Pak J Bot 47:1765–1771

Guo DL, Zhao HL, Li Q, Zhang GH, Jiang JF, Liu CH, Yu YH (2019) Genome-wide association study of berry-related traits in grape [*Vitis vinifera* L.] based on genotyping-by-sequencing markers. Hortic Res 6:11

Herzog K, Wind R, Töpfer R, Herzog K, Wind R, Töpfer R (2015) Impedance of the grape berry cuticle as a novel phenotypic trait to estimate resistance to *Botrytis cinerea*. Sensors 15:12498–12512

Hoffmann S, Di Gaspero G, Kovács L, Howard S, Kiss E, Galbács Z, Testolin R, Kozma P (2008) Resistance to *Erysiphe necator* in the grapevine 'Kishmish vatkana' is controlled by a single locus through restriction of hyphal growth. Theor Appl Genet 116:427–438

Holland JB (2015) MAGIC maize: a new resource for plant genetics. Genome Biol 16:163

Houel C, Chatbanyong R, Doligez A, Rienth M, Foria S, Luchaire N, Roux C, Adivèze A, Lopez G, Farnos M, Pellegrino A, This P, Romieu C, Torregrosa L (2015) Identification of stable QTLs for vegetative and reproductive traits in the microvine (*Vitis vinifera* L.) using the 18 K Infinium chip. BMC Plant Biol 15:205

Houle D, Govindaraju DR, Omholt S (2010) Phenomics: the next challenge. Nat Rev Genet 11:855–866

Huang Y, Bertrand Y, Guiraud J, Vialet S, Launay A, Cheynier V, Terrier N, This P (2013) Plant science expression QTL mapping in grapevine—revisiting the genetic determinism of grape skin colour. Plant Sci 207:18–24

Huang Y, Vialet S, Guiraud J, Torregrosa L, Bertrand Y, Cheynier V, This P, Terrier N (2014) VvMYBrep, a negative regulator of proanthocyanidin accumulation in grape berry, identified through expression quantitative locus mapping. New Phytol 201:795–809

Huang YF, Doligez A, Fournier-Level A, Le Cunff L, Bertrand Y, Canaguier A, Morel C, Miralles V, Veran F, Souquet JM, Cheynier V, Terrier N, This P (2012) Dissecting genetic architecture of grape proanthocyanidin composition through quantitative trait locus mapping. BMC Plant Biol 12:30

Hung H-Y, Shannon LM, Tian F, Bradbury PJ, Chen C, Flint-Garcia SA, McMullen MD, Ware D, Buckler ES, Doebley JF, Holland JB (2012) ZmCCT and the genetic basis of day-length adaptation underlying the post domestication spread of maize. Proc Natl Acad Sci 109:1913–1921

Hyma KE, Barba P, Wang M, Londo JP, Acharya CB, Mitchell SE, Sun Q, Reisch B, Cadle-Davidson L (2015) Heterozygous mapping strategy (HetMappS) for high resolution genotyping-by-sequencing markers: a case study in grapevine. PLoS ONE 10:1–31

Iakovidis M, Teixeira PJPL, Exposito-Alonso M, Cowper MG, Law TF, Liu Q, Vu MC, Dang TM, Corwin JA, Weigel D, Dangl JL, Grant SR (2016) Effector-triggered immune response in *Arabidopsis thaliana* is a quantitative trait. Genetics 204:337–353

Imazio S, Maghradze D, de Lorenzis G, Bacilieri R, Laucou V, This P, Scienza A, Failla O (2013) From the cradle of grapevine domestication: Molecular overview and description of Georgian grapevine (*Vitis vinifera* L.) germplasm. Tree Genet Genomes 9:641–658

Jaillon O, Aury JM, Noel B, Policriti A, Clepet C, Casagrande A, Choisne N, Aubourg S, Vitulo N, Jubin C, Vezzi A, Legeai F, Hugueney P, Dasilva C, Horner D, Mica E, Jublot D, Poulain J, Bruyère C, Billault A, Segurens B, Gouyvenoux M, Ugarte E, Cattonaro F, Anthouard V, Vico V, Del Fabbro C, Alaux M, Di Gaspero G, Dumas V, Felice N, Paillard S, Juman I, Moroldo M, Scalabrin S, Canaguier A, Le Clainche I, Malacrida G, Durand E, Pesole G, Laucou V, Chatelet P, Merdinoglu D, Delledonne M, Pezzotti M, Lecharny A, Scarpelli C, Artiguenave F, Pè ME, Valle G, Morgante M, Caboche M, Adam-Blondon AF, Weissenbach J, Quétier F, Wincker P (2007) The grapevine genome sequence suggests ancestral hexaploidization in major angiosperm phyla. Nature 449:463–467

Jansen J, De Jong AG, Van Oijen JW (2001) Constructing dense genetic linkage maps. Theor Appl Genet 102:1113–1122

Jonas E, De Koning D (2013) Does genomic selection have a future in plant breeding? Trends Biotechnol 31:497–504

Jones CJ, Edwards KJ, Castaglione S, Winfield MO, Sala F, Van De Wiel C, Bredemeijer G, Vosman B, Matthes M, Daly A, Brettschneider R, Bettini P, Buiatti M, Maestri E, Malcevschi A, Marmiroli N, Aert R, Volckaert G, Rueda J, Linacero R, Vazquez A, Karp A (1997) Reproducibility testing of RAPD, AFLP and SSR markers in plants by a network of European laboratories. Mol Breed 3:381–390

Jones ES, Sullivan H, Bhattramakki D, Smith JSC (2007) A comparison of simple sequence repeat and single nucleotide polymorphism marker technologies for the genotypic analysis of maize (*Zea mays* L.). Theor Appl Genet 115:361–371

Kayesh E, Zhang YY, Liu GS, Bilkish N, Sun X, Leng XP, Fang JG (2013) Development of highly polymorphic EST-SSR markers and segregation in F_1 hybrid population of *Vitis vinifera* L. Genet Mol Res 12:3871–3878

Khowaja FS, Norton GJ, Courtois B, Price AH (2009) Improved resolution in the position of drought-related QTLs in a single mapping population of rice by meta-analysis. BMC Genom 10:276

Kicherer A, Herzog K, Bendel N, Klück HC, Backhaus A, Wieland M, Rose JC, Klingbeil L, Läbe T, Hohl C, Petry W, Kuhlmann H, Seiffert U, Töpfer R (2017a) Phenoliner: a new field phenotyping platform for grapevine research. Sensors 17:1625

Kicherer A, Herzog K, Pflanz M, Wieland M, Rüger P, Kecke S, Kuhlmann H, Töpfer R (2015) An automated field phenotyping pipeline for application in grapevine research. Sensors 15:4823–4836

Kicherer A, Klodt M, Sharifzadeh S, Cremers D, Töpfer R, Herzog K (2017b) Automatic image-based determination of pruning mass as a determinant for yield potential in grapevine management and breeding. Aust J Grape Wine Res 23:120–124

Kim C, Seol Y, Lee D-J, Lee J-H, Lee T, Park D (2014) RiceQTLPro: an integrated database for quantitative trait loci marker mapping in rice plant. Bioinformation 10:664–666

Kobayashi S, Goto-Yamamoto N, Hirochika H (2004) Retrotransposon-induced mutations in grape skin color. Science 304:982

Kono A, Ban Y, Mitani N, Fujii H, Sato S, Suzaki K, Azuma A, Onoue N, Sato A (2018) Development of SSR markers linked to QTL reducing leaf hair density and grapevine downy mildew resistance in *Vitis vinifera*. Mol Breed 38:138

Korte A, Vilhjálmsson BJ, Segura V, Platt A, Long Q, Nordborg M (2012) A mixed-model approach for genome-wide association studies of correlated traits in structured populations. Nat Genet 44:1066–1071

Kosambi DD (1944) The estimation of map distances from recombination values. Ann Eugenics 12:172–175

Krivanek AF, Riaz S, Walker MA (2006) Identification and molecular mapping of PdR1, a primary resistance gene to Pierce's disease in *Vitis*. Theor Appl Genet 112:1125–1131

Labra M, Imazio S, Grassi F, Rossoni M, Sala F (2004) Vine-1 retrotransposon-based sequence-specific amplified polymorphism for *Vitis vinifera* L. genotyping. Plant Breed 123:180–185

Lacombe T, Boursiquot JM, Laucou V, Di Vecchi-Staraz M, Péros JP, This P (2013) Large-scale parentage analysis in an extended set of grapevine cultivars (*Vitis vinifera* L.). Theor Appl Genet 126:401–414

Lander ES, Green P, Abrahamson J, Barlow H, Daly M, Lincoln S, Newsbury L (1987) MAPMAKER: An interactive computer program for constructing genetic maps of experimental and natural populations. Genomics 1:174–181

Laucou V, Launay A, Bacilieri R, Lacombe T, Adam-Blondon AF, Bérard A, Chauveau A, De Andrés MT, Hausmann L, Ibáñez J, Le Paslier MC, Maghradze D, Martinez-Zapater J, Maul E, Ponnaiah M, Töpfer R, Péros JP, Boursiquot JM (2018) Extended diversity analysis of cultivated grapevine *Vitis vinifera* with 10 K genome-wide SNPs. PLoS ONE 13:1–27

Le Paslier M-C, Choisne N, Bacilieri R, Bounon R, Boursiquot J-MB, Brunel D, Di Gaspero G, Hausmann L, Lacombe T, Laucou V LA, Martinez-Zapater J, Morgante M, Raj P, PonnaiahM Q, Scalabrin S, Torres-Perez R (2013). The GrapeReSeq 18 k *Vitis* genotyping chip. In: IX international symposium on grapevine physiology and biotechnology. International society for horticultural science, Abstract Book, p 123

Li M, Liu X, Bradbury P, Yu J, Zhang Y-M, Todhunter RJ, Buckler ES, Zhang Z (2014) Enrichment of statistical power for genome-wide association studies. BMC Biol 12:73

Lijavetzky D, Cabezas J, Ibáñez A, Rodríguez V, Martínez-Zapater JM (2007) High throughput SNP discovery and genotyping in grapevine (*Vitis vinifera* L.) by combining a re-sequencing approach and SNPlex technology. BMC Genom 8:424

Lin H, Leng H, Guo Y, Kondo S, Zhao Y, Shi G, Guo X (2019) QTLs and candidate genes for downy mildew resistance conferred by interspecific grape (*V. vinifera* L. × *V. amurensis* Rupr.) crossing. Sci Hortic 244: 200–207

Lodhi MA, Ye G-N, Weeden NF, Reisch BI, Daly MJ (1995) A molecular marker based linkage map of *Vitis*. Genome 38:786–794

Lowe KM, Riaz S, Walker MA (2009) Variation in recombination rates across *Vitis* species. Tree Genet Genomes 5:71–80

Malacarne G, Costantini L, Coller E, Battilana J, Velasco R, Vrhovsek U, Grando MS, Moser C (2015) Regulation of flavonol content and composition in (Syrah × Pinot Noir) mature grapes: Integration of transcriptional profiling and metabolic quantitative trait locus analyses. J Exp Bot 66:4441–4453

Mammadov J, Aggarwal R, Buyyarapu R, Kumpatla S (2012) SNP markers and their impact on plant breeding. Int J Plant Genom 2012:728398

Marguerit E, Boury C, Manicki A, Donnart M, Butterlin G, Némorin A, Wiedemann-Merdinoglu S, Merdinoglu D, Ollat N, Decroocq S (2009) Genetic dissection of sex determinism, inflorescence morphology and downy mildew resistance in grapevine. Theor Appl Genet 118:1261–1278

Marrano A, Birolo G, Prazzoli ML, Lorenzi S, Valle G, Grando MS (2017) SNP-discovery by RAD-sequencing in a germplasm collection of wild and cultivated grapevines (*V. vinifera* L.). PLoS ONE 12:1–19

Marrano A, Micheletti D, Lorenzi S, Neale D, Grando MS (2018) Genomic signatures of different adaptations to environmental stimuli between wild and cultivated *Vitis vinifera* L. Hortic Res 5:34

Martinez AK, Soriano JM, Tuberosa R, Koumproglou R, Jahrmann T, Salvi S (2016) Yield QTLome distribution correlates with gene density in maize. Plant Sci 242:300–309

Maul E, Sudharma KN, Kecke S, Marx G, Müller C, Audeguin L, Boselli M, Boursiquot JM, Bucchetti B, Cabello F, Carraro R, Crespan M, De Andrés MT, Eiras Dias J, Ekhvaia J, Gaforio L, Gardiman M, Grando S, Gyropoulos D, Jandurova O, Kiss E, Kontic J, Kozma P, Lacombe T, Laucou V, Legrand D, Maghradze D, Marinoni D, Maletic E, Moreira F, Muñoz-Organero G, Nakhutsrishvili G, Pejic I, Peterlunger E, Pitsoli D, Pospisilova D, Preiner D, Raimondi S, Regner F, Savin G, Savvides S, Schneider A, Sereno C, Simon S, Staraz M, Zulini L, Bacilieri R, This P (2012) The European *vitis* database (www.eu-vitis.de)—a technical innovation through an online uploading and interactive modification system E. Vitis J Grapevine Res 51:79–85

Mejía N, Gebauer M, Muñoz L, Hewstone N, Muñoz C, Hinrichsen P (2007) Identification of QTLs for seedlessness, berry size, and ripening date in a seedless × seedless table grape progeny. Am J Enol Vitic 58:499–507

Mejía N, Soto B, Guerrero M, Casanueva X, Houel C, de los Ángeles Miccono M, Ramos R, Le Cunff L, Boursiquot JM, Hinrichsen P, Adam-Blondon AF (2011) Molecular, genetic and transcriptional evidence for a role of VvAGL11 in stenospermocarpic seedlessness in grapevine. BMC Plant Biol 11:57

Mercati F, De Lorenzis G, Brancadoro L, Lupini A, Abenavoli MR, Barbagallo MG, Di Lorenzo R, Scienza A, Sunseri F (2016) High-throughput 18 K SNP array to assess genetic variability of the main grapevine cultivars from Sicily. Tree Genet Genomes 12:59

Merdinoglu D, Wiedeman-Merdinoglu S, Coste P, Dumas V, Haetty S, Butterlin G, Greif C (2003) Genetic analysis of downy mildew resistance derived from *Muscadinia rotundifolia*. Acta Hortic 603:451–456

Michelmore R (1995) Molecular approaches to manipulation of disease resistance genes. Annu Rev Phytopathol 33:393–427

Michelmore RW, Paran I, Kesseli RV (1991) Identification of markers linked to disease-resistance genes by bulked segregant analysis: a rapid method to detect markers in specific genomic regions by using segregating populations. Proc Natl Acad Sci USA 88:9828–9832

Migicovsky Z, Sawler J, Gardner KM, Aradhya MK, Prins BH, Schwaninger HR, Bustamante CD, Buckler ES, Zhong G-Y, Brown PJ, Myles S (2017) Patterns of genomic and phenomic diversity in wine and table grapes. Hortic Res 4:17035

Migliaro D, Crespan M, Muñoz-Organero G, Velasco R, Moser C, Vezzulli S (2014) Structural dynamics at the berry colour locus in *Vitis vinifera* L. somatic variants. Aust J Grape Wine Res 20:485–495

Migliaro D, De Nardi B, Vezzulli S, Crespan M (2017) An upgraded core set of 11 SSR markers for grapevine cultivar identification: the case of berry color mutants. Am J Enol Vitic 68:496–498

Miller AJ, Matasci N, Schwaninger H, Aradhya MK, Prins B, Zhong GY, Simon C, Buckler ES, Myles S (2013) *Vitis* phylogenomics: hybridization intensities from a SNP array outperform genotype calls. PLoS ONE 8:e78680

Minio A, Lin J, Gaut BS, Cantu D (2017) How single molecule real-time sequencing and haplotype phasing have enabled reference-grade diploid genome assembly of wine grapes. Front Plant Sci 8:826

Minio A, Massonnet M, Figueroa-Balderas R, Castro A, Cantu D (2019) Diploid genome assembly of the wine grape carménère. G3 Genes Genomes Genet. https://doi.org/10.1534/g3.119.400030

Moreira FM, Madini A, Marino R, Zulini L, Stefanini M, Velasco R, Kozma P, Grando MS (2011) Genetic linkage maps of two interspecific grape crosses (*Vitis* spp.) used to localize quantitative trait loci for downy mildew resistance. Tree Genet Genomes 7:153–167

Moroldo M, Paillard S, Marconi R, Fabrice L, Canaguier A, Cruaud C, De Berardinis V, Guichard C, Brunaud V, Le Clainche I, Scalabrin S, Testolin R, Di Gaspero G, Morgante M, Adam-Blondon AF (2008) A physical map of the heterozygous grapevine "Cabernet Sauvignon" allows mapping candidate genes for disease resistance. BMC Plant Biol 8:1–14

Myles S, Boyko AR, Owens CL, Brown PJ, Grassi F, Aradhya MK, Prins B, Reynolds A, Chia J-M, Ware D, Bustamante CD, Buckler ES (2011) Genetic structure and domestication history of the grape. Proc Natl Acad Sci 108:3530–3535

Myles S, Chia JM, Hurwitz B, Simon C, Zhong GY, Buckler E, Ware D (2010) Rapid genomic characterization of the genus *Vitis*. PLoS ONE 5:e8219

Myles S, Mahanil S, Harriman J, Gardner KM, Franklin JL, Reisch BI, Ramming DW, Owens CL, Li L, Buckler ES, Cadle-Davidson L (2015) Genetic mapping in grapevine using SNP microarray intensity values. Mol Breed 35:88

Myles S, Peiffer J, Brown PJ, Ersoz ES, Zhang Z, Costich DE, Buckler ES (2009) Association mapping: critical considerations shift from genotyping to experimental design. Plant Cell Online 21:2194–2202

Ni J, Pujar A, Youens-Clark K, Yap I, Jaiswal P, Tecle I, Tung CW, Ren L, Spooner W, Wei X, Avraham S, Ware D, Stein L, McCouch S (2009) Gramene QTL database: development, content and applications. Database 2009

Nicolas SD, Péros JP, Lacombe T, Launay A, Le Paslier MC, Bérard A, Mangin B, Valière S, Martins F, Le Cunff L, Laucou V, Bacilieri R, Dereeper A, Chatelet P, This P, Doligez A (2016) Genetic diversity, linkage disequilibrium and power of a large grapevine (*Vitis vinifera* L) diversity panel newly designed for association studies. BMC Plant Biol 16:74

Nicolè S, Barcaccia G, Erickson DL, Kress JW, Lucchin M (2013) The coding region of the UFGT gene is a source of diagnostic SNP markers that allow single-locus DNA genotyping for the assessment of cultivar identity and ancestry in grapevine (*Vitis vinifera* L.). BMC Res Notes 6:502

Nijveen H, Ligterink W, Keurentjes JJB, Loudet O, Long J, Sterken MG, Prins P, Hilhorst HW, de Ridder D, Kammenga JE, Snoek BL (2017) AraQTL—workbench and archive for systems genetics in Arabidopsis thaliana. Plant J 89:1225–1235

Ochssner I, Hausmann L, Toepfer R (2016) *Rpvl4*, a new genetic source for *Plasmopara viticola* resistance conferred by *Vitis cinerea*. Vitis J Grapevine Res 55:79–81

Oerke E-C, Herzog K, Toepfer R (2016) Hyperspectral phenotyping of the reaction of grapevine genotypes to *Plasmopara viticola*. J Exp Bot 67:5529–5543

Ortiz JM, Martín JP, Borrego J, Chávez J, Rodríguez I, Muñoz G, Cabello F (2004) Molecular and morphological characterization of a *Vitis* gene bank for the establishment of a base collection. Genet Resour Crop Evol 51:403–409

Owens CL (2003) SNP detection and genotyping in *Vitis*. Acta Hortic 603:139–140

Pap D, Riaz S, Dry IB, Jermakow A, Tenscher AC, Cantu D, Oláh R, Walker MA (2016) Identification of two novel powdery mildew resistance loci, *Ren6* and *Ren7*, from the wild Chinese grape species *Vitis* piasezkii. BMC Plant Biol 16:170

Pauquet J, Bouquet A, This P, Adam-Blondon AF (2001) Establishment of a local map of AFLP markers around the powdery mildew resistance gene Run1 in grapevine and assessment of their usefulness for marker assisted selection. Theor Appl Genet 103:1201–1210

Peressotti E, Dolzani C, Poles L, Banchi E, Stefanini M, Salamini F, Velasco R, Vezzulli S, Riaz S, Walker MA, Reisch BI, Van De Weg WE, Bink MCAM (2015) A first pedigree-based analysis (PBA) approach for the dissection of disease resistance traits in grapevine hybrids. Acta Hortic 1082:113–122

Pindo M, Vezzulli S, Coppola G, Cartwright DA, Zharkikh A, Velasco R, Troggio M (2008) SNP high-throughput screening in grapevine using the SNPlex genotyping system. BMC Plant Biol 8:1–6

Polesani M, Desario F, Ferrarini A, Zamboni A, Pezzotti M, Kortekamp A, Polverari A (2008) cDNA-AFLP analysis of plant and pathogen genes expressed in grapevine infected with Plasmopara viticola. BMC Genom 9:142

Pollefeys P, Bousquet J (2003) Molecular genetic diversity of the French–American grapevine hybrids cultivated in North America. Genome 46:1037–1048

Qu X, Lu J, Lamikanra O, Science V, Florida A (1996) Genetic diversity in muscadine and american bunch grapes based on randomly amplified polymorphic DNA (RAPD) analysis. J Am Soc Hortic Sci 121:1020–1023

Razi M, Darvishzadeh R, Amiri ME, Doulati-Banehd H, Martínez-Gómez P (2019) Molecular characterization of a diverse Iranian table grapevine germplasm using REMAP markers: population structure, linkage disequilibrium and association mapping of berry yield and quality traits. Biologia 74:173–185

Rex F, Fechter I, Hausmann L, Töpfer R (2014) QTL mapping of black rot (*Guignardia bidwellii*) resistance in the grapevine rootstock 'börner' (*V. riparia* Gm183 × *V. cinerea* Arnold). Theor Appl Genet 127:1667–1677

Reyna N, Sneller CH (2001) Evaluation of marker-assisted introgression of yield QTL alleles into adapted soybean. Crop Sci 41:1317

Riahi L, Zoghlami N, Dereeper A, Laucou V, Mliki A, This P (2013) Single nucleotide polymorphism and haplotype diversity of the gene NAC4 in grapevine. Ind Crop Prod 43:718–724

Riaz S, Tenscher AC, Ramming DW, Walker MA (2011) Using a limited mapping strategy to identify major QTLs for resistance to grapevine powdery mildew (*Erysiphe necator*) and their use in marker-assisted breeding. Theor Appl Genet 122:1059–1073

Richter R, Gabriel D, Rist F, Topfer R, Zyprian E (2019) Identification of co-located QTLs and genomic regions affecting grapevine cluster architecture. Theor Appl Genet 132:1159–1177

Ritter E, Gebhardt C, Salamini F (1990) Estimation of recombination frequencies and construction of RFLP linkage maps in plants from crosses between heterozygous parents. Genetics 125:645–654

Ritter E, Salamini F (1996) The calculation of recombination frequencies in crosses of allogamous plant species with applications to linkage mapping. Genet Res 67:55–65

Roach MJ, Johnson DL, Bohlmann J, van Vuuren HJJ, Jones SJM, Pretorius IS, Schmidt SA, Borneman AR (2018) Population sequencing reveals clonal diversity and ancestral inbreeding in the grapevine cultivar Chardonnay. PLoS Genet 14:e1007807

Ronin Y, Mester D, Minkov D, Belotserkovski R, Jackson BN, Schnable PS, Aluru S, Korol A (2012) Two-phase analysis in consensus genetic mapping. G3 Genes Genom Genet 2:537–549

Rose J, Kicherer A, Wieland M, Klingbeil L, Töpfer R, Kuhlmann H (2016) Towards automated large-scale 3D phenotyping of vineyards under field conditions. Sensors 16:2136

Royo C, Torres-Pérez R, Mauri N, Diestro N, Cabezas JA, Marchal C, Lacombe T, Ibáñez J, Tornel M, Carreño J, Martínez-Zapater JM, Carbonell-Bejerano P (2018) The major origin of seedless grapes is associated with a missense mutation in the MADS-Box Gene VviAGL11. Plant Physiol 177:1234–1253

Said JI, Knapka JA, Song M (2015a) Cotton QTLdb: a cotton QTL database for QTL analysis, visualization, and comparison between *Gossypium hirsutum* and *G. hirsutum* × *G. barbadense* populations. Mol Genet Genom 290:1615–1625

Said JI, Song M, Wang H, Lin Z, Zhang X, Fang DD, Zhang J (2015b) A comparative meta-analysis of QTL between intraspecific *Gossypium hirsutum* and interspecific *G. hirsutum* × *G. barbadense* populations. Mol Genet Genomics 290:1003–1025

Salmaso M, Faes G, Segala C, Stefanini M, Salakhutdinov L, Zyprian E, Toepfer R, Grando MS, Velasco R (2004) Genome diversity and gene haplotypes in the grapevine (*Vitis vinifera* L.), as revealed by single nucleotide polymorphisms. Mol Breed 14:385–395

Salmaso M, Malacarne G, Troggio M, Faes G, Stefanini M, Grando MS, Velasco R (2008) A grapevine (*Vitis vinifera* L.) genetic map integrating the position of 139 expressed genes. Theor Appl Genet 116:1129–1143

Salvi S, Corneti S, Bellotti M, Carraro N, Sanguineti MC, Castelletti S, Tuberosa R (2011) Genetic dissection of maize phenology using an intraspecific introgression library. BMC Plant Biol 11:4

Salvi S, Sponza G, Morgante M, Tomes D, Niu X, Fengler KA, Meeley R, Ananiev EV, Svitashev S, Bruggemann E, Li B, Hainey CF, Radovic S, Zaina G, Rafalski J-A, Tingey SV, Miao G-H, Phillips RL, Tuberosa R (2007) Conserved noncoding genomic sequences associated with a flowering-time quantitative trait locus in maize. Proc Natl Acad Sci 104:11376–11381

Salvi S, Tuberosa R (2015) The crop QTLome comes of age. Curr Opin Biotechnol 32:179–185

Sapkota S, Chen LL, Yang S, Hyma KE, Cadle-Davidson L, Hwang CF (2019) Construction of a high-density linkage map and QTL detection of downy mildew resistance in *Vitis aestivalis*-derived 'Norton'. Theor Appl Genet 132:137–147

Schellenbaum P, Mohler V, Wenzel G, Walter B (2008) Variation in DNA methylation patterns of grapevine somaclones (*Vitis vinifera* L.). BMC Plant Biol 8:78

Schwander F, Eibach R, Fechter I, Hausmann L, Zyprian E, Töpfer R (2012) *Rpv10*: A new locus from the Asian *Vitis* gene pool for pyramiding downy mildew resistance loci in grapevine. Theor Appl Genet 124:163–176

Segura V, Vilhjálmsson BJ, Platt A, Korte A, Seren Ü, Long Q, Nordborg M (2012) An efficient multi-locus mixed-model approach for genome-wide association studies in structured populations. Nat Genet 44:825–830

Sensi E, Vignani R, Rohde W, Biricolti S (1996) Characterization of genetic biodiversity with *Vitis vinifera* L. Sangiovese and Colorino genotypes by AFLP and ISTR DNA marker technology. Vitis 35:183–188

Smita S, Lenka SK, Katiyar A, Jaiswal P, Preece J, Bansal KC (2011) QlicRice: a web interface for abiotic stress responsive QTL and loci interaction channels in rice. Database 2011:1–9

Smith HM, Smith BP, Morales NB, Moskwa S, Clingeleffer PR, Thomas MR (2018a) SNP markers tightly linked to root knot nematode resistance in grapevine (*Vitis cinerea*) identified by a genotyping-by-sequencing approach followed by Sequenom MassARRAY validation. PLoS ONE 13:1–27

Smith HM, Clarke CW, Smith BP, Carmody BM, Thomas MR, Clingeleffer PR, Powell KS (2018b) Genetic identification of SNP markers linked to a new grape phylloxera resistant locus in *Vitis cinerea* for marker-assisted selection. BMC Plant Biol 18:360

Stam P (1993) Construction of integrated genetic linkage maps by means of a new computer package: Join Map. Plant J 3:739–744

Stange M, Utz HF, Schrag TA, Melchinger AE, Würschum T (2013) High-density genotyping: an overkill for QTL mapping? Lessons learned from a case study in maize and simulations. Theor Appl Genet 126:2563–2574

Subden RE, Krizus A, Lougheed SC, Carey K (1987) Isozyme characterization of *Vitis* species and some cultivars. Am J Enol Vitic 38:76–181

Sunseri F, Lupini A, Mauceri A, De Lorenzis G, Araniti F, Brancadoro L, Dattola A, Gullo G, Zappia R, Mercati F (2018) Single nucleotide polymorphism profiles reveal an admixture genetic structure of grapevine germplasm from Calabria, Italy, uncovering its key role for the diversification of cultivars in the Mediterranean Basin. Aust J Grape Wine Res 24:345–359

Tanksley SD, Nelson JC (1996) Advanced backcross QTL analysis: a method for the simultaneous discovery and transfer of valuable QTLs from unadapted germplasm into elite breeding lines. Theor Appl Genet 92:191–203

Tautz D, Renz M (1984) Simple sequences are ubiquitous repetitive components of eukaryotic genomes. Nucleic Acids Res 12:4127–4138

Tecle IY, Menda N, Buels RM, Van Der Knaap E, Mueller LA (2010) solQTL: a tool for QTL analysis, visualization and linking to genomes at SGN database. BMC Bioinform 11:525

Teh SL, Fresnedo-ramírez J, Clark MD, Gadoury DM, Sun Q, Cadle-davidson L, Luby JJ (2017) Genetic dissection of powdery mildew resistance in interspecific half-sib grapevine families using SNP-based maps. Mol Breed 37:1–16

Tello-Ruiz MK, Stein J, Wei S, Preece J, Olson A, Naithani S, Amarasinghe V, Dharmawardhana P, Jiao Y, Mulvaney J, Kumari S, Chougule K, Elser J, Wang B, Thomason J, Bolser DM, Kerhornou A, Walts B, Fonseca NA, Huerta L, Keays M, Tang YA, Parkinson H, Fabregat A, McKay S, Weiser J, D'Eustachio P, Stein L, Petryszak R, Kersey PJ, Jaiswal P, Ware D (2016) Gramene 2016: comparative plant genomics and pathway resources. Nucleic Acids Res 44:1133–1140

Tello J, Aguirrezábal R, Hernáiz S, Larreina B, Montemayor MI, Vaquero E, Ibáñez J (2015a) Multicultivar and multivariate study of the natural variation for grapevine bunch compactness. Aust J Grape Wine Res 21:277–289

Tello J, Montemayor MI, Forneck A, Ibáñez J (2018) A new image-based tool for the high throughput phenotyping of pollen viability: evaluation of inter- and intra-cultivar diversity in grapevine. Plant Methods 14:1–17

Tello J, Torres-Pérez R, Grimplet J, Carbonell-Bejerano P, Martínez-Zapater JM, Ibáñez J (2015b) Polymorphisms and minihaplotypes in the VvNAC26 gene associate with berry size variation in grapevine. BMC Plant Biol 15:253

This P, Cuisset C, Boursiquot JM (1997) Development of stable RAPD markers for the identification of grapevine rootstocks and the analysis of genetic relationships. Am J Enol Vitic 48:492–501

This P, Jung A, Boccacci P, Borrego J, Botta R, Costantini L, Crespan M, Dangl GS, Eisenheld C, Ferreira-Monteiro F, Grando S, Ibáñez J, Lacombe T, Laucou V, Magalhães R, Meredith CP, Milani N, Peterlunger E, Regner F, Zulini L, Maul E (2004) Development of a standard set of microsatellite reference alleles for identification of grape cultivars. Theor Appl Genet 109:1448–1458

This P, Lacombe T, Cadle-Davidson M, Owens CL (2007) Wine grape (*Vitis vinifera* L.) color associates with allelic variation in the domestication gene VvmybA1. Theor Appl Genet 114:723–730

Thomas MR, Scott NS (1993) Microsatellite repeats in grapevine reveal DNA polymorphisms when analysed as sequence-tagged sites (STSs). Theor Appl Genet 86:985–990

Thongjuea S, Ruanjaichon V, Bruskiewich R, Vanavichit A (2009) RiceGeneThresher: A web-based application for mining genes underlying QTL in rice genome. Nucleic Acids Res 37:996–1000

Töpfer R, Hausmann L, Eibach R (2011) Molecular breeding. In: Adam-Blondon A-F, Martínez-Zapater JM, Kole C (eds) Genetics, genomics and breeding of grapes. CRC Press, Boca Raton, pp 160–185

Troggio M, Malacarne G, Coppola G, Segala C, Cartwright D, Massimo P, Stefanini M, Mank R, Moroldo M, Morgante M, Grando SM, Velasco R (2007) A dense single-nucleotide polymorphism based genetic linkage map of grapevine (*Vitis vinifera* L.) anchoring pinot noir bacterial artificial chromosome contigs. Genet Soc Am 176:2637–2650

Troggio M, Malacarne G, Vezzulli S, Faes G, Salmaso M, Velasco R (2008) Comparison of different methods for SNP detection in grapevine. Vitis J Grapevine Res 47:21–30

van Heerden CJ, Burger P, Vermeulen A, Prins R (2014) Detection of downy and powdery mildew resistance QTL in a 'Regent' × 'RedGlobe' population. Euphytica 200:281–295

Van Ooijen J (2006) JoinMap® 4.0: software for the calculation of genetic linkage maps in experimental population. Kyazma BV

Varoquaux F, Blanvillain R, Delseny M, Gallois P (2000) Less is better: new approaches for seedless fruit production. Trends Biotechnol 18:233–242

Varshney RK, Nayak SN, May GD, Jackson SA (2009) Next-generation sequencing technologies and their implications for crop genetics and breeding. Trends Biotechnol 27:522–530

Velasco R, Zharkikh A, Troggio M, Cartwright DA, Cestaro A, Pruss D, Pindo M, FitzGerald LM, Vezzulli S, Reid J, Malacarne G, Iliev D, Coppola G, Wardell B, Micheletti D, Macalma T, Facci M, Mitchell JT, Perazzolli M, Eldredge G, Gatto P, Oyzerski R, Moretto M, Gutin N, Stefanini M, Chen Y, Segala C, Davenport C, Dematté L, Mraz A, Battilana J, Stormo K, Costa F, Tao Q, Si-Ammour A, Harkins T, Lackey A, Perbost C, Taillon B, Stella A, Solovyev V, Fawcett JA, Sterck L, Vandepoele K, Grando SM, Toppo S, Moser C, Lanchbury J, Bogden R, Skolnick M, Sgaremella V, Bhatnagar SK, Fontana P, Gutin A, Van de Peer Y, Salamini F, Viola R (2007) A high quality draft consensus sequence of the genome of a heterozygous grapevine variety. PLoS ONE 2:e1326

Venuti S, Copetti D, Foria S, Falginella L, Hoffmann S, Bellin D, Cindrić P, Kozma P, Scalabrin S, Morgante M, Testolin R, Di Gaspero G (2013) Historical Introgression of the downy mildew resistance gene Rpv12 from the Asian species Vitis amurensis into grapevine varieties. PLoS ONE 8:e61228

Veyrieras J-B, Goffinet B, Charcosset A (2007) MetaQTL: a package of new computational methods for the meta-analysis of QTL mapping experiments. BMC Bioinform 8:49

Vezzulli S, Malacarne G, Masuero D, Vecchione A, Dolzani C, Goremykin V, Haile Mehari Z, Banchi E, Velasco R, Stefanini M, Vrhovsek U, Zulini L, Franceschi P, Moser C (2019) The *Rpv3-3* haplotype and stilbenoid induction mediate downy mildew resistance in a grapevine interspecific population. Front Plant Sci 10:234

Vezzulli S, Micheletti D, Riaz S, Pindo M, Viola R, This P, Walker MA, Troggio M, Velasco R (2008a) A SNP transferability survey within the genus *Vitis*. BMC Plant Biol 8:128

Vezzulli S, Troggio M, Coppola G, Jermakow A, Cartwright D, Zharkikh A, Stefanini M, Grando MS, Viola R, Adam-Blondon AF, Thomas M, This P, Velasco R (2008b) A reference integrated map for cultivated grapevine (*Vitis vinifera* L.) from three crosses, based on 283 SSR and 501 SNP-based markers. Theor Appl Genet 117:499–511

Viana AP, Riaz S, Walker MA (2013) Genetic dissection of agronomic traits within a segregating population of breeding table grapes. Genet Mol Res 12:951–964

Vidal JR, Moreno S, Masa A, Ortiz JM (1998) Study ofthe genetic homogeneity of Albarino (*Vitis vinifera* L.) growing in Galicia (Spain) using isozyme and RAPD markers. Vitis 37:145–146

Vos P, Hogers R, Bleeker M, Reijans M, Van De Lee T, Hornes M, Frijters A, Pot J, Peleman J, Kuiper M, Zabeau M (1995) AFLP: a new technique for DNA fingerprinting. Nucleic Acids Res 23:4407–4414

Walker AR, Lee E, Robinson SP (2006) Two new grape cultivars, bud sports of Cabernet Sauvignon bearing pale-coloured berries, are the result of deletion of two regulatory genes of the berry colour locus. Plant Mol Biol 62:623–635

Walker AR, Lee E, Bogs J, McDavid DAJ, Thomas MR, Robinson SP (2007) White grapes arose through the mutation of two similar and adjacent regulatory genes. Plant J 49:772–785

Ware DH, Jaiswal P, Ni J, Yap IV, Pan X, Clark KY, Teytelman L, Schmidt SC, Zhao W, Chang K, Cartinhour S, Stein LD, Mccouch SR (2002) Update, a tool for grass genomics. Plant Physiol 130:1606–1613

Welter LJ, Göktürk-Baydar N, Akkurt M, Maul E, Eibach R, Töpfer R, Zyprian EM (2007) Genetic mapping and localization of quantitative trait loci affecting fungal disease resistance and leaf morphology in grapevine (*Vitis vinifera* L). Mol Breed 20:359–374

Williams JG, Kubelik AR, Livak KJ, Rafalski JA, Tingey SV (1990) DNA polymorphisms amplified by arbitrary primers are useful as genetic markers. Nucleic Acids Res 18:6531–6535

Winter P, Kahl G (1995) Molecular marker technologies for plant improvement. World J Microbiol Biotechnol 11:438–448

Wu Y, Bhat PR, Close TJ, Lonardi S (2008) Efficient and accurate construction of genetic linkage maps from the minimum spanning tree of a graph. PLoS Genet 4: e1000212

Xu H, Wilson DJ, Arulsekar S, Bakalinsky AT (1995) Sequence-specific polymerase chain-reaction markers derived from randomly amplified polymorphic DNA markers for fingerprinting grape (*Vitis*) rootstocks. J Am Soc Hortic Sci 120:714–720

Xu K, Riaz S, Roncoroni NC, Jin Y, Hu R, Zhou R, Walker MA (2008) Genetic and QTL analysis of resistance to *Xiphinema index* in a grapevine cross. Theor Appl Genet 116:305–311

Xu Y (2010) Molecular breeding tools: markers and maps. In: Molecular plant breeding. CABI, Wallingford, pp 21–58

Yang S, Fresnedo-Ramírez J, Sun Q, Manns DC, Sacks GL, Mansfield AK, Luby JJ, Londo JP, Reisch BI, Cadle-Davidson LE, Fennell AY (2016a) Next generation mapping of enological traits in an F2 interspecific grapevine hybrid family. PLoS ONE 11:1–19

Yang S, Fresnedo-Ramírez J, Wang M, Cote L, Schweitzer P, Barba P, Takacs EM, Clark M, Luby J, Manns DC, Sacks G, Mansfield AK, Londo J, Fennell A, Gadoury D, Reisch B, Cadle-Davidson L, Sun Q (2016b) A next-generation marker genotyping platform (AmpSeq) in heterozygous crops: a case study for marker-assisted selection in grapevine. Hortic Res 3:16002

Yu J, Pressoir G, Briggs WH, Vroh Bi I, Yamasaki M, Doebley JF, McMullen MD, Gaut BS, Nielsen DM, Holland JB, Kresovich S, Buckler ES (2006) A unified mixed-model method for association mapping that accounts for multiple levels of relatedness. Nat Genet 38:203–208

Zabeau M, Vos P (1993) Selective restriction fragment amplification: a general method for DNA fingerprinting. Publication 0 534 858 Al. European Patent Office, Munich

Zamir D (2013) Where have all the crop phenotypes gone? PLoS Biol 11:1–4

Zeng H, Luo L, Zhang W, Zhou J, Li Z, Liu H, Zhu T, Feng X, Zhong Y (2007) PlantQTL-GE: A database system for identifying candidate genes in rice and Arabidopsis by gene expression and QTL information. Nucleic Acids Res 35:879–882

Zhang H, Fan X, Zhang Y, Jiang J, Liu C (2017) Identification of favorable SNP alleles and candidate genes for seedlessness in *Vitis vinifera* L. using genome-wide association mapping. Euphytica 213:1–13

Zhang J, Hausmann L, Eibach R, Welter LJ, Töpfer R, Zyprian EM (2009) A framework map from grapevine V3125 (*Vitis vinifera* "Schiava grossa" × 'Riesling') × rootstock cultivar "Börner" (*Vitis riparia* × *Vitis cinerea*) to localize genetic determinants of phylloxera root resistance. Theor Appl Genet 119:1039–1051

Zhang L, Marguerit E, Rossdeutsch L, Ollat N, Gambetta GA (2016) The influence of grapevine rootstocks on scion growth and drought resistance. Theor Exp Plant Physiol 28:143–157

Zhang Z, Ersoz E, Lai C-Q, Todhunter RJ, Tiwari HK, Gore MA, Bradbury PJ, Yu J, Arnett DK, Ordovas JM, Buckler ES (2010) Mixed linear model approach adapted for genome-wide association studies. Nat Genet 42:355–360

Zhao YH, Guo YS, Lin H, Liu ZD, Ma HF, Guo XW, Li K, Yang XX, Niu ZZ, Shi GG (2015) Quantitative trait locus analysis of grape weight and soluble solid content. Genet Mol Res 14:9872–9881

Zhao YH, Su K, Guo YH, Ma HF, Guo XW (2016) Molecular genetic map construction and QTL analysis of fruit maturation period in grapevine. Genet Mol Res 15:1–10

Zyprian E, Ochßner I, Schwander F, Šimon S, Hausmann L, Bonow-Rex M, Moreno-Sanz P, Grando MS, Wiedemann-Merdinoglu S, Merdinoglu D, Eibach R, Töpfer R (2016) Quantitative trait loci affecting pathogen resistance and ripening of grapevines. Mol Genet Genom 291:1573–1594

Zyprian E, Šimon S, Schwander F, Töpfer R (2015) Efficiency of single nucleotide polymorphisms to improve a genetic map of complex pedigree grapevines. Vitis J Grapevine Res 54:29–32

Status and Prospects of Systems Biology in Grapevine Research

8

José Tomás Matus, Valentino Ruggieri, Francisco José Romero, Marco Moretto and Darren C. J. Wong

Abstract

The cultivated grapevine, *Vitis vinifera* L., has gathered a vast amount of omics data throughout the last two decades, driving the imperative use of computational resources for its analysis and integration. Molecular systems biology arises from this need allowing to model and predict the emergence of phenotypes or responses in biological systems. Beyond single omics networks, integrative approaches associate the molecular components of an organism and combine them into higher order networks to model dynamic behaviors. Application of network-based methods in multi-omics data is providing additional resources to address important questions regarding grapevine fruit quality and composition. Here, we review the recent history of systems biology in this species. We highlight the most relevant aspects of the discipline and describe important integrative studies that have helped in the global understanding of how this species responds to the environment and how it triggers the fruit ripening developmental program. We also highlight the latest resources that are available for the grapevine community to exploit and take advantage of all the omics data that is being generated.

This work was supported by Grant PGC2018-099449-A-I00 and by the Ramón y Cajal program grant RYC-2017-23645, both awarded to J.T. M. from the Ministerio de Ciencia, Innovación y Universidades (MCIU, Spain), Agencia Estatal de Investigación (AEI, Spain), and Fondo Europeo de Desarrollo Regional (FEDER, European Union).

J. T. Matus (✉)
Institute for Integrative Systems Biology, I2SysBio (Universitat de València - CSIC), 46908 Paterna, Valencia, Spain
e-mail: tomas.matus@uv.es

V. Ruggieri
Center for Research in Agricultural Genomics (CRAG), CSIC-IRTA-UAB-UB, 08193 Barcelona, Spain
e-mail: valentino.ruggieri@irta.cat

V. Ruggieri
IRTA (Institut de Recerca i Tecnologia Agroalimentàries), Barcelona, Spain

F. J. Romero
Plant Development Unit, Instituto de Bioquímica Vegetal y Fotosíntesis, IBVF (Universidad de Sevilla - CSIC), Seville, Spain
e-mail: fran@us.es

F. J. Romero
Departamento de Ciencias de la Computación e Inteligencia Artificial, Universidad de Sevilla, Seville, Spain

M. Moretto
Unit of Computational Biology, Research and Innovation Centre, Fondazione Edmund Mach, 38010 San Michele all'Adige, Italy
e-mail: marco.moretto@fmach.it

D. C. J. Wong
Ecology and Evolution, Research School of Biology, Australian National University, Acton, ACT 2601, Australia
e-mail: wongdcj@gmail.com

D. Cantu and M. A. Walker (eds.), *The Grape Genome*, Compendium of Plant Genomes, https://doi.org/10.1007/978-3-030-18601-2_8

8.1 Introduction

Genes and their products perform complex cellular tasks that are essential for all living organisms. At the molecular level, they are organized as modules forming part of large networks. Within these high-order associations, genes/proteins that are functionally related interact, regulate each other, or form part of a metabolic pathway. The functional characterization of these molecules through forward and reverse genetic analyses has allowed the dissection of their networks and their involvement in diverse cellular processes. In the last decade, however, a massively promoted approach to asset the whole comprehension of a network from a global perspective has been the integration of several types of omics data.

The rise of next-generation sequencing (NGS) technologies has led to an expansion in the amount of genomic/transcriptomic data required to be stored and processed. In addition, technologies covering proteomics and other types of omics are rapidly increasing the amount of data being produced. Scientists are now racing to develop efficient data analysis algorithms, user-friendly tools, and software applications, and establishing extensive hardware infrastructure for answering different questions of modern life science. It is hypothesized that the larger the amount of omics data being generated for a species the easier for its integration, engendering more robust and reliable analyses.

The grapevine (*Vitis vinifera* L.) has become a "model" system for studying non-climacteric fleshy fruits. The increasing amount of genomics data being continuously generated within the grapevine community, after the first grape genome of the inbred line PN40024 was sequenced and released in 2007, has certainly helped in this nomination. The PN40024 genome, currently on its second assembly (12X.v2) and its third annotation (VCost.v3) comprises to date 33,568 genes (Canaguier et al. 2017). With the purpose of providing biological meaning to this remarkable amount of data, several initiatives have been introduced for describing genes within their biological context (Grimplet et al. 2009a), including not only in vivo functional characterizations but also in silico analyses such as co-expression networks and other integrative approaches (reviewed by Wong and Matus 2017).

With the commitment of consenting the efficient exploitation of *Vitis* biological resources and understanding the genetic and molecular basis of all processes in this species, the International Grapevine Genome Program (IGGP; www.vitaceae.org) is currently developing the GrapeIS system. This is an integrated set of interfaces supporting advanced data modeling, rich semantic integration and the next generation of data mining tools linking genotypes to phenotypes (Adam-Blondon et al. 2016). Within the same framework, the recently launched INTEGRAPE consortium (COST Action-mediated) aims to integrate data at different levels to maximize the power of omics and establish a manageable and open data platform. The initiatives mentioned here share the use of FAIR principles that ensure data are Findable, Accessible, Interoperable, and Reusable (Wilkinson et al. 2016). The establishment of solid integrative data platforms is compulsory to make available interoperable grapevine datasets and tools. The application of systems biology methods has arisen to fulfil this purpose. Here, we provide a brief review of the fundamentals of systems biology and the history of applying integrative omics methods in grapevine research. The best-known programming scripts/packages and web-based resources for the analysis and interpretation of omics-generated data will also be described. Before examining the state of the art, a list of terms commonly used in the field of Systems Biology is presented in Box 8.1.

Box 8.1 Glossary of terms

ATAC-seq: The technology that applies high-throughput sequencing to assay for transposase-accessible regions in the genome effectively analyzing chromatin accessibility.

Big Data/Data Science: An emerging discipline that combines computer science and statistics to analyze massive amounts of data with the goal of answering specific and practical questions of a phenomenon under study.

ChIP-seq: The technology that couples chromatin immunoprecipitation (ChIP) with high-throughput sequencing to analyze protein-DNA interactions.

Cistromics: The omics technology that analyses the cistrome or the complete set of binding sites of a given transcription factor to the DNA under specific conditions.

Community network: Network built from as few as three input networks, diminishing the limitations of each individual method. Edges supported by a higher number of methods are more reliable.

DAP-seq: The technology that couples in vitro expression of affinity-purified transcription factors with high-throughput sequencing of a genomic DNA library in order to analyze protein-DNA interactions.

Epicistromics: The omics technology that studies the epicistrome or the complete set of genomic locations occupied by nucleosomes carrying histones with distinct posttranslational modifications under specific conditions.

Gene co-expression network (GCN): A undirected network typically built from transcriptomic data such as RNA-seq or microarray data where nodes represent genes and edges are drawn between two nodes when the corresponding genes are significantly co-expressed under the analyzed conditions.

High-performance computing: The use of supercomputers and parallel computational architectures to massively process information in order to solve complex problems.

High-throughput sequencing (HTS): Techniques that sequence massive amounts of DNA in an automatic and parallel manner. High-throughput in omics is referenced to the use of automation equipment to address biological questions that are otherwise unattainable using conventional methods.

MNase-seq: The technique that applies high-throughput sequencing to the DNA protected by nucleosomes during micrococcal nuclease digestion to effectively identify nucleosome positioning.

Molecular systems biology: An emerging discipline at the intersection between molecular biology, mathematics/statistics and computer science that integrates massive amounts of omics data with the final goal of generating predictive models of biological systems focusing on biomolecular interactions rather than on isolated molecular components.

Network: A model of a system where nodes represent the system components and edges between nodes indicate an interaction between the corresponding components. Networks can be directed or undirected depending on whether or not there exists a directionality in the interactions between the system components. Networks can be weighted when numerical values are associated with edges in order to capture specific features of the corresponding interactions.

Next-Generation Sequencing (NGS): A term to describe a collection of genetic sequencing techniques that improve upon the original Sanger sequencing process. This technique utilizes DNA sequencing technologies that are capable of processing multiple sequences in parallel. Also known as massively parallel sequencing, deep sequencing or high-throughput sequencing (HTS).

Omics technologies: Techniques that detect and quantify massive amounts of molecules of a specific type from a sample.

Regulon: Group of non-contiguous genes that are regulated as a unit, generally controlled by the same regulatory gene that expresses a protein acting as a repressor or activator.

RNA-seq: The application of high-throughput sequencing to the cDNA corresponding to the entire set of transcripts in a sample. This technology allows researchers to detect and estimate the abundance of transcripts (coding and non-coding) in a sample, also including alternative splicing variants.

Transcriptional network: A directed network typically built from cistromic data corresponding to multiple transcription factors where nodes represent genes and an edge is drawn from gene_i to gene_j when gene_i codifies for a transcription factor that directly binds to the promoter of gene_j. Weights can be associated with edges to represent if the binding of the transcription factor has an activating, repressing or neutral effect over the transcription of a target gene.

Transcriptomics: The omics technology that focuses on the analysis of the transcriptome or the complete set of transcripts expressed from the genome under specific conditions.

8.2 From Elements to Relations: Overview of Plant Systems Biology

Systems biology is a computational, mathematical, and biology-based interdisciplinary field that focuses on complex interactions within biological systems. Its foundation outcomes from amending the general (Von Bertalanffy 1968) and living (Miller 1978) system theories and aims to elucidate biological phenomena applying a systemic view of interactions between molecular entities instead of describing their individual composition or function (Mesarovic 1968). By addressing the cell as a network of genes, their products, and their interactions, the latter defined as network motifs or patterns, it is feasible to study the structural design principles of living organisms. Distant networks that perform similar tasks all share similar types of recurring patterns of interconnections, thus motifs define universal classes of networks (Milo et al. 2002). From this and other studies, it was suggested that structures of different networks were governed by the same principles. This new paradigm is embodied within the Oltvai and Barabási life's complexity pyramid, here updated and revisited to include systems biology advancements (Fig. 8.1). In the model, cell components arrange themselves in persistent patterns and these in turn form modules with discrete cellular functions. Finally, these modules are hierarchically organized, defining the cell's large-scale functional organization.

Historically, reductionist studies in plants have been aimed for identifying the individual components associated with the occurrence of certain phenotypes. Although this approach has been massively adopted in the last 50 years, successfully producing extensive repertoires of plant molecular components, it begun to lose its effectiveness at the beginning of the current century when it became apparent that the majority of phenotypes were produced by complex orchestrations involving myriads of molecular components, many of which were redundant among them. This scenario became more apparent with the development of the so-called omics technologies that provide an accurate molecular snapshot of the biological processes under study by detecting and quantifying the repertoire of molecules that are present (Yuan et al. 2008). Hence, research in molecular biology is gradually shifting towards a holistic perspective, integrating the individual "omics" datasets, to gain biologically meaningful aspects of plant systems (Sheth and Thaker 2014).

The recent development of high-throughput DNA sequencing, genomics and transcriptomics have pushed these methodologies to become so far, the best-established mature and reliable techniques to characterize molecular systems (Bolger et al. 2018). Specifically, RNA-seq, the high-throughput sequencing of the cDNA corresponding to the entire set of transcripts in a sample, is applied to identify and estimate transcript abundance including different isoforms

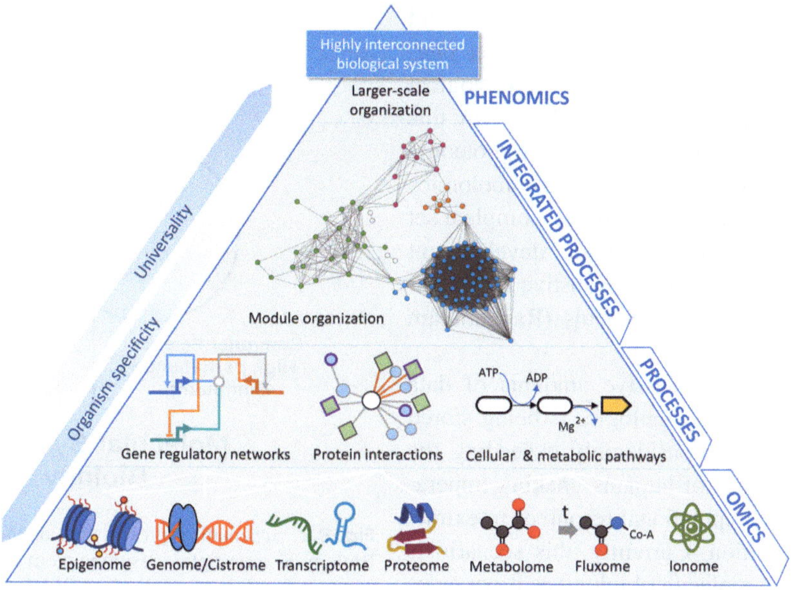

Fig. 8.1 The Oltvai and Barabási's pyramid of life reviewed by systems biology approaches. The complexity of a biological system can be represented by several layers of functional organization. Starting from the cell's building blocks; the life biomolecules, these are responsible for the genetic information to be stored, processed and finally executed in several developmental programs or in response to the environment. Genes and their epigenetic marks, transcripts, proteins and their modifications, metabolites and their fluxes and even ions can be collectively characterized and quantified through omics. The huge amount of data acquired from these technologies can only be handled with intensive bioinformatics. At the second level, biomolecules form gene regulatory and protein-interacting motifs and subcellular signaling/metabolic pathways, all of them with the inherent capacity of impacting each other. As these biological processes are tightly connected (e.g., a set of genes, proteins and metabolites being activated in response to a pathogen) they are organized in functional modules. Complex biological processes can be studied from a "multi-omics" perspective thanks to the recent improvements in genome-wide techniques and systems biology methods. Modules can be studied by integrative systems biology tools but can be further organized in higher hierarchical multidimensional structures. Larger-scale modules are also dynamic in time and translate into phenotypes. In recent efforts, modeling algorithms have been applied to largely annotate phenotypes (i.e., "phenomics"). Computational biology has supported an adequate data management, efficient data analysis, and user-friendly software applications to study biological systems at each of these levels. Although the individual components are unique to a given organism, the topologic properties of networks are surprisingly similar (Adapted from Oltvai and Barabási 2002)

produced by alternative splicing as well as to analyze differential gene expression between specific conditions (Martin et al. 2013; see Chap. 13). The main molecular mechanisms controlling gene expression, namely the interactions between transcription factors and DNA (recently named "the cistrome"), and the different posttranslational modifications of histones associated with the DNA (epicistrome) are routinely characterized using techniques such as ChIP-seq; the combination of chromatin immunoprecipitation with the high-throughput sequencing of the purified DNA (Chen et al. 2017). DAP-seq is a technique based on high-throughput sequencing that studies the cistrome based on the in vitro expression of affinity-purified transcription factors (Bartlett et al. 2017). Finally, MNase-seq, DNase-seq, and ATAC-seq are techniques used to study nucleosome positioning and chromatin accessibility that have been shown to highly influence gene expression (Pajoro et al. 2014; Sullivan et al. 2015; Pass et al. 2017; Bajic et al. 2018).

Despite the clear methodological and analytical advantages of performing genomics studies compared to other omics, it has been demonstrated that the sole use of genomics and transcriptomics is not sufficient to predict phenotypes from the molecular state of biological processes (Papatheodorou et al. 2015). In this respect, proteomics (the analysis of the proteome or the entire set of proteins), and metabolomics (the study of the metabolome or the complete set of metabolites) are currently under development aiming at providing a more exhaustive molecular description of biological systems (Ramalingam et al. 2015).

At this point, the massive amounts of data generated by omics technologies is being stored in public databases considerably exceeding the analytical capacities of humans, making imperative the use of computational resources to extract relevant information. Currently, this scenario is not exclusive to molecular biology as it pervades science in a more general context by inducing the emergence of the so call Big Data or Data Science. This is a discipline that combines high-performance computing, such as the use of computational clusters, with sophisticated statistical methods, in order to answer specific questions of phenomena under analysis (Carmichael and Marron 2018). In molecular biology, this has promoted the development of "Molecular Systems Biology". This emerging discipline lays at the intersection between molecular biology, computer science, and mathematics/statistics (Fig. 8.2). The main methodology in molecular systems biology pertains to the generation of omics data and their integration with already existing data freely available in public databases. This massive amount of data is integrated and analyzed typically using multivariate statistical methods implemented with high-performance computing. Specifically, molecular systems biology pursuits the development of computational/mathematical models of the interactions among the molecular components of the systems responsible for an observed phenotype rather than focusing on the functioning of the isolated individual components. Here, the ultimate goal relates to the generation of tools that allow to model and predict the emergence of specific phenotypes or responses in biological systems (Sheth and Thaker 2014). Commonly,

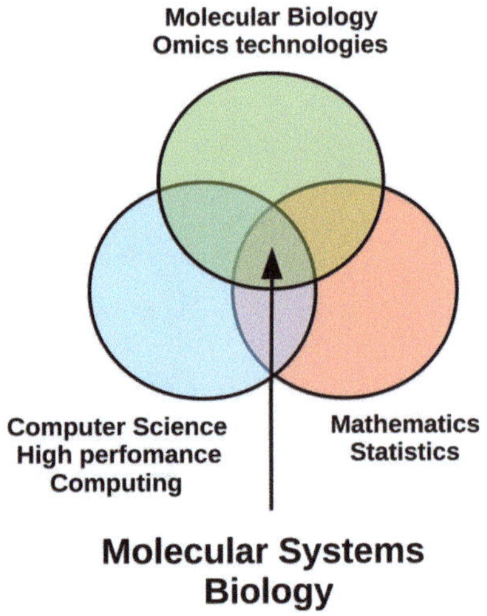

Fig. 8.2 Schematic representation of Molecular Systems Biology as a discipline resulting from the overlapping of computational, mathematical, and biological explorations

systems of differential equations are used as the modeling structure to achieve this goal. Nonetheless, network science is emerging as a central paradigm in molecular systems biology as an effective modeling framework (Li et al. 2015).

In the context of network science, a network is a graph whose nodes represent the molecular entities of the system and a directed or undirected edge is drawn between two nodes to specify the interaction between the corresponding molecular components. A numerical value termed weight can be incorporated in the edges to capture the strength of the represented interaction. Topological studies of a network, such as the analysis of free-scale properties, can identify relevant nodes called hubs that are highly connected in the network and play key roles in network robustness and dynamics. Other topological parameters such as "node transitivity", "betweenness" and "eccentricity" are especially suitable to identify relevant molecular components of the biological system under analysis. Clustering techniques and community analysis are used to unravel the underlying structure of networks and are applicable in molecular systems biology to identify molecular modules that function with a certain level of separation from the rest of the system (Aoki et al. 2007). Finally, network motif analysis or the identification of non-random subgraphs can shed light on the building blocks that occur recurrently in biological systems (Defoort et al. 2018).

Two types of gene networks are intensively used in molecular systems biology; gene co-expression networks and transcriptional networks. Gene co-expression networks are normally constructed based on a compendium of microarray and only recently, RNA-seq data sets. These are undirected networks where nodes represent genes and undirected edges are drawn between nodes to represent co-expression relationships between the corresponding genes. Transcriptional networks are constructed from ChIP-seq data corresponding to sets of different transcription factors binding to the genome. These are directed networks where nodes represent genes and a directed edge is drawn from gene_i to gene_j, where gene_i codifies for a transcription factor that binds to the promoter of gene_j. Transcriptional networks can be further refined by adding RNA-seq data corresponding to mutants or overexpressors of the transcription factors previously analyzed using ChIP-seq. According to this, weights can be associated with edges to represent an activating, repressing or neutral effect of the binding of the transcription factor to the promoter of the target gene.

8.3 A Decade Conducting Grapevine Omics. What's Yet to Come

Genomics resources for *Vitis* species have increased promptly within the last fifteen years, beginning with the sequencing of expressed sequence tags (ESTs) (Da Silva et al. 2005; Moser et al. 2005). These resources have permitted to quantitatively assess the grape transcriptome by aiding the development of cDNA and oligonucleotide microarrays (Terrier et al. 2005; Waters et al. 2005). Quantitative data acquisition through microarray analysis permitted large-scale mRNA profiling studies of gene expression to unravel the most important events of berry development and ripening. However, it was not but after the concomitant release of the *V. vinifera* PN40024 genome sequence (Jaillon et al. 2007; Velasco et al. 2007) that a burst of new transcriptomic technologies emerged for this species. In the Affymetrix Grape GeneChip Genome Array, approximately one-third of the expected genes are represented. This platform was largely used for tissue-specific mRNA expression profiling in grape berry tissues (Grimplet et al. 2007; Deluc et al. 2007) and responses to abiotic stresses (Tattersall et al. 2007; Cramer et al. 2007) and compatible viral diseases (Vega et al. 2011), where all the produced data were collected and unified in the PLEX database (PLEXdb, http://www.plexdb.org; Wise et al. 2007). The microarray Nimblegen platform was developed soon after (Fasoli et al. 2012; http://ddlab.sci.univr.it/FunctionalGenomics/), with an array representing more than 98% of the genes predicted in the 12xV1 grapevine genome annotation (090918 Vitus vinifera exp HX12 chip, with approximately 29,549 denoted genes). To date, this platform has generated the largest amount of transcriptomic data for this species (1605 experiments until July 2018). All developed arrays in *Vitis* can be found in ArrayExpress EMBL-EBI; https://www.ebi.ac.uk/arrayexpress/).

Although in situ oligonucleotide arrays are still widely used for gene expression profiling in grapevine, a rapid development of new nucleic acid technologies have been largely adopted for genomic, transcriptomic and metagenomic studies in grapevine in the last years (Fig. 8.3a). A variety of NGS technologies, including the 454 (Roche) (Margulies et al. 2005), the Genome Analyzer/Hiseq (Illumina Solexa) (Bennett et al. 2005) and the SOLiD (Life Technologies), as well as newer platforms such as Helioscope (Helicos) (Milos 2008), PacBio RS and Sequel (Pacific Bioscience) (Eid et al. 2009), Oxford Nanopore Technologies for single molecular sequencing and Ion Torrent (Life Technologies), based on a semiconductor chip (Rothberg et al. 2011), are available. Thanks to high-throughput and cost-efficient capabilities of these technologies, an unprecedented amount of data has been generated and a huge amount of genomic and

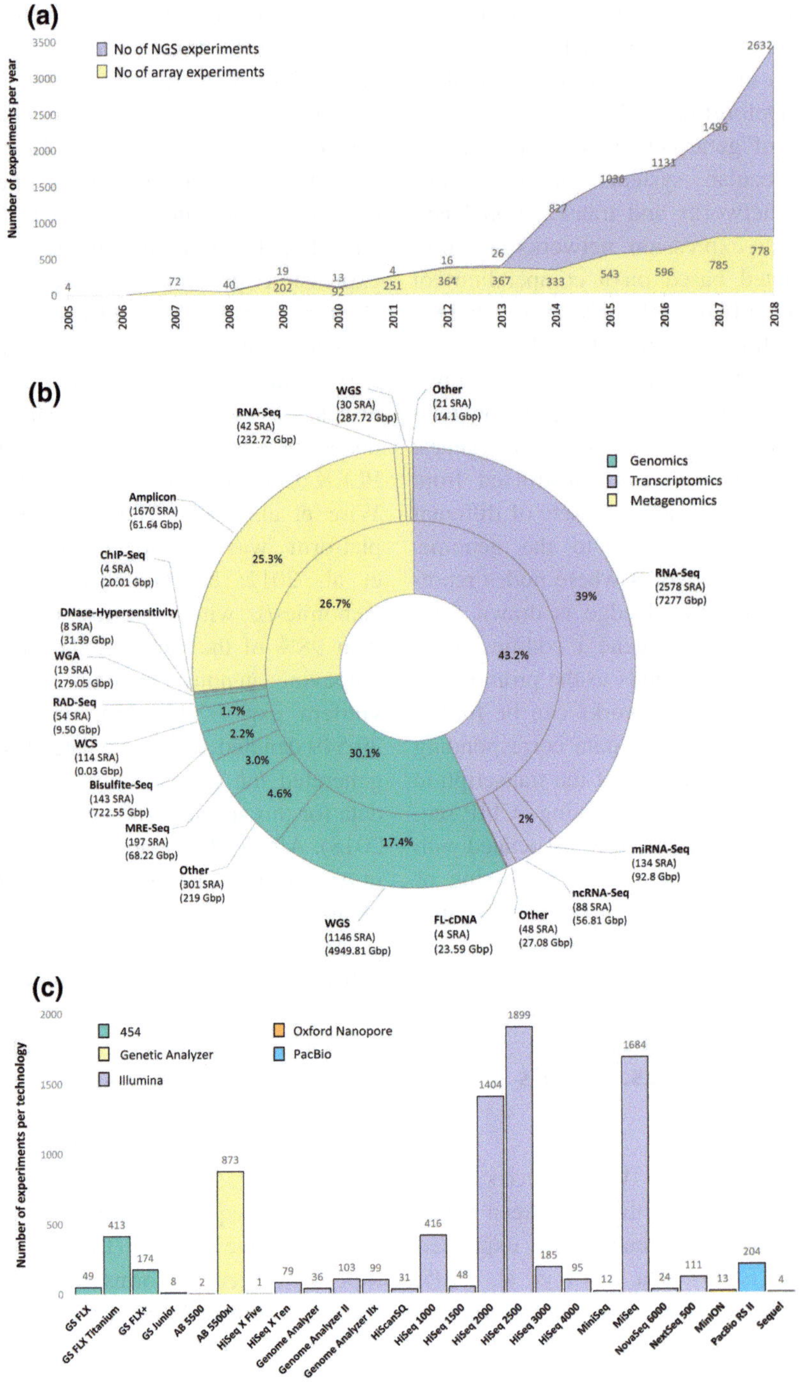
(a)
Number of experiments per year
No of NGS experiments
No of array experiments
3500
3000
2500
2000
1500
1000
500
0
4
72
40
19
13
4
16
26
827
1036
1131
1496
2632
202
92
251
364
367
333
543
596
785
778
2005
2006
2007
2008
2009
2010
2011
2012
2013
2014
2015
2016
2017
2018
(b)
Genomics
Transcriptomics
Metagenomics
WGS (30 SRA) (287.72 Gbp)
Other (21 SRA) (14.1 Gbp)
RNA-Seq (42 SRA) (232.72 Gbp)
Amplicon (1670 SRA) (61.64 Gbp)
ChIP-Seq (4 SRA) (20.01 Gbp)
DNase-Hypersensitivity (8 SRA) (31.39 Gbp)
WGA (19 SRA) (279.05 Gbp)
RAD-Seq (54 SRA) (9.50 Gbp)
WCS (114 SRA) (0.03 Gbp)
Bisulfite-Seq (143 SRA) (722.55 Gbp)
MRE-Seq (197 SRA) (68.22 Gbp)
Other (301 SRA) (219 Gbp)
WGS (1146 SRA) (4949.81 Gbp)
FL-cDNA (4 SRA) (23.59 Gbp)
Other (48 SRA) (27.08 Gbp)
ncRNA-Seq (88 SRA) (56.81 Gbp)
miRNA-Seq (134 SRA) (92.8 Gbp)
RNA-Seq (2578 SRA) (7277 Gbp)
25.3%
26.7%
39%
43.2%
1.7%
2.2%
3.0%
4.6%
30.1%
17.4%
2%
(c)
Number of experiments per technology
454
Genetic Analyzer
Illumina
Oxford Nanopore
PacBio
2000
1500
1000
500
0
49
413
174
8
2
873
1
79
36
103
99
31
416
48
1404
1899
185
95
12
1684
24
111
13
204
4
GS FLX
GS FLX Titanium
GS FLX+
GS Junior
AB 5500
AB 5500xl
HiSeq X Five
HiSeq X Ten
Genome Analyzer
Genome Analyzer II
Genome Analyzer IIx
HiScanSQ
HiSeq 1000
HiSeq 1500
HiSeq 2000
HiSeq 2500
HiSeq 3000
HiSeq 4000
MiniSeq
MiSeq
NovaSeq 6000
NextSeq 500
MinION
PacBio RS II
Sequel

◀ **Fig. 8.3** Next-generation sequencing and array data available for grapevine. Next-generation sequencing and oligonucleotide array have represented two relevant genome-scale methodologies for grapevine studies. The data presented were retrieved from the Sequence Read Archive (SRA) (https://www.ncbi.nlm.nih.gov/sra) and Gene Expression Omnibus (GEO) NCBI repositories (https://www.ncbi.nlm.nih.gov/gds/) as of December 2018, by using a keyword search "Vitis" or "Grapevine". **a** Timeline of grapevine experiments performed since 2005 according to the methodology used (in situ oligonucleotide array or NGS). **b** Number and distribution of grapevine experiments from high-throughput sequencing technologies. The inner-circle represents the distribution according to the library layer (Genomics, Transcriptomics, Metagenomics) while the outer circle is according to the library strategy used (e.g., RNA-seq, Chip-seq, etc.). For each outer section, the number of experiments (SRA) and the Giga base pair of data (Gbp) were also reported. **c** Distribution of the NGS platforms used, including Roche 454 GS System, Illumina Genome Analyzer, Applied Biosystems SOLiD System, Helicos Heliscope, Pacific Biosciences SMRT

transcriptomic data has accumulated exponentially in *Vitis* species (Fig. 8.3b, c).

The combination of high-throughput sequencing technologies and the grapevine PN40024 genome (Jaillon et al. 2007) has facilitated comprehensive sequence analysis in diverse grapevine germplasms (Table 8.1). Cultivars with different agronomic and oenological characteristics have been re-sequenced to identify genetic differences underlying the distinct phenotypes (Da Silva et al. 2014; Di Genova et al. 2014; Cardone et al. 2016; Chin et al. 2016, Minio et al. 2017, 2019; Roach et al. 2018; see Chap. 05) and comprehensive inventories of sequence variations were generated (Mercenaro et al. 2017; Zhou et al. 2017; Liang et al. 2019). On the other hand, transcriptome sequencing using NGS technologies has been widely used to detect gene expression in grapevines (see Chap. 13), including fruit (e.g. Zenoni et al. 2010), leaves (e.g. Liu et al. 2012), flowers (e.g. Domingos et al. 2016), in response to different biotic and abiotic stresses (e.g. Cheng et al. 2015; Blanco-Ulate et al. 2015; Amrine et al. 2015; Tillett et al. 2011) or to describe the expression of specific transcription factors (e.g. Sweetman et al. 2012). Other grape researchers have used high-throughput expression to examine the phenotypic plasticity of cv. "Corvina" berries at various developmental stages (Dal Santo et al. 2013). Despite its primary objective is to characterize expression profile, RNAseq technologies have been also used to identify differential splicing activity and single nucleotide polymorphisms (Zenoni et al. 2010; Vitulo et al. 2014) as well as identifying and profiling long non-coding RNAs (Vitulo et al. 2014; Harris et al. 2017).

Since grapevine naturally hosts a reservoir of microorganisms that interact with the plant and affect both the qualitative and quantitative scale of wine production (Martins et al. 2013; Zarraonaindia et al. 2015), grape metagenomics studies also are assuming an increasing resonance in the grape scientific community. Recently, high-throughput technologies have been used to characterize bacterial communities of different grapevine plant portions, such as leaves and berries (Leveau and Tech 2010), to assess the microbial communities of soils (Zarraonaindia et al. 2015; Burns et al. 2015, 2016) and to survey the associations involving grapevine microbiota, fermentation and wine chemical composition (Bokulich et al. 2014, 2016).

Despite the study of epigenetic marks (e.g., histone posttranslational modifications and DNA methylation) are known to influence gene expression and largely affect the phenotype of plants, there are still scarce epigenomic data and related resources available for grapevine. Nonetheless, Fortes and Gallusci (2017) recently proposed this species as an essential perennial woody plant model for such studies due to the impact of epigenetic modifications on agricultural traits, and also because epigenetic marks may serve as an interface between the environment and the genome (reviewed by Fabres et al. 2017; see Chap. 9). Very recently, Xie et al. (2017) used methylation-sensitive amplified polymorphisms (MSAPs) to find global patterns of DNA methylation and explored the genetic and epigenetic diversity of a single cultivar across 22 vineyards located in six different wine sub-regions.

Table 8.1 Number of SRA experiments (No. of SRA) and Gbp of data produced (Gbp of data) for grapevine cultivars according to the type of the library source (genomic or transcriptomic)

Cultivar	Genomic		Transcriptomic	
	No of SRA	Gbp of data	No of SRA	Gbp of data
Cabernet Sauvignon	6	166.59	393	805.27
Barossa Shiraz	197	68.22		
Pinot noir	15	44.16	115	341.59
Chardonnay	95	2544.81	34	48.67
Merlot	2	0.00	74	277.67
Carmenere	63	147.27		
Muscat table			54	508.32
Pinot Meunier	4	31.89	48	137.02
Thompson Seedless	3	10.63	49	174.05
Sangiovese	3	0.01	47	61.49
Sauvignon blanc	2	0.01	35	199.50
Tempranillo			36	143.73
Riesling	2	34.24	31	51.64
Cabernet Franc	2	0.01	28	85.89
Tocai friulano			30	35.50
Barbera	1	0.01	19	47.36
Kyoho			20	176.67
Semillon	3	8.61	16	40.41
Vermentino	4	12.84	12	39.80
Gaglioppo			15	50.45
Garganega	2	0.01	12	38.85
Primitivo di Manduria	2	18.16	12	42.64
Tannat	2	65.20	11	79.96
Carignan			12	39.14
Glera			12	42.68
Koshu			12	31.01
Moscatel Galego			12	50.10
Moscato bianco			12	43.17
Muscat Hamburg	3	0.96	9	13.42
All other cultivars	1224	4557.52	1344	4340.34
ND	958	3905.94	1269	3983.17

ND information not available in SRA archive

Proteomics resources have also arisen in the last decade, despite at a much lower rate. While at the beginning most of these studies used two-dimensional gel analysis and focused on berry metabolism coupled to abiotic stress responses (Vincent et al. 2007; Jellouli et al. 2008; Grimplet et al. 2009b), high-resolution techniques have also been applied to grape such as iTRAQ (Lucker et al. 2009), or much more recently, 2DE gels coupled to liquid chromatography with electrospray ionization (LC-ESI-MS/MS; Negri et al. 2015), or nanoLC ESI LTQ-Orbitrap tandem mass spectrometry (Wang et al. 2017; Kambiranda et al. 2018).

Targeted and untargeted metabolome studies have unquestionably increased within grapevine research, benefiting from a variety of tools such as massive high-performance liquid chromatography (HPLC) and gas chromatography (GC) being applied for sample separation while tandem mass spectrometry (MS) and nuclear magnetic resonance (NMR) being developed for the identification and quantification of metabolites. Solid-phase and micro solid-phase extractions (SPE and SPME), followed by GC-MS methods have been used for volatile composition studies (Savoi et al. 2016; Duchêne et al. 2017). Ultra-High-Performance Liquid Chromatography (UHPLC) coupled to triple quadrupole (QqQ) TQD mass spectrometry analysis was recently used for determining polyphenomic composition (phenylpropanoid-specific omics) and its cultivar-dependent changes in response to drought (Pinasseau et al. 2017). Also, Vondras et al. (2017) recently performed untargeted HPLC-MS to quantify amino acids, sugars, organic acids, and phenylpropanoids to compare the different ripening progressions of berries in a single cluster, while Blanco-Ulate et al. (2015) and Negri et al. (2017) studied the effect of *Botrytis cinerea* noble rot infection in the metabolome of ripening berries and postharvest withered berries, respectively, by using reversed-phase HPLC coupled to ESI mass spectrometer.

Despite metabolomics analyses are rapidly increasing in *Vitis*, metabolism must be understood as a dynamic process. Fluxomics recognizes this complexity in metabolic systems and seeks to determine the rates of metabolic reactions (Winter and Krömer 2013). With the purpose of describing how metabolic fluxes determine cellular phenotypes, Soubeyrand et al. (2018) performed targeted metabolomics and enzyme activity measurements in grape cell cultures at different time-points of nitrogen limitation in order to construct a constraint-based model (by comparing maps of metabolic fluxes in the two contrasted situations) to identify the metabolic drivers of anthocyanin accumulation under high carbon-to-nitrogen ratios.

Within the cell's functions, the transport of essential and beneficial nutrients allows all basic processes to be performed efficiently. In grapevines, ion content profiles can reflect the mineral composition of soils and therefore they can describe certain components of a *terroir*. Pii et al. (2017) studied the ionomics profile of berries grown in different areas to try to discriminate their geographical origin. By applying multi-elemental inductively-coupled plasma-mass spectrometry (ICP-MS), the authors found that rare earth elements were the best chemical descriptors.

Recent attempts for identifying transcription factor binding landscapes have been initiated and deposited in public repositories, despite no publications have yet been produced. Additional efforts are still needed to map protein-DNA and protein-protein interactions at a large scale. Also, DNAse I hypersensitivity mapping could be useful to identify pioneering transcription factors controlling grape and wine quality traits.

8.4 From Single Omics to Integrative Data Analysis

Within single omics studies, the interactions between molecules can be represented in networks, where nodes (genes, proteins, metabolites, etc.) are connected by edges that convey any type of association (e.g., relying in abundance or expression levels). In the case of gene co-expression networks (GCNs), edges represent similar gene expression behaviors, while in genome-wide transcription factor binding studies (e.g., ChIP-seq) edges represent direct target-regulator relationships. In protein-protein interaction networks, edges describe physically interacting protein pairs identified from techniques such as high-throughput yeast two-hybrid screens.

Beyond single omics networks, integrative approaches associate the molecular components of an organism and combine them into higher order networks to model dynamic behaviors. The principle is based on the fact that despite individual functions of a single network may be undetermined, its biological role can sometimes be inferred through association with other networks. Integrated/combined networks provide a more complete information of a certain biological processes as they include two or more omics' layers. In the case of combining several networks of the same type into a community network, this can also be beneficial to effectively reveal discrepancies between individual networks while stressing common associations across individual networks (Proost and Mutwil 2016). Networks of experimental evidence can be integrated by superimposing the nodes from individual networks. However, an appropriate integrative method requires biological data to be normalized, standardized, modeled and visualized in order to build an integrated model (Fig. 8.4). Data modeling requires special attention as this analysis involves generalization and simplification steps with several assumptions (Yuan et al. 2008).

The first task to perform during the integration of different multidimensional omics data consists in matching the features within each omics, as they measure diverse types of molecules and the correspondence between them is not always straight forward. For instance, a single gene can produce several transcripts with different alternative splicing. Similarly, a single transcript can give rise to multiple proteins through different posttranslational processes, making it difficult to associate genes, transcripts, and proteins when measured by genomics, transcriptomics and proteomics techniques. Moreover, cistromics and epicistromics measure transcription factor binding and occupancy of nucleosomes carrying distinct histone modifications in specific genomics regions. The association of these regions to target or regulated genes is not trivial. This problem can be tackled using different software packages such as *RGmatch* (Furió-Tarí et al. 2016), *PeakAnalyzer* or *PeakAnnotator* (Salmon-Divon et al. 2010).

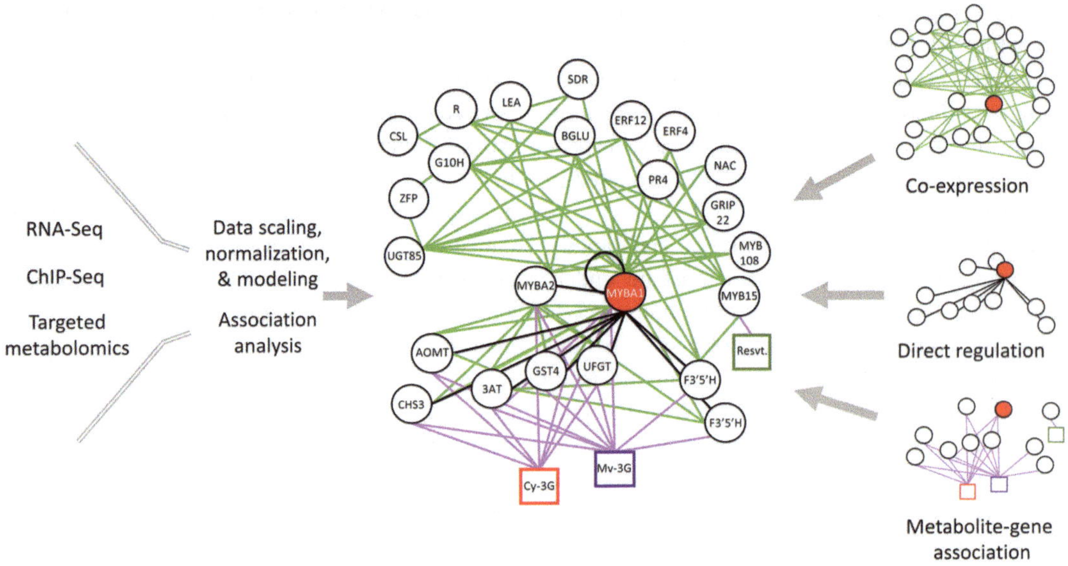

Fig. 8.4 Methods for building integrative network models. Different omics technologies generate data with diverging formats (e.g., numerical scales) and therefore are considered as multidimensional. A hypothetical regulatory network for the berry color locus was used to illustrate how gene co-expression, transcription factor binding, and metabolic data can be integrated to generate a composite network. These can be generated by applying scaling and normalization algorithms to all omics datasets (at the left) or by superposing independently-produced networks (on the right). The main anthocyanin regulator MYBA1 is centered in the network. Its co-expressed genes were taken from previous gene GCN analyses (Wong et al. 2016). Direct regulation examples are taken from experimental evidence (e.g., Matus et al. 2017). Cyanidin or malvidin-related derivatives (di or tri-hydroxylated anthocyanins) are represented by "Cy-3G" and "Mv-3G", respectively. Resvt: the stilbene resveratrol

Additional challenges faced during multi-omics data integration are represented by the heterogeneity of the different data sets. Data from each omics is measured using different units whose typical ranges vary in several orders of magnitude. This can potentially affect data analysis and is typically solved using scaling and normalization techniques. Given the wide spectrum of possible normalization techniques, it is necessary to apply as many as possible and asses their performance in order to choose the most appropriate technique for the data sets under study. The R package *Normalyzer* can be applied in this pre-processing of the data (Chawade et al. 2014).

Once data pre-processing is completed and prior to the actual multi-omics integration, some exploratory analyses need to be conducted over the individual data sets. Due to the high dimensionality of omics data typically these analyses consist in techniques able to reduce complexity in order to extract relevant information. Principal Component Analysis (PCA) constitutes the most widely used projection method in this step. PCA is a multivariate analysis technique whose final goal is to reduce the dimensionality of a large multivariate data set. Here a set of new uncorrelated or orthogonal variables are computed as linear combinations or rotations of the original ones. These new variables are called principal components and they are defined in such a way that they are sorted according to the percentage of explained variability from the original data under the constrain of being orthogonal or uncorrelated. In this way, typically, the first two or three principal components are sufficient to capture most of the variability of the original data and therefore, a projection comprising only these principal components are further considered in the analysis. Graphical representations of the selected principal components are then used to assess the quality of data replicates, uncover problems raised during sample collection (e.g., batch effects) or to unveil underlying structure in the data by applying clustering techniques. Several R packages are available to perform this step such as *factorMineR* (Lê et al. 2008) and *made4* (Culhane et al. 2005), among other methods. For instance, a clear example of data integration in grapevine was conducted by Blanco-Ulate et al. (2015) by using Multiple Factor Analysis (MFA), where four types of quantitative variables were considered: metabolome data, RNA-Seq data from grape and the fungi *Botrytis cinerea*, and *B. cinerea* biomass measurements.

Finally, multi-omics data integration is carried out. Normally, two different goals exist when integrating different omics. On one hand, researchers may be interested on exploratory analysis to identify the underlying relationship between two omics data sets. On the other hand, researchers may treat one of the omics data set as *response variables* that need to be predicted from another explanatory omics data set (considered as *predictors*). Here we discuss two statistical methods that exemplify these two goals. In both cases the input consists of two numerical matrices, $X_{n \times p}$ and $Y_{n \times q}$, that can be generated using two different omics technologies that detect and quantify p and q as different molecules from the same set of n samples.

Canonical Correlation Analysis (CCA) This is an example of exploratory analysis that generates rotations or linear combinations, U and V, of the original data, X and Y, under the constrains of maximizing the correlation $\mathrm{cor}(U_i, V_i)$ with $i = 1, \ldots, \min(p, q)$ and being uncorrelated or orthogonal. These are called canonical variates. Finally, like in any projection technique, only the two or three first canonical variates are considered to capture most of the correlation between the initial data X and Y. Several R packages are available to carry out this methodology such as *CCA* (González et al. 2008) and *mixOmics* (Rohart et al. 2017).

(Sparse) Partial Least Square regression(s) PLS is an example of a multi-omics integration technique in which researchers aim at predicting one omics data set (or physiological data) from another one. In a similar fashion to CCA, rotations U and V of the original data are performed by maximizing the covariance. Projections retaining only two or three components are then considered to perform linear regression. To assess the predictive power of the developed model, cross-validation is commonly applied. In classical PLS regression, all the original variables from *X* and *Y* are included in the rotation or linear combination making intractable the extraction of relevant information from the developed model. In order to tackle this, the sparse variant of PLS regression (sPLS; González et al. 2012) was introduced by using penalization terms based on the marginal contribution of each variable to the predictive power of the model in such way that some coefficient shrinks to zero removing the corresponding variable. This efficiently implements a feature selection technique. Graphical representations such as correlation circle plots, relevance networks and clustered image maps can be generated to facilitate the understanding and interpretation of the constructed model. The R packages *pls* (Mevik and Wehrens 2007) and *mixOmics* (Rohart et al. 2017) implement the necessary functions to apply this methodology.

8.5 Recent Experiences in Grapevine Systems Biology

Throughout the last years, several attempts for representing large biological data in networks have been conducted for elucidating the multi-layered organization of biological processes in grapevine. In this species, integrated network analyses have been mostly adopted to predict gene functions or to contribute in the study of the regulatory mechanisms that control berry composition and development, trigger defense responses to biotic and abiotic stresses or that are influenced by the *terroir* (reviewed by Wong and Matus 2017; Fabres et al. 2017). Some research efforts have defined composite networks of genes and secondary metabolites for characterizing fruit ripening processes in red and white-skinned cultivars (Massonnet et al. 2017; Palumbo et al. 2014; Zamboni et al. 2010), whereas others have constructed gene co-expression networks to describe late stages of ripening (Ghan et al. 2017) or characterize transcriptional regulators related to development, metabolism or stress responses (Loyola et al. 2016; Wong et al. 2016; Sun et al. 2018). Processes involving the rewiring of berry metabolite-transcriptional networks under environmental perturbations such as drought (Savoi et al. 2016, 2017) and elevated light exposure (du Plessis et al. 2017) have also been described. Proteomic/metabolomic composite networks (Wang et al. 2017) and those integrating genome-wide analyses of promoter regulatory elements (Wong et al. 2017) have also been generated. The integration of all these data in multilayered networks has allowed building complex maps of molecular regulation and interaction. Some relevant cases will be covered in this section.

8.5.1 Identifying Molecular Hubs Controlling Light and Cold Response Pathways

The advent and continued adoption of high-throughput transcriptome profiling platforms in grapevine research has led to the vast expansion of transcriptome datasets representing a wide range of experimental conditions (e.g., specific tissue/organ and its associated developmental series, stress—abiotic and biotic, vineyard management strategies, etc.). Although each dataset has been generated to address specific goals of its overarching study, together, individual datasets can be compiled into large expression databases to mine for novel biological insights including, but not limited to, comparative transcriptomics between grapevine and other plants, gene co-expression network analysis and functional assignment of genes, and the discovery of condition-specific *cis*-regulatory motifs (reviewed in Serin et al. 2016).

Genes involved in the same processes might share similar gene expression dynamics across an extensive collection of experiments. This relation, explained by the "guilt by association" principle (Wolfe et al. 2005), is fundamental to infer the roles of uncharacterized genes in co-expression networks. Transcription factors (TFs) comprise a suitable case of study for addressing the behavior of modules in GCNs as they exhibit plethora of protein-protein and protein-DNA interactions, shaping complex regulatory networks responsible for most developmental process. Such is the case of ELONGATED HYPOCOTYL 5 (HY5) and HY5 HOMOLOGUE (HYH), two bZIP master photomorphogenic orchestrators involved in developmental processes responsive to light environmental conditions. Loyola et al. (2016) combined microarray and RNA-Seq co-expression data with a genome-wide binding site promoter inspection to identify HY5 and HYH community gene co-expression and *cis*-regulatory sub-networks in grapevine. Search of potential gene targets identified a preferential regulation of photosynthetic-related processes, heat-shock and DNA/protein repair processes, and regulation of the flavonol biosynthetic pathway. This study was crucial for describing the molecular mechanisms explaining the high radiation adaptive mechanisms that grapevines possess (reviewed by Matus 2016).

Gene co-expression networks have also been integrated with transcription factor binding data to address grape responses to low temperature, in relation to the role of a MYB-like regulator termed AcQUIred tolerance to LOw temperatures (AQUILO; Sun et al. 2018). Here, the authors performed a multi-species GCN, incorporating gene co-expression analysis and in silico TFBS data from grape, with co-expression (associated to the heterologous overexpression of AQUILO) and DAP-seq data in Arabidopsis. The relevance of this study came from the finding that AQUILO was tightly associated with the raffinose family of oligosaccharides (RFOs), a connection that was later validated by quantifying these osmoprotectant molecules in cold-treated grape AQUILO-overexpressing calli.

8.5.2 Regulation of Phenylpropanoid Metabolism

Presently, the most widely adopted methodology to identify candidate transcriptional factors (TFs) involved in secondary metabolism pathways in grapevine involves the inference of function via sequence homology with functionally characterized proteins from model plants (for example, see Hichri et al. 2010; Cavallini et al. 2015; Matus et al. 2017). However, in the recent years, many of these regulators have been prioritized by using gene co-expression network analyses. For example, the putative functions of 134 grapevine R2R3-MYB genes were inferred based on their top 100 co-expressed genes (Wong et al. 2016). This study revealed that GCNs of many R2R3-MYB TFs (46 genes) were enriched with secondary metabolism-related functions. Demonstrating the power of such method is the ability to recover expected relationships between structural pathway genes and their known transcriptional regulators. For example, this was demonstrated with the frequent co-expression of large suites of *STILBENE SYNTHASE* genes (*STSs*) with VviMYB14 and VviMYB15, two R2R3-MYB TFs involved in the regulation of *STS* (Höll et al. 2013). Similar inferences were accounted for VviMYB13, a close homolog of VviMYB14 and VviMYB15, therefore, suggested as involved in the regulation of tissue- and stress-specific *STS* expression (Wong et al. 2016). Two recent studies have also used *STS* genes as "guides" to identify co-expressed TFs in both condition-specific (Wong and Matus 2017) and -independent contexts (Vannozzi et al. 2018). A berry-specific GCN encompassing five red cultivars across four key berry developmental stages revealed novel roles for AP2/ERF and WRKY TFs in the regulation of *STSs*. TFs of the latter two families were not only frequently co-expressed with *STSs* but were also enriched for their respective TF binding sites (TFBS) in the promoters of many *STSs*. Recent studies have now demonstrated that VviWRKY24 and VviWRKY03 are additional players in the regulation of *STSs* at various hierarchies—acting as singular effector or in synergy with VviMYB14 to activate STSs (Vannozzi et al. 2018).

The integration of non-coding RNA network analysis to existing condition-specific GCNs has also been presented to unravel the regulation of phenylpropanoid and flavonoid biosynthesis during berry development and ripening (Wong and Matus 2017). One of the key findings from this initiative was the discovery of long non-coding RNAs (lncRNAs) that were not only strongly correlated with key structural pathway genes but were also located in close proximity to their co-expressed gene). The lncRNA VIT_210s0042n00100, present in close proximity with all nine *VviSTSs* of chromosome 10 presented consistent co-expression with all of them. Another case represents one predicted lncRNA (VIT_203s0180n00020) that is linked to VviGT2 through strong co-expression and co-location. This gene encodes an enzyme putatively involved in hydroxycinnamic ester biosynthesis and proanthocyanidin galloylation (Khater et al. 2012).

GCN approaches may reveal additional layers and deconvolute the complexities of secondary metabolic pathway regulation in grapevine. Indeed, in a first study of its kind, Zhang et al. (2018) demonstrated that multiple lncRNAs, named LNC1 and LNC2, were involved in the regulation of anthocyanin biosynthesis in fruits of sea buckthorns (*Hippophae sp.*) by serving as endogenous target mimics (eTM) of miR156a and miR828, respectively. Functional studies confirmed that silencing of LNC1 and LNC2, led to the induction and repression of anthocyanin biosynthetic pathway gene expression and anthocyanin levels in fruits, respectively, validating the integrated lncRNA-miRNA-mRNA network prediction.

8.5.3 The Fight Club Goes Dry: Networks Related to Grape Berry Ripening in Response to Drought

To understand the molecular mechanisms underpinning berry development and ripening at greater detail, recent efforts have focused on understanding the transcriptome dynamics in multiple cultivars across the entire process of berry development and ripening. A study by Massonnet et al. (2017) represented the first monumental study to catalogue the genome-wide transcriptional profile of ten Italian grapevine varieties at four critical stages of berry development, all being cultivated in a single vineyard. In less than a handful of studies, network-based approaches have been applied to identify genes potentially involved in critical developmental stage transitions. Such cases often complement the findings from the widely adopted differential expression analysis but are also pivotal in revealing novel genes and relationships that were otherwise unattainable from traditional differential expression methods. For example, berry-specific gene co-expression network analysis encompassing immature-to-mature transitions has been particularly insightful in revealing groups of genes with distinct topological properties that can be classified into "party", "date" (see Han et al. 2004 for details), or "fight-club" hubs (Palumbo et al. 2014). Genes that belong to the "fight-club" hubs, in particular, were often negatively correlated with their interacting partners in gene co-expression networks, and those who do, were inferred as biologically relevant "switches" fulfilling negative regulatory roles in the transition of major developmental phases such as ripening. Although the identity of these major switches was first documented in red grapevine varieties, recent research has now ascertained several common but also reveal variety (red and white-skinned)-specific switch genes (Massonnet et al. 2017). From a total of 271 berry-specific switch genes identified to date, 131 genes were in common in both varieties while 81 and 50 genes were specific to all white and red varieties, respectively. A large proportion of these "switches" encode for transcription factors (31 genes), followed by genes involved in stress responses (31 genes), carbohydrate metabolism (22 genes), signaling (20 genes), secondary metabolism (20 genes), and cell wall metabolism (18 genes), among others (Massonnet et al. 2017).

Recent works have provided evidence for the involvement of multiple stress regulons—both ABA-dependent and ABA-independent (reviewed in Nakashima et al. 2014)—in the berry ripening program (Savoi et al. 2017). Certain TF families (e.g., NAC, bZIP, AP2/ERF) that share co-expression with downstream water deficit stress-responsive genes may be required to orchestrate the balance between the progression of berry development and stress-associated transcriptional regulation. Further analysis of gene co-expression and gene-metabolite co-response networks of the berry subjected to water deficit stress across critical berry development and ripening phases revealed several distinct modules that were congruently induced by ripening and water deficit stress (Savoi et al. 2016, 2017). Here, metabolome and transcriptome integrated network-based analysis revealed close associations between the expression behaviors of module members (especially the activation of multiple signal transduction pathways) and the dynamics of key central and specialized metabolites involved in the drought response (e.g., proline, branched-chain amino acids, phenylpropanoids, anthocyanins, and free volatile organic compounds). For example, the grapevine homologue of Arabidopsis ERF1, a key regulatory component of the jasmonate and ethylene signaling network (Cheng et al. 2013), whose expression was congruently induced by ripening and water deficit stress, was also identified to be a common berry "switch" gene. While its precise regulatory role remains to be elucidated, integrated network analysis positioned ERF1 as a putative regulator of proline and anthocyanin accumulation in the berry (Savoi et al. 2017). *VviERF1* was significantly co-expressed with pyrroline-5-carboxylate synthase (*P5CS*) and *VviMYBA2*, the key structural gene of proline biosynthesis and a key regulatory gene of anthocyanin biosynthesis in the berry, respectively, and shared significant correlation with various anthocyanin compounds. The presence of potential AP2/ERF TFBS (i.e., DRE and GCC-box) situated within the promoter region of P5CS and MYBA2 further reinforce its involvement as a regulator of berry composition during ripening and water deficit stress.

8.5.4 Non-coding RNA Networks Within Grape-Fungi Pathosystems

Grapevine diseases caused by biotic agents can be devastating for the wine and table grape industries. Among fungal-related disorders, grape trunk diseases together with downey and powdery mildew are among the most important pathologies, causing significant economic losses in vineyards practically all over the world. The symptoms of downey mildew, caused by *Plasmopara viticola*, are quite detrimental, as for instance, as soon as fruits become infected, berries completely dry out. The *Vitis* spp.—*P. viticola* association is of great interest as this oomycete is an obligate biotroph and relies entirely on the host to complete its life cycle (i.e., needs to keep its host cells alive before sporulation; Grenville-Briggs and van West 2005), and also because North American *Vitis* species are naturally resistant (Polesani et al. 2010). In order to model this complex pathosystem, Brilli et al. (2018) performed a multi-omics and multi-species functional genomic study. The authors sequenced and assembled the draft genome of *P. viticola*, identifying the lost metabolic features responsible for its total dependence on the grape host, and further studied the fungus transcriptome changes occurring during the infection process, identifying a protein triggering immunity in the resistant *V. riparia*. The most striking results from this study arise from the small RNA sequencing (sRNA-Seq) analysis in control and infected plants at different times after the infection, combined with genome-wide degradome (or parallel analysis of RNA ends) analyses in both the plant and the oomycete. As a result, a large number of sRNA-mediated cleavages exclusively occurred in infected tissues, where sRNAs produced by *P. viticola* triggered cleavage of grapevine genes while sRNAs processed from grapevine transcripts targeted the fungus mRNAs, unveiling a bi-directional RNA silencing network mediated by non-coding RNAs shuffling between the pathogen and its host (Brilli et al. 2018). As more pathogen genomes become available, a broader understanding of pathosystems and their dynamics will be achieved, especially regarding the roles of secreted effectors in interfering plant immune recognition (reviewed by Dalio et al. 2018).

Grape pathogen responses have been recently studied by addressing potential interactions of transcription factors and cis-regulatory element (CRE), and also by constructing gene co-expression networks (GCNs) of plant gene families related with defense. Wong et al. (2017) performed a genome-wide analysis of known plant CREs in all grape predicted protein-coding gene promoters, constructing an integrated CRE-driven network. Numerous CRE-driven modules inferred from using condition-dependent GCNs suggested important roles in pathogen stress responses. For example, *GCC*-core sub-modules were contained in many genes that were highly induced in berries and leaves infected with fungi such as *Botrytis cinerea* and *Erysiphe necator*. Finally, gene co-expression networks of the ATL protein family showed that many of these E3 ubiquitin ligases were induced in grapevine–pathogen interactions including *P. viticola* and necrotrophic fungi (Wong et al. 2018).

8.6 Resources

Next-generation sequencing as well as traditional Sanger sequencing methods are of great significance in unraveling the complexity of plant genomes. These are constantly generating a copious volume of sequence data to be analyzed, annotated and stored, thus creating a revolutionary demand for resources and tools to manage and handle these necessities (Basantani et al. 2017). Here we present a brief compilation of web resources that are either specific for grape or encompass a variety of plant species including *Vitis* species (Table 8.2).

At least two grape-specific platforms have been effectively used to study the extent of gene regulatory networks: the ViTis Co-expression DataBase (VTCdb; Wong et al. 2013) and VESPUCCI (Moretto et al. 2016a). These resources have played an important role in determining the roles of genes related to photomophogenic responses and secondary metabolism in targeted functional studies (Loyola et al. 2016; Malacarne et al. 2016). Integration of multi-omics datasets (i.e., gene expression, metabolite, and protein profiles), mapping of data onto relevant molecular networks, and the visualization of the dynamic interactions between the various molecular classes are also the first few steps when performing any systems biology experiments. Tools such as Cytoscape (Shannon et al. 2003) have been specially designed for this task and have been largely adopted by the grape research community to visualize and analyze complex networks. In addition, one ongoing Initiative in grapevine, VitisNet (Grimplet et al. 2009a), serves as a resource for manually curated functional gene annotation and provides a wide range of manually curated pathway-level molecular networks (over 240 categories) as templates for grapevine systems biology experiments.

The increasing release of plant genomes provided unseen opportunities and challenges for comparative genomics resources. Indeed, different genomics multi-species platforms also exist constituting relevant hubs to exploit omics data in grape. For instance, recent examples include the fruitENCODE platform (http://www.epigenome.cuhk.edu.hk/encode.html) that provides a comprehensive repository oriented to shed light on the genetic and epigenetic basis of fruit ripening in climacteric and non-climacteric species. Multi-species GCNs allowing comparative co-expression analysis are also now available for many plants including grapes (Table 8.2). Resources such as ATTED-II (http://atted.jp/) are among the most popular, providing the opportunity to query microarray and RNA-seq GCNs using the "guide" gene approach. ATTED-II also allows assessments of co-expression conservation of co-expressed genes across different plant lineages (Obayashi et al. 2018). The Plant Omics Data Center (PODC; http://plantomics.mind.meiji.ac.jp/podc/) is a NGS-derived gene expression network

Table 8.2 Online resources useful for gene network mining in grapevine

DB name	Type	Species	Datatypes	Features	Query examples:	Website
ATTED-II	GCN	Multi-species (9)	Co-expression (Microarray, RNA-seq)	Grape GCN were constructed using RNA-seq and Affymetrix micro arrays. Similarity metric = MR	1. "Guide" gene lists 2. Comparative analysis of CEG rankings across multiple species	http://atted.jp/
CoNekT	GCN	Multi-species (7)	Co-expression (RNA-seq)	Grape GCN were constructed using RNA-seq. Similarity metric = HRR	1. "Guide" gene lists 2. Comparative analysis of CEG rankings across multiple species	https://conekt.sbs.ntu.edu.sg/
AraNet/AraNetv2	Integrated (CFN)	Multi-species (29)	19 datatypes (e.g., co-expression, domain co-occurrence, genomic neighborhood of orthologs, protein-protein interactions, phylogenetic profile).	Grapevine gene function were inferred using any combination of the associated datatypes. Some include orthology-based projections from model plant species (i.e., Arabidopsis).	1. "Guide" gene lists	http://www.inetbio.org/aranet
PODC	Integrated (CFN)	Multi-species (11)	Co-expression (RNA-seq), natural language processing-based curation	Grape GCN were constructed using RNA-seq. Similarity metric = PCC and Distance in Correspondence Analysis (DCA)	1. "Guide" gene lists	http://plantomics.mind.meiji.ac.jp/podc/
COP	GCN	Multi-species (8)	Co-expression (Microarray)	Grape GCN were constructed using Affymetrix microarray data. Similarity metric = Cosine correlation (CC). Not recommended for grapevine, but fine for Arabidopsis	1. "Guide" gene lists	http://webs2.kazusa.or.jp/kagiana/cop0911/
PLANEX	GCN	Multi-species (8)	Co-expression (Microarray)	Grape GCN were constructed using Affymetrix microarray data. Similarity metric = PCC. Not recommended for grapevine, but fine for Arabidopsis	1. "Guide" gene lists	http://planex.plantbioinformatics.org/
ePlant	Vis.	Grape	Gene expression	Interactive grapevine gene atlas expression browser	1. "Guide" gene lists	http://bar.utoronto.ca/efp_grape/cgi-bin/efpWeb.cgi

(continued)

Table 8.2 (continued)

DB name	Type	Species	Datatypes	Features	Query examples:	Website
PlantReg Map	GRN	Multi-species (132 species)	CHIP-seq, DAP-seq, PBM, literature curation	Grapevine TF binding sites were inferred using orthology-based projections from model plant species (i.e., Arabidopsis). Genome-wide TFBS analysis of grapevine promoters	1. "Guide" gene lists to query downstream target genes of input gene (i.e., TFs) 2. "Guide" gene lists to query upstream regulators (TFs) of input genes	http://plantregmap.cbi.pku.edu.cn/network.php
VitisNet	Vis.	Grape	Manually curated molecular networks encompassing 247 distinct biological processes	Allows the visualization of multi-omics datasets (i.e., genes, proteins or metabolites) simultaneously on these molecular networks	1. Downloaded networks can be imported into Cytoscape for further multi-omics datasets visualization	https://www.sdstate.edu/vitisnet-molecular-networks-grapevine
STRING	Integrated (CFN)	Multi-species (2,031 species plants and animals)	8 datatypes (e.g., gene neighborhood, gene co-occurrence, textmining, co-expression, protein homology)	Grapevine gene function were inferred using any combination of the associated datatypes. Some include orthology-based projections from model plant and non-plant species. Similar to AraNet	1. "Guide" gene lists	https://string-db.org/
VTCdb	GCN	Grape	Co-expression (Microarray, RNA-seq)	Grape GCN were constructed using RNA-seq and Nimblegen arrays. Similarity metric = MR, HRR, PCC	1. "Guide" gene lists 2. Biological processes of interests	http://vtcdb.adelaide.edu.au/Home.aspx
Vespucci	GCN	Grape	Co-expression (Microarray, RNA-seq)	Grape GCN were constructed using RNA-seq and multiple microarray platforms. Similarity metric = PCC. Includes an exploratory tool to analyze expression of genes across 1608 manually curated (vocabulary-controlled) experimental conditions	1. "Guide" gene lists	http://vespucci.colombos.fmach.it/

(continued)

Table 8.2 (continued)

DB name	Type	Species	Datatypes	Features	Query examples:	Website
grape_sRNA_atlas	miRNA	Grape	miRNA (RNA-seq)	Grape miRNA-target (gene) networks were constructed using a comprehensive miRNA catalogue (both known and novel) and in silico target prediction analysis. miRNA expression browser available	1. miRNA query	https://mpss.danforthcenter.org/dbs/index.php?SITE=grape_sRNA_atlas
BIOWINE	miRNA	Grape	miRNA (RNA-seq)	Grape miRNA-target (gene) networks were constructed using in silico target prediction analysis	1. miRNA query 2. Biological processes of interests	https://alpha.dmi.unict.it/biowine/

For multi-species DB, only grapevine-specific features are highlighted. *GRN* genome-wide transcriptional regulatory interaction network, *Vis.* visualization repository aimed at integrating large-scale omics resources for a broad range of species (Ohyanagi et al. 2015). Such resources may be used in conjunction with existing grapevine-specific co-expression platforms to build community GCNs or to gain additional insights into the evolutionary context of conserved and/or species-specific co-expressed genes relationship.

Additional multi-species platforms gathering grape's omics and mainly aimed at comparative studies include Ensembl Plants (http://plants.ensembl.org) (Bolser et al. 2016), Phytozome (https://phytozome.jgi.doe.gov/pz/portal.html) (Goodstein et al. 2012), PlantGDB (http://www.plantgdb.org/) (Duvick et al. 2008), and AraNet v2 (http://www.inetbio.org/aranet) (Lee et al. 2015). These integrative resources encompassing genome-scale information (genome sequence, gene models, functional annotation, polymorphic loci, expression) offer a variety of sequence analysis tools and web services. Example of integrative platforms also come from other species including both model (Araport, Solgenomics) and non-model (Melonomics, Ginseng Genome Database) plants. A common feature underlying these resources rely on the use of customized instances of JBrowse (Buels et al. 2016), a fast and full-featured genome browser built with JavaScript and HTML5. Thanks to its speed, scalability, and versatility this platform supports complex interactive queries on large track sets representing a suitable and solid mean to handle omics data in a genomic context. In addition, a variety of analysis functions can readily be added using the plugin framework (e.g., visualization of whole-genome bisulfite sequencing data, glyphs for variants and GWAS data, small RNA visualization, etc.). Very recently, a JBrowse (v. 1.11.5) was set up to visualize and give access to some omics data in the *Vitis vinifera* 12X.v2 PN40024 assembly (https://urgi.versailles.inra.fr/jbrowse/gmod_jbrowse/?data=myData/Vitis/data_gff) (Canaguier et al. 2017). The platform hosts 11 annotations tracks, including the different releases of the grapevine genome annotations (CRIBI v1, CRIBI v2, Genoscope, Cost v3, etc.), automated and manual curated transposable elements annotations and manual curated gene family sets. In addition, 12 tracks highlighting the variants coming from re-sequencing experiments are also present in the platform, which could help in the identification of useful markers for applied research purposes.

8.6.1 VESPUCCI and NES^{2}RA as Grape-Oriented Resources

Exploring shifts in gene expression as response to different experimental conditions has become commonplace while transcriptomic experiments are being performed on a daily basis. Public available gene expression datasets, however, conceal most of their true potential since they are meant to answer to a specific biological question and aren't considered in the light of a wider context. Within transcriptomics, we have witnessed a major shift in data production with the advent of high-throughput sequencing technologies. Despite nowadays Illumina sequencing is the de facto standard for RNA-seq experiments, microarrays are still extensively used and, more importantly, constitute a wealth of public information available to be explored.

With the advent of systems biology approaches in grapevine research, data integration arises as a leading aspect to take advantage of such rich sources of information (Gligorijevic and Nataša 2015). Different methods have been proposed to carry out the task of effectively integrating gene expression data and can be usually divided in two categories: (1) direct integration and (2) meta-analysis. Direct integration (Rung and Brazma 2013) considers the sample-level measurements within each study and merges them into a single data set. The latter approach (Garrett-Mayer et al. 2008), instead, integrates gene expression analysis combining information from several data sources defining confidence levels for each study individually (without a general scheme) and is commonly used to integrate conclusions coming from different studies.

One of the platforms used for data integration in transcriptomics is COLOMBOS (Moretto et al. 2016a), originally named as a COLlection Of Microarrays for Bacterial OrganismS, which was developed for three bacterial species (*Escherichia coli*, *Bacillus subtilis*, and *Salmonella enterica* serovar Typhimurium) and later updated with others prokaryotic species and also including RNA-seq technology. The implementation of the COLOMBOS framework to the *Vitis* species led to the development of VESPUCCI (Moretto et al. 2016b) (Vitis Expression Studies Platform Using COLOMBOS Compendia Instances), an integrated gene expression database for grapevine that originally included 1500 samples at the time of its first release and now has doubled in size including most of publicly available transcriptomic data.

Both VESPUCCI and COLOMBOS fall under the direct integration methodology. Their approach to data integration is unique in the sense of directly combining gene expression information from different technological platforms and experiments, without the need for batch-normalization since it calculates log-ratios for contrasts, i.e., samples being compared that come from the same experiment and platform combination (a "batch"). This results in crossing out a high proportion of batch-related variation (Luo et al. 2010). While gathering a large amount of data is made easy for model organisms like *E. coli* (due to the abundant number of experiments available), for non-model species the situation is different as only fewer experiments are usually performed. In this case, the importance of transcriptomics data integration is even more significant as an adequate magnitude of data is needed to be able to draw valid and general conclusions. In this sense, working with plant species highlighted the need for the authors to significantly rethink some aspects of the data acquisition and annotation process. The creation of a gene expression compendium using COLOMBOS technology is facilitated by the use of COMMAND (Moretto et al. 2019), a web-based application used to download, collect and manage gene expression data from public databases, but it is still mainly a manual effort. The peculiarity and complexity of plant transcriptomes and experimental designs in plant biology require the ability to manage how probes (for microarray) and short read sequences (for RNA-seq) are mapped and thus assigned to genes. The concept of "measurable transcript" was also used to account for some technical limitations that prevent the possibility to precisely distinguish among genes with high sequence similarity.

In VESPUCCI, data and experiment-related information (meta-data) are collected and curated starting from raw intensities (for microarrays) and raw sequence reads (for RNA-Seq). A robust normalization method and a quality control procedure are performed to allow the direct comparison of gene expression values across different experimental conditions (Engelen et al. 2011). This results in a single coherent gene expression matrix in which each row represents a gene and each column represents a "sample contrast". Sample contrasts measure the difference (in log scale) between a test and a reference condition, both which are designed a priori by curators during the compendium creation process. The expression data itself is a matrix of log-ratios (base 2), so that positive values represent up-regulation, and negative values represent down-regulation of a gene in the test sample compared to the reference sample. VESPUCCI's main goal is to gather together as many expression data as possible to explore patterns of co-expression across several experimental conditions and to provide a high-quality gene expression database to be used for downstream analysis. The creation of a co-expressed genes cluster (known as module) is performed similarly to a BLAST (Camacho et al. 2009) search in which the users can look for expression values for a given set of conditions but using expression correlation instead of sequence similarity to score the best matches. Modules can be modified in several ways in order to highlight the behavior of the genes of interest and to analyze (anti)co-expression patterns.

Considering that gene expressions are represented as relative values, it is fundamental to extensively annotate samples with various sorts of meta-data to ensure that valid biological conclusions can be drawn from the exploration of the compendium. One of VESPUCCI's biggest effort and most notable feature is the manual curation and quality check of samples. Each sample has been annotated by curators using controlled vocabularies to ensure both human readability and computational tractability. To completely fulfill the properties of the FAIR (Findable Accessible Interoperable Reusable) principles (Wilkinson et al. 2016), VESPUCCI is undergoing a constant renovation to exploit standards and bio-ontologies for data annotation. Finally, the interface is the other pivotal point towards seamless integration with other services and tools and has been designed to adapt to users' needs, as well as to simplify the implementation of other tools on top of it. One example of such means is the NES^{2}RA algorithm (Asnicar et al. 2018), a mining tool for transcriptomic data used to expand a known local gene network (LGN) by finding new related genes. This method has been applied to the grapevine transcriptomic dataset using VESPUCCI as data source to expand LGNs related to the secondary metabolic pathways for anthocyanin and stilbenoid synthesis and signaling networks related to the hormones abscisic acid and ethylene (Malacarne et al. 2018). Compared to Pearson correlation, NES^{2}RA LGNs show less edges as it removes less significant interactions, due to noisy or redundant information. This allows to reduce the complexity of the network and focus on the network topology and the most likely gene interactions. NES^{2}RA is computationally demanding and relies on the BOINC platform that distributes supercomputation tasks among computers made available by the volunteers participating in the gene@home project.

Besides the importance of having a single point of access to easily check what is already available in terms of transcriptomic experiments in grapevine and, of course, the possibility to empower data analysis with thousands of integrated samples, the development of VESPUCCI has led to few considerations about the importance of correctly annotating experiments, extrapolable to all types of resources. Building the compendium itself was the most time-consuming step, as curators devoted their time and ongoing effort to describe sample conditions and their key descriptors, after carefully reading the experiment descriptions as well as scientific papers. The importance of early annotation of experiments as soon as (or even before) data are available is also underrated. It is often considered as an annoying request to fulfill before the publication, while it should be treated as an integral part of the experimental design with the same importance as notes and protocols written in lab notebooks have.

8.7 Final Remarks

The accuracy of molecular systems biology relies on efficient methods that handle, analyze and visualize large omics data sets. However, it has become evident that the use of a single omics technology is not sufficient to develop predictive models, which in turn is the ultimate goal of this new discipline. Accordingly, the multiple use of technologies such as transcriptomics, cistromics, epicistromics, proteomics, and metabolomics, over the same samples or biological conditions has started to be a central methodology in plant molecular systems biology. Multi-omics network modeling has proven to be a successful advance for unraveling the structure of biological processes in plants, as it allows identifying the key components and interactions for system regulation. Conversely, networks frequently require assumptions for data modeling, and since their methods may rely on the existing knowledge regarding the components and interactions of a system, they can evolve to more exactly represent a biological system. Thus, data should be interpreted carefully while these approaches can be complemented by reductionist methods. Notwithstanding these limitations, the use of these methodologies in grapevine research has provided novel perspectives for interpreting omics data and has already challenged the analysis of the large amount of data that are being generated for this species.

References

Adam-Blondon AF, Alaux M, Pommier C, Cantu D, Cheng ZM, Cramer GR, Davies C, Delrot S, Deluc L, Di Gaspero G, Grimplet J, Fennell A, Londo JP, Kersey P, Mattivi F, Naithani S, Neveu P, Nikolski M, Pezzotti M, Reisch BI, Töpfer R, Vivier MA, Ware D, Quesneville H (2016) Towards an open grapevine information system. Hortic Res 3:16056

Amrine KC, Blanco-Ulate B, Riaz S, Pap D, Jones L, Figueroa-Balderas R, Walker MA, Cantu D (2015) Comparative transcriptomics of Central Asian *Vitis vinifera* accessions reveals distinct defense strategies against powdery mildew. Hortic Res 2:15037

Aoki K, Ogata Y, Shibata D (2007) Approaches for extracting practical information from gene co-expression networks in plant biology. Plant Cell Physiol 48:381–390

Asnicar F, Masera L, Coller E, Gallo C, Sella N, Tolio T, Morettin P et al (2018) NES^{2}RA: network expansion by stratified variable subsetting and ranking aggregation. Int J High Perform Comput Appl 32(3):380–392

Bajic M, Maher KA, Deal RB (2018) Identification of open chromatin regions in plant genomes using ATAC-Seq. Methods Mol Biol 1675:183–201

Bartlett A, O'Malley RC, Huang SC, Galli M, Nery JR, Gallavotti A, Ecker JR (2017) Mapping genome-wide transcription-factor binding sites using DAP-seq. Nat Protoc 12:1659–1672

Basantani MK, Divya Gupta, Rajesh Mehrotra, Sandhya Mehrotra, Swati Vaish, Anjali Singh (2017) An update on bioinformatics resources for plant genomics research. Curr Plant Biol 11–12:33–40

Bennett S, Barnes C, Cox A et al (2005) Toward the 1,000 dollars human genome. Pharmacogenomics 6:373–382

Blanco-Ulate B, Amrine KC, Collins TS, Rivero RM, Vicente AR, Morales-Cruz A, Doyle CL, Ye Z, Allen G, Heymann H, Ebeler SE, Cantu D (2015) Developmental and metabolic plasticity of white-skinned grape berries in response to botrytis cinerea during noble rot. Plant Physiol 169(4):2422–2443

Bokulich NA, Thorngate JH, Richardson PM, Mills DA (2014) Microbial biogeography of wine grapes is conditioned by cultivar, vintage, and climate. Proc Natl Acad Sci 111(1):E139–E148

Bokulich NA, Collins TS, Masarweh C, Allen G, Heymann H, Ebeler SE et al (2016) Associations among wine grape microbiome, metabolome, and fermentation behavior suggest microbial contribution to regional wine characteristics. mBio 7(3):e00631-16

Bolger M, Arsova B, Usadel B (2018) Plant genome and transcriptome annotations: from misconceptions to simple solutions. Brief Bioinform 19(3):437–449

Bolser D, Staines DM, Pritchard E, Kersey P (2016) Ensembl plants: integrating tools for visualizing, mining, and analyzing plant genomics data. Methods Mol Biol 1374:115–140

Brilli M, Asquini E, Moser M, Bianchedi PL, Perazzolli M, Si-Ammour A (2018) A multi-omics study of the grapevine-downy mildew (*Plasmopara viticola*) pathosystem unveils a complex protein coding- and noncoding-based arms race during infection. Sci Rep 8(1):757

Buels R, Yao E, Diesh CM, Hayes RD, Munoz-Torres M, Helt G, Goodstein DM, Elsik CG, Lewis SE, Stein L, Holmes IH (2016) Jbrowse: a dynamic web platform for genome visualization and analysis. Genome Biol 17:66

Burns KN, Kluepfel DA, Strauss SL, Bokulich NA, Cantu D, Steenwerth KL (2015) Vineyard soil bacterial diversity and composition revealed by 16S rRNA genes: differentiation by geographic features. Soil Biol Biochem 91:232–247

Burns KN, Bokulich NA, Cantu D, Greenhut RF, Kluepfel DA, O'Geen AT, Strauss SL, Steenwerth KL (2016) Vineyard soil bacterial diversity and composition revealed by 16S rRNA genes: differentiation by vineyard management. Soil Biol Biochem 103:337–348. https://doi.org/10.1016/j.soilbio.2016.09.007

Camacho C, Coulouris G, Avagyan V, Ma N, Papadopoulos J, Bealer K, Madden TL (2009) BLAST + : architecture and applications. BMC Bioinformatics 10:421

Canaguier A, Grimplet J, Di Gaspero G, Scalabrin S, Duchêne E, Choisne N, Mohellibi N, Guichard C, Rombauts S, Le Clainche I, Bérard A, Chauveau A, Bounon R, Rustenholz C, Morgante M, Le Paslier MC, Brunel D, Adam-Blondon AF (2017) A new version of the grapevine reference genome assembly (12X.v2) and of its annotation (VCost.v3). Genomics Data 14:56–62

Cardone MF, D'Addabbo P, Alkan C, Bergamini C, Catacchio CR, Anaclerio F et al (2016) Inter-varietal structural variation in grapevine genomes. Plant J 88:648–661

Carmichael I, Marron JS (2018) Data science vs. statistics: two cultures? Jpn J Stat Data Sci 1(1):117–138

Cavallini E, Matus JT, Finezzo L, Zenoni S, Loyola R, Guzzo F, Schlechter R, Ageorges A, Arce-Johnson P, Tornielli GB (2015) The phenylpropanoid pathway is controlled at different branches by a set of R2R3-MYB C2 repressors in grapevine. Plant Physiol 167(4):1448–1470

Chawade A, Alexandersson E, Levander F (2014) Normalyzer: a tool for rapid evaluation of normalization methods for omics data sets. J Proteome Res 13:3114–3120

Chen X, Bhadauria V, Ma B (2017) ChIP-Seq: a powerful tool for studying protein-DNA interactions in plants. Curr Issues Mol Biol 27:171–180

Cheng C, Jiao C, Singer SD, Gao M, Xu X, Zhou Y et al (2015) Gibberellin-induced changes in the transcriptome of grapevine (*Vitis labrusca* × *V. vinifera*) cv. Kyoho flowers. BMC Genom 16:128

Cheng MC, Liao PM, Kuo WW, Lin TP (2013) The Arabidopsis ETHYLENE RESPONSE FACTOR1 regulates abiotic stress-responsive gene expression by binding to different cis-acting elements in response to different stress signals. Plant Physiol 162(3):1566–1582

Chin CS, Peluso P, Sedlazeck FJ, Nattestad M, Concepcion GT, Clum A, Dunn C, O'Malley R, Figueroa-Balderas R, Morales-Cruz A, Cramer GR, Delledonne M, Luo C, Ecker JR, Cantu D, Rank DR, Schatz MC (2016) Phased diploid genome assembly with single-molecule real-time sequencing. Nat Methods 13(12):1050–1054

Cramer G, Ergul A, Grimplet J, Tillett R, Tattersall E et al (2007) Water and salinity stress in grapevines: early and late changes in transcript and metabolite profiles. Funct Integr Genomics 7:111–134

Culhane AC, Thioulouse J, Perrière G, Higgins DG (2005) MADE4: an R package for multivariate analysis of gene expression data. Bioinformatics 21:2789–2790

Da Silva C, Zamperin G, Ferrarini A, Minio A, Dal Molin A, Venturini L et al (2014) The high polyphenol content of grapevine cultivar tannat berries is conferred primarily by genes that are not shared with the reference genome. Plant Cell 25:4777–4788

Da Silva FG, Iandolino A, Al-Kayal F, Bohlmann MC, Cushman MA et al (2005) Characterizing the grape transcriptome. Analysis of expressed sequence tags from multiple Vitis species and development of a compendium of gene expression during berry development. Plant Physiol 139:574–597

Dal Santo S, Tornielli GB, Zenoni S, Fasoli M, Farina L, Anesi A et al (2013) The plasticity of the grapevine berry transcriptome. Genome Biol 14:r54

Dalio RJD, Herlihy J, Oliveira TS, McDowell JM, Machado M (2018) Effector biology in focus: a primer for computational prediction and functional characterization. Mol Plant Microbe Interact 31(1):22–33

Defoort J, Van de Peer Y, Vermeirssen V (2018) Function, dynamics and evolution of network motif modules in integrated gene regulatory networks of worm and plant. Nucleic Acids Res 46(13):6480-6503. https://doi.org/10.1093/nar/gky468

Deluc LG, Grimplet J, Wheatley MD, Tillett RL, Quilici DR et al (2007) Transcriptomic and metabolite analyses of cabernet sauvignon grape berry development. BMC Genom 8:429

Di Genova A, Almeida AM, Muñoz-Espinoza C, Vizoso P, Travisany D, Moraga C et al (2014) Whole genome comparison between table and wine grapes reveals a comprehensive catalog of structural variants. BMC Plant Biol 14:7. https://doi.org/10.1186/1471-2229-14-7

Domingos S, Fino J, Paulo OS, Oliveira CM, Goulao LF (2016) Molecular candidates for early-stage flower-to-fruit transition in stenospermocarpic table grape (*Vitis vinifera* L.) inflorescences ascribed by differential transcriptome and metabolome profiles. Plant Sci 244:40–56

du Plessis K, Young PR, Eyéghé-Bickong HA, Vivier MA (2017) The transcriptional responses and metabolic consequences of acclimation to elevated light exposure in grapevine berries. Front Plant Sci 8:1261

Duchêne E, Butterlin G, Claudel P, Dumas V, Jaegli N, Hugueney P, Arnold G, Merdinoglu D (2017) Genetic determinism of the 'Muscat' flavour in grapevine (*Vitis vinifera* L.) cultivars. Acta Hortic 1157: 87–92

Duvick Jon et al (2008) PlantGDB: a resource for comparative plant genomics. Nucleic Acids Res 36 (Database issue):D959–D965

Eid J, Fehr A, Gray J et al (2009) Real-time DNA sequencing from single polymerase molecules. Science 323(5910):133–138

Engelen K, Fu Q, Meysman P, Sánchez-Rodríguez A, De Smet R, Lemmens K, Fierro AC, Marchal K (2011) COLOMBOS: access port for cross-platform bacterial expression compendia. PLoS ONE 6(7):e20938

Fabres PJ, Collins C, Cavagnaro TR, Rodríguez López CM (2017) A concise review on multi-omics data integration for terroir analysis in *Vitis vinifera*. Front Plant Sci 8:1065

Fasoli M, Dal Santo S, Zenoni S, Tornielli GB, Farina L, Zamboni A, Porceddu A, Venturini L, Bicego M, Murino V, Ferrarini A, Delledonne M, Pezzotti M (2012) The grapevine expression atlas reveals a deep transcriptome shift driving the entire plant into a maturation program. Plant Cell 24(9):3489–3505

Fortes AM, Gallusci P (2017) Plant stress responses and phenotypic plasticity in the epigenomics era: perspectives on the grapevine scenario, a model for perennial crop plants. Front Plant Sci 8:82

Furió-Tarí P, Consea A, Tarazona S (2016) RGmatch: matching genomic regions to proximal genes in omics data integration. BMC Bioinformatics 17:427

Garrett-Mayer E, Parmigiani G, Zhong X, Cope L, Gabrielson E (2008) Cross-study validation and combined analysis of gene expression microarray data. Biostatistics 9(2):333–354

Ghan R, Petereit J, Tillett RL, Schlauch KA, Toubiana D, Fait A, Cramer GR (2017) The common transcriptional subnetworks of the grape berry skin in the late stages of ripening. BMC Plant Biol 17(1):94

Gligorijević V, Nataša P (2015) Methods for biological data integration: perspectives and challenges. J R Soc Interface 12(112):20150571

Goodstein DM, Shu S, Howson R, Neupane R, Hayes RD, Fazo J, Mitros T, Dirks W, Hellsten U, Putnam N, Rokhsar DS (2012) Phytozome: a comparative platform for green plant genomics. Nucleic Acids Res 40(Database issue):D1178–D1186. https://doi.org/10.1093/nar/gkr944

González I, Déjean S, Martin P, Baccini A (2008) CCA: an R package to extend canonical correlation analysis. J Stat Softw 23:12

González I, Cao KA, Davis MJ, Déjean S (2012) Visualising associations between paired 'omic' data sets. BioData Min 5(19):1–23

Grenville-Briggs LJ, van West P (2005) The biotrophic stages of oomycete-plant interactions. Adv Appl Microbiol 57:217–243

Grimplet J, Cramer GR, Dickerson JA, Mathiason K, Van Hemert J, Fennell AY (2009a) VitisNet: "Omics" integration through grapevine molecular networks. PLoS ONE 4(12):e8365

Grimplet J, Deluc LG, Tillett RL, Wheatley MD, Schlauch KA et al (2007) Tissue-specific mRNA expression profiling in grape berry tissues. BMC Genom 8:187

Grimplet J, Wheatley MD, Jouira HB, Deluc LG, Cramer GR, Cushman JC (2009b) Proteomic and selected metabolite analysis of grape berry tissues under well-watered and water-deficit stress conditions. Proteomics 9(9):2503–2528

Han JD, Bertin N, Hao T, Goldberg DS, Berriz GF, Zhang LV, Dupuy D, Walhout AJ, Cusick ME, Roth FP, Vidal M (2004) Evidence for dynamically organized modularity in the yeast protein–protein interaction network. Nature 430(6995):88–93

Harris ZN, Kovacs LG, Londo JP (2017) RNA-seq-based genome annotation and identification of long-noncoding RNAs in the grapevine cultivar 'Riesling'. BMC Genom 18(1):937

Hichri I, Heppel SC, Pillet J, Léon C, Czemmel S, Delrot S, Lauvergeat V, Bogs J (2010) The basic helix-loop-helix transcription factor MYC1 is involved in the regulation of the flavonoid biosynthesis pathway in grapevine. Mol Plant 3(3):509–523

Höll J, Vannozzi A, Czemmel S, D'Onofrio C, Walker AR, Rausch T, Lucchin M, Boss PK, Dry IB, Bogs J (2013) The R2R3-MYB transcription factors MYB14 and MYB15 regulate stilbene biosynthesis in *Vitis vinifera*. Plant Cell 25 (10):4135–4149

Jaillon O, Aury J-M, Noel B, Policriti A, Clepet C et al (2007) The grapevine genome sequence suggests ancestral hexaploidization in major angiosperm phyla. Nature 449:463–467

Jellouli N, Jouira BH, Skouri H, Ghorbel A, Gourgouri A et al (2008) Proteomic analysis of Tunisian grapevine cultivar Razegui under salt stress. J Plant Physiol 165:471–481

Kambiranda D, Basha SM, Singh R, Snowden J, Mercer R (2018) Proteome profile of American hybrid grape cv. Blanc du Bois during ripening reveals proteins associated with flavor volatiles and ethylene production. Proteomics 18(8):e1700305

Khater F, Fournand D, Vialet S, Meudec E, Cheynier V, Terrier N (2012) Identification and functional characterization of cDNAs coding for hydroxybenzoate/hydroxycinnamate glucosyltransferases co-expressed with genes related to proanthocyanidin biosynthesis. J Exp Bot 63(3):1201–1214

Lê S, Josse J, Husson F (2008) FactoMineR: an R package for multivariate analysis. J Stat Softw 25:1

Lee T, Yang S, Kim E, Ko Y, Hwang S, Shin J, Shim JE, Shim H, Kim H, Kim C, Lee I (2015) Aranet V2: an improved database of co-functional gene networks for the study of *Arabidopsis thaliana* and 27 other nonmodel plant species. Nucleic Acids Res 43: D996–D1002

Leveau JH, Tech JJ (2010) Grapevine microbiomics: bacterial diversity on grape leaves and berries revealed by high-throughput sequence analysis of 16S rRNA amplicons. Int Symp Biol Control Postharvest Dis Chall Oppor 905:31–42

Li Y, Pearl SA, Jackson SA (2015) Gene networks in plant biology: approaches in reconstruction and analysis. Trends Plant Sci 20:664–675

Liang Z, Duan S, Sheng J, Zhu S, Ni X, Shao J, Liu C, Nick P, Du F, Fan P, Mao R, Zhu Y, Deng W, Yang M, Huang H, Liu Y, Ding Y, Liu X, Jiang J, Zhu Y, Li S, He X, Chen W, Dong Y (2019) Whole-genome resequencing of 472 Vitis accessions for grapevine diversity and demographic history analyses. Nat Commun 10(1):1190

Liu G, Wang J, Cramer G, Dai Z, Duan W, Xu H et al (2012) Transcriptomic analysis of grape (*Vitis vinifera* L.) leaves during and after recovery from heat stress. BMC Plant Biol 12:174

Loyola R, Herrera D, Mas A, Wong DCJ, Höll J, Cavallini E et al (2016) The photomorphogenic factors UV-B RECEPTOR 1, ELONGATED HYPOCOTYL 5, and HY5 HOMOLOGUE are part of the UV-B signalling pathway in grapevine and mediate flavonol accumulation in response to the environment. J Exp Bot 67(18):5429–5445

Lucker J, Laszczak M, Smith D, Lund ST (2009) Generation of a predicted protein database from EST data and application to I-TRAQ analyses in grape (*Vitis vinifera* cv. Cabernet Sauvignon) berries at ripening initiation. BMC Genom 10:50

Luo J, Schumacher M, Scherer A, Sanoudou D, Megherbi D, Davison T, Shi T et al (2010) A comparison of batch effect removal methods for enhancement of prediction performance using MAQC-II microarray gene expression data. Pharmacogenomics J 10(4):278–291

Malacarne G, Coller E, Czemmel S, Vrhovsek U, Engelen K, Goremykin V et al (2016) The grapevine VvibZIPC22 transcription factor is involved in the regulation of flavonoid biosynthesis. J Exp Bot 67:3509–3522

Malacarne G, Pilati S, Valentini S, Asnicar F, Moretto M, Sonego P, Masera L, Cavecchia V, Blanzieri E, Moser C (2018) Discovering causal relationships in grapevine expression data to expand gene networks. a case study: four networks related to climate change. Front Plant Sci 9:1385. https://doi.org/10.3389/fpls.2018.01385. eCollection 2018

Margulies M, Egholm M, Altman WE et al (2005) Genome sequencing in microfabricated high-density picolitre reactors. Nature 437:376–380

Martin LBB, Zhangjun F, Giovannoni JJ, Rose JKC (2013) Catalyzing plant science research with RNA-seq. Front Plant Sci 4:66

Martins G, Lauga B, Miot-Sertier C, Mercier A, Lonvaud A, Soulas ML et al (2013) Characterization of epiphytic bacterial communities from grapes, leaves, bark and soil of grapevine plants grown, and their relations. PLoS ONE 8(8):e73013

Massonnet M, Fasoli M, Tornielli GB, Altieri M, Sandri M, Zuccolotto P, Paci P, Gardiman M, Zenoni S, Pezzotti M (2017) Ripening transcriptomic program in red and white grapevine varieties correlates with berry skin anthocyanin accumulation. Plant Physiol 174 (4):2376–2396. https://doi.org/10.1104/pp.17.00311

Matus JT, Cavallini E, Loyola R, Höll J, Finezzo L, Dal Santo S, Vialet S, Commisso M, Roman F, Schubert A, Alcalde JA, Bogs J, Ageorges A, Tornielli GB, Arce-Johnson P (2017) A group of grapevine MYBA transcription factors located in chromosome 14 control anthocyanin synthesis in vegetative organs with different specificities compared with the berry color locus. Plant J 91(2):220–236

Matus JT (2016) Transcriptomic and metabolomic networks in the grape berry illustrate that it takes more than flavonoids to fight against ultraviolet radiation. Front Plant Sci 7:1337

Mercenaro L et al (2017) Sequence polymorphisms and structural variations among four grapevine (*Vitis vinifera* L.) cultivars representing Sardinian agriculture. Front Plant Sci 8:1279

Mesarovic MD (1968) Systems theory and biology—view of a theoretician. In: Mesarovic MD (ed) Systems theory and biology. Springer, New York, pp 59–87

Mevik B, Wehrens R (2007) The pls package: principal component and partial least squares regression in R. J Stat Softw 18:2

Miller JG (1978) Living systems. Mcgraw-Hill, New York

Milo R, Shen-Orr S, Itzkovitz S, Kashtan N, Chklovskii D, Alon U (2002) Network motifs: simple building blocks of complex networks. Science 298 (5594):824–827

Milos P (2008) Helicos BioSciences. Pharmacogenomics 9:477–480

Minio A, Massonnet M, Figueroa-Balderas R, Castro A, Cantu D (2019) Diploid genome assembly of the wine grape carménère. G3 (Bethesda) 9(5):1331–1337. https://doi.org/10.1534/g3.119.400030

Minio A, Lin J, Gaut BS, Cantu D (2017) How single molecule real-time sequencing and haplotype phasing have enabled reference-grade diploid genome assembly of wine grapes. Front Plant Sci 8:826

Moretto Marco, Sonego Paolo, Dierckxsens Nicolas, Brilli Matteo, Bianco Luca, Ledezma-Tejeida Daniela, Gama-Castro Socorro et al (2016a) COLOMBOS v3.0: leveraging gene expression compendia for cross-species analyses. Nucleic Acids Res 44(D1): D620–D623

Moretto M, Sonego P, Pilati S, Malacarne G, Costantini L, Grzeskowiak L et al (2016b) VESPUCCI: exploring patterns of gene expression in grapevine. Front Plant Sci 7:633

Moretto Marco, Sonego Paolo, Villaseñor-Altamirano Ana B, Engelen Kristof (2019) First step toward gene expression data integration: transcriptomic data acquisition with COMMAND. BMC Bioinformatics 20 (1):54

Moser C, Segala C, Fontana P, Salakhudtinov I, Gatto P et al (2005) Comparative analysis of expressed sequence tags from different organs of *Vitis vinifera* L. Funct Integr Genomics 5:208–217

Nakashima K, Yamaguchi-Shinozaki K, Shinozaki K (2014) The transcriptional regulatory network in the drought response and its crosstalk in abiotic stress responses including drought, cold, and heat. Front Plant Sci 5:170

Negri AS, Prinsi B, Failla O, Scienza A, Espen L (2015) Proteomic and metabolic traits of grape exocarp to explain different anthocyanin concentrations of the cultivars. Front Plant Sci 6:603

Negri S, Lovato A, Boscaini F, Salvetti E, Torriani S, Commisso M, Danzi R, Ugliano M, Polverari A, Tornielli GB, Guzzo F (2017) The induction of noble rot (*Botrytis cinerea*) infection during postharvest withering changes the metabolome of grapevine berries (*Vitis vinifera* L., cv. Garganega). Front Plant Sci 8:1002

Obayashi T, Aoki Y, Tadaka S, Kagaya Y, Kinoshita K (2018) ATTED-II in 2018: a plant coexpression database based on investigation of the statistical property of the mutual rank index. Plant Cell Physiol 59(1):e3

Ohyanagi H, Takano T, Terashima S, Kobayashi M, Kanno M, Morimoto K, Kanegae H, Sasaki Y, Saito M, Asano S, Ozaki S, Kudo T, Yokoyama K, Aya K, Suwabe K, Suzuki G, Aoki K, Kubo Y, Watanabe M, Matsuoka M, Yano K (2015) Plant Omics Data Center: An integrated web repository for interspecies gene expression networks with NLP-based curation. Plant Cell Physiol 56:e9

Oltvai ZN, Barabási AL (2002) Systems biology: life's complexity pyramid. Science 298(5594):763–764

Pajoro A, Madrigal P, Muiño JM, Matus JT, Jin J, Mecchia MA, Debernardi JM, Palatnik JF, Balazadeh S, Arif M, Ó'Maoiléidigh DS, Wellmer F, Krajewski P, Riechmann JL, Angenent GC, Kaufmann K (2014) Dynamics of chromatin accessibility and gene regulation by MADS-domain transcription factors in flower development. Genome Biol 15(3): R41

Palumbo MC, Zenoni S, Fasoli M, Massonnet M, Farina L, Castiglione F et al (2014) Integrated network analysis identifies fight-club nodes as a class of hubs encompassing key putative switch genes that induce major transcriptome reprogramming during grapevine development. Plant Cell 26(12):4617–4635

Papatheodorou I, Oellrich A, Smedley D (2015) Linking gene expression to phenotypes via pathway information. J Biomed Semant 6:17

Pass DA, Sornay E, Marchbank A, Crawford MR, Paszkiewicz K, Kent NA, Murray JAH (2017) Genome-wide chromatin mapping with size resolution reveals a dynamic sub-nucleosomal landscape in Arabidopsis. PLoS Genet 13:e1006988

Pii Y, Zamboni A, Dal Santo S, Pezzotti M, Varanini Z, Pandolfini T (2017) Prospect on Ionomic Signatures for the Classification of Grapevine Berries According to Their Geographical Origin. Front Plant Sci 8:640

Pinasseau L, Vallverdú-Queralt A, Verbaere A, Roques M, Meudec E, Le Cunff L, Péros JP, Ageorges A, Sommerer N, Boulet JC, Terrier N, Cheynier V (2017) Cultivar diversity of grape skin polyphenol composition and changes in response to drought investigated by LC-MS based metabolomics. Front Plant Sci 8:1826

Polesani M, Bortesi L, Ferrarini A, Zamboni A, Fasoli M, Zadra C, Lovato A, Pezzotti M, Delledonne M, Polverari A (2010) General and species-specific transcriptional responses to downy mildew infection in a susceptible (*Vitis vinifera*) and a resistant (*V. riparia*) grapevine species. BMC Genom 11:117

Proost S, Mutwil M (2016) Tools of the trade: studying molecular networks in plants. Curr Opin Plant Biol 30:130–140

Ramalingam A, Kudapa H, Pazhamala LT, Weckwerth W, Varshney RK (2015) Proteomics and metabolomics: two emerging areas for legume improvement. Front Plant Sci 6:1116

Roach MJ, Johnson DL, Bohlmann J, van Vuuren HJJ, Jones SJM, Pretorius IS, Schmidt SA, Borneman AR (2018) Population sequencing reveals clonal diversity and ancestral inbreeding in the grapevine cultivar Chardonnay. PLoS Genet 14(11):e1007807

Rohart F, Gautier B, Singh A, Lê Cao KA (2017) mixOmics: An R package for omics feature selection and multiple data integration. PLoS Comput Biol 13: e1005752

Rothberg JM, Hinz W, Rearick TM et al (2011) An integrated semiconductor device enabling non-optical genome sequencing. Nature 475(7356):348–352

Rung Johan, Brazma Alvis (2013) Reuse of public genome-wide gene expression data. Nat Rev Genet 14(2):89–99

Salmon-Divon M, Dvinge H, Tommoja K, Bertone P (2010) PeakAnalyzer: genome-wide annotation of chromatin binding and modification loci. BMC Bioinformatics 11:415

Savoi S, Wong DC, Arapitsas P, Miculan M, Bucchetti B, Peterlunger E, Fait A, Mattivi F, Castellarin SD (2016) Transcriptome and metabolite profiling reveals that prolonged drought modulates the phenylpropanoid and terpenoid pathway in white grapes (*Vitis vinifera* L.). BMC Plant Biol 16:67

Savoi S, Wong DCJ, Degu A, Herrera JC, Bucchetti B, Peterlunger E, Fait A, Mattivi F, Castellarin SD (2017) Multi-omics and integrated network analyses reveal new insights into the systems relationships between metabolites, structural genes, and transcriptional regulators in developing grape berries (*Vitis vinifera* L.) exposed to water deficit. Front Plant Sci 8:1124

Serin EA, Nijveen H, Hilhorst HW, Ligterink W (2016) Learning from co-expression networks: possibilities and challenges. Front Plant Sci 7:444

Shannon P, Markiel A, Ozier O, Baliga NS, Wang JT, Ramage D, Amin N, Schwikowski B, Ideker T (2003) Cytoscape: a software environment for integrated models of biomolecular interaction networks. Genome Res 13(11):2498–2504

Sheth BP, Thaker VS (2014) Plant systems biology: insights, advances and challenges. Planta 240:33–54

Soubeyrand E, Colombié S, Beauvoit B, Dai Z, Cluzet S, Hilbert G, Renaud C, Maneta-Peyret L, Dieuaide-Noubhani M, Mérillon JM, Gibon Y, Delrot S, Gomès E (2018) Constraint-based modeling highlights cell energy, redox status and α-ketoglutarate availability as metabolic drivers for anthocyanin accumulation in grape cells under nitrogen limitation. Front Plant Sci 9:421

Sullivan AM, Kerry LB, Sandstrom R, Stamatoyannpoulos JA, Queitsch C (2015) DNase I hypersensitivity mapping, genomic footprinting, and transcription factor networks in plants. Curr Plant Biol 3–4:40–47

Sun X, Matus JT, Wong DCJ, Wang Z, Chai F, Zhang L, Fang T, Zhao L, Wang Y, Han Y, Wang Q, Li S, Liang Z, Xin H (2018) The GARP/MYB-related grape transcription factor AQUILO improves cold tolerance and promotes the accumulation of raffinose family oligosaccharides. J Exp Bot 69(7):1749–1764

Sweetman C, Wong DC, Ford CM, Drew DP (2012) Transcriptome analysis at four developmental stages of grape berry (*Vitis vinifera* cv. Shiraz) provides insights into regulated and coordinated gene expression. BMC Genom 13:691

Tattersall E, Grimplet J, Deluc L, Wheatley M, Vincent D et al (2007) Transcript abundance profiles reveal larger and more complex responses of grapevine to chilling compared to osmotic and salinity stress. Funct Integr Genomics 7:317–333

Terrier N, Glissant D, Grimplet J, Barrieu F, Abbal P et al (2005) Isogene specific oligo arrays reveal multifaceted changes in gene expression during grape berry (*Vitis vinifera* L.) development. Planta 222:832–847

Tillett RL, Ergül A, Albion RL, Schlauch KA, Cramer GR, Cushman JC (2011) Identification of tissue-specific, abiotic stress-responsive gene expression patterns in wine grape (*Vitis vinifera* L.) based on curation and mining of large-scale EST data sets. BMC Plant Biol 11:86

Vannozzi A, Wong DCJ, Höll J, Hmmam I, Matus JT, Bogs J, Ziegler T, Dry I, Barcaccia G, Lucchin M (2018) Combinatorial regulation of stilbene synthase genes by WRKY and MYB transcription factors in grapevine (*Vitis vinifera* L.). Plant Cell Physiol 59 (5):1043–1059

Vega A, Gutiérrez RA, Peña-Neira A, Cramer GR, Arce-Johnson P (2011) Compatible GLRaV-3 viral infections affect berry ripening decreasing sugar accumulation and anthocyanin biosynthesis in *Vitis vinifera*. Plant Mol Biol 77(3):261–274

Velasco R, Zharkikh A, Troggio M, Cartwright DA, Cestaro A et al (2007) A high quality draft consensus sequence of the genome of a heterozygous grapevine variety. PLoS ONE 2:1326

Vincent D, Ergul A, Bohlman MC, Tattersall EA, Tillett RL et al (2007) Proteomic analysis reveals differences between *Vitis vinifera* L. cv. Chardonnay and cv. Cabernet Sauvignon and their responses to water deficit and salinity. J Exp Bot 58:1873–1892

Vitulo N, Forcato C, Carpinelli EC, Telatin A, Campagna D, D'Angelo M, Zimbello R, Corso M, Vannozzi A, Bonghi C, Lucchin M, Valle G (2014) A deep survey of alternative splicing in grape reveals changes in the splicing machinery related to tissue, stress condition and genotype. BMC Plant Biol 14:99

Von Bertalanffy L (1968) General system theory. Foundations, development, applications. George Braziller, New York

Vondras AM, Commisso M, Guzzo F, Deluc LG (2017) Metabolite profiling reveals developmental inequalities in pinot noir berry tissues late in ripening. Front Plant Sci 8:1108

Wang L, Sun X, Weiszmann J, Weckwerth W (2017) System-level and granger network analysis of integrated proteomic and metabolomic dynamics identifies key points of grape berry development at the interface of primary and secondary metabolism. Front Plant Sci 8:1066

Waters DL, Holton TA, Ablett EM, Lee LS, Henry RJ (2005) cDNA microarray analysis of developing grape (*Vitis vinifera* cv. Shiraz) berry skin. Funct Integr Genomics 5:40–58

Wilkinson MD, Dumontier M, Aalbersberg IJ, Appleton G, Axton M, Baak A, Blomberg N et al (2016) The FAIR Guiding Principles for scientific data management and stewardship. Sci Data 3:160018

Winter G, Krömer JO (2013) Fluxomics–connecting 'omics analysis and phenotypes'. Environ Microbiol 15(7):1901–1916

Wise RP, Caldo RA, Hong L, Shen L, Cannon E, Dickerson JA (2007) BarleyBase/PLEXdb. Methods Mol Biol 406:347–363

Wolfe CJ, Kohane IS, Butte AJ (2005) Systematic survey reveals general applicability of "guilt-by-association" within gene coexpression networks. BMC Bioinformatics 6:227

Wong DCJ, Matus JT (2017) constructing integrated networks for identifying new secondary metabolic pathway regulators in grapevine: recent applications and future opportunities. Front Plant Sci 8:505

Wong DCJ, Zhang L, Merlin I, Castellarin SD, Gambetta GA (2018) Structure and transcriptional regulation of the major intrinsic protein gene family in grapevine. BMC Genom 19(1):248

Wong DCJ, Lopez-Gutierrez R, Gambetta GA, Castellarin SD (2017) Genome-wide analysis of cis-regulatory element structure and discovery of motif-driven gene co-expression networks in grapevine. DNA Res 24(3):311–326

Wong DCJ, Schlechter R, Vannozzi A, Höll J, Hmmam I, Bogs J et al (2016) A systems-oriented analysis of the grapevine R2R3-MYB transcription factor family uncovers new insights into the regulation of stilbene accumulation. DNA Res 23:451–466

Wong DCJ, Sweetman C, Drew DP, Ford CM (2013) VTCdb: a gene co-expression database for the crop species *Vitis vinifera* (grapevine). BMC Genom 14:882

Xie H, Konate M, Sai N, Tesfamicael KG, Cavagnaro T, Gilliham M, Breen J, Metcalfe A, Stephen JR, De Bei R, Collins C, Lopez CMR (2017) Global DNA methylation patterns can play a role in defining terroir in grapevine (*Vitis vinifera* cv. Shiraz). Front Plant Sci 8:1860

Yuan JS, Galbraith DW, Dai SY, Griffin P, Stewart N (2008) Plant system biology comes of age. Trends Plant Sci 13(4):165–171

Zamboni A, Di Carli M, Guzzo F, Stocchero M, Zenoni S, Ferrarini A et al (2010) Identification of putative stage-specific grapevine berry biomarkers and omics data integration into networks. Plant Physiol 154:1439–1459

Zarraonaindia I, Owens SM, Weisenhorn P, West K, Hampton-Marcell J, Lax S et al (2015) The soil microbiome influences grapevine-associated microbiota. MBio 6(2):e02527-14

Zenoni S, Ferrarini A, Giacomelli E, Xumerle L, Fasoli M, Malerba G et al (2010) Characterization of transcriptional complexity during berry development in *Vitis vinifera* using RNA-Seq. Plant Physiol 152 (4):1787–1795

Zhang G, Chen D, Zhang T, Duan A, Zhang J, He C (2018) Transcriptomic and functional analyses unveil the role of long non-coding RNAs in anthocyanin biosynthesis during sea buckthorn fruit ripening. DNA Res 25(5), 465–476. https://doi.org/10.1093/dnares/dsy017

Zhou Y, Massonnet M, Sanjak JS, Cantu D, Gaut BS (2017) Evolutionary genomics of grape (*Vitis vinifera* ssp. vinifera) domestication. Proc Natl Acad Sci USA 114(44):11715–11720

Epigenetic Regulation in Fleshy Fruit: Perspective for Grape Berry Development and Ripening

9

Junhua Kong, Margot Berger, Amélie Colling, Linda Stammitti, Emeline Teyssier and Philippe Gallusci

Abstract

Epigenetic regulation mainly refers to histone post-translational modifications and DNA methylation, which are critical to plant gene regulation and contribute to the development of plants and to their response to the environment. Recent molecular and epigenomic studies have shown that epigenetic regulations play critical roles in tomato fruit development and ripening, the current model for climacteric fruit. This led to a new model of ripening control where active DNA demethylation plays a central role being necessary to the induction of several genes that control fruit ripening. Whether this is a general model applying to all type of fruit, including non-climacteric fruit for which grape berry stands as a general model, is an open question that requires investigating the genome-wide variations of epigenetic marks during fruit development and ripening in many different species. Finally, the potential roles of epigenetic regulations in grapevine, a perennial, grafted, and clonally propagated plant, are discussed.

Junhua Kong and Margot Berger contributed equally to this work

J. Kong · M. Berger · A. Colling · L. Stammitti · E. Teyssier · P. Gallusci (✉)
UMR EGFV, Université de Bordeaux, Institut national de la Recherche Agronomique, Institut des Sciences de la Vigne et du Vin, 210 Chemin de Leysotte, CS 50008, 33882 Villenave-d'Ornon, France
e-mail: philippe.gallusci@inra.fr

J. Kong
e-mail: junhua.kong@inra.fr

M. Berger
e-mail: margot.berger@inra.fr

A. Colling
e-mail: amelie.colling@inra.fr

L. Stammitti
e-mail: linda.stammitti@inra.fr

E. Teyssier
e-mail: emeline.teyssier@inra.fr

9.1 Introduction: Relevance of Epigenetic Regulations in Plants

In eukaryotes, DNA is tightly associated with histones to form the chromatin, a highly dynamic structure that plays critical roles in genome functioning. Chromatin is made of elementary units called nucleosomes that are composed of octamers of the core histones (H2A, H2B, H3, and H4) around which 147 bp of DNA is rolled up. Nucleosomes are separated by a 50-bp-long linker DNA that interacts with histone H1. Traditionally, two distinct chromatin states have been described: the highly condensed heterochromatin, which is considered as inactive, and euchromatin which corresponds to a less condensed and

D. Cantu and M. A. Walker (eds.), *The Grape Genome*, Compendium of Plant Genomes, https://doi.org/10.1007/978-3-030-18601-2_9

transcriptionally active chromatin state. Indeed, dynamic changes on chromatin play critical roles in gene regulation and have therefore been the subject of intensive studies over the last decades both in animals and in plants (Exner and Hennig 2008; Zheng and Liu 2019).

Epigenetics was initially defined as "the branch of biology which studies the causal interactions between genes and their products which bring the phenotype into being" (Waddington 1942). Epigenetics now refers to "the study of changes in gene function that are mitotically and/or meiotically heritable and that do not entail change in DNA sequence" (Wu and Morris 2001). Epigenetic regulations are mediated by the so-called epigenetic marks that include the methylation of the cytosines on the 5th carbon (5-methylCytosine, 5mC) as well as several histone post-translational modifications (HPTMs), but also involve small RNAs and histone variants (Law and Jacobsen 2010; Maeji and Nishimura 2018; Rothbart and Strahl 2014). Both types of marks contribute to defining specific chromatin states and consequent gene expression patterns that can be maintained after cell division during tissue and organ development (Birnbaum and Roudier 2017; Eichten et al. 2014; Pikaard and Scheid 2014).

Epigenetic modifications are now emerging as crucial players controlling various aspects of plant development, such as for example transitions between developmental phases (Trindade et al. 2017), plant reproduction (Wang and Köhler 2017), root (Kawakatsu et al. 2016), seed (Kawakatsu et al. 2017), and fruit development (Gallusci et al. 2016; Giovannoni et al. 2017). It also participates in the response of plants to environmental stresses (Chinnusamy and Zhu 2009; Crisp et al. 2016).

In this chapter, we will mainly focus on the role of epigenetic regulations in fleshy fruit, an organ of primary importance for plants as it insures seed dispersal and for humankind, because fleshy fruits are an important source of nutrients in human nutrition (Klee and Giovannoni 2011) and provide raw material for products of high economical value such as wine. Studies in tomato, grape, strawberry, and others have now shown that the development and ripening of fleshy fruit rely on the establishment and maintenance of differential gene expression patterns (Alba 2005; Osorio et al. 2011) and complex regulatory pathways that involve both genetic and hormonal controls critical at these developmental phases (Osorio et al. 2013). However, several studies have now shown that both DNA methylation and histone PTMs also regulate fruit development and ripening (Bucher et al. 2018; Gallusci et al. 2016; Giovannoni et al. 2017) indicating that epigenetic regulations require to be considered as well. Most of these studies have been performed on tomato, the model plant for climacteric fruit. However, tomato fruit presents specific developmental and physiological features including high endoreduplication levels and a monophasic growth curve. Therefore, it remains unclear whether similar mechanisms are operating in other fruits with different characteristics, such as grape, the model for non-climacteric fruit.

Here, we summarize the current knowledge of epigenetic mechanisms in plants and present the most recent studies highlighting the role of epigenetic regulations in fruit development and ripening. As a conclusion, we discuss the specificity of grape as a grafted perennial plant that is clonally propagated and develops non-climacteric fruit.

9.2 Fleshy Fruit Development and Ripening: Specificities of Grape Berries

Fruit is an organ specific to angiosperms designed for seed protection and dispersal that has long been considered essential in the human diet because it contains fibers, vitamins, carbohydrates, and antioxidants that are essential to humans (Klee and Giovannoni 2011; Seymour et al. 2013). Most fruits develop from ovaries, although accessory tissues, for example the receptacle in strawberry, may be used as well (Seymour et al. 2013). The development of fleshy fruit is in most cases initiated by fertilization and is characterized by two main steps that precede fruit ripening: (1) a cell division phase which is initiated shortly after pollination and followed by (2) a cell expansion phase that is responsible for the increase in fruit size (Gillaspy et al. 1993). In contrast to dry fruits that undergo lignification, fleshy fruit enters a complex ripening process characterized by extensive metabolic modifications such as soluble sugar accumulation, cell wall degradation, and synthesis of a wide range of secondary compounds of high nutritional value such as carotenoids or anthocyanins, and several vitamins. In most cases, fruit ripening results in significant changes in fruit appearance, including fruit color modifications and fruit softening (Lee et al. 2012; Seymour et al. 2013).

Among fleshy fruit, grape berry presents specific developmental features. In contrast to most fruits that present a typical simple sigmoid growth curve, grape berry growth follows a double sigmoid curve as fruit size increases both before and after the induction of ripening (Conde et al. 2007; Serrano et al. 2017). The first increase in berry size starts shortly after fruit set and is due to cell division and subsequent cell expansion. It is characterized by organic acid accumulation in vacuoles and the synthesis of tannins and hydroxycinnamates. The berry size stops to increase during the so-called lag phase that precedes the "véraison stage," which is characterized by berry softening, ABA synthesis, and initiation of sugar and anthocyanin accumulation (Castellarin et al. 2015). Following, grape berry size increases again due to additional cell expansion events in the mesocarp. This second growth phase, which occurs during ripening, is characterized by important metabolic changes that include the accumulation of glucose and fructose along with a decrease in organic acid levels, berry softening, and the synthesis of precursors of various aromatic compounds including terpenes, isoprenoids, esters, and thiols.

Fleshy fruits have been classified based on the physiological mechanisms that control the induction of ripening. Climacteric fruits for which tomato stands as a model (Giovannoni et al. 2017) are characterized by an intense respiratory burst associated with ethylene synthesis that precedes fruit ripening induction. This contrasts with non-climacteric fruits such as grape and strawberry, for which no specific physiological parameter that marks the initiation of ripening has been identified (Bapat et al. 2010), even if hormones, including ethylene and ABA, are now known to have important roles in the ripening of this type of fruit (Fortes et al. 2015). Genetic control of ripening has also been demonstrated for climacteric fruit, mainly in the tomato model, and several mutations affecting essential regulators of ripening have been described in this plant (Gapper et al. 2014; Bucher et al. 2018; Gallusci et al. 2016). The recent discovery that epigenetic regulators are major players in the control of fruit development, ripening, and senescence has deeply changed the proposed models describing the regulation of fruit development and raises the question of the general function of such mechanisms in all types of fruit. So far, most studies indicate that epigenetic regulations may be important in other types of fruit.

9.3 Epigenetic Mechanisms

Epigenetic regulations are based on two main mechanisms, histone post-translational modifications (HPTMs) and DNA methylation, and also include additional processes such as short interfering RNAs (siRNAs) synthesis and specific histone isoforms, called histone variants. These mechanisms have been the subject of many recent reviews (see, e.g., Maeji and Nishimura 2018) and will be only summarized here with a focus on the most recent findings.

9.3.1 Histone Post-translational Modifications

The mechanisms responsible for histone post-translational modifications (HPTMs) are conserved in plants and animals (Feng and Jacobsen 2011; Fuchs et al. 2006). The following part presents these conserved mechanisms using examples taken from plant models (except when data were obtained from animal models only) and discusses a few differences discriminating plants from animals.

9.3.1.1 Numerous Histone Post-translational Modifications and Histone Variants Contribute to the Epigenetic Information

All histones are subjected to a wide variety of post-translational modifications that include methylation, acetylation, phosphorylation, ubiquitylation, sumoylation, and ADP ribosylation (Bannister and Kouzarides 2011; Berger 2007; Feng and Jacobsen 2011; Jenuwein and Allis 2001). These modifications affect various amino acids at different positions. The nucleosomal histones are mostly modified at their NH2 terminus which protrudes out of the nucleosome. In addition, histone H2A, histone H3, and histone H1 are encoded by small gene families, allowing the production of different isoforms usually referred to as histone variants that bear specific roles and may be subjected to differential post-translational modifications (Jiang and Berger 2017; Talbert and Henikoff 2017). Importantly, most histone marks are found both in plants and in animals, but the same histone mark can have a different distribution and physiological function in different organisms. A striking example is the mark H3K9me3 which is mostly associated with heterochromatin in organisms ranging from fission yeast to humans (Becker et al. 2016), but it is typically found in euchromatin in Arabidopsis (Roudier et al. 2011).

Histone modifications and histone variants control several processes linked to genome function, such as DNA replication, DNA repair, DNA recombination, and transcriptional activation/inactivation (Vergara and Gutierrez 2017). Most studies have focused on their function in gene expression, which relies on two main mechanisms (Bannister and Kouzarides 2011; Berger 2007; Engelhorn et al. 2014). First HPTMs, like histone acetylation, neutralize the positive charge of histones and weaken the interaction between histones and the negatively charged DNA molecule leading to an increased DNA accessibility to the transcriptional machinery. Recent data based on a multiscale computational study have shown that histone lysine acetylation also unfolds chromatin by decreasing tail availability for inter-nucleosome interactions, which are important for the chromatin fiber compaction (Collepardo-Guevara et al. 2015). In addition, HPTMs are recognized by a diverse set of effector proteins, also called histone readers, which participate in the control of gene expression, for example chromatin remodeling proteins or transcriptional regulators. Hence, a large array of protein domains has been characterized, which recognize and bind to specific histone modifications. Some of the HPTM readers are directly responsible for a specific functional outcome such as the DNA methyltransferase CMT3 which recognizes H3K9me2 (Du et al. 2012; Lindroth et al. 2004) and is responsible for CHG methylation (Lindroth et al. 2001). Alternatively, HPTM readers can act through their interaction with effector proteins. For example, the Arabidopsis MORF-related gene (MRG) group proteins, MRG1 and MRG2, recognize the H3K4me3/H3K36me3 marks on the *FLOWERING LOCUS T* (FT) promoter; this interaction favors the activation of *FT* transcription through a physical interaction between MRG1/MRG2 and the transcription factor CONSTANS (Bu et al. 2014). Because they rely on a number of different protein partners, such mechanisms can be precisely controlled. Finally, recent data suggest that HPTMs play a role in the 3D organization of genomic DNA, contributing to the formation of specific nuclear territories, characterized by precise expression output (Liu et al. 2016; Rodriguez-Granados et al. 2016; Veluchamy et al. 2016).

9.3.1.2 The Genome-Wide Distribution of HPTMs Shapes the Epigenetic Landscape

The recent development of genome-wide analysis of epigenetic mark distribution has shown that histone PTMs together with DNA methylation (see below) can form specific combinations that define genome territories with either active or repressive chromatin states in multiple organisms from metazoa (Baker 2011) to plants, including rice (Li et al. 2008), Arabidopsis (Luo et al. 2013; Roudier et al. 2011; Sequeira-Mendes et al. 2014; Wang et al. 2015), and barley (Baker et al. 2015). These studies allowed the identification of a finite number of chromatin states along chromosomes, characterized by distinct sets of epigenetic marks. Interestingly, genomic elements are often distinguished by specific chromatin states. For example, in Arabidopsis, silent heterochromatin is associated with H3.1, H3K9me2, H3K27me1, and 5mC, and the transcriptional start site (TSS) of many actively transcribed genes with a combination of H2Bub, H3K36me3, and H3K4me3. Alternatively, repressed genes present in euchromatic regions are associated with H3K27me3 within a nucleosome context enriched in H3.1 (Roudier et al. 2011; Sequeira-Mendes et al. 2014).

Interestingly, some genes are associated with both active and repressive marks, as illustrated by the chromatin state 2 defined by Sequeira-Mendes et al. (2014), where H3K4me2 and H3K27me2 coexist. Such bivalent chromatin states have been described at genes coding for important developmental regulators such as AGAMOUS (Saleh et al. 2007) or floral integrators (Qian et al. 2018) and could be necessary for fine-tuning gene expression.

9.3.1.3 HPTMs Dynamic Is Controlled by Specific Enzymes

Active and repressive histone marks are established and removed by specific enzymes referred to as HPTM writers and erasers, respectively. The level of each HPTM is therefore determined in a dynamic fashion, by the relative abundance/activity of its specific writer(s) and eraser(s) (Fig. 9.1). Although HPTMs are reversible marks, their stability is variable. For example, histone acetylation is a very dynamic epigenetic mark. The estimation of H3 and H4 acetylation turnover rates in human cells revealed very short half-lives (Zheng et al. 2013), with 12 histone sites displaying half-life below one hour (Weinert et al. 2018). As a consequence, modification of histone acetylation status could be essential when rapid changes in gene expression are required, for example in response to environmental stimuli (Barth and Imhof 2010). On the contrary, H3K27me3 was initially considered a very stable epigenetic mark that was conserved through cell division perpetuating the stable repressive state of the chromatin at specific loci. Consequently, H3K27me3 is considered a major determinant of cell identity, although it is now clearly established that this mark can be actively removed by the Jumonji-type of histone demethylases (Chen et al. 2011; Liu et al. 2010; Xiao et al. 2016).

Many genes coding for HPTM writers and erasers have been identified and functionally characterized in Arabidopsis (Fig. 9.1). Most studies have focused on histone methylation and acetylation, so that other HPTMs, such as histone phosphorylation or sumoylation, have been overlooked. Over the past decade, functional analyses of writers and erasers have also been conducted in a few other models and crop species, like tomato (Boureau et al. 2016; How Kit et al. 2010), rice (Jiang et al. 2018a, b; Li et al. 2014; Liu et al. 2017; Zheng et al. 2015), *Brassica napus* (Jiang et al. 2018c), poplar (Fan et al. 2018), wheat (Liu et al. 2018), and maize (Forestan et al. 2018; Rossi et al. 2007). These studies are mainly based on the characterization of genes presenting homologies with those originally identified in Arabidopsis. As shown in Fig. 9.1, each histone mark is set up by a specific set of enzymes, which are frequently specialized in the addition of a precise number of modifications. For example, whereas ARABIDOPSIS TRITHORAX-RELATED PROTEIN 5 (ATRX5) and ATRX6 of the trithorax group are responsible for the addition of one methyl group at histone 3 lysine 27 (H3K27me1) (Jacob et al. 2009). Enhancer of Zeste proteins from the Polycomb group family are part of the Polycomb Repressive Complex 2 (PRC2) and are in charge of the addition of 2 and 3 methyl groups at the same residue (H3K27me3) (Liu et al. 2010; Fig. 9.1).

In addition, most writers and erasers function as multiprotein complexes. As mentioned above, the Enhancer of Zeste (E(z)) proteins which catalyze the H3K27 trimethylation is part of the PRC2 complex. PRC2s contain three additional core proteins, a protein of the Suppressor of

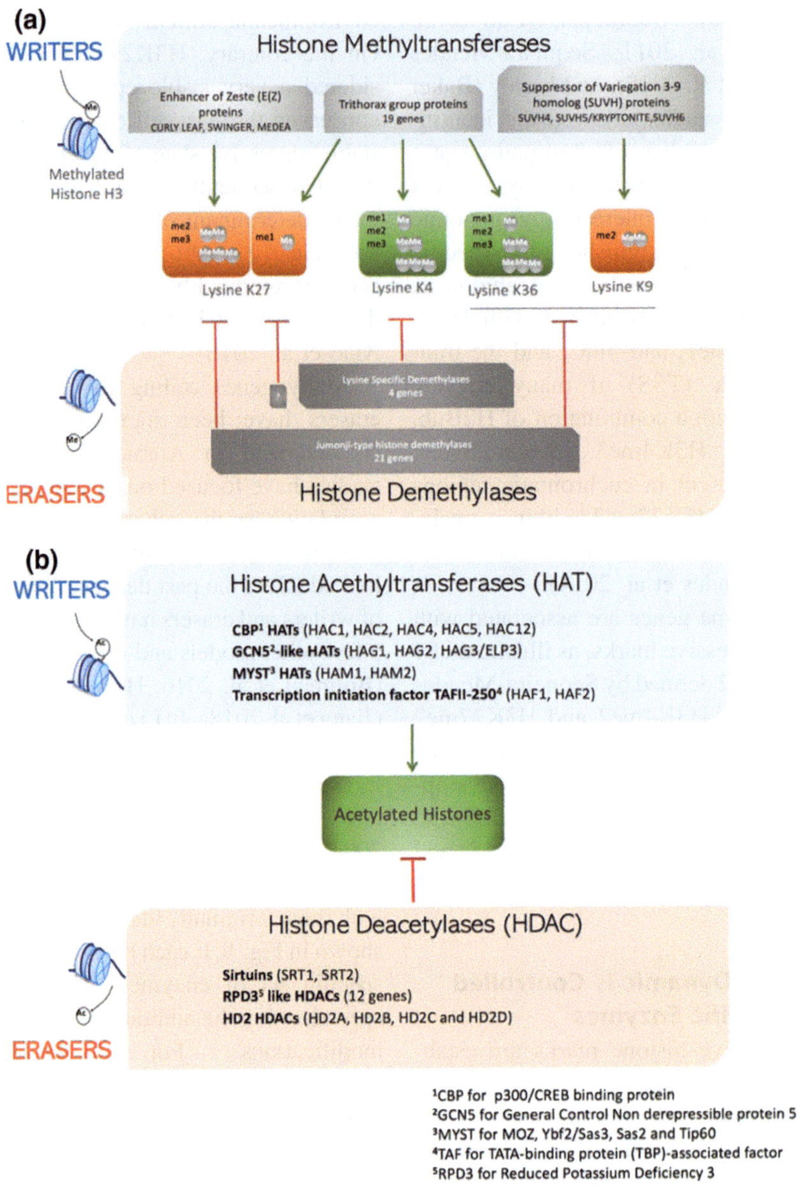

Fig. 9.1 Histone H3 major post-translational modifications and corresponding enzymes. **a** Proteins responsible for histone H3 methylation/demethylation. Depending on the modified lysine residue (lysine K4, K9, K27, or K36), different protein families are involved. Moreover, different proteins may be required depending on the number of methyl residues added/eliminated, as reviewed in Liu et al. (2010); Chen et al. (2011); and (Xiao et al. 2016). **b** Proteins responsible for histone acetylation and deacetylation. For each type of regulators, the number of genes found in the Arabidopsis genome is specified. In a few cases, the name of these genes is indicated. Of note, for gene families which include a large number of genes, such as the trithorax group proteins, only a few genes have been functionally characterized. The transcriptional state (active or inactive) mainly associated with each HPTM is indicated using the following color code: active in green/inactive in red

Zeste 12 (Su (z)12) family, a protein of the Extra Sex Comb (ESC) family, and a Multicopy Suppressor of IRA 1 (MSI) protein. The four PRC2 core proteins are necessary for PRC2 to function in vivo (Schubert et al. 2005), but only the E(z) protein harbors the methyltransferase catalytic domain (the so-called SET domain). Many histone deacetylases (HDACs) have also been shown to associate with other proteins to form multi-subunit complexes, suggesting that they function cooperatively with other epigenetic regulators and in association with transcription factors (for recent results, see Hung et al. 2018; Kim et al. 2016; and Yu et al. 2017).

Another important common trait of writers and erasers in plants is that they are both encoded by multigene families leading to the production of multiple isoforms that controls each histone PTM. In Arabidopsis, for example, the E(z) proteins are encoded by three genes, respectively, *CURLY LEAF* (*CLF*), *SWINGER* (*SWN*), and *MEDEA* (*MEA*). Hence, a variety of PRC2 complexes are produced, which act in a redundant manner and/or at distinct developmental transitions during the life cycle (Chanvivattana et al. 2004; Derkacheva and Hennig 2013; Kinoshita et al. 2001; Mozgova and Hennig 2015).

9.3.1.4 A Diversity of Mechanisms Is Involved in the Targeting of Histone Writers/Erasers

The molecular mechanisms responsible for the recruitment of the epigenetic writers and erasers to their specific target loci have been a long-standing question. Recent data suggest that different mechanisms may be involved (Deng et al. 2018). Although this does not appear as a general feature, some enzymes responsible for histone mark editing contain DNA-binding domains, which participate in their recruitment at specific DNA consensus sequences. As an example, relative of early flowering, also known as Jumonji domain-containing protein 12 (JMJ12), which specifically demethylates H3K27me3 (Lu et al. 2011), recognizes a CTCTGYTY motif through its four Cys2His2 zinc fingers (Cui et al. 2016; Li et al. 2016). A second and more general mechanism involves transcription factors and corepressors, which can recruit epigenetic regulators either through direct protein–protein interactions or because they are partners in the same multi-subunit complexes (Vachon et al. 2018). This has been demonstrated for a number of different epigenetic regulators including PcG proteins (Questa et al. 2016; Roy et al. 2018; Xiao et al. 2017; Yuan et al. 2016; Zhou et al. 2018), Jumonji domain-containing histone demethylases (Cheng et al. 2018b; Hou et al. 2014; Ning et al. 2015; Zhang et al. 2015), and HDACs (Cheng et al. 2018c; Tang et al. 2016a, 2017). In addition, transcription factor binding at specific gene regulatory regions can induce the displacement of writers/erasers from their target loci, as demonstrates at least in two plant studies (Luo et al. 2018; Sun et al. 2014). Non-coding RNAs are also involved in the targeting of HPTM regulators. Two long non-coding RNAs play a role in the repression of FLOWERING LOCUS C (FLC) expression by PcG proteins (Heo and Sung 2011; Kim et al. 2017; Kim and Sung 2017), participating in their recruitment through an uncharacterized mechanism (Kim et al. 2017). Also, an intronic non-coding RNA was shown to be necessary for the CLF-dependent repression of AGAMOUS (Wu et al. 2018). Whether this mechanism is more general remains to be demonstrated. Finally, a few epigenetic regulators are recruited through their interaction with other epigenetic marks, or of histone variants, thereby generating specific epigenetic mark combinations. For example, according to the canonical model, PRC1 complexes are recruited to PcG target genes through the recognition of H3K27me3, leading to the addition of the H2Ub marks at the same loci and to the stable repression of the corresponding genes (Del Prete et al. 2015).

Altogether, these mechanisms ensure that writers and erasers are recruited only at specific loci at specific times. In addition, HPTM editing can be controlled through the regulation of the production of the writers/erasers and of their enzymatic activity.

9.3.1.5 Regulation of HPTM Remodeling

A few epigenetic regulators are expressed at specific developmental stages or in response to precise environmental changes. For example, MEDEA, an E(z) coding for an H3K27me3 methyltransferase, is specifically expressed in the female gametophyte, in the endosperm or in response to an infection by a pathogen (Chaudhury et al. 1997; Roy et al. 2018; Spillane et al. 2000; Yadegari et al. 2000). Another example is the histone demethylase JMJ30, whose expression oscillates with a circadian rhythm and plays a role in the regulation of the period length (Jones et al. 2010; Lu et al. 2011). Hence, as a first regulation level, cells can control the timing of epigenetic changes by a tight regulation of the synthesis of the epigenetic writers/erasers, at least in some specific cases. In addition, epigenetic regulators can be post-translationally regulated through direct protein–protein interactions. For example, the activity of the histone deacetylase HDA6 has been shown to be regulated by phosphorylation (Yu et al. 2017), the activity of histone methyltransferase ATX1 by O-GlcNacylation (Xing et al. 2018), and the activity of the histone methyltransferase CLF by an F-box protein responsible for protein ubiquitylation before their degradation through the ubiquitin–26S proteasome (Woong et al. 2011). Moreover, as described above (Sect. 9.3.1.4), histone modifiers can also be controlled by transcription factors through a regulation of their recruitment and/or eviction to/from their target sites. On top of that, an increasing number of data suggest that HPTM is under metabolic control (for a review, see: Shen et al. 2016). Indeed, several regulators use metabolites as substrate or cofactor: for example, histone acetyltransferases, which necessitate acetyl-coA, and histone methyltransferases, which depend on S-adenosyl methionine availability.

As described in the above paragraph, our knowledge about the mechanisms underlying gene expression regulation through HPTM is rapidly growing, revealing a tight cross talk between histone modifiers, chromatin remodeling complexes, and the transcription machinery (Ojolo et al. 2018). In addition, multiple histone-related epigenetic regulators may be required in a highly coordinated manner for the proper control of gene expression, as it has been demonstrated for *FLOWERING LOCUS C* (*FLC*) coding for a central floral repressor in Arabidopsis (Berry and Dean 2015; Fletcher 2017; Hepworth and Dean 2015; Whittaker and Dean 2017). In addition, HPTMs do not act alone, but in combination with DNA methylation. Several data suggest a functional coupling between histone and DNA methylation, including the aforementioned interaction between H3K9me2 and the DNA methyltransferase CMT3 (for reviews: Du et al. 2015; Torres and Fujimori 2015).

9.3.2 DNA Methylation

DNA methylation is an important and a highly conserved epigenetic mark that has been studied in detail in fungi, animals, and plants and plays fundamental roles in genome functioning and protection. It refers to the transfer of a methyl group to the fifth position of the cytosine ring of nuclear genomic DNA to form 5 methylcytosine. In contrast to mammalian where DNA methylation mainly occurs at CG sites, in plants genomic DNA can be methylated in all cytosine sequence contexts, including the symmetrical CG, CHG motives, and the non-symmetrical CHH motif (which H represents A, T, or C) (He et al. 2011; Law and Jacobsen 2010). Each sequence context requires different mechanisms for establishment and maintenance of DNA methylation (Fig. 9.2).

9.3.2.1 Mechanisms of DNA Methylation in Plants

The mechanisms that control both initiation and maintenance of DNA methylation have received much attention in Arabidopsis (Matzke et al. 2015; Matzke and Mosher 2014; Law and Jacobsen 2010), although studies have also been performed in crop plants including corn, rice, and tomato (Chodavarapu et al. 2012; Corem et al. 2018; Eichten et al. 2013; Fu et al. 2018a; Hu et al. 2014; Li et al. 2012). DNA replication is a semiconservative process that leads to the formation of hemi-methylated DNA molecules. During replication, only non-methylated cytosines are incorporated in the newly synthesized DNA strand. Cells have therefore developed specific mechanisms to fully re-establish DNA methylation patterns. In mammalian, this is insured by the enzyme, Dnmt1, that recognizes

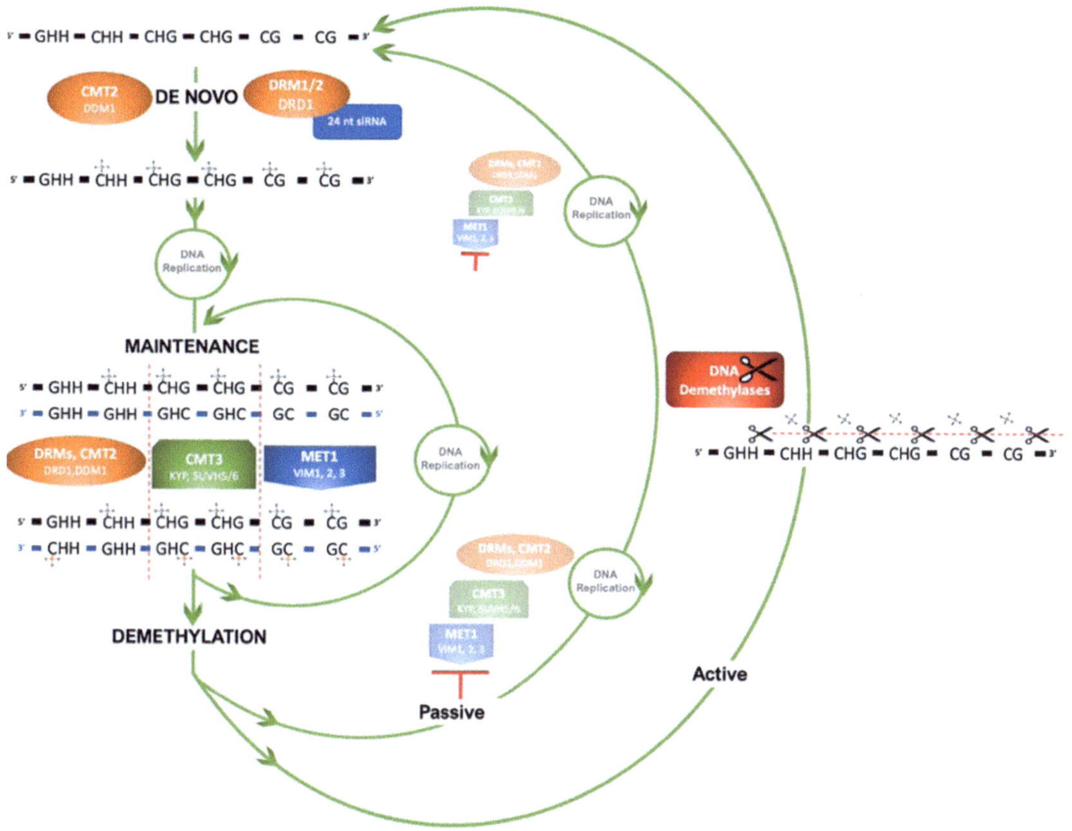

Fig. 9.2 Mechanisms of de novo and maintenance of DNA methylation in plants. DNA methyltransferases and demethylases are involved in 5mC de novo methylation, maintenance of methylation, and demethylation in higher plants. Names of enzymes are those identified in the Arabidopsis model. De novo DNA methylation is set up by the RNA-directed DNA methylation (RdDM) pathway involving the DRM1/2 methyltransferases, DRD1, and 24nt-long small RNAs, and by the chromomethylase CMT2 with DDM1 in the CHH sequence context at heterochromatic regions (Zemach et al. 2013). After replication, newly produced DNA is hemi-methylated at CG and CHG symmetrical sites, but at the non-symmetrical CHH sites only one of the two newly synthesized DNA molecules is not methylated. Maintenance of methylation in the CG context depends on MET1 and VIM1, 2, and 3, and maintenance in the CHG context is catalyzed by CMT3. CHH maintenance of methylation depends both on the RdDM pathway and on CMT2 activity. Both CMTs are dependent on histone methylation mediated by KYP and SUVH5 and 6. DNA demethylation can occur passively in a replication-dependent way, when the methylation machinery is not or poorly active. 5mC cytosine can be actively removed by DNA glycosylase/lyase, also called DNA demethylase, independently from DNA replication. Newly synthesized DNA strands are colored in deep blue. Shaded figures represent enzymes showing reduced activity. Enzyme names are from Arabidopsis. DRM1/2, CMT2/3 (chromomethylase2/3), MET1 (cytosine DNA methyltransferase 1), VIM1-3 (variant in methylation1-3), KYP/SUVH4 [KYP/Su-(var)3-9 homolog 4], SUVH5/6 [Su-(var)3-9 homolog 5/6], DRD1 (defective in rna-directed DNA methylation), DDM1 (decrease in DNA methylation), and 24nt siRNA (24 nucleotide small interfering RNAs)

hemi-methylated DNA template at CG motives (Law and Jacobsen 2010). In plants, different mechanisms that are specific to each of the sequence contexts for DNA methylation have been identified that fulfill these tasks (Fig. 9.2). The DNA methyltransferase 1 (MET1), which is orthologous to the mammalian Dnmt1 (Achour et al. 2008; Sharif et al. 2007), is recruited to hemi-methylated DNA by VIM1 and 2 (variant in methylation 1 and 2) and insures the maintenance of methylation at CG sites (Vongs et al. 1993). Both VIM1 and 2 proteins contain an

SRA (SET- and RING-associated) domain that mediates their binding to hemi-methylated DNA (Kim et al. 2014; Woo et al. 2008). The CHG methylation is maintained by plant-specific DNA methyltransferases, namely the chromomethylases (CMTs), that include CMT3 in Arabidopsis (Bartee et al. 2001; Bewick et al. 2017; Jackson et al. 2002) and its maize homolog ZMET2 (Du et al. 2012; Papa et al. 2001). CMTs contain a BAH domain (bromo-adjacent homology) and a chromodomain (chromatin organization modifier) that is required to their binding to histone H3 when dimethylated on lysine K9 (H3K9me2). Genome-wide analysis of CMT3 distribution has shown that it co-localizes with H3K9me2, an interaction that seems necessary for CMT3 activity in vivo (Bernatavichute et al. 2008; Du et al. 2012). Based on the current model, CMT3 and ZMET2 are recruited to their targets following binding to H3K9me2, which is set up by suppressor of variegation homolog 4 (SUVH4)/KRYPTONITE (KYP), SUVH5, and SUVH6 (Bartee et al. 2001; Du et al. 2014; Gouil and Baulcombe 2016; Jackson et al. 2002). Consistent with this view, mutations impairing SUVH4/KRYP present a dramatic reduction in both H3K9me2 and CHG methylation levels (Jackson et al. 2002; Malagnac et al. 2002). As SUVH4/KRYP contains an SRA domain that allows its recruitment to methylated DNA, it is thought that CMTs and KRYP are working in a regulatory loop to maintain CHG methylation (Du et al. 2014). Finally, CHG methylation and H3K9me2 interactions are further highlighted by the study of the *ibm1* mutant (increase in bonsai methylation) that shows an increased level of both H3K9me and CHG methylation at active genes (Miura et al. 2009). The *IBM1* gene encodes a Jumonji type of histone demethylase necessary to eliminate H3K9me2 at genes, thereby preventing CHG methylation and insuring an active chromatin state (Inagaki et al. 2010; Saze et al. 2008). Recently in Arabidopsis, CMT2 was shown to maintain CHH and CHG methylation in large heterochromatin peri-centromeric regions enriched in large transposons (TEs) (Zemach et al. 2013), most likely via its interaction with the H3K9me2 histone PTMs (Stroud et al. 2014).

Maintenance of methylation at CHH sites and initiation of DNA methylation at non-methylated sites irrespective to the sequence context are both catalyzed by a third class of DNA methyltransferases, the domain rearranged methyltransferases (DRMs; reviewed in Law and Jacobsen 2010). These enzymes are directed to their target loci by 24 nt small interfering RNA (siRNA) in a process named RNA-directed DNA methylation (RdDM; Matzke et al. 2015). The synthesis of these small RNAs has been studied in great details in Arabidopsis over the last decades and will not be discussed here as several recent reviews are available (Matzke et al. 2015; Matzke and Mosher 2014; Wendte and Pikaard 2017).

9.3.2.2 DNA Demethylation

Although DNA methylation is considered as a stable epigenetic mark, reprogramming of DNA methylation patterns has been observed in various plant tissues and at specific developmental stages (Li et al. 2018). DNA methylation can be either actively removed or passively lost (Fig. 9.2; Law and Jacobsen 2010). Passive demethylation occurs after DNA replication when non-methylated cytosines incorporated in the newly synthesized DNA strand cannot be methylated because the DNA methylation machinery is not operating. This results in a rapid and non-specific dilution of methylation as observed in *met1* and other mutants affected in methylation control that presented a general decrease in DNA methylation levels (Cokus et al. 2008; Stroud et al. 2013). In contrast, active demethylation can specifically eliminate methylated cytosines at specific loci. Active demethylation has been observed during endosperm development and imprinting (Bauer and Fischer 2011; Choi et al. 2002; Hsieh et al. 2009; Schoft et al. 2011), gametophyte and gamete development (Park et al. 2016), tomato fruit ripening (Liu et al. 2015), and for the establishment of a successful symbiosis with *Bradyrhizobium* in Medicago (Satgé et al. 2016). Plant active DNA demethylation is catalyzed by bifunctional enzymes, the DNA glycosylase/lyases (DNA GLs) initially identified in Arabidopsis. The Arabidopsis genome contains four genes encoding DNA GLs: *REPRESSOR OF SILENCING 1* (*ROS1*), *DEMETER* (*DME*), and two *DEMETER-like* (*DML*) genes, *DML2* and *DML3*; (Choi et al. 2002; Gong et al. 2002; Ortega-Galisteo et al. 2008; Penterman et al. 2007). ROS1 and DME display in vitro nicking activity on methylated DNA consistent with their DNA GL activity; DNA demethylation requires cytosine removal, a process that involves the cleavage of the DNA backbone at the site of cytosine removal mediated by the AP lyase activity of ROS1 and subsequent reparation by an unknown mechanism which likely involves a putative polynucleotide kinase, a DNA polymerase, and a DNA ligase (Li et al. 2018). This results in the removal and replacement of methylated cytosines via a pathway related to base excision repair (BER; Agius et al. 2006).

Studies in Arabidopsis have suggested that multiple factors may lead the DNA demethylases to their targets (Li et al. 2018). These include ROS3 (Zheng et al. 2008), ROS4, a histone acetyltransferase, also known as IDM1 (increase in DNA methylation 1) (Qian et al. 2012), the methyl-CpG-binding protein 7 (MBD7; Lang et al. 2015), the Harbinger transposon-derived protein 1 and 2 (HDP1 and 2; Duan et al. 2017), and other partners (Li et al. 2018) that cooperate to address ROS1 to its target loci. In addition, expression of *DML* genes seems to be tightly controlled in plants. Indeed, *ROS1*, *DML2*, and *DML3* gene expressions are rather ubiquitous in Arabidopsis (Ortega-Galisteo et al. 2008; Penterman et al. 2007) as is the expression of the tomato *ROS1* orthologous genes, *SlDML1* and *SlDML2* (Liu et al. 2015). However, some of the *DML* genes display distinct patterns of expression and have been recruited for specific developmental functions. This is the case for *DEMETER* (*DME*) gene in Arabidopsis and related species. *DME* is specifically expressed in the central cell of the megagametophyte, which restricts DME activity to this cell type. Another example is the *SlDML2* gene that in addition of its general expression in young plant tissues together with *SlDML1* is the only tomato *DML* gene strongly overexpressed at the onset of fruit ripening, which correlates with its role in the induction of fruit ripening (Liu et al. 2015). Recent evidence also indicates that DNA methylation levels may also participate in controlling *DML* gene expression. This was suggested following the observation that expression of the *ROS1* gene is repressed in the Arabidopsis met1 or RdDM mutants, which are characterized by a hypomethylated genome (Mathieu et al. 2007). More recently, the *ROS1* promoter was shown to contain a 39 bp DNA methylation monitoring sequence (MEMS) that acts like a "methylstat" able to sense DNA methylation level and control *ROS1* expression, thereby maintaining a dynamic balance between DNA methylation and active DNA demethylation (Lei et al. 2015; Williams et al. 2015).

9.3.2.3 DNA Methylation Distribution in Plants

The development of genome-wide strategies to analyze DNA methylation such as methylated DNA immunoprecipitation sequencing (MeDIP-seq) or whole-genome bisulfite sequencing (WGBS; Beck and Rakyan 2008; Kim et al. 2014; Yong et al. 2016) has allowed determining the distribution of DNA methylation in several eukaryotes. Among these two methods, WGBS is considered the golden standard method as it allows unraveling the position of methylated cytosines at one base resolution and therefore provides the most precise view of the distribution of 5mC in eukaryote genomes (Yong et al. 2016). In plants, the description of the genome-wide distribution of methylated cytosines was first reported in Arabidopsis (Cokus et al. 2008; Zhang et al. 2006; Zilberman et al. 2007). An increasing number of plant methylomes has now been investigated (Niederhuth et al. 2016), including crops such as rice (Li et al. 2012), maize (Eichten et al. 2013), and tomato (Zhong et al. 2013). Results indicate that DNA methylation levels vary significantly between species irrespective of the sequence contexts although in most cases similar rules seem to apply (Niederhuth et al. 2016). In plants, CG methylation is the highest in all species tested and can vary up to threefold between species: The lowest mCG content was found in Arabidopsis (circa 30%; Niederhuth et al. 2016) and the highest in *Beta vulgaris* (circa 90%; Niederhuth et al. 2016). In the plant species analyzed, mCHG and mCHH contents were found at lower levels than CG methylation and ranged between 9.3 and 81.2% and between 1.4 and 18.8%, respectively, and the highest levels being found in *Beta vulgaris* in each case. The range of methylation variations in these two contexts is therefore much higher than the one observed for the CG context. When considering the distribution of mC within genomes, various studies have shown that the centromeric and peri-centromeric regions of chromosomes that are enriched in transposable elements (TEs) and tandem repeats are the most heavily methylated (Cokus et al. 2008; Lister et al. 2008; Seymour et al. 2014), although some variations between plant species were observed (Niederhuth et al. 2016). High methylation levels at TEs are consistent with 5mC being of primary importance in the control of their activity and are thought to inhibit their transcription (Cui and Cao 2014).

The distribution of DNA methylation differs in genes and TEs, and presents common features between plant species. First, early work on Arabidopsis showed that only 5% of the genes were methylated within their promoter region (Zhang et al. 2006). However, these studies were performed using mixture of tissues, which makes difficult to determine the precise number of genes with methylated promoters and the relation with gene expression. Since that time, other studies have analyzed organ-specific DNA methylation patterns in relation to gene expression profiles demonstrating an inverse correlation between DNA methylation in promoters and gene expression. For example, analysis of DNA methylation during soybean seed development and maturation has allowed identifying 40, 66, and 2136 genes with changes in DNA methylation levels in the CG, CHG, and CHH contexts, respectively. Most of the genes with differentially methylated regions in the CHH context showed a negative correlation between methylation and expression levels (An et al. 2017). Similarly in tomato fruit, low methylation levels at promoters of a subset of ripening-induced genes have been correlated with gene expression (Lang et al. 2017; Liu et al. 2015; Zhong et al. 2013). Thus, promoter methylation is likely associated with the repression of gene expression although recent evidence suggests that the opposite is also possible (Lang et al. 2017).

The body of genes was also shown to be methylated, but only in the CG context, a process called gene body methylation (GbM). GbM seems conserved in plants and affects orthologous genes between species (Takuno and Gaut 2011); depletion of CHG and CHH methylation in gene bodies suggests that these two types of methylation are antagonist to transcription elongation whereas CG methylation is not (Coleman-Derr and Zilberman 2012; Feng et al. 2010; Takuno and Gaut 2011; Zemach et al. 2010; Zilberman et al. 2007). For now, the function of GbM is not understood. In Arabidopsis, more than 20% of the genes harbor GbM, corresponding in general to genes that are moderately expressed and constitutively active (Zhang et al. 2006; Zilberman et al. 2007). However, some plants have lost GbM methylation, suggesting it either is not required for plant viability or can be compensated by other mechanisms (Bewick and Schmitz 2017). Such situations remain rare, which suggests that GbM plays an important function in plants, still to be discovered. Interestingly, in Arabidopsis GbM seems to partially depend on latitude, which may reflect an adaptive function to the environment (Dubin et al. 2015). In addition to GbM, CHG and CHH methylations can also be found in the body of genes. CHG genes are usually expressed at low levels, below all genes, and those with CHH methylation, also called RdDM genes, are not expressed (Niederhuth et al. 2016; Bewick and Schmitz 2017).

The recent literature we have summarized here clearly shows that the function of DNA methylation in plants is complex and depends on both the sequence context and the localization.

9.4 Epigenetic Regulations in Fleshy Fruit

9.4.1 Evidence that HPTMs Are Essential to Fleshy Fruit Development

As mentioned above, HPTMs are critical to many plant development processes, and recent evidence indicates that these epigenetic marks are essential during fruit development and ripening (Bucher et al. 2018; Gallusci et al. 2016). Genes encoding histone deacetylases (HDACs), histone acetyltransferases (HATs), histone methyl transferases (HMTs), and histone demethylases (HDMs) have been identified in several fleshy fruit species such as apple (Janssen et al. 2008), banana (Fu et al. 2018a, b), sweet orange (Xu et al. 2015), strawberry (Gu et al. 2016), and tomato (Cigliano et al. 2013; Zhao et al. 2015). Studies have shown that some of the genes encoding histone modifiers are preferentially expressed in fruit, with stage-specific expression patterns that depend on both fruit species and HPTM modifiers. In grapevine, genome-wide analysis has revealed 33 gene-encoding proteins containing a SET domain, 10 *PRC2* genes, and 7 and 13 genes coding for putative HATs and HDAC, respectively. Some of these genes show expression patterns consistent with a possible involvement in grape berry development and ripening (Almada et al. 2011; Aquea et al. 2010; Aquea et al. 2011). Overall, these observations suggest that the corresponding proteins are recruited for the control of fruit development, ripening, and abscission in fleshy fruit species. Although not in grapevine, evidence of their role in fruit development was provided by loss and gain of function in tomato (for recent reviews: Bucher et al. 2018; Gallusci et al. 2016; Giovannoni et al. 2017).

Early studies have analyzed the tomato's high pigment mutants (*hp1*, *hp2*) which present increased carotenoid content in fruits. The corresponding tomato genes encode two subunits of an ubiquitin ligase complex, DDB1 and DET1, respectively (Tang et al. 2016b). In human, this complex is known to target histone proteins for ubiquitination in response to DNA damages (Hu et al. 2004; Wang et al. 2006). In tomato, by impeding light signal transduction by preventing the ubiquitination of H2B histones (Benvenuto et al. 2002; Lieberman et al. 2004), these mutations may affect the transcriptional repression of genes involved in the production of carotenoids and other compound/s, therefore generating the enhanced pigmented fruit phenotype. More recently, silencing studies were conducted in tomato on different components of the histone modifier complex PRC2 (Polycomb repressive complex 2). They targeted genes encoding the enhancer of zeste EZ1 and EZ2 proteins (Boureau et al. 2016; How Kit et al. 2010) and the FIE protein (Fertilization-Independent Endosperm Development; Liu et al. 2012). These studies revealed the roles of these genes during flower formation and early fruit development (reviewed in: Bucher et al. 2018; Gallusci et al. 2016). In a more recent work, impairment of MSI1 (multi-suppressor of IRA 1), a putative component of the tomato PRC2s, was shown to affect fruit ripening (Liu et al. 2016). However, MSI1 is also a member of the CAF-1 complex involved in chromatin assembly (Henning et al. 2005). As none of the other PRC2 components affect fruit ripening when repressed, it is possible that the effect on ripening is due to impairment of the CAF-1 complex activity rather than to the inhibition of PRC2 activity. Indeed, chromatin assembly activity might be of primary importance in tomato fruit due to the high endoreduplication level (Teyssier et al. 2008). Finally, other studies have shown that the control of histone acetylation is also important to fine-tune induction of ripening. For example, plants with reduced activity of various HDACs present delayed carotenoid accumulation and ripening (Guo et al. 2017a, b) or an opposite effect on both processes (Guo et al. 2018).

Evidence of the role of HPTMs in fruit was also provided in the frame of the fruit ENCODE project that aimed at analyzing the evolution of fleshy fruit ripening control in angiosperms. Combined genetic and epigenetic approaches were implemented on 13 different fruit species including (1) climacteric fruit species (tomato, apple, pear, banana, melon, papaya, and peach), (2) non-climacteric fruit species (grape, strawberry, cucumber, and watermelon), and (3) dry fruit species (Arabidopsis and rice; Lü et al. 2018). The project generated multidimensional dataset based on transcriptomic DNA methylation and histone PTMs with a focus on H3K27me3 and H3K4me3 profiles to decipher genetic and epigenetic events controlling fruit ripening (Lü et al. 2018). In this context, researchers focused on key molecular players involved in ethylene-dependent ripening circuits in climacteric fruit and their orthologues in non-climacteric and dry fruit. Although global- and locus-specific DNA methylation changes were observed in all fruit species during ripening induction, DNA demethylation was suggested to be only required for tomato ripening. However, these conclusions were based on correlative studies without functional foundation and are not consistent with the recent demonstration that in addition to tomato fruit ripening (see below; Lang et al. 2017; Liu et al. 2015), strawberry fruit ripening and sweet orange fruit ripening are also under DNA methylation control although different mechanisms are operating (Cheng et al. 2018a; Huang et al. 2019). In contrast, Lü et al. (2018) suggested that, instead of DNA methylation, the repressive mark H3K27me3 may play a conserved—and maybe central—role in regulating fruit ripening, although its precise function and importance may vary between fruit species. Indeed, for a few ripening-related genes, a correlation was found between their induction during ripening and the removal of H3K27m3 in several fruit species, therefore suggesting an ancestral inherited role for this mark in angiosperm fruit ripening (Lü et al. 2018). Interestingly, a recent study indicates that H3K27me3 may be involved in the control of methoxypyrazines (MPs) accumulation in grape berries, a compound known to contribute to the herbaceous characters in wine (Battilana et al. 2017). MPs biosynthesis depends on the expression of the *VvOMT3* gene which encodes a protein controlling the final and key step of this biosynthetic pathway in grape. However, MP accumulation is variety dependent. For example, berries from Cabernet Sauvignon accumulate MPs, but those of the Pinot Meunier-derived dwarf do not. A recent study has shown the mark H3K27me3 is abundant at the *VvOMT3* locus in Pinot Meunier dwarf but not in Cabernet Sauvignon berries (Battilana et al. 2017), suggesting that H3K27me3 inhibits *VvOMT3* gene expression resulting in the inhibition of MP biosynthesis. Although these results are consistent with an important role of H3K27me3 in fruit ripening control, this mark does not seem to be critical for ripening in all fleshy fruit species shown in tomato (Boureau et al. 2016; How Kit et al. 2010; Liu et al. 2012).

The characterization of PRC2 mutants or of mutants affected in the removal of the H3K27me3 mark will now be necessary to better assess the importance of this epigenetic mark in modulating the epigenetic landscape and its consequences on gene expression and fruit phenotypes

9.4.2 DNA Methylation Role in Fruit Development and Shape

So far, very few studies have investigated the possible role of epigenetic mechanisms in the control of organogenesis and early development of fruit. However, a few examples show that DNA methylation is likely part of the regulatory networks that control fruit shape and size. One recent example is provided by the analysis of the mantled phenotype in oil palm (*Elaeis guineensis*) that was identified in plants generated by somatic embryogenesis (Rival et al. 1998). Oil palm plants with the mantled phenotype are characterized by the development of flowers with carpeloid structures in place of male organs leading to the formation of a fruit with various phenotypes ranging from normal-looking fruits to very small fruits (Dussert et al. 2000). This phenotype was recently shown to be caused by the hypo-methylation of a Karma-like LINE retrotransposon located within an intron of the *DEFICIENS* (*DEF*) gene. Normal fruits develop when the Karma retrotransposon is methylated, whereas its hypo-methylation leads to the mantled phenotype due to the inhibition of *DEF* splicing (Ong-Abdullah et al. 2015). For tomato, impairing DNA demethylases does not only inhibit ripening (see Sect. 9.4.2.1), but also alter flower and fruit shape. In particular, fruit presented a significant increase in the number of locules that resulted from an increased number of carpels formed during flower development (Liu et al. 2015). However, it is still unclear whether this effect is a direct or indirect consequence of a deficient demethylation process.

A final example comes from the analysis of apple fruit development using two double haploid apple varieties with fruit, whose size correlates with the number of cells in the fruit (Daccord et al. 2017). While these two varieties have genomes that only differ by a limited number of single-nucleotide polymorphisms (SNPs), 294 differentially methylated regions (DMRs) were identified in proximity to genes that could be involved in fruit growth and development. The causal relationship between these DMRs and difference in fruit size is still elusive (Daccord et al. 2017).

9.4.2.1 Evidence that DNA Methylation Is Critical to Fruit Ripening

DNA methylation changes were first documented in tomato decades ago by Hadfield et al. (1993), who showed that genes induced at the onset of fruit ripening had changes in their methylation state. Since that time, the description of the *Colorless Non-Ripening* (*Cnr*) epimutation provided compelling evidence that DNA methylation control is essential to fruit ripening (Manning et al. 2006). Fruits of the *Cnr* epimutant are characterized by a severe reduction in ethylene production, an inhibition of fruit softening, and a lack of carotenoid synthesis and accumulation (Thompson et al. 1999). The *Cnr* epimutant phenotype is very stable, and reverting sectors were only observed on 3 individual fruits on independent plants from more than 3000 plants. This allowed the positional cloning of the *CNR* locus that was shown to contain only one gene differentially regulated between *Cnr* and WT fruits, yet without any sequence differences between both genetic backgrounds (Manning et al. 2006). This gene, which encodes a SQUAMOSA promoter-binding protein-like (SBP-box/SPL) transcription factor, presented a 286-bp-long hyper-methylated region located 2.3 kb upstream of the TSS. Hyper-methylation was only found in the *Cnr* background and resulted in *CNR* gene repression and blocking of fruit ripening (Manning et al. 2006). Additional evidence that methylation upstream of the promoter was responsible for the repression of the *CNR* gene was provided using virus-induced gene silencing (VIGS) to repress the expression of the tomato *CMT3* gene in the *Cnr* background that allowed reversing the *Cnr* phenotype to WT, whereas the same approach using *MET1* or the *DRM* genes had much weaker effects (Chen et al. 2015). This approach was sufficient to reduce methylation at the CHG sites located in the hyper-methylated 286-bp region of the *CNR* promoter and to increase the expression of *CNR* indicating that the methylation of *CNR* gene in the *Cnr* background requires the functional maintenance of methylation machinery. Hence, maintenance of methylation at the *Cnr* locus is necessary for the somatic stability of the epimutation (Chen et al. 2015). Since the description of *Cnr*, other studies have led to the identification of epialleles in tomato. They include the demonstration that variations in vitamin E content of tomato fruit are determined in part by the methylation level of the promoter region of *VTE3*, a gene which encodes a 2-methyl-6-phytylquinol methyltransferase, responsible for an essential step in tocopherol biosynthesis (Quadrana et al. 2014). Methylation variations were observed between tomato accessions that were correlated with changes in *VTE3* gene expression and fruit vitamin E content. Additional epialleles were also identified in the progeny of crossings between M82, a commercial tomato accession, and *Solanum penellii*, a wild tomato relative (Gouil and Baulcombe 2018). However, the stability of the newly generated epialleles was not established in this case. Epialleles that determine the color of the skin were also found in apple and pear (El-Sharkawy et al. 2015; Telias et al. 2011; Wang et al. 2013). In both cases, hyper-methylation of the promoter region of *MYB10* was associated with repression of the gene and of anthocyanin biosynthesis in the skin.

9.4.2.2 DNA Methylation Reprogramming in Fleshy Fruit

Analysis of the global DNA methylation level at different stages of tomato fruit development indicated that the total content in 5mC decreased in the pericarp of tomato fruit from 29.9% at the breaker stage to 20.1% at the red ripe stage (Teyssier et al. 2008). This decrease in DNA methylation level was confirmed by WGBS of the tomato fruit genomic DNA at four developmental stages, namely immature green, breaker, turning, and fully ripe fruit of WT plants and also at two stages in the *Cnr* and *ripening inhibitor* (*rin*) mutant genetic backgrounds, both impaired in the ripening process (Zhong et al. 2013). Results indicated that in addition to a decrease in methylation level at CG sites observed in TEs-rich regions, DNA methylation was also reduced at the promoters of genes that are induced during fruit ripening, including gene-encoding proteins with important role in this process, such as the *CNR*, the *RIN*, or the *NOR* genes (Reviewed in: Bucher et al. 2018; Gallusci et al. 2016; Giovannoni et al. 2017). Noteworthy, CHH methylation is high in tomato (11% in ripe fruit, 13% in non-ripe fruit, and 8.3% in leaves; Zhong et al. 2013) as compared to previously described CHH methylation levels in Arabidopsis (1.5%; Cokus et al. 2008) and in other plants (Niederhuth et al. 2016), and was found higher in fruit (Zhong et al. 2013).

With the aim to investigate the mechanisms underlying the loss of genomic DNA methylation occurring at the onset of fruit ripening, Liu et al. (2015) have identified four tomato genes encoding putative DNA demethylase. One of them, *SlDML2*, was strongly upregulated at the onset of ripening, simultaneously to the decrease in DNA methylation. Inhibition of *SlDML2* gene expression using RNAi and VIGS strategies (Liu et al. 2015) or by CRISPR-Cas9-mediated mutagenesis (Lang et al. 2017) indicated that *SlDML2* is an absolute requirement for tomato fruit ripening to occur. Ripening inhibition was associated with the repression of genes encoding the *RIN*, *NOR*, and *CNR* transcription factors that play a major role in the induction of tomato fruit ripening (Lang et al. 2017; Liu et al. 2015). Of note, the promoter region of these transcription factors is normally demethylated during fruit ripening, whereas loss of *SlDML2* function was associated with the absence of demethylation and gene repression. A similar situation was observed at 600 ripening-induced genes that failed to be expressed and remained hyper-methylated in their promoter region. Interestingly, 598 other hyper-methylated genes normally repressed during the ripening of wild-type tomato fruit maintained their expression level in the mutant background (Lang et al. 2017), suggesting that DNA methylation is also associated with gene expression in tomato fruit.

It was recently suggested in the frame of a fruit ENCODE project that DNA demethylation might not be a general process controlling fleshy fruit ripening and dry fruit maturation, in contrast to H3K27me3 (Lü et al. 2018). However, recent works indicate that DNA methylation control is likely important in other fruits as well. The description of the strawberry fruit methylome indicates that fruit genomic DNA becomes massively demethylated during the ripening process (Cheng et al. 2018b), as observed in tomato (Teyssier et al. 2008; Zhong et al. 2013). Demethylated regions were enriched at a large subset of genes induced during ripening suggesting a direct link with the expression of ripening-induced genes, consistent with the demonstration that the treatment of strawberry fruit with a demethylating agent accelerates fruit ripening (Cheng et al. 2018b). Interestingly, in strawberry, no demethylase-encoding gene could be identified that was involved in the loss of methylation. Decrease in methylation was rather associated with repression of the RdDM pathway that could in turn lead to demethylation at specific loci (Cheng et al. 2018b). In a more recent study, Huang et al. (2019) analyzed the changes in genomic DNA methylation in the skin of orange fruit and demonstrated a general increase in DNA methylation along with fruit development and ripening. Inhibition of methylation by means of azacytidine, a demethylating agent, resulted in delayed ripening indicating that increase in DNA methylation is necessary for orange fruit ripening to occur (Huang et al. 2019). Taken together, these results highlight the general importance of DNA methylation control in fleshy fruit, even though it becomes clear that a diversity of mechanisms is operating depending on the plant species under study (Fig. 9.3).

9.5 Interaction Between Hormones and Epigenetic Regulations in Fleshy Fruit Development and Ripening

Other important regulatory pathways, including hormones and transcription factors, control fruit ripening. Their complex interactions with chromatin-based regulations need to be investigated. Several recent works have illustrated that hormone signaling may involve an epigenetic component (Yamamuro et al. 2016), but very few studies have addressed this question in fruit so far (Lü et al. 2018; Zuo et al. 2018).

Fruit set is known to be under hormonal control, and a diversity of hormones plays a critical role in this process (see Chap. 12). They include auxins, gibberellic acids, or cytokinins that can promote parthenocarpic fruit development when applied alone, although their combined action appears much more efficient in both dry and fleshy fruits (for recent reviews: Joldersma and Liu 2018; Kumar et al. 2014). The involvement of epigenetic regulation during this developmental phase is still poorly studied. At present, evidence is mounting that PRC2 complexes might be involved in this process as illustrated by the elongation of fruit in the absence of fertilization in Arabidopsis PRC2 mutants (Goodrich et al. 1997) and parthenocarpy in tomato (Liu et al. 2012). However, it is not clear whether PRC2s control fruit elongation directly or through auxin signaling. Consistent with the latter, it has been shown that genes involved in auxin biosynthesis or signaling are enriched in the H3K27me3 repressive mark, which is established by PRC2s (Lafos et al. 2011). In addition, *met1* mutants show an elongation of fruit without pollination, suggesting that maintenance of DNA methylation is necessary to prevent fruit development in the absence of fertilization (FitzGerald et al. 2008). In this case, interaction with hormonal regulations has not been yet investigated, even though interplay between PRC2 and DNA methylations has been suggested in the megagametophyte of Arabidopsis developing flowers. Therefore, auxins, DNA methylation, and histone marks could control the induction of seed and fruit development in a coordinate manner.

The role of hormones varies between fruit types, with ethylene being the major player in climacteric fruit, whereas ABA appears to have a more prominent role in non-climacteric fruit (McAtee et al. 2013) including grapevine (Fortes et al. 2015). Yet, the relationship between hormonal and epigenetic regulations in fruit ripening control is still poorly understood. As far as histone PTMs are concerned, a recent study performed in banana has shown that the ethylene response factor11 (MaERF11), a negative regulator of banana fruit ripening, may recruit the MaHDA1 HDAC at the promoters of the *MaEXP2*, *MaEXP7*, *MaEXP8*, and *MaACO1* genes in immature green fruit (Han et al. 2016). This would result in deacetylation and repression of these genes, before ripening induction, an effect that would be relieved by the massive synthesis of ethylene occurring at the onset of ripening (Han et al. 2016). HDACs were also suggested to interact with ethylene to regulate gene expression involved in longan fruit senescence (Kuang et al. 2011). There is, however, stronger evidence that ethylene and DNA methylation interact to control fruit ripening, at least in the tomato (Liu et al. 2015), where genes involved in ethylene biosynthesis are misregulated in *Sldml2* mutants (Lang et al. 2017). Inversely, tomato plants affected in ethylene signal transduction were shown to have deeply modified fruit methylation patterns, consistent with a loop regulation between DNA methylation/demethylation and ethylene biosynthesis in tomato fruit (Zuo et al. 2018).

ABA is thought to play a much more prominent role in the control of ripening of non-climacteric fruit (McAtee et al. 2013). In strawberry, some of the ABA biosynthetic genes

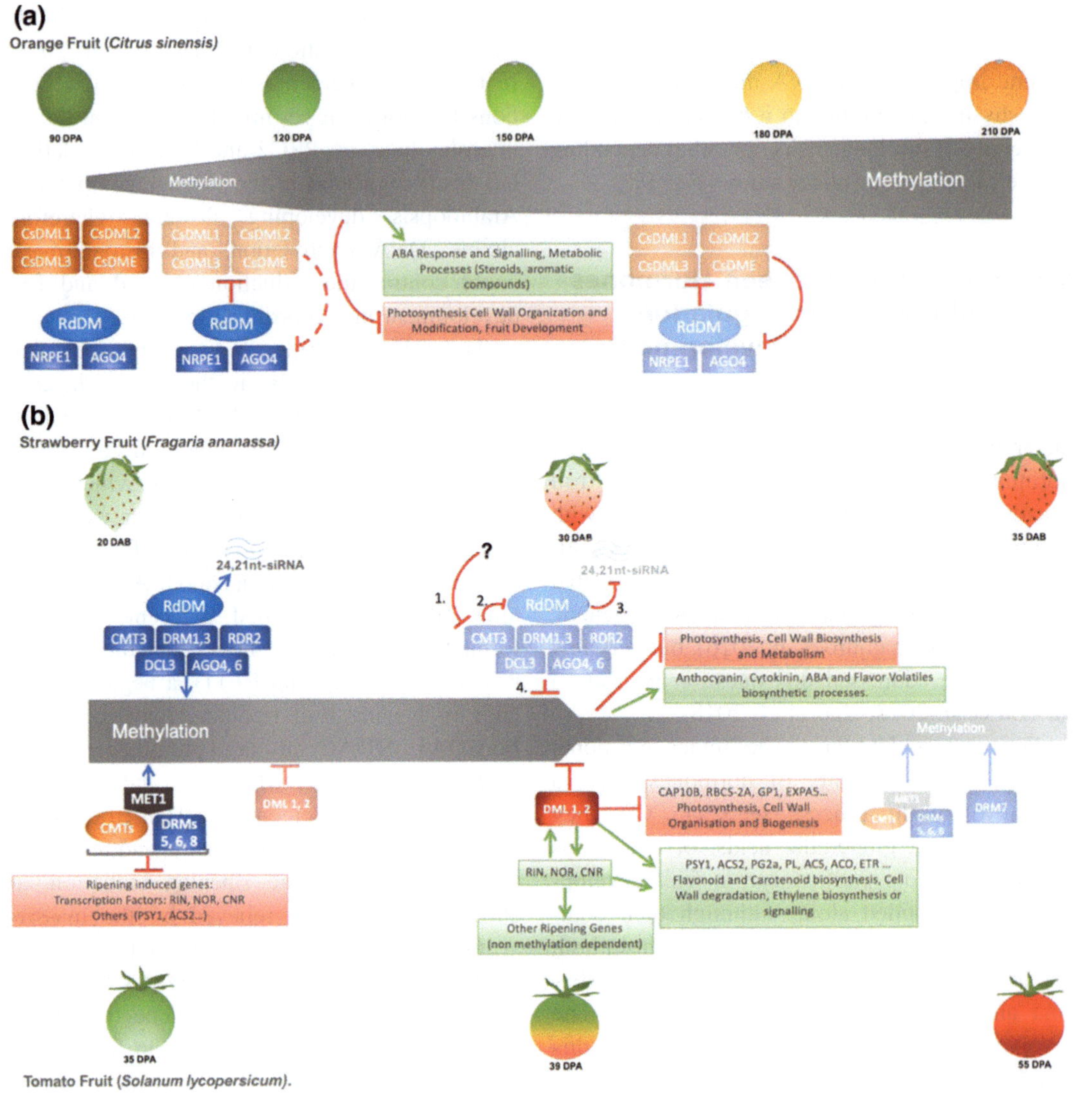

Fig. 9.3 Putative roles of genomic DNA methylation in fleshy fruit. **a** Function of DNA methylation in sweet orange fruit: Genomic DNA methylation increases from 13% of total cytosine in 90 dpa old sweet orange fruit to 14.5% in 210 DPA old fruit. Increase in DNA methylation is correlated with the gradual decrease in the expression of DNA demethylase (DML) genes and of genes involved in the RNA-directed DNA methylation pathway (*NRPE1*, *AGO4*). Ripening-associated hyper-methylated regions were associated with hundreds of genes normally expressed at early stages of fruit development, as those involved in photosynthesis, but also with the induction of several genes involved in orange fruit ripening. Results suggest that DNA methylation is critical to ripening of sweet orange fruit, as confirmed by the ripening inhibitory effect of azacytidine, an inhibitor of genomic DNA methylation. Up- and down-regulated processes shown in the figure are, respectively, associated with DEGs correlated with hyper-DMR (gain of methylation during ripening). **b** Function of DNA demethylation in strawberry fruit and in tomato fruit: genomic DNA methylation in young strawberry immature fruit is 7.5% and decreases during fruit ripening. Loss of methylation occurs at genes involved in the ripening process (anthocyanin accumulation, secondary compound synthesis, etc.), suggesting that demethylation is necessary for ripening induction. Consistent with this view, fruit treatment with azacytidine results in early ripening. Reduction of methylation was correlated with the reduction of the expression of genes involved in the RdDM pathway and with reduced accumulation of short interfering RNAs of 24 nt. In contrast, DNA demethylase-encoding genes are not induced. Genomic DNA methylation decreases from 30% of total cytosine in young immature fruit to 20% in red ripe fruit (Teyssier et al. 2008). Decrease in DNA methylation correlates with up-regulation of *SlDML2*, one of the tomato DNA demethylases. Genes encoding RIN, NOR, CNR transcription factors that control fruit ripening

and other genes encoding enzymes necessary to ripening (phytoene synthase 1, polygalacturonase, etc.) have hyper-methylated promoters and are repressed in immature green tomato fruit (Lang et al. 2017; Liu et al. 2015). Some of the genes involved in photosynthesis are expressed in young fruit even though their promoter is methylated at this stage (Lang et al. 2017). Reduction of DNA methylation that occurs at the onset of fruit ripening necessitates the expression of the *SlDML2* gene (Liu et al. 2015) and correlates with the reduced expression of genes involved maintenance of DNA methylation (Teyssier et al. 2008). Demethylation occurs in the promoter region of many of the genes encoding the CNR, RIN, and NOR transcription factors, as well as of genes involved in carotenoid (phytoene synthase 1), ethylene synthesis (ACC synthase 2), and cell wall metabolism (polygalacturonase, etc.), among others, and is associated with their expression and fruit ripening (Lang et al. 2017; Liu et al. 2015). For some genes (*CAP10*, *RBCS*, etc.) demethylation was correlated with gene repression (Lang et al. 2017). SlMET1 (cytosine DNA methyltransferase 1), CMT (chromomethylase), DRM (domain, rearranged methyltransferase), DML (DEMETER-like demethylase), PSY1 (phytoene synthase 1), ACS2 (ACC synthase 2), RIN (ripening inhibitor), NOR (non-ripening), CNR (colorless non-ripening). Genes in boxes with intense colors (orange, blue, or gray) are strongly expressed. Those in boxes with pale colors are weakly expressed. Green arrows indicate activation, and red bars repression. Repressed processes and genes are indicated in red, and those activated in green

are hypomethylated in their promoter region and present an enhanced expression during ripening (Cheng et al. 2018b). However, there is no evidence of a causal interaction between ABA synthesis and transduction signal and variations in DNA methylation at these genes.

9.6 Conclusions: Specific Aspect of Epigenetic Regulations in Grapevine

The importance of epigenetic regulations has been demonstrated in Arabidopsis, for which a plethora of mutants have been generated that affect the regulation of DNA methylation and histone PTMs and were used to illustrate the prominent roles of epigenetic regulations in plant development and adaptation to stresses. However, it is becoming clear that although epigenetic mechanisms have been conserved within the plant kingdom, they have been recruited for a diversity of developmental processes that may vary between species. In addition, different epigenetic mechanisms may fulfill similar physiological functions in different plants. An example is provided by the function of the DNA demethylase SlDML2 that mediates the active demethylation of tomato fruit genomic DNA, a process necessary to tomato fruit ripening (Liu et al. 2015), whereas in strawberry ripening specific DNA demethylation is controlled by inhibition of de novo methylation through the RdDM pathway (Cheng et al. 2018b), and in some other cases such as sweet orange there is no massive demethylation during fruit ripening (Huang et al. 2019).

Noteworthy, recent works also indicate that epigenetic regulations may have much stronger impacts on plant phenotypes and gene expression in crops than in the model plant Arabidopsis (Gallusci et al. 2016; Mirouze and Vitte 2014). A diversity of reasons may contribute to this observation including the lower methylation level and transposon content of Arabidopsis as compared to most crops (Lee and Kim 2014), and differences in genome organization, for example the distance between genes and transposons (Niederhuth et al. 2016). Genome analysis has shown that the grapevine contains more transposons than Arabidopsis (Jaillon et al. 2007). The most striking difference between the two species is the alternation in grapevines of regions with high and low gene density along chromosomes, together with the high density of transposons nearby genes and within introns. In addition to possible impact on gene expression, higher transposon density increases the probability that their mobility will generate variants due to loss of gene function. Indeed, genetic variations due to transposons that are inserted within or in the vicinity of genes have been described in grape and other plants (Hirsch and Springer 2017; Lijavetzky et al. 2006; This et al. 2007; Verriès et al. 2000). The most striking example is the white color of grape berries that has been shown due to the insertion of the *GRET1* transposon in the promoter region of *MybA1* in berry skin cells (Kobayashi et al. 2004;

Lijavetzky et al. 2006). Hence, the control of transposon mobility is likely to be an important issue in grapevine even more because it is a perennial plant that is clonally propagated, which allows maintaining somatic variations in a population.

As far as fruit is concerned, several studies have already highlighted the relevance of epigenetic regulations in fruit crops. Whereas DNA methylation was shown to play important roles in tomato, strawberry, and orange fruit during ripening (Cheng et al. 2018b; Huang et al. 2019; Liu et al. 2015), histone PTMs are also likely important at various phases of fleshy fruit development (Gallusci et al. 2016; Lü et al. 2018). So far, evidence of the role of both types of epigenetic marks in grape berries, and in many other fruit crops, awaits demonstration. The combination of high-throughput sequencing associated with chromatin immunoprecipitation or with bisulfite treatment of DNA will undoubtedly shed light on the dynamics of epigenetic marks in fruit, as illustrated in the fruit ENCODE project (Lü et al. 2018), but such approaches remain correlative in nature and will require to be completed by functional analysis of the corresponding genes. In grapevine, generation of loss of function variants is hampered by the difficulty to generate RNAi lines and CRISPR-Cas9 mutations due to the limited efficiency of plant transformation/regeneration processes (see Chap. 16). So far, in silico analyses, conducted on grapevine, have identified candidate genes involved in the control of epigenetic marks (see Sect. 9.4.1). Many of these genes are differentially expressed in grape berries (Almada et al. 2011; Aquea et al. 2010, 2011), suggesting that histone PTMs—and more globally, chromatin remodeling—could play a key role in grapefruit development and ripening. However, ChIP analysis would be necessary to determine the variations of histone mark distribution. Similarly, expression analysis of genes involved in the control of DNA methylation associated with the genome-wide description of DNA methylation changes would be necessary to assess the potential role of DNA methylation in fruit. Noteworthy, given the clear metabolic differences observed between the skin and the pulp, such studies should be performed in each tissue separately. The final demonstration of the role of epigenetic marks in grape berries will require studying the effects of mutations affecting genes that encode histone writers and erasers, as well as enzymes involved in DNA methylation control. Pharmacological approaches using specific drugs interfering with these epigenetic processes could also provide alternative strategies to study the function of epigenetic marks in grape berries (Baubec et al. 2009; Finnegan et al. 2018; Griffin et al. 2016).

In addition to the specificity of grape berry development and ripening, grapevine development and propagation strategies present features that may emphasize the impact of epigenetic regulations on plant phenotypes. First, grapevine is a clonally propagated plant, which contributes to limit its genetic diversity and subsequent phenotypic variations, although both human selection and naturally occurring mutations contribute to the phenotypic diversity (Ferreira et al. 2018). As far as natural clonal propagation is concerned, epigenetic processes are likely contributing to the adaptation of plants to their local environment and may provide selective advantage (Verhoeven and Preite 2014). In line with this idea, a recent study has shown that plants of the mangrove species *Laguncularia racemosa*, have little genetic differences, but possess important DNA methylation differences, suggesting that epigenetic variation in natural plant populations may have an important role in the adaptation to different environments (Lira-Medeiros et al. 2010). Additional evidence of the role of epigenetic processes in clonally propagated plants is provided by the analysis of the transgenerational memory of stresses in white clover (González et al. 2016, 2018). Results indicate that among the various stresses applied to the parental plants, drought-generated transgenerational effects in clonally propagated offspring were transmitted concomitantly to DNA methylation changes and maintained during several clonal offspring generations. So far, there was no causal relationship demonstrated between DNA methylation changes and transgenerational effects in these studies, but results suggest a possible link between both types of event. As far as grape is concerned, such studies have not been performed and it is unknown whether genetically identical clones may be epigenetically different.

In addition, clones of the same origin may become with time epigenetically different. Indeed, environmental conditions do impact the epigenetic status of plants as epigenetic mechanisms are essential to plant responses to both non-biotic and biotic stresses (Gutzat and Scheid 2012; Kinoshita and Seki 2014; Lämke and Bäurle 2017). However, the stability and maintenance of stress-induced epigenetic modifications have been a matter for debate in annual plants (Crisp et al. 2016). As far as perennial plants are concerned, new epigenetic imprints generated by environmental conditions could accumulate over the years and be maintained in the meristem, thereby generating specific epigenetic status for the plants depending on their location and environment (Lafon-Placette et al. 2018; Raj et al. 2011). Hence, genetically identical clones could become epigenetically distinct based on their growing location. The recent demonstration of important changes in methylation patterns that seem to depend on the grapevine growing region is consistent with this idea, although clones of the same origin were not used in this study (Xie et al. 2017).

In addition to stresses (Fortes and Gallusci 2017), climate changes have important consequences on grapevine phenology: it has been shown that timing of budburst and flowering as well as fruit quality are impacted by global warming (Van Leeuwen and Darriet 2016). The relevance of epigenetic-based processes involved in the adaptation of grape plants to these environmental constraints is so far unclear. However, budburst was shown to be under methylation control in poplar, active demethylation being involved in the induction of bud opening after winter (Conde et al. 2017). Whether epigenetic mechanisms exist in grapevines that control budburst is still unknown, recent studies have suggested that PcG proteins might be involved in the control of bud break and flowering (Almada et al. 2011), a function that would be reminiscent to the epigenetic control of vernalization in Arabidopsis. Indeed, a better understanding of the role of chromatin-based regulations in the control of developmental stages during the annual life cycle of grape may provide new strategies to modify grapevine phenology and improve adaptation of this important fruit crop to climate changes.

A very important additional specific feature that differentiates grapevines from other plants is that since the second half of the nineteenth century, grapevines are mostly grown grafted on rootstocks, to protect the plant from Phylloxera and other soilborne pests and diseases (Ollat et al. 2017; see Chap. 14). Grafting does not correspond to the simple juxtaposition of two organisms: the two associated graft partners, rootstock and scion, actively interact with each other. Hence, grafting is known to induce phenotypic changes in the scion and in the rootstock and to improve scion growth potential and fruit yield and quality (Albacete et al. 2015; Kyriacou et al. 2017; Warschefsky et al. 2016). Hetero-grafting (association of a scion and a rootstock with different genotypes) was shown in some cases to generate inheritable sporadic phenotypic changes in the scion, affecting diverse developmental processes including fruit growth and ripening (Hirata 1980; Taller et al. 1998; Yagishita 1961). Although the molecular bases for graft-dependent phenotypic variations are obviously multiple including hormonal, proteins, and mRNA exchange (Albacete et al. 2015; Gregory et al. 2013; Ollat et al. 2017), recent data suggest that epigenetic mechanisms could be among them (Berger et al. 2018). Indeed, several reports indicate that hetero-grafting induces changes in DNA methylation patterns in the scion in different species including Arabidopsis (Lewsey et al. 2016), Hevea (Uthup et al. 2018), solanaceous (Wu et al. 2013), and cucurbitaceous (Avramidou et al. 2015; Xanthopoulou et al. 2019) crops. Moreover, part of these modifications was shown to be inheritable (Wu et al. 2013). These epigenetic changes could induce phenotypic variations, although no example of such functional relationship has been demonstrated yet. Interestingly, mechanistic studies performed in Arabidopsis and in different Solanaceae species have revealed a molecular mechanism which is responsible for the production of epi-variants in grafted plants: small RNAs produced in the scion can induce de novo methylation in the rootstock (Bai et al. 2011; Kasai et al. 2016; Melnyk et al. 2011) and vice versa (Bai et al. 2011). Such epigenetic modifications were shown to occur at loci with homologous sequences to the exchanged small RNAs. When these loci correspond to gene regulatory regions, they can impact gene expression, hence plant phenotype. Whether such graft-dependent mechanisms also exist in grapevine and could generate stable phenotypic diversity remains to be determined. As a conclusion, whereas genetics is a driving force in shaping the phenotypic diversity of grape plants, epigenetics is likely providing an additional layer of variability that could impact grape development and adaptation to environment, and generate stable phenotypical variants.

References

Achour M, Jacq X, Rondé P, Alhosin M, Charlot C, Chataigneau T et al (2008) The interaction of the SRA domain of ICBP90 with a novel domain of DNMT1 is involved in the regulation of VEGF gene expression. Oncogene. https://doi.org/10.1038/sj.onc.1210855

Agius F, Kapoor A, Zhu J-K (2006) Role of the Arabidopsis DNA glycosylase/lyase ROS1 in active DNA demethylation. Proc Natl Acad Sci 103 (31):11796–11801. https://doi.org/10.1073/pnas.0603563103

Alba R (2005) Transcriptome and selected metabolite analyses reveal multiple points of ethylene control during tomato fruit development. Plant Cell 17 (11):2954–2965. https://doi.org/10.1105/tpc.105.036053

Albacete A, Martínez-Andújar C, Martínez-Pérez A, Thompson AJ, Dodd IC, Pérez-Alfocea F (2015) Unravelling rootstock × scion interactions to improve food security. J Exp Bot 66(8):2211–2226. https://doi.org/10.1093/jxb/erv027

Almada R, Cabrera N, Casaretto JA, Peña-Cortés H, Ruiz-Lara S, Villanueva EG (2011) Epigenetic repressor-like genes are differentially regulated during grapevine (*Vitis vinifera* L.) development. Plant Cell Rep 30(10):1959

Aquea F, Timmermann T, Arce-Johnson P (2010) Analysis of histone acetyltransferase and deacetylase families of *Vitis vinifera*. Plant Physiol Biochem 48 (2–3):194–199

Aquea F, Vega A, Timmermann T, Poupin MJ, Arce-Johnson P (2011) Genome-wide analysis of the SET DOMAIN GROUP family in grapevine. Plant Cell Rep 30(6):1087–1097

An YC, Goettel W, Han Q, Bartels A, Liu Z, Xiao W (2017) Dynamic changes of genome-wide DNA methylation during soybean seed development. Sci Rep 7(12263):1–14

Avramidou E, Kapazoglou A, Aravanopoulos FA, Xanthopoulou A, Ganopoulos I, Tsaballa A et al (2015) Global DNA methylation changes in Cucurbitaceae inter-species grafting. Crop Breed Appl Biotechnol 15 (2):112–116

Bai S, Kasai A, Yamada K, Li T, Harada T (2011) A mobile signal transported over a long distance induces systemic transcriptional gene silencing in a grafted partner. J Exp Bot 62(13):4561–4570

Baker K, Dhillon T, Colas I, Cook N, Milne I, Milne L et al (2015) Chromatin state analysis of the barley epigenome reveals a higher-order structure defined by H3K27me1 and H3K27me3 abundance. Plant J 84 (1):111–124

Baker M (2011) Making sense of chromatin states. Nature Publishing Group, London

Bannister AJ, Kouzarides T (2011) Regulation of chromatin by histone modifications. Cell Res 21(3):381

Bapat VA, Trivedi PK, Ghosh A, Sane VA, Ganapathi TR, Nath P (2010) Ripening of fleshy fruit: molecular insight and the role of ethylene. Biotechnol Adv 28 (1):94–107

Bartee L, Malagnac F, Bender J (2001) Arabidopsis cmt3 chromomethylase mutations block non-CG methylation and silencing of an endogenous gene. Genes Dev 15(14):1753–1758. https://doi.org/10.1101/gad.905701

Barth TK, Imhof A (2010) Fast signals and slow marks: the dynamics of histone modifications. Trends Biochem Sci 35(11):618–626

Battilana J, Dunlevy JD, Boss PK (2017) Histone modifications at the grapevine VvOMT3 locus, which encodes an enzyme responsible for methoxypyrazine production in the berry. Funct Plant Biol 44(7):655–664

Baubec T, Pecinka A, Rozhon W, Mittelsten Scheid O (2009) Effective, homogeneous and transient interference with cytosine methylation in plant genomic DNA by zebularine. Plant J 57(3):542–554

Bauer MJ, Fischer RL (2011) Genome demethylation and imprinting in the endosperm. Curr Opin Plant Biol 14 (2):162–167

Beck S, Rakyan VK (2008) The methylome: approaches for global DNA methylation profiling. Trends Genet 24(5):231–237

Becker JS, Nicetto D, Zaret KS (2016) H3K9me3-dependent heterochromatin: barrier to cell fate changes. Trends Genet 32(1):29–41

Benvenuto G, Formiggini F, Laflamme P, Malakhov M, Bowler C (2002) The photomorphogenesis regulator DET1 binds the amino-terminal tail of histone H2B in a nucleosome context. Curr Biol 12(17):1529–1534

Berger MMJ, Gallusci P, Teyssier E (2018) Chapter seven —roles of epigenetic mechanisms in grafting and possible applications. Plant Epigenetics Coming Age Breed Appl 88:203–246

Berger SL (2007) The complex language of chromatin regulation during transcription. Nature 447(7143):407

Bernatavichute YV, Zhang X, Cokus S, Pellegrini M, Jacobsen SE (2008) Genome-wide association of histone H3 lysine nine methylation with CHG DNA methylation in *Arabidopsis thaliana*. PLoS ONE 3(9): e3156

Berry S, Dean C (2015) Environmental perception and epigenetic memory: mechanistic insight through FLC. Plant J 83(1):133–148

Bewick AJ, Niederhuth CE, Ji L, Rohr NA, Griffin PT, Leebens-Mack J, Schmitz RJ (2017) The evolution of CHROMOMETHYLASES and gene body DNA methylation in plants. Genome Biol 18(1):65

Bewick AJ, Schmitz RJ (2017) Gene body DNA methylation in plants. Curr Opin Plant Biol 36:103–110

Birnbaum KD, Roudier F (2017) Epigenetic memory and cell fate reprogramming in plants. Regeneration 4 (1):15–20

Boureau L, How Kit A, Teyssier E, Drevensek S, Rainieri M, Joubès J et al (2016) A CURLY LEAF homologue controls both vegetative and reproductive development of tomato plants. Plant Mol Biol 90(4–5):485–501

Bu Z, Yu Y, Li Z, Liu Y, Jiang W, Huang Y, Dong A-W (2014) Regulation of Arabidopsis flowering by the histone mark readers MRG1/2 via interaction with CONSTANS to modulate FT expression. PLoS Genet 10(9):e1004617

Bucher E, Kong J, Teyssier E, Gallusci P (2018) Chapter ten—epigenetic regulations of fleshy fruit development and ripening and their potential applications to breeding strategies. In: Mirouze M, Bucher E, Gallusci BR (eds) Plant epigenetics coming of age for breeding applications, vol 88, pp 327–360. https://doi.org/10.1016/bs.abr.2018.09.015

Castellarin SD, Gambetta GA, Wada H, Krasnow MN, Cramer GR, Peterlunger E et al (2015) Characterization of major ripening events during softening in grape: turgor, sugar accumulation, abscisic acid metabolism, colour development, and their relationship with growth. J Exp Bot 67(3):709–722

Chanvivattana Y, Bishopp A, Schubert D, Stock C, Moon Y-H, Sung ZR, Goodrich J (2004) Interaction of Polycomb-group proteins controlling flowering in Arabidopsis. Development 131(21):5263–5276

Chaudhury AM, Ming L, Miller C, Craig S, Dennis ES, Peacock WJ (1997) Fertilization-independent seed development in *Arabidopsis thaliana*. Proc Natl Acad Sci 94(8):4223–4228. https://doi.org/10.1073/pnas.94.8.4223

Chen W, Kong J, Qin C, Yu S, Tan J, Chen Y et al (2015) Requirement of CHROMOMETHYLASE3 for somatic inheritance of the spontaneous tomato epimutation Colourless non-ripening. Sci Rep 5:9192. https://doi.org/10.1038/srep09192

Chen X, Hu Y, Zhou D-X (2011) Epigenetic gene regulation by plant Jumonji group of histone demethylase. Biochim Biophys Acta: BBA Gene Regul Mech 1809(8):421–426

Cheng J, Niu Q, Zhang B, Chen K, Yang R, Zhu J-K et al (2018a) Downregulation of RdDM during strawberry fruit ripening. Genome Biol 19(1):212

Cheng S, Tan F, Lu Y, Liu X, Li T, Yuan W et al (2018b) WOX11 recruits a histone H3K27me3 demethylase to promote gene expression during shoot development in rice. Nucleic Acids Res 46(5):2356–2369

Cheng X, Zhang S, Tao W, Zhang X, Liu J, Sun J et al (2018c) INDETERMINATE SPIKELET1 recruits histone deacetylase and a transcriptional repression complex to regulate rice salt tolerance. Plant Physiol 1:1. https://doi.org/10.1104/pp.18.00324

Chinnusamy V, Zhu JK (2009) Epigenetic regulation of stress responses in plants. Curr Opin Plant Biol 12 (2):133–139. https://doi.org/10.1016/j.pbi.2008.12.006

Chodavarapu RK, Feng S, Ding B, Simon SA, Lopez D, Jia Y et al (2012) Transcriptome and methylome interactions in rice hybrids. Proc Natl Acad Sci 109 (30):12040–12045

Choi Y, Gehring M, Johnson L, Hannon M, Harada JJ, Goldberg RB et al (2002) DEMETER, a DNA glycosylase domain protein, is required for endosperm gene imprinting and seed viability in Arabidopsis. Cell 110(1):33–42

Cigliano RA, Sanseverino W, Cremona G, Ercolano MR, Conicella C, Consiglio FM (2013) Genome-wide analysis of histone modifiers in tomato: gaining an insight into their developmental roles. BMC Genom 14(1):57

Cokus SJ, Feng S, Zhang X, Chen Z, Merriman B, Haudenschild CD et al (2008) Shotgun bisulphite sequencing of the Arabidopsis genome reveals DNA methylation patterning. Nature 452(7184):215

Coleman-Derr D, Zilberman D (2012) Deposition of histone variant H2A. Z within gene bodies regulates responsive genes. PLoS Genet 8(10):e1002988

Collepardo-Guevara R, Portella G, Vendruscolo M, Frenkel D, Schlick T, Orozco M (2015) Chromatin unfolding by epigenetic modifications explained by dramatic impairment of internucleosome interactions: a multiscale computational study. J Am Chem Soc 137 (32):10205–10215

Conde C, Silva P, Fontes N, Dias AT, Sousa RM, Agasse MJ, Delrot A, Gerós H (2007) Biochemical changes throughout grape berry development and fruit and wine quality. Food 1:1–22

Conde D, Le Gac AL, Perales M, Dervinis C, Kirst M, Maury S et al (2017) Chilling-responsive DEMETER-LIKE DNA demethylase mediates in poplar bud break. Plant, Cell Environ 40(10):2236–2249. https://doi.org/10.1111/pce.13019

Corem S, Doron-Faigenboim A, Jouffroy O, Maumus F, Arazi T, Bouché N (2018) Redistribution of CHH methylation and small interfering RNAs across the genome of tomato ddm1 mutants. Plant Cell 30 (7):1628–1644

Crisp PA, Ganguly D, Eichten SR, Borevitz JO, Pogson BJ (2016) Reconsidering plant memory:

intersections between stress recovery, RNA turnover, and epigenetics. Sci Adv 2(2):e1501340

Cui X, Lu F, Qiu Q, Zhou B, Gu L, Zhang S et al (2016) REF6 recognizes a specific DNA sequence to demethylate H3K27me3 and regulate organ boundary formation in Arabidopsis. Nat Genet 48(6):694

Cui X, Cao X (2014) Epigenetic regulation and functional exaptation of transposable elements in higher plants. Curr Opin Plant Biol 21:83–88

Daccord N, Celton JM, Linsmith G, Becker C, Choisne N, Schijlen E et al (2017) High-quality de novo assembly of the apple genome and methylome dynamics of early fruit development. Nat Genet 49(7):1099–1106. https://doi.org/10.1038/ng.3886

del Prete S, Mikulski P, Schubert D, Gaudin V (2015) One, two, three: Polycomb proteins hit all dimensions of gene regulation. Genes 6(3):520–542

Deng X, Qiu Q, He K, Cao X (2018) The seekers: how epigenetic modifying enzymes find their hidden genomic targets in Arabidopsis. Curr Opin Plant Biol 45:75–81

Derkacheva M, Hennig L (2013) Variations on a theme: Polycomb group proteins in plants. J Exp Bot 65 (10):2769–2784

Du J, Johnson LM, Groth M, Feng S, Hale CJ, Li S et al (2014) Mechanism of DNA methylation-directed histone methylation by KRYPTONITE. Mol Cell 55 (3):495–504

Du J, Johnson LM, Jacobsen SE, Patel DJ (2015) DNA methylation pathways and their crosstalk with histone methylation. Nat Rev Mol Cell Biol 16(9):519

Du J, Zhong X, Bernatavichute YV, Stroud H, Feng S, Caro E et al (2012) Dual binding of chromomethylase domains to H3K9me2-containing nucleosomes directs DNA methylation in plants. Cell 151(1):167–180

Duan C-G, Wang X, Xie S, Pan L, Miki D, Tang K et al (2017) A pair of transposon-derived proteins function in a histone acetyltransferase complex for active DNA demethylation. Cell Res 27(2):226–240. https://doi.org/10.1038/cr.2016.147

Dubin MJ, Zhang P, Meng D, Remigereau M-S, Osborne EJ, Paolo Casale F et al (2015) DNA methylation in Arabidopsis has a genetic basis and shows evidence of local adaptation. ELife 4:e05255–e05255. https://doi.org/10.7554/eLife.05255

Dussert S, Verdeil JL, Jaligot E, Beulé T, Rival A (2000) Somaclonal variation in oil palm (*Elaeis guineensis* Jacq.): the DNA methylation hypothesis. Plant Cell Rep 19(7):684–690. https://doi.org/10.1007/s002999900177

Eichten SR, Briskine R, Song J, Li Q, Swanson-Wagner R, Hermanson PJ et al (2013) Epigenetic and genetic influences on DNA methylation variation in maize populations. Plant Cell 25(8):2783–2797. https://doi.org/10.1105/tpc.113.114793

Eichten SR, Schmitz RJ, Springer NM (2014) Epigenetics: beyond chromatin modifications and complex genetic regulation. Plant Physiol 165(3):933–947

El-Sharkawy I, Liang D, Xu K (2015) Transcriptome analysis of an apple (*Malus* × *domestica*) yellow fruit somatic mutation identifies a gene network module highly associated with anthocyanin and epigenetic regulation. J Exp Bot 66(22):7359–7376

Engelhorn J, Blanvillain R, Carles CC (2014) Gene activation and cell fate control in plants: a chromatin perspective. Cell Mol Life Sci 71(16):3119–3137

Exner V, Hennig L (2008) Chromatin rearrangements in development. Curr Opin Plant Biol 11(1):64–69

Fan D, Wang X, Tang X, Ye X, Ren S, Wang D, Luo K (2018) Histone H3K9 demethylase JMJ25 epigenetically modulates anthocyanin biosynthesis in poplar. Plant J 96(6):1121–1136

Feng S, Bostick M, Sadler KC, Cokus SJ, Strauss SH, Jain J et al (2010) Conservation and divergence of methylation patterning in plants and animals. Proc Natl Acad Sci 107(19):8689–8694. https://doi.org/10.1073/pnas.1002720107

Feng S, Jacobsen SE (2011) Epigenetic modifications in plants: an evolutionary perspective. Curr Opin Plant Biol 14(2):179–186

Ferreira V, Pinto-Carnide O, Arroyo-García R, Castro I (2018) Berry color variation in grapevine as a source of diversity. Plant Physiol Biochem132:696–707

Finnegan EJ, Ford B, Wallace X, Pettolino F, Griffin PT, Schmitz RJ et al (2018) Zebularine treatment is associated with deletion of FT-B1 leading to an increase in spikelet number in bread wheat. Plant, Cell Environ 41(6):1346–1360

FitzGerald J, Luo M, Chaudhury A, Berger F (2008) DNA methylation causes predominant maternal controls of plant embryo growth. PLoS ONE 3(5):e2298

Fletcher JC (2017) State of the art: trxG factor regulation of post-embryonic plant development. Front Plant Sci 8:1925

Forestan C, Farinati S, Rouster J, Lassagne H, Lauria M, Ferro ND, Varotto S (2018) Control of maize vegetative and reproductive. Genetics 208:1443–1466. https://doi.org/10.1534/genetics.117.300625/-/DC1.1

Fortes AM, Gallusci P (2017) Plant stress responses and phenotypic plasticity in the epigenomics era: perspectives on the grapevine scenario, a model for perennial crop plants. Front Plant Sci 8:82

Fortes A, Teixeira R, Agudelo-Romero P (2015) Complex interplay of hormonal signals during grape berry ripening. Molecules 20(5):9326–9343

Fu C-C, Han Y-C, Guo Y-F, Kuang J-F, Chen J-Y, Shan W, Lu W-J (2018a) Differential expression of histone deacetylases during banana ripening and identification of MaHDA6 in regulating ripening-associated genes. Postharvest Biol Technol 141:24–32

Fu F-F, Dawe RK, Gent JI (2018b) Loss of RNA-directed DNA methylation in maize chromomethylase and DDM1-type nucleosome remodeler mutants. Plant Cell 30(7):1617–1627

Fuchs J, Demidov D, Houben A, Schubert I (2006) Chromosomal histone modification patterns—from conservation to diversity. Trends Plant Sci 11 (4):199–208. https://doi.org/10.1016/j.tplants.2006.02.008

Gallusci P, Hodgman C, Teyssier E, Seymour GB (2016) DNA methylation and chromatin regulation during fleshy fruit development and ripening. Front Plant Sci 7:807

Gapper NE, Giovannoni JJ, Watkins CB (2014) Understanding development and ripening of fruit crops in an "omics" era. Horticu Res 1:14034. https://doi.org/10.1038/hortres.2014.34

Gillaspy G, Ben-david H, Gruissem W, Darwin C (1993) Fruits: a developmental perspective. Plant Cell 5 (October):1439–1451

Giovannoni J, Nguyen C, Ampofo B, Zhong S, Fei Z (2017) The epigenome and transcriptional dynamics of fruit ripening. Annu Rev Plant Biol 68:61–84

Gong Z, Morales-Ruiz T, Ariza RR, Roldán-Arjona T, David L, Zhu J-K (2002) ROS1, a repressor of transcriptional gene silencing in Arabidopsis, encodes a DNA glycosylase/lyase. Cell 111(6):803–814

González APR, Chrtek J, Dobrev PI, Dumalasová V, Fehrer J, Mráz P, Latzel V (2016) Stress-induced memory alters growth of clonal offspring of white clover (*Trifolium repens*). Am J Bot 103(9): 1567–1574

González APR, Preite V, Verhoeven KJF, Latzel V (2018) Transgenerational effects and epigenetic memory in the clonal plant *Trifolium repens*. Front Plant Sci 9:1677. https://doi.org/10.3389/fpls.2018.01677

Goodrich J, Puangsomlee P, Martin M, Long D, Meyerowitz EM, Coupland G (1997) A Polycomb-group gene regulates homeotic gene expression in Arabidopsis. Nature 386(6620):44

Gouil Q, Baulcombe DC (2016) DNA methylation signatures of the plant chromomethyltransferases. PLoS Genet 12(12):e1006526

Gouil Q, Baulcombe DC (2018) Paramutation-like features of multiple natural epialleles in tomato. BMC Genom 19(1):203

Gregory PJ, Atkinson CJ, Bengough AG, Else MA, Fernández-Fernández F, Harrison RJ, Schmidt S (2013) Contributions of roots and rootstocks to sustainable, intensified crop production. J Exp Bot 64(5):1209–1222

Griffin PT, Niederhuth CE, Schmitz RJ (2016) A comparative analysis of 5-azacytidine- and zebularine-induced DNA demethylation. G3 6(9):2773–2780. https://doi.org/10.1534/g3.116.030262

Gu T, Han Y, Huang R, McAvoy RJ, Li Y (2016) Identification and characterization of histone lysine methylation modifiers in Fragaria vesca. Sci Rep 6:23581

Guo J-E, Hu Z, Li F, Zhang L, Yu X, Tang B, Chen G (2017a) Silencing of histone deacetylase SlHDT3 delays fruit ripening and suppresses carotenoid accumulation in tomato. Plant Sci 265:29–38

Guo J-E, Hu Z, Yu X, Li A, Li F, Wang Y et al (2018) A histone deacetylase gene, SlHDA3, acts as a negative regulator of fruit ripening and carotenoid accumulation. Plant Cell Rep 37(1):125–135

Guo J-E, Hu Z, Zhu M, Li F, Zhu Z, Lu Y, Chen G (2017b) The tomato histone deacetylase SlHDA1 contributes to the repression of fruit ripening and carotenoid accumulation. Sci Rep 7(1):7930

Gutzat R, Scheid OM (2012) Epigenetic responses to stress: triple defense? Curr Opin Plant Biol 15(5): 568–573

Hadfield KA, Dandekar AM, Romani RJ (1993) Demethylation of ripening specific genes in tomato fruit. Plant Sci 92(1):13–18

Han Y-C, Kuang J-F, Chen J-Y, Liu X-C, Xiao Y-Y, Fu C-C et al (2016) Banana transcription factor MaERF11 recruits histone deacetylase MaHDA1 and represses the expression of MaACO1 and expansins during fruit ripening. Plant Physiol 171(2):1070–1084

He XJ, Chen T, Zhu JK (2011) Regulation and function of DNA methylation in plants and animals. Cell Res 21 (3):442–465. https://doi.org/10.1038/cr.2011.23

Hennig L, Bouveret R, Gruissem W (2005) MSI1-like proteins: an escort service for chromatin assembly and remodeling complexes. Trends Cell Biol 15(6):295–302

Heo JB, Sung S (2011) Vernalization-mediated epigenetic silencing by a long intronic noncoding RNA. Science 331(6013):76–79

Hepworth J, Dean C (2015) Flowering Locus C's lessons: conserved chromatin switches underpinning developmental timing and adaptation. Plant Physiol 168 (4):1237–1245

Hirata Y (1980) Graft-induced in tomato changes in skin and flesh color (*Lycopersicon esculentum* Mill.). Soc Hortic Sci 49(2):211–216

Hirsch CD, Springer NM (2017) Transposable element influences on gene expression in plants. Biochim Biophys Acta (BBA) Gene Regul Mech 1860(1): 157–165

Hou X, Zhou J, Liu C, Liu L, Shen L, Yu H (2014) Nuclear factor Y-mediated H3K27me3 demethylation of the SOC1 locus orchestrates flowering responses of Arabidopsis. Nat Commun 5:4601

How Kit A, Boureau L, Stammitti-Bert L, Rolin D, Teyssier E, Gallusci P (2010) Functional analysis of SlEZ1 a tomato Enhancer of zeste (E (z)) gene demonstrates a role in flower development. Plant Mol Biol 74(3):201–213

Hsieh T-F, Ibarra CA, Silva P, Zemach A, Eshed-Williams L, Fischer RL, Zilberman D (2009) Genome-wide demethylation of Arabidopsis endosperm. Science 324(5933):1451–1454

Hu J, McCall CM, Ohta T, Xiong Y (2004) Targeted ubiquitination of CDT1 by the DDB1-CUL4A-ROC1 ligase in response to DNA damage. Nat Cell Biol 6 (10):1003

Hu L, Li N, Xu C, Zhong S, Lin X, Yang J et al (2014) Mutation of a major CG methylase in rice causes genome-wide hypomethylation, dysregulated genome expression, and seedling lethality. Proc Natl Acad Sci 111(29):10642–10647. https://doi.org/10.1073/pnas.1410761111

Huang H, Liu R, Niu Q, Tang K, Zhang B, Zhang H et al (2019) Global increase in DNA methylation during orange fruit development and ripening. Proc Natl Acad Sci 116(4):1430–1436

Hung FY, Chen FF, Li C, Chen C, Lai YC, Chen JH et al (2018) The Arabidopsis LDL1/2-HDA6 histone modification complex is functionally associated with CCA1/LHY in regulation of circadian clock genes. Nucleic Acids Res 46(20):10669–10681. https://doi.org/10.1093/nar/gky749

Inagaki S, Miura-Kamio A, Nakamura Y, Lu F, Cui X, Cao X et al (2010) Autocatalytic differentiation of epigenetic modifications within the Arabidopsis genome. EMBO J 29(20):3496–3506

Jackson JP, Lindroth AM, Cao X, Jacobsen SE (2002) Control of CpNpG DNA methylation by the KRYPTONITE histone H3 methyltransferase. Nature 416 (6880):556

Jacob Y, Feng S, LeBlanc CA, Bernatavichute YV, Stroud H, Cokus S et al (2009) ATXR5 and ATXR6 are H3K27 monomethyltransferases required for chromatin structure and gene silencing. Nat Struct Mol Biol 16(7):763

Jaillon O, Aury J-M, Noel B, Policriti A, Clepet C, Casagrande A et al (2007) The grapevine genome sequence suggests ancestral hexaploidization in major angiosperm phyla. Nature 449(7161):463–467. https://doi.org/10.1038/nature06148

Janssen BJ, Thodey K, Schaffer RJ, Alba R, Balakrishnan L, Bishop R et al (2008) Global gene expression analysis of apple fruit development from the floral bud to ripe fruit. BMC Plant Biol 8(1):16

Jenuwein T, Allis CD (2001) Translating the histone code. Science 293(5532):1074–1080

Jiang D, Berger F (2017) Histone variants in plant transcriptional regulation. Biochim Biophys Acta (BBA) Gene Regul Mech 1860(1):123–130

Jiang L, Li D, Jin L, Ruan Y, Shen WH, Liu C (2018a) Histone lysine methyltransferases BnaSDG8.A and BnaSDG8.C are involved in the floral transition in *Brassica napus*. Plant J 95(4):672–685. https://doi.org/10.1111/tpj.13978

Jiang P, Wang S, Jiang H, Cheng B, Wu K, Ding Y (2018b) The COMPASS-like complex promotes flowering and panicle branching in rice. Plant Physiol 176 (4):2761–2771

Jiang P, Wang S, Zheng H, Li H, Zhang F, Su Y et al (2018c) SIP 1 participates in regulation of flowering time in rice by recruiting OsTrx1 to Ehd1. New Phytol 219(1):422–435

Joldersma D, Liu Z (2018) The making of virgin fruit: the molecular and genetic basis of parthenocarpy. Oxford University Press, Oxford

Jones MA, Covington MF, DiTacchio L, Vollmers C, Panda S, Harmer SL (2010) Jumonji domain protein JMJD5 functions in both the plant and human circadian systems. Proc Natl Acad Sci 107 (50):21623–21628. https://doi.org/10.1073/pnas.1014204108

Kasai A, Bai S, Hojo H, Harada T (2016) Epigenome editing of potato by grafting using transgenic tobacco as siRNA donor. PLoS ONE 11(8):e0161729

Kawakatsu T, Nery JR, Castanon R, Ecker JR (2017) Dynamic DNA methylation reconfiguration during seed development and germination. Genome Biol 18 (1):171

Kawakatsu T, Stuart T, Valdes M, Breakfield N, Schmitz RJ, Nery JR et al (2016) Unique cell-type-specific patterns of DNA methylation in the root meristem. Nat Plants 2(5):16058

Kim D-H, Xi Y, Sung S (2017) Modular function of long noncoding RNA, COLDAIR, in the vernalization response. PLoS Genet 13(7):e1006939

Kim DH, Sung S (2017) Vernalization-triggered intragenic chromatin loop formation by long noncoding RNAs. Dev Cell 40(3):302–312.e4. https://doi.org/10.1016/j.devcel.2016.12.021

Kim J, Kim JH, Richards EJ, Chung KM, Woo HR (2014) Arabidopsis VIM proteins regulate epigenetic silencing by modulating DNA methylation and histone modification in cooperation with MET1. Mol Plant 7 (9):1470–1485

Kim YJ, Wang R, Gao L, Li D, Xu C, Mang H et al (2016) POWERDRESS and HDA9 interact and promote histone H3 deacetylation at specific genomic sites in Arabidopsis. Proc Natl Acad Sci 113 (51):14858–14863

Kinoshita T, Harada J, Goldberg R, Fischer R (2001) Polycomb repression of flowering during early plant development. Proc Natl Acad Sci USA 98(24):14156–14161. https://doi.org/10.1073/pnas.241507798

Kinoshita T, Seki M (2014) Epigenetic memory for stress response and adaptation in plants. Plant Cell Physiol 55(11):1859–1863

Klee HJ, Giovannoni JJ (2011) Genetics and control of tomato fruit ripening and quality attributes. Annu Rev Genet 45:41–59

Kobayashi S, Goto-Yamamoto N, Hirochika H (2004) Retrotransposon-induced mutations in grape skin color. Science 304(5673):982

Kuang J, Chen J, Luo M, Wu K, Sun W, Jiang Y, Lu W (2011) Histone deacetylase HD2 interacts with ERF1 and is involved in longan fruit senescence. J Exp Bot 63(1):441–454

Kumar R, Khurana A, Sharma AK (2014) Role of plant hormones and their interplay in development and ripening of fleshy fruits. J Exp Bot 65(16):4561–4575

Kyriacou MC, Rouphael Y, Colla G, Zrenner R, Schwarz D (2017) Vegetable grafting: the implications of a growing agronomic imperative for vegetable fruit quality and nutritive value. Front Plant Sci 8:741

Lafon-Placette C, Le Gac AL, Chauveau D, Segura V, Delaunay A, Lesage-Descauses MC et al (2018) Changes in the epigenome and transcriptome of the poplar shoot apical meristem in response to water availability affect preferentially hormone pathways. J Exp Bot 69(3):537–551. https://doi.org/10.1093/jxb/erx409

Lafos M, Kroll P, Hohenstatt ML, Thorpe FL, Clarenz O, Schubert D (2011) Dynamic regulation of H3K27 trimethylation during Arabidopsis differentiation. PLoS Genet 7(4):e1002040

Lämke J, Bäurle I (2017) Epigenetic and chromatin-based mechanisms in environmental stress adaptation and stress memory in plants. Genome Biol 18(1):124

Lang Z, Lei M, Wang X, Tang K, Miki D, Zhang H et al (2015) The methyl-CpG-binding protein MBD7 facilitates active DNA demethylation to limit DNA hyper-methylation and transcriptional gene silencing. Mol Cell 57(6):971–983. https://doi.org/10.1016/j.molcel.2015.01.009

Lang Z, Wang Y, Tang K, Tang D, Datsenka T, Cheng J et al (2017) Critical roles of DNA demethylation in the activation of ripening-induced genes and inhibition of ripening-repressed genes in tomato fruit. Proc Natl Acad Sci 114(22):E4511–E4519

Law JA, Jacobsen SE (2010) Establishing, maintaining and modifying DNA methylation patterns in plants and animals. Nat Rev Genet 11(3):204

Lee JM, Joung J, McQuinn R, Chung M, Fei Z, Tieman D et al (2012) Combined transcriptome, genetic diversity and metabolite profiling in tomato fruit reveals that the ethylene response factor SlERF6 plays an important role in ripening and carotenoid accumulation. Plant J 70(2):191–204

Lee S-I, Kim N-S (2014) Transposable elements and genome size variations in plants. Genomics Inform 12 (3):87

Lei M, Zhang H, Julian R, Tang K, Xie S, Zhu J-K (2015) Regulatory link between DNA methylation and active demethylation in Arabidopsis. Proc Natl Acad Sci 112 (11):3553–3557

Lewsey MG, Hardcastle TJ, Melnyk CW, Molnar A, Valli A, Urich MA et al (2016) Mobile small RNAs regulate genome-wide DNA methylation. Proc Natl Acad Sci 113(6):E801–E810

Li C, Gu L, Gao L, Chen C, Wei CQ, Qiu Q et al (2016) Concerted genomic targeting of H3K27 demethylase REF6 and chromatin-remodeling ATPase BRM in Arabidopsis. Nat Genet 48(6):687–693. https://doi.org/10.1038/ng.3555

Li S, Zhou B, Peng X, Kuang Q, Huang X, Yao J et al (2014) OsFIE2 plays an essential role in the regulation of rice vegetative and reproductive development. New Phytol 201(1):66–79

Li X, Wang X, He K, Ma Y, Su N, He H et al (2008) High-resolution mapping of epigenetic modifications of the rice genome uncovers interplay between DNA methylation, histone methylation, and gene expression. Plant Cell 20(2):259–276. https://doi.org/10.1105/tpc.107.056879

Li X, Zhu J, Hu F, Ge S, Ye M, Xiang H et al (2012) Single-base resolution maps of cultivated and wild rice methylomes and regulatory roles of DNA methylation in plant gene expression. BMC Genom 13(1):300. https://doi.org/10.1186/1471-2164-13-300

Li Y, Kumar S, Qian W (2018) Active DNA demethylation: mechanism and role in plant development. Plant Cell Rep 37(1):77–85

Lieberman M, Segev O, Gilboa N, Lalazar A, Levin I (2004) The tomato homolog of the gene encoding UV-damaged DNA binding protein 1 (DDB1) underlined as the gene that causes the high pigment-1 mutant phenotype. Theor Appl Genet 108 (8):1574–1581

Lijavetzky D, Ruiz-García L, Cabezas JA, De Andrés MT, Bravo G, Ibáñez A et al (2006) Molecular genetics of berry colour variation in table grape. Mol Genet Genomics 276(5):427–435

Lindroth AM, Cao X, Jackson JP, Zilberman D, McCallum CM, Henikoff S, Jacobsen SE (2001) Requirement of CHROMOMETHYLASE3 for maintenance of CpXpG methylation. Science 292(5524):2077 LP–2080 LP

Lindroth AM, Shultis D, Jasencakova Z, Fuchs J, Johnson L, Schubert D et al (2004) Dual histone H3 methylation marks at lysines 9 and 27 required for interaction with CHROMOMETHYLASE3. EMBO J 23(21):4286–4296

Lira-Medeiros CF, Parisod C, Fernandes RA, Mata CS, Cardoso MA, Ferreira PCG (2010) Epigenetic variation in mangrove plants occurring in contrasting natural environment. PLoS ONE 5(4):e10326

Lister R, O'Malley RC, Tonti-Filippini J, Gregory BD, Berry CC, Millar AH, Ecker JR (2008) Highly integrated single-base resolution maps of the epigenome in Arabidopsis. Cell 133(3):523–536

Liu C, Lu F, Cui X, Cao X (2010) Histone methylation in higher plants. Annu Rev Plant Biol 61:395–420

Liu D-D, Dong Q-L, Fang M-J, Chen K-Q, Hao Y-J (2012) Ectopic expression of an apple apomixis-related gene MhFIE induces co-suppression and results in abnormal vegetative and reproductive development in tomato. J Plant Physiol 169(18): 1866–1873

Liu D-D, Zhou L-J, Fang M-J, Dong Q-L, An X-H, You C-X, Hao Y-J (2016) Polycomb-group protein SlMSI1 represses the expression of fruit-ripening genes to prolong shelf life in tomato. Sci Rep 6:31806

Liu J, Zhi P, Wang X, Fan Q, Chang C (2018) Wheat WD40-repeat protein TaHOS15 functions in a histone deacetylase complex to fine-tune defense responses to *Blumeria graminis* f. sp. tritici. J Exp Bot 70(1):255–268

Liu K, Yu Y, Dong A, Shen W (2017) SET DOMAIN GROUP701 encodes a H3K4-methytransferase and regulates multiple key processes of rice plant development. New Phytol 215(2):609–623

Liu R, How-Kit A, Stammitti L, Teyssier E, Rolin D, Mortain-Bertrand A et al (2015) A DEMETER-like DNA demethylase governs tomato fruit ripening. Proc Natl Acad Sci 112(34):10804–10809. https://doi.org/10.1073/pnas.1503362112

Lü P, Yu S, Zhu N, Chen Y-R, Zhou B, Pan Y et al (2018) Genome encode analyses reveal the basis of convergent evolution of fleshy fruit ripening. Nat Plants 4(10):784–791. https://doi.org/10.1038/s41477-018-0249-z

Lu SX, Knowles SM, Webb CJ, Celaya RB, Cha C, Siu JP, Tobin EM (2011) The Jumonji C domain-containing protein JMJ30 regulates period length in the Arabidopsis circadian clock. Plant

Physiol 155(2):906–915. https://doi.org/10.1104/pp.110.167015

Luo C, Sidote DJ, Zhang Y, Kerstetter RA, Michael TP, Lam E (2013) Integrative analysis of chromatin states in Arabidopsis identified potential regulatory mechanisms for natural antisense transcript production. Plant J 73(1):77–90

Luo X, Gao Z, Wang Y, Chen Z, Zhang W, Huang J et al (2018) The NUCLEAR FACTOR-CONSTANS complex antagonizes Polycomb repression to de-repress FLOWERING LOCUS T expression in response to inductive long days in Arabidopsis. Plant J 95(1):17–29. https://doi.org/10.1111/tpj.13926

Maeji H, Nishimura T (2018) Chapter two-epigenetic mechanisms in plants. https://doi.org/10.1016/bs.abr.2018.09.014

Malagnac F, Bartee L, Bender J (2002) An Arabidopsis SET domain protein required for maintenance but not establishment of DNA methylation. EMBO J 21 (24):6842–6852

Manning K, Tör M, Poole M, Hong Y, Thompson AJ, King GJ et al (2006) A naturally occurring epigenetic mutation in a gene encoding an SBP-box transcription factor inhibits tomato fruit ripening. Nat Genet 38 (8):948

Mathieu O, Reinders J, Čaikovski M, Smathajitt C, Paszkowski J (2007) Transgenerational stability of the Arabidopsis epigenome is coordinated by CG methylation. Cell 130(5):851–862

Matzke MA, Kanno T, Matzke AJM (2015) RNA-directed DNA methylation: the evolution of a complex epigenetic pathway in flowering plants. Annu Rev Plant Biol 66:243–267

Matzke MA, Mosher RA (2014) RNA-directed DNA methylation: an epigenetic pathway of increasing complexity. Nat Rev Genet 15(6):394

McAtee P, Karim S, Schaffer RJ, David K (2013) A dynamic interplay between phytohormones is required for fruit development, maturation, and ripening. Front Plant Sci 4:79

Melnyk CW, Molnar A, Bassett A, Baulcombe DC (2011) Mobile 24 nt small RNAs direct transcriptional gene silencing in the root meristems of *Arabidopsis thaliana*. Curr Biol 21(19):1678–1683

Mirouze M, Vitte C (2014) Transposable elements, a treasure trove to decipher epigenetic variation: insights from Arabidopsis and crop epigenomes. J Exp Bot 65 (10):2801–2812

Miura A, Nakamura M, Inagaki S, Kobayashi A, Saze H, Kakutani T (2009) An Arabidopsis jmjC domain protein protects transcribed genes from DNA methylation at CHG sites. EMBO J 28(8):1078–1086

Mozgova I, Hennig L (2015) The polycomb group protein regulatory network. Annu Rev Plant Biol 66:269–296

Niederhuth CE, Bewick AJ, Ji L, Alabady MS, Do Kim K, Li Q et al (2016) Widespread natural variation of DNA methylation within angiosperms. Genome Biol 17(1):194

Ning YQ, Ma ZY, Huang HW, Mo H, Zhao TT, Li L et al (2015) Two novel NAC transcription factors regulate gene expression and flowering time by associating with the histone demethylase JMJ14. Nucleic Acids Res 43(3):1469–1484. https://doi.org/10.1093/nar/gku1382

Ojolo SP, Cao S, Priyadarshani S, Li W, Yan M, Aslam M et al (2018) Regulation of plant growth and development: a review from a chromatin remodeling perspective. Front Plant Sci 9:1232. https://doi.org/10.3389/fpls.2018.01232

Ollat N, Cookson SJ, Lauvergeat V, Marguerit E, Barrieu F, Gambetta G et al (2017) Grapevine roots: the dark side. Acta Hortic 1188:213–226. https://doi.org/10.17660/ActaHortic.2017.1188.28

Ong-Abdullah M, Ordway JM, Jiang N, Ooi SE, Kok SY, Sarpan N et al (2015) Loss of Karma transposon methylation underlies the mantled somaclonal variant of oil palm. Nature 525(7570):533–537. https://doi.org/10.1038/nature15365

Ortega-Galisteo AP, Morales-Ruiz T, Ariza RR, Roldán-Arjona T (2008) Arabidopsis DEMETER-LIKE proteins DML2 and DML3 are required for appropriate distribution of DNA methylation marks. Plant Mol Biol 67(6):671–681

Osorio S, Alba R, Damasceno CMB, Lopez-Casado G, Lohse M, Zanor MI et al (2011) Systems biology of tomato fruit development: combined transcript, protein, and metabolite analysis of tomato transcription factor (nor, rin) and ethylene receptor (Nr) mutants reveals novel regulatory interactions. Plant Physiol 157(1):405–425

Osorio S, Scossa F, Fernie A (2013) Molecular regulation of fruit ripening. Front Plant Sci 4:198

Papa CM, Springer NM, Muszynski MG, Meeley R, Kaeppler SM (2001) Maize chromomethylase Zea methyltransferase2 is required for CpNpG methylation. Plant Cell 13(8):1919–1928

Park K, Kim MY, Vickers M, Park J-S, Hyun Y, Okamoto T et al (2016) DNA demethylation is initiated in the central cells of Arabidopsis and rice. Proc Natl Acad Sci 113(52):15138 LP–15143 LP. https://doi.org/10.1073/pnas.1619047114

Penterman J, Zilberman D, Huh JH, Ballinger T, Henikoff S, Fischer RL (2007) DNA demethylation in the Arabidopsis genome. Proc Natl Acad Sci 104 (16):6752–6757. https://doi.org/10.1073/pnas.0701861104

Pikaard CS, Scheid OM (2014) Epigenetic regulation in plants. Cold Spring Harb Perspect Biol 6(12):a019315

Qian S, Lv X, Scheid RN, Lu L, Yang Z, Chen W et al (2018) Dual recognition of H3K4me3 and H3K27me3 by a plant histone reader SHL. Nat Commun 9(1): 2425

Qian W, Miki D, Zhang H, Liu Y, Zhang X, Tang K et al (2012) Arabidopsis. 336(June): 1445–1448

Quadrana L, Almeida J, Asís R, Duffy T, Dominguez PG, Bermúdez L et al (2014) Natural occurring epialleles

determine vitamin E accumulation in tomato fruits. Nat Commun 5:4027
Questa JI, Song J, Geraldo N, An H (2016) The sequence specific transcriptional repressor VAL1 triggers Polycomb silencing at FLC. Science 1(6298):1–5
Raj S, Bräutigam K, Hamanishi ET, Wilkins O, Thomas BR, Schroeder W et al (2011) Clone history shapes Populus drought responses. Proc Natl Acad Sci 108(30):12521–12526
Rival A, Tregear J, Verdeil JL, Richaud F, Beulé T, Duval Y et al (1998) Molecular search for MRNA and genomic markers of the oil palm "mantled" somaclonal variation. Acta Hort 461:165–172. https://doi.org/10.17660/ActaHortic.1998.461.16
Rodriguez-Granados NY, Ramirez-Prado JS, Veluchamy A, Latrasse D, Raynaud C, Crespi M et al (2016) Put your 3D glasses on: plant chromatin is on show. J Exp Bot 67(11):3205–3221
Rossi V, Locatelli S, Varotto S, Donn G, Pirona R, Henderson DA et al (2007) Maize histone deacetylase hda101 Is involved in plant development, gene transcription, and sequence-specific modulation of histone modification of genes and repeats. Plant Cell 19(4):1145–1162. https://doi.org/10.1105/tpc.106.042549
Rothbart SB, Strahl BD (2014) Interpreting the language of histone and DNA modifications. Biochim Biophys Acta (BBA) Gene Regul Mech 1839(8):627–643
Roudier F, Ahmed I, Bérard C, Sarazin A, Mary-Huard T, Cortijo S et al (2011) Integrative epigenomic mapping defines four main chromatin states in Arabidopsis. EMBO J 30(10):1928–1938. https://doi.org/10.1038/emboj.2011.103
Roy S, Gupta P, Rajabhoj MP, Maruthachalam R, Nandi AK (2018) The polycomb-group repressor MEDEA attenuates pathogen defense. Plant Physiol 177(4):1728–1742
Saleh A, Al-Abdallat A, Ndamukong I, Alvarez-Venegas R, Avramova Z (2007) The Arabidopsis homologs of trithorax (ATX1) and enhancer of zeste (CLF) establish "bivalent chromatin marks" at the silent AGAMOUS locus. Nucleic Acids Res 35(18):6290–6296. https://doi.org/10.1093/nar/gkm464
Satgé C, Moreau S, Sallet E, Lefort G, Auriac M-C, Remblière C et al (2016) Reprogramming of DNA methylation is critical for nodule development in Medicago truncatula. Nat Plants 2:16166
Saze H, Shiraishi A, Miura A, Kakutani T (2008) Control of genic DNA methylation by a jmjC domain-containing protein in *Arabidopsis thaliana*. Science 319(5862):462–465
Schoft VK, Chumak N, Choi Y, Hannon M, Garcia-Aguilar M, Machlicova A et al (2011) Function of the DEMETER DNA glycosylase in the *Arabidopsis thaliana* male gametophyte. Proc Natl Acad Sci 108(19):8042–8047
Schubert D, Clarenz O, Goodrich J (2005) Epigenetic control of plant development by Polycomb-group proteins. Curr Opin Plant Biol 8(5):553–561
Sequeira-Mendes J, Araguez I, Peiro R, Mendez-Giraldez R, Zhang X, Jacobsen SE et al (2014) The functional topography of the arabidopsis genome is organized in a reduced number of linear motifs of chromatin states. Plant Cell 26(6):2351–2366. https://doi.org/10.1105/tpc.114.124578
Serrano A, Espinoza C, Armijo G, Inostroza-Blancheteau C, Poblete E, Meyer-Regueiro C et al (2017) Omics approaches for understanding grapevine berry development: regulatory networks associated with endogenous processes and environmental responses. Front Plant Sci 8:1486
Seymour DK, Koenig D, Hagmann J, Becker C, Weigel D (2014) Evolution of DNA methylation patterns in the Brassicaceae is driven by differences in genome organization. PLoS Genet 10(11):e1004785
Seymour GB, Østergaard L, Chapman NH, Knapp S, Martin C (2013) Fruit development and ripening. Annu Rev Plant Biol 64:219–241
Sharif J, Muto M, Takebayashi S, Suetake I, Iwamatsu A, Endo TA et al (2007) The SRA protein Np95 mediates epigenetic inheritance by recruiting Dnmt1 to methylated DNA. Nature 450(7171):908
Shen Y, Issakidis-Bourguet E, Zhou D-X (2016) Perspectives on the interactions between metabolism, redox, and epigenetics in plants. J Exp Bot 67 (18):5291–5300
Spillane C, MacDougall C, Stock C, Köhler C, Vielle-Calzada JP, Nunes SM et al (2000) Interaction of the Arabidopsis polycomb group proteins FIE and MEA mediates their common phenotypes. Curr Biol 10(23):1535–1538
Stroud H, Do T, Du J, Zhong X, Feng S, Johnson L et al (2014) Non-CG methylation patterns shape the epigenetic landscape in Arabidopsis. Nat Struct Mol Biol 21 (1):64
Stroud H, Greenberg MVC, Feng S, Bernatavichute YV, Jacobsen SE (2013) Comprehensive analysis of silencing mutants reveals complex regulation of the Arabidopsis methylome. Cell 152(1–2):352–364
Sun B, Looi L-S, Guo S, He Z, Gan E-S, Huang J et al (2014) Timing mechanism dependent on cell division is invoked by Polycomb eviction in plant stem cells. Science 343(6170):1248559
Takuno S, Gaut BS (2011) Body-methylated genes in *Arabidopsis thaliana* are functionally important and evolve slowly. Mol Biol Evol 29(1):219–227
Talbert PB, Henikoff S (2017) Histone variants on the move: substrates for chromatin dynamics. Nat Rev Mol Cell Biol 18(2):115
Taller J, Hirata Y, Yagishita N, Kita M, Ogata S (1998) Graft-induced genetic changes and th inheritance of several characteristics in pepper. Theor Appl Genet 97:705–713
Tang N, Ma S, Zong W, Yang N, Lv Y, Yan C et al (2016a) MODD mediates deactivation and degradation of OsbZIP46 to negatively regulate ABA signaling and drought resistance in rice. Plant Cell 28 (9):2161–2177

Tang X, Miao M, Niu X, Zhang D, Cao X, Jin X et al (2016b) Ubiquitin-conjugated degradation of golden 2-like transcription factor is mediated by CUL4-DDB1-based E3 ligase complex in tomato. New Phytol 209(3):1028–1039. https://doi.org/10.1111/nph.13635

Tang Y, Liu X, Liu X, Li Y, Wu K, Hou X (2017) Arabidopsis NF-YCs mediate the light-controlled hypocotyl elongation via modulating histone acetylation. Mol Plant 10(2):260–273

Telias A, Lin-Wang K, Stevenson DE, Cooney JM, Hellens RP, Allan AC et al (2011) Apple skin patterning is associated with differential expression of MYB10. BMC Plant Biol 11(1):93

Teyssier E, Bernacchia G, Maury S, How Kit A, Stammitti-Bert L, Rolin D, Gallusci P (2008) Tissue dependent variations of DNA methylation and endoreduplication levels during tomato fruit development and ripening. Planta 228(3):391

This P, Lacombe T, Cadle-Davidson M, Owens CL (2007) Wine grape (*Vitis vinifera* L.) color associates with allelic variation in the domestication gene VvmybA1. Theor Appl Genet 114(4):723–730

Thompson AJ, Tor M, Barry CS, Vrebalov J, Orfila C, Jarvis MC et al (1999) Molecular and genetic characterization of a novel pleiotropic tomato-ripening mutant. Plant Physiol 120(2):383–390

Torres IO, Fujimori DG (2015) Functional coupling between writers, erasers and readers of histone and DNA methylation. Curr Opin Struct Biol 35:68–75

Trindade I, Schubert D, Gaudin V (2017) Epigenetic regulation of phase transitions in *Arabidopsis thaliana*. In: Rajewsky N, Jurga S, Barciszewski J (ed) Plant epigenetics. Springer, Berlin, pp 359–383

Uthup TK, Karumamkandathil R, Ravindran M, Saha T (2018) Heterografting induced DNA methylation polymorphisms in *Hevea brasiliensis*. Planta 248 (3):579–589

Vachon G, Engelhorn J, Carles CC (2018) Interactions between transcription factors and chromatin regulators in the control of flower development. Oxford University Press, Oxford

van Leeuwen C, Darriet P (2016) The impact of climate change on viticulture and wine quality. J Wine Econ 11(1):150–167

Veluchamy A, Jégu T, Ariel F, Latrasse D, Mariappan KG, Kim S-K et al (2016) LHP1 regulates H3K27me3 spreading and shapes the three-dimensional conformation of the Arabidopsis genome. PLoS ONE 11(7):e0158936

Vergara Z, Gutierrez C (2017) Emerging roles of chromatin in the maintenance of genome organization and function in plants. Genome Biol 18(1):96

Verhoeven KJF, Preite V (2014) Epigenetic variation in asexually reproducing organisms. Evolution 68 (3):644–655

Verriès C, Bès C, This P, Tesnière C (2000) Cloning and characterization of Vine-1, a LTR-retrotransposon-like element in *Vitis vinifera* L., and other Vitis species. In: Genome, vol 43. https://doi.org/10.1139/g99-139

Vongs A, Kakutani T, Martienssen RA, Richards EJ (1993) *Arabidopsis thaliana* DNA methylation mutants. Science 260(5116):1926 LP–1928 LP. https://doi.org/10.1126/science.8316832

Waddington CH (1942) The epigenotype. Endeavour 1:18–20

Wang C, Liu C, Roqueiro D, Grimm D, Schwab R, Becker C et al (2015) Genome-wide analysis of local chromatin packing in *Arabidopsis thaliana*. Genome Res 25(2):246–256

Wang G, Köhler C (2017) Epigenetic processes in flowering plant reproduction. J Exp Bot 68(4):797–807

Wang H, Zhai L, Xu J, Joo H-Y, Jackson S, Erdjument-Bromage H et al (2006) Histone H3 and H4 ubiquitylation by the CUL4-DDB-ROC1 ubiquitin ligase facilitates cellular response to DNA damage. Mol Cell 22(3):383–394

Wang Z, Meng D, Wang A, Li T, Jiang S, Cong P, Li T (2013) The methylation of the PcMYB10 promoter is associated with green-skinned sport in Max Red Bartlett pear. Plant Physiol 162(2):885–896

Warschefsky EJ, Klein LL, Frank MH, Chitwood DH, Londo JP, von Wettberg EJB, Miller AJ (2016) Rootstocks: diversity, domestication, and impacts on shoot phenotypes. Trends Plant Sci 21(5):418–437

Weinert BT, Narita T, Satpathy S, Srinivasan B, Hansen BK, Schölz C et al (2018) Time-resolved analysis reveals rapid dynamics and broad scope of the CBP/p300 acetylome. Cell 174(1):231–244

Wendte JM, Pikaard CS (2017) The RNAs of RNA-directed DNA methylation. Biochim Biophys Acta (BBA) Gene Regul Mech 1860(1):140–148

Whittaker C, Dean C (2017) The FLC locus: a platform for discoveries in epigenetics and adaptation. Annu Rev Cell Dev Biol 33:555–575

Williams BP, Pignatta D, Henikoff S, Gehring M (2015) Methylation-sensitive expression of a DNA demethylase gene serves as an epigenetic rheostat. PLoS Genet 11(3):e1005142

Woo HR, Dittmer TA, Richards EJ (2008) Three SRA-domain methylcytosine-binding proteins cooperate to maintain global CpG methylation and epigenetic silencing in Arabidopsis. PLoS Genet 4(8): e1000156

Woong C, Roh H, Vi T, Do Y, Fischer RL, Seob J et al (2011) An E3 ligase complex regulates SET-domain polycomb group protein activity in *Arabidopsis thaliana*. PNAS 108(19):8036–8041. https://doi.org/10.1073/pnas.1104232108

Wu C, Morris JR (2001) Genes, genetics, and epigenetics: a correspondence. Science 293(5532):1103 LP–1105 LP. https://doi.org/10.1126/science.293.5532.1103

Wu HW, Deng S, Xu H, Mao HZ, Liu J, Niu QW et al (2018) A noncoding RNA transcribed from the AGAMOUS (AG) second intron binds to CURLY LEAF and represses AG expression in leaves. New Phytol 219(4):1480–1491. https://doi.org/10.1111/nph.15231

Wu R, Wang X, Lin Y, Ma Y, Liu G, Yu X et al (2013) Inter-species grafting caused extensive and heritable alterations of DNA methylation in Solanaceae plants. PLoS ONE 8(4):e61995

Xanthopoulou A, Tsaballa A, Ganopoulos I, Kapazoglou A, Avramidou E, Aravanopoulos FA et al (2019) Intra-species grafting induces epigenetic and metabolic changes accompanied by alterations in fruit size and shape of *Cucurbita pepo* L. Plant Growth Regul 87(1):93–108

Xiao J, Jin R, Yu X, Shen M, Wagner JD, Pai A et al (2017) Cis and trans determinants of epigenetic silencing by Polycomb repressive complex 2 in Arabidopsis. Nat Genet 49(10):1546–1552. https://doi.org/10.1038/ng.3937

Xiao J, Lee U-S, Wagner D (2016) Tug of war: adding and removing histone lysine methylation in Arabidopsis. Curr Opin Plant Biol 34:41–53

Xie H, Konate M, Sai N, Tesfamicael KG, Cavagnaro T, Gilliham M et al (2017) Global DNA methylation patterns can play a role in defining terroir in grapevine (*Vitis vinifera* cv. Shiraz). Front Plant Sci 8:1860

Xing L, Liu Y, Xu S, Xiao J, Wang B, Deng H et al (2018) Arabidopsis O-GlcNAc transferase SEC activates histone methyltransferase ATX1 to regulate flowering. EMBO J 37(19):e98115

Xu J, Xu H, Liu Y, Wang X, Xu Q, Deng X (2015) Genome-wide identification of sweet orange (*Citrus sinensis*) histone modification gene families and their expression analysis during the fruit development and fruit-blue mold infection process. Front Plant Sci 6:607

Yadegari R, Kinoshita T, Lotan O, Cohen G, Katz A, Choi Y et al (2000) Mutations in the FIE and MEA genes that encode interacting Polycomb proteins cause parent-of-origin effects on seed development by distinct mechanisms. Plant Cell 12(12):2367–2382. https://doi.org/10.1105/tpc.12.12.2367

Yagishita N (1961) Studies on graft hybrids of *Capsicun annuum* L. II. Variation in fruit shape caused by grafting for three generations and the effects in the progeny. Bot Mag Tokyo 74:480–489

Yamamuro C, Zhu J-K, Yang Z (2016) Epigenetic modifications and plant hormone action. Mol Plant 9 (1):57–70

Yong W-S, Hsu F-M, Chen P-Y (2016) Profiling genome-wide DNA methylation. Epigenetics Chromatin 9(1):26

Yu C-W, Tai R, Wang S-C, Yang P, Luo M, Yang S et al (2017) HISTONE DEACETYLASE6 acts in concert with histone methyltransferases SUVH4, SUVH5, and SUVH6 to regulate transposon silencing. Plant Cell 29 (8):1970–1983

Yuan W, Luo X, Li Z, Yang W, Wang Y, Liu R et al (2016) A cis cold memory element and a trans epigenome reader mediate Polycomb silencing of FLC by vernalization in Arabidopsis. Nat Genet 48(12):1527

Zemach A, Kim MY, Hsieh P-H, Coleman-Derr D, Eshed-Williams L, Thao K et al (2013) The Arabidopsis nucleosome remodeler DDM1 allows DNA methyltransferases to access H1-containing heterochromatin. Cell 153(1):193–205

Zemach A, McDaniel IE, Silva P, Zilberman D (2010) Genome-wide evolutionary analysis of eukaryotic DNA methylation. Science 328(5980):916–919

Zhang S, Zhou B, Kang Y, Cui X, Liu A, Deleris A et al (2015) C-terminal domains of histone demethylase JMJ14 interact with a pair of NAC transcription factors to mediate specific chromatin association. Cell Discov 1:15003

Zhang X, Yazaki J, Sundaresan A, Cokus S, Chan SW-L, Chen H et al (2006) Genome-wide high-resolution mapping and functional analysis of DNA methylation in Arabidopsis. Cell 126(6):1189–1201

Zhao H, Zhao K, Wang J, Chen X, Chen Z, Cai R, Xiang Y (2015) Comprehensive analysis of Dicer-like, Argonaute, and RNA-dependent RNA polymerase gene families in grapevine (*Vitis vinifera*). J Plant Growth Regul 34(1):108–121

Zheng M, Wang Y, Wang Y, Wang C, Ren Y, Lv J et al (2015) DEFORMED FLORAL ORGAN1 (DFO1) regulates floral organ identity by epigenetically repressing the expression of OsMADS58 in rice (*Oryza sativa*). New Phytol 206(4):1476–1490

Zheng X, Pontes O, Zhu J, Miki D, Zhang F, Li W-X et al (2008) ROS3 is an RNA-binding protein required for DNA demethylation in Arabidopsis. Nature 455 (7217):1259

Zheng Y, Liu X (2019) Chromatin organization in plant and animal stem cell maintenance. Plant Sci 281:173–179. https://doi.org/10.1016/j.plantsci.2018.12.026

Zheng Y, Thomas PM, Kelleher NL (2013) Measurement of acetylation turnover at distinct lysines in human histones identifies long-lived acetylation sites. Nat Commun 4:2203

Zhong S, Fei Z, Chen Y-R, Zheng Y, Huang M, Vrebalov J et al (2013) Single-base resolution methylomes of tomato fruit development reveal epigenome modifications associated with ripening. Nat Biotechnol 31(2):154

Zhou Y, Wang Y, Krause K, Yang T, Dongus JA, Zhang Y, Turck F (2018) Telobox motifs recruit CLF/SWN–PRC2 for H3K27me3 deposition via TRB factors in Arabidopsis. Nat Genet 50(5):638

Zilberman D, Gehring M, Tran RK, Ballinger T, Henikoff S (2007) Genome-wide analysis of *Arabidopsis thaliana* DNA methylation uncovers an interdependence between methylation and transcription. Nat Genet 39(1):61

Zuo J, Wang Y, Zhu B, Luo Y, Wang Q, Gao L (2018) Comparative analysis of DNA methylation reveals specific regulations on ethylene pathway in tomato fruit. Genes 9(5):266

From Phenotyping to Phenomics: Present and Future Approaches in Grape Trait Analysis to Inform Grape Gene Function

10

Lance Cadle-Davidson, Jason Londo, Dani Martinez, Surya Sapkota and Ben Gutierrez

Abstract

Phenotyping in grapevines is the assessment of qualitative and quantitative traits including growth, development, tolerance, resistance, architecture, physiology, chemistry, ecology, and yield. Traditionally, phenotyping techniques relied on measurement of visual, chemical, physiological, or other characteristics by experts, often at low-throughput. The use of standardized OIV or phenological descriptors and scales to phenotype grapevine traits has provided a good foundation for international adoption of phenotyping standards and cross-comparison of results. However, many of these descriptors are subjective, fail to capture complete trait variation, or may not be relevant to some studies. Phenomics, the future of phenotyping, brings opportunities and challenges in increased throughput, objectivity, precision, dynamic measures, and integration that demand new approaches for standardization, data management, and analysis. Here, with a focus on large-scale genetic studies, such as QTL mapping, we describe current phenotyping approaches and their limitations and introduce some future opportunities in phenomics, including the promotion of FAIR data principles of Findability, Accessibility, Interoperability, and Reusability.

All authors made equal contributions to this chapter. Mention of trade names or commercial products is solely for the purpose of providing specific information and does not imply recommendation or endorsement by the U.S. Department of Agriculture (USDA). USDA is an equal opportunity provider and employer.

L. Cadle-Davidson (✉) · J. Londo · B. Gutierrez
USDA-ARS Plant Genetic Resources Unit, Geneva, NY, USA
e-mail: Lance.CadleDavidson@usda.gov

J. Londo
e-mail: Jason.Londo@usda.gov

B. Gutierrez
e-mail: Ben.Gutierrez@usda.gov

D. Martinez · S. Sapkota
School of Integrative Plant Sciences, Geneva, NY, USA
e-mail: dm676@cornell.edu

S. Sapkota
e-mail: sds322@cornell.edu

10.1 Introduction

Until relatively recently, the past decade or so, the scientific bottleneck for advancing knowledge in grapevine lay primarily in the high cost of genetic analysis. Specifically, the complex, heterozygous, and high diversity nature of the grapevine genome reduced researchers' ability to make rapid associations between genes and phenotypes. While it was common to perform QTL analysis using classical molecular markers

D. Cantu and M. A. Walker (eds.), *The Grape Genome*, Compendium of Plant Genomes, https://doi.org/10.1007/978-3-030-18601-2_10

such as SSR loci, the traits that could be analyzed tended to be those with strong effects (Dalbó et al. 2000; Doligez et al. 2002, 2006; Fischer et al. 2004; Blasi et al. 2011). Fine mapping of these traits was seldom attempted as the marker association itself was sufficient for breeding applications. However, with the successful sequencing of the grape genome (Jaillon et al. 2007; Velasco et al. 2007) and the rapid decreases in costs associated with development of new molecular markers and genome tools, a flip of the bottleneck has occurred. The phenotyping bottleneck now limits rapid progress in advancing grapevine research.

One common thread that has become apparent in grapevine phenotyping is the challenge associated with consistent data collection and analysis as it relates to each particular phenotype. As studies are conducted in parallel by different laboratories or data are compared between field and greenhouse conditions, it is essential that phenotyping be consistent and reflects biology. Simply understanding what and how to phenotype are the predominant obstacles for developing a high-throughput (HT) method. When you try to reduce something complex into something that is simple and fast, how do you avoid measuring in error and keep it relevant to commercial grape production? Some traits are more amenable to HT methods by leveraging associations between whole vine and sampled vine aspects, such as disease resistance (as highlighted below). Other traits are simply easier to collect HT data based on their long history of correlation in grapevine, such as pruning weights and vine vigor (Kicherer et al. 2017b). Rapid developments for these types of traits may provide the initial push needed to conceptualize more complex methods.

Several phenotypes currently cannot be conducted in a HT manner. For example, physiological studies require expensive equipment with long calibration times in order to collect meaningful data (e.g., water use efficiency, WUE). Despite substantial literature detailing the complex interactions of the vine with water availability, rapid methods of assessing vine status in a way that informs viticulture remain elusive. For physiological traits that remain too complex or are so influenced by environment that they pose a challenge to emulate vineyard conditions, efforts need to be spent defining the critical aspects of the trait.

In this chapter, we focus primarily on HT phenotyping, identifying methods, and strategies that have been successfully applied in genetic experiments involving larger germplasm sets (species collections, breeding/mapping populations, and mutagenesis) rather than focused sets (transgenes and gene editing) and envisioning technologies that may be applied in these same scenarios. The goals of such HT phenotyping studies typically include germplasm characterization or selection, QTL analysis, and/or gene discovery. In some cases, like table grape fruit quality and viticultural treatments (e.g., gibberellic acid applications) are embedded in the experimental design and may enable scientists to study viticultural treatment effects. And there are many opportunities for the phenotypic analyses themselves to become assays used in viticultural management as we currently see for fruit chemistry traits.

In spite of all the successes referenced in this chapter, a major challenge for the grapevine community exists regarding the study of genotype by environment interactions (GxE). Addressing the challenges of GxE requires, among other things, standardization and careful documentation of phenotyping protocols, an effort that is gaining international traction. Several collaborative, international projects and networks on phenotyping (transplant, regional and international plant phenotyping networks, and ELIXIR-EXCELERATE) have developed resources for standardized phenotyping. One noteworthy effort is Minimum Information About a Plant Phenotyping Experiment (www.miappe.org), which outlines suggested and required attributes for metadata description of experiments. Standardization and careful documentation promote improved data stewardship and makes data re-usable for purposes beyond those initially envisioned or beyond current resources. To this end, a set of FAIR principles (Findability, Accessibility, Interoperability, and Reusability) have been developed (Wilkinson et al. 2016). The vision for FAIR as it relates to the grape community has begun to be organized through a global grape information system (GrapeIS) organized by the International Grapevine Genome Program (IGGP; www.vitaceae.org), and its success will depend upon the active participation of those in the grape community generating, analyzing, and publishing data.

In the following sections, rather than organize traits into silos, we organize the phenotypes by how they are collected (visual, physiological, chemical, and molecular), and provide exemplary traits for each, focusing on phenotyping challenges and future opportunities.

10.2 Visual Ratings

Sensorial traits characterize the oldest and the most intuitive phenotypes. Imagine hunters and gatherers accustomed to seeing and collecting small clusters of wild female grapes and one day by chance they find large and perfect-flowered clusters. Visually observing this key domestication trait, they took a note to return and care for these. This visual cue of selection for large clusters has been repeatedly followed by selection for other visual traits such as berry color, size, and shape, as well as other sensorial traits like seed trace, texture, flavor, and aroma. In this section, we focus on phenotypes that are visually rated as a starting point using a comprehensive set of traits described by the OIV phenotypic scales (OIV 2018), which were developed primarily for the standardized description of grapevine varieties and species.

10.2.1 Challenges with Standardizing Visual Phenotyping Methods and Scales

The OIV phenotypic scales represent well over 100 traits that are measured by visual assessment, including diverse traits such as ampelographic measures, abiotic and biotic susceptibility, berry and cluster measures, and phenology. The primary approach has been to categorize and provide examples in order to standardize data collection around the world. While an excellent tool for its intended purpose, there are some widely acknowledged weaknesses to this system when applied to phenotyping for genetic analyses. Many of the described phenotypes are continuously variable traits that lose information upon categorization. Further, many of the 9-point OIV scales only have five defined categories, so for most users they functionally become 5-point scales and the limited resolution of the scale determines limited resolution of knowledge gained. The expression or biology of traits may differ greatly among grape germplasm, necessitating a different phenotyping method to capture the trait biology within the germplasm studied. Finally, during the preparation of this chapter, a second edition of OIV descriptors was published in 2018 (OIV 2018) and we as authors were not using the same editions. It quickly became apparent that for the goal of standardizing, it will be important for the international community to use and reference this enhanced second edition.

10.2.1.1 Visual Phenotyping of Powdery Mildew

The degree of care in designing, executing, and describing powdery mildew foliar resistance phenotyping experiments are highly variable but recent studies have improved attention to detail. Table 10.1 shows some phenotyping studies in which sufficient detail was provided to interpret the experimental design. Most studies (12 of 23) rated disease on a categorical scale following natural infection, often attempting to relate their ratings to the scale OIV455. Several of these studies rated disease progression over time and found that the significance of QTL changed over time (Pap et al. 2016; Zendler et al. 2017), with some QTL being undetectable if the wrong time point was selected (Barba et al. 2014; Cadle-Davidson et al. 2016). Time series ratings provide the added opportunity for area under the disease progress curve (AUDPC) analysis (Teh et al. 2017). Studies in other pathosystems have emphasized the subjectivity and imprecision of visual ratings (Sherwood et al. 1983) as well as the importance of rater subjectivity (Poland and Nelson 2011). Fortunately, QTL were consistent across raters, even if inconsistency in ratings affected the magnitude of those QTL effects (Poland and Nelson 2011). Recently, efforts to develop and standardize controlled inoculation in vitro have shown promise for detection of moderate or minor QTL like *REN2* and *REN9* that may not be detected in vineyard evaluations (Cadle-Davidson et al. 2016; Zendler et al. 2017).

For loci that have been genetically mapped using multiple phenotypes, the degree of QTL significance can provide some insights into which phenotyping methods best explain the genetics of resistance. For *REN1*, single isolate in vitro inoculations reproducibly generated higher LOD scores than vineyard and greenhouse ratings (Hoffmann et al. 2008; Cadle-Davidson et al. 2016). For *REN3/REN9*, which was only mapped using vineyard data, better results were obtained when replicated progeny vines were analyzed (Zendler et al. 2017). Finally, for *REN6* and *REN7*, careful phenotyping using several methods indicated that visual ratings after controlled inoculation in a greenhouse or on detached leaves reflected the genetics better than vineyard evaluations or qPCR-based quantitation of fungal growth (Pap et al. 2016). Based on the studies, presented in Table 10.1, for vineyard evaluations we recommend repeated measures over the course of the growing season on replicated vines. Further, if resources are available, single isolate inoculation of detached leaves or disks appears to be the current best method for detecting minor or moderate QTL and may be the most relevant for fine mapping and characterization of candidate genes.

10.2.1.2 Visual Phenotyping of Fruit Clusters

Big berries, long berries, many berries, open cluster architecture—in the world of genetic improvement for fruit clusters, there are dozens of breeding goals, each with different challenges in visual phenotyping and often with trade-offs. Further, many visual fruit traits are of specific importance to table grapes, for which gibberellic acid or other treatments may interact with genetic effects. The OIV system has categorical descriptors for several visual phenotypes related to fruit (cluster dimensions and density, berry size, shape, and color), but most genetic and genomic studies choose to quantify phenotypes with greater precision and objectivity.

Components of cluster architecture have been carefully defined in several studies (Shavrukov et al. 2004; Correa et al. 2014) and recently reviewed in detail (Tello and Ibáñez 2018). These use standard tools to measure lengths (caliper), angles (protractor), volumes (beaker), weights (balance), and counts (fingers and toes?). Within the context of QTL analysis, careful measurement of 23 such parameters identified QTL of moderate effect, explaining 13–24% of the phenotypic variance (Correa et al. 2014). Simply weighing berries has indicated QTL in multiple studies (Zhao et al. 2015; Ban et al. 2016). Similarly, in a candidate gene analysis of a diversity panel, simple weighing and measuring showed that berry length, width, volume, and weight were significantly predicted by alleles of *VvNAC26* (Tello et al. 2015).

Given the success of simple phenotyping methods applied in these genetic studies, the question arises, what are the limitations of these methods, and what changes are needed? Without a doubt the biggest limitation of manual measurements is labor. Using calipers to measure each rachis internode or weighing or measuring a 30-berry subsample represents a labor-intensive process that requires removing berries from the stem and significant time when applied to hundreds of genotypes with any level of replication. Thus, several research programs are pursuing the tasks of imaging for computer vision either in controlled environments or in the vineyard (discussed below in Sects. 10.2.2 and 10.6).

10.2.1.3 Visual Phenotyping of Phenology

Grape phenology is the study of cyclic developmental processes, especially in relation to the

Table 10.1 Varied methods and response variables reported for powdery mildew resistance phenotyping

Environment	Inoculum	Locus	Response variable[a]	Observations (years)[b]
Vineyard	Natural	*REN3*	OIV455	1 (5)
Vineyard	Natural	*REN8*	OIV455	1 (6)
Vineyard	Natural	*SEN1*	OIV455	5 and 2 (2)
Vineyard	Natural	*REN3*	OIV455 and 7-point	8 and 10 (2)
Vineyard	Natural	*REN10*	OIV455 and 7-point	8 and 10 (2)
Vineyard	Natural	*REN3/9*	OIV455, but 5-point	2 (2), 2 replicate vines
Vineyard	Natural	*REN4*	6-point version of OIV455	1 (1)
Vineyard	Natural	*RUN2.1*	6-point version of OIV455	1 (1)
Vineyard	Natural	*RUN2.2*	6-point version of OIV455	1 (1)
Vineyard	Natural	*REN1*	4-point rating	1–2 (2)
Vineyard	Natural	*REN4*	4-point rating	1 (3)
Vineyard	Natural	*SEN1*	4-point rating	1 (1)
Vineyard	Natural and Mixed isolate	*REN6/7*	4-point rating	3–4 (2)
Greenhouse	Natural	*REN1*	OIV455	3 (2)
Greenhouse	Natural	*REN4*	Foliar incidence and severity	1 (1), 2 locations
Greenhouse	Natural	*RUN2.1*	Foliar incidence and severity	2 (1), 2 locations
Greenhouse	Single isolate	*REN6/7*	5-point scale	1 (1), 2 raters, 3–4 replicate vines
Petri dish	Mixed isolate	*REN9*	Necrosis, 7-9 dpi microscopy	1 (1)
Petri dish	Single isolate	*REN6/7*	5-point modified OIV455, 14–15 dpi	1 (1), 2 raters
Petri dish	Single isolate	*REN6/7*	qPCR, 14–15 dpi	1 (1)
Petri dish	Single isolate	*REN5*	6-point ratings, coverage, and sporulation, 7 dpi	1 (2)
Agar tray	Single isolate	*REN1*	Hyphal transects, 8–9 dpi	1–2 (3)
Agar tray	Single isolate	*REN2*	Hyphal transects, sporulation, 8 dpi	1–2 (2)

[a]The IPGRI rating scale OIV455 describes odd number foliar ratings on a 1–9 scale. For in vitro inoculations, the days post inoculation (dpi) of data collection is provided
[b]The number of observations in each year is shown, along with the number of years in which observations were made in parentheses, followed by additional efforts in replicated plants and/or raters

growing season and climate. Grape phenology growth stages are commonly defined by the BBCH scale or by the modified Eichorn-Lorenz scale (Coombe 1995), which describes 47 stages such as winter bud, budburst, flowering, berries harvest-ripe, and end of leaf fall. Most geneticists and breeders are interested in the relative timing of these stages, which are determined by visually monitoring vines over time.

One challenge is phenological heterogeneity across environments, within a vine, and even within organs of a vine. For example, bloom is asynchronous at sites with warmer winters, such as South Australia, with basal clusters on distal shoots (away from the trunk) being 4–6 days delayed within the same vine (Gadoury 2015). While phenological heterogeneity at warm-winter sites is easily observed at bloom, the effects can be detected at most stages of development. This and other environmental effects on phenology are likely reflected in QTL studies of phenological processes. For example, QTL were significant in only one environment for the periods from inflorescence appearance to 50% flowering and from flowering to véraison in a Picovine × Ugni Blanc flb family (Houel et al. 2015). Fechter et al. (2014) combined two approaches to effectively address the uncertainty generated by phenology GxE: (1) they categorized relative flower date with the 5-point OIV-302 scale to account for year-to-year shifts in bloom date and used median values of the phenotypic categories across years and (2) they used independent mapping families within the same experimental design for validation of the key QTL involved in the traits studied. Interestingly, for growth cessation response to photoperiod from *V. riparia*, in the absence of inducing temperatures a QTL on LG13 explained more than 80% of phenotypic variation, but in the presence of inducing temperatures a QTL on LG11 also explained nearly all the phenotypic variation (Garris et al. 2009). But not all phenological studies are so environmentally fickle. In another study, the date of véraison and duration between bloom and véraison both indicated a major QTL named *Ver1* on Chr16, which explained up to 70% of the phenotypic variance observed with early onset of véraison from GF. GA-47-42 (Zyprian et al. 2016).

To overcome the effects of environment on physiological processes, Vivin et al. (2017) suggested that models integrating physiological processes with their genetic control may aid in the genetic characterization of complex traits with genotype × environment interactions. Climate change significantly affects phenotyping of phenology, both as a challenge for long-term applicability of current or past results and as a justification for the importance of studying phenology in preparation for expected climate change. Over the past 50 years, several phenological stages (budbreak, flowering, and véraison) have shifted significantly earlier in France and are forecast to push even earlier in future decades (García de Cortázar-Atauri et al. 2017). In Italy, over the past four decades harvest has shifted 25–40 days earlier for four widely planted grape cultivars (Koch and Oehl 2018).

10.2.2 Opportunities from Computer Vision for Phenotyping

Phenotyping methodologies in plant science are usually based on detecting, extracting, and quantifying observable features from biological samples. Those physical features can be detected visually by human sight or by optical instruments and electronic sensors. As described above, most of this phenotyping work has been done by human experts. However, this becomes infeasible in experiments with a large number of samples requiring a huge amount of time and human resources. Furthermore, the differences in the subjective reasoning among human experts can add error to the statistical analysis of the results.

The automation of phenotyping methodologies by using latest approaches from information technologies (IT) has been gaining interest among researchers. The main goal is to achieve a more robust, objective, and faster method than human perception and decision making by applying computer-based procedures such as computer vision to emulate human sight. The effectiveness from first approaches of such methods has motivated researchers to search for ambitious implementations capable of handling large numbers of samples per experiment that provide more solid and reliable results. This concept is popularly known as high-throughput (HT) phenotyping.

Of the numerous approaches developed to-date, here, we provide a few examples relevant to the above visual ratings. Automated phenotyping has been used for the characterization of plant diseases, and one of the first automated HT phenotyping microscopy systems focused on characterizing susceptibility to the barley powdery mildew, *Blumeria graminis* f. sp. *hordeii* (*Bgh*) (Ihlow et al. 2008). More recently, hyperspectral cameras have been used for the detection of *Bgh* (Kuska et al. 2015). Unfortunately, enough dissimilarities exist in the morphology of *Bgh* and *E. necator*, as well as in host morphology, that computer vision tools developed for *Bgh* may not be directly transferable to *E. necator*.

The laborious task of fruit phenotyping inspired numerous groups to pursue computer vision approaches. In controlled environments, imaging systems can capture 2D and 3D representations of grape clusters for accurate computer vision analysis of cluster length, width, elongation, and volume (Tello et al. 2016). A smartphone phenotyping application was developed capable of performing fast estimation of the number of grapes detected from smartphone-captured images of clusters (Aquino et al. 2018). In the vineyard, robot systems can be used for phenotyping, which enables upgraded automation capabilities such as unsupervised operation. A field-phenotyping enabled mobile robot was developed capable of performing multiple field tasks based on image acquisition such as berry size and color measurement (Kicherer et al. 2015). Furthermore, they continued working on this concept and developed a new phenotyping platform called Phenoliner, which extends the capabilities of the previous platform with an improved vehicle carrier based on a grapevine harvester, extra sensors, and a 3D reconstruction procedure for grape vine modeling (Kicherer et al. 2017a). Such a phenotyping platform could be envisioned to house additional phenotyping tools and strategies from other efforts, such as dynamic imaging of plant growth responses to soil water deficit (Granier et al. 2006) or statistical analysis of leaf features (Failmezger et al. 2018).

The above computer vision approaches generate big data for which some technologies have recently been applied to improve and extend the decision-making procedures on many AI-based applications including newer HT phenotyping approaches (Coppens et al. 2017). The identification of patterns, features, and/or noise exclusion in images or other data types from huge amounts of data collections (and its management) are the challenging fronts on these newer automated phenotyping implementations. Singh et al. (2016) proposed to use machine learning tools in a big data scheme for plant stress phenotyping in order to be able to process large amount of different gathered sensor data in different timepoints. The conclusions from this work highlight the feasibility of such methods and encourage researchers to apply these popular artificial intelligence approaches in order to speedup data processing as well as to increase result accuracy rates.

Information technologies are a focus of the new advances in phenotyping methodologies, which are mainly based on computationally capturing and processing data. The main goal in this field is to maximize phenotyping throughput and accuracy in order to generate consistent results for better genetic prediction. Such goals can only be achieved by applying HT automated phenotyping methodologies based on IT in many experimental designs, due to large sample sizes and genetic variability in input samples that are infeasible to address with manual procedures.

10.3 Physiological Measurements

Accurate HT phenotyping of physiological processes is necessary for the future sustainability and productivity of grapevine production. Numerous studies have modeled the projected effects of changing climate parameters on current and future viticultural regions (Jones et al. 2005; Luedeling 2012; Mozell and Thach 2014; Duchene 2016; Wolkovich et al. 2017). While predictions vary, there is little doubt that a number of climate-related factors will make cultivating grapevines more challenging in the future. Elevated average temperatures threaten to advance ripening stages of grapes to the hottest points of the summer season, degrading color attributes, reducing photosynthetic capacity, and ultimately reducing harvest quality (Webb et al. 2007; Mira de Orduña 2010; Salazar Parra et al. 2010; Sweetman et al. 2014). Increased variation in water availability through alternating drought and deluge conditions threaten to impact aspects of yield and berry integrity (Schultz 2000; Webb et al. 2008). Warming spring temperatures have already resulted in phenological shifts in grapevine (Tomasi et al. 2011) and predictions of shifting of earlier budburst may expose vines to greater frost risk (Zapata et al. 2017; Leolini et al. 2018). To address this, new viticultural areas could be developed at higher latitudes; however, despite extended growing seasons, higher latitudes will continue to be faced with freezing winter temperatures (Caffarra and Eccel 2011; Luedeling 2012; Mosedale et al. 2015).

10.3.1 Challenges with Standardizing Physiological Phenotyping Methods and Scales

Due to complexity of physiology and the strong interaction of environment on physiological traits, the present state of phenotyping physiological traits and adaptation to environment is one of low-throughput. One of the key issues with development of HT methods for assessing grape physiology is the interaction between plant plasticity and changing environment as it affects the relevance of a given phenotype. Most physiological traits (e.g., WUE, salt tolerance, cold tolerance, and photosynthetic capacity) are not understood with enough depth, or require too much time for measurements, for rigorous phenotypes to be measured in a HT method. The primary way researchers have negotiated this challenge is to isolate the plants under controlled conditions to somehow standardize the phenotype (e.g., potted greenhouse plant assays). This approach can be used to phenotype some components of physiology, for example, understanding how stomatal conductance or leaf water status differs under drought simulation (Toumi et al. 2008; Salazar-Parra et al. 2015). These studies may also be large enough to assess many different cultivars or rootstocks at once, or more than one stresser (Serra et al. 2014). However, the results are seldom tested or replicated under field conditions. Ultimately the phenotype we measure has to have a real-world impact for it to be of use in grapevine production or in the breeding of new cultivars. The full list of studies examining aspects of physiological responses of grapevine are too long to review in this chapter, but we have attempted to capture the breadth of studies below.

Phenotypic variation among cultivars and among species have been observed for water use efficiency (WUE, as reviewed in Flexas et al. 2010; Tomás et al. 2014), regulation of stomatal conductance (Costa et al. 2012; Pou et al. 2012; Coupel-Ledru et al. 2014; Duursma et al. 2018), cavitation resistance and embolism repair (Lovisolo and Tramontini 2010), root suberization (Barrios-Masias et al. 2015), stomatal density (Boso et al. 2016), ABA sensitivity (Rossdeutsch et al. 2016), isohydric versus anisohydric behavior (Lavoie-Lamoureux et al. 2017), chloride/sodium exclusion (Henderson et al. 2014, 2017), photosynthetic capacity and control of respiration (Pellegrini et al. 2015; Coupel-Ledru et al. 2016), and low temperature and lethal temperatures (Ferguson et al. 2014; Londo and Kovaleski 2017). What is abundantly clear in reviewing these studies is that physiological traits are almost never amenable to HT methods. Which of these traits can be measured quickly enough for use in large populations? Additionally, each of the phenotypes represents just a small aspect of the overall trait; are those small aspects relevant to whole vine or field responses? It is essential to develop highly robust phenotypes that are vineyard relevant in order to most efficiently move toward functional and genetic elucidation of physiological traits. Listed below are a few nice studies where some aspects of HT phenotyping were conducted. All are forced to reduce the complexity of physiology to a few factors in order to conduct the assessments. This is not specifically a critique of those studies or their methods, but instead an open acknowledgement to the long and complicated road ahead for the development of HT physiological phenotyping.

10.3.1.1 Drought and Water Relations

Enhancing grapevine drought tolerance is a major goal for grapevine researchers. Climate change predictions indicate that future precipitation patterns will become more erratic. Despite being moderately drought adapted, grapevine still requires irrigation over much of its current production area to maintain yield and quality. Competition with other crops and other societal demands may also increase the challenges in the future for securing sufficient water for irrigated production. One example of a potential HT phenotype for drought/water relations leveraged leaf dehydration rate as a proxy for grapevine's relative drought resistance (Hopper et al. 2014). The study used water loss from detached grapevine leaves as well as leaf responses to ABA applications to assess relative stomatal control and compared results with reports of the different genotypes' drought tolerance. The study demonstrated clear differences in water loss resistance among the genotypes and also demonstrated a major effect of the developmental stage of the leaves tested. How do the results of this study then translate to large field studies? Is water loss from detached leaves a good proxy for field-based drought resistance? Comparing well-established measures of WUE in the field between leaf level measures and whole canopy measures typically demonstrates that there is a general lack of correlation between these two scales in grapevine (Medrano et al. 2015). The next step is to investigate if this HT concept can translate to differences in drought response in mapping families or field grown vines.

As another example, QTL mapping of hydric behavior (isohydric vs anisohydric) in a cross between the cultivars Syrah (anisohydric) and Grenache (isohydric) were able to identify QTL for leaf water potential in potted plants (Coupel-Ledru et al. 2014). The study made use of potted vines with high-tech control of irrigation level, controlled lighting, humidity, and temperature and is one of the few examples where a physiological trait has been examined in a mapping population. Data was reproducible between years but correlations between traits were low, suggesting additional knowledge is needed about the master control differences between hydric behaviors. In further studies of this population, the potential to breed for increased transpiration efficiency was demonstrated through reducing night-time water loss from respiration (Coupel-Ledru et al. 2016). QTL were identified for both night-time and day-time loss of water from plants, but higher variation was noted during day-time sampling. Genotypes with reduced loss of water at night maintained higher growth rates under water withholding conditions. This trait appears to localize to different QTL from those associated with day-time stomatal response. While these phenotypes were not conducted in the field, the results raise the possibility that breeders could select on different aspects of water control to increase overall WUE.

10.3.1.2 Temperature Response

Understanding the complexity of the grapevine response to high or low temperatures is complicated by the thermal variation that occurs seasonally and annually. Selecting phenotypes that may be adaptive for a future of temperature extremes are essential to maintaining crop harvests and post-harvest product quality. Heat stress interacts with drought stress, but also directly impacts photosynthetic capacity and impacts berry color and flavor development, resulting in poor quality (Chuine et al. 2004; Greer and Weedon 2013). To examine genetic variation in heat tolerance, detached leaves of 47 grapevine genotypes were exposed to high temperatures (47 °C) and assayed for oxygen evolution rate, chlorophyll *a* fluorescence and ion leakage as a proxy for cell membrane damage (Xu et al. 2014a). The authors note that most wild grapevine species were more resistant to damage than *V. vinifera* and that these assays may be an indirect measure of field-based heat tolerance. Further, measuring the changes in chlorophyll *a* fluorescence was more sensitive and convenient for evaluating these traits, but the question remains if these traits capture vine level heat resistance. Use of the dwarf grapevine breeding tool "microvine" (Chaïb et al. 2010) offers some potential for accelerating phenotyping of physiological stress responses. The effects of elevated heat treatments on berry development and photosynthesis revealed the important role of carbon balance in heat tolerance and berry quality (Torregrosa et al. 2017). Phenotyping was conducted on the microvine samples to identify traits that varied by temperature treatment (dry matter and biomass, gene expression, and berry ripening parameters), and QTL analysis in a microvine × "Ugni Blanc" population uncovered genetic regions tied to these phenotypes. Like all studies, it remains to be seen if these phenotypes capture the complexity of high heat impacts under vineyard conditions.

At the other end of the thermal spectrum is low-temperature stress and damage. Freeze damage typically occurs either as frost damage on exposed leaves and inflorescence tissues or as reduced ability of dormant tissues to survive winter. Only two studies examining frost resistance of green tissue by visual assessment of damage have been described and each simulated frost (Fuller and Telli 1999; Londo et al. 2018). Very little variation in this trait was observed but screening must be expanded to determine if trait variation exists. Winter survival is a complex mix of traits with various organ specific processes including dormancy induction (Garris et al. 2009), changes in bud supercooling ability (Mills et al. 2006; Londo and Kovaleski 2017; Shellie et al. 2018), and chilling requirement (Dokoozlian 1999; Londo and Johnson 2014). Variation in dormancy levels is frequently assessed in grapevine via forcing assays to determine chilling requirement; dormant cane material is placed in warm conditions and the time needed to observe budburst is recorded (Fila et al. 2012; Londo and Johnson 2014). Typically, these forcing assay studies take weeks to see the phenological indicator of budburst. While it is possible that forcing assays could be done at a mapping population level, the long duration needed for the phenotype to manifest makes it decidedly low-throughput. The most common phenotypes for freeze resistance for green growing and woody tissues are ion leakage, a proxy for cell membrane integrity. Tissue samples are placed into glycol baths at various freezing temperatures, frozen, and evaluated for increases in ions within a wash solution (Ershadi et al. 2015; Gale and Moyer 2017). For leaf ion leakage assays, no studies could be found examining genetic or phenotypic variation within grapevine, though this method is used in many other horticultural species (Lindén et al. 2000; Morin et al. 2007; Pagter and Williams 2011; Väinölä et al. 1997). The most common phenotype for dormant bud tissues is evaluation of the low-temperature exotherm using differential thermal analysis (Mills et al. 2006; Ferguson et al. 2011, 2014; Londo and Kovaleski 2017), whereby lethal temperatures are evaluated by the failure of the supercooling mechanism of suppressing freezing. Assessments of phenotypic variation for low-temperature exotherms have described various levels of cold hardiness between cultivars and wild species. These studies have demonstrated the large impact of winter temperature variation on expression of the phenotype (Ferguson et al. 2014; Dami et al. 2016; Londo and Kovaleski 2017). As a result, no QTL studies to date have been conducted to examine LTE variation, despite several breeding programs using this method for plant selection.

10.3.1.3 Root Behavior

While not explicitly "physiology", rooting behavior is another phenotyping area where many different methods are in development. Root phenotyping also contributes to our understanding of other soil-based physiological stresses like nutrient deficiency, salt tolerance, and of course, drought response (Yıldırım et al. 2018). A number of root traits have been suggested as targets for phenotyping including root system size and the ratio of root to shoot mass, total root length and surface area, fine root attributes, and root regrowth aspects (as reviewed in Comas et al. 2013). Traditional phenotyping in grapevine involves using trenches to examine rooting patterns in the soil, or mini-rhizotron systems to examine branching patterns in semi-natural conditions. Studies have focused on root angle aspects to describe differences in predicted drought resistance, namely that deeper-rooted genotypes will be more resistant. Root trait differences between rootstocks often correlate with demonstrated differences as it relates to drought response (Yıldırım et al. 2018; Sucu et al. 2018), though drought response is usually tested under greenhouse conditions. Differences in root system architecture correlated with differences in drought resistance and suggested that phenotypes like root length (primary and lateral) as well as root area and number have potential for HT phenotyping. However, root pattern data suggest that soil compactness and type in the field are major contributors to the actual rooting pattern under vineyard conditions (Smart et al. 2006). Root growth patterns clearly have an exploitable genetic component as different wild grapevines inhabit very different soil types (Callen et al. 2016), an aspect leveraged by rootstock breeding programs. Though HT root phenotyping has a long way to go, methods to visualize root growth in field soil with ground penetrating radar or in soil analogs with CT scanning (Atkinson et al. 2018) may be one way to scale up phenotyping efforts designed to understand what root traits contribute to vine success.

10.3.2 Opportunities with Future Technologies Applied to Physiological Measures

What is the future of phenotyping as it relates to physiological traits? The key is: (1) defining keystone traits that translate from HT methods to field relevant performance, (2) leverage the power of clonal reproduction to replicate experimental vineyards across varied environments, or (3) combine these strategies. To begin with, classical mapping family approaches could be used to identify QTL, much like the studies mentioned above. These populations could be replicated across environments and in particular, across stresses. Including and acknowledging environmental variation in the design of the study (E. Duchene, pers. comm) by interspersing control cultivars could also offer the possibility for finer detection of QTL. Smart vineyards could then build on this concept of capturing environmental variation and determining the GxE component of physiological traits (Kustas et al. 2018). Sensors could be placed in the soil (Adamchuk et al. 2004), in and near the canopy (Taylor et al. 2017), interfacing data from stationary sensors with aerial drones (Bellvert et al. 2014; Anastasiou et al. 2018) and if at all possible, in the vines themselves (Pagay et al. 2014). Coordination between research groups, domestic and international, to replicate studies could truly leverage the range of environmental variation to better our understanding of grape physiology. The implementation of drones and analysis from aerial images has already demonstrated the potential for precision viticulture (Bates et al. 2018). What is missing is the application of these technologies in a discovery stage of research, rather than solely in established production fields. Despite many challenges associated with phenotyping physiological traits, the sheer number of researchers committed to increasing our collective knowledge of how physiology works, and what phenotypes are most relevant, bodes well for the development of true HT phenotyping in grape.

10.4 Chemical Analysis

From the quality of wine to the consumer appeal of table grapes, and to biotic and abiotic stress responses, fruit metabolites play an intricate role in viticulture and in the economic value of the grape industries. Understanding the primary and secondary metabolites, particularly their impact on fruit quality, will serve to meet industry standards and can facilitate development of new cultivars. As a trait, metabolite composition, or metabotype, is highly informative; fruit and vine metabolites can differentiate wild species and hybrids from *Vitis vinifera* (De Rosso et al. 2014) and distinguish between cultivars and wines (Versari et al. 2014; Crupi et al. 2015; Billet et al. 2018). Studies in grapes range from analysis of targeted metabolites to the metabolome, analysis of all metabolites and their fluctuations in a given tissue (as reviewed by (Jorge et al. 2016). Standard techniques vary from low-input, low-resolution methods, such as refractometry (°Brix) and spectrophotometry (pigmentation), to high-resolution chromatography–mass spectrometry.

10.4.1 Challenges with Standardizing Chemical Phenotyping Methods and Scales

Though largely determined by genetics, metabolites are influenced by environmental and developmental factors. Sampling time is one of the most important considerations, as broad changes in primary and secondary metabolites occur from early development to maturity and post-harvest (Zamboni et al. 2010). This can be difficult to standardize between cultivars (genotypes) due to phenological variation (Wolkovich et al. 2017). Disease status can negatively impact fruit metabolites, viruses can delay fruit ripening and decrease sugars and pigmentation (Vega et al. 2011), and pests and pathogens can alter flavor (Hall et al. 2018; Schueuermann et al. 2019). However, certain infections, such as noble rot, can improve the flavor and composition of grape berries (Blanco-Ulate et al. 2015). Abiotic factors, including drought stress and light (Sun et al. 2017; Pinasseau et al. 2017) and horticultural practices (Wang et al. 2018) also contribute to changes in fruit metabolites. Additionally, sample type, including juice, wine, berries, seeds, and vegetative tissues, may require special treatments prior to analysis, making cross application of methods challenging. Within sample types, additional variation exists, such as differences between peel and flesh of berries, and the metabolic changes associated with fermentation in wine. Sample preparation varies for targeted (hypothesis-driven) and untargeted (discovery-based) approaches. Certain techniques required additional sample preparation, such as derivatization of non-volatile metabolites for gas chromatography (Cevallos-Cevallos et al. 2009).

10.4.1.1 Anthocyanins and Tannins

Anthocyanins and tannins are phenolic compounds that contribute to the quality of grapes and wine. Anthocyanins are the red and blue pigments found in grapes, and tannins are bitter, astringent metabolites providing mouth feel for wine and stabilize wine color (Bautista-Ortín et al. 2016). Both contribute to the nutritional value of grapes and wine. In grapes, anthocyanins exist as aglycone or glycosylated forms. Tannins are polymers of the procyanidins epicatechin, catechin, and epicatechin-3-*O*-gallate, and are characterized by the mean degree of polymerization (mDP). Extraction protocols for these compounds, whether for industry purposes (e.g., wine making) or research, impact quality and reproducibility, giving rise to the tongue-in-cheek term "extractomics." There are many methods to extract these compounds from berries (Liang et al. 2012; Koyama et al. 2017), and wine and juice (Tang et al. 2018; Sommer and Cohen 2018), with no standard protocols. Developmental stage is critical for sampling. Extractable tannins decrease during ripening berries caused by increased tannin binding to the cell wall (Bindon et al. 2014; Bautista-Ortín et al. 2016). Pulsed electric field treatments may facilitate extraction of bound tannins (Delsart et al. 2012). Genetic diversity in wild grapes and hybrids may also impact extraction and quantification. Wild *Vitis* and hybrids have distinct tannin and anthocyanin profiles from *V. vinifera*. Co-elution of tannins and anthocyanins in wild grapes can result in an incomplete measurement of tannin content (Koyama et al. 2017). In hybrid grapes, increased quantities of pathogenesis-related proteins bind to tannins and impact berry tannin extraction and exogenous tannin retention in wine (Springer and Sacks 2014; Springer et al. 2016).

10.4.2 Opportunities with Future Technologies

Phenotyping platforms for chemical analysis continue to improve, through technical improvements (GS/MS, NMR) and accessibility and ease of use (Parpinello et al. 2013; Pinelli et al. 2018). New techniques can reduce sample preparation time and complexity and are amenable to HT phenotyping. For example, direct analysis in real time mass spectrometry (DART-MS) can be used to rapidly measure compounds in their native state or in combination with separation methods. Jastrzembski et al. (2017) combined solid-phase microextraction with DART-MS to measure trace volatiles in grapes in ~30 s/sample, compared to ~30 min/sample for GC-MS.

Most promising is the integration of metabolic data with other –omics data. More extensive metabolite profiling can be paired with genomic, proteomic, and transcriptomic data to give powerful view of grape physiology. Additionally, multivariate analysis can better integrate disparate datasets collect from various phenotyping approaches. Utilization of this approach can lead to new understanding of developmental processes, including chemical changes associated with disease, fruit development, post-harvest treatments, and vinification.

10.5 Molecular Tools

During the last 20 years, there has been significant improvement in classical sequencing and molecular profiling of plant genomes. These techniques have provided valuable tools to select desirable alleles at many loci through marker assisted selection and more recently genomic selection (Furbank 2009; Tester and Langridge 2010; Furbank and Tester 2011). Phenotypic prediction is challenging due to the large number of genes and gene products that contribute simultaneously to phenotypes under influence from complex and changeable environments. Therefore, a robust and reliable phenotyping system is necessary to overcome these shortcomings (Rahaman et al. 2015). "Phenomics," a new branch of biological sciences, could respond to the functional analysis, merging the gap between two (Houle et al. 2010). While many phenotypes of interest represent the endpoints of gene expression, the molecular and genetic signals themselves could be considered valuable phenotypes to be measured and mapped.

10.5.1 Challenges with Standardizing Molecular Phenotyping Methods and Scales

Challenges remain in understanding the majority of transcripts, proteins, and protein families involved in stress response, despite generating a tremendous amount of data from the study of traits such as fruit biochemical pathways and defense pathways in response to both abiotic and biotic stresses in grape cultivars. The limited availability of genome sequences of different cultivars limits the characterization of species- or cultivar-specific transcript and protein sequences (see Chap. 5). Similarly, a major challenge remains in inferring biological meaning from these data with a majority of sequences lacking annotated function (Delaunois et al. 2013; George et al. 2015).

10.5.1.1 Detection, Diagnosis, and Quantification of Plant Pathogens

In recent years, molecular techniques for phenotyping diseases have been well established in grapevines (Table 10.2). Unlike conventional methods that rely on visual symptoms, isolation, and/or culturing, pathogens can be detected using molecular techniques, such as enzyme-linked immunosorbent assays (ELISA), DNA/RNA probes, or polymerase chain amplification of nucleic acids including via quantitative PCR or reverse-transcription PCR (McCartney et al. 2003; Donoso and Valenzuela 2018). The availability of extensive DNA and RNA sequence information greatly benefits most techniques for molecular detection and diagnostics of plant pathogens. ELISA-based assays, which upon fluorescence or other visible chemical reaction confirms the presence of disease, have been standard techniques in detecting viruses, fungi, and other microbes (Sankaran et al. 2010; Boonham et al. 2014), and are quick, cheap, and available for on-site testing without need of specially-trained personnel. However, immunological procedures rely on antibody-based recognition of antigens produced by the pathogen, which may not be available for all pathogens of interest.

PCR-based assays target sequences from the pathogen for amplification and detection (Ward et al. 2004), and species-specific primers have provided a powerful tool for pathogen identification. These primers usually target regions of

Table 10.2 A survey of representative molecular techniques deployed to assess diverse phenotypes in grapevines

Phenotypes	Type	Molecular method	Reference
Biotic stress	Powdery mildew	PCR, HPLC	Brewer et al. (2011), Frenkel et al. (2012)
	Downy mildew	Proteomics, metabolomics	Palmieri et al. (2012), Chitarrini et al. (2017), Negrel et al. (2018)
	Botrytis	Nested PCR-RFLP, qPCR	Cadle-Davidson (2008), Saito et al. (2009)
	Trunk diseases	Transcriptomics	Spagnolo et al. (2012)
	Bacteria response	PCR, ELISA, proteomics	Minsavage et al. (1994), Katam et al. (2015)
	Virus	RT-PCR, PCR	Dubiela et al. (2013)
	Pest response	Chemical fingerprinting	Benheim et al. (2011)
Abiotic stress	Water deficit	Proteomics	Grimplet et al. (2009)
	Salt stress	Proteomics	Jellouli et al. (2008)
	High temperature and heat shock	PCR	Liu et al. (2012)
	Herbicide	Proteomics	Castro et al. (2005)
	Cold storage	Proteomics	Yuan et al. (2014)
	Dormancy	HPLC	George et al. (2018)
	Freeze shock	Transcriptomics	Tattersall et al. (2007), Xin et al. (2013), Xu et al. (2014b), Londo et al. (2018)
	Cold tolerance	RT-PCR	Hou et al. (2018)
	UV stress	RT-PCR	Schoedl et al. (2013)
Physiology and development	Post-harvest withering	Proteomics	Di Carli et al. (2011)
	Dormancy Induction	Transcriptomics	Fennell et al. (2015)
	Berry flesh development	Proteomics	Martínez-Esteso et al. (2011)
	Berry development	Transcriptomic, proteomic and metabolomics	Zamboni et al. (2010)
	Berry anthocyanins	Mass spectrometry	Picariello et al. (2014)
	Fermentation and yeast	PAGE, PCR, and RT-PCR, proteomics	Marks et al. (2003), Blein-Nicolas et al. (2013)
	Seedlessness	PCR	Lahogue et al. (1998)
	Variety identification	Proteomics	Povero et al. (2010)
	Leaf metabolome	LC–MS	Marti et al. (2014)

ribosomal RNA genes exhibiting sufficient diversity among taxa, such as the internal transcribed spacer regions, ITS1 and/or ITS2, in fungi. In addition, using nested PCR and multiple primer pairs (multiplexing) can increase the specificity of differentiating related pathogens within a short time interval (Alaniz et al. 2009). The next iteration is characterization of the phytobiome via metagenomics to study the composition and expression profiles of microbial communities in and around the grapevines. While most phytobiomes change drastically in response to environment, there appears to be a genetic basis to phytobiome as a phenotype. Such studies can generate details on biological and metabolic processes in grape-microbe interactions (Zarraonaindia et al. 2015; Alaimo et al. 2017).

10.5.1.2 Biomarkers for Phenology and Berry Development

Extensive studies have been conducted to characterize phenology and berry development in grapevines using molecular tools. At various stages of flower and berry development, (Wang et al. 2014) determined an expression profile of nine genes, three of which (*VvAP1*, *VvAP3,* and *VvFLC*) accurately predicted grapevine phenology. They used this "genetic phenology" to guide urea fertilizer timing and suggested that gene expression could be used in accurate diagnosis or pre-diagnosis of the corresponding phenophases for viticultural treatments. In another study, the grapevine R2R3-MYB transcription factor *VvMYBF1* was shown to regulate flavonol synthesis in the developing berries, with high expression during flower development and in skins of ripening barriers correlating with accumulation of flavonols (Czemmel et al. 2009).

Proteomic tools can be applied to study protein expression, function, and interactions while characterizing biological functions with possible applications as biomarkers. Traditionally, two-dimensional electrophoresis (2-DE) has been extensively used to study grapevine proteomics to examine defense and stress responses (Spagnolo et al. 2012; Delaunois et al. 2013). With the release of the grape genome sequence, traditional approaches are being replaced by shotgun proteomics techniques including iTRAQ and TMT, which have been widely used to study physiological responses to fungal infections, heat stress, and ripening events (Kambiranda et al. 2014; Liu et al. 2014). Along these lines, proteomic responses have been studied in numerous systems, such as: developing and ripening berries responding to various stresses (Negri et al. 2008; Kambiranda et al. 2014); grapevine stems in response to *Xylella fastidiosa* (Yang et al. 2011); and resistance induced against downy mildew by *Trichoderma harzianum* (Palmieri et al. 2012). While each story seems straightforward when succinctly summarized, the phenome of an organism is complex, dynamic, and conditional, often determined by external responses, adding complexity to pinpoint single time point expression (Pendergrass et al. 2015). Thus, more complex, dynamic analyses may be desirable in some situations.

10.5.1.3 Systems Biology and Expression QTL

Recently, systems biology has provided information about the interaction of genes, proteins, and metabolites through integration of omics data (see Chap. 8). As technologies for generating omics data improve in sensitivity, resolution, accuracy, depth, and speed, databases and data analysis pipelines must keep pace. For example, next generation sequencing technologies have simplified simultaneously obtaining transcriptome-wide expression profiles and genome-wide marker data, which have created an opportunity for expression QTL (eQTL) studies. At the simplest, eQTL can analyze expression of a single gene as the response variable. In analyzing expression of *VvUFGT* by reverse-transcription quantitative PCR (RT-qPCR) in the family Syrah × Grenache, a cis-eQTL in *VvUFGT* explained 20% of expression variance and a trans-eQTL at the *VvMYBA* locus explained 35% (Huang et al. 2013). Building on this, five proanthocyanidins also measured by RT-qPCR indicated 21 eQTLs, of which only four were previously known (Huang et al. 2014). In studying the *Rda1* locus for resistance to *Diaporthe ampelina*, an eQTL approach based on RNASeq analysis of a subset of recombinants identified 16 candidate genes, and the *Rda1* locus predicted expression of 6 of those genes, including two NB-LRR genes (Barba et al. 2018).

Integration of multiple omics data may be justified by the modest correlations between gene and protein expression levels: for example, only one-third of proteins identified in mature berries were significantly correlated with their RNASeq transcript abundance (Ghan et al. 2015). This poses a complex problem: If proteomic and metabolic data lie downstream from RNA, closer to the phenotype, but our assays do not access all expressed proteins and metabolites, how do we gain clearer understanding of phenotypes from more complex data having more assumptions, comparisons, strengths, and weaknesses? It becomes a challenge not just to analyze systems biology data, but also to visualize and provide meaningful interpretation.

10.5.2 Opportunities with Future Technologies

Although molecular methods have been developing rapidly and constantly, their application as phenotypes has been limited. In part, this may be due to their indirect relationship to the selective trait of interest (e.g., quantitation of pathogen DNA vs quantitation of visible disease) and due to the more accurate measures obtained by direct observation. For instance, when using both qPCR and disease ratings to discover *REN6* and *REN7*, while both methods detected the loci, disease ratings explained more phenotypic variance and had higher LOD scores (Pap et al. 2016). Molecular phenotypes have a bright future where their throughput is significantly higher than other phenotypes while retaining trait correlation, or where they create opportunities not otherwise possible, such as in eQTL studies and other systems biology approaches.

10.6 Information Technologies in Agriculture (Precision Agriculture and Big Data)

In recent decades, IT has had an important role in agriculture management as a response to the continuous increase of production demands and the need to fulfill stricter quality control requirements from governmental institutions. The precision agriculture concept was adapted in the 1990s by Lowenberg-DeBoer and Boehlje (1996) in order to reference the implementation of new IT to traditional agriculture science. The challenge of precision agriculture is to maximize crop production and product quality, while optimizing returns on economic investments. Therefore, precision agriculture follows a sustainable agriculture scheme by means of strategy adaptation based on collected data. The feasibility of applying these methods for monitoring large field areas and its capability of handling environmental variance (Santesteban et al. 2013) has been demonstrated in the scientific literature during the last decades.

Precision agriculture methods are based on gathering a wide scope of different field metrics, markers, and heuristics which are accurately georeferenced by means of distribution maps usually generated from airborne or satellite imagery. The popularization of small unmanned aerial systems (UAS) has been fostering the development of new mapping approaches for precision agriculture (Zhang and Kovacs 2012) as well as making final implementations more affordable for researchers, companies, and farmers. In addition, local-operating ground mobile robots with automated task capabilities can generate or use data from precision agriculture methods. For example, autonomous tractors use the global positioning system (GPS) for localization and visual sensors for obstacle detection (Moorehead et al. 2012). In addition, the simultaneous localization and mapping (SLAM) method, a well-known relative localization method in mobile robotics, was implemented for the navigation system of automated agricultural machinery as an effective solution in areas that often experience GPS signal losses (Auat Cheein et al. 2011).

Computer vision methods in precision agriculture are commonly used to obtain visual quantifiable data. However, the application of such methods for field experimentation usually involves an exponential increase of difficulty due to added variables from uncontrolled environmental factors (e.g., lighting, weather, and terrain variations). As a result, there is not a wide scope of robust computer vision systems for agriculture, and most of those solutions work under specific, restricted conditions. However, even with these limitations, much progress has been made. For example, fruit yield estimations are considered important information for producers and breeders, and its automation has been popularly addressed in the literature. One such image analysis method estimates total fruit load from mango trees by means of taking high-resolution photos (Payne et al. 2013). Similarly, immature citrus fruit can be detected by means of applying color filtering, illumination enhancement, watershed transform, and texture feature extraction (Zhao et al. 2016), and orange citrus can be counted based on color and watershed segmentations (Dorj et al. 2017). Individual red grape counting for yield estimation was developed to work autonomously at night by using artificial lighting (Font et al. 2014b). Automated harvesting was also addressed by using similar fruit detection mechanisms and a robotic manipulator to collect citrus fruit within a computationally estimated 3D plane (Mehta and Burks 2014). Although IR-based depth cameras are not suitable for outdoor operation, other depth sensor systems can be used such as stereovision cameras for fruit harvesting (Font et al. 2014a). Moreover, fruit detectability can be also a challenging problem when dealing with complex plant structures where fruits are often covered by leaves. This problem can be addressed by assessing fruit detectability from different camera viewpoints, emphasizing the importance of setting a balance between computational cost and effectiveness (Hemming et al. 2014).

Literature on precision agriculture presents many important contributions to the technical challenges of using computational devices and sensors for data gathering and information fusion. However, many researchers consider that there is a lack of research and application on the next step, the decision making (Lindblom et al. 2017). All the collected data is almost useless without an effective information management plan capable of generating final conclusions that define effective strategies to be applied. Moreover, the capability of deployment and implementation of such strategies are also considered as a critical point since farmers and field managers are often reticent to the adaption of such modern techniques. Sonka (2016) referenced big data as the evolution of the precision agriculture concept regarding the new advances on artificial intelligence which are becoming powerful and are gaining presence in our society. Bronson and Knezevic (2016) share a similar perspective on the application of big data technologies in food and agriculture; moreover, they highlight the need of encouraging society to get updated for such rising technologies.

10.7 Conclusions

In this chapter, we provided a perspective on current phenotyping approaches as they relate to large-scale genetic studies, considering first how phenotypes are perceived and measured and then giving examples of specific traits, to encourage a broad perspective on phenotyping. With increased access to genomic data at reduced cost, we see a phenomics revolution in process, which promises to bring improved precision, objectivity, reproducibility, and throughput. However, with new approaches come challenges and standardization and data management and analysis will be critical to the success of phenomics in each application. The other key is the need to validate phenotypes within the context of viticulture and breeding. We recommend a mindset of breaking traits down to measurable phenotype components, then building models back up to vineyard validation and application.

References

Adamchuk VI, Hummel JW, Morgan MT, Upadhyaya SK (2004) On-the-go soil sensors for precision agriculture. Comput Electron Agric 44:71–91

Alaimo S, Marceca GP, Giugno R et al (2017) Current knowledge and computational techniques for grapevine meta-omics analysis. Front Plant Sci 8:2241

Alaniz S, Armengol J, García-Jiménez J et al (2009) A multiplex PCR system for the specific detection of *Cylindrocarpon liriodendri*, *C. macrodidymum*, and *C. pauciseptatum* from Grapevine. Plant Dis 93:821–825

Anastasiou E, Balafoutis A, Darra N et al (2018) Satellite and proximal sensing to estimate the yield and quality of table grapes. Collect FAO Agric 8:94

Aquino A, Barrio I, Diago M-P et al (2018) vitisBerry: an Android-smartphone application to early evaluate the number of grapevine berries by means of image analysis. Comput Electron Agric 148:19–28

Atkinson JA, Pound MP, Bennett MJ, Wells DM (2018) Uncovering the hidden half of plants using new advances in root phenotyping. Curr Opin Biotechnol 55:1–8

Auat Cheein F, Steiner G, Perez Paina G, Carelli R (2011) Optimized EIF-SLAM algorithm for precision agriculture mapping based on stems detection. Comput Electron Agric 78:195–207

Ban Y, Mitani N, Sato A et al (2016) Genetic dissection of quantitative trait loci for berry traits in interspecific hybrid grape (*Vitis labruscana* × *Vitis vinifera*). Euphytica 211:295–310

Barba P, Cadle-Davidson L, Harriman J et al (2014) Grapevine powdery mildew resistance and susceptibility loci identified on a high-resolution SNP map. Theor Appl Genet 127:73–84

Barba P, Lillis J, Luce RS et al (2018) Two dominant loci determine resistance to Phomopsis cane lesions in F1 families of hybrid grapevines. Theor Appl Genet 131:1173–1189

Barrios-Masias FH, Knipfer T, McElrone AJ (2015) Differential responses of grapevine rootstocks to water stress are associated with adjustments in fine root hydraulic physiology and suberization. J Exp Bot 66:6069–6078

Bates T, Dresser J, Eckstrom R, et al (2018) Variable-rate mechanical crop adjustment for crop load balance in "Concord" vineyards. In: 2018 IoT vertical and topical summit on agriculture—Tuscany (IOT Tuscany), pp 1–4

Bautista-Ortín AB, Martínez-Hernández A, Ruiz-García Y et al (2016) Anthocyanins influence tannin–cell wall interactions. Food Chem 206:239–248

Bellvert J, Zarco-Tejada PJ, Girona J, Fereres E (2014) Mapping crop water stress index in a "Pinot-noir" vineyard: comparing ground measurements with thermal remote sensing imagery from an unmanned aerial vehicle. Precis Agric 15:361–376

Benheim D, Rochfort S, Ezernieks V et al (2011) Early detection of grape phylloxera (*Daktulosphaira vitifoliae* Fitch) infestation through identification of chemical biomarkers. Acta Hortic 904:17–24

Billet K, Houillé B, Dugé de Bernonville T et al (2018) Field-based metabolomics of *Vitis vinifera* L. stems provides new insights for genotype discrimination and polyphenol metabolism structuring. Front Plant Sci 9:798

Bindon KA, Madani SH, Pendleton P et al (2014) Factors affecting skin tannin extractability in ripening grapes. J Agric Food Chem 62:1130–1141

Blanco-Ulate B, Amrine KCH, Collins TS et al (2015) Developmental and metabolic plasticity of white-skinned grape berries in response to *Botrytis cinerea* during noble rot. Plant Physiol 169:2422–2443

Blasi P, Blanc S, Wiedemann-Merdinoglu S et al (2011) Construction of a reference linkage map of *Vitis amurensis* and genetic mapping of Rpv8, a locus conferring resistance to grapevine downy mildew. Theor Appl Genet 123:43–53

Blein-Nicolas M, Albertin W, Valot B et al (2013) Yeast proteome variations reveal different adaptive responses to grape must fermentation. Mol Biol Evol 30:1368–1383

Boonham N, Kreuze J, Winter S et al (2014) Methods in virus diagnostics: from ELISA to next generation sequencing. Virus Res 186:20–31

Boso S, Gago P, Alonso-Villaverde V et al (2016) Density and size of stomata in the leaves of different hybrids (*Vitis* sp.) and *Vitis vinifera* varieties. Vitis. https://doi.org/10.5073/vitis.2016.55.17-22

Brewer MT, Cadle-Davidson L, Cortesi P et al (2011) Identification and structure of the mating-type locus and development of PCR-based markers for mating type in powdery mildew fungi. Fungal Genet Biol 48:704–713

Bronson K, Knezevic I (2016) Big Data in food and agriculture. Big Data Soc 3:2053951716648174

Cadle-Davidson L (2008) Monitoring pathogenesis of natural *Botrytis cinerea* infections in developing grape berries. Am J Enol Vitic 59:387–395

Cadle-Davidson L, Gadoury D, Fresnedo-Ramírez J et al (2016) Lessons from a phenotyping center revealed by the genome-guided mapping of powdery mildew resistance loci. Phytopathology 106:1159–1169

Caffarra A, Eccel E (2011) Projecting the impacts of climate change on the phenology of grapevine in a mountain area: effects of climate change on grape phenology. Aust J Grape Wine Res 17:52–61

Callen ST, Klein LL, Miller AJ (2016) Climatic niche characterization of 13 North American *Vitis* species. Am J Enol Vitic 67:339–349

Castro AJ, Carapito C, Zorn N et al (2005) Proteomic analysis of grapevine (*Vitis vinifera* L.) tissues subjected to herbicide stress. J Exp Bot 56:2783–2795

Cevallos-Cevallos JM, Reyes-De-Corcuera JI, Etxeberria E et al (2009) Metabolomic analysis in food science: a review. Trends Food Sci Technol 20: 557–566

Chaïb J, Torregrosa L, Mackenzie D et al (2010) The grape microvine—a model system for rapid forward and reverse genetics of grapevines: *Grape microvines*. Plant J 62:1083–1092

Chitarrini G, Soini E, Riccadonna S et al (2017) Identification of biomarkers for defense response to *Plasmopara viticola* in a resistant grape variety. Front Plant Sci 8:1524

Chuine I, Yiou P, Viovy N et al (2004) Historical phenology: grape ripening as a past climate indicator. Nature 432:289–290

Comas LH, Becker SR, Cruz VMV et al (2013) Root traits contributing to plant productivity under drought. Front Plant Sci 4:442

Coombe BG (1995) Growth stages of the grapevine: adoption of a system for identifying grapevine growth stages. Aust J Grape Wine Res 1:104–110

Coppens F, Wuyts N, Inzé D, Dhondt S (2017) Unlocking the potential of plant phenotyping data through

integration and data-driven approaches. Curr Opin Syst Biol 4:58–63
Correa J, Mamani M, Muñoz-Espinoza C et al (2014) Heritability and identification of QTLs and underlying candidate genes associated with the architecture of the grapevine cluster (*Vitis vinifera* L.). Theor Appl Genet 127:1143–1162
Costa JM, Ortuño MF, Lopes CM, Chaves MM (2012) Grapevine varieties exhibiting differences in stomatal response to water deficit. Funct Plant Biol 39:179–189
Coupel-Ledru A, Lebon É, Christophe A et al (2014) Genetic variation in a grapevine progeny (*Vitis vinifera* L. cvs Grenache × Syrah) reveals inconsistencies between maintenance of daytime leaf water potential and response of transpiration rate under drought. J Exp Bot 65:6205–6218
Coupel-Ledru A, Lebon E, Christophe A et al (2016) Reduced nighttime transpiration is a relevant breeding target for high water-use efficiency in grapevine. Proc Natl Acad Sci USA 113:8963–8968
Crupi P, Bergamini C, Perniola R et al (2015) A chemometric approach to identify the grape cultivar employed to produce nutraceutical fruit juice. Eur Food Res Technol 241:487–496
Czemmel S, Stracke R, Weisshaar B et al (2009) The grapevine R2R3-MYB transcription factor VvMYBF1 regulates flavonol synthesis in developing grape berries. Plant Physiol 151:1513–1530
Dalbó MA, Ye GN, Weeden NF et al (2000) A gene controlling sex in grapevines placed on a molecular marker-based genetic map. Genome 43:333–340
Dami IE, Li S, Zhang Y (2016) Evaluation of primary bud freezing tolerance of twenty-three winegrape cultivars new to the Eastern United States. Am J Enol Vitic 67:139–145
De Rosso M, Tonidandel L, Larcher R et al (2014) Identification of new flavonols in hybrid grapes by combined liquid chromatography–mass spectrometry approaches. Food Chem 163:244–251
Delaunois B, Colby T, Belloy N et al (2013) Large-scale proteomic analysis of the grapevine leaf apoplastic fluid reveals mainly stress-related proteins and cell wall modifying enzymes. BMC Plant Biol 13:24
Delsart C, Ghidossi R, Poupot C et al (2012) Enhanced extraction of phenolic compounds from Merlot grapes by pulsed electric field treatment. Am J Enol Vitic 63:205–211
Di Carli M, Zamboni A, Pè ME et al (2011) Two-dimensional differential in gel electrophoresis (2D-DIGE) analysis of grape berry proteome during postharvest withering. J Proteome Res 10:429–446
Dokoozlian NK (1999) Chilling temperature and duration interact on the Budbreak of 'Perlette' grapevine cuttings. HortScience 34:1–3
Doligez A, Bouquet A, Danglot Y et al (2002) Genetic mapping of grapevine (*Vitis vinifera* L.) applied to the detection of QTLs for seedlessness and berry weight. Theor Appl Genet 105:780–795
Doligez A, Adam-Blondon AF, Cipriani G et al (2006) An integrated SSR map of grapevine based on five mapping populations. Theor Appl Genet 113:369–382
Donoso A, Valenzuela S (2018) In-field molecular diagnosis of plant pathogens: recent trends and future perspectives. Plant Pathol 67:1451–1461
Dorj U-O, Lee M, Yun S-S (2017) An yield estimation in citrus orchards via fruit detection and counting using image processing. Comput Electron Agric 140:103–112
Dubiela CR, Fajardo TVM, Souto ER et al (2013) Simultaneous detection of Brazilian isolates of grapevine viruses by TaqMan real-time RT-PCR. Trop Plant Pathol 38:158–165
Duchene E (2016) How can grapevine genetics contribute to the adaptation to climate change? OENO One. https://doi.org/10.20870/oeno-one.2016.50.3.98
Duursma RA, Blackman CJ, Lopéz R et al (2018) On the minimum leaf conductance: its role in models of plant water use, and ecological and environmental controls. New Phytol. https://doi.org/10.1111/nph.15395
Ershadi A, Karimi R, Mahdei KN (2015) Freezing tolerance and its relationship with soluble carbohydrates, proline and water content in 12 grapevine cultivars. Acta Physiol Plant 38:2
Failmezger H, Lempe J, Khadem N et al (2018) MowJoe: a method for automated-high throughput dissected leaf phenotyping. Plant Methods 14:27
Fechter I, Hausmann L, Zyprian E et al (2014) QTL analysis of flowering time and ripening traits suggests an impact of a genomic region on linkage group 1 in *Vitis*. Theor Appl Genet 127:1857–1872
Fennell AY, Schlauch KA, Gouthu S et al (2015) Short day transcriptomic programming during induction of dormancy in grapevine. Front Plant Sci 6:834
Ferguson JC, Tarara JM, Mills LJ et al (2011) Dynamic thermal time model of cold hardiness for dormant grapevine buds. Ann Bot 107:389–396
Ferguson JC, Moyer MM, Mills LJ et al (2014) Modeling dormant bud cold hardiness and budbreak in twenty-three *Vitis* genotypes reveals variation by region of origin. Am J Enol Vitic 65:59–71
Fila G, Di Lena B, Gardiman M et al (2012) Calibration and validation of grapevine budburst models using growth-room experiments as data source. Agric For Meteorol 160:69–79
Fischer BM, Salakhutdinov I, Akkurt M et al (2004) Quantitative trait locus analysis of fungal disease resistance factors on a molecular map of grapevine. Theor Appl Genet 108:501–515
Flexas J, Galmés J, Gallé A et al (2010) Improving water use efficiency in grapevines: potential physiological targets for biotechnological improvement. Aust J Grape Wine Res 16:106–121
Font D, Pallejà T, Tresanchez M et al (2014a) A proposal for automatic fruit harvesting by combining a low cost stereovision camera and a robotic arm. Sensors 14:11557–11579

Font D, Palleja T, Tresanchez M et al (2014b) Counting red grapes in vineyards by detecting specular spherical reflection peaks in RGB images obtained at night with artificial illumination. Comput Electron Agric 108: 105–111

Frenkel O, Portillo I, Brewer MT et al (2012) Development of microsatellite markers from the transcriptome of *Erysiphe necator* for analysing population structure in North America and Europe: polymorphic markers from the *Erysiphe necator* transcriptome. Plant Pathol 61:106–119

Fuller MP, Telli G (1999) An investigation of the frost hardiness of grapevine (*Vitis vinifera*) during bud break. Ann Appl Biol 135:589–595

Furbank RT (2009) Foreword: plant phenomics: from gene to form and function. Funct Plant Biol 36:v–vi

Furbank RT, Tester M (2011) Phenomics–technologies to relieve the phenotyping bottleneck. Trends Plant Sci 16:635–644

Gadoury DM (2015) Climate, asynchronous phenology, ontogenic resistance, and the risk of disease in deciduous fruit crops. IOBC-WPRS Bull 110:15–24

Gale EJ, Moyer MM (2017) Cold hardiness of *Vitis vinifera* roots. Am J Enol Vitic 68:468–477

García de Cortázar-Atauri I, Duchêne E, Destrac-Irvine A et al (2017) Grapevine phenology in France: from past observations to future evolutions in the context of climate change. OENO One 51:115

Garris A, Clark L, Owens C et al (2009) Mapping of photoperiod-induced growth cessation in the wild grape *Vitis riparia*. J Am Soc Hortic Sci 134:261–272

George IS, Pascovici D, Mirzaei M, Haynes PA (2015) Quantitative proteomic analysis of cabernet sauvignon grape cells exposed to thermal stresses reveals alterations in sugar and phenylpropanoid metabolism. Proteomics 15:3048–3060

George IS, Fennell AY, Haynes PA (2018) Shotgun proteomic analysis of photoperiod regulated dormancy induction in grapevine. J Proteom 187:13–24

Ghan R, Van Sluyter SC, Hochberg U et al (2015) Five omic technologies are concordant in differentiating the biochemical characteristics of the berries of five grapevine (*Vitis vinifera* L.) cultivars. BMC Genom 16:946

Granier C, Aguirrezabal L, Chenu K et al (2006) PHENOPSIS, an automated platform for reproducible phenotyping of plant responses to soil water deficit in Arabidopsis thaliana permitted the identification of an accession with low sensitivity to soil water deficit. New Phytol 169:623–635

Greer DH, Weedon MM (2013) The impact of high temperatures on *Vitis vinifera* cv. Semillon grapevine performance and berry ripening. Front Plant Sci 4:491

Grimplet J, Wheatley MD, Jouira HB et al (2009) Proteomic and selected metabolite analysis of grape berry tissues under well-watered and water-deficit stress conditions. Proteomics 9:2503–2528

Hall ME, Loeb GM, Cadle-Davidson L et al (2018) Grape sour rot: a four-way interaction involving the host, yeast, acetic acid bacteria, and insects. Phytopathology. https://doi.org/10.1094/phyto-03-18-0098-r

Hemming J, Ruizendaal J, Hofstee JW, van Henten EJ (2014) Fruit detectability analysis for different camera positions in sweet-pepper. Sensors 14:6032–6044

Henderson SW, Baumann U, Blackmore DH et al (2014) Shoot chloride exclusion and salt tolerance in grapevine is associated with differential ion transporter expression in roots. BMC Plant Biol 14:273

Henderson SW, Dunlevy JD, Wu Y et al (2017) Functional differences in transport properties of natural HKT1;1 variants influence shoot Na + exclusion in grapevine rootstocks. New Phytol. https://doi.org/10.1111/nph.14888

Hoffmann S, Di Gaspero G, Kovács L et al (2008) Resistance to *Erysiphe necator* in the grapevine "Kishmish vatkana" is controlled by a single locus through restriction of hyphal growth. Theor Appl Genet 116:427–438

Hopper DW, Ghan R, Cramer GR (2014) A rapid dehydration leaf assay reveals stomatal response differences in grapevine genotypes. Hortic Res 1:2

Hou L, Zhang G, Zhao F et al (2018) VvBAP1 is involved in cold tolerance in *Vitis vinifera* L. Front Plant Sci 9:726

Houel C, Chatbanyong R, Doligez A et al (2015) Identification of stable QTLs for vegetative and reproductive traits in the microvine (*Vitis vinifera* L.) using the 18 K Infinium chip. BMC Plant Biol 15:205

Houle D, Govindaraju DR, Omholt S (2010) Phenomics: the next challenge. Nat Rev Genet 11:855–866

Huang Y-F, Bertrand Y, Guiraud J-L et al (2013) Expression QTL mapping in grapevine—revisiting the genetic determinism of grape skin colour. Plant Sci 207:18–24

Huang Y-F, Vialet S, Guiraud J-L et al (2014) A negative MYB regulator of proanthocyanidin accumulation, identified through expression quantitative locus mapping in the grape berry. New Phytol 201:795–809

Ihlow A, Schweizer P, Seiffert U (2008) A high-throughput screening system for barley/powdery mildew interactions based on automated analysis of light micrographs. BMC Plant Biol 8:6

Jaillon O, Aury J-M, Noel B et al (2007) The grapevine genome sequence suggests ancestral hexaploidization in major angiosperm phyla. Nature 449:463–467

Jastrzembski JA, Bee MY, Sacks GL (2017) Trace-level volatile quantitation by direct analysis in real time mass spectrometry following headspace extraction: optimization and validation in grapes. J Agric Food Chem 65:9353–9359

Jellouli N, Ben Jouira H, Skouri H et al (2008) Proteomic analysis of Tunisian grapevine cultivar Razegui under salt stress. J Plant Physiol 165:471–481

Jones GV, White MA, Cooper OR, Storchmann K (2005) Climate change and global wine quality. Clim Change 73:319–343

Jorge TF, Rodrigues JA, Caldana C et al (2016) Mass spectrometry-based plant metabolomics: metabolite responses to abiotic stress. Mass Spectrom Rev 35:620–649

Kambiranda D, Katam R, Basha SM, Siebert S (2014) iTRAQ-based quantitative proteomics of developing and ripening muscadine grape berry. J Proteome Res 13:555–569

Katam R, Chibanguza K, Latinwo LM, Smith D (2015) Proteome biomarkers in xylem reveal pierce's disease tolerance in grape. J Proteom Bioinform 8:217–224

Kicherer A, Herzog K, Pflanz M et al (2015) An automated field phenotyping pipeline for application in grapevine research. Sensors 15:4823–4836

Kicherer A, Herzog K, Bendel N et al (2017a) Phenoliner: a new field phenotyping platform for grapevine research. Sensors 17:1625. https://doi.org/10.3390/s17071625

Kicherer A, Klodt M, Sharifzadeh S et al (2017b) Automatic image-based determination of pruning mass as a determinant for yield potential in grapevine management and breeding: image-based automated estimation of pruning mass. Aust J Grape Wine Res 23:120–124

Koch B, Oehl F (2018) Climate change favors grapevine production in temperate zones. AS 09:247–263

Koyama K, Kamigakiuchi H, Iwashita K et al (2017) Polyphenolic diversity and characterization in the red-purple berries of East Asian wild *Vitis* species. Phytochemistry 134:78–86

Kuska M, Wahabzada M, Leucker M et al (2015) Hyperspectral phenotyping on the microscopic scale: towards automated characterization of plant–pathogen interactions. Plant Methods 11:28

Kustas WP, Anderson MC, Alfieri JG et al (2018) The grape remote sensing atmospheric profile and evapotranspiration experiment (GRAPEX). Bull Am Meteorol Soc. https://doi.org/10.1175/bams-d-16-0244.1

Lahogue F, This P, Bouquet A (1998) Identification of a codominant scar marker linked to the seedlessness character in grapevine. Theor Appl Genet 97:950–959

Lavoie-Lamoureux A, Sacco D, Risse P-A, Lovisolo C (2017) Factors influencing stomatal conductance in response to water availability in grapevine: a meta-analysis. Physiol Plant 159:468–482

Leolini L, Moriondo M, Fila G et al (2018) Late spring frost impacts on future grapevine distribution in Europe. Field Crops Res 222:197–208

Liang Z, Yang Y, Cheng L, Zhong G-Y (2012) Polyphenolic composition and content in the ripe berries of wild *Vitis* species. Food Chem 132:730–738

Lindblom J, Lundström C, Ljung M, Jonsson A (2017) Promoting sustainable intensification in precision agriculture: review of decision support systems development and strategies. Precis Agric 18:309–331

Lindén L, Palonen P, Lindén M (2000) Relating freeze-induced electrolyte leakage measurements to lethal temperature in red raspberry. J Am Soc Hortic Sci 125:429–435

Liu G-T, Wang J-F, Cramer G et al (2012) Transcriptomic analysis of grape (*Vitis vinifera* L.) leaves during and after recovery from heat stress. BMC Plant Biol 12:174

Liu G-T, Ma L, Duan W et al (2014) Differential proteomic analysis of grapevine leaves by iTRAQ reveals responses to heat stress and subsequent recovery. BMC Plant Biol 14:110

Londo JP, Johnson LM (2014) Variation in the chilling requirement and budburst rate of wild *Vitis* species. Environ Exp Bot 106:138–147

Londo JP, Kovaleski AP (2017) Characterization of wild North American grapevine cold hardiness using differential thermal analysis. Am J Enol Vitic 68:203–212

Londo JP, Kovaleski AP, Lillis JA (2018) Divergence in the transcriptional landscape between low temperature and freeze shock in cultivated grapevine (*Vitis vinifera*). Hortic Res 5:10

Lovisolo C, Tramontini S (2010) Methods for assessment of hydraulic conductance and embolism extent in grapevine organs. In: Delrot S, Medrano H, Or E, Bavaresco L, Grando S (eds) Methodologies and results in grapevine research. Springer, Dordrecht, pp 71–85

Lowenberg-DeBoer J, Boehlje M (1996) Revolution, evolution or dead-end: economic perspectives on precision agriculture. In: Robert PC, Rust RH and Larson WE (eds) Precision agriculture. American Society of Agronomy, Crop Science Society of America, Soil Science Society of America, Madison, pp 923–944

Luedeling E (2012) Climate change impacts on winter chill for temperate fruit and nut production: a review. Sci Hortic 144:218–229

Marks VD, van der Merwe GK, van Vuuren HJJ (2003) Transcriptional profiling of wine yeast in fermenting grape juice: regulatory effect of diammonium phosphate. FEMS Yeast Res 3:269–287

Marti G, Schnee S, Andrey Y et al (2014) Study of leaf metabolome modifications induced by UV-C radiations in representative *Vitis*, *Cissus* and *Cannabis* species by LC–MS based metabolomics and antioxidant assays. Molecules 19:14004–14021

Martínez-Esteso MJ, Sellés-Marchart S, Lijavetzky D et al (2011) A DIGE-based quantitative proteomic analysis of grape berry flesh development and ripening reveals key events in sugar and organic acid metabolism. J Exp Bot 62:2521–2569

McCartney HA, Foster SJ, Fraaije BA, Ward E (2003) Molecular diagnostics for fungal plant pathogens. Pest Manag Sci 59:129–142

Medrano H, Tomás M, Martorell S, Flexas J, Hernández E, Rosselló J, Pou A, Escalona JM, Bota J (2015) From leaf to whole-plant water use efficiency (WUE) in complex canopies: limitations of leaf WUE as a selection target. Crop J 3(3):220–228

Mehta SS, Burks TF (2014) Vision-based control of robotic manipulator for citrus harvesting. Comput Electron Agric 102:146–158

Mills LJ, Ferguson JC, Keller M (2006) Cold-hardiness evaluation of grapevine buds and cane tissues. Am J Enol Vitic 57:194–200

Minsavage GV, Thompson CM, Hopkins DL et al (1994) Development of a polymerase chain reaction protocol for detection of *Xylella fastidiosa* in plant tissue. Phytopathology 84:456–461

Mira de Orduña R (2010) Climate change associated effects on grape and wine quality and production. Food Res Int 43:1844–1855

Moorehead SJ, Wellington CK, Gilmore BJ, Vallespi C (2012) Automating orchards: a system of autonomous tractors for orchard maintenance. In: Proceedings of the IEEE international conference of intelligent robots and systems; workshop on agricultural robots

Morin X, Améglio T, Ahas R et al (2007) Variation in cold hardiness and carbohydrate concentration from dormancy induction to bud burst among provenances of three European oak species. Tree Physiol 27:817–825

Mosedale JR, Wilson RJ, Maclean IMD (2015) Climate change and crop exposure to adverse weather: changes to frost risk and grapevine flowering conditions. PLoS ONE 10:e0141218

Mozell MR, Thach L (2014) The impact of climate change on the global wine industry: challenges & solutions. Wine Econ Policy 3:81–89

Negrel L, Halter D, Wiedemann-Merdinoglu S et al (2018) Identification of lipid markers of *Plasmopara viticola* infection in grapevine using a non-targeted metabolomic approach. Front Plant Sci 9:360

Negri AS, Prinsi B, Rossoni M et al (2008) Proteome changes in the skin of the grape cultivar Barbera among different stages of ripening. BMC Genom 9:378

OIV (2018) OIV descriptor list for grape varieties and *Vitis* species. In: The International Organization of Vine and Wine, 2nd edn. http://www.oiv.int/public/medias/2274/code-2e-edition-finale.pdf

Pagay V, Santiago M, Sessoms DA et al (2014) A microtensiometer capable of measuring water potentials below −10 MPa. Lab Chip 14:2806–2817

Pagter M, Williams M (2011) Frost dehardening and rehardening of *Hydrangea macrophylla* stems and buds. HortScience 46:1121–1126

Palmieri MC, Perazzolli M, Matafora V et al (2012) Proteomic analysis of grapevine resistance induced by *Trichoderma harzianum* T39 reveals specific defence pathways activated against downy mildew. J Exp Bot 63:6237–6251

Pap D, Riaz S, Dry IB et al (2016) Identification of two novel powdery mildew resistance loci, Ren6 and Ren7, from the wild Chinese grape species *Vitis piasezkii*. BMC Plant Biol 16:170

Parpinello GP, Nunziatini G, Rombolà AD et al (2013) Relationship between sensory and NIR spectroscopy in consumer preference of table grape (cv Italia). Postharvest Biol Technol 83:47–53

Payne AB, Walsh KB, Subedi PP, Jarvis D (2013) Estimation of mango crop yield using image analysis—segmentation method. Comput Electron Agric 91:57–64

Pellegrini E, Campanella A, Paolocci M et al (2015) Functional leaf traits and diurnal dynamics of photosynthetic parameters predict the behavior of grapevine varieties towards ozone. PLoS ONE 10:e0135056

Pendergrass SA, Verma A, Okula A et al (2015) Phenome-wide association studies: embracing complexity for discovery. Hum Hered 79:111–123

Picariello G, Ferranti P, Garro G et al (2014) Profiling of anthocyanins for the taxonomic assessment of ancient purebred *V. vinifera* red grape varieties. Food Chem 146:15–22

Pinasseau L, Vallverdú-Queralt A, Verbaere A et al (2017) Cultivar diversity of grape skin polyphenol composition and changes in response to drought investigated by LC–MS based metabolomics. Front Plant Sci 8:1826

Pinelli P, Romani A, Fierini E, Agati G (2018) Prediction models for assessing anthocyanins in grape berries by fluorescence sensors: dependence on cultivar, site and growing season. Food Chem 244:213–223

Poland JA, Nelson RJ (2011) In the eye of the beholder: the effect of rater variability and different rating scales on QTL mapping. Phytopathology 101(2):290–298

Pou A, Medrano H, Tomàs M et al (2012) Anisohydric behaviour in grapevines results in better performance under moderate water stress and recovery than isohydric behaviour. Plant Soil 359:335–349

Povero G, Papale M, Gesualdo L et al (2010) Identification of grapevine cultivar biomarkers using surface-enhanced laser desorption and ionization (SELDI-TOF-MS). Am J Enol Vitic 61:492–497

Rahaman MM, Chen D, Gillani Z et al (2015) Advanced phenotyping and phenotype data analysis for the study of plant growth and development. Front Plant Sci 6:619

Rossdeutsch L, Edwards E, Cookson SJ et al (2016) ABA-mediated responses to water deficit separate grapevine genotypes by their genetic background. BMC Plant Biol 16:91

Saito S, Suzuki S, Takayanagi T (2009) Nested PCR-RFLP is a high-speed method to detect fungicide-resistant *Botrytis cinerea* at an early growth stage of grapes. Pest Manag Sci 65:197–204

Salazar Parra C, Aguirreolea J, Sánchez-Díaz M et al (2010) Effects of climate change scenarios on Tempranillo grapevine (*Vitis vinifera* L.) ripening: response to a combination of elevated CO2 and temperature, and moderate drought. Plant Soil 337:179–191

Salazar-Parra C, Aranjuelo I, Pascual I et al (2015) Carbon balance, partitioning and photosynthetic acclimation in fruit-bearing grapevine (*Vitis vinifera* L. cv. Tempranillo) grown under simulated climate change (elevated CO2, elevated temperature and moderate drought) scenarios in temperature gradient greenhouses. J Plant Physiol 174:97–109

Sankaran S, Mishra A, Ehsani R, Davis C (2010) A review of advanced techniques for detecting plant diseases. Comput Electron Agric 72:1–13

Santesteban LG, Guillaume S, Royo JB, Tisseyre B (2013) Are precision agriculture tools and methods relevant at the whole-vineyard scale? Precis Agric 14:2–17

Schoedl K, Schuhmacher R, Forneck A (2013) Correlating physiological parameters with biomarkers for UV-B stress indicators in leaves of grapevine cultivars Pinot noir and Riesling. J Agric Sci 151:189–200

Schueuermann C, Steel CC, Blackman JW et al (2019) A GC–MS untargeted metabolomics approach for the classification of chemical differences in grape juices based on fungal pathogen. Food Chem 270:375–384

Schultz H (2000) Climate change and viticulture: a European perspective on climatology, carbon dioxide and UV-B effects. Aust J Grape Wine Res 6:2–12

Serra I, Strever A, Myburgh PA, Deloire A (2014) Review: the interaction between rootstocks and cultivars (*Vitis vinifera* L.) to enhance drought tolerance in grapevine: Rootstocks to enhance drought tolerance in grapevine. Aust J Grape Wine Res 20:1–14

Shavrukov YN, Dry IB, Thomas MR (2004) Inflorescence and bunch architecture development in *Vitis vinifera* L. Aust J Grape Wine Res 10:116–124

Shellie K, Kovaleski AP, Londo JP (2018) Water deficit severity during berry development alters timing of dormancy transitions in wine grape cultivar Malbec. Sci Hortic 232:226–230

Sherwood RT, Berg CC, Hoover MR, Zeiders KE (1983) Illusions in visual assessment of Stagonospora leaf spot of orchardgrass. Phytopathology 73:173–177

Singh A, Ganapathysubramanian B, Singh AK, Sarkar S (2016) Machine learning for high-throughput stress phenotyping in plants. Trends Plant Sci 21:110–124

Smart DR, Schwass E, Lakso A, Morano L (2006) Grapevine rooting patterns: a comprehensive analysis and a review. Am J Enol Vitic 57:89–104

Sommer S, Cohen S (2018) Comparison of different extraction methods to predict anthocyanin concentration and color characteristics of red wines. Fermentation 4:39

Sonka ST (2016) Big data: fueling the next evolution of agricultural innovation. J Innov Manag 4:114–136

Spagnolo A, Magnin-Robert M, Alayi TD et al (2012) Physiological changes in green stems of *Vitis vinifera* L. cv. Chardonnay in response to esca proper and apoplexy revealed by proteomic and transcriptomic analyses. J Proteome Res 11:461–475

Springer LF, Sacks GL (2014) Protein-precipitable tannin in wines from *Vitis vinifera* and interspecific hybrid grapes (*Vitis* ssp.): differences in concentration, extractability, and cell wall binding. J Agric Food Chem 62:7515–7523

Springer LF, Sherwood RW, Sacks GL (2016) Pathogenesis-related proteins limit the retention of condensed tannin additions to red wines. J Agric Food Chem 64:1309–1317

Sucu S, Yağcı A, Yıldırım K (2018) Changes in morphological, physiological traits and enzyme activity of grafted and ungrafted grapevine rootstocks under drought stress. Erwerbs-Obstbau 60:127–136

Sun R-Z, Cheng G, Li Q et al (2017) Light-induced variation in phenolic compounds in Cabernet Sauvignon grapes (*Vitis vinifera* L) involves extensive transcriptome reprogramming of biosynthetic enzymes, transcription factors, and phytohormonal regulators. Front Plant Sci 8:547

Sweetman C, Sadras VO, Hancock RD et al (2014) Metabolic effects of elevated temperature on organic acid degradation in ripening *Vitis vinifera* fruit. J Exp Bot 65:5975–5988

Tang X, Wang Y, Han J et al (2018) Separation, purification of anthocyanin and vitis linn polysaccharide from grape juice by the two-step extraction and dialysis. J Food Process Preserv 42:e13344

Tattersall EAR, Grimplet J, DeLuc L et al (2007) Transcript abundance profiles reveal larger and more complex responses of grapevine to chilling compared to osmotic and salinity stress. Funct Integr Genom 7:317–333

Taylor JA, Link K, Taft T et al (2017) A protocol to map vine size in commercial single high-wire trellis vineyards using "off-the-shelf" proximal canopy-sensing systems. Catal Discov Pract 1:35–47

Teh SL, Fresnedo-Ramírez J, Clark MD et al (2017) Genetic dissection of powdery mildew resistance in interspecific half-sib grapevine families using SNP-based maps. Mol Breed 37:1

Tello J, Ibáñez J (2018) What do we know about grapevine bunch compactness? A state-of-the-art review: review on bunch compactness. Aust J Grape Wine Res 24:6–23

Tello J, Torres-Pérez R, Grimplet J et al (2015) Polymorphisms and minihaplotypes in the VvNAC26 gene associate with berry size variation in grapevine. BMC Plant Biol 15:253

Tello J, Cubero S, Blasco J et al (2016) Application of 2D and 3D image technologies to characterise morphological attributes of grapevine clusters. J Sci Food Agric 96:4575–4583

Tester M, Langridge P (2010) Breeding technologies to increase crop production in a changing world. Science 327:818–822

Tomás M, Medrano H, Escalona JM et al (2014) Variability of water use efficiency in grapevines. Environ Exp Bot 103:148–157

Tomasi D, Jones GV, Giust M et al (2011) Grapevine phenology and climate change: relationships and trends in the Veneto region of Italy for 1964–2009. Am J Enol Vitic 62:329–339

Torregrosa L, Bigard A, Doligez A et al (2017) Developmental, molecular and genetic studies on grapevine response to temperature open breeding strategies for adaptation to warming. OENO One 51:155

Toumi I, Gargouri M, Nouairi I et al (2008) Water stress induced changes in the leaf lipid composition of four grapevine genotypes with different drought tolerance. Biol Plant 52:161–164

Väinölä A, McNamara S, Pellett H (1997) Stem and flower bud hardiness of deciduous azaleas. J Environ Hortic 15:45–50

Vega A, Gutiérrez RA, Peña-Neira A et al (2011) Compatible GLRaV-3 viral infections affect berry ripening decreasing sugar accumulation and anthocyanin biosynthesis in *Vitis vinifera*. Plant Mol Biol 77:261–274

Velasco R, Zharkikh A, Troggio M et al (2007) A high quality draft consensus sequence of the genome of a heterozygous grapevine variety. PLoS ONE 2:e1326

Versari A, Laurie VF, Ricci A et al (2014) Progress in authentication, typification and traceability of grapes and wines by chemometric approaches. Food Res Int 60:2–18

Vivin P, Lebon É, Dai Z et al (2017) Combining ecophysiological models and genetic analysis: a promising way to dissect complex adaptive traits in grapevine. OENO One 51:181–189

Wang C, Han J, Shangguan L et al (2014) Depiction of grapevine phenology by gene expression information and a test of its workability in guiding fertilization. Plant Mol Biol Rep 32:1070–1084

Wang Y, He Y-N, Chen W-K et al (2018) Effects of cluster thinning on vine photosynthesis, berry ripeness and flavonoid composition of Cabernet Sauvignon. Food Chem 248:101–110

Ward E, Foster SJ, Fraaije BA, Mccartney HA (2004) Plant pathogen diagnostics: immunological and nucleic acid-based approaches. Ann Appl Biol 145:1–16

Webb LB, Whetton PH, Barlow EWR (2007) Modelled impact of future climate change on the phenology of winegrapes in Australia. Aust J Grape Wine Res 13:165–175

Webb LB, Whetton PH, Barlow EWR (2008) Climate change and winegrape quality in Australia. Clim Res 36:99–111

Wilkinson MD, Dumontier M, Aalbersberg IJJ et al (2016) The FAIR Guiding Principles for scientific data management and stewardship. Sci Data 3:160018

Wolkovich EM, Burge DO, Walker MA, Nicholas KA (2017) Phenological diversity provides opportunities for climate change adaptation in winegrapes. J Ecol 105:905–912

Xin H, Zhu W, Wang L et al (2013) Genome wide transcriptional profile analysis of *Vitis amurensis* and *Vitis vinifera* in response to cold stress. PLoS ONE 8: e58740

Xu H, Liu G, Liu G et al (2014a) Comparison of investigation methods of heat injury in grapevine (Vitis) and assessment to heat tolerance in different cultivars and species. BMC Plant Biol 14:156

Xu W, Li R, Zhang N et al (2014b) Transcriptome profiling of *Vitis amurensis*, an extremely cold-tolerant Chinese wild *Vitis* species, reveals candidate genes and events that potentially connected to cold stress. Plant Mol Biol 86:527–541

Yang L, Lin H, Takahashi Y et al (2011) Proteomic analysis of grapevine stem in response to *Xylella fastidiosa* inoculation. Physiol Mol Plant Pathol 75:90–99

Yıldırım K, Yağcı A, Sucu S, Tunç S (2018) Responses of grapevine rootstocks to drought through altered root system architecture and root transcriptomic regulations. Plant Physiol Biochem 127:256–268

Yuan X, Wu Z, Li H et al (2014) Biochemical and proteomic analysis of "Kyoho" grape (*Vitis labruscana*) berries during cold storage. Postharvest Biol Technol 88:79–87

Zamboni A, Di Carli M, Guzzo F et al (2010) Identification of putative stage-specific grapevine berry biomarkers and omics data integration into networks. Plant Physiol 154:1439–1459

Zapata D, Salazar-Gutierrez M, Chaves B et al (2017) Predicting key phenological stages for 17 grapevine cultivars (*Vitis vinifera* L.). Am J Enol Vitic 68:60–72

Zarraonaindia I, Owens SM, Weisenhorn P et al (2015) The soil microbiome influences grapevine-associated microbiota. MBio 6:e02527-14. https://doi.org/10.1128/mbio.02527-14

Zendler D, Schneider P, Töpfer R, Zyprian E (2017) Fine mapping of Ren3 reveals two loci mediating hypersensitive response against *Erysiphe necator* in grapevine. Euphytica 213:68

Zhang C, Kovacs JM (2012) The application of small unmanned aerial systems for precision agriculture: a review. Precis Agric 13:693–712

Zhao YH, Guo YS, Lin H et al (2015) Quantitative trait locus analysis of grape weight and soluble solid content. Genet Mol Res 14:9872–9881

Zhao C, Lee WS, He D (2016) Immature green citrus detection based on colour feature and sum of absolute transformed difference (SATD) using colour images in the citrus grove. Comput Electron Agric 124:243–253

Zyprian E, Ochßner I, Schwander F et al (2016) Quantitative trait loci affecting pathogen resistance and ripening of grapevines. Mol Genet Genom 291:1573–1594

Response and Recovery of Grapevine to Water Deficit: From Genes to Physiology

11

Silvina Dayer, Idan Reingwirtz, Andrew J. McElrone and Gregory A. Gambetta

Abstract

Grapevine is a crop of global economic importance which is often cultivated in dry Mediterranean climates. In the context of climatic change, periods of drought could increase and become more intense. Growers will face increasing pressure to increase irrigation efficiently and/or adopt new grapevine varieties with increased drought resistance and water use efficiency. Adapting viticulture to these challenges requires an improved understanding of how grapevines behave under drought to enable sustainable management strategies and develop new varieties and rootstocks. This chapter summarizes our current understanding of the changes in physiology, signaling, metabolism, and gene expression that mediate grapevine's response and adaptation to drought.

11.1 Introduction

Water scarcity, which occurs when demands exceed supplies, threatens crop production in dry growing regions across the globe. Changing climatic conditions could exacerbate this situation, as more intense and prolonged drought events are predicted for many regions (IPCC 2014). Grapevines are a high-value crop in many parts of the world and are commonly grown in Mediterranean-like regions with long, dry summers making them prone to extended periods of drought. Unusually prolonged droughts (even considered mega-droughts) have recently wreaked havoc on grape growers in Australia, California, and Chile (Thrupp et al. 2008; Abare 2008; Garreaud et al. 2017). With warmer winters, regions in the western USA are also dealing with less snowpack accumulation in mountain ranges, which provide surface runoff that supplies irrigation water (Mote et al. 2008). Growers in regions that rely on irrigation have recently faced restricted water allocations as they compete with demands from urban, industrial, and

S. Dayer
INRA, Institut des Sciences de la Vigne et du Vin, UMR 1287, Ecophysiologie et Génomique Fonctionnelle de la Vigne, 33140 Villenave d'Ornon, France
e-mail: silvina.dayer@inra.fr

I. Reingwirtz · A. J. McElrone
Department of Viticulture and Enology, University of California, Davis, CA 95616, USA
e-mail: ireingwirtz@ucdavis.edu

A. J. McElrone
e-mail: ajmcelrone@ucdavis.edu

A. J. McElrone
USDA-ARS, Davis, CA 95616, USA

G. A. Gambetta (✉)
Bordeaux Science Agro, Institut des Sciences de la Vigne et du Vin, UMR 1287, Ecophysiologie et Génomique Fonctionnelle de la Vigne, 33140 Villenave D'Ornon, France
e-mail: gregory.gambetta@agro-bordeaux.fr

D. Cantu and M. A. Walker (eds.), *The Grape Genome*,
Compendium of Plant Genomes, https://doi.org/10.1007/978-3-030-18601-2_11

conservation sectors. Given that agriculture water use dominates total water use in these regions (i.e., ~80% of the total in some regions), growers will face increasing pressure to use water more efficiently. This requires an improved understanding of grapevine behavior under drought to enable growers to manage deficit irrigation strategies while respecting the vine's stress thresholds.

Wine grapes in many parts of the world are traditionally grown without supplemental irrigation. This tradition still holds in many regions (e.g., France, Spain, Italy), while many other growing regions (and other grape commodities, e.g., table grapes, juice, and raisins) rely on irrigation to improve vine yields and avoid drought-induced vine mortality. Even under conditions where irrigation is applied, growers often deliberately impose a water deficit particularly for premium wine grape production and to facilitate earlier harvests and time to market for table grapes. This is often accomplished using regulated deficit irrigation, where less water is applied than that needed to match the evapotranspiration demands of the vineyard. This results in soil water depletion over time and increased water stress in the vines particularly if the deficit coincides with increased atmospheric demand during the hottest portion of the growing season. Deficit irrigation applied at the right time and right intensity helps to control vegetative growth, reduce humidity, and allow adequate light penetration in the fruiting zone (Keller 2015). Maximizing water use efficiency in vineyards requires adequate understanding of the physiological constraints imposed by water deficits so stress thresholds can be approached without long-lasting detrimental effects that prevent fruit ripening or bud fruitfulness in future growing seasons.

The vast majority (>95%) of water absorbed by grapevine root systems is transported directly to the canopy and lost to the atmosphere via transpiration. Water exits the leaves through the stomata, where it is exchanged for CO_2 needed for photosynthesis. Water that remains within grapevines is used for maintaining cell turgor, building and expanding new cells, translocating nutrients and sugars, providing evaporative cooling, and facilitating gas exchange (Keller 2015). Under drought, these physiological processes can be largely disrupted, but the timing and degree of these disruptions vary across these processes. Mild stress in grapevines occurs when leaf water potential (Ψ_{leaf}) ranges from approximately −0.8 MPa to −1.1 MPa, while moderate stress is often characterized when Ψ_{leaf} is −1.2 to −1.4 MPa. Severely stress grapevines exhibit Ψ_{leaf} below −1.6 MPa.

11.1.1 Overview of Grapevine Response to Drought: From Mild to Severe Stress

Growth and expansion of tissues are one of the most sensitive indicators of drought-induced water stress in grapevines. Non-stressed vines that are actively growing usually have long tendrils that extend past the shoot tip. Under mild water stress, turgor and relative water content start to decrease in grapevine cells, which results in reduced cell division and expansion. One of the earliest signals of drought stress is reduced shoot tip and tendril growth. At this same time, plants reduce cell wall synthesis and protein production needed to drive cellular metabolism (Hsiao 1973).

As soil water content continues to decrease and water stress increases, abscisic acid (ABA), a plant hormone and key water stress response signal, is produced and combines with turgor loss to initiate stomatal closure under mild–moderate drought stress. This leads to initial reductions in photosynthesis due to substrate limitation (i.e., CO_2). ABA production also impacts other key physiological processes at the molecular level including cellular osmotic adjustment, regulation of aquaporin activity, and antagonizing auxin to inhibit cell loosening/expansion.

As grapevines approach moderate water stress, shoot growth and leaf expansion cease completely (Schultz and Matthews 1988). Decreased canopy size and photosynthetic capacity lead to less carbon export to sinks and thus depletion of reserves from storage sites in woody organs (Holzapfel et al. 2010). This process is likely associated with altered transport processes in the phloem.

Root growth decreases under water stress due to lost cell turgor and increases penetration resistance of drying soils (Bengough et al. 2011), but the reduction in root growth is generally less severe than that of the canopy (likely associated with higher expansin protein activity and osmotic regulation of root tips; During and Dry 1995), thus leading to higher root:shoot ratios under drought stress. Grapevine root respiration is known to decrease with soil water deficit, and a loss of membrane integrity leads to root dieback under severe drought stress. Both responses are likely associated with lacuna formation in the cortex of grapevine fine roots that also reduces the hydraulic conductivity and precedes root shrinkage and xylem embolism in these organs (Cuneo et al. 2016).

Under moderate to more severe stress shoot tips will dry up and fall off and reduce the apical dominance within the shoot. This response likely induces a hormonal signal down the shoot triggering responses in older leaves on the shoot. Leaves change angle and orient themselves parallel to the sun's rays, thus reducing incident radiation and heat load as evaporative cooling associated with transpiration is lost. Moderate to severe water stress limits photosynthesis via damage to various components integral to light harvesting, electron transport, and carbon fixation by photosynthetic enzymes. Delays in ripening, reduced bud fruitfulness, reduced winter hardiness, and even sudden vine collapse can eventually occur at this stage.

11.2 Regulating Water Use Under Drought

11.2.1 Stomatal Regulation

Leaf gas exchange in vascular plants is facilitated by stomata, tiny pores at the leaf surface each encompassed by a pair of adjacent guard cells. Changes in the turgor of the guard cells allow the plant to open and close the stomata, regulating the trade-off between carbon uptake and water loss (Buckley and Mott 2002). Stomatal closure initiates during the early stages of drought stress. Plants close the stomata to avoid excessive water loss, and consequently, xylem tensions that could trigger cavitation. The physical mechanism by which g_s and water potential are coordinated is complex and poorly understood because the stomata are responding to a spectrum of factors at any moment, from intercellular signaling to a wide range of environmental factors (Hetherington and Woodward 2003). Stomatal regulation can result directly from hydraulic signals, i.e., changes in the local water status of (or around) the guard cells (Fig. 11.1). These changes in water status can result from osmotic changes within the guard cells themselves, and/or through changes in the water potential gradient resulting from the hydraulic conductance of the pathway. At the same time, biochemical factors (e.g., ABA discussed below) mediate stomatal response to water deficit and can trigger stomatal closure even in the absence of changes in leaf water potential (Christmann et al. 2007).

Hydraulic and chemical signals have been extensively studied, but their relative contribution to stomatal regulation remains under debate. In some species and experimental conditions, one signal may dominate (e.g., Comstock 2002; Ahmadi et al. 2009). Experiments assessing the stomatal response to ABA in basal lineages such as ferns and lycophytes indicated that these plants use only passive hydraulic mechanisms for stomatal regulation (Brodribb and McAdam 2011; McAdam and Brodribb 2012). Further examination of this response was studied by comparing the stomatal responses to vapor pressure deficit (VPD) under a wider group of phylogenetic representative species including ABA-sensitive stomata (angiosperms) and ABA-insensitive stomata (ferns and conifers, Brodribb and McAdam 2011; McAdam et al. 2016a, b). These studies observed that only angiosperms are able to rapidly increase foliar ABA levels during a VPD transition (from low to high levels) to regulate stomatal closure, while minimal changes in foliar ABA levels were

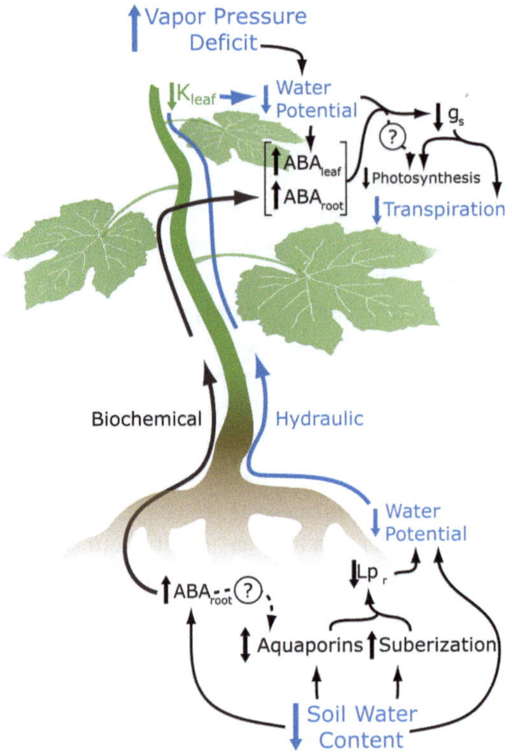

Fig. 11.1 Summary of grapevine whole plant integration under drought. The two pathways modulating stomatal conductance, transpiration, and photosynthesis are biochemical (black) and hydraulic (blue). Biochemical signaling results from the production and sensing of chemical signals (e.g., ABA), either locally in leaves and/or via the long-distance transport from roots to leaves. Hydraulic signals likely originate through the integration of decreases in root (Lp_r) and leaf (K_{leaf}) hydraulic conductance resulting in decreases in water potential that impact stomatal conductance (g_s)

observed for ferns and conifers (McAdam and Brodribb 2015, 2016). Recent studies in grapevine showed that the hydraulic control was dominant during the early phases of water stress, while chemical signals seemed to have an additive effect involved in the long-term maintenance of stomatal closure under prolonged water stress (Tombesi et al. 2015). Thus, in grapevine an integrated system that includes both types of signals seems to be more likely than a control based on either chemical or hydraulic signaling alone (Fig. 11.1) (Tardieu and Davies 1993; Peccoux et al. 2017).

At a molecular level, many proteins that regulate stomatal responses to the environment have been identified. The ERECTA transcription factor family, putative leucine-rich repeat receptor-like kinases have been related to the perception of water stress signals across the cell membranes in Arabidopsis (Masle et al. 2005). ERECTA coordinates transpiration and photosynthesis, and as such is regarded as a transpiration efficiency gene (Reynolds and Tuberosa 2008). On the other hand, several proteins located in the plasma membrane and tonoplast of guard cells, including channels and carriers, are also known to be involved in the regulation of stomatal movements (Chaves et al. 2011; Costa et al. 2015). For instance, aquaporins (membrane water channels) play an important role in stomatal regulation by facilitating the exchange of water across membranes (Chaumont and Tyerman 2014). Experiments on grapevine showed that the leaf hydraulic conductance decreased by about 30% under water stress concomitantly with a decrease of expression of some aquaporin isoforms (Pou et al. 2013). In that study, positive correlations were observed between stomatal conductance (g_s), leaf hydraulic conductance, and leaf aquaporin expression and activity, suggesting a contribution of aquaporins in regulating vine water use at the leaf level. Similarly, experiments on field-grown Chasselas grapevines growing under different radiation and irrigation regimes revealed that short-term changes in the hydraulic conductivity of the petioles were explained largely by changes in the leaf g_s and the expression of aquaporins (Dayer et al. 2017a).

11.2.2 ABA as a Key Regulator of Stomatal Conductance

ABA is a plant growth regulator involved in various physiological processes that include positive or negative roles depending on the plant conditions. For example, when ABA is at low concentration under non-stressful conditions, it has been shown to be essential for vegetative growth in several organs (e.g., primary root growth) (Sharp et al. 2000), but when ABA accumulates under drought it reduces growth and inhibits stomatal opening. ABA can be synthesized in all cells and organs, including guard cells, and thus plays an important role in regulating gas exchange via stomatal closure as water stress increases (Munns and Cramer 1996; Boursiac et al. 2013). The 9-*cis*-epoxycarotenoid dioxygenase (NCED) genes catalyze the first step in ABA biosynthesis and represent the rate-limiting step in Arabidopsis and presumably many other plant species (Endo et al. 2008). In *V. vinifera,* the *VviNCED1* and *VviNCED2* genes are linked to ABA synthesis and were shown to be up-regulated during water deficit (Speirs et al. 2013; Rossdeutsch et al. 2016). The expression of other genes in the NCED family (*VviNCED3, VviNCED5,* and *VviNCED6*) varies across three different genotypes of *Vitis;* although the relative contributions of these different isogenes in the control of ABA biosynthesis, it is not entirely clear (Hopper et al. 2016).

Under water deficit roots and shoots synthesize ABA and there have been conflicting views on the relative contribution of root and leaf derived ABA in stomatal regulation (Davies and Zhang 1991; Tardieu and Simonneau 1998; Dodd 2005). ABA content in roots is well correlated with both soil moisture and root-relative water content in many plant species. At the molecular level, Speirs et al. (2013) reported that the expression of the ABA biosynthesis genes *VviNCED1* and *VviNCED2* were activated in roots, but not in leaves, in response to water deficit, suggesting that roots could link stomatal response to soil moisture status. On the other hand, leaf cells are known to synthesize ABA (Cutler and Krochko 1999) when their water status is affected by local environmental conditions such as high VPD so one would expect that the same would be true for changes in water status brought about by soil water deficits (via hydraulic signals). In fact, there is an increasing number of studies that suggest leaf derived ABA is the dominant regulator of stomata. Reciprocal grafting studies in tomato showed that changes in apoplastic ABA levels in leaves were responsible for stomatal closure, and that ABA production by roots was not required to trigger the response (Holbrook et al. 2002). In Arabidopsis, Christmann et al. (2007) demonstrated that changes in turgor pressure of leaf mesophyll cells occurred within minutes of root-induced osmotic stress and elicited activation of ABA biosynthesis in shoots, putatively signaling stomatal closure. In grapevine, the source of xylem sap ABA was suggested to originate from the leaf rather than the roots due to the abundance of leaf ABA and the increased expression of *VviNCED1* and another ABA biosynthetic gene *VviZEP* in the leaves during the day (Soar et al. 2006). Interestingly, shoot derived ABA likely influences root physiology as well. In angiosperms, ABA levels in the roots, as well as root growth, were influenced by ABA synthesized in the leaves rather than sourced from the roots (McAdam et al. 2016a). Although the importance and role of root-sourced ABA are still controversial some of the conflicting observations may be due to differences in the intensity and speed of the development of water deficit under experimental conditions.

ABA biosynthesis and its subsequent regulation of stomata are complex. In Arabidopsis, the ABA biosynthesis core signal network involves at least 138 proteins and over 500 interactions (Lumba et al. 2014). In the absence of ABA, the central ABA signaling 2C protein phosphatases (PP2C) inhibit the activity of serine/threonine protein kinases (SnRKs) and downstream ABA signaling (Fig. 11.2). When ABA is present, the PYR/PYL/RCAR protein family of ABA receptors (Ma et al. 2009; Park et al. 2009) bind ABA increasing their interaction with the PP2Cs. This interaction disrupts the PP2C–SnRK interaction, thus liberating the SnRKs to activate downstream ABA responses. In grapevine, studies have characterized how the expression of some of these signaling components changes in response to drought. The PP2Cs, *VviHAI1* and *VviAHG3*, and the SnRK, *VviOST1* (ortholog of OST1 from rice), increase in *Vitis* leaves under water

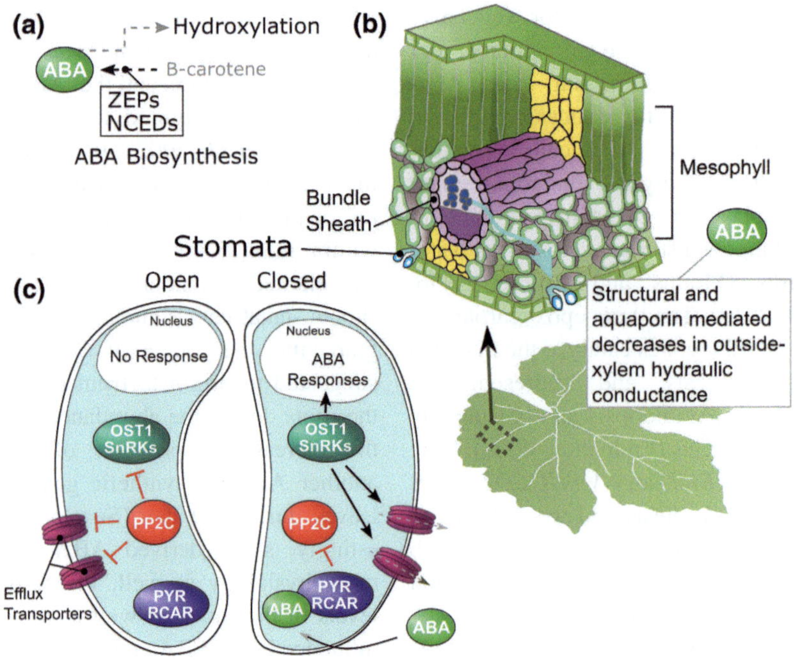

Fig. 11.2 On overview of ABA biosynthesis, signaling, and its role in mediating changes in leaf hydraulic conductance and stomatal regulation during drought. **a** ABA is biosynthesized from B-carotene (*not all steps are shown*) with the zeaxanthin epoxidase (ZEP) and 9-*cis*-epoxycarotenoid dioxygenase (NCED) proteins catalyzing the rate-limiting steps. ABA is catabolized via hydroxylation. **b** In response to drought, ABA is thought to mediate decreases in outside-xylem hydraulic conductance in the leaf lamina. **c** In the stomatal guard cells, ABA signaling mediates stomatal closure. Under well-watered conditions, the 2C protein phosphatases (PP2C) inhibit the activity of serine/threonine-protein kinases (SnRKs) and downstream ABA responses. When ABA is present, it binds to the PYR/PYL/RCAR receptors which disrupt the PP2C-SnRK interaction, thus liberating the SnRKs to activate downstream ABA responses

deficit (Hopper et al. 2016). In addition, the abundance of *VviABI5* and *VviABF2,* ABA-responsive transcription factors which are targeted by *VviOST1*, also increases during water deficit (Haider et al. 2017). Two receptors involved in ABA perception, *VviRCAR5* and *VviRCAR6,* were downregulated in leaves and roots. Both genes are putative negative regulators of *VviPP2C4* and *VviPP2C9* and showed higher expression under water deficit (Boneh et al. 2012; Rossdeutsch et al. 2016).

Recent studies have identified the role of protein phosphorylation in ABA-induced stomatal closure that involves kinases and phosphatases (Zhang et al. 2014). However, more effort should be focused on revealing the protein abundance and phosphorylation status of these proteins to complete our understanding in the plant response to stress.

Changes in the pH of xylem sap commonly observed under drought stress can be an important component of root-to-shoot signaling and may act synergistically with ABA. The potential effects of pH have been outlined elsewhere (Wilkinson 1999) and include (1) changes in ABA metabolism resulting in increased leaf ABA concentration; (2) direct effects on leaf water status that could alter guard cell turgor or sensitivity to leaf ABA concentrations; (3) direct effects on ion fluxes through the guard cell plasma membrane; and (4) an increase of ABA concentration in the apoplast surrounding guard cells.

11.2.3 The Influence of Root and Leaf Hydraulic Conductance on Plant Water Use

In addition to stomatal regulation, the hydraulic conductance of leaves and roots also contributes to the regulation of plant water use. Under water deficit, hydraulic conductance decreases sharply in fine roots and leaves. This drop-in hydraulic conductance plays an important role in protecting grapevines from more severe levels of water stress that can result in embolism and mortality (see Sect. 11.4).

While water deficit tends to decrease root hydraulic conductance (Vandeleur et al. 2009) contrasting results have been obtained for ABA applications (Gambetta et al. 2017). An increase in the root hydraulic conductance regulated by aquaporins was observed in maize mutants overexpressing ABA (Parent et al. 2009). In contrast, other studies have observed only a transient increase in root hydraulic conductance (Hose et al. 2000), or even no effect (Wan and Zwiazek 2001; Aroca et al. 2003), in response to ABA applications. The increase in root hydraulic conductance by ABA has been interpreted as a mechanism to improve the water supply to the shoot, decreasing the water potential gradient along the flow pathway under soil or atmospheric water stress (Kudoyarova et al. 2011; Pantin et al. 2013).

Diurnal changes in root hydraulic conductance have also been observed under well-watered conditions concomitantly with changes in shoot transpiration (Vandeleur et al. 2009). In general, these variations correlate with the transcript abundance of aquaporins in roots suggesting that aquaporins facilitate water transport across roots to meet the transpirational demand of the shoots (Sakurai-Ishikawa et al. 2011; Laur and Hacke 2013; Vandeleur et al. 2014). Gene expression studies in various plant species have reported contrasting responses of aquaporin expression to water stress. Experiments using mercuric chloride demonstrated a decrease in aquaporin activity in water-stressed dessert plants and *Populus* sp. seedlings (Martre et al. 2001; Siemens and Zwiazek 2003; North et al. 2004). In grapevine, Gambetta et al. (2012) observed differences in root hydraulic conductance between low and high vigor conferring rootstocks that corresponded to differences in the expression and activity of aquaporins.

On the other hand, a great variation in the apparent sensitivity of leaf hydraulic conductance to xylem ABA concentration has been reported (Correia et al. 1995). For instance, a large variability in leaf hydraulic conductance sensitivity to exogenous ABA was observed between different grapevine genotypes (Coupel-Ledru et al. 2017). Those authors found that ABA accumulation in the xylem sap of intact grapevine plants was highly dependent on the genotype, suggesting variability in ABA biosynthesis capacity or catabolism. This observation was further confirmed in nine grapevine genotypes where ABA-mediated responses to water deficit separate the genotypes by their genetic background (Rossdeutsch et al. 2016). Thus, stomatal regulation likely results from the complex integration of guard cell osmotic pressure, leaf water status and hydraulic conductance, and root-to-shoot controls.

Under water deficit, the increase in ABA concentration in roots and leaves is coincident with decreases in hydraulic conductance. In leaves, studies showed that xylem fed-ABA decreases the leaf hydraulic conductivity by decreasing water permeability in the vascular bundle sheath cells (Shatil-Cohen et al. 2011). Pantin et al. (2013) further demonstrated that vascular ABA decreased the leaf hydraulic conductance putatively by inactivating bundle sheath aquaporins, indicating that ABA indirectly impacts guard cell water relations through these changes in leaf hydraulics. These results led the authors to suggest that ABA regulates stomata via an additional indirect mechanism, whereby reduced water permeability within leaf vascular tissues results in local changes in water potential that are sensed by guard cells (Fig. 11.2) (Pantin et al. 2013). These decreases in the hydraulic conductance in the pathway between the xylem and the stomata (i.e., the outside-xylem pathway) occur across species and contribute to stomatal closure and protection from more severe stress

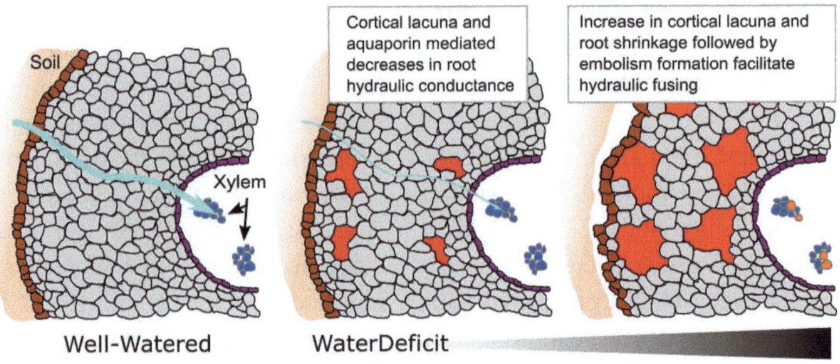

Fig. 11.3 Responses of grapevine fine roots to drought. Water transport (light blue arrow) from the soil across the root cortex (gray) into the xylem decreases under water deficit. This decrease in water uptake is first mediated by decreases in hydraulic conductance which occurs as a result of structural changes (e.g., lacuna formation in red) and aquaporin mediated decreases. As stress increase, the lacunas expand and the root shrinks largely disconnecting the root from the soil, a process referred to as hydraulic fusing. Eventually, if the stress becomes severe, enough xylem vessels embolize (orange vessels)

levels (Scoffoni et al. 2017a, b); similar findings have been found recently in grapevine leaves (Albuquerque et al. unpublished data). The same is true in fine roots where water deficit leads to sharp decreases in hydraulic conductance which occurs as a result of structural changes (Cuneo et al. 2016) and aquaporin mediated decreases (Fig. 11.3).

11.2.4 Cultivar Sensitivity to ABA: The Iso/Aniso Debate

In some species, g_s appears to regulate plant water status so tightly that leaf water potential does not vary significantly (Tardieu and Davies 1993; Saliendra et al. 1995). Plants that present this conservative response under drought have been classified as "isohydric". In contrast, plants that have a less strict stomatal control, exhibiting more negative water potentials under drought have been classified as "anisohydric" (Tardieu and Simonneau 1998; Soar et al. 2006). This broad classification assumes that genotype fixes a plant's behavior somewhere in between these two theoretical extremes; however, it is widely recognized that this is not always the case (Chaves et al. 2010; Domec and Johnson 2012). For instance, contrasting studies are plentiful in the literature demonstrating the same grapevine variety can exhibit different behaviors depending on the growing conditions (e.g., field grown versus potted plants; Medrano et al. 2003; Sousa et al. 2006; Lovisolo et al. 2010; Charrier et al. 2018). This classification has also been used to describe the underlying mechanisms of drought-induced changes in plant physiology such as root and leaf hydraulic conductance (Schultz 2003; Vandeleur et al. 2009), nighttime g_s (Cirelli et al. 2015), vulnerability to cavitation (Hukin et al. 2005), and plant mortality (McDowell et al. 2008). For instance, the degree of iso/anisohydric behavior has been explained by the differential expression of root aquaporins in two grapevine genotypes (Grenache and Syrah; Vandeleur et al. 2009). In that study, both varieties show increased root suberization under water stress, thus reducing the total hydraulic conductance of the root system, but only cv. Chardonnay (the more drought-sensitive, anisohydric) seemed to partially compensate for this decrease through increased expression of the grape aquaporin *VvPIP1;1*.

Differences in stomatal response to drought might be partially determined by genetic differences in the capacity to produce ABA. Only part of this variation is under heritable control since leaf developmental stage and environmental preconditioning exert a large influence on the stomatal response to drought (Chaves et al.

2010). In grapevine, different *Vitis* genotypes exhibiting different levels of drought adaptation differ in key steps involved in ABA metabolism and signaling; both under well-watered conditions and in response to water deficit (Rossdeutsch et al. 2016).

11.2.5 Other Hormone Pathways: Ethylene, GABA

Even though ABA signaling is seen as the main pathway for stomatal regulation, chemical signals other than ABA have been proposed (Christmann et al. 2007; Wilkinson et al. 2007) including the nonprotein amino acid γ-aminobutyric acid (GABA; Serraj et al. 1998). Rapid accumulation of GABA was identified in plant tissues upon exposure to drought (Serraj et al. 1998). In water deficit studies in Arabidopsis, GABA accumulation was observed to be stress-specific and its accumulation induced stomatal closure (Mekonnen et al. 2016). In addition, other studies have identified specific plant transporter proteins (e.g., aluminum-activated malate transporter) that are modulated by GABA and affect diverse aspects of the drought response (Ramesh et al. 2015).

Ethylene could be another important factor under water deficit. A precursor of ethylene 1-aminocyclopropane-1-carboxylic acid (ACC) that moves in the xylem from root to shoots has been observed to increase in water-stressed grapevines (Haider et al. 2017). A role for ethylene under drought was demonstrated by the use of ACC oxidase (ACO, which catalyzes the conversion of ACC into ethylene) antisense lines in tomato (Sobeih et al. 2004). In these plants, ethylene evolution was much lower than normal under both well-watered and drought conditions. Under water deficit, the stomatal response in the ACO antisense plants was the same as the wild type, but a decrease in leaf growth was measured in wild type, but not ACO antisense plants. ACC synthase (ACS) is the rate-limiting enzyme in the biosynthesis of ethylene and dehydrated leaves of Cabernet Sauvignon exhibited increases in the expression of *VviACS7*, *VviACS4*, and *VviACS8-like* (Hopper et al. 2016).

In Arabidopsis, the ethylene response factors (ERFs) are considered integrators of hormone pathways, and ERF5 and ERF6 play a crucial role in leaf growth as response to dehydration (Dubois et al. 2013). Hopper et al. (2016) observed an increase in *VviERF6-like* in *Vitis vinifera* cv. Cabernet Sauvignon leaves under water stress. Equally, the ethylene receptors *VviETR2*, *VviERS2*, and *VviERS1* are all increased under water deficit. The WRKY gene family is also known to affect the ethylene signaling. In Arabidopsis, *AtWRKY40* is regulated by members of the APETALA 2/ethylene-responsive element binding factor (AP2/ERF) transcription factor family (Koyama et al. 2013), and in some grape genotypes the grape orthologue, *VviWRKY40*, is up-regulated under water deficit along with AP2/ERF transcription factors (Hopper et al. 2016). Genes from the ERF family, *VviERF9*, *VviERF055*, *VviERF022*, and *VviERF128* showed increased expression under water deficit (Hopper et al. 2016). *VviERF055* is homologous to an ERF transcription factor in *Arabidopsis*, the translucent green (TG), which is thought to increase drought tolerance by binding to aquaporin promoters. These coordinated changes in gene expression suggest a role for ethylene and ethylene signaling in the drought response, but more research on this topic is needed.

Stomata play a key role in plant adaptation to the environment, as they regulate the trade-off between water and CO_2 and modeling is an effective tool to investigate the integration, simulation, and prediction of environmental effects on stomatal regulation (Zhu et al. 2017, 2018). However, models could be improved by incorporating a more nuanced understanding of additional chemical signals. For example, hydrogen peroxide is an important reactive oxygen species (ROS) molecule involved in guard cell functioning and more specifically in the guard cell ABA-signaling network (Schroeder et al. 2001). Including the concentration of hydrogen peroxide in plant, models may provide an essential and complementary link between g_s, photosynthesis, and ABA (Damour et al. 2010).

11.3 Photosynthesis and the Effect of Drought

Decreases in carbon fixation observed in grapevines subjected to water stress is initially due to stomatal closure (see above; Chaves 1991; Flexas et al. 2004) as evidenced by a close correlation between g_s and photosynthesis (Naor and Wample 1994; Flexas et al. 2002), by full recovery of photosynthesis when exposing the leaves to saturating amounts of CO_2 (Cornic 2000), and by increasing instantaneous water use efficiency (i.e., the ratio of photosynthesis to transpiration) under these conditions (Cornic and Fresneau 2002). Further decreases in photosynthetic carbon assimilation under water stress are associated with other biochemical processes such as photophosphorylation and reduced activity of RuBisCO (Tezara et al. 1999). Significant disruption of the photosynthetic machinery occurs under severe stress that can often coincide with high light and high-temperature conditions that exacerbate the damage.

11.3.1 Diffusive Versus Metabolic Limitations to Photosynthesis

This diffusive limitation to CO_2 is not only imposed by the stomata but also by the pathway from the substomatal cavity into mesophyll cells and sites of carboxylation in the chloroplasts (Perez-Martin et al. 2009). Conductance of CO_2 into mesophyll cells (g_m) can impose a significant limitation on photosynthesis (Centritto et al. 2003; Flexas et al. 2007). It was proposed that aquaporins and carbonic anhydrase play an important role in regulating g_m (Flexas et al. 2006; Kawase et al. 2013), and recent work showed that most of the variations observed in g_s and g_m in olive leaves was explained by two leaf aquaporins and the expression of carbonic anhydrase had a significant effect on g_m under water-stressed conditions (Perez-Martin et al. 2014).

When water stress becomes severe alterations of photosynthetic metabolism occur, such as decreases in ATP production, ribulose-1,5-biphosphate RuBP regeneration, and RuBisCO activity (Chaves 1991; Cornic 2000; Flexas et al. 2004). Primary events of photosynthesis such as the electron transport rate are very resilient to drought, and changes in the efficiency of photosystem II (PSII) do not occur until photosynthesis becomes very low (g_s below 0.05 mol H_2O m^{-2} s^{-1}; Flexas et al. 2002, 2004; Medrano et al. 2002). At this level of severe water stress photosynthesis does not recover upon re-watering (Quick et al. 1992), indicating that non-stomatal inhibition is dominant. As g_s decreases, further RuBisCO activity steeply declines (Bota et al. 2002, 2004; Flexas et al. 2002; Maroco et al. 2002). Thus, RuBisCO has been proved to be highly stable and resistant to water stress.

Similar to RuBisCO, the key carbon and nitrogen metabolic enzymes sucrose-phosphate synthase and nitrate reductase are also highly stable under water stress (Flexas et al. 2004). By contrast, less attention has been paid to other enzymes involved in the regeneration of RuBP in the Calvin cycle, and there is still lack of knowledge regarding their regulation under drought, particularly for grapevine.

11.3.2 Sugar Signaling Metabolism and Osmotic Adjustment

Carbohydrates have different roles in the plant, from energy storage compounds to metabolic signaling molecules. There is evidence that an increase of sugars in the guard cells under water stress may determine the stomatal sensitivity to ABA (Wilkinson and Davies 2002). In general, soluble sugars tend to be maintained or even increased under water stress despite a lower carbon assimilation rate. This is possible mainly because other processes such as growth and sucrose transport to sink tissues are inhibited. In contrast, the concentration of starch decreases under drought (Chaves 1991; Dayer et al. 2016). In addition, sugars seem to favor the expression of genes related to biosynthesis and storage of reserves (e.g., starch) and repress those associated with photosynthesis and remobilization of sugars (Ho 2001). Some evidence has been provided that water deficit and other related abiotic stresses affect the expression of sugar transporter genes. For instance, in Arabidopsis transcript accumulation of the tonoplast monosaccharide transporters was increased in response to drought treatment (Wormit et al. 2006). In grapevine, water stress increased the gene expression for sucrose transporters known to code for mesophyll cell proteins in leaves without affecting the transcript abundance for the phloem loading protein (Pastenes et al. 2014). In addition, water stress may inhibit important functions of vacuolar invertase-mediated sucrose hydrolysis and osmotic potential modulation (Andersen 2002). Studies in grapevine observed that water stress induction of *VvGIN2* gene encoding a putative vacuolar invertase may contribute to the increase of cell osmotic potential in response to water deficit that helps maintain basic metabolic functions (Medici et al. 2014).

Grapevines have the ability to support growth and productivity under water deficit through osmotic adjustment (Schultz and Matthews 1993; Patakas and Nortsakis 1999). The accumulation of osmolytes in leaves is attributable to a variety of small molecules with both metabolomic and transcriptomic studies highlighting the accumulation of sugar and amino acids (Hochberg et al. 2013; Medici et al. 2014; Haider et al. 2017). For instance, Patakas et al. (2002) demonstrated the importance of organic solute and ion accumulation under water stress in grapevines. Proline metabolism is a common osmoprotectant across plant species and is among the three most responsive amino acids that change in response to water deficit, increasing as much as two to three times in *V. vinifera* leaves (Cramer et al. 2007). Haider et al. (2017) reported an increase in proline levels during water deficit as well. Increase in proline results from an increase in delta 1-pyrroline-5-carboxylate synthetase (P5CS) abundance, a biosynthetic enzyme that initiates the proline pathway (Cramer et al. 2007). Another important enzyme in proline metabolism is proline dehydrogenase (PDH) whose expression also increases as a result of water deficit (Peng et al. 1996). In *Vitis*, *PDH*, *P5CS*, and other genes involved in proline metabolism were up-regulated under water deficit (Haider et al. 2017). This osmotic adjustment may have long-term effects on grapevine performance under drought. For example, vines that have undergone successive water deficits are able to maintain slightly higher levels of g_s, which are thought to result in part from osmotic adjustment (Hochberg et al. 2017a). Two rootstocks (M4 and 101-14) that differ in their drought resistance exhibited differences in their ability to osmotically adjust with the more drought-resistant rootstock (M4) accumulating greater concentrations of sugars, amino acids, and osmotin like-proteins in response to drought (Prinsi et al. 2018).

Some transcript factors are involved in osmoprotection change in response to abiotic stress. For example, fructose bisphosphate aldolase and galactinol synthase experienced an increase in transcript abundance at an early stage of water deficit in grapevines (Cramer et al. 2007). In *Poplar*, genes encoding sucrose synthase, galactinol synthase, and raffinose synthase were all increased under water deficits (Shatil-Cohen et al. 2011). Similarly, genes encoding galactinol and raffinose synthases were similarly up-regulated in loblolly pine under drought stress (Lorenz et al. 2011).

11.3.3 Photosynthetic Pigments and Antioxidant Defense

Water stress reduces the tissue concentration of photosynthetic pigments such as chlorophylls and carotenoids (Poormohammad Kiani et al. 2008), primarily through the production of reactive oxygen species in the thylakoids (Niyogi 1999; Reddy et al. 2004). Carotenoids, in addition to their function as accessory pigments, play an important function as antioxidants protecting and sustaining photochemical processes (Havaux 1998). Carotenoids form a key part of the plant antioxidant defense system but are very susceptible to oxidative destruction. β-Carotene, presents in the chloroplasts of all green plants, is exclusively bound to the core complexes of PSI and PSII (Havaux 1998). A major protective role of β-carotene in photosynthetic tissue may be through direct quenching of triplet chlorophyll, which prevents the generation of singlet oxygen and protects from oxidative damage (Farooq et al. 2009), which becomes increasingly important under severe water stress conditions.

11.3.4 Photoinhibition and Oxidative Stress

Under field conditions, plants are normally exposed to different stresses simultaneously, such as water deficit, high temperatures and radiation regimes, and high VPD. Under well-watered conditions, most of the light absorbed by the leaves is used for photosynthesis and photorespiration processes. However, in situations where stomata close (e.g., water deficit) the combination of high irradiance with low CO_2 availability cause the plant to absorb an excess of radiant energy that has the potential to damage the photosynthetic apparatus. Under these conditions, the leaves experience a transient decrease of the photochemical efficiency of PSII in a process called photoinhibition, which is a form of non-photochemical quenching (Gamon and Pearcy 1990; Baker 2008). Photoinhibition is most commonly equated with photodamage, a long-term depression of quantum efficiency due to damage to the photosynthetic apparatus as a result of excess photosynthetic photon flux density (Walters and Horton 1993). Chronic photoinhibition may be considered as a depression of photosynthetic efficiency from which the plant does not recover after 3–4 days in shade (Greer and Laing 1992). To avoid this damage, plants can prevent this excess of light absorption by either adjusting their leaf angles to the sun, losing the chlorophyll content, or diverting the absorbed light to different processes such as thermal dissipation (Demmig-Adams and Adams 2006). Thermal dissipation is a very important non-radiative process that can dissipate >75% of the light energy absorbed by the leaves (Niyogi 1999). The xanthophyll cycle plays a primordial role in the thermal dissipation process (Demmig-Adams and Adams 2006) and also a direct action as antioxidant by increasing the tolerance of the thylakoid membrane to lipid peroxidation (Niyogi 1999).

When the leaf cannot keep pace between the light energy absorbed and thermal dissipation of this energy, the production of highly reactive molecules is exacerbated. These molecules are referred to as reactive oxygen species (ROS) and are generated mainly in the chloroplast and may lead to an oxidative damage (e.g., photooxidation) of the photosynthetic apparatus if the plant is not efficient in scavenging these molecules (Niyogi 1999). Some of the ROS molecules reported in the literature include hydrogen peroxide (H_2O_2), superoxide and hydroxyl radicals and singlet oxygen (O_2^-). Reactive oxygen species are also essential signaling molecules that mediate ABA-induced stomatal closure and ABA-induced inhibition of stomatal opening (Yan et al. 2007). Among all ROS, hydrogen peroxide emerges as one of the most important considering its role in guard cell functioning and more specifically in the guard cell ABA-signaling network (Schroeder et al. 2001; Wang and Song 2008). In addition, Gunes et al. (2006) showed that grapevine leaves can generate O_2^- and H_2O_2 in response to boron excess, which may happen under water deficit as well.

The balance between ROS synthesis and scavenging depends on the rate and duration of the water stress (Lawlor and Tezara 2009). For example, when the water stress develops rapidly over days under high light, ROS damage is observed (Demmig-Adams and Adams 2006). Detoxification mechanisms consume reducing power and form water and include reactions with reduced compounds such as ascorbate and glutathione (Mittler 2002; Asada 2006). Interestingly, increased ROS production along with the high redox state of the electron membrane chain under water stress, induce the expression of genes coding for components of energy-dissipating and regulation systems in the chloroplasts, allowing acclimation to stress conditions (Pfannschmidt et al. 2003). In *Vitis*, genes associated with ROS increased when exposed to water deficit (Cramer et al. 2013). Genes involved in ROS detoxification such as phospholipid hydroperoxide glutathione peroxidase (TC45235, 048646), gamma-glutamylcysteine synthetase, and NADPH glutathione reductase showed increases in their gene expression under water deficit (Cramer et al. 2007). Photorespiratory enzymes of the glyoxysome/peroxisome participate in water stress signal and in oxygen free-radical metabolism (Corpas et al. 2001; Moreno et al. 2005). Cramer et al. (2007) showed that several of these enzymes increased their transcript abundance in grapevines during water deficit. GABA transaminase subunit isozyme 1 is an enzyme in the "GABA shunt" pathway, which is known for its role in defense against ROS (Bouché et al. 2003; Fiorani et al. 2005; Umbach et al. 2005). Cramer et al. (2007) showed an increase in grapevine GABA transaminase transcript abundance in response to water deficit.

11.3.5 Membrane Stability

Cell and organelle membranes are one of the first receptors of stress, and they can protect the cell through modifications affecting both stress perception and rigidity of the cell structure. Quantitative changes in the membrane lipids, such as unsaturation level of phospholipids and glycolipids, affect membrane fluidity and as a consequence the activity of membrane-bound proteins (Quartacci et al. 2002). Drought causes alterations in membrane fluidity, and membrane stability is commonly used as a physiological index for the evaluation of resistance to drought tolerance (Premachandra et al. 1990). In addition, cell membranes are susceptible to damage from ROS produced via the metabolism of the cell, and/or as a result of stress (Koca et al. 2006), and the interaction between ROS and cell membranes produces lipid peroxides that can be used as a stress indicator. Because ROS species are produced in the chloroplasts, chloroplast membranes are particularly susceptible to oxidative stress.

A decrease in cellular volume caused by membrane disruption increases the cytoplasmic compounds, and the chances of molecular interactions that can cause protein denaturation and membrane fusion (Farooq et al. 2009). A broad range of compounds has been identified that can prevent such adverse molecular interactions. Some of these include proline, glutamate, glycine betaine, mannitol, sorbitol, polyols, trehalose, sucrose, fructans, macromolecules (Hoekstra et al. 2001). Such responses have not been addressed in grapevines.

11.4 Extreme Drought and Long-Term Productivity

11.4.1 Hydraulic Fusing and Embolism

Under severe water deficits, grapevines have more drastic responses such as petiole embolism, leaf shedding, and in severe cases stem embolism. However, the vulnerability of grapevine organs to embolism is not equal with grapevine petioles and leaves being significantly more vulnerable to embolism than stems (Hochberg et al. 2016, 2017b; Charrier et al. 2016). This phenomenon is referred to as "vulnerability segmentation" or "hydraulic fusing". First put forth by Zimmermann (1983), segmentation (or fusing) results when an increased vulnerability to embolism in distal organs such as petioles, leaves, and/or fine roots prevents embolism in perennial organs such as stems and trunks. Studies suggest that grapevine leaves and petioles have a P_{50} (i.e., the pressure at which there is 50% loss of hydraulic conductance via embolism) ranging from −1.0 to −2.0 MPa (Hochberg et al. 2016, 2017b; Charrier et al. 2016) while stems have a P_{50} ranging from approximately −2.0 to −3.0 MPa (Choat et al. 2010; Brodersen et al. 2013; Charrier et al. 2018). Grapevine stems become less and less vulnerable through the season and this likely increases the segmentation between leaves/petioles and stems (Charrier et al. 2018).

Equally, roots could also be more vulnerable to embolism to protect the vine against more negative water potentials (Lovisolo and Schubert 2006; Lovisolo et al. 2008). More recent results using noninvasive methods corroborated these results and demonstrated that xylem of grapevine fine roots had a P_{50} similar to that of leaves (−1.8 MPa) (Cuneo et al. 2016). It was also recently discovered that grapevine fine roots subjected to drought stress form lacuna prior to root shrinkage and embolism formation. Together, these responses likely result in fine roots becoming hydraulically disconnected from the drying soil (Cuneo et al. 2016).

Hydraulic fusing in grapevine leads to premature leaf senescence and leaf shedding (Hochberg et al. 2017b), and the progression of leaf mortality mirrors increases in leaf and petiole embolism (Charrier et al. 2018). Together with other mechanisms (e.g., in roots), these responses appear to isolate drought-induced damage of the xylem systems to expendable plant parts other than stems and trunks (Charrier et al. 2018). Stem embolism is extremely detrimental to the plant, and significant levels are typically fatal (from 50 to 90% loss of conductivity depending on species; Brodribb and Cochard 2009; Urli et al. 2013; Li et al. 2015) so its prevention and/or repair (discussed below) are likely critical. In general, leaf shedding represents a move toward dormancy helping deciduous plants such as grapevine escape severe levels of water deficit (Zhao et al. 2017; Volaire 2018). Although this "abandon the current season and wait it out" strategy may be effective for long-term survival, it would have severely negative effects on current season productivity in an agricultural setting.

11.4.2 Recovery and Repair

Drought stress responses such as reduced growth and/or stomatal closure are largely reversible over a short time frame. Stomatal conductance recovers rapidly when grapevines are re-watered while under moderate levels of water deficit (Hochberg et al. 2017a; Dayer et al. 2017b). However, this recovery time lengthens as the severity of the stress experienced by the vine increases (Charrier et al. 2018). Other responses such as leaf shedding can only be reversed over longer time frames. The repair (i.e., refilling) of embolized xylem vessels can take place over both short (hours to days) and long (over winter) time frames (Brodersen and McElrone 2013). Although embolism repair has been the subject of debate because of methodological artifacts leading to false conclusions (Torres-Ruiz et al. 2015), the increasing use of noninvasive imaging, especially X-ray microCT, now provides a much more robust means to examine embolism repair in situ (Brodersen et al. 2010; Knipfer et al. 2016; Hochberg et al. 2017b). Studies using these technologies confirm that grapevines are not as susceptible to embolism as previously thought and thus routine cycles of embolism formation and repair do not appear to occur on a daily basis during the growing season.

The mechanisms involved in embolism repair are still largely based on speculation. Root pressure has traditionally been invoked as a cornerstone mechanism in xylem repair across many species including grapevine (Sperry 1993; Tibbetts and Ewers 2000; Isnard and Silk 2009). MicroCT studies have associated grapevine embolism repair with root pressure (Knipfer et al. 2015; Charrier et al. 2016), and *Vitis* species differing in their ability to produce root pressure under drought exhibited corresponding abilities to refill embolized xylem vessels (Knipfer et al. 2015). Certainly overwintering in grapevine, with the significant amount of root pressure produced in spring, should facilitate significant embolism repair.

In the absence of root pressure, embolism repair is thought to involve solute loading into embolized vessels from adjacent living xylem parenchyma thus creating an osmotic driving force to facilitate vessel refilling (Brodersen and McElrone 2013). Using microCT, Brodersen et al. (2010) illustrated that vessel refilling in grapevine was achieved by water influx from the xylem parenchyma manifesting as droplets that expand until the vessels is filled. The orientation of this refilling was most often associated with ray tissues suggesting a possible role for carbohydrates in the process. Studies in other species also invoke the role of carbohydrates in the production of the osmotic gradients that could potentially drive the refilling process (Salleo et al. 2009; Nardini et al. 2011). However, it should be pointed out that refiling has only been observed in potted grapevines where soil is uniformly saturated when re-watered, a case that is almost always absent under field conditions.

There are currently no functional studies that unequivocally identify any molecular mechanism involved in embolism repair; however, numerous attempts have been made to correlate changes in gene expression with drought recovery in xylem associated tissues. Transcriptomic studies in Poplar during recovery from water deficit highlight an induction of genes involved in transport, including aquaporins and ion transporters, and carbon metabolism (Secchi and Zwieniecki 2010; Secchi et al. 2011). These findings correspond with the hypothesized mechanisms discussed above. In grapevine, a study by Chitarra et al. (2014) revealed similar changes in targeted drought, aquaporin, and carbon-related genes. Studies that combine function analyses of putative proteins involved in the repair process with noninvasive, real-time visualization of refilling are required to make firm conclusions regarding the molecular mechanisms involved in embolism repair.

11.4.3 Carry Over Effects

Since water deficits are commonly applied in viticulture, there are questions regarding their effects on crop performance over the long-term; to what extent do repeated seasonal water deficits have carry over effects on growth and/or yield, and to what extent can grapevines recover from both moderate and more severe water deficits? Some drought stress responses such as reduced growth and/or stomatal closure are largely reversible over a short time frame (i.e., within season) while others such as leaf shedding can only be reversed by overwintering.

Water deficits clearly decrease vigor and yields in the current season and sometimes can lead to carry over effects that reduce yields in the following season through negatively impacting bud fertility (Buttrose 1974; Williams and Matthews 1990). However, this appears to be dependent on the crop load (Dayer et al. 2013) suggesting an important impact of source–sink relationships and carbohydrate reserves. Several recent leaf removal studies effects on grape berry composition and starch reserves were only observed in treatments that severely reduced the source–sink ratio suggesting grapevines largely compensate for these changes (Bobeica et al. 2015; Silva et al. 2017). The compensatory capacity of grape berries to maintain normal ripening (i.e., sugar accumulation) seems especially high (Pellegrino et al. 2014). At the molecular level, Silva et al. (2017) demonstrated compensatory changes in woody tissues that increased sink strength via the upregulation of *VvSusy*, a key regulator of starch synthesis, and an increase in acid invertase activity when the source was limiting. Similar changes may be expected under water deficit where stomatal closure and decreased photosynthesis equally limit source production (discussed below).

11.5 Conclusions

Recent advances in grapevine have demonstrated that a large number of genes are involved in plant drought responses. There is strong evidence that ABA plays a key role in various aspects of metabolism in the overall response. The identification of genes that lead to the stress-induced production of ABA and the perception of this signal are important in understanding stomatal regulation under mild water deficit. However, further work is required to fully elucidate the signal transduction and transcriptional regulation of these genes under stress conditions, especially at the protein level.

Further studies are essential to determine the molecular basis of altered carbon assimilation and transport of sugars within the plant. For instance, there is still lack of knowledge about the enzymes involved in the regeneration of RuBP in the Calvin cycle and their regulation under drought, particularly for grapevine. Dormancy and the redistribution of carbon stores from season to season are also poorly understood although they likely have a cornerstone role in growth and productivity over the lifespan of a vineyard.

Often drought is accompanied by other environmental stresses such as high temperatures and high VPD that also result in oxidative stress. And like drought, scavenging of the reactive oxygen species, cell membrane stability, expression of aquaporins, and osmotic adjustment are some of the protective mechanisms that allow plants to cope with these stresses as well. Research has advanced in the identification of redox signals (e.g., hydrogen peroxide) that may regulate the energy balance of the leaf involving the expression of several genes that are linked to photosynthesis and other metabolic pathways. It is critical to understand how these different stress response pathways are integrated in grapevine and other plants.

References

Abare (2008) Drought and reduced financial returns affect wine grape growers in the Murray Valley and Barossa regions. http://www.abareconomics.com/corporate/media/2008_releases/16dec_08.html

Ahmadi SH, Andersen MN, Poulsen RT et al (2009) A quantitative approach to developing more mechanistic gas exchange models for field grown potato: a new insight into chemical and hydraulic signalling. Agric For Meteorol 149:1541–1551. https://doi.org/10.1016/j.agrformet.2009.04.009

Andersen MN (2002) Soluble invertase expression is an early target of drought stress during the critical, abortion-sensitive phase of young ovary development in maize. Plant Physiol. https://doi.org/10.1104/pp.005637

Aroca R, Vernieri P, Irigoyen JJ et al (2003) Involvement of abscisic acid in leaf and root of maize (*Zea mays* L.) in avoiding chilling-induced water stress. Plant Sci 165:671–679. https://doi.org/10.1016/S0168-9452(03)00257-7

Asada K (2006) Production and scavenging of reactive oxygen species in chloroplasts and their functions. Plant Physiol. https://doi.org/10.1104/pp.106.082040

Baker NR (2008) Chlorophyll fluorescence: a probe of photosynthesis in vivo. Annu Rev Plant Biol. https://doi.org/10.1146/annurev.arplant.59.032607.092759

Bengough AG, McKenzie BM, Hallett PD, Valentine TA (2011) Root elongation, water stress, and mechanical impedance: a review of limiting stresses and beneficial root tip traits. J Exp Bot 62:59–68. https://doi.org/10.1093/jxb/erq350

Bobeica N, Poni S, Hilbert G et al (2015) Differential responses of sugar, organic acids and anthocyanins to source-sink modulation in Cabernet Sauvignon and *Sangiovese grapevines*. Front Plant Sci 6:382. https://doi.org/10.3389/fpls.2015.00382

Boneh U, Biton I, Zheng C et al (2012) Characterization of potential ABA receptors in *Vitis vinifera*. Plant Cell Rep 31:311–321

Bota J, Flexas J, Keys AJ et al (2002) CO2/O2 specificity factor of ribulose-1,5-bisphosphate carboxylase/oxygenase in grapevines (*Vitis vinifera* L.): first in vitro determination and comparison to in vivo estimations. Vitis 41:163–168

Bota J, Medrano H, Flexas J (2004) Is photosynthesis limited by decreased Rubisco activity and RuBP content under progressive water stress? New Phytol. https://doi.org/10.1111/j.1469-8137.2004.01056.x

Bouché N, Fait A, Bouchez D et al (2003) Mitochondrial succinic-semialdehyde dehydrogenase of the γ-aminobutyrate shunt is required to restrict levels of reactive oxygen intermediates in plants. Proc Natl Acad Sci 100:6843 LP–6848

Boursiac Y, Léran S, Corratgé-Faillie C et al (2013) ABA transport and transporters. Trends Plant Sci 18:325–333. https://doi.org/10.1016/J.TPLANTS.2013.01.007

Brodersen CR, McElrone AJ (2013) Maintenance of xylem network transport capacity: a review of embolism repair in vascular plants. Front Plant Sci 4:108. https://doi.org/10.3389/fpls.2013.00108

Brodersen CR, McElrone AJ, Choat B et al (2010) The dynamics of embolism repair in xylem. In vivo visualizations using high-resolution computed tomography. Plant Physiol 154:1088–1095. https://doi.org/10.1104/pp.110.162396

Brodersen CR, McElrone AJ, Choat B et al (2013) In vivo visualizations of drought-induced embolism spread in *Vitis vinifera*. Plant Physiol 161:1820–1829. https://doi.org/10.1104/pp.112.212712

Brodribb TJ, Cochard H (2009) Hydraulic failure defines the recovery and point of death in water-stressed conifers. Plant Physiol 149:575–584. https://doi.org/10.1104/pp.108.129783

Brodribb TJ, McAdam SAM (2011) Passive origins of stomatal control in vascular plants. Science (80–) 331:582–585. https://doi.org/10.1126/science.1197985

Buckley TN, Mott KA (2002) Dynamics of stomatal water relations during the humidity response:

implications of two hypothetical mechanisms. Plant Cell Environ 25:407–419. https://doi.org/10.1046/j.0016-8025.2001.00820.x
Buttrose MS (1974) Fruitfulness in grapevines: effects of water stress. Vitis 12:299–305
Centritto M, Loreto F, Chartzoulakis K (2003) The use of low [CO2] to estimate diffusional and non-diffusional limitations of photosynthetic capacity of salt-stressed olive saplings. Plant Cell Environ. https://doi.org/10.1046/j.1365-3040.2003.00993.x
Charrier G, Torres-Ruiz JM, Badel E et al (2016) Evidence for hydraulic vulnerability segmentation and lack of xylem refilling under tension. Plant Physiol 172:1657–1668. https://doi.org/10.1104/pp.16.01079
Charrier G, Delzon S, Domec J-C et al (2018) Drought will not leave your glass empty: low risk of hydraulic failure revealed by long-term drought observations in world's top wine regions. Sci Adv 4:eaao6969. https://doi.org/10.1126/sciadv.aao6969
Chaumont F, Tyerman SD (2014) Aquaporins: highly regulated channels controlling plant water relations. Plant Physiol 164:1600–1618. https://doi.org/10.1104/pp.113.233791
Chaves MM (1991) Effects of water deficits on carbon assimilation. J Exp Bot 42:1–16
Chaves MM, Zarrouk O, Francisco R et al (2010) Grapevine under deficit irrigation: hints from physiological and molecular data. Ann Bot 105:661–676. https://doi.org/10.1093/aob/mcq030
Chaves MM, Miguel Costa J, Madeira Saibo NJ (2011) Recent advances in photosynthesis under drought and salinity. Adv Bot Res 57:49–104. https://doi.org/10.1016/B978-0-12-387692-8.00003-5
Chitarra W, Balestrini R, Vitali M et al (2014) Gene expression in vessel-associated cells upon xylem embolism repair in *Vitis vinifera* L. petioles. Planta 239:887–899. https://doi.org/10.1007/s00425-013-2017-7
Choat B, Drayton WM, Brodersen CR et al (2010) Measurement of vulnerability to water stress-induced cavitation in grapevine: a comparison of four techniques applied to a long-vesseled species. Plant Cell Environ. https://doi.org/10.1111/j.1365-3040.2010.02160.x
Christmann A, Weiler EW, Steudle E, Grill E (2007) A hydraulic signal in root-to-shoot signalling of water shortage. Plant J 52:167–174. https://doi.org/10.1111/j.1365-313X.2007.03234.x
Cirelli D, Equiza MA, Lieffers VJ, Tyree MT (2015) Populus species from diverse habitats maintain high night-time conductance under drought. Tree Physiol 36:229–242. https://doi.org/10.1093/treephys/tpv092
Comstock JP (2002) Hydraulic and chemical signalling in the control of stomatal conductance and transpiration. J Exp Bot 53:195–200
Cornic G (2000) Drought stress inhibits photosynthesis by decreasing stomatal aperture—not by affecting ATP synthesis. Trends Plant Sci 5(5):187–188. https://doi.org/10.1016/S1360-1385(00)01625-3
Cornic G, Fresneau C (2002) Photosynthetic carbon reduction and carbon oxidation cycles are the main electron sinks for photosystem II activity during a mild drought. Ann Bot. https://doi.org/10.1093/aob/mcf064
Corpas FJ, Barroso JB, del Río LA (2001) Peroxisomes as a source of reactive oxygen species and nitric oxide signal molecules in plant cells. Trends Plant Sci 6:145–150. https://doi.org/10.1016/S1360-1385(01)01898-2
Correia MJ, Pereira JS, Chaves MM et al (1995) ABA xylem concentrations determine maximum daily leaf conductance of field-grown *Vitis vinifera* L. plants. Plant Cell Environ 18:511–521. https://doi.org/10.1111/j.1365-3040.1995.tb00551.x
Costa JM, Monnet F, Jannaud D et al (2015) Open all night long: the dark side of stomatal control. Plant Physiol 167:289–294. https://doi.org/10.1104/pp.114.253369
Coupel-Ledru A, Tyerman SD, Masclef D et al (2017) Abscisic acid down-regulates hydraulic conductance of grapevine leaves in isohydric genotypes only. Plant Physiol 175:1121–1134. https://doi.org/10.1104/pp.17.00698
Cramer GR, Ergül A, Grimplet J et al (2007) Water and salinity stress in grapevines: early and late changes in transcript and metabolite profiles. Funct Integr Genom 7:111–134
Cramer GR, Van Sluyter SC, Hopper DW et al (2013) Proteomic analysis indicates massive changes in metabolism prior to the inhibition of growth and photosynthesis of grapevine (*Vitis vinifera* L.) in response to water deficit. BMC Plant Biol 13:49
Cuneo IF, Knipfer T, Brodersen CR, McElrone AJ (2016) Mechanical failure of fine root cortical cells initiates plant hydraulic decline during drought. Plant Physiol 172:1669–1678. https://doi.org/10.1104/pp.16.00923
Cutler AJ, Krochko JE (1999) Formation and breakdown of ABA. Trends Plant Sci 4:472–477. https://doi.org/10.1016/S1360-1385(99)01497-1
Damour G, Simonneau T, Cochard H, Urban L (2010) An overview of models of stomatal conductance at the leaf level. Plant Cell Environ 33:1419–1438. https://doi.org/10.1111/j.1365-3040.2010.02181.x
Davies WJ, Zhang J (1991) Root signals and the regulation of growth and development of plants in drying soil. Annu Rev Plant Physiol Plant Mol Biol 42:55–76. https://doi.org/10.1146/annurev.pp.42.060191.000415
Dayer S, Prieto JA, Galat E, Perez Peña J (2013) Carbohydrate reserve status of Malbec grapevines after several years of regulated deficit irrigation and crop load regulation. Aust J Grape Wine Res 19:422–430. https://doi.org/10.1111/ajgw.12044
Dayer S, Prieto JA, Galat E, Peña JP (2016) Leaf carbohydrate metabolism in Malbec grapevines: combined effects of regulated deficit irrigation and crop load. Aust J Grape Wine Res. https://doi.org/10.1111/ajgw.12180

Dayer S, Peña JP, Gindro K et al (2017a) Changes in leaf stomatal conductance, petiole hydraulics and vessel morphology in grapevine (*Vitis vinifera* cv. Chasselas) under different light and irrigation regimes. Funct Plant Biol 44:679. https://doi.org/10.1071/fp16041

Dayer S, Tyerman SD, Garnett T, Pagay V (2017b) Relationship between hydraulic and stomatal conductance and its regulation by root and leaf aquaporins under progressive water stress and recovery and exogenous application of ABA in *Vitis vinifera* L. 'Syrah'. Acta Hortic. https://doi.org/10.17660/actahortic.2017.1188.29

Demmig-Adams B, Adams III WW (2006) Photoprotection in an ecological context: the remarkable complexity of thermal energy dissipation. New Phytol 172 (1):11–21. https://doi.org/10.1111/j.1469-8137.2006.01835.x

Dodd IC (2005) Root-to-shoot signalling: assessing the roles of "up" in the up and down world of long-distance signalling in planta. Plant Soil 274:251–270. https://doi.org/10.1007/s11104-004-0966-0

Domec JC, Johnson DM (2012) Does homeostasis or disturbance of homeostasis in minimum leaf water potential explain the isohydric versus anisohydric behavior of *Vitis vinifera* L. cultivars? Tree Physiol 32:245–248. https://doi.org/10.1093/treephys/tps013

Dubois M, Skirycz A, Claeys H et al (2013) ETHYLENE RESPONSE FACTOR6 acts as a central regulator of leaf growth under water-limiting conditions in Arabidopsis. Plant Physiol 162:319 LP–332

During H, Dry PR (1995) Osmoregulation in water stressed roots: responses of leaf conductance and photosynthesis. Vitis 34:15–17

Endo A, Sawada Y, Takahashi H et al (2008) Drought induction of arabidopsis 9-*cis*-epoxycarotenoid dioxygenase occurs in vascular parenchyma cells. Plant Physiol 147:1984 LP–1993

Farooq M, Wahid A, Kobayashi N et al (2009) Plant drought stress: effects, mechanisms and management. Agron Sustain Dev. https://doi.org/10.1051/agro:2008021

Fiorani F, Umbach AL, Siedow JN (2005) The alternative oxidase of plant mitochondria is involved in the acclimation of shoot growth at low temperature. A study of arabidopsis AOX1a; transgenic plants. Plant Physiol 139:1795 LP–1805

Flexas J, Bota J, Escalona JM et al (2002) Effects of drought on photosynthesis in grapevines under field conditions: an evaluation of stomatal and mesophyll limitations. Funct Plant Biol. https://doi.org/10.1071/pp01119

Flexas J, Bota J, Cifre J et al (2004) Understanding down-regulation of photosynthesis under water stress: future prospects and searching for physiological tools for irrigation management. Ann Appl Biol. https://doi.org/10.1111/j.1744-7348.2004.tb00343.x

Flexas J, Ribas-Carbó M, Hanson DT et al (2006) Tobacco aquaporin NtAQP1 is involved in mesophyll conductance to CO2 in vivo. Plant J 48:427–439. https://doi.org/10.1111/j.1365-313X.2006.02879.x

Flexas J, Diaz-Espejo A, Galmés J et al (2007) Rapid variations of mesophyll conductance in response to changes in CO2 concentration around leaves. Plant Cell Environ. https://doi.org/10.1111/j.1365-3040.2007.01700.x

Gambetta GA, Manuck CM, Drucker ST et al (2012) The relationship between root hydraulics and scion vigour across Vitis rootstocks: what role do root aquaporins play? J Exp Bot 63:6445–6455. https://doi.org/10.1093/jxb/ers312

Gambetta GA, Knipfer T, Fricke W, Mcelrone AJ (2017) Aquaporins and root water uptake. In: Chaumont F, Tyermaneds SD (eds) Plant aquaporins: from transport to signaling. Springer, Cham, pp 133–153. https://doi.org/10.1007/978-3-319-49395-4_6

Gamon JA, Pearcy RW (1990) Photoinhibition in *Vitis californica*: the role of temperature during high-light treatment. Plant Physiol 92(2):487–494. https://doi.org/10.1104/pp.92.2.487

Garreaud RD et al (2017) The 2010–2015 megadrought in central Chile. Hydrol Earth Syst Sci 21:6307–6327. https://doi.org/10.5194/hess-21-6307-2017

Greer DH, Laing WA (1992) Photoinhibition of photosynthesis in intact kiwifruit (*Actinidia deliciosa*) leaves: changes in susceptibility to photoinhibition and recovery during the growth season

Gunes A, Soylemezoglu G, Inal A et al (2006) Antioxidant and stomatal responses of grapevine (*Vitis vinifera* L.) to boron toxicity. Sci Hortic (Amsterdam) 110:279–284. https://doi.org/10.1016/J.SCIENTA.2006.07.014

Haider MS, Zhang C, Kurjogi MM et al (2017) Insights into grapevine defense response against drought as revealed by biochemical, physiological and RNA-Seq analysis. Sci Rep 7:13134. https://doi.org/10.1038/s41598-017-13464-3

Havaux M (1998) Carotenoids as membrane stabilizers in chloroplasts. Trends Plant Sci 3(4):147–151. https://doi.org/10.1016/S1360-1385(98)01200-X

Hetherington AM, Woodward FI (2003) The role of stomata in sensing and driving environmental change. Nature 424:901–908. https://doi.org/10.1038/nature01843

Ho S-L (2001) Sugar coordinately and differentially regulates growth- and stress-related gene expression via a complex signal transduction network and multiple control mechanisms. Plant Physiol. https://doi.org/10.1104/pp.125.2.877

Hochberg U, Degu A, Toubiana D et al (2013) Metabolite profiling and network analysis reveal coordinated changes in grapevine water stress response. BMC Plant Biol 13:184. https://doi.org/10.1186/1471-2229-13-184

Hochberg U, Albuquerque C, Rachmilevitch S et al (2016) Grapevine petioles are more sensitive to drought induced embolism than stems: evidence from *in vivo* MRI and microcomputed tomography observations of hydraulic vulnerability segmentation. Plant Cell Environ 39:1886–1894. https://doi.org/10.1111/pce.12688

Hochberg U, Bonel AG, David-Schwartz R et al (2017a) Grapevine acclimation to water deficit: the adjustment of stomatal and hydraulic conductance differs from petiole embolism vulnerability. Planta 245:1091–1104. https://doi.org/10.1007/s00425-017-2662-3

Hochberg U, Windt CW, Ponomarenko A et al (2017b) Stomatal closure, basal leaf embolism, and shedding protect the hydraulic integrity of grape stems. Plant Physiol 174:764–775. https://doi.org/10.1104/pp.16.01816

Hoekstra FA, Golovina EA, Buitink J (2001) Mechanisms of plant desiccation tolerance. Trends Plant Sci. https://doi.org/10.1016/s1360-1385(01)02052-0

Holbrook NM, Shashidhar VR, James RA, Munns R (2002) Stomatal control in tomato with ABA-deficient roots: response of grafted plants to soil drying. J Exp Bot 53:1503–1514

Holzapfel BP, Smith JP, Field SK, Hardie WJ (2010) Dynamics of carbohydrate reserves in cultivated grapevines. In: Janick J (ed) Horticultural reviews, vol 37. https://doi.org/10.1002/9780470543672.ch3

Hopper DW, Ghan R, Schlauch KA, Cramer GR (2016) Transcriptomic network analyses of leaf dehydration responses identify highly connected ABA and ethylene signaling hubs in three grapevine species differing in drought tolerance. BMC Plant Biol 16:118. https://doi.org/10.1186/s12870-016-0804-6

Hose E, Steudle E, Hartung W (2000) Abscisic acid and hydraulic conductivity of maize roots: a study using cell- and root-pressure probes. Planta 211:874–882. https://doi.org/10.1007/s004250000412

Hsiao TC (1973) Plant responses to water stress. Annu Rev Plant Physiol 24:519–570

Hukin D, Cochard H, Dreyer E et al (2005) Cavitation vulnerability in roots and shoots: does *Populus euphratica* Oliv., a poplar from arid areas of Central Asia, differ from other poplar species? J Exp Bot 56:2003–2010. https://doi.org/10.1093/jxb/eri198

IPCC (2014) Intergovernmental Panel on Climate Change- Fifth Assessment Report (AR5): The Synthesis Report

Isnard S, Silk WK (2009) Moving with climbing plants from Charles Darwin's time into the 21st century. Am J Bot 96:1205–1221. https://doi.org/10.3732/ajb.0900045

Kawase M, Hanba YT, Katsuhara M (2013) The photosynthetic response of tobacco plants overexpressing ice plant aquaporin McMIPB to a soil water deficit and high vapor pressure deficit. J Plant Res. https://doi.org/10.1007/s10265-013-0548-4

Keller M (2015) The science of grapevines. Elsevier Inc., Amsterdam. https://doi.org/10.1016/C2013-0-06797-7

Knipfer T, Eustis A, Brodersen CR et al (2015) Grapevine species from varied native habitats exhibit differences in embolism formation/repair associated with leaf gas exchange and root pressure. Plant Cell Environ 38:1503–1513. https://doi.org/10.1111/pce.12497

Knipfer T, Cuneo IF, Brodersen CR, McElrone AJ (2016) In situ visualization of the dynamics in xylem embolism formation and removal in the absence of root pressure: a study on excised grapevine stems. Plant Physiol 171:1024–1036. https://doi.org/10.1104/pp.16.00136

Koca H, Ozdemir F, Turkan I (2006) Effect of salt stress on lipid peroxidation and superoxide dismutase and peroxidase activities of *Lycopersicon esculentum* and *L. pennellii*. Biol Plant. https://doi.org/10.1007/s10535-006-0121-2

Koyama T, Nii H, Mitsuda N et al (2013) A regulatory cascade involving class II ETHYLENE RESPONSE FACTOR transcriptional repressors operates in the progression of leaf senescence. Plant Physiol 162:991 LP–1005

Kudoyarova G, Veselova S, Hartung W et al (2011) Involvement of root ABA and hydraulic conductivity in the control of water relations in wheat plants exposed to increased evaporative demand. Planta 233:87–94. https://doi.org/10.1007/s00425-010-1286-7

Laur J, Hacke UG (2013) Transpirational demand affects aquaporin expression in poplar roots. J Exp Bot 64:2283–2293. https://doi.org/10.1093/jxb/ert096

Lawlor DW, Tezara W (2009) Causes of decreased photosynthetic rate and metabolic capacity in water-deficient leaf cells: a critical evaluation of mechanisms and integration of processes. Ann Bot 103:561–579. https://doi.org/10.1093/aob/mcn244

Li S, Feifel M, Karimi Z et al (2015) Leaf gas exchange performance and the lethal water potential of five European species during drought. Tree Physiol 36: tpv117. https://doi.org/10.1093/treephys/tpv117

Lorenz WW, Alba R, Yu Y-S et al (2011) Microarray analysis and scale-free gene networks identify candidate regulators in drought-stressed roots of loblolly pine (*P. taeda* L.). BMC Genom 12:264. https://doi.org/10.1186/1471-2164-12-264

Lovisolo C, Schubert A (2006) Mercury hinders recovery of shoot hydraulic conductivity during grapevine rehydration: evidence from a whole-plant approach. New Phytol 172:469–478. https://doi.org/10.1111/j.1469-8137.2006.01852.x

Lovisolo C, Perrone I, Hartung W, Schubert A (2008) An abscisic acid-related reduced transpiration promotes gradual embolism repair when grapevines are rehydrated after drought. New Phytol 180:642–651. https://doi.org/10.1111/j.1469-8137.2008.02592.x

Lovisolo C, Perrone I, Carra A et al (2010) Drought-induced changes in development and function of grapevine (*Vitis* spp.) organs and in their hydraulic and non-hydraulic interactions at the whole-plant level: a physiological and molecular update. Funct Plant Biol 37:98–116. https://doi.org/10.1071/FP09191

Lumba S, Toh S, Handfield L-F et al (2014) A mesoscale abscisic acid hormone interactome reveals a dynamic signaling landscape in Arabidopsis. Dev Cell 29:360–372. https://doi.org/10.1016/J.DEVCEL.2014.04.004

Ma Y, Szostkiewicz I, Korte A et al (2009) Regulators of PP2C phosphatase activity function as abscisic acid sensors. Science (80–) 324:1064 LP–1068

Maroco JP, Rodrigues ML, Lopes C, Chaves MM (2002) Limitations to leaf photosynthesis in field-grown grapevine under drought—metabolic and modelling approaches. Funct Plant Biol 29:451–459. https://doi.org/10.1071/PP01040

Martre P, Cochard H, Durand J-L (2001) Hydraulic architecture and water flow in growing grass tillers (*Festuca arundinacea* Schreb.). Plant Cell Environ 24:65–76. https://doi.org/10.1046/j.1365-3040.2001.00657.x

Masle J, Gilmore SR, Farquhar GD (2005) The ERECTA gene regulates plant transpiration efficiency in Arabidopsis. Nature 436:866–870. https://doi.org/10.1038/nature03835

McAdam SAM, Brodribb TJ (2012) Stomatal innovation and the rise of seed plants. Ecol Lett 15:1–8. https://doi.org/10.1111/j.1461-0248.2011.01700.x

McAdam SAM, Brodribb TJ (2015) The evolution of mechanisms driving the stomatal response to vapor pressure deficit. Plant Physiol 167:833–843. https://doi.org/10.1104/pp.114.252940

McAdam SAM, Brodribb TJ (2016) Linking turgor with ABA biosynthesis: implications for stomatal responses to vapor pressure deficit across land plants. Plant Physiol 171:2008–2016. https://doi.org/10.1104/pp.16.00380

McAdam SAM, Brodribb TJ, Ross JJ (2016a) Shoot-derived abscisic acid promotes root growth. Plant Cell Environ 39:652–659. https://doi.org/10.1111/pce.12669

McAdam SAM, Sussmilch FC, Brodribb TJ (2016b) Stomatal responses to vapour pressure deficit are regulated by high speed gene expression in angiosperms. Plant Cell Environ 39:485–491. https://doi.org/10.1111/pce.12633

McDowell N, Pockman WT, Allen CD et al (2008) Mechanisms of plant survival and mortality during drought: why do some plants survive while others succumb to drought? New Phytol 178:719–739. https://doi.org/10.1111/j.1469-8137.2008.02436.x

Medici A, Laloi M, Atanassova R (2014) Profiling of sugar transporter genes in grapevine coping with water deficit. FEBS Lett 588:3989–3997. https://doi.org/10.1016/J.FEBSLET.2014.09.016

Medrano H, Escalona JM, Bota J et al (2002) Regulation of photosynthesis of C3 plants in response to progressive drought: Stomatal conductance as a reference parameter. Ann Bot 89:895–905. https://doi.org/10.1093/aob/mcf079

Medrano H, Escalona JM, Cifre J et al (2003) A ten year study on the physiology of two Spanish grapevine cultivars under field conditions: effects of water availability from leaf photosynthesis to grape yield and quality. Funct Plant Biol 30:607–619

Mekonnen DW, Flügge U-I, Ludewig F (2016) Gamma-aminobutyric acid depletion affects stomata closure and drought tolerance of Arabidopsis thaliana. Plant Sci 245:25–34. https://doi.org/10.1016/J.PLANTSCI.2016.01.005

Mittler R (2002) Oxidative stress, antioxidants and stress tolerance. Trends Plant Sci 7(9):405–410. https://doi.org/10.1016/S1360-1385(02)02312-9

Moreno I, Martín R, Castresana C (2005) Arabidopsis SHMT1, a serine hydroxymethyltransferase that functions in the photorespiratory pathway influences resistance to biotic and abiotic stress. Plant J 41:451–463. https://doi.org/10.1111/j.1365-313X.2004.02311.x

Mote P, Hamlet A, Salathe E (2008) Has spring snowpack declined in the Washington Cascades? Hydrol Earth Syst Sci 12(1):193–206

Munns R, Cramer GR (1996) Is coordination of leaf and root growth mediated by abscisic acid? Opinion. Plant Soil 185:33–49. https://doi.org/10.1007/BF02257563

Naor A, Wample RL (1994) Gas exchange and water relations of field-grown concord (*Vitis labruscana* Bailey) grapevines. Am J Enol Vitic 45:333–337

Nardini A, Lo Gullo MA, Salleo S (2011) Refilling embolized xylem conduits: Is it a matter of phloem unloading? Plant Sci 180:604–611. https://doi.org/10.1016/j.plantsci.2010.12.011

Niyogi KK (1999) Photoprotection revisited: genetic and molecular approaches. Annu Rev Plant Biol 50 (1):333–359. https://doi.org/10.1146/annurev.arplant.50.1.333

North GB, Martre P, Nobel PS (2004) Aquaporins account for variations in hydraulic conductance for metabolically active root regions of Agave deserti in wet, dry, and rewetted soil. Plant Cell Environ 27:219–228. https://doi.org/10.1111/j.1365-3040.2003.01137.x

Pantin F, Monnet F, Jannaud D et al (2013) Rapid report the dual effect of abscisic acid on stomata. New Phytol 197:65–72. https://doi.org/10.1111/nph.12013

Parent B, Hachez C, Redondo E et al (2009) Drought and abscisic acid effects on aquaporin content translate into changes in hydraulic conductivity and leaf growth rate: a trans-scale approach. Plant Physiol 149:2000–2012. https://doi.org/10.1104/pp.108.130682

Park S-Y, Fung P, Nishimura N et al (2009) Abscisic acid inhibits type 2C protein phosphatases via the PYR/PYL family of START proteins. Science (80–) 324:1068 LP–1071

Pastenes C, Villalobos L, Ríos N et al (2014) Carbon partitioning to berries in water stressed grapevines: the role of active transport in leaves and fruits. Environ Exp Bot 107:154–166. https://doi.org/10.1016/j.envexpbot.2014.06.009

Patakas A, Nikolaou N, Zioziou E et al (2002) The role of organic solute and ion accumulation in osmotic adjustment in drought-stressed grapevines. Plant Sci 163:361–367

Peccoux A, Loveys B, Zhu J et al (2017) Dissecting the rootstock control of scion transpiration using model-assisted analyses in grapevine. Tree Physiol. https://doi.org/10.1093/treephys/tpx153

Pellegrino A, Clingeleffer P, Cooley N, Walker R (2014) Management practices impact vine carbohydrate status

to a greater extent than vine productivity. Front Plant Sci 5:283. https://doi.org/10.3389/fpls.2014.00283
Peng Z, Lu Q, Verma DPS (1996) Reciprocal regulation of delta 1-pyrroline-5-carboxylate synthetase and proline dehydrogenase genes controls proline levels during and after osmotic stress in plants. Mol Gen Genet MGG 253:334–341. https://doi.org/10.1007/PL00008600
Perez-Martin A, Flexas J, Ribas-Carbó M et al (2009) Interactive effects of soil water deficit and air vapour pressure deficit on mesophyll conductance to CO2 in *Vitis vinifera* and *Olea europaea*. J Exp Bot 60:2391–2405. https://doi.org/10.1093/jxb/erp145
Perez-Martin A, Michelazzo C, Torres-Ruiz JM et al (2014) Regulation of photosynthesis and stomatal and mesophyll conductance under water stress and recovery in olive trees: correlation with gene expression of carbonic anhydrase and aquaporins. J Exp Bot. https://doi.org/10.1093/jxb/eru160
Pfannschmidt T, Schütze K, Fey V et al (2003) Chloroplast redox control of nuclear gene expression—a new class of plastid signals in interorganellar communication. Antioxid Redox Signal. https://doi.org/10.1089/152308603321223586
Poormohammad Kiani S, Maury P, Sarrafi A, Grieu P (2008) QTL analysis of chlorophyll fluorescence parameters in sunflower (*Helianthus annuus* L.) under well-watered and water-stressed conditions. Plant Sci. https://doi.org/10.1016/j.plantsci.2008.06.002
Pou A, Medrano H, Flexas J, Tyerman SD (2013) A putative role for TIP and PIP aquaporins in dynamics of leaf hydraulic and stomatal conductances in grapevine under water stress and re-watering. Plant Cell Environ 36:828–843. https://doi.org/10.1111/pce.12019
Premachandra GS, Saneoka H, Fujita K, Ogata S (1990) Cell membrane stability and leaf water relations as affected by phosphorus nutrition under water stress in maize. Soil Sci Plant Nutr. https://doi.org/10.1080/00380768.1990.10416803
Prinsi B, Negri AS, Failla O et al (2018) Root proteomic and metabolic analyses reveal specific responses to drought stress in differently tolerant grapevine rootstocks. BMC Plant Biol 18:126. https://doi.org/10.1186/s12870-018-1343-0
Quartacci MF, Glisić O, Stevanović B, Navari-Izzo F (2002) Plasma membrane lipids in the resurrection plant *Ramonda serbica* following dehydration and rehydration. J Exp Bot 53:2159–2166
Quick WP, Chaves MM, Wendler R, David M, Rodrigues ML, Passaharinho JA, Pereira JS, Adcock MD, Leegood RC, Stitt M (1992) The effect of water stress on photosynthetic carbon metabolism in four species grown under field conditions. Plant Cell Environ 15(1):25–35. https://doi.org/10.1111/j.1365-3040.1992.tb01455.x
Ramesh SA, Tyerman SD, Xu B et al (2015) GABA signalling modulates plant growth by directly regulating the activity of plant-specific anion transporters. Nat Commun 6:1–10. https://doi.org/10.1038/ncomms8879
Reddy AR, Chaitanya KV, Vivekanandan M (2004) Drought-induced responses of photosynthesis and antioxidant metabolism in higher plants. J Plant Physiol 161(11):1189–1202.https://doi.org/10.1016/j.jplph.2004.01.013
Reynolds M, Tuberosa R (2008) Translational research impacting on crop productivity in drought-prone environments. Curr Opin Plant Biol 11:171–179. https://doi.org/10.1016/j.pbi.2008.02.005
Rossdeutsch L, Edwards E, Cookson SJ et al (2016) ABA-mediated responses to water deficit separate grapevine genotypes by their genetic background. BMC Plant Biol 16:91. https://doi.org/10.1186/s12870-016-0778-4
Sakurai-Ishikawa J, Murai-Hatano M, Hayashi H et al (2011) Transpiration from shoots triggers diurnal changes in root aquaporin expression. Plant Cell Environ 34:1150–1163. https://doi.org/10.1111/j.1365-3040.2011.02313.x
Saliendra NZ, Sperry J, Comstock JP (1995) Influence of leaf water status on stomatal response to humidity, hydraulic conductance, and soil drought in *Betula occidentalis*. Planta 196:357–366
Salleo S, Trifilò P, Esposito S et al (2009) Starch-to-sugar conversion in wood parenchyma of field-growing *Laurus nobilis* plants: a component of the signal pathway for embolism repair? Funct Plant Biol 36:815. https://doi.org/10.1071/FP09103
Schroeder JI, Allen GJ, Hugouvieux V et al (2001) Guard cell signal transduction. Annu Rev Plant Physiol Plant Mol Biol 52:627–658. https://doi.org/10.1146/annurev.arplant.52.1.627
Schultz HR (2003) Differences in hydraulic architecture account for near- isohydric and anisohydric behaviour of two eld-grown. Plant Cell Environ 26:1393–1406. https://doi.org/10.1046/j.1365-3040.2003.01064.x
Schultz HR, Matthews MA (1988) Vegetative growth distribution during water deficits in *Vitis vinifera* L. Aust J Plant Physiol 15:641–656
Schultz HR, Matthews MA (1993) Growth, osmotic adjustment, and cell-wall mechanics of expanding grape leaves during water deficits. Crop Sci 33:287. https://doi.org/10.2135/cropsci1993.0011183X003300020015x
Scoffoni C, Albuquerque C, Brodersen CR et al (2017a) Outside-xylem vulnerability, not xylem embolism, controls leaf hydraulic decline during dehydration. Plant Physiol 173:1197–1210. https://doi.org/10.1104/pp.16.01643
Scoffoni C, Sack L, Ort D (2017b) The causes and consequences of leaf hydraulic decline with dehydration. J Exp Bot 68:4479–4496. https://doi.org/10.1093/jxb/erx252
Secchi F, Zwieniecki MA (2010) Patterns of PIP gene expression in *Populus trichocarpa* during recovery from xylem embolism suggest a major role for the PIP1 aquaporin subfamily as moderators of refilling

process. Plant Cell Environ 33:1285–1297. https://doi.org/10.1111/j.1365-3040.2010.02147.x
Secchi F, Gilbert ME, Zwieniecki MA (2011) Transcriptome response to embolism formation in stems of *Populus trichocarpa* provides insight into signaling and the biology of refilling. Plant Physiol 157:1419–1429. https://doi.org/10.1104/pp.111.185124
Serraj R, Shelp BJ, Sinclair TR (1998) Accumulation of gamma-aminobutyric acid in nodulated soybean in response to drought stress. Physiol Plant 102:79–86. https://doi.org/10.1034/j.1399-3054.1998.1020111.x
Sharp RE, LeNoble ME, Else MA et al (2000) Endogenous ABA maintains shoot growth in tomato independently of effects on plant water balance: evidence for an interaction with ethylene. J Exp Bot 51:1575–1584. https://doi.org/10.1093/jexbot/51.350.1575
Shatil-Cohen A, Attia Z, Moshelion M (2011) Bundle-sheath cell regulation of xylem-mesophyll water transport via aquaporins under drought stress: a target of xylem-borne ABA? Plant J 67:72–80. https://doi.org/10.1111/j.1365-313X.2011.04576.x
Siemens JA, Zwiazek JJ (2003) Effects of water deficit stress and recovery on the root water relations of trembling aspen (*Populus tremuloides*) seedlings. Plant Sci 165:113–120. https://doi.org/10.1016/S0168-9452(03)00149-3
Silva A, Noronha H, Dai Z et al (2017) Low source–sink ratio reduces reserve starch in grapevine woody canes and modulates sugar transport and metabolism at transcriptional and enzyme activity levels. Planta 246:525–535. https://doi.org/10.1007/s00425-017-2708-6
Soar CJ, Spei J, Maffei SM et al (2006) Grape vine varieties Shiraz and Grenache differ in their stomatal response to VPD: apparent links with ABA physiology and gene expression in leaf tissue. Aust J Grape Wine Res 12:2–12. https://doi.org/10.1111/j.1755-0238.2006.tb00038.x
Sobeih WY, Dodd IC, Bacon MA et al (2004) Long-distance signals regulating stomatal conductance and leaf growth in tomato (*Lycopersicon esculentum*) plants subjected to partial root-zone drying. J Exp Bot 55:2353–2363. https://doi.org/10.1093/jxb/erh204
Sousa TA, Oliveira MT, Pereira JM (2006) Physiological indicators of plant water status of irrigated and non-irrigated grapevines grown in a low rainfall area of Portugal. Plant Soil 282:127–134. https://doi.org/10.1007/s11104-005-5374-6
Speirs J, Binney A, Collins M et al (2013) Expression of ABA synthesis and metabolism genes under different irrigation strategies and atmospheric VPDs is associated with stomatal conductance in grapevine (*Vitis vinifera* L. cv Cabernet Sauvignon). J Exp Bot 64:1907–1916. https://doi.org/10.1093/jxb/ert052
Sperry J (1993) Winter xylem embolism and spring recovery in *Betula cordifolia*, *Fagus grandifolia*, *Abies balsamea* and *Picea rubens*. In: Borghetti M, Grace J, Raschi A (eds) Water transport in plants under climatic stress. Cambridge University Press, Cambridge, pp 86–98
Tardieu F, Davies WJ (1993) Integration of hydraulic and chemical signalling in the control of stomatal conductance and water status of droughted plants. Plant Cell Environ 16:341–349. https://doi.org/10.1111/j.1365-3040.1993.tb00880.x
Tardieu F, Simonneau T (1998) Variability among species of stomatal control under fluctuating soil water status and evaporative demand: modelling isohydric and anisohydric behaviours. J Exp Bot 49:419–432. https://doi.org/10.1093/jxb/49.Special_Issue.419
Tezara W, Mitchell VJ, Driscoll SD, Lawlor DW (1999) Water stress inhibits plant photosynthesis by decreasing coupling factor and ATP. Nature 401:914–917. https://doi.org/10.1038/44842
Thrupp LA, Browde J, Francioni L, Jordan A (eds) (2008) Reducing risks through sustainable winegrowing: a grower's guide. California Sustainable Winegrowing Alliance
Tibbetts TJ, Ewers FW (2000) Root pressure and specific conductivity in temperate lianas: exotic *Celastrus orbiculatus* (Celastraceae) vs. native *Vitis riparia* (Vitaceae). Am J Bot 87:1272–1278
Tombesi S, Nardini A, Frioni T et al (2015) Stomatal closure is induced by hydraulic signals and maintained by ABA in drought-stressed grapevine. Sci Rep 5:12449. https://doi.org/10.1038/srep12449
Torres-Ruiz JM, Jansen S, Choat B et al (2015) Direct x-ray microtomography observation confirms the induction of embolism upon xylem cutting under tension. Plant Physiol 167:40–43. https://doi.org/10.1104/pp.114.249706
Umbach AL, Fiorani F, Siedow JN (2005) Characterization of transformed arabidopsis with altered alternative oxidase levels and analysis of effects on reactive oxygen species in tissue. Plant Physiol 139:1806 LP–1820
Urli M, Porte AJ, Cochard H et al (2013) Xylem embolism threshold for catastrophic hydraulic failure in angiosperm trees. Tree Physiol 33:672–683. https://doi.org/10.1093/treephys/tpt030
Vandeleur RK, Mayo G, Shelden MC et al (2009) The role of plasma membrane intrinsic protein aquaporins in water transport through roots: diurnal and drought stress responses reveal different strategies between isohydric and anisohydric cultivars of grapevine. Plant Physiol 149:445–460. https://doi.org/10.1104/pp.108.128645
Vandeleur RK, Sullivan W, Athman A et al (2014) Rapid shoot-to-root signalling regulates root hydraulic conductance via aquaporins. Plant Cell Environ 37:520–538. https://doi.org/10.1111/pce.12175
Volaire F (2018) A unified framework of plant adaptive strategies to drought: crossing scales and disciplines. Glob Chang Biol. https://doi.org/10.1111/gcb.14062
Walters RG, Horton P (1993) Theoretical assessment of alternative mechanisms for non-photochemical quenching of PS II fluorescence in barley leaves. Photosynth Res. https://doi.org/10.1007/bf00016277
Wan X, Zwiazek JJ (2001) Root water flow and leaf stomatal conductance in aspen (Populus tremuloides)

seedlings treated with abscisic acid. Planta 213:741–747. https://doi.org/10.1007/s004250100547

Wang P, Song CP (2008) Guard-cell signalling for hydrogen peroxide and abscisic acid. New Phytol 178(4):703–718. https://doi.org/10.1111/j.1469-8137.2008.02431.x

Wilkinson S (1999) PH as a stress signal. J Plant Growth Regul 28:87–99. https://doi.org/10.1023/A:1006203715640

Wilkinson S, Bacon MA, Davies WJ (2007) Nitrate signalling to stomata and growing leaves: interactions with soil drying, ABA, and xylem sap pH in maize. J Exp Bot 58:1705–1716. https://doi.org/10.1093/jxb/erm021

Wilkinson S, Davies WJ (2002) ABA-based chemical signalling: the co-ordination of responses to stress in plants. Plant Cell Environ 25(2):195–210. https://doi.org/10.1046/j.0016-8025.2001.00824.x

Williams LE, Matthews MA (1990) Grapevine. In: Stewart BA, Nielsen DR (eds) Irrigation of agricultural crops. ASA-CSSA-SSSA, Madison, pp 1019–1055

Wormit A, Trentmann O, Feifer I et al (2006) Molecular identification and physiological characterization of a novel monosaccharide transporter from arabidopsis involved in vacuolar sugar transport. Plant Cell Online 18:3476–3490. https://doi.org/10.1105/tpc.106.047290

Yan J, Tsuichihara N, Etoh T, Iwai S (2007) Reactive oxygen species and nitric oxide are involved in ABA inhibition of stomatal opening. Plant Cell Environ. https://doi.org/10.1111/j.1365-3040.2007.01711.x

Zhang T, Chen S, Harmon AC (2014) Protein phosphorylation in stomatal movement. Plant Signal Behav 9: e972845. https://doi.org/10.4161/15592316.2014.972845

Zhao Y, Gao J, Im Kim J et al (2017) Control of plant water use by ABA induction of senescence and dormancy: an overlooked lesson from evolution. Plant Cell Physiol 58:1319–1327. https://doi.org/10.1093/pcp/pcx086

Zhu J, Dai Z, Vivin P et al (2017) A 3-D functional–structural grapevine model that couples the dynamics of water transport with leaf gas exchange. Ann Bot. https://doi.org/10.1093/aob/mcx141

Zhu J, Génard M, Poni S, Gambetta GA, Vivin P, Vercambre G, Trought MCT, Ollat N, Delrot S, Dai Z (2018) Modelling grape growth in relation to whole-plant carbon and water fluxes. J Exp Bot 70:2505–2521. https://doi.org/10.1093/jxb/ery367

Zimmerman MH (1983) Xylem structure and the ascent of sap. Springer, New York, 143p

The Genomics of Grape Berry Ripening

12

Rachele Falchi, Darren C. J. Wong, Yifan Yan, Stefania Savoi, Gregory A. Gambetta and Simone D. Castellarin

Abstract

Because of their economic and cultural importance, grapes are arguably the most studied fruit crop and are considered a model system for research on non-climacteric fruits. The sequencing of the grapevine genome has led to major discoveries that have increased our understanding of the molecular regulation of fruit ripening and berry metabolism, and how the environment and viticultural practices affect berry physiology. This chapter reviews the most recent studies on the molecular and metabolic pathways associated with grape berry ripening including the pathways involved in berry growth and softening, and sugar, organic acid, phenolic, and aroma accumulation. The role of hormones and hormone crosstalk, as well as a compendium of the most recent research on transcription factors (TFs) and non-coding RNAs are presented.

R. Falchi
Department of Agricultural, Food, Environmental and Animal Sciences, University of Udine, Via Delle Scienze 206, 33100 Udine, Italy
e-mail: rachele.falchi@uniud.it

R. Falchi · G. A. Gambetta
Bordeaux Science Agro, Institut des Sciences de la Vigne et du Vin, Ecophysiologie et Génomique Fonctionnelle de la Vigne, UMR 1287, 33140 Villenave d'Ornon, France
e-mail: gregory.gambetta@agro-bordeaux.fr

D. C. J. Wong · Y. Yan · S. D. Castellarin (✉)
Wine Research Centre, The University of British Columbia, 2205 East Mall, Vancouver, BC V6T 1Z4, Canada
e-mail: simone.castellarin@ubc.ca

D. C. J. Wong
e-mail: wongdcj@gmail.com

Y. Yan
e-mail: evelineyan@foxmail.com

S. Savoi
AGAP, CIRAD, INRA, Montpellier SupAgro, University of Montpellier, 2 Place Pierre Viala, 34060 Montpellier, France
e-mail: savoi.stefania@gmail.com

12.1 Introduction: General Physiological Aspects of Ripening

Grape berry growth follows a double-sigmoid pattern where two rapid phases of growth are interrupted by "lag" during which there is little or no growth (Matthews and Shackel 2005). The first growth stage (I) begins at flowering (i.e., anthesis) and continues until the lag stage (II), while the start of the final growth stage (III) is coincident with the onset of ripening, or veraison (Fig. 12.1). Stage I growth results from both cell division and cell expansion, but stage III growth results exclusively from expansion (Coombe 1976; Ojeda et al. 1999). The transition from stage II to stage III is abrupt (i.e., veraison) in

D. Cantu and M. A. Walker (eds.), *The Grape Genome*, Compendium of Plant Genomes, https://doi.org/10.1007/978-3-030-18601-2_12

Fig. 12.1 *Zinfandel grape (Vitis vinifera L.) clusters at the onset of ripening (i.e., veraison).* The timing of veraison is heterogeneous among berries of the same cluster and clusters of the same vine. In the picture, some berries have just begun ripening (light pink), whereas others are still green

individual berries. In viticulture, veraison is regarded as a critical moment because, in addition to the resumption of growth, numerous ripening processes begin, including softening, rapid sugar accumulation, and most conspicuously a change in color in red grape varieties.

Ripening is a critical stage for determining grape and wine quality and has major implications for the economic value of the crop. The grape berry is a non-climacteric fruit, which means that ripening is not related to, or modulated by, a burst of respiration and ethylene as in climacteric fruits such as tomato or apple (Coombe 1976; Gapper et al. 2013). In fact, the onset of ripening was originally thought to be a coordinated process where a multitude of physiological changes (softening, sugar accumulation, increase in ABA, and color development) were coincident and preceded the resumption of growth by several days (Coombe and Bishop 1980; Coombe 1992). More recently, studies have delimited the earliest events at the onset of ripening: softening, the associated decreases in cell turgor, and increases in ABA concentration (Thomas et al. 2006; Wada et al. 2009; Castellarin et al. 2016). Increases in sugar concentration and color development appear to occur only later, when the firmness of the berry has already decreased dramatically and the ABA concentration has further increased (Castellarin et al. 2016). Besides ABA, other hormones such as brassinosteroids and ethylene are involved in the ripening process, as well as sugars, which affect the synthesis of anthocyanins (Symons et al. 2006; Hayes et al. 2007; Chervin et al. 2008; Davies and Böttcher 2009; Dai et al. 2013). Auxins—normally accumulated at early stages of berry development—act as negative modulators of the ripening process, and their deactivation is necessary for ripening to begin (Böttcher et al. 2010, 2012a; Gouthu and Deluc 2015).

Sugars are one of the major metabolites that accumulate in the grape berry during ripening. Other compounds that accumulate during ripening are flavonols, which protect the berry from UV light, anthocyanins which determine the pink/red/blue coloration of red grape varieties, and several volatile organic compounds (VOCs), such as norisoprenoids, monoterpenes, thiols, or their conjugated precursors (Adams 2006; Teixeira et al. 2013; Robinson et al. 2014a, b). These VOCs determine the aroma of grapes, juices, and wines, particularly when chemical changes associated with acid and enzymatic modifications of conjugated precursors occur during fermentation and wine aging.

Many key compounds for fruit and wine quality are synthesized before veraison and normally decrease in concentration during the ripening period. This is the case for organic acids, hydroxycinnamates, tannins, and methoxypyrazines. The two major organic acids accumulated in the grape berry, tartaric and malic acid (Kliewer 1966; Kliewer et al. 1967; Shiraishi et al. 2010), strongly affect juice and wine pH and contribute to the quality (freshness and sourness notes) and longevity of wine. Phenolic compounds such as hydroxycinnamates and tannins confer bitterness and astringency to juices and wines (Teixeira et al. 2013). Finally, methoxypyrazines impart the sensory characteristics of bell pepper, asparagus, or pea to grapes and wines. These aromas can be perceived as good or bad depending on variety and wine style (Robinson et al. 2014a, b).

12.2 Berry Growth and Softening

12.2.1 Cell Division and Expansion

Final berry size dictates in large part yield, and thus genetic and molecular studies focused on understanding the mechanisms controlling rates of cell division and expansion are of agronomic interest. Transcriptomic studies highlight the transition from cell division driven growth, during early stage I, to cell expansion driven growth, later during stage I and stage III (Deluc et al. 2007). To date, very few cornerstone regulators of grape berry size have been identified. The fleshless berry (*flb*) mutation, originally a somatic variant and later used in crosses, exhibits profound effects on fruit set and/or fruit size depending on the meristem cell layers affected (Fernandez et al. 2006a, b). Follow-up studies identified that the mutation results from mis-expression of a PISTILLATA-like MADS-box transcription factor, *VviPI* (Fernandez et al. 2013). Chialva et al. (2016) identified three potential genes involved in cell division during stage I. Members of the grape AP2/ERF transcription factor family, AINTEGUMENTA (ANT) and AINTEGUMENTA-like (AIL), were differentially expressed across different genotypes that varied in ovary size and cell number. One candidate, in particular, *VviANT1*, co-localizes with previously identified QTLs for berry size in both table and wine grapes (Doligez et al. 2002; Cabezas et al. 2006; Chialva et al. 2016).

Later in stage I, and during stage III, berry growth results from cell expansion. Cell expansion is driven by cell turgor pressure, and the rate of expansion is determined by cell wall extensibility (i.e., the yield threshold; Cosgrove 2005). Therefore, expansive growth will be modulated through a combination of processes that affect turgor, such as solute accumulation, and processes that affect cell wall extensibility and involve cell wall modifying enzymes (Matthews and Shackel 2005). During stage I, there is evidence that both processes indeed contribute to growth. Water deficits reduce berry growth, resulting largely from decreases in berry turgor pressure (Thomas et al. 2006). At the same time, expression analyses during stage I across table grape genotypes with contrasting rates of growth highlighted differences in many genes encoding cell wall modifying enzymes (Muñoz-Espinoza et al. 2016).

Grape berry cell turgor is high during stage I, but decreases during stage II, and reaches very low levels at the onset of ripening (Thomas et al. 2006; Wada et al. 2009; Castellarin et al. 2016). This decrease in turgor prior to the onset of ripening is thought to contribute to softening (discussed below), but it creates a conundrum regarding the resumption of growth that occurs at the same time. Extremely low turgor requires a corresponding decrease in the cell wall yield threshold in order for rapid expansive growth to resume. In fact, numerous studies have concluded that the resumption of growth at the onset of ripening corresponds to the upregulation of many genes encoding cell wall modifying enzymes (Nunan et al. 2001; Deluc et al. 2007; Schlosser et al. 2008; Castellarin et al. 2016). Nicolas et al. (2013) identified a basic helix–loop–helix transcription factor, VviCEB1, that positively regulates grape berry size through enhanced cell expansion, and its action was confirmed through ectopic expression in *Arabidopsis* and tobacco (Lim et al. 2018). *VviCEB1* overexpression led to the induction of numerous genes encoding cell wall modification enzymes, which suggests a possible role for these enzymes in changing the yield threshold to modulate cell expansion (Nicolas et al. 2013). During berry development, *VviCEB1* expression increases throughout stage I, peaks at the onset of ripening, and remains high during stage III, consistent with the period of expansive berry growth.

Stage III berry growth is peculiar because grape berries are largely buffered hydraulically from the parent plant (Matthews and Shackel 2005; Thomas et al. 2006). The traditional view, that this hydraulic buffering was a result of a physical disconnection of the xylem, has been refuted (Keller et al. 2006), although the buffering does involve decreases in hydraulic conductivity (Choat et al. 2009; Knipfer et al. 2015). The membrane water channel proteins, aquaporins, may contribute to these decreases in berry hydraulic conductivity; however, the regulation of this gene family during ripening is complex (Choat et al. 2009; Wong et al. 2018). The extent to which aquaporins mediate berry growth remains unknown, but it is fair to speculate that they play a role in berry growth via their effects on berry water relations (Tyerman et al. 2012).

12.2.2 Softening: Decreases in Turgor and Changes in Cell Wall Composition

Berry softening occurs approximately 10 days prior to the onset of ripening and represents one of the earliest detectable changes in berry physiology leading to veraison (Wada et al. 2008; Matthews et al. 2009; Castellarin et al. 2016). Softening is thought to result from the same two compatible mechanisms as growth does decreases in cell turgor (introduced above) and changes in the structure of cell walls (Brummell and Harpster 2001; Gapper et al. 2013).

Interestingly, both of these mechanisms have links with abscisic acid (ABA), one of the key hormones regulating the onset of ripening in grape (Gambetta et al. 2010; Castellarin et al. 2016; Pilati et al. 2017) and other fruits (Leng et al. 2014). The decrease in turgor associated with softening in grape corresponds to increases in ABA, and both precede the increase in sugar concentration at the onset of ripening (Castellarin et al. 2016). The decrease in turgor results from the accumulation of solutes, mostly malate and sugars, in the apoplast of the berry (Wada et al. 2008, 2009). This accumulation of solutes in the berry apoplast may result from apoplastic sucrose unloading from the phloem and an upregulation of acid invertases, which ABA stimulates (Pan et al. 2005; Zhang et al. 2006; Koyama et al. 2010).

Many genes encoding cell wall modification enzymes are up-regulated during softening in grape, including many members of the expansin and pectin methylesterase gene families, among others (Dal Santo et al. 2013; Castellarin et al. 2016; Fasoli et al. 2016). In addition, cell wall modification enzymes are thought to contribute to postharvest changes in fruit texture and quality (Brummell and Harpster 2001), and this is consistent with findings in grape where many genes encoding cell wall modification enzymes continue to be up-regulated late into ripening and throughout the postharvest period (Castellarin et al. 2016; Zenoni et al. 2016). The master regulators of these increases are still unknown, but ABA has been shown to up-regulate cell wall modification enzymes, including expansins and pectin methylesterases, in tomato (Sun et al. 2012). Increases in *VviCEB1* expression (discussed above) correspond to softening, and along with *VviCEB1*'s induction of genes encoding cell wall modification enzymes, one can speculate a role for *VviCEB1* in softening as well (Nicolas et al. 2013).

12.3 Berry Composition

Grape composition determines grape, juice, and wine sensorial attributes. It changes dramatically during fruit ripening and is strongly affected by the genotype, the environment, and the viticultural practices applied in the vineyard. The complex regulation of the physiological and metabolic pathways that determine grape composition, as well as the modulation of these pathways by the environment or viticultural practices, have been intensively investigated during recent years.

12.3.1 Sugars

Sugars play an important role in shaping berry sensory properties, in determining alcohol concentration after fermentation, and as precursors for the synthesis of organic acids, phenolics, and aroma compounds (Dai et al. 2011). *Vitis vinifera* berries accumulate large amounts of sugars, predominantly glucose and fructose (in equal concentrations) with only a trace amount of sucrose (Hawker et al. 1976; Liu et al. 2006; Shiraishi et al. 2010). Grapevine varieties exhibit an impressively large range of sugar concentrations at maturity. For example, Kliewer et al. (1967) compared 78 table and wine grape varieties and found that total soluble solids of the berry juice—a good representation of berry sugar concentration—varied at harvest from 18.5 to 28.2 °Brix.

In plants, sugars are synthesized in the cytoplasm of the leaf mesophyll cells and transported, in the form of sucrose, via phloem into other parts of the plant (Cheng et al. 2018). In the grape berry, sucrose is then hydrolyzed by invertases and stored in the vacuole in the form of glucose and fructose. At the onset of berry ripening or just before, sugar loading into the berry from the phloem shifts from a symplastic to an apoplastic pathway (Zhang et al. 2006). The latter requires at least two transporters—one secreting sugars from sieve elements/companion cells, the other mediating reuptake into the adjacent sink cells (Lalonde et al. 2004). Sugar transport across membranes is mainly mediated by the proton-coupled sucrose transporters (SUTs, the disaccharide transporters) and hexose transporters (HTs, the monosaccharide transporters), together with several other subfamilies of monosaccharide transporters. Acidic invertases (AI), located in the vacuole or cell wall, and neutral invertases (NI), located in the cytoplasm, are the two major classes of sucrose metabolic enzymes contributing to hexose accumulation in grape berry. Although the vacuolar invertases are considered important for sugar accumulation, the expression of the genes encoding these enzymes precedes the onset of hexose accumulation by some weeks; therefore, the synthesis of these enzymes cannot be considered a trigger for sugar accumulation in grape berry (Davies and Robinson 1996).

SUTs are essential for sucrose translocation in plants (Lalonde et al. 2004). Four genes encoding sucrose transporters have been identified in grapevine, namely *VviSUC11/VviSUT1*, *VviSUC12*, *VviSUC27*, and *VviSUT2*. *VviSUC11* and *VviSUC12* are high affinity sucrose transporters (Ageorges et al. 2000; Manning et al. 2001; Afoufa-Bastien et al. 2010), and *VviSUC27* is a low affinity sucrose transporter that has a very similar structure to *VviSUT2* (Zhang et al. 2007). *VviSUC11* and *VviSUC12* expressions have been detected in all organs. The weakest expression for both genes was observed in berries at fruit set (Afoufa-Bastien et al. 2010), but a significant upregulation was observed during ripening (Lecourieux et al. 2014). Afoufa-Bastien et al. (2010) suggest that *VviSUC12* either might be involved in phloem unloading or in sucrose import into the berry, and that *VviSUC11* might control sucrose uptake into berry vacuoles. In contrast, *VviSUT27* transcript amounts significantly decrease during ripening (Davies et al. 1999), which suggests a different physiological function for this transporter. On the other hand, *VviSUC27* transcripts have been detected at a high level in petioles, stems, and tendrils, and less abundantly in young leaves, mature leaves, and roots (Afoufa-Bastien et al. 2010). The "Sugars Will Eventually be Exported Transporter" (SWEET) proteins are a newly identified family of sugar efflux transporters (Chen 2014). SWEETs are integral membrane proteins and function as a prerequisite for *SUT1*-mediated phloem loading (Chen et al. 2012). There are 17 *SWEET* genes, with different expression levels among vegetative and reproductive organs, identified in grapevine. Generally, most *VviSWEET* genes are more highly expressed in the berry, and their expression level increases throughout berry ripening (Chong et al. 2014).

HTs in grapevine are encoded by a multigene family, of which five members (*VviHT1-5*) are well studied (Tanner and Caspari 1996; Zhang et al. 2007; Agasse et al. 2009), and 17 were identified more recently (*VvHT8-24*) (Afoufa-Bastien et al. 2010). *VviHT1* is expressed mainly in grape berry (Fillion et al. 1999), and its transcription greatly increases during leaf development. *VviHT3* and *VviHT5* are expressed in both mature leaves and grape berries, though *VviHT5* has a much lower expression level than *VviHT3*. *VviHT4*, whose function is restricted to glucose, is also expressed in grape berries (Hayes et al. 2007). *VvHT1*, *VvHT2*, and particularly *VvHT3* are highly expressed at all stages of berry development, with transcriptional patterns consistent with the shift from a symplastic to an apoplastic phloem unloading pathway that occurs prior to veraison (Lecourieux et al. 2014). A gene named *VviHT8*, which has a high similarity to *VviHT1*, was identified as a molecular target for the selection of grapes with improved sugar accumulation (Xin et al. 2013).

Other monosaccharide transporters present in the grapevine genome include tonoplast monosaccharide transporters (*VviTMTs*), polyol/monosaccharide transporters (*VviPMTs*), glucose transporters (*VviGlcTs*), and ERD6-like transporters (Afoufa-Bastien et al. 2010).

12.3.2 Organic Acids

Tartaric acid and malic acid are the major organic acids in grapevine. Most of the tartrate and malate in immature berries originate from glucose and fructose (Hardy 1968). Tartaric and malic acid accumulate in berry cell vacuoles before veraison. Unlike many other fruits, grape berries do not contain large amounts of citrate. During ripening, the concentration of tartaric acid remains stable, but the concentration decreases through a dilution effect determined by cell expansion (Dai et al. 2011; Regalado et al. 2013). Malic acid also decreases in concentration during ripening, but in contrast to tartrate, most of this decrease is due to degradation, use in respiration, and conversion into sugars (Sweetman et al. 2009).

Tartaric acid is synthesized from L-ascorbic acid (vitamin C). L-idonate dehydrogenase (L-IdnDH) is responsible for catalyzing the proposed rate-limiting step, the oxidization of L-idonic acid to 5-keto-gluconic acid (DeBolt et al. 2006; Cholet et al. 2016), and is the only known enzyme to be involved in tartaric acid accumulation (DeBolt et al. 2006). The sudden increase of tartaric acid during stage I is paralleled by *VviL-IdnDH* gene expression and translation (Grimplet et al. 2007; Wen et al. 2010; Cholet et al. 2016). There are three different isoforms of *VviL-IdnDH* genes: two of them are specifically expressed in young berries, and the third increases during berry ripening (Sweetman et al. 2012).

The accumulation of malate before the onset of ripening is thought to be mainly due to its de novo synthesis in berries (Sweetman et al. 2009). Malic acid is produced from phosphoenolpyruvate (PEP) through the activity of different enzymes: phosphoenolpyruvate carboxylase (PEPC), malate dehydrogenase (MDH) (Givan 1999; Sweetman et al. 2012), malic enzyme (ME) (Sweetman et al. 2012), and fumarase (FUM) (Shangguan et al. 2015). There are two *VviPEPC*s, one *VviMDH*s, and two *VviFUM*s identified in grapevine (Shangguan et al. 2015).

The cytoplasmic MDH and the mitochondrial ME appear to be key enzymes for malic acid synthesis, since the decrease in expression of their codifying genes correlates to decreases in malate concentration during ripening (Sweetman et al. 2012).

MDH enzymes catalyze the reversible conversion of oxaloacetate into malate; therefore, the possible decrease of oxaloacetate in mature berries caused by altered expression of *VviPEPC* and *VviPEPCK* could influence malate degradation by shifting the function of MDH enzymes towards malate catabolism (Sweetman et al. 2012). Since the catabolism of malate can only occur when the acid is accessible to metabolic enzymes outside the vacuole, the compartmentation of malate may also influence the rates of its degradation during berry development. For this reason, the decrease of malate could also be attributed partly to the down-regulation of the genes encoding the tonoplast dicarboxylate transporters (*VviTDTs*) (Sweetman et al. 2009, 2012), which are responsible for the transport of malate into vacuoles. Moreover, the decrease in acid content during grape ripening has been mainly associated with mitochondrial malate oxidation (Regalado et al. 2013). Three mitochondrial dicarboxylate/tricarboxylate carriers (*VviDTC1–VviDTC3*) have been characterized in *Vitis vinifera*. *VviDTC1* is able to transport all the dicarboxylates/tricarboxylates of the TCA cycle, with the exception of fumarate, and exhibits high specificity for malate. The expression of *VviDTC2* and *VviDTC3* transcripts is strongly enhanced in the mesocarp at the onset of ripening, which suggests that their role in the transport of malate into mitochondria might be critical (Regalado et al. 2013).

12.3.3 Phenolics

Phenolics are synthesized from phenylalanine via the phenylpropanoid, flavonoid, and stilbenoid pathways. The phenylpropanoid pathway leads to the production of *p*-coumaryl-CoA from phenylalanine, which involves enzymes such as phenylalanine ammonia lyase (PAL), cinnamate-4-hydroxylase (C4H), and 4-coumarate-CoA ligase (4CL). *p*-Coumaryl-CoA and malonyl-CoA are the substrates of both chalcone synthase (CHS) and stilbene synthase (STS), which catalyze the first steps of the flavonoid and stilbenoid pathway, respectively.

Hydroxycinnamic acids, such as *p*-coumaric, caffeic, and ferulic acid and their esterified forms coutaric, caftaric, and fertaric acid are the major phenolic acids in the berry. Their synthesis occurs before veraison via modifications of the intermediates of the phenylpropanoid pathway catalyzed by caffeic acid 3-*O*-metyltransferase (COMT) and caffeoyl-CoA 3-*O*-methyltransferase (CCoAOMT). Recently, two TFs, VviMYB4a and VviMYB4b, have been characterized as negative regulators of phenylpropanoid genes and hydrocinnamic acid synthesis (Cavallini et al. 2015).

Stilbenoids (e.g., *cis*- and *trans*-resveratrol, piceatannol, *cis*- and *trans*-piceid, astringin, pallidol, and α-, β-, γ-, δ-, ε-viniferin) are mostly accumulated from veraison onward (Gatto et al. 2008) and are strongly modulated by both biotic and abiotic factors (Vannozzi et al. 2012; Savoi et al. 2017). Forty-five stilbene synthases are found in the grapevine genome, with at least 33 encoding full-length proteins. This gene family arose from multiple events of tandem and segmental duplications (Vannozzi et al. 2012). Recent large-scale transcriptomic analysis has shown that the expression of many *VviSTSs* changes during fruit development and ripening (Massonnet et al. 2017). In red berry varieties, induction of *VviSTSs* is particularly pronounced during the late stages of ripening. The two R2R3 MYB transcription factors, VviMYB14 and VviMYB15 (Höll et al. 2013), which are known to regulate stilbene biosynthesis, also share similar expression profiles. Nonetheless, among the many TFs proposed to regulate this pathway (Wong et al. 2016b; Vannozzi et al. 2018), two WRKY TFs, VviWRKY24 and VviWRKY03, participate at different levels of *VviSTS* regulation—via direct activation of *VviSTSs* or synergistic action with MYB TFs to regulate *VviSTS*s.

The flavonoid pathway leads to the production of flavonols, flavan-3-ols, and anthocyanins. The modulation of the pathway during berry development and under environmental stresses has been largely investigated in grapevine (Teixeira et al. 2013; Kuhn et al. 2013). Most of the genes of the flavonoid pathway are present in low copy numbers except for those encoding the flavonoid-3′,5′-hydroxylases (F3′5′H s). Flavonoid-3′-hydroxylases (F3′Hs) and F3′5′Hs divide the pathway into two major branches, whose compounds are either di-hydroxylated or tri-hydroxylated. In most plants, F3′5′H genes are present in low copy numbers, but a proliferation of the F3′5′Hs has occurred in the grapevine genome and given rise to 15 paralogs within 650 kb (Falginella et al. 2010). Most *VviF3′5′Hs* are predominantly expressed in berries, and differences in cis-regulatory sequences of promoter regions are paralleled by temporal specialization of gene transcription during fruit ripening and in berry tissues (Falginella et al. 2010, 2012).

Flavonol synthases (*FLSs*) are key enzymes for the synthesis of berry flavonols such as kaempferol, quercetin, myricetin, isorhamnetin, laricitrin, and syringetin (Downey et al. 2004). The expression of the FLSs is well known to be under the control of a light-induced transcription factor (VviMYBF1/VviMYB12) (Czemmel et al. 2009). Two recent studies now show that three additional bZIP TFs, VviHY5, VviHYH, and VvibZIPC22 (Malacarne et al. 2015; Loyola et al. 2016), are involved in the regulation of flavonol synthases and flavonol accumulation in the berry. VviMYBF1 was shown to be part of a regulatory cascade of VviHY5/HYH that potentially involves positive feedback regulation (Loyola et al. 2016; Czemmel et al. 2017). Flavonols are normally glycosylated (as glucosides, galactosides, rhamnosides, rutinosides, and glucuronides) and the flavonol-3-*O*-glycosyltransferases (VviGT3-5-6) and flavonol-3-*O*-rhamnosyltransferase (VviRhaT1) responsible for this glycosylation have been recently characterized in grapevine (Ono et al. 2010; Czemmel et al. 2017).

Flavan-3-ols are produced via the activity of leucoanthocyanidin reductases (LAR1-2) or an anthocyanidin reductase (ANR) (Bogs et al. 2005). Their synthesis is promoted from anthesis to veraison and is regulated by transcription factors of the MYB family. In particular, *VviLAR1* and *VviANR* are under the control of VviMYBPA1 and VviMYBPA2 (Bogs et al. 2007; Terrier et al. 2009), whereas *VviLAR2* is under the control of VviMYBPAR (Koyama et al. 2014). The monomeric flavan-3-ols accumulated in grape, (+)-catechin, (−)-epicatechin, (−)-epicatechin-3-*O*-gallate, (+)-gallocatechin and (−)-epigallocatechin, differ according to stereochemistry, level of hydroxylation, and acylation by gallic acid (Mattivi et al. 2009). Until now, the mechanisms involved in either polymerization into tannins, galloylation, and transport into the vacuoles have not yet been well understood (Zhao et al. 2010). However, a QTL study revealed different genetic determinisms for PA composition in seeds and skin, including PA total content, PA building blocks, degree of polymerization, and ratio between building blocks (Huang et al. 2012). Three annotated glycosyltransferases (VviGT1-3) were described to be putatively involved in the galloylation of proanthocyanidins and the production of hydroxycinnamic esters (Khater et al. 2012), and two specific transporters of proanthocyanidin were identified (VviPAMATE1-2) (Pérez-Díaz et al. 2014).

Anthocyanins are responsible for the pigmentation of the grape berries. They are synthetized in the epidermis and hypodermis cells from veraison onward and then stored in the vacuole. *Teinturier* varieties, such as Alicante Bouschet, also accumulate anthocyanin in the flesh (Castellarin et al. 2011; Falginella et al. 2012). In *Vitis vinifera*, anthocyanins are glycosylated at the 3′ position by the addition of a glucose moiety through the activity of the enzyme UDP-glucose, flavonoid-3-*O*-glucosyltransferase (UFGT). Both di-hydroxylated and tri-hydroxylated anthocyanins are synthetized by VviUFGT. The *O*-methyltransferases (VviAOMT1-3) methylate cyanidin-3-*O*-glucoside and delphinidin-3-*O*-glucoside into peonidin-3-*O*-glucoside, petunidin-3-*O*-glucoside, and malvidin-3-*O*-glucoside (Fournier-Level et al. 2011). Moreover, anthocyanins can also be acylated at the 6″ position of the glucose, which produces 3-*O*-6″-acetyl-, 3-*O*-6″-coumaroyl- and 3-*O*-6″-caffeoyl-monoglucosides and, recently, an anthocyanin-3-*O*-glucoside-6″-*O*-acyltransferase was characterized (Vvi3AT) (Rinaldo et al. 2015).

The MYBA1-A2 TFs are crucial genetic determinants of berry color (Walker et al. 2007). Recent studies show that additional members of the MYBA cluster, VviMYBA6 and VviMYBA7, have the capacity to influence fruit anthocyanin pigmentation and composition under severe environmental conditions (i.e., UV-B) during veraison (Czemmel et al. 2017). Anthocyanin-acylglucosides are translocated into the vacuole by MATE-type transporters localized in the tonoplast (VviAnthoMATE1-3) (Gomez et al. 2009), whereas the glycosylated anthocyanins are translocated via a glutathione-dependent, ATP-binding cassette (ABC) protein (VviABCC1) (Francisco et al. 2013).

Furthermore, a recent QTL study identified a set of new candidate genes for the regulation of anthocyanin variation among cultivars (Costantini et al. 2015).

Overall, the synthesis of hydroxycinnamic acids, stilbenes, flavonols, flavan-3-ols, and anthocyanins is spatiotemporally separated during grape berry development and ripening and tightly regulated by positive and/or negative regulators. Besides the TFs described above, two (VviMYB5a-b) are general regulators of the flavonoid pathway and, in particular, modulate the expression profile of several flavonoid genes (*VviCHI*, *VviF3′5′H*, *VviLDOX*, *VviLAR*, and *VviANR*) during berry development and ripening (Lauvergeat et al. 2006; Cavallini et al. 2015). Recently, two TFs (VviMYBC2-L1 and L3) were characterized as repressors of both proanthocyanidin and anthocyanin biosynthesis (Huang et al. 2014; Cavallini et al. 2015). Moreover, a bHLH (VviMYC1) interacts with VviMYB5a-b, VviMYBPA1, and VviMYBA1-A2 in the transcriptional control of proanthocyanidin and anthocyanins biosynthesis in grapevine (Hichri et al. 2010).

12.3.4 Volatile Organic Compounds

Terpenes are a major class of volatiles in grapes and strongly affect the aroma of grapes and wines of several varieties. The sesquiterpenes and monoterpenes accumulate in the berry before and after veraison, respectively. Two independent pathways produce terpenes in plants: (1) the plastidial 2C-methyl-erythritol-4-phosphate (MEP) pathway, which is the predominant pathway for monoterpenes (C_{10}) and diterpenes (C_{20}), and (2) the cytosolic mevalonate (MVA) pathway, which is the primary pathway for sesquiterpenes (C_{15}) (Bohlmann and Keeling 2008).

The major monoterpenes produced in grapes are linalool, geraniol, nerol, citronellol, hotrienol, α-terpineol, and rose oxides (Matarese et al. 2014); these compounds confer flowery and fruity notes to wines (Robinson et al. 2014a; Siebert et al. 2018). Sesquiterpenes have a minor impact on grape and wine aroma because usually their concentrations are below the olfactory threshold. The most studied sesquiterpene is rotundone, which gives peppery character in some red and white varieties (Siebert et al. 2008; Wood et al. 2008; Mattivi et al. 2011; Caputi et al. 2011). Recently, key genes (*VviGuaS*, *VviTPS24*, *VviSTO2*) involved in rotundone biosynthesis were identified (Drew et al. 2015; Takase et al. 2015).

Among the several structural genes of the MEP pathway, 1-deoxy-xylulose 5-phosphate synthase (*VviDXS*) was identified as a key modulator of total monoterpene content in grapevine (Battilana et al. 2009, 2011). Terpene synthases (TPSs) control monoterpene or sesquiterpene production (Martin et al. 2010; Matarese et al. 2013, 2014). Interestingly, in the genome of *Vitis vinifera* there are 69 putative terpene synthases, 39 of them functionally characterized (Martin et al. 2010). Generally, TPSs are divided into seven clades: TPS-a, TPS-b, TPS-c, TPS-d, TPS-e/f, TPS-g, and TPS-h (Chen et al. 2011). The TPS-a clade (30 genes) contains mostly sesquiterpene and possibly diterpene synthases, whereas the TPS-b clade (19 genes) and TPS-g clade (17 genes) consist mostly of monoterpene synthases. TPS-c (2 genes) and TPS-e/f (1 gene) clades contain plant hormone metabolism genes that are typically represented with a single gene copy in plant genomes. No full-length TPS-d and TPS-h were found in grapevine (Martin et al. 2010). Recently, several genes, such as nudix hydroxylase, vesicle-associated proteins, ABCG transporters, glutathione S-transferases, and amino acid permeases have been proposed as candidate genes for regulating the monoterpene biosynthesis and accumulation in the berry (Costantini et al. 2017). Moreover, positive correlation between aroma production and ERF TFs indicates that ethylene signaling could be a factor in affecting the final terpene content (Cramer et al. 2014). In addition, a major role of jasmonic acid and methyljasmonate has been hypothesized for the regulation of terpene biosynthesis in grapes (Savoi et al. 2016; D'Onofrio et al. 2018).

Most monoterpenes and sesquiterpenes are present in grapevine as non-volatile terpene glycosides. In grapevine, only three monoterpenol glycosyltransferases have been characterized, *VviGT7-14-15* (Bönisch et al. 2014a, b; Li et al. 2017) and the cytochrome P450 CYP76F14, which catalyzes the conversion of linalool to (E)-8-carboxylinalool, which, during wine fermentation, generates a wine lactone, a key odorant of Gewurztraminer wines (Ilc et al. 2017).

Other terpenoids synthesized in the berry before ripening are the carotenoids, which are pigments contributing to light harvesting and to protecting the photosynthetic apparatus from photooxidation (Rodríguez-Concepción and Boronat 2002). The genes involved in their biosynthetic pathway were recently identified in grapevine (Young et al. 2012). Carotenoids can be cleaved via other carotenoid cleavage dioxygenases (VviCCD1a/b, VvCCD4a/b/c) (Lashbrooke et al. 2013) to form volatile flavor and aroma-related compounds, such as the C_{13}-norisoprenoids β-ionone and β-damascenone, which contribute to floral and fruity aromas. The majority of them are glycosylated in grape (Robinson et al. 2014a).

The unsaturated C_{18} fatty acids linoleic acid and linolenic acid are the precursors of other volatile organic compounds such as C_6-aldehydes and alcohols like hexanal and hexanol (Kalua and Boss 2009). They are formed by the activity of lipoxygenases (*VviLOX*) (Podolyan et al. 2010), hydroperoxide lyase (*VviHPL1-2*) (Zhu et al. 2012), and *(3Z)-(2E)* enal isomerase and alcohol dehydrogenase (VviADH) (Kalua and Boss 2009). Their synthesis occurs mainly pre-veraison (Kalua and Boss 2009), and they are responsible of green-grassy aromas even though, considering their detection threshold, they rarely contribute to the herbaceous character of juices and wines (Robinson et al. 2014a).

Methoxypyrazines like 3-isobutyl-2-methoxypyrazineare (IBMP), 3-isopropyl-2-methoxypyrazine (IPMP) are extremely volatile compounds accumulated before veraison. They contribute to the specific green-herbaceous aroma of some wines such as Sauvignon blanc, Cabernet Sauvignon, Cabernet Franc, and Merlot. Their biosynthesis starts with an adicarbonyl addition to the amino acid leucine or valine for IBMP and IPMP, respectively, followed by methoxylation reactions to form the final methoxypyrazines. Four O-methyltransferases (VviOMT1-4) have been identified in grape, with VviOMT3 having a major role in IBMP production (Dunlevy et al. 2010; Guillaumie et al. 2013).

Finally, thiols confer typical aromatic features to some varieties such as Sauvignon blanc. The thiols in grape are normally accumulated during ripening in a non-volatile form, bounded to S-cysteine or S-glutathione via the VviGST3 and VviGST4 activity (Kobayashi et al. 2011). These compounds are released during and after fermentation, conferring to wines many desired properties and sometimes off-flavors, depending on the concentration (Peña-Gallego et al. 2012).

12.4 Hormonal Regulation of Berry Ripening

Several hormones participate in the control of grape ripening. Genomic and high throughput technologies have been essential in characterizing the crosstalk between hormones and the expression of associated downstream genes (McAtee et al. 2013; Fortes et al. 2015) (Fig. 12.2).

12.4.1 Auxins

Several studies have established that IAA decline is associated with the initiation of ripening, both in climacteric fruit and in non-climacteric fruit such as grapes (Böttcher et al. 2011; Fortes et al. 2015). Auxin treatments retard sugar and anthocyanin accumulation and prevent the decrease in acidity and chlorophyll concentration, but also cause a delay in the usual ripening-associated increase in the levels of abscisic acid (ABA), by altering gene expression in grape berry (Davies et al. 1997; Ziliotto et al. 2012).

Gouthu and Deluc (2015) showed that the timing of ripening initiation is related to an auxin

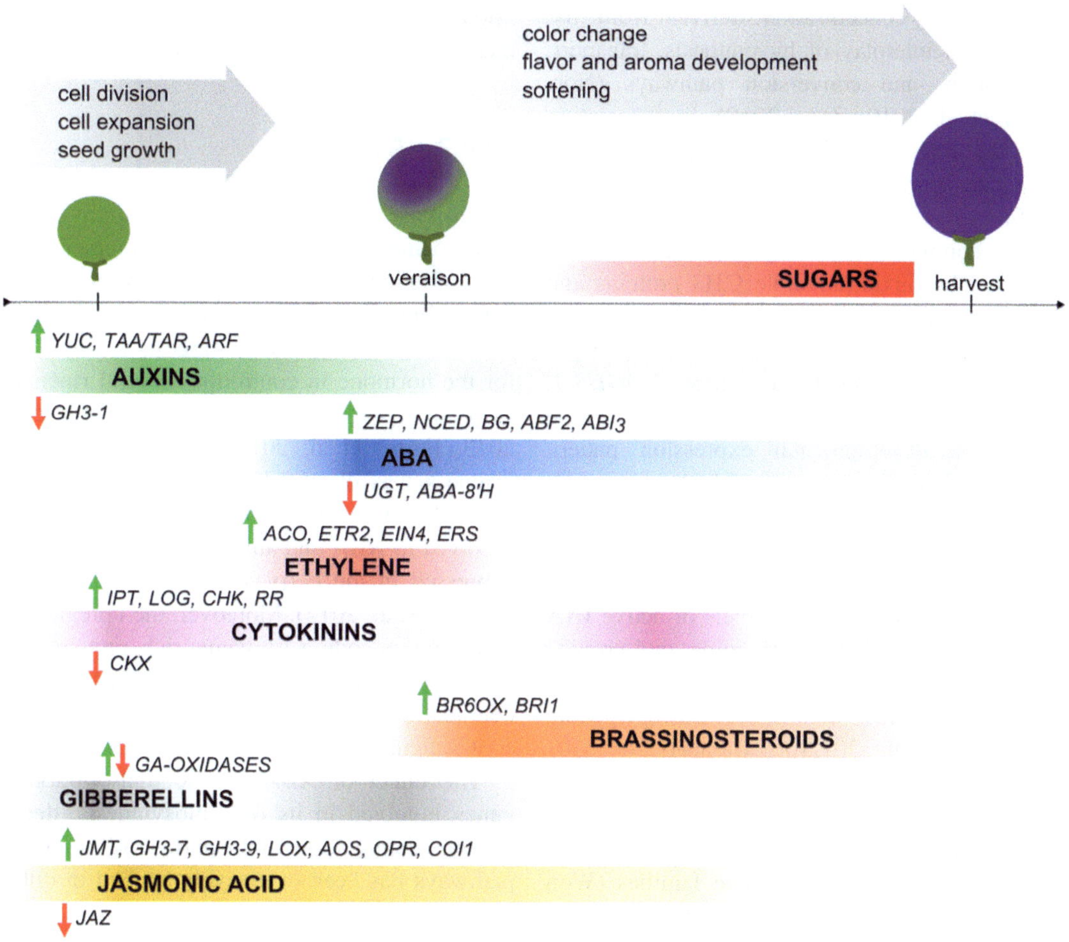

Fig. 12.2 *Hormone dynamics during berry development and ripening*. Several studies have shown that increases in auxin, cytokinin, gibberellin, and jasmonic acid occur during the first phases of fruit growth (Stage I); brassinosteroids, ethylene, and ABA are mainly involved in physiological changes related to berry ripening (Stage III). The up- and down-regulation of the main biosynthetic/catabolic and associated downstream signaling genes are reported for each different hormone. In detail, gene names are abbreviated as follows: TRYPTOPHAN AMINOTRANSFERASE OF ARABIDOPSIS1/TRYPTOPHAN AMINOTRANSFERASE RELATED (*TAA/TAR*); YUCCA (*YUC*); auxin response factors (*ARF*); IAA-amido synthetase (*GH3-1*); 9-cis-epoxy-carotenoid dioxygenase (*NCED*); zeaxanthin epoxidase (*ZEP*); β-glucosidases (*BG*); transcription factors ABA insensitive (*ABI3*); ABRE-binding factors (*ABF*); UDP-glucosyltransferases (*UGT*); ABA 8'-hydroxylase (*ABA-8'H*); ACC oxidase (*ACO*); ethylene receptors (*ETR2*, *EIN4*, *ERS*); Adenosine phosphate-isopentenyltransferase (*IPT*); phosphoribohydrolase "Lonely guy" (*LOG*); cytokinin histidine kinase (*CHK*) receptors; response regulators (*RR*); cytokinin oxidase/dehydrogenase (*CKX*); brassinosteroid 6-oxidase gene (*BR6OX*); BR receptors (*BRI1*); GA-oxidases; *S*-adenosyl-L-methionine:jasmonic acid carboxyl methyltransferase (*JMT*); JA-amido synthetases (*GH3-7* and *GH3-9*); lipoxygenase (*LOX*); allene oxide synthase (*AOS*); 12-oxophytodienoate reductase (*OPR*), CORONATINE INSENSITIVE 1 (*COI1*) jasmonate receptor; jasmonate ZIM domain (*JAZ*)

signal and is linked to the relative seed content in berries. In a recent study that compared the berry physiology and composition to the whole genome gene expression analyzed by RNA-seq, a potential role of auxin and its conjugates in determining asynchrony between berries of different sizes was suggested (Wong et al. 2016a). Moreover, it was shown that the tight control of the hormone concentration derives from the coordinated interplay of biosynthesis, transport, degradation, and conversion pathways (Normanly et al. 2010; Zhao 2010), in association with the fine regulation of the pool of IAA conjugates during grape ripening (Fortes et al. 2015).

The conjugation of IAA to amino acids is catalyzed by auxin-inducible GH3 proteins and provides a negative feedback loop to control auxin homoeostasis (Böttcher et al. 2010). A putative IAA-amido synthetase gene, *VviGH3-1*, was identified in grape berries. This gene displays a developmental expression pattern consistent with the increase of IAA-conjugates, which in turn is coupled to several ripening-associated processes in the berry. Indeed, the increasing levels of IAA-aspartate in grapes might be linked to the low levels of active IAA that were observed during ripening, and provide evidence for a possible mechanism for the maintenance of low auxin levels during ripening (Böttcher et al. 2012b). Members of both the TRYPTOPHAN AMINOTRANSFERASE OF ARABIDOPSIS1/TRYPTOPHAN AMINOTRANSFERASE RELATED (*TAA1/TAR*) and YUCCA (*YUC*) gene families (Won et al. 2011), involved in the two-step pathway of auxin biosynthesis, are also expressed in developing berries. Recent transcriptomic analyses revealed a consistency between TAA/TAR and YUC transcripts' evolution and auxin accumulation during berry development and ripening (Wong et al. 2016a).

Auxins' effects are mediated by early response genes, such as Aux/IAA, GH3, and SAUR family members. Several putative auxin response elements (AuxREs) have been identified, and it has been demonstrated that the conserved motif TGTCTG is responsible for the binding of the auxin response factors (ARFs) that confer specificity to auxin response through the selection of target genes, i.e., transcription factors (Hayashi 2012; Li et al. 2016). Nineteen *VviARF* genes, categorized into four groups (Classes 1, 2, 3 and 4) have been identified. Most *VviARFs* display the highest transcript levels in the berry, suggesting that they may play important roles in the regulation of grape berry maturation processes (Wan et al. 2014).

12.4.2 ABA

An increase in free ABA levels around veraison accompanies sugar accumulation, pigmentation, and softening (Deluc et al. 2007; Wheeler et al. 2009; Sun et al. 2010; Gambetta et al. 2010; Pilati et al. 2017), which suggests a major role for the hormone in controlling several ripening-associated processes in grape berry (Kuhn et al. 2013; Fortes et al. 2015). A decrease in fruit firmness was observed by transforming tomato with the *Vitis* transcription factor *VvABF2*, involved in ABA and abiotic stress signaling and expressed in the berry at the onset of ripening (Nicolas et al. 2014). Moreover, the upregulation of a gene encoding a glycine-rich protein, possibly involved in cell wall biogenesis and degradation, confirms a role for the hormone in fruit softening (Rattanakon et al. 2016).

The effect of ABA on the transcription of genes involved in its own biosynthesis, degradation, conjugation, transport, and signaling pathways has been extensively studied in different organs of grapevine (Rattanakon et al. 2016; Pilati et al. 2017). These studies highlighted that a small amount of ABA can trigger a positive feedback regulation of genes involved in ABA biosynthesis, including a significant up-regulation of *VviABI3* (transcription factor involved in ABA responsiveness) during the lag phase, which further supports the regulatory role of ABA in grape ripening (Rattanakon et al. 2016).

ABA biosynthesis comprises crucial steps catalyzed by 9-cis-epoxy-carotenoid dioxygenase (VviNCED) and zeaxanthin epoxidase (VviZEP). The genes codifying for those proteins are up-regulated around veraison. Conversely, *ABA 8′-hydroxylase* (*VviABA-8′H*), which regulates ABA catabolism, is down-regulated at the same stage (Deluc et al. 2007; Fortes et al. 2015). Moreover, the activity of cytosolic UDP-glucosyltransferases (VviUGTs), which conjugate ABA to form the ABA-glucose ester, and the activity of β-glucosidases (VviBGs), which release ABA from the above conjugated form, further control ABA levels in the berry tissues (Owen et al. 2009).

Higher accumulation of anthocyanins has been observed in the skin of berries treated with ABA (Wheeler et al. 2009; Gambetta et al. 2010). This is consistent with the increased expression of anthocyanins' biosynthetic genes *VviCHI*, *VviF3H*, *VviDFR*, and *VviUFGT*, and of the related transcription factors *VviMYBA1* and *VviMYBA2* (Koyama et al. 2010). ABA is also a key modulator of water stress responses, and water deficit promotes ripening and color accumulation in grape berries (Castellarin and Di Gaspero 2007; Herrera and Castellarin 2016; Savoi et al. 2017); however, several studies have shown that under water deficit, ABA is not the only signal for color development, and sugars and other stimuli may co-regulate the metabolic response of the berry (Gambetta et al. 2010; Ferrandino and Lovisolo 2014; Pilati et al. 2017). Supporting this hypothesis, Pilati et al. (2017) analyzed berry skin transcriptional modulation by RNA-seq, and observed that ABA treatment by itself did not induce anthocyanins' biosynthetic genes.

In addition to the regulation of secondary metabolism, ABA may be able to hasten the initiation of sugar accumulation when applied before veraison by stimulating the uptake and storage of sugars in berries (Davies and Böttcher 2009; Fortes et al. 2015). The link between ABA and sugar metabolism is also supported by a study demonstrating that ABA increased the activity of both soluble and cell wall acid invertases in berry discs (Pan et al. 2005).

12.4.3 Other Hormones

12.4.3.1 Ethylene

The role of ethylene in regulating berry ripening was usually considered negligible (Sun et al. 2010; Muñoz-Robredo et al. 2013). However, ethylene can alter the progression of ripening. For example, the application of an ethylene-releasing compound (2-chloroethylphosphonic acid, 2-CEPA) delayed ripening when applied early in berry development, and treatments with an inhibitor of ethylene biosynthesis, aminoethoxyvinylglycine (AVG), advanced ripening (Böttcher et al. 2013). However, the response to CEPA and AVG clearly changed during berry development, and this was speculated to be due to the different sensitivity of the ethylene biosynthesis and perception pathways to exogenous ethylene at different times (Böttcher et al. 2013). Interestingly, CEPA application at veraison generated an increase in the concentration of anthocyanin in Cabernet Sauvignon berries, with a concomitant increase in expression of genes such as *VviCHS*, *VviF3H*, and *VviUFGT* (El-Kereamy et al. 2003).

Ethylene also promotes berry size, stimulating the expression of several genes encoding aquaporins, polygalacturonases, xyloglucan endotransglycosylase, cellulose synthases, and expansins (Chervin et al. 2008). Ethylene is perceived by transmembrane-receptor proteins, belonging to the EThylene Receptor (ETR) family, localized in the endoplasmic reticulum. Chervin and Deluc (2010) analyzed the transcript abundance of several ethylene receptors (*VviETR2*, *VviEIN4*, *VviERS*) and transcription factors (*VviEIN3* and *VviMADS4*) across berry development and the impact of the ethylene inhibitor 1-MCP on their expression. Recently, a phylogenetic analysis performed on ETRs and related proteins, in both climacteric and non-climacteric fruits, pointed out that both classes share many aspects of ethylene perception and signaling during fruit ripening. Moreover, grape, as non-climacteric fruit, exhibits an earlier expression peak of four ETRs, concomitant with the onset of sugar accumulation (Chen et al. 2018). One gene coding for ACC oxidase (*VviACO*) was found to increase its expression at the early stages of berry development (Deluc et al. 2007), with a peak around veraison; a similar observation, together with the increase of ethylene levels, was related to the beginning of fruitlet abscission in Chardonnay berries (Hilt and Bessis 2003). Recently, the expression of genes involved in the ethylene signaling pathway, as well as ethylene transcription factors with recognized roles in leaf senescence, were found to increase during the late stages of ripening of Cabernet Sauvignon, which suggests that ethylene may play a bigger role than expected in regulating grape berry ripening (Cramer et al. 2014).

12.4.3.2 Cytokinins

Although previous studies reported that cytokinins do not participate in ripening in grapevine (Inaba et al. 1976), more recently some studies have highlighted the importance of this hormone both at the pre- and post-veraison stages (Böttcher et al. 2015; Pilati et al. 2017). Grapevine orthologues of five *Arabidopsis* gene families involved in cytokinin metabolism and signaling were identified, and their expression patterns were analyzed in developing berries. Genes regulating cytokinin biosynthesis (*VviIPTs*), activation (*VviLOGs*), perception (*VviCHKs*), and signaling (*VviRRs*) were found to be expressed in all stages of berry development and most significantly just before and after veraison, and during this time the expression of genes involved in cytokinin degradation (*VviCKXs*) progressively decrease (Böttcher et al. 2015).

12.4.3.3 Brassinosteroids

Expression analysis of genes encoding brassinosteroid (BR) biosynthetic enzymes or BR receptors (i.e., *VviBRI1*) during berry development revealed transcript accumulation patterns consistent with the dramatic increase in endogenous BR levels observed at the onset of fruit ripening (Symons et al. 2006). It has been shown that levels of castasterone, the bioactive BR, and its precursor 6-deoxo-castasterone increase at veraison and remain high during ripening in Cabernet Sauvignon berries due to the upregulation of a brassinosteroid 6-oxidase gene *(VviBR6OX)* (Symons et al. 2006). The application of exogenous brassinosteroid increases the total anthocyanin content in berries, and the full coloration of grapes occurred earlier in BR-treated samples, with increased expression of anthocyanin biosynthetic genes (i.e., *VviF3H*, *VviF3'5'H*, *VviDFR*, *VviANS*, *VviUFGT*) (Luan et al. 2013; Serrano et al. 2017). In addition, the involvement of BR in sugar unloading into the berry has been recently demonstrated. Exogenous treatment of Cabernet Sauvignon berries with BR (24-epibrassinolide) increases the soluble sugar content by enhancing the activities of enzymes related to sugar unloading, including neutral and acidic invertases and sucrose synthase, and up-regulating the expression of sucrose transporter genes (Xu et al. 2015).

12.4.3.4 Gibberellins

The involvement of gibberellins (GAs), produced in the seeds, in grape berry development and size determination is well known (Coombe 1960). GAs peak early during stage I (Davies and Böttcher 2009), and increase again at the initiation of stage III (Pérez et al. 2000).

A comprehensive annotation and characterization of GA-oxidases (GAox)—involved in GAs biosynthesis and deactivation—has been performed in grapevine (Giacomelli et al. 2013). The authors propose that the pool of bioactive GAs is controlled by the stage- and tissue-specific regulation of GA oxidase, and *VviGA3ox1* and *VviGA2ox4* transcripts are significantly up-regulated at fruit set.

RNA-seq analysis of "Centennial Seedless" berries treated with GAs after flowering showed an increased expression of xyloglucan endotransglycosylase (*VviXET*) genes, which participate in cell wall expansion. A crosstalk between GAs, ABA, and ethylene during berry enlargement period has also been reported, and GA3-application induces gene expression changes in plant hormone metabolism and signaling pathways (Chai et al. 2014). Moreover, GAs' soaking of cv. Kyoho clusters strongly hastens berry coloration, which allows the hypothesis of a role for the hormone in regulating anthocyanin biosynthesis (Cheng et al. 2015). In the same study, a large number of the identified differentially expressed genes were involved in GA biosynthetic and signaling pathways. Zhang et al. (2014) provided new insights into the crosstalk mechanism of GAs and glucose hexokinase-dependent signaling during grape berry sugar accumulation, and hypothesized that GAs might regulate the expression of invertase and sucrose synthase genes in order to maintain intracellular sugar levels and normal cell metabolism.

12.4.3.5 Sugars

Notably, besides their role as a metabolic substrate, sugars directly or indirectly control a wide range of processes, including photosynthesis, sugar transport itself, phenylpropanoid metabolism, cell wall metabolism, auxin homeostasis, and ultimately berry growth and ripening (Smeekens et al. 2010). The sugar-dependent regulation of anthocyanin pathway and of biotic/abiotic stress responses has been extensively reviewed by Lecourieux et al. (2014). Interaction between sugar and ABA signaling pathways likely plays a pivotal role in ripening, which is suggested by the parallel increase of sugars and ABA in the berries at veraison (Gambetta et al. 2010; Lecourieux et al. 2014). Interestingly, both sucrose and ABA were able to increase *VviSK1*—a gene encoding a protein kinase with sugar signaling function—expression in grape cell suspensions, which underlines the tight interaction between sugars and hormone signaling pathways (Smeekens 2000; Finkelstein and Gibson 2002; León and Sheen 2003).

12.4.3.6 Jasmonic Acid

The plant hormone jasmonic acid (JA) is crucial for stress responses in plants, but its role in fruit development and ripening is becoming increasingly clear. In non-climacteric fruits such as grape, the jasmonate levels are high at early developmental stages, decreasing to lower values at the onset of ripening (Kondo and Fukuda 2001; Fortes et al. 2011, 2015). Conjugation of JA to isoleucine (JA-Ile) is a critical step in the JA signaling pathway since only JA-Ile is recognized by the jasmonate receptor. The conjugation reaction is catalyzed by JA-amido synthetases, belonging to the family of GH3 proteins. Böttcher et al. (2015) report that the transcriptional profiles of two grapevine GH3 genes, *VviGH3-7* and *VviGH3-9*, support a primary role for JA signaling in fruit set and cell division, but do not justify JA's involvement in the ripening process.

Methyl jasmonate (MeJA) also plays an important role in signal transduction processes that regulate the synthesis of secondary metabolites (Pauwels et al. 2009); grapevine plants and cell cultures respond to MeJA with an increase in aroma compounds or stilbene levels (D'Onofrio et al. 2009; Almagro et al. 2014; D'Onofrio et al. 2018; Portu et al. 2018). The gene coding for S-adenosyl-L-methionine:jasmonic acid carboxyl methyltransferase (JMT), putatively involved in volatile methyl jasmonate synthesis, was down-regulated in ripe fruits of three grape varieties. On the other hand, the gene coding for the jasmonate ZIM domain (JAZ) containing protein 8, a repressor of jasmonic acid signaling, has been identified as a putative positive marker of ripening (Agudelo-Romero et al. 2013). Treatments with MeJA increase the transcription levels of several ripening-related genes, such as color-related genes (i.e., *VviPAL1*, *VviDFR*, *VviCHI*, *VviF3H*, *VviGST*, *VviCHS*, and *VviUFGT*), softening-related genes (i.e., *VviPG*, *VviPL*, *VviPE*, *VviCell*, *VviEG1*, and *VviXTH1*), and aroma-related genes (i.e.,*VviEcar*, *VviQR*, and *VviEGS*). Moreover, jasmonic acid positively regulated its biosynthesis pathway genes such as lipoxygenase (*LOX*), allene oxide synthase (*AOS*), 12-oxophytodienoate reductase (*OPR*), and signal pathway genes such as *VviCOI1* and *VviJMT*. In addition, the overexpression of grape jasmonic acid receptor *VviCOI1* in strawberry fruit accelerated the fruit ripening process (Jia et al. 2016).

12.5 Molecular Regulators of Fruit Ripening

Transcription factors (TFs) regulate the spatial and temporal expression of genes by specific binding to cis-regulatory elements (CREs or "motifs") present in the promoter region of genes. In plants, as many as 58 TF families have been described (Jin et al. 2016), of which many play essential roles in biological processes, including fleshy fruit development, ripening, and regulation of fruit quality/composition (Karlova et al. 2014). A plethora of TFs involved in ripening have been discovered using tomato, a climacteric fruit species, as the model species for understanding fruit ripening. For example, the MADS-box (e.g., RIPENING-INHIBITOR, RIN; FRUITFULL, FUL1 and FUL2), SBP (e.g., COLORLESS NON-RIPENING, CNR; TOMATO AGAMOUS-LIKE1, TAGL1), NAC (e.g., NON-RIPENING, NOR; NAC4), HD-Zip homeobox (HB1), and AP2/ERF (e.g., APETALA2a) TFs are among the many widely known regulators of ripening. Moreover, TFs involved in hormone response and signaling such as AP2/ERFs (e.g., ERF1, ERF6) and ARF (e.g., ARF2) are also implicated in fruit ripening and participate in the regulation of ripening-associated phenotypic traits such as flavonoid/anthocyanin biosynthesis, sugar accumulation, and softening.

While much is known about the regulation of climacteric fruit ripening, our understanding of the TFs involved in ripening remains limited for non-climacteric fruit. The roles of some TFs involved in tomato development and ripening have been elucidated also in grapevine. For example, the MADS-box TF SEPALLATA (VviSEP4) may fulfil similar functions to RIN in grapes, as revealed by its ability to partially complement the non-ripening phenotype of *RIN* mutants (Mellway and Lund 2013).

A grapevine bZIP TF, namely, ABSCISIC ACID RESPONSE ELEMENT-BINDING FACTOR2 (VviABF2), was shown to play a direct role in the ABA-dependent berry ripening processes (Nicolas et al. 2014). Regulatory networks encompassing ABA responses were either enhanced and/or altered by VviABF2, which led to enhanced sensitivity to ABA. In addition, the role of VviABF2 in the regulation of ripening-associated processes such as the biosynthesis of phenolic metabolites was also demonstrated in tomato and grapevine. The lack of MADS-box TF participation together with the enrichment of TFs (i.e., bZIP, AP2/ERF, R2R3-MYB, and NAC) in the ABA signaling network during berry ripening (Pilati et al. 2017) suggest that grapevine MADS-box TFs do not play a key role in overall ripening regulation in grapevine. This is also supported by a strong enrichment of cis-regulatory motifs bound by bZIP and NAC TFs and the lack of MADS-box TF motifs in the promoters of ABA-modulated genes in the berry (Pilati et al. 2017). Nonetheless, other TFs such as VviERF045 (AP2/ERF) (Leida et al. 2016) and VviCEB1 (bHLH) (Nicolas et al. 2013) have been implicated in the control of ripening. For example, genes involved in wax metabolism, cell expansion, defense, and phenylpropanoid/flavonoid metabolism are potential targets of VviERF045, while VviCEB1 may stimulate cell expansion through the activation of auxin metabolism, auxin signaling, and multiple cell expansion related genes.

Beyond these few cases, the function of the vast majority of TFs remains to be elucidated. To facilitate the discovery of fruit-associated TF functions, adoption of multi-omics approaches (i.e., transcriptome, metabolome), the application of network-based approaches to analyze the omics data, and subsequent network integration across different domains could be particularly useful (reviewed in Wong and Matus 2017). For example, gene co-expression network analysis of a large accession of berry cultivars during fruit development and ripening has been performed to identify putative regulators of berry developmental and ripening (Palumbo et al. 2014; Massonnet et al. 2017). Not surprisingly, many of these putative genes encode TFs that belong to AP2/ERF, MYB, NAC, and WRKY families. Independent studies were also able to link several of these ripening-related TFs to their potential roles during berry ripening using gene-metabolite co-response networks (Savoi et al. 2017). For example, VviERF1 and VviNAC33, two common berry TFs (Massonnet et al. 2017), are potentially related to the regulation of proline biosynthesis in the berry, given their strong coordinated regulation with pyrroline-5-carboxylate synthase (P5CS), the gene encoding enzyme involved in proline biosynthesis, and with proline content in the berry. Similarly, NACs such as VviNAC13 and VviNAC33 are potentially new candidate regulators for anthocyanin compounds that exhibit tight association with several anthocyanin biosynthetic gene and metabolite profiles (Savoi et al. 2017).

Such approaches can also be used to infer the regulatory candidates involved in the regulation of fruit-associated volatiles (e.g., terpenes), one of the least understood components of berry ripening. For example, Savoi et al. (2016) highlighted one promising regulatory candidate (VviMYB24) for monoterpene biosynthesis, given its strong gene-metabolite co-response profile with several TPS and monoterpene (e.g., linalool, nerol, α-terpineol) abundance in the fruit during ripening and under an abiotic stress such as drought (Fig. 12.3).

Notwithstanding the crucial roles fulfilled by various TFs during ripening, new evidence supporting the involvement of regulatory non-coding RNA classes, especially micro RNA (miRNA) and long non-coding RNA (lncRNA), in the regulation of fruit ripening and composition have been described. Although it is possible to infer the function of miRNAs in fruits through comprehensive miRNA expression profiling during development and ripening and performing in silico target prediction analysis (Gao et al. 2015; Xin et al. 2015; Zeng et al. 2015; Belli Kullan et al. 2015), the first and only study to date demonstrating a direct role for miRNAs in overall ripening regulation and fruit softening investigated the tomato miR157 and miRNA156 (Chen et al. 2015). Tomato miR156 impacts fruit softening especially at the late stages of ripening but contributes little to overall ripening regulation (Chen et al. 2015). Interestingly, miR156 sequences are highly conserved in plants, including grapevine (Belli Kullan et al. 2015). Like its tomato counterpart, grapevine miR156 also exhibits ripening-associated expression, and it has been postulated to induce ripening via the regulation of multiple SPL (Squamosa Promoter binding Like protein) and anthocyanin pathway genes (Belli Kullan et al. 2015).

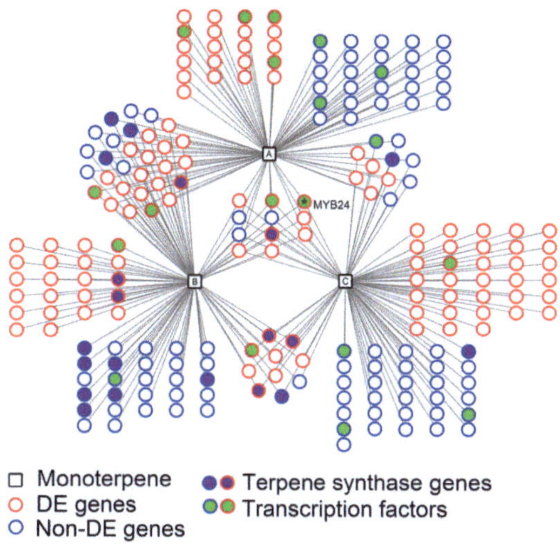

Fig. 12.3 *Predicted gene-metabolite networks related to nerol (A), α-terpineol (B), and linalool (C) accumulation in grape berries during development.* Genes and metabolites are represented by circle and square nodes, respectively. Edges represent associations ($P < 0.001$) between transcripts and metabolites. Node borders in red represent genes that are modulated (differentially expressed, DE) under drought. Purple and green nodes identify terpene synthase genes and transcription factors, respectively. The network was re-designed from Savoi et al. (2016)

Compared to miRNAs, lncRNAs are an emerging class of RNA species that are operationally defined as non-coding transcripts, greater than 200nt in length. The advent of sequencing technologies has led to the discovery of thousands of lncRNAs in both model (Liu et al. 2015) and non-model fruit crops such as tomato (Wang et al. 2018), grapevine (Vitulo et al. 2014; Harris et al. 2017), kiwi (Tang et al. 2016), and sea buckthorn (Zhang et al. 2018); however, for the vast majority of these crops, the functions of lncRNAs remain unknown. Only a small fraction of these have been validated experimentally (Liu et al. 2015). lncRNAs are known to possess tissue- and developmental stage-specific expression in plants and these properties also manifests in the fruit (Tang et al. 2016; Zhang et al. 2018; Wang et al. 2018). Only recently their role in the regulation of fruit ripening and composition was confirmed. For example, using a combination of lncRNA-miRNA-mRNA network and functional analysis, LNC1 and LNC2 were shown to be negative and positive regulators, respectively, of anthocyanin in sea buckhorn fruits.

While novel lncRNAs continue to be discovered in grapevines (Vitulo et al. 2014; Harris et al. 2017), very little work has been done to profile their expression during ripening and/or to infer their potential regulatory role in the fruit. To date, this was done only to understand the complex regulation of phenylpropanoid and flavonoid biosynthesis in the grape berry (Wong and Matus 2017). Using integrated lncRNA-miRNA-mRNA network analysis (as in Zhang et al. 2018), several lncRNAs identified showed strong co-regulated expression and co-location with key structural pathway genes. Notable examples include one lncRNA (VIT_210s0042n00100) that is situated within close proximity of nine *VviSTSs*. The expression pattern of the lncRNA closely mirrored the ripening-associated expression of the nine *VviSTSs*. Similarly, one predicted lncRNA (VIT_203s0180n00020) was co-located and closely mirrored the expression of *VviGT2*, a gene potentially involved in the production of hydroxycinnamic esters and proanthocyanidins galloylation (Khater et al. 2012). Such initiatives have provided a glimpse into the potential large-scale regulatory function of lncRNAs on the regulation of fruit composition during development and ripening.

12.6 Conclusion

Taken together, all these studies and information indicate the complex feedback and multifaceted regulation of grape berry ripening. The long-standing interest in grapevine production has led to a good knowledge in this field, but a large number of research questions, many of which have crucial practical implications, still need to be answered. New insights about the control of berry metabolism and ripening will be gained by clearly assigning functions to key regulators of these processes. This is challenging and will require innovative functional genomic approaches; in this regard, new-generation sequencing and emerging genome editing technologies, currently being developed for grapevine, could provide important contributions to our understanding.

References

Adams DO (2006) Phenolics and ripening in grape berries. Am J Enol Vitic 57:249–256

Afoufa-Bastien D, Medici A, Jeauffre J, Coutos-Thévenot P, Lemoine R, Atanassova R et al (2010) The Vitis vinifera sugar transporter gene family: phylogenetic overview and macroarray expression profiling. BMC Plant Biol 10:245. https://doi.org/10.1186/1471-2229-10-245

Agasse A, Vignault C, Kappel C, Conde C, Gerós H, Delrot S. (2009) Sugar transport & sugar sensing in grape. In: Grapevine molecular physiology & biotechnology. Springer, Dordrecht, pp 105–139

Ageorges A, Issaly N, Picaud S, Delrot S, Romieu C (2000) Identification and functional expression in yeast of a grape berry sucrose carrier. Plant Physiol Biochem 38:177–185. https://doi.org/10.1016/S0981-9428(00)00730-0

Agudelo-Romero P, Erban A, Sousa L, Pais MS, Kopka J, Fortes AM (2013) Search for transcriptional and metabolic markers of grape pre-ripening and ripening and insights into specific aroma development in three portuguese cultivars. PLoS ONE 8:e60422. https://doi.org/10.1371/journal.pone.0060422

Almagro L, Carbonell-Bejerano P, Belchí-Navarro S, Bru R, Martínez-Zapater JM, Lijavetzky D, Pedreño MA (2014) Dissecting the transcriptional response to elicitors in *Vitis vinifera* cells. PLoS ONE 9:e109777. https://doi.org/10.1371/journal.pone.0109777

Battilana J, Costantini L, Emanuelli F, Sevini F, Segala C, Moser S, Velasco R, Versini G, Grando MS (2009) The 1-deoxy-D-xylulose 5-phosphate synthase gene co-localizes with a major QTL affecting monoterpene content in grapevine. Theor Appl Genet 118:653–669. https://doi.org/10.1007/s00122-008-0927-8

Battilana J, Emanuelli F, Gambino G, Gribaudo I, Gasperi F, Boss PK, Grando MS (2011) Functional effect of grapevine 1-deoxy-D-xylulose 5-phosphate synthase substitution K284 N on Muscat flavour formation. J Exp Bot 62:5497–5508. https://doi.org/10.1093/jxb/err231

Belli Kullan J, Lopes Paim Pinto D, Bertolini E, Fasoli M, Zenoni S, Tornielli GB, Pezzotti M, Meyers BC, Farina L, Pè ME, Mica E (2015) miRVine: a microRNA expression atlas of grapevine based on small RNA sequencing. BMC Genom 16:393. https://doi.org/10.1186/s12864-015-1610-5

Bogs J, Downey MO, Harvey JS, Ashton AR, Tanner GJ, Robinson SP (2005) Proanthocyanidin synthesis and expression of genes encoding leucoanthocyanidin reductase and anthocyanidin reductase in developing grape berries and grapevine leaves. Plant Physiol 139:652–663. https://doi.org/10.1104/pp.105.064238

Bogs J, Jaffé FW, Takos AM, Walker AR, Robinson SP (2007) The grapevine transcription factor VvMYBPA1 regulates proanthocyanidin synthesis during fruit development. Plant Physiol 143:1347–1361. https://doi.org/10.1104/pp.106.093203

Bohlmann J, Keeling CI (2008) Terpenoid biomaterials. Plant J 54:656–669. https://doi.org/10.1111/j.1365-313X.2008.03449.x

Bönisch F, Frotscher J, Stanitzek S, Rühl E, Wüst M, Bitz O, Schwab W (2014a) Activity-based profiling of a physiologic aglycone library reveals sugar acceptor promiscuity of family 1 UDP-glucosyltransferases from grape. Plant Physiol 166:23–39. https://doi.org/10.1104/pp.114.242578

Bönisch F, Frotscher J, Stanitzek S, Rühl E, Wüst M, Bitz O, Schwab W (2014b) A UDP-glucose: monoterpenol glucosyltransferase adds to the chemical diversity of the grapevine metabolome. Plant Physiol 165:561–581. https://doi.org/10.1104/pp.113.232470

Böttcher C, Boss PK, Davies C (2012a) Delaying Riesling grape berry ripening with a synthetic auxin affects malic acid metabolism and sugar accumulation, and alters wine sensory characters. Funct Plant Biol 39:745. https://doi.org/10.1071/FP12132

Böttcher C, Boss PK, Davies C (2011) Acyl substrate preferences of an IAA-amido synthetase account for variations in grape (*Vitis vinifera* L.) berry ripening caused by different auxinic compounds indicating the importance of auxin conjugation in plant development. J Exp Bot 62:4267–4280. https://doi.org/10.1093/jxb/err134

Böttcher C, Burbidge CA, Boss PK, Davies C (2015) Changes in transcription of cytokinin metabolism and signalling genes in grape (*Vitis vinifera* L.) berries are associated with the ripening-related increase in isopentenyladenine. BMC Plant Biol 15:223. https://doi.org/10.1186/s12870-015-0611-5

Böttcher C, Dennis EG, Booker GW, Polyak SW, Boss PK, Davies C (2012b) A novel tool for studying auxin-metabolism: the inhibition of grapevine indole-3-acetic acid-amido synthetases by a reaction intermediate analogue. PLoS ONE 7:e37632. https://doi.org/10.1371/journal.pone.0037632

Böttcher C, Harvey KE, Boss PK, Davies C (2013) Ripening of grape berries can be advanced or delayed by reagents that either reduce or increase ethylene levels. Funct Plant Biol 40:566. https://doi.org/10.1071/FP12347

Böttcher C, Keyzers RA, Boss PK, Davies C (2010) Sequestration of auxin by the indole-3-acetic acid-amido synthetase GH3-1 in grape berry (*Vitis vinifera* L.) and the proposed role of auxin conjugation during ripening. J Exp Bot 61:3615–3625. https://doi.org/10.1093/jxb/erq174

Brummell DA, Harpster MH (2001) Cell wall metabolism in fruit softening and quality and its manipulation in transgenic plants. Plant Mol Biol 47:311–340

Cabezas JA, Cervera MT, Ruiz-García L, Carreño J, Martínez-Zapater JM (2006) A genetic analysis of seed and berry weight in grapevine. Genome 49:1572–1585. https://doi.org/10.1139/g06-122

Caputi L, Carlin S, Ghiglieno I, Stefanini M, Valenti L, Vrhovsek U, Mattivi F (2011) Relationship of changes in rotundone content during grape ripening and winemaking to manipulation of the 'peppery' character of wine. J Agric Food Chem 59:5565–5571. https://doi.org/10.1021/jf200786u

Castellarin SD, Di Gaspero G (2007) Transcriptional control of anthocyanin biosynthetic genes in extreme phenotypes for berry pigmentation of naturally occurring grapevines. BMC Plant Biol 7:46. https://doi.org/10.1186/1471-2229-7-46

Castellarin SD, Gambetta GA, Wada H, Krasnow MN, Cramer GR, Peterlunger E, Shackel KA, Matthews MA (2016) Characterization of major ripening events during softening in grape: turgor, sugar accumulation, abscisic acid metabolism, colour development, and their relationship with growth. J Exp Bot 67:709–722. https://doi.org/10.1093/jxb/erv483

Castellarin SD, Gambetta GA, Wada H, Shackel KA, Matthews MA (2011) Fruit ripening in *Vitis vinifera*: spatiotemporal relationships among turgor, sugar accumulation, and anthocyanin biosynthesis. J Exp Bot 62:4345–4354. https://doi.org/10.1093/jxb/err150

Cavallini E, Matus JT, Finezzo L, Zenoni S, Loyola R, Guzzo F, Schlechter R, Ageorges A, Arce-Johnson P, Tornielli GB (2015) The phenylpropanoid pathway is controlled at different branches by a set of R2R3-MYB C2 repressors in grapevine. Plant Physiol 167:1448–1470. https://doi.org/10.1104/pp.114.256172

Chai L, Li Y, Chen S, Perl A, Zhao F, Ma H (2014) RNA sequencing reveals high resolution expression change of major plant hormone pathway genes after young seedless grape berries treated with gibberellin. Plant Sci 229:215–224. https://doi.org/10.1016/j.plantsci.2014.09.010

Chen F, Tholl D, Bohlmann J, Pichersky E (2011) The family of terpene synthases in plants: a mid-size family of genes for specialized metabolism that is highly diversified throughout the kingdom. Plant J 66:212–229. https://doi.org/10.1111/j.1365-313X.2011.04520.x

Chen L-Q (2014) SWEET sugar transporters for phloem transport and pathogen nutrition. New Phytol 201:1150–1155

Chen L-Q, Qu X-Q, Hou B-H, Sosso D, Osorio S, Fernie AR et al (2012) Sucrose efflux mediated by SWEET proteins as a key step for phloem transport. Science (80-) 335:207–211. https://doi.org/10.1126/science.1213351

Chen W, Kong J, Lai T, Manning K, Wu C, Wang Y, Qin C, Li B, Yu Z, Zhang X, He M, Zhang P, Gu M, Yang X, Mahammed A, Li C, Osman T, Shi N, Wang H, Jackson S, Liu Y, Gallusci P (2015) Tuning LeSPL-CNR expression by SlymiR157 affects tomato fruit ripening. Sci Rep 5:7852. https://doi.org/10.1038/srep07852

Chen Y, Grimplet J, David K, Castellarin SD, Terol J, Wong DCJ, Luo Z, Schaffer R, Celton J-M, Talon M, Gambetta GA, Chervin C (2018) Ethylene receptors and related proteins in climacteric and non-climacteric fruits. Plant Sci 276:63–72. https://doi.org/10.1016/J.PLANTSCI.2018.07.012

Cheng C, Jiao C, Singer SD, Gao M, Xu X, Zhou Y, Li Z, Fei Z, Wang Y, Wang X (2015) Gibberellin-induced changes in the transcriptome of grapevine (*Vitis labrusca* × *V. vinifera*) cv. Kyoho flowers. BMC Genom 16:128. https://doi.org/10.1186/s12864-015-1324-8

Cheng J, Wen S, Xiao S, Lu B, Ma M, Bie Z (2018) Overexpression of the tonoplast sugar transporter CmTST2 in melon fruit increases sugar accumulation. J Exp Bot 69:511–523. https://doi.org/10.1093/jxb/erx440

Chervin C, Deluc L (2010) Ethylene signalling receptors and transcription factors over the grape berry development: Gene expression profiling. Vitis J Grapevine Res 49:129–136. https://doi.org/10.1016/j.ssi.2010.01.014

Chervin C, Tira-umphon A, Terrier N, Zouine M, Severac D, Roustan J-P (2008) Stimulation of the grape berry expansion by ethylene and effects on related gene transcripts, over the ripening phase. Physiol Plant 134:534–546. https://doi.org/10.1111/j.1399-3054.2008.01158.x

Chialva C, Eichler E, Grissi C, Muñoz C, Gomez-Talquenca S, Martínez-Zapater JM, Lijavetzky D (2016) Expression of grapevine AINTEGUMENTA-like genes is associated with variation in ovary and berry size. Plant Mol Biol 91:67–80. https://doi.org/10.1007/s11103-016-0443-1

Choat B, Gambetta GA, Shackel KA, Matthews MA (2009) Vascular function in grape berries across development and its relevance to apparent hydraulic isolation. Plant Physiol 151:1677–1687. https://doi.org/10.1104/pp.109.143172

Cholet C, Claverol S, Claisse O, Rabot A, Osowsky A, Dumot V, Ferrari G, Gény L (2016) Tartaric acid pathways in *Vitis vinifera* L. (cv. Ugni blanc): a comparative study of two vintages with contrasted climatic conditions. BMC Plant Biol 16:144. https://doi.org/10.1186/s12870-016-0833-1

Chong J, Piron M-C, Meyer S, Merdinoglu D, Bertsch C, Mestre P (2014) The SWEET family of sugar transporters in grapevine: VvSWEET4 is involved in the interaction with *Botrytis cinerea*. J Exp Bot 65:6589–6601. https://doi.org/10.1093/jxb/eru375

Coombe B, Bishop G (1980) Development of the grape berry. II Changes in diameter and deformability during veraison. Aust J Agric Res 31:499–509

Coombe BG (1976) The development of fleshy fruit. Ann Rev Plant Physiol 27:507–528

Coombe BG (1992) Research on development and ripening of the grape berry. Am J Enol Vitic 43:101–110

Coombe BG (1960) Relationship of growth and development to changes in sugars, auxins, and gibberellins in fruit of seeded and seedless varieties of *Vitis vinifera*. Plant Physiol 35:241–250

Cosgrove DJ (2005) Growth of the plant cell wall. Nat Rev Mol Cell Biol 6:850–861. https://doi.org/10.1038/nrm1746

Costantini L, Kappel CD, Trenti M, Battilana J, Emanuelli F, Sordo M, Moretto M, Camps C, Larcher R, Delrot S, Grando MS (2017) Drawing links from transcriptome to metabolites: the evolution of aroma in the ripening berry of Moscato Bianco (*Vitis vinifera* L.). Front Plant Sci. https://doi.org/10.3389/fpls.2017.00780

Costantini L, Malacarne G, Lorenzi S, Troggio M, Mattivi F, Moser C, Grando MS (2015) New candidate genes for the fine regulation of the colour of grapes. J Exp Bot 66:4427–4440. https://doi.org/10.1093/jxb/erv159

Cramer GR, Ghan R, Schlauch KA, Tillett RL, Heymann H, Ferrarini A, Delledonne M, Zenoni S, Fasoli M, Pezzotti M (2014) Transcriptomic analysis of the late stages of grapevine (*Vitis vinifera* cv. Cabernet Sauvignon) berry ripening reveals significant induction of ethylene signaling and flavor pathways in the skin. BMC Plant Biol 14:370. https://doi.org/10.1186/s12870-014-0370-8

Czemmel S, Höll J, Loyola R, Arce-Johnson P, Alcalde JA, Matus JT, Bogs J (2017) Transcriptome-wide identification of novel uv-b- and light modulated flavonol pathway genes controlled by VviMYBF1. Front Plant Sci 8:1084. https://doi.org/10.3389/fpls.2017.01084

Czemmel S, Stracke R, Weisshaar B, Cordon N, Harris NN, Walker AR, Robinson SP, Bogs J (2009) The grapevine R2R3-MYB transcription factor VvMYBF1 regulates flavonol synthesis in developing grape berries. Plant Physiol 151:1513–1530. https://doi.org/10.1104/pp.109.142059

D'Onofrio C, Matarese F, Cuzzola A (2018) Effect of methyl jasmonate on the aroma of Sangiovese grapes and wines. Food Chem 242:352–361. https://doi.org/10.1016/J.FOODCHEM.2017.09.084

D'Onofrio C, Cox A, Davies C, Boss PK (2009) Induction of secondary metabolism in grape cell cultures by jasmonates. Funct Plant Biol 36:323. https://doi.org/10.1071/FP08280

Dai ZW, Leon C, Feil R et al (2013) Metabolic profiling reveals coordinated switches in primary carbohydrate metabolism in grape berry (*Vitis vinifera* L.), a non-climacteric fleshy fruit. J Exp Bot 64:1345–1355. https://doi.org/10.1093/jxb/ers396

Dai ZW, Ollat N, Gomes E, Decroocq S, Tandonnet J-P, Bordenave L, Pieri P, Hilbert G, Kappel C, van Leeuwen C, Vivin P, Delrot S (2011) Ecophysiological, genetic, and molecular causes of variation in grape berry weight and composition: a review. Ann Rev Plant Physiol 62:413–425. https://doi.org/10.5344/ajev.2011.10116

Dal Santo S, Vannozzi A, Tornielli GB, Fasoli M, Venturini L, Pezzotti M, Zenoni S (2013) Genome-wide analysis of the expansin gene superfamily reveals grapevine-specific structural and functional characteristics. PLoS ONE 8:e62206. https://doi.org/10.1371/journal.pone.0062206

Davies C, Boss PK, Robinson SP (1997) Treatment of Grape Berries, a Nonclimacteric Fruit with a Synthetic Auxin, Retards Ripening and Alters the Expression of Developmentally Regulated Genes. Plant Physiol 115:1155–1161. https://doi.org/10.1104/PP.115.3.1155

Davies C, Böttcher C (2009) Hormonal control of grape berry ripening. In: Roubelakis-Angelakis KA (ed) Grapevine molecular physiology & biotechnology. Springer, Dordrecht, pp 229–261

Davies C, Robinson SP (1996) Sugar accumulation in grape berries. Cloning of two putative vacuolar invertase cDNAs and their expression in grapevine tissues. Plant Physiol 111:275–283
Davies C, Wolf T, Robinson SP (1999) Three putative sucrose transporters are differentially expressed in grapevine tissues. Plant Sci 147:93–100. https://doi.org/10.1016/S0168-9452(99)00059-X
DeBolt S, Cook DR, Ford CM (2006) L-Tartaric acid synthesis from vitamin C in higher plants. Proc Natl Acad Sci 103:5608–5613. https://doi.org/10.1073/pnas.0510864103
Deluc LG, Grimplet J, Wheatley MD, Tillett RL, Quilici DR, Osborne C, Schooley DA, Schlauch KA, Cushman JC, Cramer GR (2007) Transcriptomic and metabolite analyses of Cabernet Sauvignon grape berry development. BMC Genom 8:429. https://doi.org/10.1186/1471-2164-8-429
Doligez A, Bouquet A, Danglot Y, Lahogue F, Riaz S, Meredith C, Edwards K, This P (2002) Genetic mapping of grapevine (*Vitis vinifera* L.) applied to the detection of QTLs for seedlessness and berry weight. Theor Appl Genet 105:780–795. https://doi.org/10.1007/s00122-002-0951-z
Downey MO, Harvey JS, Robinson SP (2004) The effect of bunch shading on berry development and flavonoid accumulation in Shiraz grapes. Aust J Grape Wine Res 10:55–73. https://doi.org/10.1111/j.1755-0238.2004.tb00008.x
Drew DP, Andersen TB, Sweetman C, Møller BL, Ford C, Simonsen HT (2015) Two key polymorphisms in a newly discovered allele of the *Vitis vinifera* TPS24 gene are responsible for the production of the rotundone precursor α-guaiene. J Exp Bot. https://doi.org/10.1093/jxb/erv491
Dunlevy JD, Soole KL, Perkins MV, Dennis EG, Keyzers RA, Kalua CM, Boss PK (2010) Two O-methyltransferases involved in the biosynthesis of methoxypyrazines: grape-derived aroma compounds important to wine flavour. Plant Mol Biol 74:77–89. https://doi.org/10.1007/s11103-010-9655-y
El-Kereamy A, Chervin C, Roustan J-P, Cheynier V, Souquet J-M, Moutounet M, Raynal J, Ford C, Latche A, Pech J-C, Bouzayen M (2003) Exogenous ethylene stimulates the long-term expression of genes related to anthocyanin biosynthesis in grape berries. Physiol Plant 119:175–182. https://doi.org/10.1034/j.1399-3054.2003.00165.x
Falginella L, Castellarin SD, Testolin R, Gambetta GA, Morgante M, Di Gaspero G (2010) Expansion and subfunctionalisation of flavonoid 3′,5′-hydroxylases in the grapevine lineage. BMC Genom 11:562. https://doi.org/10.1186/1471-2164-11-562
Falginella L, Di Gaspero G, Castellarin SD (2012) Expression of flavonoid genes in the red grape berry of "Alicante Bouschet" varies with the histological distribution of anthocyanins and their chemical composition. Planta 236:1037–1051. https://doi.org/10.1007/s00425-012-1658-2
Fasoli M, Dell'Anna R, Dal Santo S, Balestrini R, Sanson A, Pezzotti M, Monti F, Zenoni S (2016) Pectins, hemicelluloses and celluloses show specific dynamics in the internal and external surfaces of grape berry skin during ripening. Plant Cell Physiol 57:1332–1349. https://doi.org/10.1093/pcp/pcw080
Fernandez L, Chaïb J, Martinez-Zapater J-M, Thomas MR, Torregrosa L (2013) Mis-expression of a *PISTILLATA-* like MADS box gene prevents fruit development in grapevine. Plant J 73:918–928. https://doi.org/10.1111/tpj.12083
Fernandez L, Doligez A, Lopez G, Thomas MR, Bouquet A, Torregrosa L (2006a) Somatic chimerism, genetic inheritance, and mapping of the *fleshless berry* (*flb*) mutation in grapevine (*Vitis vinifera* L.). Genome 49:721–728. https://doi.org/10.1139/g06-034
Fernandez L, Romieu C, Moing A, Bouquet A, Maucourt M, Thomas MR, Torregrosa L (2006b) The grapevine fleshless berry mutation. A unique genotype to investigate differences between fleshy and nonfleshy fruit. Plant Physiol 140:537–547. https://doi.org/10.1104/pp.105.067488
Ferrandino A, Lovisolo C (2014) Abiotic stress effects on grapevine (*Vitis vinifera* L.): focus on abscisic acid-mediated consequences on secondary metabolism and berry quality. Environ Exp Bot 103:138–147. https://doi.org/10.1016/J.ENVEXPBOT.2013.10.012
Fillion L, Ageorges A, Picaud S et al (1999) Cloning and expression of a hexose transporter gene expressed during the ripening of grape berry. Plant Physiol 120:1083–1094. https://doi.org/10.1104/pp.120.4.1083
Finkelstein RR, Gibson SI (2002) ABA and sugar interactions regulating development: cross-talk or voices in a crowd? Curr Opin Plant Biol 5:26–32
Fortes A, Teixeira R, Agudelo-Romero P (2015) Complex interplay of hormonal signals during grape berry ripening. Molecules 20:9326–9343. https://doi.org/10.3390/molecules20059326
Fortes AM, Agudelo-Romero P, Silva MS, Ali K, Sousa L, Maltese F, Choi YH, Grimplet J, Martinez-Zapater JM, Verpoorte R (2011) Transcript and metabolite analysis in Trincadeira cultivar reveals novel information regarding the dynamics of grape ripening. BMC Plant Biol 11:149–183
Fournier-Level A, Hugueney P, Verriès C et al (2011) Genetic mechanisms underlying the methylation level of anthocyanins in grape (*Vitis vinifera* L.). BMC Plant Biol 11:179. https://doi.org/10.1186/1471-2229-11-179
Francisco RM, Regalado A, Ageorges A, Burla BJ, Bassin B, Eisenach C, Zarrouk O, Vialet S, Marlin T, Chaves MM, Martinoia E, Nagy R (2013) ABCC1, an ATP binding cassette protein from grape berry, transports anthocyanidin 3-*O*-glucosides. Plant Cell 25:1840–1854. https://doi.org/10.1105/tpc.112.102152
Gambetta GA, Matthews MA, Shaghasi TH, McElrone AJ, Castellarin SD (2010) Sugar and abscisic acid signaling orthologs are activated at the

onset of ripening in grape. Planta 232:219–234. https://doi.org/10.1007/s00425-010-1165-2

Gao C, Ju Z, Cao D, Zhai B, Qin G, Zhu H, Fu D, Luo Y, Zhu B (2015) MicroRNA profiling analysis throughout tomato fruit development and ripening reveals potential regulatory role of RIN on microRNAs accumulation. Plant Biotechnol J 13:370–382. https://doi.org/10.1111/pbi.12297

Gapper NE, McQuinn RP, Giovannoni JJ (2013) Molecular and genetic regulation of fruit ripening. Plant Mol Biol 82:575–591. https://doi.org/10.1007/s11103-013-0050-3

Gatto P, Vrhovsek U, Muth J, Segala C, Romualdi C, Fontana P, Pruefer D, Stefanini M, Moser C, Mattivi F, Velasco R (2008) Ripening and genotype control stilbene accumulation in healthy grapes. J Agric Food Chem 56:11773–11785. https://doi.org/10.1021/jf8017707

Giacomelli L, Rota-Stabelli O, Masuero D, Acheampong AK, Moretto M, Caputi L, Vrhovsek U, Moser C (2013) Gibberellin metabolism in *Vitis vinifera* L. during bloom and fruit-set: functional characterization and evolution of grapevine gibberellin oxidases. J Exp Bot 64:4403–4419. https://doi.org/10.1093/jxb/ert251

Givan CV (1999) Evolving concepts in plant glcolysis: two centuries of progress. Biol Rev 74:277–309

Gomez C, Terrier N, Torregrosa L, Vialet S, Fournier-Level A, Verriès C, Souquet JM, Mazauric JP, Klein M, Cheynier V, Ageorges A (2009) Grapevine MATE-type proteins act as vacuolar H + -dependent acylated anthocyanin transporters. Plant Physiol 150:402–415. https://doi.org/10.1104/pp.109.135624

Gouthu S, Deluc LG (2015) Timing of ripening initiation in grape berries and its relationship to seed content and pericarp auxin levels. BMC Plant Biol 15:46. https://doi.org/10.1186/s12870-015-0440-6

Grimplet J, Deluc L, Tillett R, Wheatley M, Schlauch K, Cramer G, Cushman J (2007) Tissue-specific mRNA expression profiling in grape berry tissues. BMC Genom 8:187

Guillaumie S, Ilg A, Rety S, Brette M, Trossat-Magnin C, Decroocq S, Leon C, Keime C, Ye T, Baltenweck-Guyot R, Claudel P, Bordenave L, Vanbrabant S, Duchene E, Delrot S, Darriet P, Hugueney P, Gomes E (2013) Genetic analysis of the biosynthesis of 2-methoxy-3-isobutylpyrazine, a major grape-derived aroma compound impacting wine quality. Plant Physiol 162:604–615. https://doi.org/10.1104/pp.113.218313

Hardy PJ (1968) Metabolism of sugars and organic acids in immature grape berries. Plant Physiol 43:224–228. https://doi.org/10.1104/pp.43.2.224

Harris ZN, Kovacs LG, Londo JP (2017) RNA-seq-based genome annotation and identification of long-noncoding RNAs in the grapevine cultivar 'Riesling'. BMC Genom 18:937. https://doi.org/10.1186/s12864-017-4346-6

Hawker JS, Ruffner HP, Walker RR (1976) The sucrose content of some Australian grapes. Am J Enol Vitic 27:125–129

Hayashi K (2012) The interaction and integration of auxin signaling components. Plant Cell Physiol 53:965–975. https://doi.org/10.1093/pcp/pcs035

Hayes MA, Davies C, Dry IB (2007) Isolation, functional characterization, and expression analysis of grapevine (*Vitis vinifera* L.) hexose transporters: differential roles in sink and source tissues. J Exp Bot 58:1985–1997. https://doi.org/10.1093/jxb/erm061

Herrera JC, Castellarin SD (2016) Preveraison water deficit accelerates berry color change in merlot grapevines. Am J Enol Vitic 67:356–360. https://doi.org/10.5344/ajev.2016.15083

Hichri I, Heppel SC, Pillet J, Léon C, Czemmel S, Delrot S, Lauvergeat V, Bogs J (2010) The basic helix-loop-helix transcription factor MYC1 is involved in the regulation of the flavonoid biosynthesis pathway in grapevine. Mol Plant 3:509–523. https://doi.org/10.1093/mp/ssp118

Hilt C, Bessis R (2003) Abscission of grapevine fruitlets in relation to ethylene biosynthesis. Vitis 42:1–3

Höll J, Vannozzi A, Czemmel S, D'Onofrio C, Walker AR, Rausch T, Lucchin M, Boss PK, Dry IB, Bogs J (2013) The R2R3-MYB transcription factors MYB14 and MYB15 regulate stilbene biosynthesis in *Vitis vinifera*. Plant Cell 25:4135–4149. https://doi.org/10.1105/tpc.113.117127

Huang Y-F, Doligez A, Fournier-Level A et al (2012) Dissecting genetic architecture of grape proanthocyanidin composition through quantitative trait locus mapping. BMC Plant Biol 12:30. https://doi.org/10.1186/1471-2229-12-30

Huang Y-F, Doligez A, Fournier-Level A, Le Cunff L, Bertrand Y, Canaguier A, Morel C, Miralles V, Veran F, Souquet J-M, Cheynier V, Terrier N, This P (2014) A negative MYB regulator of proanthocyanidin accumulation, identified through expression quantitative locus mapping in the grape berry. New Phytol 201:795–809. https://doi.org/10.1111/nph.12557

Ilc T, Halter D, Miesch L, Lauvoisard F, Kriegshauser L, Ilg A, Baltenweck R, Hugueney P, Werck-Reichhart D, Duchêne E, Navrot N (2017) A grapevine cytochrome P450 generates the precursor of wine lactone, a key odorant in wine. New Phytol 213:264–274. https://doi.org/10.1111/nph.14139

Inaba A, Ishida M, Sobajima Y (1976) Changes in endogenous hormone concentrations during berry development in relation to the ripening of delaware grapes. J Jpn Soc Hortic Sci 45:245–252. https://doi.org/10.2503/jjshs.45.245

Jia H, Zhang C, Pervaiz T, Zhao P, Liu Z, Wang B, Wang C, Zhang L, Fang J, Qian J (2016) Jasmonic acid involves in grape fruit ripening and resistant against *Botrytis cinerea*. Funct Integr Genomics 16:79–94. https://doi.org/10.1007/s10142-015-0468-6

Jin J, Tian F, Yang D-C, Meng Y-Q, Kong L, Luo J, Gao G (2016) PlantTFDB 4.0: toward a central hub for transcription factors and regulatory interactions in plants. Nucleic Acids Res 45:gkw982. https://doi.org/10.1093/nar/gkw982

Kalua CM, Boss PK (2009) Evolution of volatile compounds during the development of cabernet sauvignon grapes (*Vitis vinifera* L.). J Agric Food Chem 57:3818–3830. https://doi.org/10.1021/jf803471n

Karlova R, Chapman N, David K, Angenent GC, Seymour GB, De Maagd RA (2014) Transcriptional control of fleshy fruit development and ripening. J Exp Bot 65:4527–4541. https://doi.org/10.1093/jxb/eru316

Keller M, Smith JP, Bondada BR (2006) Ripening grape berries remain hydraulically connected to the shoot. J Exp Bot 57:2577–2587. https://doi.org/10.1093/jxb/erl020

Khater F, Fournand D, Vialet S, Meudec E, Cheynier V, Terrier N (2012) Identification and functional characterization of cDNAs coding for hydroxybenzoate/hydroxycinnamate glucosyltransferases co-expressed with genes related to proanthocyanidin biosynthesis. J Exp Bot 63:1201–1214. https://doi.org/10.1093/jxb/err340

Kliewer WM (1966) Sugars and organic acids of *Vitis vinifera*. Plant Physiol 41:923–931

Kliewer WM, Howarth L, Omori M (1967) Concentrations of tartaric acid and malic acids and their salts in *Vitis vinifera* grapes. Am J Enol Vitic 18:42–54

Knipfer T, Fei J, Gambetta GA, McElrone AJ, Shackel KA, Matthews MA (2015) Water transport properties of the grape pedicel during fruit development: insights into xylem anatomy and function using microtomography. Plant Physiol 168:1590–1602. https://doi.org/10.1104/pp.15.00031

Kobayashi H, Takase H, Suzuki Y, Tanzawa F, Takata R, Fujita K, Kohno M, Mochizuki M, Suzuki S, Konno T (2011) Environmental stress enhances biosynthesis of flavor precursors, *S*-3-(hexan-1-ol)-glutathione and *S*-3-(hexan-1-ol)-L-cysteine, in grapevine through glutathione *S*-transferase activation. J Exp Bot 62:1325–1336. https://doi.org/10.1093/jxb/erq376

Kondo S, Fukuda K (2001) Changes of jasmonates in grape berries and their possible roles in fruit development. Sci Hortic (Amsterdam) 91:275–288. https://doi.org/10.1016/S0304-4238(01)00271-0

Koyama K, Numata M, Nakajima I, Goto-Yamamoto N, Matsumura H, Tanaka N (2014) Functional characterization of a new grapevine MYB transcription factor and regulation of proanthocyanidin biosynthesis in grapes. J Exp Bot 65:4433–4449. https://doi.org/10.1093/jxb/eru213

Koyama K, Sadamatsu K, Goto-Yamamoto N (2010) Abscisic acid stimulated ripening and gene expression in berry skins of the Cabernet Sauvignon grape. Funct Integr Genomics 10:367–381. https://doi.org/10.1007/s10142-009-0145-8

Kuhn N, Guan L, Dai ZW, Wu B-H, Lauvergeat V, Gomès E, Li S-H, Godoy F, Arce-Johnson P, Delrot S (2013) Berry ripening: recently heard through the grapevine. J Exp Bot 65:4543–4559. https://doi.org/10.1093/jxb/ert395

Lalonde S, Wipf D, Frommer WB (2004) Transport mechanisms for organic forms of carbon and nitrogen between source and sink. Ann Rev Plant Biol 55:341–372. https://doi.org/10.1146/annurev.arplant.55.031903.141758

Lashbrooke JG, Young PR, Dockrall SJ, Vasanth K, Vivier MA (2013) Functional characterisation of three members of the *Vitis vinifera* L. carotenoid cleavage dioxygenase gene family. BMC Plant Biol 13:156. https://doi.org/10.1186/1471-2229-13-156

Lauvergeat V, Decendit A, Richard T, Deluc L (2006) Characterization of a grapevine R2R3-MYB transcription factor that regulates the phenylpropanoid pathway. Plant Physiol 140:499–511. https://doi.org/10.1104/pp.105.067231.ered

Lecourieux F, Kappel C, Lecourieux D, Serrano A, Torres E, Arce-Johnson P, Delrot S (2014) An update on sugar transport and signalling in grapevine. J Exp Bot 65:821–832. https://doi.org/10.1093/jxb/ert394

Leida C, Rì AD, Costa LD, Gómez MD, Pompili V, Sonego P, Engelen K, Masuero D, Ríos G, Moser C (2016) Insights into the Role of the Berry-Specific Ethylene Responsive Factor VviERF045. Frontiers in Plant Science 7:1793. https://doi.org/10.3389/fpls.2016.01793

Leng P, Yuan B, Guo Y (2014) The role of abscisic acid in fruit ripening and responses to abiotic stress. J Exp Bot 65:4577–4588

León P, Sheen J (2003) Sugar and hormone connections. Trends Plant Sci 8:110–116. https://doi.org/10.1016/S1360-1385(03)00011-6

Li S-B, Xie Z-Z, Hu C-G, Zhang J-Z (2016) A review of auxin response factors (ARFs) in plants. Front Plant Sci 7:47. https://doi.org/10.3389/fpls.2016.00047

Li X-Y, Wen Y-Q, Meng N, Qian X, Pan Q-H (2017) Monoterpenyl glycosyltransferases differentially contribute to production of monoterpenyl glycosides in two aromatic *Vitis vinifera* varieties. Front Plant Sci 8:1–13. https://doi.org/10.3389/fpls.2017.01226

Lim SD, Yim WC, Liu D, Hu R, Yang X, Cushman JC (2018) A *Vitis vinifera* basic helix-loop-helix transcription factor enhances plant cell size, vegetative biomass and reproductive yield. Plant Biotechnol J 1:15. https://doi.org/10.1111/pbi.12898

Liu H-F, Wu B-H, Fan P-G, Li S-H, Li L-S (2006) Sugar and acid concentrations in 98 grape cultivars analyzed by principal component analysis. J Sci Food Agric 86:1526–1536. https://doi.org/10.1002/jsfa.2541

Liu J, Wang H, Chua N-H (2015) Long noncoding RNA transcriptome of plants. Plant Biotechnol J 13:319–328. https://doi.org/10.1111/pbi.12336

Loyola R, Herrera D, Mas A, Wong DCJ, Höll J, Cavallini E, Amato A, Azuma A, Ziegler T, Aquea F, Castellarin SD, Bogs J, Tornielli GB, Peña-Neira A, Czemmel S, Alcalde JA, Matus JT, Arce-Johnson P (2016) The photomorphogenic factors UV-B RECEPTOR 1, ELONGATED HYPOCOTYL 5, and

HY5 HOMOLOGUE are part of the UV-B signalling pathway in grapevine and mediate flavonol accumulation in response to the environment. J Exp Bot 67:5429–5445. https://doi.org/10.1093/jxb/erw307

Luan LY, Zhang ZW, Xi ZM, Huo SS, Ma LN (2013) Brassinosteroids regulate anthocyanin biosynthesis in the ripening of grape berries. S Afr J Enol Vitic 34:196–203

Malacarne G, Costantini L, Coller E, Battilana J, Velasco R, Vrhovsek U, Grando MS, Moser C (2015) Regulation of flavonol content and composition in (Syrah × Pinot Noir) mature grapes: integration of transcriptional profiling and metabolic quantitative trait locus analyses. J Exp Bot 66:4441–4453. https://doi.org/10.1093/jxb/erv243

Manning K, Davies C, Bowen HC, White PJ (2001) Functional characterization of two ripening-related sucrose transporters from grape berries. Ann Bot 87:125–129. https://doi.org/10.1006/anbo.2000.1316

Martin DM, Aubourg S, Schouwey MB, Daviet L, Schalk M, Toub O, Lund ST, Bohlmann J (2010) Functional annotation, genome organization and phylogeny of the grapevine (*Vitis vinifera*) terpene synthase gene family based on genome assembly, flcdna cloning, and enzyme assays. BMC Plant Biol 10:226. https://doi.org/10.1186/1471-2229-10-226

Massonnet M, Fasoli M, Tornielli GB, Altieri M, Sandri M, Zuccolotto P, Paci P, Gardiman M, Zenoni S, Pezzotti M (2017) Ripening transcriptomic program in red and white grapevine varieties correlates with berry skin anthocyanin accumulation. Plant Physiol 174:2376–2396. https://doi.org/10.1104/pp.17.00311

Matarese F, Cuzzola A, Scalabrelli G, D'Onofrio C (2014) Expression of terpene synthase genes associated with the formation of volatiles in different organs of *Vitis vinifera*. Phytochemistry 105:12–24. https://doi.org/10.1016/j.phytochem.2014.06.007

Matarese F, Scalabrelli G, Onofrio CD (2013) Analysis of the expression of terpene synthase genes in relation to aroma content in *Vitis vinifera* L. flowers, developing berries and other tissues and implications for viticultural practices. Funct Plant Biol 40:552–565

Matthews MA, Thomas TR, Shackel KA (2009) Fruit ripening in *Vitis vinifera* L.: possible relation of veraison to turgor and berry softening. Aust J Grape Wine Res 15:278–283. https://doi.org/10.1111/j.1755-0238.2009.00060.x

Matthews MA, Shackel KA (2005) Growth and water transport in fleshy fruit. In: Holbrook NM, Zwieniecki MA (eds) Vascular transport in plants. Elsevier, pp 181–197

Mattivi F, Caputi L, Carlin S, Lanza T, Minozzi M, Nanni D, Valenti L, Vrhovsek U (2011) Effective analysis of rotundone at below-threshold levels in red and white wines using solid-phase microextraction gas chromatography/tandem mass spectrometry. Rapid Commun Mass Spectrom 25:483–488. https://doi.org/10.1002/rcm.4881

Mattivi F, Vrhovsek U, Masuero D, Trainotti D (2009) Differences in the amount and structure of extractable skin and seed tannins amongst red grape varieties. Aust J Grape Wine Res 15:27–35. https://doi.org/10.1111/j.1755-0238.2008.00027.x

McAtee P, Karim S, Schaffer R, David K (2013) A dynamic interplay between phytohormones is required for fruit development, maturation, and ripening. Front Plant Sci 4:79. https://doi.org/10.3389/fpls.2013.00079

Mellway RD, Lund ST (2013) Interaction analysis of grapevine MIKCc-type MADS transcription factors and heterologous expression of putative véraison regulators in tomato. J Plant Physiol 170:1424–1433. https://doi.org/10.1016/j.jplph.2013.05.010

Muñoz-Espinoza C, Di Genova A, Correa J, Silva R, Maass A, González-Agüero M, Orellana A, Hinrichsen P (2016) Transcriptome profiling of grapevine seedless segregants during berry development reveals candidate genes associated with berry weight. BMC Plant Biol 16:104. https://doi.org/10.1186/s12870-016-0789-1

Muñoz-Robredo P, Gudenschwager O, Chervin C, Campos-Vargas R, González-Agüero M, Defilippi BG (2013) Study on differential expression of 1-aminocyclopropane-1-carboxylic acid oxidase genes in table grape cv. Thompson Seedless. Postharvest Biol Technol 76:163–169. https://doi.org/10.1016/J.POSTHARVBIO.2012.10.006

Nicolas P, Lecourieux D, Gomès E, Delrot S, Lecourieux F (2013) The grape berry-specific basic helix-loop-helix transcription factor VvCEB1 affects cell size. J Exp Bot 64:991–1003. https://doi.org/10.1093/jxb/ers374

Nicolas P, Lecourieux D, Kappel C, Cluzet S, Cramer G, Delrot S, Lecourieux F (2014) The basic leucine zipper transcription factor ABSCISIC ACID RESPONSE ELEMENT-BINDING FACTOR2 is an important transcriptional regulator of abscisic acid-dependent grape berry ripening processes. Plant Physiol 164:365–383. https://doi.org/10.1104/pp.113.231977

Normanly J, Slovin JP, Cohen JD (2010) Auxin biosynthesis and metabolism. In: Davies PJ (ed) Plant hormones. Springer, Dordrecht, pp 36–62

Nunan KJ, Davies C, Robinson SP, Fincher GB (2001) Expression patterns of cell wall-modifying enzymes during grape berry development. Planta 214:257–264

Ojeda H, Deloire A, Carbonneau A, Ageorges A, Romieu C (1999) Berry development of grapevines: Relations between the growth of berries and their DNA content indicate cell multiplication and enlargement. Vitis 38:145–150

Ono E, Homma Y, Horikawa M, Kunikane-Doi S, Imai H, Takahashi S, Kawai Y, Ishiguro M, Fukui Y, Nakayama T (2010) Functional differentiation of the glycosyltransferases that contribute to the chemical diversity of bioactive flavonol glycosides in grapevines (Vitis vinifera). Plant Cell 22:2856–2871. https://doi.org/10.1105/tpc.110.074625

Owen SJ, Lafond MD, Bowen P, Bogdanoff C, Usher K, Abrams SR (2009) Profiles of abscisic acid and its catabolites in developing Merlot grape (*Vitis vinifera*) berries. Am J Enol Vitic 60:277–284

Palumbo MC, Zenoni S, Fasoli M, Massonnet M, Farina L, Castiglione F, Pezzotti M, Paci P (2014) Integrated network analysis identifies fight-club nodes as a class of hubs encompassing key putative switch genes that induce major transcriptome reprogramming during grapevine development. Plant Cell 26:4617–4635. https://doi.org/10.1105/tpc.114.133710

Pan Q-H, Li M-J, Peng C-C, Zhang N, Zou X, Zou K-Q, Wang X-L, Yu X-C, Wang X-F, Zhang D-P (2005) Abscisic acid activates acid invertases in developing grape berry. Physiol Plant 125:157–170. https://doi.org/10.1111/j.1399-3054.2005.00552.x

Pauwels L, Inzé D, Goossens A (2009) Jasmonate-inducible gene: what does it mean? Trends Plant Sci 14:87–91. https://doi.org/10.1016/J.TPLANTS.2008.11.005

Peña-Gallego A, Hernández-Orte P, Cacho J, Ferreira V (2012) S-Cysteinylated and S-glutathionylated thiol precursors in grapes. A review. Food Chem 131:1–13. https://doi.org/10.1016/j.foodchem.2011.07.079

Pérez-Díaz R, Ryngajllo M, Pérez-Díaz J, Peña-Cortés H, Casaretto JA, González-Villanueva E, Ruiz-Lara S (2014) VvMATE1 and VvMATE2 encode putative proanthocyanidin transporters expressed during berry development in *Vitis vinifera* L. Plant Cell Rep 33:1147–1159. https://doi.org/10.1007/s00299-014-1604-9

Pérez FJ, Viani C, Retamales J (2000) Bioactive gibberellins in seeded and seedless grapes: identification and changes in content during berry development. Am J Enol Vitic 51:315–318

Pilati S, Bagagli G, Sonego P, Moretto M, Brazzale D, Castorina G, Simoni L, Tonelli C, Guella G, Engelen K, Galbiati M, Moser C (2017) Abscisic acid is a major regulator of grape berry ripening onset: new insights into ABA signaling network. Front Plant Sci 8:1093. https://doi.org/10.3389/fpls.2017.01093

Podolyan A, White J, Jordan B, Winefield C (2010) Identification of the lipoxygenase gene family from *Vitis vinifera* and biochemical characterisation of two 13-lipoxygenases expressed in grape berries of Sauvignon Blanc. Funct Plant Biol 37:767–784

Portu J, López R, Santamaría P, Garde-Cerdán T (2018) Methyl jasmonate treatment to increase grape and wine phenolic content in Tempranillo and Graciano varieties during two growing seasons. Sci Hortic (Amsterdam) 240:378–386. https://doi.org/10.1016/J.SCIENTA.2018.06.019

Rattanakon S, Ghan R, Gambetta GA, Deluc LG, Schlauch KA, Cramer GR (2016) Abscisic acid transcriptomic signaling varies with grapevine organ. BMC Plant Biol 16:72. https://doi.org/10.1186/s12870-016-0763-y

Regalado A, Pierri CL, Bitetto M, Laera VL, Pimentel C, Francisco R, Passarinho J, Chaves MM, Agrimi G (2013) Characterization of mitochondrial dicarboxylate/tricarboxylate transporters from grape berries. Planta 237:693–703. https://doi.org/10.1007/s00425-012-1786-8

Rinaldo A, Cavallini E, Jia Y, Moss SMA, McDavid DAJ, Hooper LC, Robinson SP, Tornielli GB, Zenoni S, Ford CM, Boss PK, Walker AR (2015) A grapevine anthocyanin acyltransferase, transcriptionally regulated by VvMYBA, can produce most acylated anthocyanins present in grape skins. Plant Physiol 169:1897–1916. https://doi.org/10.1104/pp.15.01255

Robinson AL, Boss PK, Solomon PS, Trengove RD, Heymann H, Ebeler SE (2014a) Origins of grape and wine aroma. Part 1". Chemical components and viticultural impacts. Am J Enol Vitic 65:1–24. https://doi.org/10.5344/ajev.2013.12070

Robinson AL, Boss PK, Solomon PS, Trengove RD, Heymann H, Ebeler SE (2014b) "Origins of grape and wine aroma. Part 2". Chemical and sensory analysis. Am J Enol Vitic 65:25–42. https://doi.org/10.5344/ajev.2013.13106

Rodríguez-Concepción M, Boronat A (2002) Elucidation of the methylerythritol phosphate pathway for isoprenoid biosynthesis in bacteria and plastids. a metabolic milestone achieved through genomics. Plant Physiol 130:1079–1089. https://doi.org/10.1104/pp.007138

Savoi S, Wong DCJ, Arapitsas P, Miculan M, Bucchetti B, Peterlunger E, Fait A, Mattivi F, Castellarin SD (2016) Transcriptome and metabolite profiling reveals that prolonged drought modulates the phenylpropanoid and terpenoid pathway in white grapes (*Vitis vinifera* L.). BMC Plant Biol 16:67. https://doi.org/10.1186/s12870-016-0760-1

Savoi S, Wong DCJ, Degu A, Herrera JC, Bucchetti B, Peterlunger E, Fait A, Mattivi F, Castellarin SD (2017) Multi-omics and integrated network analyses reveal new insights into the systems relationships between metabolites, structural genes, and transcriptional regulators in developing grape berries (*Vitis vinifera* L.) exposed to water deficit. Front Plant Sci 8:1–19. https://doi.org/10.3389/fpls.2017.01124

Schlosser J, Olsson N, Weis M, Reid K, Peng F, Lund S, Bowen P (2008) Cellular expansion and gene expression in the developing grape (*Vitis vinifera* L.). Protoplasma 232:255–265. https://doi.org/10.1007/s00709-008-0280-9

Serrano A, Espinoza C, Armijo G, Inostroza-Blancheteau C, Poblete E, Meyer-Regueiro C, Arce A, Parada F, Santibáñez C, Arce-Johnson P (2017) Omics approaches for understanding grapevine berry development: regulatory networks associated with endogenous processes and environmental responses. Front Plant Sci 8:1486. https://doi.org/10.3389/fpls.2017.01486

Shangguan L, Sun X, Zhang C, Mu Q, Leng X, Fang J (2015) Genome identification and analysis of genes encoding the key enzymes involved in organic acid biosynthesis pathway in apple, grape, and sweet orange. Sci Hortic (Amsterdam) 185:22–28. https://doi.org/10.1016/J.SCIENTA.2015.01.012

Shiraishi M, Fujishima H, Chijiwa H (2010) Evaluation of table grape genetic resources for sugar, organic acid, and amino acid composition of berries. Euphytica 174:1–13. https://doi.org/10.1007/s10681-009-0084-4

Siebert TE, Barter SR, de Barros Lopes MA, Herderich MJ, Francis IL (2018) Investigation of 'stone fruit' aroma in Chardonnay, Viognier and botrytis Semillon wines. Food Chem 256:286–296. https://doi.org/10.1016/j.foodchem.2018.02.115

Siebert TE, Wood C, Elsey GM, Pollnitz AP (2008) Determination of rotundone, the pepper aroma impact compound, in grapes and wine. J Agric Food Chem 56:3745–3748. https://doi.org/10.1021/jf800184t

Smeekens S (2000) Sugar-induced signal transduction in plants. Ann Rev Plant Physiol Plant Mol Biol 51:49–81. https://doi.org/10.1146/annurev.arplant.51.1.49

Smeekens S, Ma J, Hanson J, Rolland F (2010) Sugar signals and molecular networks controlling plant growth. Curr Opin Plant Biol 13:273–278. https://doi.org/10.1016/j.pbi.2009.12.002

Sun L, Sun Y, Zhang M, Wang L, Ren J, Cui M, Wang Y, Ji K, Li P, Li Q, Chen P, Dai S, Duan C, Wu Y, Leng P (2012) Suppression of 9-cis-epoxycarotenoid dioxygenase, which encodes a key enzyme in abscisic acid biosynthesis, alters fruit texture in transgenic tomato. Plant Physiol 158:283–298. https://doi.org/10.1104/pp.111.186866

Sun L, Zhang M, Ren J, Qi J, Zhang G, Leng P (2010) Reciprocity between abscisic acid and ethylene at the onset of berry ripening and after harvest. BMC Plant Biol 10:257–268

Sweetman C, Deluc LG, Cramer GR, Ford CM, Soole KL (2009) Regulation of malate metabolism in grape berry and other developing fruits. Phytochemistry 70:1329–1344. https://doi.org/10.1016/j.phytochem.2009.08.006

Sweetman C, Wong DC, Ford CM, Drew DP (2012) Transcriptome analysis at four developmental stages of grape berry (*Vitis vinifera* cv. Shiraz) provides insights into regulated and coordinated gene expression. BMC Genom 13:691. https://doi.org/10.1186/1471-2164-13-691

Symons GM, Davies C, Shavrukov Y, Dry IB, Reid JB, Thomas MR (2006) Grapes on steroids. Brassinosteroids are involved in grape berry ripening. Plant Physiol 140:150–158. https://doi.org/10.1104/pp.105.070706

Takase H, Sasaki K, Shinmori H, Shinohara A, Mochizuki C, Kobayashi H, Ikoma G, Saito H, Matsuo H, Suzuki S, Takata R (2015) Cytochrome P450 CYP71BE5 in grapevine (*Vitis vinifera*) catalyzes the formation of the spicy aroma compound (−)-rotundone. J Exp Bot 67:787–798. https://doi.org/10.1093/jxb/erv496

Tang W, Zheng Y, Dong J, Yu J, Yue J, Liu F, Guo X, Huang S, Wisniewski M, Sun J, Niu X, Ding J, Liu J, Fei Z, Liu Y (2016) Comprehensive transcriptome profiling reveals long noncoding rna expression and alternative splicing regulation during fruit development and ripening in kiwifruit (*Actinidia chinensis*). Front Plant Sci 7:335. https://doi.org/10.3389/fpls.2016.00335

Tanner W, Caspari T (1996) MEMBRANE TRANSPORT CARRIERS. Ann Rev Plant Physiol Plant Mol Biol 47:595–626. https://doi.org/10.1146/annurev.arplant.47.1.595

Teixeira A, Eiras-Dias J, Castellarin SD, Gerós H (2013) Berry phenolics of grapevine under challenging environments. Int J Mol Sci 14:18711–18739. https://doi.org/10.3390/ijms140918711

Terrier N, Torregrosa L, Ageorges A, Vialet S, Verriès C, Cheynier V, Romieu C (2009) Ectopic expression of VvMybPA2 promotes proanthocyanidin biosynthesis in grapevine and suggests additional targets in the pathway. Plant Physiol 149:1028–1041. https://doi.org/10.1104/pp.108.131862

Thomas TR, Matthews MA, Shackel KA (2006) Direct in situ measurement of cell turgor in grape (*Vitis vinifera* L.) berries during development and in response to plant water deficits. Plant, Cell Environ 29:993–1001. https://doi.org/10.1111/j.1365-3040.2006.01496.x

Tyerman SD, Chaves MM, Barrieu F (2012) Water relations of the grape berry and aquaporins. In: Gerós H, Chaves MM, Delrot S (eds) The biochemistry of the grape berry. Bentham Science, Bussum, pp 3–22

Vannozzi A, Chern D, Wong J, Ho J, Hmmam I, Bogs J, Ziegler T, Dry I, Barcaccia G, Lucchin M (2018) Combinatorial regulation of stilbene synthase genes by WRKY and MYB transcription factors in grapevine (*Vitis vinifera* L.). Plant Cell Physiol 59:1043–1059. https://doi.org/10.1093/pcp/pcy045

Vannozzi A, Dry IB, Fasoli M, Zenoni S, Lucchin M (2012) Genome-wide analysis of the grapevine stilbene synthase multigenic family: genomic organization and expression profiles upon biotic and abiotic stresses. BMC Plant Biol 12:130. https://doi.org/10.1186/1471-2229-12-130

Vitulo N, Forcato C, Carpinelli E, Telatin A, Campagna D, D'Angelo M, Zimbello R, Corso M, Vannozzi A, Bonghi C, Lucchin M, Valle G (2014) A deep survey of alternative splicing in grape reveals changes in the splicing machinery related to tissue, stress condition and genotype. BMC Plant Biol 14:99. https://doi.org/10.1186/1471-2229-14-99

Wada H, Matthews MA, Shackel KA (2009) Seasonal pattern of apoplastic solute accumulation and loss of cell turgor during ripening of *Vitis vinifera* fruit under field conditions. J Exp Bot 60:1773–1781. https://doi.org/10.1093/jxb/erp050

Wada H, Shackel KA, Matthews MA (2008) Fruit ripening in *Vitis vinifera*: apoplastic solute accumulation accounts for pre-veraison turgor loss in berries. Planta 227:1351–1361. https://doi.org/10.1007/s00425-008-0707-3

Walker AR, Lee E, Bogs J, McDavid DAJ, Thomas MR, Robinson SP (2007) White grapes arose through the mutation of two similar and adjacent regulatory genes. Plant Journal 49:772–785. https://doi.org/10.1111/j.1365-313X.2006.02997.x

Wan S, Li W, Zhu Y, Liu Z, Huang W, Zhan J (2014) Genome-wide identification, characterization and expression analysis of the auxin response factor gene family in *Vitis vinifera*. Plant Cell Rep 33:1365–1375. https://doi.org/10.1007/s00299-014-1622-7

Wang M, Zhao W, Gao L, Zhao L (2018) Genome-wide profiling of long non-coding RNAs from tomato and a comparison with mRNAs associated with the regulation of fruit ripening. BMC Plant Biol 18:75. https://doi.org/10.1186/s12870-018-1300-y

Wen Y-Q, Li J-M, Zhang Z-Z, Zhang Y-F, Pan Q-H (2010) Antibody preparation, gene expression and subcellular localization of L-idonate dehydrogenase in grape berry. Biosci Biotechnol Biochem 74:2413–2417. https://doi.org/10.1271/bbb.100448

Wheeler S, Loveys B, Ford C, Davies C (2009) The relationship between the expression of abscisic acid biosynthesis genes, accumulation of abscisic acid and the promotion of *Vitis vinifera* L. berry ripening by abscisic acid. Aust J Grape Wine Res 15:195–204

Won C, Shen X, Mashiguchi K, Zheng Z, Dai X, Cheng Y, Kasahara H, Kamiya Y, Chory J, Zhao Y (2011) Conversion of tryptophan to indole-3-acetic acid by TRYPTOPHAN AMINOTRANSFERASES OF ARABIDOPSIS and YUCCAs in arabidopsis. Proc Natl Acad Sci 108:18518–18523. https://doi.org/10.1073/pnas.1108436108

Wong DCJ, Lopez Gutierrez R, Dimopoulos N, Gambetta GA, Castellarin SD (2016a) Combined physiological, transcriptome, and cis-regulatory element analyses indicate that key aspects of ripening, metabolism, and transcriptional program in grapes (*Vitis vinifera* L.) are differentially modulated accordingly to fruit size. BMC Genom 17:416. https://doi.org/10.1186/s12864-016-2660-z

Wong DCJ, Matus JT (2017) Constructing integrated networks for identifying new secondary metabolic pathway regulators in grapevine: recent applications and future opportunities. Front Plant Sci 8:505. https://doi.org/10.3389/fpls.2017.00505

Wong DCJ, Schlechter R, Vannozzi A, Höll J, Hmmam I, Bogs J, Tornielli GB, Castellarin SD, Matus JT (2016b) A systems-oriented analysis of the grapevine R2R3-MYB transcription factor family uncovers new insights into the regulation of stilbene accumulation. DNA Res 23:451–466. https://doi.org/10.1093/dnares/dsw028

Wong DCJ, Zhang L, Merlin I, Castellarin SD, Gambetta GA (2018) Structure and transcriptional regulation of the major intrinsic protein gene family in grapevine. BMC Genom 19:248. https://doi.org/10.1186/s12864-018-4638-5

Wood C, Siebert TE, Parker M, Capone DL, Elsey GM, Pollnitz AP, Eggers M, Meier M, Vössing T, Widder S, Krammer G, Sefton MA, Herderich MJ (2008) From wine to pepper: rotundone, an obscure sesquiterpene, is a potent spicy aroma compound. J Agric Food Chem 56:3738–3744. https://doi.org/10.1021/jf800183k

Xin C, Liu W, Lin Q, Zhang X, Cui P, Li F, Zhang G, Pan L, Al-Amer A, Mei H, Al-Mssallem IS, Hu S, Al-Johi HA, Yu J (2015) Profiling microRNA expression during multi-staged date palm (*Phoenix dactylifera* L.) fruit development. Genomics 105:242–251. https://doi.org/10.1016/J.YGENO.2015.01.004

Xin H, Zhang J, Zhu W, Wang N, Fang P, Han Y, Ming R, Li S (2013) The effects of artificial selection on sugar metabolism and transporter genes in grape. Tree Genet Genomes 9:1343–1349. https://doi.org/10.1007/s11295-013-0643-7

Xu F, Xi Z, Zhang H, Zhang C, Zhang Z (2015) Brassinosteroids are involved in controlling sugar unloading in *Vitis vinifera* 'Cabernet Sauvignon' berries during véraison. Plant Physiol Biochem 94:197–208. https://doi.org/10.1016/j.plaphy.2015.06.005

Young P, Lashbrooke J, Alexandersson E, Jacobson D, Moser C, Velasco R, Vivier M (2012) The genes and enzymes of the carotenoid metabolic pathway in *Vitis vinifera* L. BMC Genom 13:243

Zeng S, Liu Y, Pan L, Hayward A, Wang Y (2015) Identification and characterization of miRNAs in ripening fruit of *Lycium barbarum* L. using high-throughput sequencing. Front Plant Sci 6:778. https://doi.org/10.3389/fpls.2015.00778

Zenoni S, Fasoli M, Guzzo F, Dal Santo S, Amato A, Anesi A, Commisso M, Herderich M, Ceoldo S, Avesani L, Pezzotti M, Tornielli GB (2016) Disclosing the molecular basis of the postharvest life of berry in different grapevine genotypes. Plant Physiol 172:1821–1843. https://doi.org/10.1104/pp.16.00865

Zhang G, Chen D, Zhang T, Duan A, Zhang J, He C (2018) Transcriptomic and functional analyses unveil the role of long non-coding RNAs in anthocyanin biosynthesis during sea buckthorn fruit ripening. DNA Res. https://doi.org/10.1093/dnares/dsy017

Zhang XY, Wang XL, Wang XF, Xia GH, Pan QH, Fan RC, Wu FQ, Yu XC, Zhang DP (2006) A shift of phloem unloading from symplasmic to apoplasmic pathway is involved in developmental onset of ripening in grape berry. Plant Physiol 142:220–232. https://doi.org/10.1104/pp.106.081430

Zhang YL, Meng QY, Zhu HL, Guo Y, Gao HY, Luo YB, Lu J (2007) Functional characterization of a LAHC sucrose transporter isolated from grape berries in yeast. Plant Growth Regul 54:71–79. https://doi.org/10.1007/s10725-007-9226-7

Zhang Y, Zhen L, Tan X et al (2014) The involvement of hexokinase in the coordinated regulation of glucose and gibberellin on cell wall invertase and sucrose synthesis in grape berry. Mol Biol Rep 41:7899–7910. https://doi.org/10.1007/s11033-014-3683-7

Zhao J, Pang Y, Dixon RA (2010) The mysteries of proanthocyanidin transport and polymerization. Plant Physiol 153:437–443. https://doi.org/10.1104/pp.110.155432

Zhao Y (2010) Auxin biosynthesis and its role in plant development. Ann Rev Plant Biol 61:49–64. https://doi.org/10.1146/annurev-arplant-042809-112308

Zhu BQ, Xu XQ, Wu YW, Duan CQ, Pan QH (2012) Isolation and characterization of two hydroperoxide lyase genes from grape berries. Mol Biol Rep 39:7443–7455. https://doi.org/10.1007/s11033-012-1577-0

Ziliotto F, Corso M, Rizzini FM, Rasori A, Botton A, Bonghi C (2012) Grape berry ripening delay induced by a pre-véraison NAA treatment is paralleled by a shift in the expression pattern of auxin- and ethylene-related genes. BMC Plant Biol 12:185. https://doi.org/10.1186/1471-2229-12-185

Grape Transcriptomics and Viticulture

13

Mélanie Massonnet, Marianna Fasoli, Amanda M. Vondras, Sara Zenoni, Silvia Dal Santo, Alessandro Vannozzi, Simone D. Castellarin, Mario Pezzotti and Dario Cantu

Abstract

A major goal of viticulture is to exert control over ripening and produce fruit of reproducible yield and quality. This implies developing effective viticultural practices, breeding cultivars with improved characteristics, and requires considering the numerous variables that can influence development and ripening, like cultivar-specific traits, regional climate, and stresses. Molecular tools aid these efforts. Among them, transcriptome measurements that capture expression across the genome allow monitoring which genomic features are transcribed given the aforementioned variables. The technologies used to study the transcriptome have rapidly improved and become less expensive since the early 2000s, increasing the feasibility of developing molecular marker-driven practices. This chapter briefly reviews the history and state of transcriptomic technologies since they have been applied to grapevine, reviews the seminal publications that have used these tools, and proposes a direction for this field in the future.

13.1 Introduction

Grapevine is one of the most extensively grown fruit crops worldwide (http://www.fao.org/). Grapes are predominantly used for winemaking

M. Massonnet · A. M. Vondras · D. Cantu (✉)
Department of Viticulture and Enology, University of California Davis, One Shields Ave, Davis, CA 95616, USA
e-mail: dacantu@ucdavis.edu

M. Massonnet
e-mail: mmassonnet@ucdavis.edu

A. M. Vondras
e-mail: amvondras@ucdavis.edu

M. Fasoli
Viticulture, Chemistry and Enology, E. & J. Gallo Winery, 600 Yosemite Blvd, Modesto, CA 95354, USA
e-mail: Marianna.Fasoli@ejgallo.com

S. Zenoni · S. D. Santo · M. Pezzotti
Department of Biotechnology, University of Verona, Strada le Grazie 15, 37134 Verona, Italy
e-mail: sara.zenoni@univr.it

S. D. Santo
e-mail: silvia.dalsanto@univr.it

M. Pezzotti
e-mail: mario.pezzotti@univr.it

A. Vannozzi
Department of Agronomy, Food, Natural Resources, Animals and Environment, University of Padova, Agripolis, Viale dell'Università 16, 35020 Legnaro, Italy
e-mail: alessandro.vannozzi@unipd.it

S. D. Castellarin
Wine Research Centre, The University of British Columbia, 2329 West Mall, Vancouver, BC V6T 1Z4, Canada
e-mail: simone.castellarin@ubc.ca

D. Cantu and M. A. Walker (eds.), *The Grape Genome*, Compendium of Plant Genomes, https://doi.org/10.1007/978-3-030-18601-2_13

(56%; http://www.oiv.int/), though are also consumed fresh, pressed for juice, dried into raisins, and distilled into spirits. Wine grapes are commonly evaluated by measuring fruits' sugars, organic acids, pigments, tannins, and metabolites associated with aroma potential (Bisson 2001; Cozzolino and Dambergs 2010). What is considered optimal fruit quality varies with their purpose and the market's preferences (Poni et al. 2018). To achieve desired fruit composition, viticultural practices thought to change the balance between vegetative (photosynthetic organs) and reproductive (fruit) growth like shoot and cluster thinning, diverse rootstock–scion pairings, and planting cover-crops are used (Jackson and Lombard 1993; Matthews and Nuzzo 2007; Vaillant-Gaveau et al. 2014; Poni and Gatti 2017).

Complicating these efforts, however, are the numerous environmental factors that impact the fruit (van Leeuwen et al. 2004; Dai et al. 2011). In pursuit of an optimal product, grape growers contend with diverse abiotic and biotic stresses and the regional environmental conditions sometimes described as *terroir*. These variables influence how regions distinguish themselves and which cultivars they grow (van Leeuwen and Seguin 2006; Renouf et al. 2010; Jones et al. 2012; Anderson and Aryal 2013). Stresses like drought and disease can have devastating effects by reducing fruit quality and causing major crop losses (Madden and Wheelis 2003; Mittler 2006; Oerke 2006; Alston et al. 2013; Bock et al. 2013; Fuller et al. 2014; Suzuki et al. 2014). For grape, as in other crops, mitigating or circumventing these pressures involves developing management practices to deploy in the vineyard and developing varieties with superior traits (Duchêne et al. 2010; Østergård et al. 2009; Viers et al. 2013; Poni et al. 2018). Both strategies might be expedited with a deep understanding of the grapevine genome that whole-genome expression (transcriptome) studies provide.

Transcriptomic studies describe the genome-wide expression and co-expression dynamics during developmental processes, in response to treatments, and how these processes vary between samples (Wang et al. 2009; Lowe et al. 2017). This helps identify the metabolic and signaling pathways involved in these responses. The technology can also be used to improve the annotation of reference genomes by resolving full-length, individual transcripts and make it possible to know which gene isoforms are contextually relevant (Abdel-Ghany et al. 2016; Wang et al. 2016; Minio et al. 2019). Importantly, expression studies can aid the identification of markers for breeding and for adopting practices based on expression markers (Gramazio et al. 2016; Pandey et al. 2016; Xu et al. 2018). Finally, the biochemical uniformity of transcripts' composition makes their collective measurement relatively simpler than measuring the metabolome or proteome and yields functional information that cannot be known from the genome sequence alone.

The methods used for measuring the transcriptome have changed and improved repeatedly over the last decades. Early transcriptome studies used Sanger sequencing or a gel-based profiling method to characterize gene expression differences (Vuylsteke et al. 2007; Lowe et al. 2017). Then, the availability of hybridization-based arrays and the publication of the grape genome led to an expansion of grapevine transcriptome research (Jaillon et al. 2007; Velasco et al. 2007). Without the requirement of predefined, transcript-specific probes, RNA sequencing (RNA-seq) and isoform sequencing (Iso-Seq) are now commonly used to measure expression with higher sensitivity than earlier technologies and to annotate novel transcripts (Xu et al. 2013; Zhao et al. 2014; Minio et al. 2019).

In this chapter, we review how transcriptomic technologies have been applied historically to studying grapevine and discuss the seminal publications about the grapevine transcriptome, specifically those concerned with the impact of source-sink management, rootstocks, terroir, drought stress, and pathogens on fruit development and ripening.

13.2 History and Current State of Grapevine Transcriptomics

The first large-scale studies of gene expression in *Vitis vinifera* were performed using low-throughput Sanger sequencing of expressed sequence tags (ESTs), i.e., short transcript sequences generated from randomly selected complementary DNA (cDNA) clones (Parkinson and Blaxter 2009). The EST-based method was used to determine the transcript composition of multiple grapevine organs, including the grape berry, at different developmental stages (Ablett et al. 2000; Terrier et al. 2001; Moser et al. 2005). This costly approach generated a limited number of ESTs per sample, ranging from 100 to 2,279 sequences. Grape berry ripening was also studied using gel-based amplified fragment length polymorphism, another large-scale approach that does not rely on sequencing information (cDNA-AFLP; Venter et al. 2001; Burger and Botha 2004). Both approaches revealed differences in transcript content across grape organs and berry development, consistent with what was observed by Davies and Robinson (2000) in a study of berry transcripts that used a differential screening method to monitor 17 grape ripening-induced ("Grip") cDNAs.

The first high-throughput gene expression profiling analyses became possible with the emergence of the DNA arrays. The first hybridization-based gene expression profiling analyses on grape berry development were performed using cDNA-based arrays (Terrier et al. 2005; Waters et al. 2005). Then, the concomitant increase of publicly available DNA sequence information and rapid technological progress led to the first commercial high-density oligonucleotide array for grapevine in 2004. This array could monitor the expression of up to 14,000 *Vitis vinifera* transcripts and 1,700 transcripts from other *Vitis* species (Affymetrix GeneChip® *Vitis vinifera* Genome Array). This array was used to study tissue-specific gene expression (Grimplet et al. 2007), gene expression during the development of Cabernet Sauvignon and Pinot Noir berries (Deluc et al. 2007; Pilati et al. 2007), and the effects of heat, water, and other environmental stresses on berry ripening (Deluc et al. 2009; Mori et al. 2007).

The release of the first complete grapevine genome sequence in 2007 (Jaillon et al. 2007; Velasco et al. 2007) led to the creation of microarray platforms that qualitatively monitor grapevine transcripts. These included the CombiMatrix GrapeArray 1.2 array and the NimbleGen 12 × 135 K array which was developed using the 12X genome assembly and V1 gene prediction (Forcato 2010; Pastore et al. 2011). The NimbleGen array included 12 sub-arrays containing 135,000 60-mer oligonucleotide probes; each sub-array could detect the expression of 29,549 grapevine transcripts. This array was used to build a tissue and developmental stage-specific transcriptome atlas for the grapevine cultivar Corvina (Fasoli et al. 2012). This study revealed a deep transcriptome shift driving maturation.

Next-generation sequencing methods like RNA sequencing (RNA-seq) generate short reads of cDNA sequences that can be absolutely quantified when aligned to reference sequences and counted (Mortazavi et al. 2008). The advent of this technology was very useful for studying grapevine. The first application of RNA-seq to grapevine research was a 2010 study of Corvina berries at three stages of development (Zenoni et al. 2010). Approximately, 59 million single-ended reads between 36 and 44 base pairs (bp) long were generated and aligned to the PN40024 reference genome (Jaillon et al. 2007) and 17,324 transcripts with diverse expression levels and patterns were captured. The study reported substantial transcriptional complexity during berry development. Many studies since have used RNA-seq to show genome-wide transcriptional dynamics during grape berry ripening, identifying important genes involved in the regulation of berry development (Massonnet et al. 2017a; see Chap. 14), characterizing cultivar-specific and phenotype-associated gene expression patterns (Da Silva et al. 2013), and evaluating the transcriptional responses of different grape organs to biotic stress as well as profiling grapevine pathogen transcriptomes during infection (Amrine et al. 2015; Blanco-Ulate et al.

2015, 2017; Massonnet et al. 2017b; Brilli et al. 2018; Massonnet et al. 2018; Morales-Cruz et al. 2018). In addition to these applications, RNA-seq was used to detect novel and cultivar-specific transcripts, expressed single nucleotide polymorphisms, and splicing variants (Zenoni et al. 2010; Da Silva et al. 2013; Venturini et al. 2013; Amrine et al. 2015; Gambino et al. 2017; Minio et al. 2019).

Further technological advancements made full-length cDNA sequencing in long reads possible; the Iso-Seq method developed by Pacific Biosciences provides accurate information about alternative transcripts. This data has helped improve genome annotations and gene discovery (Clavijo et al. 2017; Li et al. 2017a, b; Minio et al. 2019). It has also been used to identify alternative transcripts that participate in various biological processes and stress responses (Cheng et al. 2017; Kim et al. 2017; Li et al. 2017a, b; Zhu et al. 2017; Minio et al. 2019). Importantly, full-length cDNA sequencing removes the necessity of a reference genome and has the potential to unlock information about cultivar-specific traits, plant defense, and plant development yet unseen.

13.3 Impact of the Viticultural Practices on the Berry Transcriptome

13.3.1 Source-Sink Management

Achieving high yield and quality is the most important objective in viticulture. Crop yield is often referred to as the amount of ripened fruit produced by a vine or vineyard and fruit quality is related to its composition, which includes sugars, acids, polyphenolics, and other metabolites. Maximizing crop yield without reducing grape quality requires optimally balancing the vine's vegetative and reproductive growth or its "source to sink" ratio. This balance can be assessed by measuring crop load, which is crop size (yield per vine or per unit of land area) relative to vine size (assessed as dormant pruning weight or leaf area). In general, a leaf area of 10–15 cm^2 is required to fully ripen 1 g of fruit and this normally results in a yield to pruning weight ratio between 5 and 10 (Kliewer and Dokoozlian 2005). If crop load is lower than this, the vine is considered undercropped or sink limited and will tend to divert more resources toward vegetative growth to the detriment of fruit quality (Kliewer and Dokoozlian 2005). Conversely, with insufficient leaf area, the vine may be unable to support ripening and is considered overcropped or source limited (Kliewer and Dokoozlian 2005). When crop size and vegetative growth are optimally balanced, grapevines produce a greater yield of high-quality fruit (Kliewer and Dokoozlian 2005). Many cultural practices are used to achieve vine balance, including defoliation and cluster thinning. The impact of these practices on the berry transcriptome has been studied and will be discussed in this section.

Defoliation involves selectively removing leaves around grape clusters. This practice reduces the leaf photosynthetic area and increases air circulation and the exposure of clusters to sunlight (Poni et al. 2006). Leaves can be removed at any time between pre-bloom and berry véraison with different consequences (Hunter et al. 1991). Pre-bloom defoliation causes a slight increase in sugar and anthocyanin levels in Sangiovese berries and defoliation at véraison can reduce anthocyanin concentration and increase the incidence of sun damage (Pastore et al. 2013). In order to determine the molecular mechanisms underlying those changes in berry composition, Pastore et al. (2013) did a transcriptomic analysis during berry ripening using a genome-wide microarray. Defoliated vines were transcriptionally different than control vines; their ripening programs were relatively delayed and photosynthetic genes were shut down relatively later than control vines. The timing of defoliation also caused disparate transcriptional effects. Structural and regulatory genes controlling anthocyanin biosynthesis were differentially expressed and accompanied differences in anthocyanins between the treatments. More recently, Zenoni et al. (2017) compared the agronomic and molecular berry responses to pre-bloom defoliation in Sangiovese and three other Italian cultivars

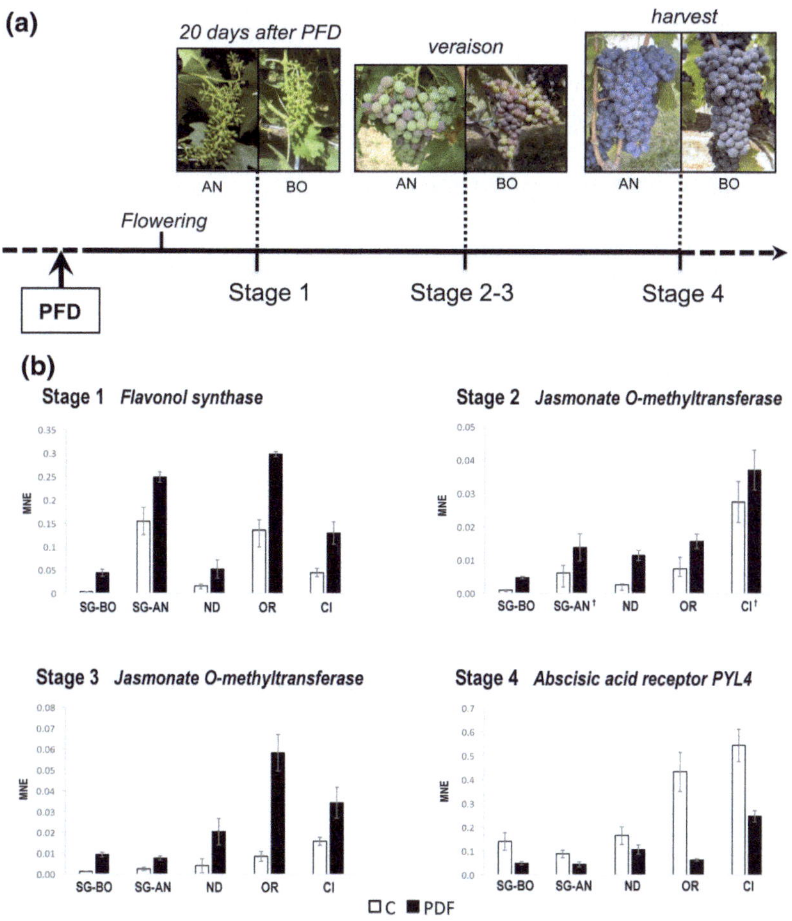

Fig. 13.1 Identification of molecular markers of pre-flowering defoliation (PFD) treatment in different genotypes and environments (Zenoni et al. 2017). **a** Schematic representation of the sampling design used. **b** Real-time qPCR analysis of PFD treatment common molecular markers in 2013. Real-time qPCR analysis of the genes encoding flavonol synthase (VIT_18s0001g03470) at Stage 1, jasmonate O-methyltransferase (VIT_18s0001g12890) at Stages 2 and 3 and abscisic acid receptor PYL4 (VIT_08s0058g00470) at Stage 4 of berry development from PFD and control (C) vines of Sangiovese at Bologna (SG-BO) and Ancona (SG-AN) sites, Nero d'Avola, (ND), Ortrugo (OR) and Ciliegiolo (CI). The mean normalized expression (MNE)-value was calculated for each sample referred to *VvUBIQUITIN1* (VIT_16s0098g01190). Bars represent means ± SE of three biological replicates. All genes in all genotypes resulted significantly modulated (t-test; *P*-value < 0.05) between C and PFD berries. *Source* Zenoni et al. (2017)

grown in different regions to identify molecular markers of early defoliation treatment independent of genotype and environment. The study used the NimbleGen 12 × 135 K microarray to evaluate the significance of the interaction between location and early defoliation on Sangiovese berry development. One hundred and twenty-five putative molecular markers associated specifically with pre-flowering defoliation were identified. Three candidates were validated across all genotypes using real-time qPCR (Fig. 13.1). These included a flavonol synthase gene, a jasmonate *O*-methyltransferase gene, and the gene encoding the ABA receptor PYL4.

Cluster thinning is another method of pursuing vine balance (Kliewer and Dokoozlian 2005; Dokoozlian 2009). This practice affects berry ripening rate (Dokoozlian and Hirschfelt 1995;

Palliotti and Cartechini 2000; Guidoni et al. 2002, 2008). The impact of cluster thinning on the berry transcriptome during ripening was first shown in Sangiovese, where it also caused an increase in berry sugars and anthocyanins (Pastore et al. 2011). Modifying the source:sink ratio via cluster thinning affects genes associated with carbohydrate metabolism and synthesis and transport of secondary products. Flavonoid and anthocyanin biosynthesis pathway genes, including a dihydroflavonol reductase (*DFR*), *VviMYBA1*, and three flavonoid glucosyltransferases, as well as anthocyanin-related transporters (e.g., the glutathione-S-transferase *VviGST4*), were up-regulated in berries from cluster-thinned vines and reflect the increase in anthocyanins observed. Non-anthocyanin, flavonoid-related genes, like a flavanone 3-hydroxylase (*F3H*), leucoanthocyanidin dioxygenase (*LDOX*), and the leucoanthocyanidin reductase *VviLAR1*, were down-regulated in berries from cluster-thinned plants.

The impact of crop load on the transcriptome and metabolome was recently studied in Pinot noir using RNA-seq (Fasoli et al. 2018a). Pinot noir grapevines at three crop load levels achieved via cluster thinning (overcropped, undercropped, and balanced) were compared throughout berry development in three consecutive vintages. The data generated from weekly sampling showed that crop load manipulation affects genes that may trigger ripening (Fasoli et al. 2018b). Genes involved in softening and other crucial components of ripening initiation responded to crop load changes. Differential expression of these genes likely influenced the whole ripening phase. Consistent with earlier reports, anthocyanins biosynthesis was higher at lower crop loads. This coincided with the up-regulation of a key enzyme in the anthocyanin biosynthetic pathway during maturation, UDP glucose:flavonoid-3-*O*-glucosyltransferase (*VviUFGT*) (Fig. 13.2).

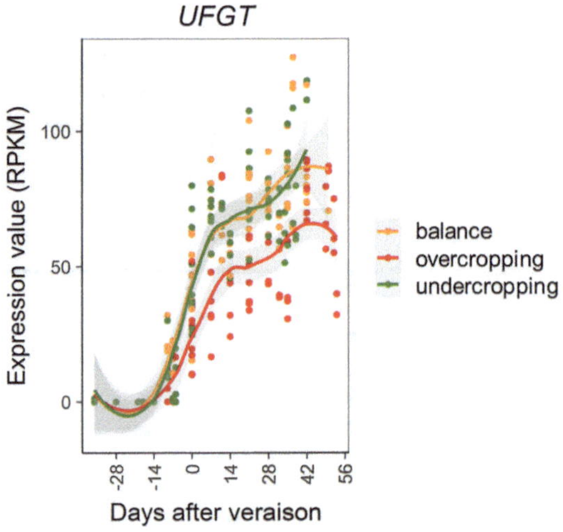

Fig. 13.2 Expression profile of *VviUFGT* (VIT_16s0039g02230) by crop load level (Fasoli et al. 2018a). Line graphs were created using data from three vintages plotted by days after véraison. Gray shading indicates 0.95 confidence level relative to the smoothed conditional means plotting method. RPKM: reads per kilobase of transcript per million mapped reads

These studies support that grape metabolism and the berry transcriptome are remarkably flexible, with treatments inducing extensive, genome-wide changes in expression during development. Their results support the potential of modifying source:sink ratios as means of optimizing grape yield and quality. Moreover, if the molecular basis of both variables is better understood, more precise vineyard management regimens can be developed and practiced.

13.3.2 Rootstock Selection

Grafting valuable grape varieties on tolerant rootstocks is a common practice in modern viticulture. More than 80% of the vineyards worldwide use grafted plants made with a *V. vinifera* scion grafted onto a rootstock of single American *Vitis* species or interspecific hybrids of *Vitis* species that combine multiple desirable traits, like *V. riparia*, *V. berlandieri*, *V. rupestris*, and *V. vinifera* (Ollat et al. 2016; see Chaps. 2 and 16). This practice was first adopted in Europe in the late nineteenth century because of the Phylloxera epidemic. Then, the replacement of the entire root system of *vinifera* varieties with non-*vinifera* species gradually became a general practice as a biological control strategy against the soil-borne pest. In addition to conferring resistance to various root pathogens, rootstocks can provide adaptation to abiotic factors including drought (Gambetta et al. 2012; Marguerit et al. 2012; Corso et al. 2015) and salinity (Fisarakis et al. 2001; Meggio et al. 2014), as well as benefits to the scion, such as regulating vine vigor and fruit quality (Walker et al. 2002, 2004; Gregory et al. 2013). Rootstocks can also differentially impact the ripening rate, likely by influencing the abundance of auxin and the expression of auxin-related genes (Corso et al. 2016).

Combining multiple favorable traits in elite rootstocks is the focus of breeding programs worldwide. Though marker-assisted selection accelerated the development of improved genotypes, our understanding of the molecular mechanisms underlying rootstock traits is still limited. Few studies focused on the transcriptomic behaviors of rootstocks in response to abiotic stress. Corso et al. (2015) used a comparative transcriptomic approach to characterize the biological processes affected by water deficit in root and leaf tissues of two rootstock genotypes with contrasting hydraulic behavior. The RNA-seq study revealed that stilbene- and flavonoid-related genes were higher expressed in roots and leaves, respectively, of the drought-tolerant rootstock during stress. Authors speculated that the induction of phenolic biosynthesis helps drought-tolerant rootstocks cope with oxidative stress associated with water deficit. The response of the grape berry transcriptome to water deficit can also be influenced by the rootstock used. For instance, expression of genes associated with jasmonic acid biosynthesis and secondary metabolism was found induced (and/or less repressed) by water limitation in Pinot noir berries from vines grafted on the drought-sensitive rootstock 125AA compared to the drought-tolerant one 110R (Berdeja et al. 2015).

Along with water uptake, rootstock genotype can also affect scion growth by its ability to cope with nutrient-limited conditions. Transcriptomics has been used to better understand nitrogen metabolism and iron deficiency tolerance (Cochetel et al. 2017; Vannozzi et al. 2017). In their study, Cochetel et al. (2017) investigated the transcriptomic responses to changes in nitrate availability in two rootstocks, *V. riparia* cv. Riparia Gloire de Montpellier and the hybrid 1103P; these rootstocks are known to have disparate effects on scion growth. Comparative transcriptomic analysis of root samples showed that the two rootstocks responded in a genotype-dependent manner to heterogeneous nitrogen availability. Interestingly, the transcriptomic response was more pronounced in the rootstock conferring lower scion vigor (Riparia Gloire de Montpellier). This suggested that the ability of a rootstock to uptake and assimilate nitrogen influenced scion vigor. Rootstock responses to iron deficiency were also studied at the transcriptomic level (Vannozzi et al. 2017). RNA-seq analysis of root apices of the hybrid 101.14 (*V. riparia* × *V. rupestris*), a commonly used grapevine rootstock susceptible to iron chlorosis, showed that many ortholog genes of the *Arabidopsis* "ferrome" were differentially expressed in roots of iron-deprived plants versus non-stressed plants. Comparison between the rootstocks 101.14 and M1, another rootstock genotype with high tolerance to iron deficiency, showed a correlation between the expression of two genes, encoding a plasma membrane H+-ATPase (*AHA2*) and an iron deficiency-inducible ferric chelate reductase (*FRO2*), and the manifestation of leaf chlorosis symptoms.

Our understanding of the molecular basis of graft compatibility, a critical factor determining the success of the union between rootstock and scion, is also still limited (Pina and Errea 2005; Lider et al. 1978; Fallot et al. 1979; Todić et al. 2005). Successful grafting is a complex process, requiring the adhesion of the two individuals, followed by a callus formation, and finally the establishment of a functional vascular system between the two grafting partners (Milien et al. 2012). In order to dissect the transcriptomic dynamics occurring at the rootstock–scion interface during grafting process, Cookson et al. (2013, 2014) studied the transcriptional changes induced at the rootstock–scion union site in both homo- and hetero-grafted plants using the NimbleGen 12 × 135 K microarray. Self-grafting (Cabernet Sauvignon onto Cabernet Sauvignon) led to the differential expression of many genes involved in secondary metabolism, wounding response (e.g., chitinases, peroxidases, germin-like proteins, and transcription factors), hormone signaling (jasmonate-related genes), and callus maintenance (LATERAL ORGAN BOUNDARIES DOMAIN (LBD) proteins) during graft union formation (Cookson et al. 2013). In contrast, grafting Cabernet Sauvignon onto two different rootstocks (Cookson et al. 2014) revealed that grafting with non-self rootstocks triggers the up-regulation of genes associated with plant stress responses at the graft interface, including genes involved in oxidative and biotic stress responses like pathogenesis-related (PR) proteins. Those results suggested that cells at the graft interface are capable of detecting the presence of the non-self-grafting partner and induce an immune-type response.

13.4 Effect of the Environment on Genome-Wide Transcriptional Dynamics During Berry Ripening

13.4.1 Impact of the "*Terroir*": The Berry Transcriptomic Plasticity

In viticulture and enology, the environmental factors that characterize a specific vineyard and impact grape and wine composition are referred to as "terroir". A first definition of this term was initially given by Seguin (1988), classifying terroir as an interactive ecosystem in a given place including climate, soil, and the vine. In a non-scientific context, the term terroir grew to include the impact of human intervention and gave rise to another term, typicity, which describes the specific qualitative properties of wines (van Leeuwen et al. 2004). Wine typicity arises from the extensive phenotypic plasticity of grapes. Plasticity refers to the ability of a single genotype to produce a range of phenotypes as a function of its environment (Bradshaw et al. 1965). Grapevines are characterized by considerable phenotypic plasticity, with the same clone showing variability within individual berries, among berries within a cluster, between clusters on a vine, and among vines in the vineyard (Dai et al. 2011). A widely accepted notion in viticulture is that different cultivated genotypes (cultivars) uniquely interact with a given environment (van Leeuwen et al. 2004). When phenotypic plasticity differs between genotypes, it is attributed to a "Genotype x Environment" (GxE) interaction (Saltz et al. 2018). Despite the scientific and commercial importance of genotype interactions with growing conditions, few studies have characterized the molecular basis of phenotypic plasticity.

The transcriptional plasticity of Corvina in different environments and in different vintages was characterized using a microarray (Dal Santo et al. 2013). The Corvina berry transcriptome is highly sensitive to growing conditions. Most of the berry transcriptome clustered according to the year of growth rather than common environmental or viticultural practices, highlighting the significant impact of climate conditions on berry ripening and fruit composition (van Leeuwen et al. 2004). Transcripts related to secondary metabolism, especially those involved in phenylpropanoid metabolism, were significantly repressed during a vintage with unfavorable climate. These results were supported by metabolomic data that confirmed the extreme sensitivity of grapes to their environment and adverse climate conditions. Considering fruit from 11 different vineyards in a single year, environmentally sensitive genes were approximately 18% of genes modulated during berry development. Gene ontology categories including "DNA/RNA metabolic process", "transcription factor activity", "transport", and "secondary metabolism" were enriched among plastic genes. Together, these results implied that GxE effects may have consequences for berry development, ripening, and wine quality.

A study of the metabolomic and transcriptomic basis of the broad phenotypic plasticity of Garganega, a white cultivar, was studied at four locations with different pedoclimates (Dal Santo et al. 2016). The study revealed many environmentally modulated genes. Moreover, transcripts commonly modulated during berry development showed extensive expression plasticity, indicated by different coefficients of variation between the four ripening stages. Plasticity in the expression of core metabolism prompted the authors to compare Garganega (white berry cultivar) and Corvina (red berry cultivar) transcriptomes, and more specifically, to compare the gene expression and metabolite accumulation in the context of the phenylpropanoid/flavonoid pathway. The levels of transcripts and metabolites varied by vineyard for Garganega berries, whereas Corvina samples were similar at all four sites, suggesting that the white cultivar was more flexible during ripening and given different environments than the red cultivar.

Another GxE study in grapevine compared berry development in two cultivars (Sangiovese and Cabernet Sauvignon) grown in three different environments over two consecutive years (Dal Santo et al. 2018). Sangiovese, a typical central Italian cultivar, differently modulated almost twofold more genes across the three locations than Cabernet Sauvignon. The lower transcriptomic plasticity of Cabernet Sauvignon, with its relatively invariable response to its environment, could have contributed to the widespread cultivation of the variety. Gene functions related to photosynthesis, the generation of energy, and central carbohydrate metabolism were overrepresented among developmental stage-specific genes unaffected by genotype or environment. Biotic stress genes were enriched among cultivar-specific genes unaffected by the environment. Importantly, vintage and location variables interacted to influence the berry transcriptome. The area of cultivation alone contributed less to variation in berry gene expression during ripening than other variables, though was associated with variably expressed secondary metabolism genes. GxE-specific gene expression was also enriched with secondary metabolism-related genes involved in the phenylpropanoid and anthocyanin biosynthetic pathways, lignin biosynthesis, and volatile metabolite production. These data suggest that location plays an important role in determining the performance of different varieties and could mediate the abundance of metabolites related to wine aroma, structure, and color.

A climate changing over time will alter the terroir of major wine-producing regions and could disrupt the conventional notion of terroir entirely (White et al. 2009). Studies that explore the link between terroir and the transcriptome during berry development and ripening will help broaden our understanding of terroir and sustain viticulture and wine production.

13.4.2 Effect of Drought Conditions on the Berry Transcriptome During Development

Many premium wine-producing areas are located in dry and warm regions where grapevines often suffer from periods of seasonal drought. Well-known examples are the Mediterranean regions, where limited summer precipitation and high evaporative demand due to high temperature frequently lead to moderate and even severe water deficits in vineyards, especially later in the growing season during the period of fruit ripening. Grapevines respond to water deficit by activating a plethora of physiological and metabolic pathways that ultimately lead to a reduction of canopy and berry growth, as well as changes in berry composition (Castellarin et al. 2007a, b; Chaves et al. 2010). In non-irrigated regions, dry vintages are often considered better (van Leeuwen et al. 2009). Accordingly, deficit irrigation, a cultural practice that involves maintaining/imposing a moderate water deficit, has been used to improve berry composition by stimulating the production of key compounds associated with wine quality (Chaves et al. 2010). Deficit irrigation can stimulate fruit ripening in red varieties (Castellarin et al. 2007a; Gambetta et al. 2010; Herrera and Castellarin 2016) and the biosynthesis of key phenolics and aromatics (Bindon et al. 2007; Castellarin et al. 2007b). For this reason, managing vine water use through the choice of plant material and/or irrigation is a major issue in viticulture. Different grapevine varieties have been described as being more or less drought-tolerant, and this definition has been largely based on differences in their hydraulic behavior, either isohydric (water-saving) or anisohydric (water-consuming; Schultz 2003; Chaves et al. 2010; Lovisolo et al. 2010; Zarrouk et al. 2016). More than 60 studies have been done since the first publication about this topic; together, this research suggests that the differences may arise as much from environmental differences as from genetic backgrounds (Herrera et al. 2017; Hochberg et al. 2018; Charrier et al. 2018). The adoption of 'omics' technologies, including genome sequencing, metabolomics, and transcriptomics, has generated valuable insights into the regulation of fruit metabolism and composition in response to water deficit.

Deluc et al. (2009) were the first to apply transcriptomics to analyze the effects of long-term, seasonal water deficit on berries of Cabernet Sauvignon and Chardonnay vines grown in California and Nevada vineyards, respectively. Authors used the Affymetrix® Vitis Genome Array for transcriptomic analyses and considered seven berry developmental stages. The study showed that water deficit affected key metabolic pathways related to berry physiology and quality, including the phenylpropanoid, abscisic acid (ABA), isoprenoid, carotenoid, amino acid, and fatty acid metabolic pathways. Drought treatment also triggered the expression modulation of genes associated with the cell response to osmotic stresses, including those involved in glutamate and proline synthesis. Interestingly, the study highlighted that the response of the transcriptome to water deficit was inconsistent between varieties. For instance, in the red grape variety Cabernet Sauvignon, water deficit strongly modulated the expression of anthocyanin-related genes (including *VviMYBA1*, *VviMYBA2*, and *VviUFGT*). In the white grape variety Chardonnay, water deficit induced the expression of flavonol synthase genes and genes associated with the production of aroma compounds, including a terpenoid synthase and a carotenoid-cleavage dioxygenase. Commonalities in the deficit response between varieties were also found. For example, water deficit had similar effects for both varieties on fatty acid metabolism, inducing the expression of lipoxygenase and hydroperoxide lyase genes that lead to the production of some aroma compounds.

More recently, two studies used transcriptomic and large-scale metabolic analyses to characterize the responses of berries from two cultivars, the white-fruited Tocai Friulano and the red-fruited Merlot, to water deficit (Savoi et al. 2016, 2017). Both studies were conducted in the same experimental vineyard and applied similar deficit irrigation treatments from early stages of berry development to harvest. Water deficit increased the concentration of phenylpropanoids, monoterpenes, and tocopherols in the white cultivar Tocai Friulano; carotenoid and flavonoid concentrations were differentially affected according to the berry developmental stage (Savoi et al. 2016). Consistently, phenylpropanoid, flavonoid, carotenoid, and terpenoid structural genes were modulated by water deficit. Given the contribution of monoterpenes to wine aroma, the authors analyzed gene and metabolite relationships, focusing on ripening-related monoterpenes induced by water stress. The gene-metabolite network included the top 100 correlated transcripts for each monoterpene. Half of the genes belonging to the network were differentially expressed under water deficit (52%), and a large proportion of the correlated genes were involved in terpenoid, lipid, and hormone metabolism (Fig. 13.3a). Interestingly, the analysis also identified a promising candidate that might regulate monoterpene biosynthetic pathways in grapevine. The transcription factor gene, *VviMYB24*, has high homology with Arabidopsis MYBs that regulate terpenoid biosynthesis. Its expression was strongly correlated with the concentration of the deficit-responsive monoterpenes and the expression of several terpene synthases modulated by water limitation. In Merlot berries, water deficit promoted the accumulation of proline, branched-chain amino acids, phenylpropanoids, anthocyanins, and free volatile organic compounds in the berry, and the increases in concentration of these compounds coincided with the regulation of key structural pathway genes (Savoi et al. 2017). A total of 447 transcription factors were modulated in the berry in response to water deficit. Gene-gene and gene-metabolite network analyses showed that water-deficit-responsive transcription factors such as bZIPs, AP2/ERFs, MYBs, and NACs could be involved in the molecular regulation of the synthesis of those metabolites. The expression of deficit-modulated bZIP (e.g., VviABF4 and VviGBF3), AP2/ERF (e.g., VviRAP2.1, VviRAP2.4, VviERF62), and NAC (e.g., VviNAC87, VviRD26) transcription factors were frequently correlated with various deficit-modulated anthocyanin structural and regulatory genes and anthocyanin contents (Fig. 13.3b). This analysis provided new insight into the regulation of the metabolic response of grape berries to water deficit and narrowed down the list of candidate genes for the functional validation of these regulators.

Understanding how varying degrees of water deficit affect berry development, grape composition, and wine quality, and how the climate and genotype interact to produce responses to water deficit remains among the major research priorities for the grape and wine industry. Transcriptomics has proven to be an invaluable tool that in combination with other 'omics' technologies has generated important insights into water deficit response.

13.5 The Application of Transcriptomics to Study Grapevine–Pathogen Interactions

Cultivated grapevines are susceptible to numerous pathogens that negatively impact plant fitness, fruit composition, and fruit yield. These pathogens include bacteria, fungi, oomycetes, and viruses (Bertsch et al. 2013; Wilcox et al. 2015; Armijo et al. 2016). Downstream of the detection of pathogens and their often destructive effects are an elaborate array of transcriptional, post-transcriptional, epigenetic, hormonal, and other responses. The outcome of the interaction between the plant and its pathogen depends on how the plant recognizes and defends itself, and whether the pathogen is able to undermine plant defense, grow, and reproduce in plant tissue (Jones and Dangl 2006). Multiple sources of genetic resistance have been identified within *Vitis* and *Muscadinia rotundifolia* for some diseases, like powdery mildew, downy mildew, and Pierce's disease (Ruel and Walker 2006; Gessler et al. 2011; Qiu et al. 2015; Buonassisi et al. 2017). Characterizing the basis of both host resistance and microbial virulence is crucial for creating sustainable cultivars through breeding programs and for developing phytosanitary strategies.

This portion of the chapter will review the studies that applied transcriptomic tools to understand the responses of grapevines to diverse pathogens, the basis of resistance, and the manifestation of symptoms, as well to investigate the virulence mechanisms of some pathogens (Table 13.1). These studies compared infected to uninfected vines, have exploited variability in the effects of disease (Camps et al. 2010), identified genes associated with compatible versus incompatible responses to different strains of the same pathogen (Li et al. 2015), and characterized the responses of resistant varieties to infection (Weng et al. 2014; Amrine et al. 2015). Moreover, experiments comparing infected to uninfected plants sometimes include hormone and/or secondary metabolite data to support their findings and demonstrate the validity of using transcriptomic technologies to predict the functional implications of disease (Vega et al. 2011; Li et al. 2015; Agudelo-Romero et al. 2015; Blanco-Ulate et al. 2015, 2017; Massonnet et al. 2017b). The application of transcriptomics to survey the vineyard metagenome, profile pathogen transcripts, and better understand virulence will also be discussed. This body of work generated valuable insights into the relationships between plant and pathogen, and the architecture and regulation of their transcriptomes.

Vineyards can host a multitude of microbes, among which fungi and oomycetes are the most numerous known pathogens that affect grapevine (Wilcox et al. 2015). However, the interaction between grapevine and only a handful of fungal pathogens has been investigated at the

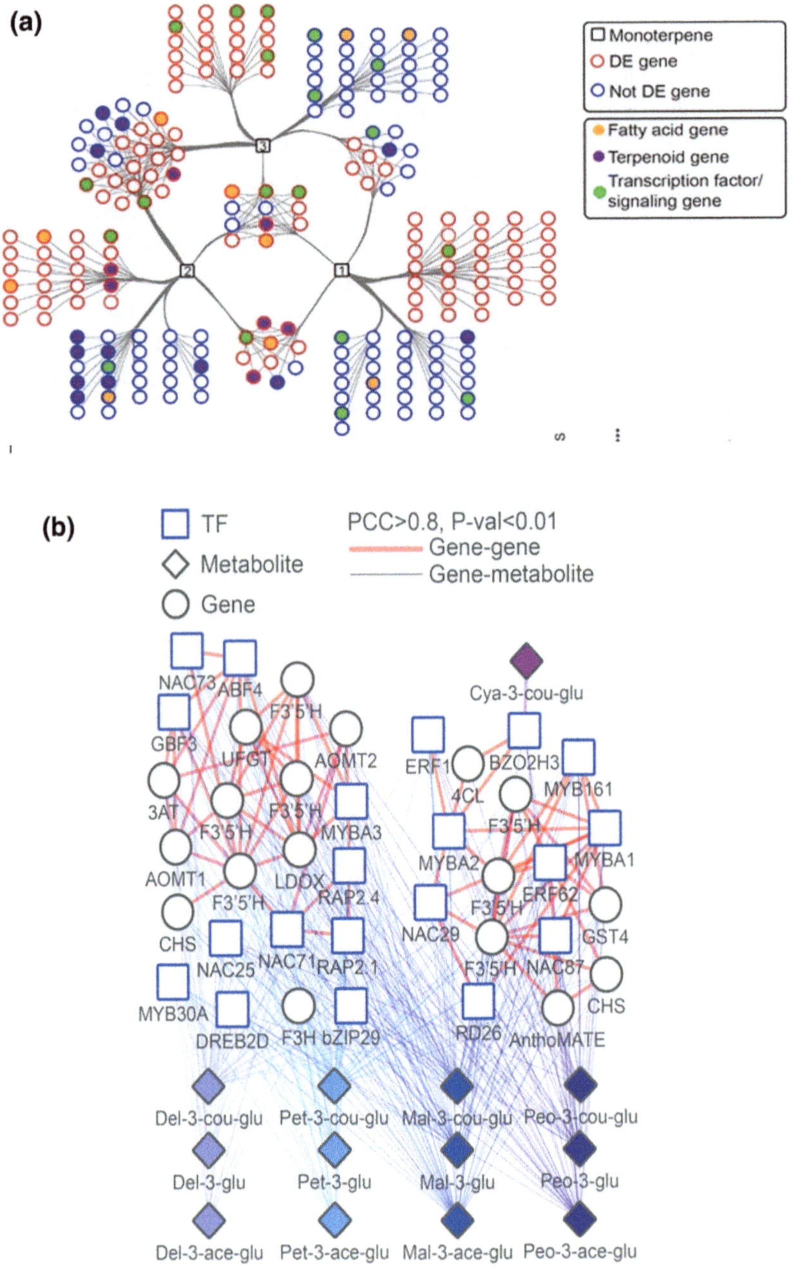

Fig. 13.3 Examples of gene-metabolite network of white and red cultivars, Tocai and Merlot, respectively, during berry development under water deficit (Savoi et al. 2016, 2017). **a** Predicted gene-metabolite network related to monoterpenes: linalool (1), α-terpineol (2), and nerol (3), in Tocai berries. Genes and metabolites are represented by circle and square nodes, respectively. Edges represent associations (P-value < 0.001) between transcripts and metabolites. The top 100 correlators for each metabolite are shown. Node borders in red represent differentially expressed (DE) transcripts. Node colors indicate the pathway of the transcripts. **b** Network representation of co-expressed genes having ripening-associated expression patterns in Merlot berries, centered on significantly correlated transcription factors (TFs) and/or structural genes with anthocyanin (pink, lavender blue, cyan, blue, and purple). Structural genes, TF genes, and metabolites are represented by circle, square, and diamond nodes, respectively. Thick edges in light red represent associations between structural genes/TFs (Pearson Correlation Coefficient (PCC) > 0.8; P-value < 0.01). Gene-metabolite associations (PCC > 0.8; P-value < 0.01) are depicted in thinner edges with colors denoting the different anthocyanin categories. *Source* Savoi et al. (2016, 2017)

Table 13.1 Transcriptomics applied to grape–pathogen interaction

Grape pathogens	Disease name	Causal agent	Site(s) of infection	Studied tissue(s)	Studied transcriptome(s)	Approach	References
Bacteria	Pierce's disease	*Xylella fastidiosa*	Xylem	Leaf	Grape	Microarray	Choi et al. (2013)
						RNA-seq	Zaini et al. (2018)
				Petiole	Grape	RNA-seq	Rapicavoli et al. (2018)
	Bois noir (GY)	*Candidatus* Phytoplasma *solani*	Phloem	Leaf midrib	Grape	Microarray	Hren et al. (2009)
	Flavescence dorée (GY)	*Candidatus* Phytoplasma *vitis*	Phloem	Leaf	Pathogen	RNA-seq	Abbà et al. (2014)
Oomycetes	Downy mildew	*Plasmopara viticola*	Any green tissues	Leaf	Grape	Microarray	Polesani et al. (2010)
						RNA-seq	Wu et al. (2010)
							Vannozzi et al. (2012)
							Li et al. (2015)
				In vitro plants, sporangia	Grape and pathogen	RNA-seq	Brilli et al. (2018)
Fungi	Powdery mildew	*Erysiphe necator*	Any green tissues	Leaf	Grape	Microarray	Fung et al. (2008)
						RNA-seq	Weng et al. (2014)
							Amrine et al. (2015)
	Bunch rot	*Botrytis cinerea*	Berry	Berry	Grape	Microarray	Agudelo-Romero et al. (2015)
					Grape and pathogen	Microarray	Kelloniemi et al. (2015)
	Noble rot	*Botrytis cinerea*	Berry	Berry	Pathogen	RNA-seq	Blanco-Ulate et al. (2014)
					Grape	RNA-seq	Blanco-Ulate et al. (2015)
	Eutypa dieback (GTD)	*Eutypa lata*	Trunk	Leaf	Grape	Microarray	Camps et al. (2010)
	Botryosphaeria dieback (GTD)	*Neofusicoccum parvum*	Trunk	Leaf	Grape	RNA-seq	Czemmel et al. (2015)
				Woody stem, leaf	Grape	RNA-seq	Massonnet et al. (2017b)
					Pathogen	RNA-seq	Massonnet et al. (2018)
	GTDs	Grapevine trunk pathogens	Trunk	Wood	Pathogens	RNA-seq	Morales-Cruz et al. (2018)

(continued)

Table 13.1 (continued)

Grape pathogens	Disease name	Causal agent	Site(s) of infection	Studied tissue(s)	Studied transcriptome(s)	Approach	References
Viruses	Rupestris stem pitting	GRSPaV	Entire plant (systemic)	Leaf, berry	Grape	Microarray	Gambino et al. (2012)
	Grapevine leafroll disease	GLRaV-3	Entire plant (systemic)	Leaf	Grape	Microarray	Espinoza et al. (2007)
				Leaf, berry	Grape	Microarray	Vega et al. (2011)
	Red blotch disease	GRBaV	Entire plant (systemic)	Berry	Grape	RNA-seq	Blanco-Ulate et al. (2017)

GY grapevine yellows, *GTD* grapevine trunk diseases, *GRSPaV* Rupestris stem-pitting virus, *GLRaV-3* Grapevine leafroll-associated virus strain 3, *GRBaV* Grapevine red blotch-associated virus transcriptomic level. Those include the oomycete *Plasmopara viticola* and the ascomycete *Erysiphe necator*, causal agents of grapevine downy (DM) and powdery (PM) mildew, respectively, the ascomycete *Botrytis cinerea*, responsible for bunch rot and noble rot, and some grapevine trunk pathogens. Profiles of vineyards' microbial diversity and understanding pathogens' infection strategies complement our understanding of grapevine biology and are the basis of understanding how they interact.

Morales-Cruz et al. (2018) characterized the metatranscriptome of the vineyard microbiome associated with trunk diseases by mapping RNA-seq reads from woody tissues to a multi-species transcriptome reference composed of common and consequential grapevine trunk fungal pathogens including *Eutypa lata*, *Neofusicoccum parvum*, *Diplodia seriata*, *Phaeoacremonium minimum*, *Phaeomoniella chlamydospora* and *Diaporthe ampelina*. Gene expression of putative virulence factors distinguished samples with different disease symptoms (Morales-Cruz et al. 2018). Grapevine trunk pathogens colonize the woody structures of the plant through wounds, often pruning wounds, causing chronic infections that compromise the translocation of water and nutrients throughout the plant and cause a variety of symptoms in growing green tissues (Bertsch et al. 2013; Gramaje et al. 2018). Using a grapevine microarray, Camps et al. (2010) compared the leaf transcriptomes of healthy, symptomatic, and asymptomatic vines infected with *E. lata* to identify genes that might prevent foliar symptoms. Asymptomatic plants uniquely up-regulated genes primarily associated with energy metabolism and the light phase of photosynthesis, possibly to help maintain chloroplast function and redox balance and circumvent the otherwise harmful effects of *E. lata* toxins.

The interaction between grapevine and another trunk pathogen, *N. parvum*, was also studied at the transcriptomic level (Massonnet et al. 2017b). Host plant leaves and stems infected with *N. parvum* underwent extensive, common, and temporally separated transcriptional changes. Woody stems, where the pathogen is localized, reacted earlier than leaves to infection. The temporal difference in response was indicated by genes related to signal perception, signal transduction, and downstream biological processes including oxidative stress, cell wall rearrangement, and cell death. The results suggested that leaves perceive similar signals as the infection site without interacting directly with the pathogen. The RNA-seq data were also used to investigate the virulence mechanisms used by *N. parvum* during the infection (Massonnet et al. 2018). Gene expression analysis showed that *N. parvum* co-expresses genes associated with secondary metabolism and plant cell wall degradation in function of the growth substrate and the stage of plant infection. Co-expressed genes were found to be physically clustered and to share common regulatory elements in their promoters, suggesting that their co-regulation might contribute to its virulence.

Transcriptomic analyses performed during *P. viticola* infection showed that the grape response involves the activation of defense-related mechanisms, including the expression of PR genes, phenylpropanoid genes, signal transduction (Mitogen-activated protein kinases (MAPKs), calmodulin-binding proteins, receptor kinases) and hormone signaling pathway genes, albeit to a lesser extent in *V. vinifera* than in the DM-resistant *V. riparia* (Polesani et al. 2010; Wu et al. 2010; Vannozzi et al. 2012). Li et al. (2015) used a DM-resistant species, *Vitis amurensis*, as a model to study the molecular determinants of compatible versus incompatible responses that result from infection with different strains of *Plasmopara viticola*. The incompatible interaction and onset of DM symptoms were associated with the up-regulation of 37 resistance genes shortly after infection, genes that participate in the reactive oxygen species and nitric oxide (ROS/NO) and phenylpropanoid biosynthesis, and MAPK and hormone signaling pathways. In contrast, the incompatible interaction involved the repression of photosynthesis and fatty acid synthesis genes. *De novo* sequencing and assembly of *P. viticola* genome, combined with transcriptome profiling of *V. vinifera* during infection identified an RxLR effector gene induced during pathogen colonization (Brilli et al. 2018). *In planta* expression of this effector triggered a strong necrotic response in *V. riparia*, though no noticeable symptoms were visible on *V. vinifera* leaves. This suggested that susceptibility of *V. vinifera* to DM might be partly due to a failure to recognize *P. viticola* virulence factors.

Infecting PM-resistant *V. pseudoreticulata* with *E. necator* revealed a strong induction of effector-triggered immunity, basal defense, systemic acquired resistance (SAR), and secondary metabolism (Weng et al. 2014). Weng et al. (2014) proposed that *V. pseudoreticulata* resistance does not include preventing host-cell penetration by the fungus, but involves the accumulation of phytoalexins, a heightened salicylic acid-related response, depressed jasmonic acid-associated response, cell wall thickening, SAR, and ROS-dependent hypersensitive responses. In another PM study, constitutively expressed and PM-inducible genes shared among resistant accessions were identified, as were 81 genes with expression linked to phenotypic differences among the most resistant accessions (Amrine et al. 2015). The study used sequenced transcripts to examine expression and the basis of variable resistance (Amrine et al. 2015).

Integrating metabolite and transcriptomic data has yielded particularly interesting insights into *Botrytis cinerea* infections, a fungus that can cause either noble rot (desirable) or bunch rot (undesirable) (Agudelo-Romero et al. 2015; Blanco-Ulate et al. 2015; Kelloniemi et al. 2015). Blanco-Ulate et al. (2015) observed an up-regulation of genes during noble rot that affects the accumulation of valuable aroma compounds contributing to the distinctiveness of botrytized wines, as well an induction of the phenylpropanoid genes coupled with the accumulation of anthocyanins, the first such observation in white-skinned berries. Authors compared their data to that published by Agudelo-Romero et al. (2015), a study of bunch rot, and found only 11.9% of noble rot-responsive genes behaving similarly during bunch rot. However, no study has simultaneously compared bunch rot, noble rot, and uninfected plants to our knowledge.

Grapevine diseases caused by bacteria include Pierce's disease, crown gall, bacterial blight, and grapevine yellows. Pierce's disease is caused by the xylem-inhabiting bacterium *Xyllela fastidiosa*, and a few studies have explored its effects and the basis of its virulence (Choi et al. 2013; Rapicavoli et al. 2018). Grapevines respond to Pierce's disease infection by up-regulating genes encoding phytoalexins, PR proteins, and proteins associated with abscisic acid- and jasmonic acid-responsive biosynthesis, and down-regulating transcripts related to photosynthesis and growth (Choi et al. 2013; Zaini et al. 2018). Like most of the Gram-negative bacteria, the *X. fastidiosa* cell envelope is composed of lipopolysaccharides that can be rapidly recognized by the plant and induce a quick oxidative burst (Erbs and Newman 2012). In their study, Rapicavoli et al. (2018) showed that the bacterium produces a long-chain O-antigen that masks the elicitor portions of the lipopolysaccharides. Transcriptomic analysis showed that the lack of the O-antigen leads to a fast plant perception of mutant *X. fastidiosa*, triggering the induction of PR genes and a salicylic acid-mediated defense pathway. The authors suggested that the long-chain O-antigen enables *X. fastidiosa* to delay the initial plant recognition, thereby allowing it to effectively subvert plant defense responses and establish itself in the host.

To date, nearly 70 different viruses have been identified as grape pathogens, accounting for ~ 25 different diseases (Martelli 2014). The most damaging viral diseases include Fanleaf degeneration, Leafroll disease, Rupestris stem-pitting disease, and Red blotch (Meng et al. 2017). An early study of *Rupestris stem-pitting virus* (GRSPaV) used the NimbleGen microarray for grapevine (Gambino et al. 2012). Responses were tissue-specific and included the induction of senescence-related genes, consistent with separate observations of leaves infected with Grapevine leafroll-associated virus 3 (GLRaV-3) (Espinoza et al. 2007). In infected leaves, signal-transducing kinases and hormone signaling were up-regulated compared to uninfected leaves, as were secondary metabolism genes associated with terpene, flavonol, and lignin biosynthesis. In berries, genes affected by the viral infection were predominantly down-regulated and many were associated with plant defense. During GLRaV-3 infection, sugar transporters and anthocyanin biosynthesis-related genes were down-regulated in fruit; Vega et al. (2011) concomitantly observed reduced sugar levels, total anthocyanins and malvidin-3-O-glucoside in GLRaV-3-infected berries. Like GLRaV-3, Grapevine red blotch-associated virus (GRBaV) has a negative effect on fruit composition (Blanco-Ulate et al. 2017). Transcriptomic, enzymatic, and metabolite data supported an alteration of hormone signaling and secondary metabolism in berries during ripening (Blanco-Ulate et al. 2017). Moreover, sequencing of virus-derived small RNAs has been shown to be a useful diagnostic tool for detecting known and novel viruses and viral genome reconstruction (Navarro et al. 2009; Pantaleo et al. 2010; Giampetruzzi et al. 2012; Czotter et al. 2018).

Transcriptomic tools have been effectively used to better understand the relationship between grapevine and its pathogens and for discovering virulence factors, assessing plant sanitary status, and discovering new pathogens. Efforts can and have been made to understand how specific virulence factors, like those identified by Morales-Cruz et al. (2018), are regulated and cause disease symptoms. Gene co-expression may also be an important determinant of virulence and should continue to be evaluated in transcriptomic studies (Massonnet et al. 2017b, 2018). Furthermore, understanding the basis of plant resistance and pathogen virulence will be essential to sustaining viticulture, particularly as the number of pathogens resistant to fungicides grows (Gubler et al. 1996; Baudoin et al. 2008; Gisi and Sierotzki 2008). Finally, we would be remiss to not mention that stress can selectively induce the expression of specific gene isoforms and alternative transcripts (Vitulo et al. 2014; Liu et al. 2016; Han et al. 2017; Jiang et al. 2017). With the advent of full-length isoform sequencing technologies, our understanding of infection will grow as we learn how pathogens alter the isoform landscape and whether particular isoforms are related to resistance.

13.6 Conclusions

This chapter described the expansive application of transcriptomics to viticulture research. Thanks to technological progress and declining sequencing costs over the last 20 years, the grape community has been able to apply global gene expression profiling to investigate many key issues of viticulture. Combined with other experimental data, like agronomical and physiological measurements, hormone and secondary metabolite information, and by integrating additional "omics" data (see Chap. 8), genome-wide transcriptional profiling has provided a deeper understanding of the effects of viticultural practices, scion–rootstock pairings, variable environmental conditions, and diverse types of stress. The implementation of novel tools, like Iso-Seq and 3'RNA-seq, should further improve and accelerate the application of transcriptomics to viticulture. In addition to facilitating gene candidate identification when combined with genetic association approaches, transcriptomics will help identify useful molecular markers that can be used to improve viticulture practices, e.g., to predict flavonoid composition or to signal the type of stress experienced by grapevines.

References

Abbà S, Galetto L, Carle P, Carrère S, Delledonne M, Foissac X, Palmano S, Veratti F, Marzachì C (2014) RNA-Seq profile of flavescence dorée phytoplasma in grapevine. BMC Genom 15(1):1088. https://doi.org/10.1186/1471-2164-15-1088

Abdel-Ghany SE, Hamilton M, Jacobi JL, Ngam P, Devitt N, Schilkey F, Ben-Hur A, Reddy AS (2016) A survey of the sorghum transcriptome using single-molecule long reads. Nat Commun 7:11706. https://doi.org/10.1038/ncomms11706

Ablett E, Seaton G, Scott K, Shelton D, Graham MW, Baverstock P, Lee LS, Henry R (2000) Analysis of grape ESTs: global gene expression patterns in leaf and berry. Plant Sci 159(1):87–95. https://doi.org/10.1016/S0168-9452(00)00335-6

Agudelo-Romero P, Erban A, Rego C, Carbonell-Bejerano P, Nascimento T, Sousa L, Martínez-Zapater JM, Kopka J, Fortes AM (2015) Transcriptome and metabolome reprogramming in *Vitis vinifera* cv. Trincadeira berries upon infection with *Botrytis cinerea*. J Exp Bot 66(7):1769–1785. https://doi.org/10.1093/jxb/eru517

Alston JM, Fuller K, Kaplan J, Tumber K (2013) The economic consequences of pierce's disease and related policy in the California Winegrape Industry. J Agric Resour Econ 38(2):269–297

Amrine KC, Blanco-Ulate B, Riaz S, Pap D, Jones L, Figueroa-Balderas R, Walker MA, Cantu D (2015) Comparative transcriptomics of Central Asian *Vitis vinifera* accessions reveals distinct defense strategies against powdery mildew. Hortic Res 2:15037. https://doi.org/10.1038/hortres.2015.37

Anderson K, Aryal NR (2013) Which winegrape varieties are grown where? A global empirical picture. University of Adelaide Press, Adelaide. https://doi.org/10.20851/winegrapes

Armijo G, Schlechter R, Agurto M, Muñoz D, Nuñez C, Arce-Johnson P (2016) Grapevine pathogenic microorganisms: understanding infection strategies and host response scenarios. Front Plant Sci 7:382. https://doi.org/10.3389/fpls.2016.00382

Baudoin A, Olaya G, Delmotte F, Colcol JF, Sierotzki H (2008) QoI resistance of *Plasmopara viticola* and *Erysiphe necator* in the mid-atlantic United States. Plant Health Progress 122:122. https://doi.org/10.1094/PHP-2008-0211-02-RS

Berdeja M, Nicolas P, Kappel C, Dai ZW, Hilbert G, Peccoux A, Lafontaine M, Ollat N, Gomès E, Delrot S (2015) Water limitation and rootstock genotype interact to alter grape berry metabolism through transcriptome reprogramming. Hortic Res 2:15012. https://doi.org/10.1038/hortres.2015.12

Bertsch C, Ramirez-Suero M, Magnin-Robert M, Larignon P, Chong J, Abou-Mansour E, Spagnolo A, Clément C, Fontaine F (2013) Grapevine trunk diseases: complex and still poorly understood. Plant Pathol 62:243–265. https://doi.org/10.1111/j.1365-3059.2012.02674.x

Bindon KA, Dry PR, Loveys BR (2007) Influence of plant water status on the production of C13-norisoprenoid precursors in *Vitis vinifera* L. Cv. Cabernet Sauvignon grape berries. J Agric Food Chem 55(11):4493–4500. https://doi.org/10.1021/jf063331p

Bisson L (2001) In search of optimal grape maturity. Pract Winery Vineyard J 23:32–43

Blanco-Ulate B, Morales-Cruz A, Amrine KC, Labavitch JM, Powell AL, Cantu D (2014) Genome-wide transcriptional profiling of *Botrytis cinerea* genes targeting plant cell walls during infections of different hosts. Front Plant Sci 5:435. https://doi.org/10.3389/fpls.2014.00435

Blanco-Ulate B, Amrine KC, Collins TS, Rivero RM, Vicente AR, Morales-Cruz A, Doyle CL, Ye Z, Allen G, Heymann H, Ebeler SE, Cantu D (2015) Developmental and metabolic plasticity of white-skinned grape berries in response to *Botrytis cinerea* during noble rot. Plant Physiol 169(4):2422–2443. https://doi.org/10.1104/pp.15.00852

Blanco-Ulate B, Hopfer H, Figueroa-Balderas R, Ye Z, Rivero RM, Albacete A, Pérez-Alfocea F, Koyama R, Anderson MM, Smith RJ, Ebeler SE, Cantu D (2017) Red blotch disease alters grape berry development and metabolism by interfering with the transcriptional and hormonal regulation of ripening. J Exp Bot 68 (5):1225–1238. https://doi.org/10.1093/jxb/erw506

Bock A, Sparks TH, Estrella N, Menzel A (2013) Climate-induced changes in grapevine yield and must sugar content in Franconia (Germany) between 1805 and 2010. PLoS ONE 8:10. https://doi.org/10.1371/journal.pone.0069015

Bradshaw AD, Caspari EW, Thoday JM (1965) Evolutionary significance of phenotypic plasticity in plants. Adv Genet 13:115–155. https://doi.org/10.1016/S0065-2660(08)60048-6

Brilli M, Asquini E, Moser M, Bianchedi PL, Perazzolli M, Si-Ammour A (2018) A multi-omics study of the grapevine-downy mildew (*Plasmopara viticola*) pathosystem unveils a complex protein coding- and noncoding-based arms race during infection. Sci Rep 8(1):757. https://doi.org/10.1038/s41598-018-19158-8

Buonassisi D, Colombo M, Migliaro D, Dolzani C, Peressotti E, Mizzotti C, Velasco R, Masiero S, Perazzolli M, Vezzulli S (2017) Breeding for grapevine downy mildew resistance: a review of "omics" approaches. Euphytica 213:103. https://doi.org/10.1007/s10681-017-1882-8

Burger AL, Botha FC (2004) Ripening-related gene expression during fruit ripening in *Vitis vinifera* L. cv. Cabernet Sauvignon and Clairette blanche. Vitis 43(2):59–63

Camps C, Kappel C, Lecomte P, Léon C, Gomès E, Coutos-Thévenot P, Delrot S (2010) A transcriptomic study of grapevine (*Vitis vinifera* cv. Cabernet-Sauvignon) interaction with the vascular ascomycete fungus Eutypa lata. J Exp Bot 61(6):1719–1737. https://doi.org/10.1093/jxb/erq040

Castellarin SD, Matthews MA, Di Gaspero G, Gambetta GA (2007a) Water deficits accelerate ripening and induce changes in gene expression regulating flavonoid biosynthesis in grape berries. Planta 227:101–112. https://doi.org/10.1007/s00425-007-0598-8

Castellarin SD, Pfeiffer A, Sivilotti P, Degan M, Peterlunger E, Di Gaspero G (2007b) Transcriptional regulation of anthocyanin biosynthesis in ripening fruits of grapevine under seasonal water deficit. Plant Cell Environ 30:1381–1399. https://doi.org/10.1111/j.1365-3040.2007.01716.x

Charrier G, Delzon S, Domec JC, Zhang L, Delmas CEL, Merlin I, Corso D, King A, Ojeda H, Ollat N, Prieto JA, Scholach T, Skinner P, van Leeuwen C, Gambetta GA (2018) Drought will not leave your glass empty: low risk of hydraulic failure revealed by long-term drought observations in world's top wine regions. Sci Adv 4:eaao6969. https://doi.org/10.1126/sciadv.aao6969

Chaves MM, Zarrouk O, Francisco R, Costa JM, Santos T, Regalado AP, Rodrigues ML, Lopes CM (2010) Grapevine under deficit irrigation: hints from physiological and molecular data. Ann Bot 105:661–676. https://doi.org/10.1093/aob/mcq030

Cheng B, Furtado A, Henry RJ (2017) Long-read sequencing of the coffee bean transcriptome reveals the diversity of full-length transcripts. Gigascience 6 (11):1–13. https://doi.org/10.1093/gigascience/gix086

Choi HK, Iandolino A, da Silva FG, Cook DR (2013) Water deficit modulates the response of *Vitis vinifera* to the Pierce's disease pathogen *Xylella fastidiosa*. Mol Plant Microbe Interact 26(6):643–657. https://doi.org/10.1094/MPMI-09-12-0217-R

Clavijo BJ, Venturini L, Schudoma C et al (2017) An improved assembly and annotation of the allo-hexaploid wheat genome identifies complete families of agronomic genes and provides genomic evidence for chromosomal translocations. Genome Res 27 (5):885–896. https://doi.org/10.1101/gr.217117.116

Cochetel N, Escudié F, Cookson SJ, Dai Z, Vivin P, Bert PF, Muñoz MS, Delrot S, Klopp C, Ollat N, Lauvergeat V (2017) Root transcriptomic responses of grafted grapevines to heterogeneous nitrogen availability depend on rootstock genotype. J Exp Bot 68 (15):4339–4355. https://doi.org/10.1093/jxb/erx224

Cookson SJ, Clemente Moreno MJ, Hevin C, Nyamba Mendome LZ, Delrot S, Trossat-Magnin C, Ollat N (2013) Graft union formation in grapevine induces transcriptional changes related to cell wall modification, wounding, hormone signalling, and secondary metabolism. J Exp Bot 64:2997–3008. https://doi.org/10.1093/jxb/ert144

Cookson SJ, Clemente Moreno MJ, Hevin C, Nyamba Mendome LZ, Delrot S, Magnin N, Trossat-Magnin C, Ollat N (2014) Heterografting with nonself rootstocks induces genes involved in stress responses at the graft interface when compared with autografted controls. J Exp Bot 65:2473–2481. https://doi.org/10.1093/jxb/eru145

Corso M, Vannozzi A, Maza E, Vitulo N, Meggio F, Pitacco A, Telatin A, D'Angelo M, Feltrin E, Negri AS, Prinsi B, Valle G, Ramina A, Bouzayen M, Bonghi C, Lucchin M (2015) Comprehensive transcript profiling of two grapevine rootstock genotypes contrasting in drought susceptibility links the phenylpropanoid pathway to enhanced tolerance. J Exp Bot 66(19):5739–5752. https://doi.org/10.1093/jxb/erv274

Corso M, Vannozzi A, Ziliotto F, Zouine M, Maza E, Nicolato T, Vitulo N, Meggio F, Valle G, Bouzayen M, Müller M, Munné-Bosch S, Lucchin M, Bonghi C (2016) Grapevine rootstocks differentially affect the rate of ripening and modulate auxin-related genes in Cabernet Sauvignon berries. Front Plant Sci 7:69. https://doi.org/10.3389/fpls.2016.00069

Cozzolino D, Dambergs RG (2010) Instrumental analysis of grape, must and wine. In: Reynolds A (ed) Managing wine quality, vol 1. Viticulture and wine quality. Woodhead Publishing, Cambridge, pp 134–161. https://doi.org/10.1533/9781845699284.2.134

Czemmel S, Galarneau ER, Travadon R, McElrone AJ, Cramer GR, Baumgartner K (2015) Genes expressed in grapevine leaves reveal latent wood infection by the fungal pathogen Neofusicoccum parvum. PLoS ONE 10(3):e0121828. https://doi.org/10.1371/journal.pone.0121828

Czotter N, Molnar J, Szabó E, Demian E, Kontra L, Baksa I, Szittya G, Kocsis L, Deak T, Bisztray G, Tusnady GE, Burgyan J, Varallyay E (2018) NGS of virus-derived small RNAs as a diagnostic method used to determine viromes of hungarian vineyards. Front Microbiol 9:122. https://doi.org/10.3389/fmicb.2018.00122

Dai ZW, Ollat N, Gomès E, Decroocq S, Tandonnet J-P, Bordenave L, Pieri P, Hilbert G, Kappel C, van Leeuwen C, Vivin P, Delrot S (2011) Ecophysiological, genetic, and molecular causes of variation in grape berry weight and composition: a review. Am J Enol Vitic 62:413–425. https://doi.org/10.5344/ajev.2011.10116

Dal Santo S, Tornielli GB, Zenoni S, Fasoli M, Farina L, Anesi A, Guzzo F, Delledonne M, Pezzotti M (2013) The plasticity of the grapevine berry transcriptome. Genome Biol 14(6):r54. https://doi.org/10.1186/gb-2013-14-6-r54

Dal Santo S, Fasoli M, Negri S, D'Incà E, Vicenzi N, Guzzo F, Tornielli GB, Pezzotti M, Zenoni S (2016) Plasticity of the berry ripening program in a white grape variety. Front Plant Sci 7:970. https://doi.org/10.3389/fpls.2016.00970

Dal Santo S, Zenoni S, Sandri M, De Lorenzis G, Magris G, De Paoli E, Di Gaspero G, Del Fabbro C, Morgante M, Brancadoro L, Grossi D, Fasoli M, Zuccolotto P, Tornielli GB, Pezzotti M (2018) Grapevine field experiments reveal the contribution of genotype, the influence of environment and the effect of their interaction (G × E) on the berry transcriptome. Plant J 93(6):1143–1159. https://doi.org/10.1111/tpj.13834

Da Silva C, Zamperin G, Ferrarini A, Minio A, Dal Molin A, Venturini L, Buson G, Tononi P, Avanzato C, Zago E, Boido E, Dellacassa E, Gaggero C, Pezzotti M, Carrau F, Delledonne M (2013) The high polyphenol content of grapevine cultivar tannat berries is conferred primarily by genes that are not shared with the reference genome. Plant Cell 25(12):4777–4788. https://doi.org/10.1105/tpc.113.118810

Davies C, Robinson SP (2000) Differential screening indicates a dramatic change in mRNA profiles during grape berry ripening. Cloning and characterization of cDNAs encoding putative cell wall and stress response proteins. Plant Physiol 122(3):803–812. https://doi.org/10.1104/pp.122.3.803

Deluc LG, Grimplet J, Wheatley MD, Tillett RL, Quilici DR, Osborne C, Schooley DA, Schlauch KA, Cushman JC, Cramer GR (2007) Transcriptomic and metabolite analyses of Cabernet Sauvignon grape berry development. BMC Genom 8:429. https://doi.org/10.1186/1471-2164-8-429

Deluc LG, Quilici DR, Decendit A, Grimplet J, Wheatley MD, Schlauch KA, Mérillon JM, Cushman JC,

Cramer GR (2009) Water deficit alters differentially metabolic pathways affecting important flavor and quality traits in grape berries of Cabernet Sauvignon and Chardonnay. BMC Genom 10:212. https://doi.org/10.1186/1471-2164-10-212

Dokoozlian NK (2009) Integrated canopy management: a twenty year evolution in California. In: Proceedings of recent advances in grapevine canopy management. Davis, California, July 16 2009, pp 43–52

Dokoozlian NK, Hirschfelt DJ (1995) The influence of cluster thinning at various stages of fruit development on flame seedless table grapes. Am J Enol Vitic 46:429–436

Duchêne E, Huard F, Dumas V, Schneider C, Merdinoglu D (2010) The challenge of adapting grapevine varieties to climate change. Clim Res 41:193–204. https://doi.org/10.3354/cr00850

Erbs G, Newman MA (2012) The role of lipopolysaccharide and peptidoglycan, two glycosylated bacterial microbe-associated molecular patterns (MAMPs), in plant innate immunity. Mol Plant Pathol 13(1):95–104. https://doi.org/10.1111/j.1364-3703.2011.00730.x

Espinoza C, Medina C, Somerville S, Arce-Johnson P (2007) Senescence-associated genes induced during compatible viral interactions with grapevine and Arabidopsis. J Exp Bot 58(12):3197–3212. https://doi.org/10.1093/jxb/erm165

Fallot J, Ruchaud C, Durquety PM, Gazeau JP (1979) The clone and its reaction to grafting. III. The transmission of incompatibility when grafting 5 BB and *Vitis vinifera*. Progrès Agricole et Viticole 96(10):211–216

Fasoli M, Dal Santo S, Zenoni S, Tornielli GB, Farina L, Zamboni A, Porceddu A, Venturini L, Bicego M, Murino V, Ferrarini A, Delledonne M, Pezzotti M (2012) The grapevine expression atlas reveals a deep transcriptome shift driving the entire plant into a maturation program. Plant Cell 24(9):3489–3505. https://doi.org/10.1105/tpc.112.100230

Fasoli M, Richter CL, Zenoni S, Bertini E, Vitulo N, Dal Santo S, Green EA, Dokoozlian NK, Pezzotti M, Tornielli GB (2018a) Unraveling the key molecular events of grape berry ripening under varying crop loads. In: XII international conference on grapevine breeding and genetics. Bordeaux, France, p 56

Fasoli M, Richter CL, Zenoni S, Bertini E, Vitulo N, Dal Santo S, Dokoozlian NK, Pezzotti M, Tornielli GB (2018b) The timing and order of the molecular events that mark the onset of berry ripening in grapevine. Plant Physiol 178(3):1187–1206. https://doi.org/10.1104/pp.18.00559

Fisarakis I, Chartzoulakis K, Stavrakas D (2001) Response of Sultana vines (V. vinifera L.) on six rootstocks to NaCl salinity exposure and recovery. Agric Water Manag 51:13–27. https://doi.org/10.1016/S0378-3774(01)00115-9

Forcato C (2010) Gene prediction and functional annotation in the *Vitis vinifera* genome. PhD Thesis, Universita' Degli Studi Di Padova, Italy

Fuller KB, Alston JM, Sambucci OS (2014) The value of powdery mildew resistance in grapes: evidence from California. Wine Econ Policy 3:90–107. https://doi.org/10.1016/j.wep.2014.09.001

Fung RW, Gonzalo M, Fekete C, Kovacs LG, He Y, Marsh E, McIntyre LM, Schachtman DP, Qiu W (2008) Powdery mildew induces defense-oriented reprogramming of the transcriptome in a susceptible but not in a resistant grapevine. Plant Physiol 146 (1):236–249. https://doi.org/10.1104/pp.107.108712

Gambetta GA, Matthews MA, Shaghasi TH, McElrone AJ, Castellarin SD (2010) Sugar and abscisic acid signaling orthologs are activated at the onset of ripening in grape. Planta 232:219–234. https://doi.org/10.1007/s00425-010-1165-2

Gambetta GA, Manuck CM, Drucker ST, Shaghasi T, Fort K, Matthews MA, Walker MA, McElrone AJ (2012) The relationship between root hydraulics and scion vigour across Vitis rootstocks: what role do root aquaporins play? J Exp Bot 63:6445–6455. https://doi.org/10.1093/jxb/ers312

Gambino G, Cuozzo D, Fasoli M, Pagliarani C, Vitali M, Boccacci P, Pezzotti M, Mannini F (2012) Co-evolution between Grapevine rupestris stem pitting-associated virus and *Vitis vinifera* L. leads to decreased defence responses and increased transcription of genes related to photosynthesis. J Exp Bot 63 (16):5919–5933. https://doi.org/10.1093/jxb/ers244

Gambino G, Dal Molin A, Boccacci P, Minio A, Chitarra W, Avanzato CG, Tononi P, Perrone I, Raimondi S, Schneider A, Pezzotti M, Mannini F, Gribaudo I, Delledonne M (2017) Whole-genome sequencing and SNV genotyping of 'Nebbiolo' (*Vitis vinifera* L.) clones. Sci Rep 7(1):17294. https://doi.org/10.1038/s41598-017-17405-y

Gessler C, Pertot I, Perazzolli M (2011) *Plasmopara viticola*: a review of knowledge on downy mildew of grapevine and effective disease management. Phytopathol Mediterr 50:3–44. https://doi.org/10.14601/Phytopathol_Mediterr-9360

Giampetruzzi A, Roumi V, Roberto R, Malossini U, Yoshikawa N, La Notte P, Terlizzi F, Credi R, Saldarelli P (2012) A new grapevine virus discovered by deep sequencing of virus- and viroid-derived small RNAs in Cv Pinot gris. Virus Res 163:262–268. https://doi.org/10.1016/j.virusres.2011.10.010

Gisi U, Sierotzki H (2008) Fungicide modes of action and resistance in downy mildews. Eur J Plant Pathol 122:157–167. https://doi.org/10.1007/s10658-008-9290-5

Gramaje D, Úrbez-Torres JR, Sosnowski MR (2018) Managing grapevine trunk diseases with respect to etiology and epidemiology: current strategies and future prospects. Plant Dis 102:12–39. https://doi.org/10.1094/PDIS-04-17-0512-FE

Gramazio P, Blanca J, Ziarsolo P, Herraiz FJ, Plazas M, Prohens J, Vilanova S (2016) Transcriptome analysis and molecular marker discovery in *Solanum incanum* and *S. aethiopicum*, two close relatives of the common eggplant (*Solanum melongena*) with interest for breeding. BMC Genom 17:300. https://doi.org/10.1186/s12864-016-2631-4

Gregory PJ, Atkinson CJ, Bengough AG, Else MA, Fernández-Fernández F, Harrison RJ, Schmidt S (2013) Contributions of roots and rootstocks to sustainable, intensified crop production. J Exp Bot 64:1209–1222. https://doi.org/10.1093/jxb/ers385

Grimplet J, Deluc LG, Tillett RL, Wheatley MD, Schlauch KA, Cramer GR, Cushman JC (2007) Tissue-specific mRNA expression profiling in grape berry tissues. BMC Genom 8:187. https://doi.org/10.1186/1471-2164-8-187

Gubler WD, Ypema HL, Ouimette DE, Bettiga LJ (1996) Occurrence of resistance in Uncinula necator to triadimefon, myclobutanil, and fenarimol in California grapevines. Plant Dis 80(8):902–909. https://doi.org/10.1094/PD-80-0902

Guidoni S, Allara P, Schubert A (2002) Effect of cluster thinning on berry skin anthocyanin composition of *Vitis vinifera* cv. Nebbiolo. Am J Enol Vitic 53:224–226

Guidoni S, Ferrandino A, Novello V (2008) Effects of seasonal and agronomical practices on skin anthocyanin profile of Nebbiolo grapes. Am J Enol Vitic 59:22–29

Han N, Ji XL, Du YP, He X, Zhao XJ, Zhai H (2017) Identification of a novel alternative splicing variant of VvPMA1 in grape root under salinity. Front Plant Sci 8:605. https://doi.org/10.3389/fpls.2017.00605

Herrera JC, Castellarin SD (2016) Preveraison water deficit accelerates berry color change in merlot grapevines. Am J Enol Vitic 67:356–360. https://doi.org/10.5344/ajev.2016.15083

Herrera JC, Hochberg U, Degu A, Sabbatini P, Lazarovitch N, Castellarin SD, Fait A, Alberti G, Peterlunger E (2017) Grape metabolic response to postveraison water deficit is affected by interseason weather variability. J Agric Food Chem 65(29):5868–5878. https://doi.org/10.1021/acs.jafc.7b01466

Hochberg U, Rockwell FE, Holbrook NM, Cochard H (2018) Iso/anisohydry: a plant-environment interaction rather than a simple hydraulic trait. Trends Plant Sci 23(2):112–120. https://doi.org/10.1016/j.tplants.2017.11.002

Hunter JJ, De Villiers OT, Watts JE (1991) The effect of partial defoliation on quality characteristics of *Vitis vinifera* L. cv. Cabernet Sauvignon grapes. II. Skin color, skin sugar and wine quality. Am J Enol Vitic 42:13–18

Hren M, Nikolić P, Rotter A, Blejec A, Terrier N, Ravnikar M, Dermastia M, Gruden K (2009) 'Bois noir' phytoplasma induces significant reprogramming of the leaf transcriptome in the field grown grapevine. BMC Genom 10:460. https://doi.org/10.1186/1471-2164-10-460

Jackson DI, Lombard PB (1993) Environmental and management practices affecting grape composition and wine quality: a review. Am J Enol Vitic 44:409–430

Jaillon O, Aury JM, Noel B, The French–Italian Public Consortium for Grapevine Genome Characterization et al (2007) The grapevine genome sequence suggests ancestral hexaploidization in major angiosperm phyla. Nature 449(7161):463–467. https://doi.org/10.1038/nature06148

Jiang J, Liu X, Liu C, Liu G, Li S, Wang L (2017) Integrating omics and alternative splicing reveals insights into grape response to high temperature. Plant Physiol 173(2):1502–1518. https://doi.org/10.1104/pp.16.01305

Jones GV, Reid R, Vilks A (2012) Climate, grapes, and wine: structure and suitability in a variable and changing climate. In: Dougherty P (ed) The geography of wine. Springer, Dordrecht, pp 109–133. https://doi.org/10.1007/978-94-007-0464-0_7

Jones JD, Dangl JL (2006) The plant immune system. Nature 444:323–329. https://doi.org/10.1038/nature05286

Kelloniemi J, Trouvelot S, Héloir MC, Simon A, Dalmais B, Frettinger P, Cimerman A, Fermaud M, Roudet J, Baulande S, Bruel C, Choquer M, Couvelard L, Duthieuw M, Ferrarini A, Flors V, Le Pêcheur P, Loisel E, Morgant G, Poussereau N, Pradier JM, Rascle C, Trdá L, Poinssot B, Viaud M (2015) Analysis of the molecular dialogue between gray mold (*Botrytis cinerea*) and grapevine (*Vitis vinifera*) reveals a clear shift in defense mechanisms during berry ripening. Mol Plant Microbe Interact 28(11):1167–1180. https://doi.org/10.1094/MPMI-02-15-0039-R

Kim MA, Rhee JS, Kim TH, Lee JS, Choi AY, Choi BS, Choi IY, Sohn YC (2017) Alternative splicing profile and sex-preferential gene expression in the female and male Pacific Abalone Haliotis discus hannai. Genes 8(3):99. https://doi.org/10.3390/genes8030099

Kliewer WM, Dokoozlian NK (2005) Leaf area/crop weight ratios of grapevines: influence on fruit composition and wine quality. Am J Enol Vitic 56:170–181

Li J, Harata-Lee Y, Denton MD, Feng Q, Rathjen JR, Qu Z, Adelson DL (2017a) Long read reference genome-free reconstruction of a full-length transcriptome from Astragalus membranaceus reveals transcript variants involved in bioactive compound biosynthesis. Cell Discov 3:17031. https://doi.org/10.1038/celldisc.2017.31

Li X, Wu J, Yin L, Zhang Y, Qu J, Lu J (2015) Comparative transcriptome analysis reveals defense-related genes and pathways against downy mildew in Vitis amurensis grapevine. Plant Physiol Biochem 95:1–14. https://doi.org/10.1016/j.plaphy.2015.06.016

Li Y, Wei W, Feng J, Luo H, Pi M, Liu Z, Kang C (2017b) Genome re-annotation of the wild strawberry Fragaria vesca using extensive Illumina- and SMRT-based RNA-seq datasets. DNA Res 25(1):61–70. https://doi.org/10.1093/dnares/dsx038

Lider LA, Ferrari NL, Bowers KW (1978) A study of longevity of graft combinations in California vineyards, with special interest in the vinifera X rupestris hybrids. Am J Enol Vitic 29:18–24

Liu J, Chen X, Liang X, Zhou X, Yang F, Liu J, He SY, Guo Z (2016) Alternative splicing of rice WRKY62 and WRKY76 transcription factor genes in pathogen defense. Plant Physiol 171(2):1427–1442. https://doi.org/10.1104/pp.15.01921

Lovisolo C, Perrone I, Carra A, Ferrandino A, Flexas J, Medrano H, Schubert A (2010) Drought-induced changes in development and function of grapevine (Vitis spp.) organs and in their hydraulic and non-hydraulic interactions at the whole-plant level: a physiological and molecular update. Funct Plant Biol 37:98–116. https://doi.org/10.1071/FP09191

Lowe R, Shirley N, Bleackley M, Dolan S, Shafee T (2017) Transcriptomics technologies. PLoS Comput Biol 13(5):e1005457. https://doi.org/10.1371/journal.pcbi.1005457

Madden LV, Wheelis M (2003) The threat of plant pathogens as weapons against U.S. crops. Annu Rev Phytopathol 41:155–176. https://doi.org/10.1146/annurev.phyto.41.121902.102839

Marguerit E, Brendel O, Lebon E, van Leeuwen C, Ollat N (2012) Rootstock control of scion transpiration and its acclimation to water deficit are controlled by different genes. New Phytol 194:416–429. https://doi.org/10.1111/j.1469-8137.2012.04059.x

Martelli GP (2014) Directory of virus and virus-like diseases of the grapevine and their agents. J Plant Pathol 96:1–136

Massonnet M, Fasoli M, Tornielli GB, Altieri M, Sandri M, Zuccolotto P, Paci P, Gardiman M, Zenoni S, Pezzotti M (2017a) Ripening transcriptomic program in red and white grapevine varieties correlates with berry skin anthocyanin accumulation. Plant Physiol 174 (4):2376–2396. https://doi.org/10.1104/pp.17.00311

Massonnet M, Figueroa-Balderas R, Galarneau ERA, Miki S, Lawrence DP, Sun Q, Wallis CM, Baumgartner K, Cantu D (2017b) Neofusicoccum parvum colonization of the grapevine woody stem triggers asynchronous host responses at the site of infection and in the leaves. Front Plant Sci 8:1117. https://doi.org/10.3389/fpls.2017.01117

Massonnet M, Morales-Cruz A, Figueroa-Balderas R, Lawrence DP, Baumgartner K, Cantu D (2018) Condition-dependent co-regulation of genomic clusters of virulence factors in the grapevine trunk pathogen Neofusicoccum parvum. Mol Plant Pathol 19(1):21–34. https://doi.org/10.1111/mpp.12491

Matthews MA, Nuzzo V (2007) Berry size and yield paradigms on grapes and wines quality. Acta Hortic. https://doi.org/10.17660/ActaHortic.2007.754.56

Meggio F, Prinsi B, Negri AS, Simone Di Lorenzo G, Lucchini G, Pitacco A, Failla O, Scienza A, Cocucci M, Espen L (2014) Biochemical and physiological responses of two grapevine rootstock genotypes to drought and salt treatments. Aust J Grape Wine Res 20:310–323. https://doi.org/10.1111/ajgw.1207

Meng B, Martelli GP, Golino DA, Fuchs M (eds) (2017) Grapevine viruses: molecular biology, diagnostics and management. Springer, Heidelberg

Milien M, Renault-Spilmont AS, Cookson SJ, Sarrazin A, Verdeil JL (2012) Visualization of the 3D structure of the graft union of grapevine using X-ray tomography. Sci Hortic 144:130–140. https://doi.org/10.1016/j.scienta.2012.06.045

Minio A, Massonnet M, Figueroa-Balderas R, Vondras AM, Blanco-Ulate B, Cantu D (2019) Iso-Seq allows genome-independent transcriptome profiling of grape berry development. G3 (Bethesda, MD). https://doi.org/10.1534/g3.118.201008

Mittler R (2006) Abiotic stress, the field environment and stress combination. Trends Plant Sci 11:15–19. https://doi.org/10.1016/j.tplants.2005.11.002

Morales-Cruz A, Allenbeck G, Figueroa-Balderas R, Ashworth VE, Lawrence DP, Travadon R, Smith RJ, Baumgartner K, Rolshausen PE, Cantu D (2018) Closed-reference metatranscriptomics enables in planta profiling of putative virulence activities in the grapevine trunk disease complex. Mol Plant Pathol 19 (2):490–503. https://doi.org/10.1111/mpp.12544

Mori K, Goto-Yamamoto N, Kitayama M, Hashizume K (2007) Loss of anthocyanins in red-wine grape under high temperature. J Exp Bot 58(8):1935–1945. https://doi.org/10.1093/jxb/erm055

Mortazavi A, Williams BA, McCue K, Schaeffer L, Wold B (2008) Mapping and quantifying mammalian transcriptomes by RNA-seq. Nat Methods 5(7):621–628. https://doi.org/10.1038/nmeth.1226

Moser C, Segala C, Fontana P, Salakhudtinov I, Gatto P, Pindo M, Zyprian E, Toepfer R, Grando MS, Velasco R (2005) Comparative analysis of expressed sequence tags from different organs of *Vitis vinifera* L. Funct Integr Genom 5(4):208–217. https://doi.org/10.1007/s10142-005-0143-4

Navarro B, Pantaleo V, Gisel A, Moxon S, Dalmay T, Bisztray G, Di Serio F, Burgyán J (2009) Deep sequencing of viroid-derived small RNAs from grapevine provides new insights on the role of RNA silencing in plant–viroid interaction. PLoS ONE 4: e7686. https://doi.org/10.1371/journal.pone.0007686

Oerke E-C (2006) Crop losses to pests. J Agric Sci 144:31–43. https://doi.org/10.1017/S0021859605005708

Ollat N, Bordenave L, Tandonnet JP, Boursiquot JM, Marguerit E (2016) Grapevine rootstocks: origins and perspectives. Acta Hort 1136:11–22. https://doi.org/10.17660/ActaHortic.2016.1136.2

Østergård H, Finckh M, Fontaine L, Goldringer I, Hoad S, Kristensen K, Lammerts van Bueren E, Mascher F, Munk L, Wolfe M (2009) Time for a shift in crop production: Embracing complexity through diversity at all levels. J Sci Food Agric 89:1439–1445. https://doi.org/10.1002/jsfa.3615

Palliotti A, Cartechini A (2000) Cluster thinning effects on yield and grape composition in different grapevine cultivars. Acta Hortic. https://doi.org/10.17660/ActaHortic.2000.512.11

Pandey MK, Roorkiwal M, Singh VK, Ramalingam A, Kudapa H, Thudi M, Chitikineni A, Rathore A, Varshney RK (2016) Emerging genomic tools for legume breeding: current status and future prospects.

Front Plant Sci 7:455. https://doi.org/10.3389/fpls.2016.00455
Pantaleo V, Saldarelli P, Miozzi L, Giampetruzzi A, Gisel A, Moxon S, Dalmay T, Bisztray G, Burgyan J (2010) Deep sequencing analysis of viral short RNAs from an infected Pinot Noir grapevine. Virology 408:49–56. https://doi.org/10.1016/j.virol.2010.09.001
Parkinson J, Blaxter M (2009) Expressed sequence tags: an overview. Methods Mol Biol 533:1–12. https://doi.org/10.1007/978-1-60327-136-3_1
Pastore C, Zenoni S, Tornielli GB, Allegro G, Dal Santo S, Valentini G, Intrieri C, Pezzotti M, Filippetti I (2011) Increasing the source/sink ratio in *Vitis vinifera* (cv Sangiovese) induces extensive transcriptome reprogramming and modifies berry ripening. BMC Genom 12:631. https://doi.org/10.1186/1471-2164-12-631
Pastore C, Zenoni S, Fasoli M, Pezzotti M, Tornielli GB, Filippetti I (2013) Selective defoliation affects plant growth, fruit transcriptional ripening program and flavonoid metabolism in grapevine. BMC Plant Biol 13:30. https://doi.org/10.1186/1471-2229-13-30
Pilati S, Perazzolli M, Malossini A, Cestaro A, Demattè L, Fontana P, Dal Ri A, Viola R, Velasco R, Moser C (2007) Genome-wide transcriptional analysis of grapevine berry ripening reveals a set of genes similarly modulated during three seasons and the occurrence of an oxidative burst at véraison. BMC Genom 8:428. https://doi.org/10.1186/1471-2164-8-428
Pina A, Errea P (2005) A review of new advances in mechanism of graft compatibility–incompatibility. Sci Hortic 106:1–11. https://doi.org/10.1016/j.scienta.2005.04.003
Polesani M, Bortesi L, Ferrarini A, Zamboni A, Fasoli M, Zadra C, Lovato A, Pezzotti M, Delledonne M, Polverari A (2010) General and species-specific transcriptional responses to downy mildew infection in a susceptible (*Vitis vinifera*) and a resistant (V. riparia) grapevine species. BMC Genom 11:117. https://doi.org/10.1186/1471-2164-11-117
Poni S, Gatti M (2017) Affecting yield components and grape composition through manipulations of the source-sink balance. Acta Hortic 1188:21–34. https://doi.org/10.17660/ActaHortic.2017.1188.4
Poni S, Casalini L, Bernizzoni F, Civardi S, Intrieri C (2006) Effects of early defoliation on shoot photosynthesis, yield components, and grape composition. Am J Enol Vitic 57:397–407
Poni S, Gatti M, Palliotti A, Dai Z, Duchêne E, Truong T-T, Ferrara G, Matarrese AMS, Gallotta A, Bellincontro A, Mencarelli F, Tombesi S (2018) Grapevine quality: a multiple choice issue. Sci Hortic 234:445–462. https://doi.org/10.1016/j.scienta.2017.12.035
Qiu W, Feechan A, Dry I (2015) Current understanding of grapevine defense mechanisms against the biotrophic fungus (*Erysiphe necator*), the causal agent of powdery mildew disease. Hortic Res 2:15020. https://doi.org/10.1038/hortres.2015.20
Rapicavoli JN, Blanco-Ulate B, Muszyński A, Figueroa-Balderas R, Morales-Cruz A, Azadi P, Dobruchowska JM, Castro C, Cantu D, Roper MC (2018) Lipopolysaccharide O-antigen delays plant innate immune recognition of *Xylella fastidiosa*. Nat Commun 9(1):390. https://doi.org/10.1038/s41467-018-02861-5
Renouf V, Tregoat O, Roby JP, van Leeuwen C (2010) Soils, rootstocks and grapevine varieties in prestigious bordeaux vineyards and their impact on yield and quality. Journal International des sciences de la vigne et du vin 44:127–134. https://doi.org/10.20870/oeno-one.2010.44.3.1471
Ruel JJ, Walker MA (2006) Resistance to Pierce's disease in muscadinia rotundifolia and other native grape species. Am J Enol Vitic 57:158–165
Saltz JB, Bell AM, Flint J, Gomulkiewicz R, Hughes KA, Keagy J (2018) Why does the magnitude of genotype-by-environment interaction vary? Ecol Evolut 8(12):6342–6353. https://doi.org/10.1002/ece3.4128
Savoi S, Wong DC, Arapitsas P, Miculan M, Bucchetti B, Peterlunger E, Fait A, Mattivi F, Castellarin SD (2016) Transcriptome and metabolite profiling reveals that prolonged drought modulates the phenylpropanoid and terpenoid pathway in white grapes (*Vitis vinifera* L.). BMC Plant Biol 16:67. https://doi.org/10.1186/s12870-016-0760-1
Savoi S, Wong DCJ, Degu A, Herrera JC, Bucchetti B, Peterlunger E, Fait A, Mattivi F, Castellarin SD (2017) Multi-omics and integrated network analyses reveal new insights into the systems relationships between metabolites, structural genes, and transcriptional regulators in developing grape berries (*Vitis vinifera* L.) exposed to water deficit. Front Plant Sci 8:1124. https://doi.org/10.3389/fpls.2017.01124
Schultz HR (2003) Differences in hydraulic architecture account for near- isohydric and anisohydric behaviour of two eld-grown. Plant Cell Environ 26:1393–1406. https://doi.org/10.1046/j.1365-3040.2003.01064.x
Seguin G (1988) Ecosystems of the great red wines produced in the maritime climate of Bordeaux. In: Fuller-Perrine L (ed) Proceedings of the symposium on maritime climate winegrowing. Department of Horticultural Sciences, Cornell University, Geneva, NY, pp 36–53
Suzuki N, Rivero RM, Shulaev V, Blumwald E, Mittler R (2014) Abiotic and biotic stress combinations. New Phytol 203(1):32–43. https://doi.org/10.1111/nph.12797
Terrier N, Ageorges A, Abbal P, Romieu C (2001) Generation of ESTs from grape berry at various developmental stages. J Plant Physiol 158(12):1575–1583. https://doi.org/10.1078/0176-1617-00566
Terrier N, Glissant D, Grimplet J, Barrieu F, Abbal P, Couture C, Ageorges A, Atanassova R, Léon C, Renaudin JP, Dédaldéchamp F, Romieu C, Delrot S, Hamdi S (2005) Isogene specific oligo arrays reveal multifaceted changes in gene expression during grape berry (*Vitis vinifera* L.) development. Planta 222(5):832–847. https://doi.org/10.1007/s00425-005-0017-y

Todić S, Bešlić Z, Kuljančić I (2005) Varying degree of grafting compatibility between cv. Chardonnay, Merlot and different grapevine rootstocks. J Cent Eur Agric 6(2):115–120

Vaillant-Gaveau N, Wojnarowiez G, Petit A-N, Jacquens L, Panigai L, Clement C, Fontaine F (2014) Relationships between carbohydrates and reproductive development in chardonnay grapevine: impact of defoliation and fruit removal treatments during four successive growing seasons. Journal International des sciences de la vigne et du vin 48(4):219–229. https://doi.org/10.20870/oeno-one.2014.48.4.1694

van Leeuwen C, Seguin G (2006) The concept of terroir in viticulture. J Wine Res 17:1–10. https://doi.org/10.1080/09571260600633135

van Leeuwen C, Friant PH, Choné X, Trégoat O, Koundouras S, Dubourdieu D (2004) The influence of climate, soil and cultivar on terroir. Am J Enol Vitic 55:207–217

van Leeuwen C, Trégoat O, Choné X, Bois B, Pernet D, Gaudillère J-P (2009) Vine water status is a key factor in grape ripening and vintage quality for red Bordeaux wine. How can it be assessed for vineyard management purposes? Journal International des sciences de la vigne et du vin 43:121. https://doi.org/10.20870/oeno-one.2009.43.3.798

Vannozzi A, Dry IB, Fasoli M, Zenoni S, Lucchin M (2012) Genome-wide analysis of the grapevine stilbene synthase multigenic family: genomic organization and expression profiles upon biotic and abiotic stresses. BMC Plant Biol 12:130. https://doi.org/10.1186/1471-2229-12-130

Vannozzi A, Donnini S, Vigani G, Corso M, Valle G, Vitulo N, Bonghi C, Zocchi G, Lucchin M (2017) Transcriptional characterization of a widely-used grapevine rootstock genotype under different iron-limited conditions. Front Plant Sci 7:1994. https://doi.org/10.3389/fpls.2016.01994

Vega A, Gutiérrez RA, Peña-Neira A, Cramer GR, Arce-Johnson P (2011) Compatible GLRaV-3 viral infections affect berry ripening decreasing sugar accumulation and anthocyanin biosynthesis in *Vitis vinifera*. Plant Mol Biol 77(3):261–274. https://doi.org/10.1007/s11103-011-9807-8

Velasco R, Zharkikh A, Troggio M et al (2007) A high quality draft consensus sequence of the genome of a heterozygous grapevine variety. PLoS ONE 2(12): e136. https://doi.org/10.1371/journal.pone.0001326

Venter M, Burger AL, Botha FC (2001) Molecular analysis of fruit ripening: the identification of differentially expressed sequences in *Vitis vinifera* using cDNA-AFLP technology. Vitis 40(4):191–196

Venturini L, Ferrarini A, Zenoni S, Tornielli GB, Fasoli M, Dal Santo S, Minio A, Buson G, Tononi P, Zago ED, Zamperin G, Bellin D, Pezzotti M, Delledonne M (2013) De novo transcriptome characterization of *Vitis vinifera* cv. Corvina unveils varietal diversity. BMC Genom 14:41. https://doi.org/10.1186/1471-2164-14-41

Viers JH, Williams JN, Nicholas KA, Barbosa O, Kotzé I, Spence L, Webb LB, Merenlender A, Reynolds M (2013) Vinecology: pairing wine with nature. Conserv Lett 6:287–299. https://doi.org/10.1111/conl.12011

Vitulo N, Forcato C, Carpinelli EC, Telatin A, Campagna D, D'Angelo M, Zimbello R, Corso M, Vannozzi A, Bonghi C, Lucchin M, Valle G (2014) A deep survey of alternative splicing in grape reveals changes in the splicing machinery related to tissue, stress condition and genotype. BMC Plant Biol 14:99. https://doi.org/10.1186/1471-2229-14-99

Vuylsteke M, Peleman JD, van Eijk MJ (2007) AFLP-based transcript profiling (cDNA-AFLP) for genome-wide expression analysis. Nat Protoc 2(6): 1399–1413. https://doi.org/10.1038/nprot.2007.174

Walker RR, Blackmore DH, Clingeleffer PR, Correll RL (2002) Rootstock effects on salt tolerance of irrigated field-grown grapevines (*Vitis vinifera* L. cv. Sultana). 1. Yield and vigour inter-relationships. Aust J Grape Wine Res 8(1):3–14. https://doi.org/10.1111/j.1755-0238.2002.tb00206.x

Walker RR, Blackmore DH, Clingeleffer PR, Correll RL (2004) Rootstock effects on salt tolerance of irrigated field-grown grapevines (*Vitis vinifera* L. cv. Sultana) 2. Ion concentrations in leaves and juice. Aust J Grape Wine Res 10:90–99. https://doi.org/10.1111/j.1755-0238.2004.tb00011.x

Wang B, Tseng E, Regulski M, Clark TA, Hon T, Jiao Y, Lu Z, Olson A, Stein JC, Ware D (2016) Unveiling the complexity of the maize transcriptome by single-molecule long-read sequencing. Nat Commun 7:11708. https://doi.org/10.1038/ncomms11708

Wang Z, Gerstein M, Snyder M (2009) RNA-Seq: a revolutionary tool for transcriptomics. Nat Rev Genet 10(1):57–63. https://doi.org/10.1038/nrg2484

Waters DLE, Holton TA, Ablet EM, Slade Lee L, Henry RJ (2005) cDNA microarray analysis of developing grape (*Vitis vinifera* cv. Shiraz) berry skin. Funct Integr Genom 5(1):40–58. https://doi.org/10.1007/s10142-004-0124-z

Weng K, Li ZQ, Liu RQ, Wang L, Wang YJ, Xu Y (2014) Transcriptome of *Erysiphe necator*-infected Vitis pseudoreticulata leaves provides insight into grapevine resistance to powdery mildew. Hortic Res 1:14049. https://doi.org/10.1038/hortres.2014.49

Wilcox WF, Gubler WD, Uyemoto JK (2015) Diseases of Grape (*Vitis vinifera* L.). https://www.apsnet.org/publications/commonnames/Pages/Grape.aspx. Accessed 14 Dec 2018

White MA, Whalen P, Jones GV (2009) Land and wine. Nat Geosci 2:82–84. https://doi.org/10.1038/ngeo429

Wu J, Zhang Y, Zhang H, Huang H, Folta KM, Lu J (2010) Whole genome wide expression profiles of Vitis amurensis grape responding to downy mildew by using Solexa sequencing technology. BMC Plant Biol 10:234. https://doi.org/10.1186/1471-2229-10-234

Xu S, Wang J, Shang H, Huang Y, Yao W, Chen B, Zhang M (2018) Transcriptomic characterization and potential marker development of contrasting sugarcane cultivars. Sci Rep 8:1683. https://doi.org/10.1038/s41598-018-19832-x

Xu X, Zhang Y, Williams J, Antoniou E, McCombie WR, Wu S, Zhu W, Davidson NO, Denoya P, Li E (2013) Parallel comparison of Illumina RNA-Seq and Affymetrix microarray platforms on transcriptomic profiles generated from 5-aza-deoxy-cytidine treated HT-29 colon cancer cells and simulated datasets. BMC Bioinform 14(Suppl 9):S1. https://doi.org/10.1186/1471-2105-14-S9-S1

Zaini PA, Nascimento R, Gouran H, Cantu D, Chakraborty S, Phu M, Goulart LR, Dandekar AM (2018) Molecular profiling of Pierce's disease outlines the response circuitry of *Vitis vinifera* to *Xylella fastidiosa* Infection. Front Plant Sci 9:771. https://doi.org/10.3389/fpls.2018.00771

Zarrouk O, Costa JM, Francisco R et al (2016) Drought and water management in Mediterranean vineyards. In: Gerós H, Chaves M, Medrano H, Delrot S (eds) Grapevine in a changing environment: a molecular and ecophysiological perspective. Wiley, Hoboken, pp 38–67

Zenoni S, Ferrarini A, Giacomelli E, Xumerle L, Fasoli M, Malerba G, Bellin D, Pezzotti M, Delledonne M (2010) Characterization of transcriptional complexity during berry development in *Vitis vinifera* using RNA-Seq. Plant Physiol 152(4):1787–1795. https://doi.org/10.1104/pp.109.149716

Zenoni S, Dal Santo S, Tornielli GB, D'Inca E, Filippetti I, Pastore C, Allegro G, Silvestroni O, Lanari V, Pisciotta A, Di Lorenzo R, Palliotti A, Tombesi S, Gatti M, Poni S (2017) Transcriptional responses to pre-flowering leaf defoliation in grapevine berry from different growing sites, years, and genotypes. Front Plant Sci 8:630. https://doi.org/10.3389/fpls.2017.00630

Zhao S, Fung-Leung W-P, Bittner A, Ngo K, Liu X (2014) Comparison of RNA-Seq and microarray in transcriptome profiling of activated T cells. PLoS ONE 9(1):e78644. https://doi.org/10.1371/journal.pone.0078644

Zhu FY, Chen MX, Ye NH, Shi L, Ma KL, Yang JF, Cao YY, Zhang Y, Yoshida T, Fernie AR, Fan GY, Wen B, Zhou R, Liu TY, Fan T, Gao B, Zhang D, Hao GF, Xiao S, Liu YG, Zhang J (2017) Proteogenomic analysis reveals alternative splicing and translation as part of the abscisic acid response in Arabidopsis seedlings. Plant J 91(3):518–533. https://doi.org/10.1111/tpj.13571

Grape Rootstock Breeding and Their Performance Based on the Wolpert Trials in California

14

Jean Catherine Dodson Peterson, Roger Duncan, Donna Hirschfelt, Chuck Ingels, Glenn McGourty, Rhonda Smith, Ed Weber, James Wolpert, Michael Anderson, Jason Benz and M. Andrew Walker

Abstract

Pest and disease pressures have traditionally driven the use of grapevine rootstocks. However, with competing demands on limited water resources and changing climate conditions, the water available for agriculture will likely continue to diminish. This inevitability has resulted in an increased interest in grape rootstock effects on scion growth, specifically yield components, as a function of parentage and in terms of drought tolerance. It has also spurred efforts to breed new rootstocks with combined pest, disease, and abiotic stress resistance. Field studies examining rootstock effect on scion growth and development have been inconsistent due to complications associated with varying soil types and weather factors unique to each site. Other factors tend to vary from site to site including trellising systems, scion cultivar selection and management practices, each of which also has a role in determining vine phenology. Due to these issues, rootstock trial data is often presented on a site-by-site basis with the objective of determining if a specific rootstock tends to yield more or less than other rootstocks on a given site. Although inherently limited by the aforementioned challenges, rootstock trails are arguably one of the best methods of providing insight into rootstock performance and scion interactions. The Wolpert rootstock trails in California were one of the more

J. C. Dodson Peterson
Wine and Viticulture, California Polytechnic State University, One Grand Avenue, San Luis Obispo, CA, USA
e-mail: jdodsonp@calpoly.edu

R. Duncan · D. Hirschfelt · C. Ingels · G. McGourty · R. Smith · E. Weber
Cooperative Extension, University of California, Oakland, CA, USA
e-mail: raduncan@ucanr.edu

D. Hirschfelt
e-mail: djhirschfelt@ucdavis.edu

C. Ingels
e-mail: caingels@ucanr.edu

G. McGourty
e-mail: gtmcgourty@ucanr.edu

R. Smith
e-mail: rhsmith@ucanr.edu

E. Weber
e-mail: eaweber@ucanr.edu

J. Wolpert · M. Anderson · J. Benz · M. A.Walker (✉)
Department of Viticulture and Enology, University of California, One Shields Ave, Davis, CA 95616, USA
e-mail: awalker@ucdavis.edu

J. Wolpert
e-mail: jawolpert@ucdavis.edu

M. Anderson
e-mail: mike.mmanders@gmail.com

J. Benz
e-mail: mlbenz@ucdavis.edu

D. Cantu and M. A. Walker (eds.), *The Grape Genome*,
Compendium of Plant Genomes, https://doi.org/10.1007/978-3-030-18601-2_14

comprehensive efforts to understand and classify rootstock performance as a function of site. General trends in performance were observed as a function of rootstock selection with regard to yield and pruning weights.

14.1 Introduction

The utilization of rootstocks in grape production is relatively recent considering the extensive history humans have cultivating grapes. The catalyst for rootstock breeding and implementation as an industry mainstay traces back to the European phylloxera crisis. Grape phylloxera (*Daktuloshpaira vitifoliae*) was imported into Europe in 1845 and quickly devastated the European winegrape industry (Campbell 2005). By 1873, it was found feeding on California vineyards (Granett et al. 1987a, b). Although it was widely assumed at the time that all Native American grapevine species would be equally resistant to phylloxera feeding, it was quickly realized that some rootstocks were more tolerant to feeding than others (Lider 1958). This prompted much of the initial rootstock research designed to quantify the resistance and viticultural attributes by genetic parentage (Ramming 2010). The initial California rootstock research centering around phylloxera tolerance was completed by the State Viticultural Commission (Doyle 1894).

Although the California State Viticultural Commission was deeply invested in phylloxera research, it was the French and Italians who initially took the lead with respect to breeding resistant rootstocks. By the end of the late nineteenth century, France was grappling with the introduction of phylloxera, powdery and downy mildew. The almost simultaneous appearance of these issues into France resulted in two distinctive, but complementary efforts by French grape breeders (Reynolds 2015). The first was the creation of rootstocks resistant to phylloxera that would allow the preservation of the well-known cultivars that were currently in production. Secondarily, French breeders established grape breeding programs designed to integrate resistance to phylloxera, powdery and downy mildew, into *V. vinifera* backgrounds, while attempting to preserve the favorable fruit quality and sensory attributes of *V. vinifera* (Reynolds 2015). Key French breeders behind these early hybridization efforts included Albert Seibel (1844–1936), Eugene Kuhlman (1858–1932), Bertille Seyve (1895–1959) and Joannes Seyve (1900–1966) to name a few. Additionally, aside from the thousands of French-American hybrids bred by the French, there were also many noteworthy rootstocks created. The majority of the initial rootstocks released for use were hybrids of *Vitis rupestris* × *V. riparia*. French expeditions to North America resulted in the collection of many *Vitis* species, but the only two that rooted well from dormant cuttings and possessed strong phylloxera resistance were *V. rupestris* and *V. riparia*. Despite being phylloxera resistant, the French soon found that these rootstock hybrids failed in the calcareous soils common in much of Europe. This eventually prompted the collection of *V. berlandieri* for the sole purpose of incorporating lime tolerance into their breeding efforts.

Equally important as the French efforts, the Italians devoted the majority of their breeding efforts to new rootstocks starting from the time phylloxera was first formally identified in Italy in 1879 (Bavaresco et al. 2015). It was the Italians who recognized the long-term benefits of capitalizing on the native American species that co-evolved with phylloxera, and as a result, they founded the first American species nursery in 1881 (Bavaresco et al. 2015). Notably, Federico Paulsen (1861–1943) began breeding grapes rootstocks (1889) with *V. berlandieri* parentage. Paulsen's effort to incorporate the lime tolerance of *V. berlandieri* was one of the first documented efforts in grape rootstock breeding that focused on traits beyond phylloxera tolerance. Paulsen also recognized the potential of the relatively expansive root systems of his *V. berlandieri* × *V. rupestris* hybrids had with respect to drought tolerance. His lime- and drought-tolerant rootstock, 1103P remains one of the leading rootstocks in use worldwide. Another early Italian grape breeder who considered genotypic attributes beyond phylloxera tolerance was Antonio Ruggeri (1859–1897). Ruggeri also focused on *V. berlandieri* × *V. rupestris* hybrids, but also created *V. berlandieri* × *V. riparia* hybrids. His 140Ru rootstock, a *V. berlandieri* × *V. rupestris* cross, is still used today for its strong drought and salt tolerance.

In 1875, phylloxera was documented in part of Hungary which is now modern-day Serbia. Zsigmond Teleki (1854–1910) quickly became the primary Hungarian rootstock expert. Teleki began by planting a variety of rootstocks/native American species grafted to traditional winegrape cultivars (Hajdu 2015). He was displeased with the performance of those trials. Unfortunately, at the time a black rot quarantine prevented the free movement of plant material. To avoid the quarantine, Teleki began breeding his own rootstocks from seeds he obtained from France. His efforts focused on selecting seedlings with phylloxera resistance and lime tolerance, as well as factors such as vigor induction and rootstock–scion graft compatibility. Károly Bakonyi continued Teleki's work in Hungary since 1970 and that work is currently directed by László Kocsis (Hajdu 2015).

German efforts at rootstock breeding also date back to the nineteenth century (Ruehl et al. 2015). Like France, Germany was searching for solutions to foliar diseases such as downy and powdery mildew. Unfortunately, German breeders were prohibited from obtaining American rootstocks and hybrids. To get around this restriction, breeders imported *V. riparia* seeds from New England selections. Challenges associated with the lack of lime tolerance in *V. riparia* temporarily stalled German rootstock breeding efforts. Eventually, it was the Teleki hybrids that made it possible for the German breeding programs to move forward (Ruehl et al. 2015). Teleki shared ten of his most promising selections with Austrian viticulturist Franz Kober. Kober fine-tuned Teleki's screening process and made selections of the most robust and vigor inducing hybrids. The partnership between the Hungarian and Austrian programs eventually expanded to include Geisenheim under the direction of Heinrich Birk (Ruehl et al. 2015). The Teleki rootstocks resulting from these partnerships include Teleki-Fuhr SO4, Teleki-Kober 5BB and Teleki 5C (Csepregi and Zilai 1955), all of which have been widely used since their development.

Grape breeding efforts in the Western USA were conducted by Harold Olmo (1931–1979) at the University of California, Davis. Olmo's research stretched beyond breeding table, raisin and winegrape cultivars, and included the *V. vinifera* × *Muscadinia rotundifolia* (VR) rootstock selection O39-16 (Walker et al. 1994) which controls fanleaf disease and its dagger nematode vector, *Xiphinema index*. Olmo's efforts formed the foundation for M. Andrew Walker's breeding program that has included classical breeding of rootstocks and wine grapes, the identification of new resistance sources and using molecular tools to accelerate breeding. In addition to Walker's Pierce's Disease (*Xylella fastidiosa*), nematode, powdery mildew and ongoing phylloxera research efforts, he has pursued drought and salt tolerance. This all-inclusive approach to rootstock breeding has made significant strides toward addressing new and ongoing pest and disease issues as well as working to clarify rootstock performance expectations. To date, Walker has released (2008) five rootstocks (GRN1–5). The GRNs are the result of a rather complex hybridization of several species including *V. rupestris*, *M. rotundifolia*, *V. rufotomentosa*, *V. champinii* and *V. riparia*. These new releases have been screened against several nematode species and have documented resistance to phylloxera. Unlike the original pest resistance trials, Walker has recognized the importance of determining and characterizing the role rootstock genetic background has on the scion performance, which is often complicated by environmental factors such as soil texture, water availability, climatic variation and management practices.

As in the Western USA, breeding programs in other parts of the world remain active as well although some have shifted focus from rootstocks to scion selection. Efforts in Germany from Geisenheim University and Geilweilerhof (part of the Julius Kuhn-Institute) are currently directed at clonal variation, genetic diversity between clones and exploring variation in vine growth factors (cluster compactness, berry size and berry chemistry factors). These programs have also been selecting clones that demonstrate higher tolerance to Botrytis bunch rot. Hungarian breeding efforts are currently led by Pál Kozma Jr., János Korbuly and Lászlo Kocsis (Hajdu 2015). Kozma has continued developing hybrids using *V. vinifera* and *M. rotundifolia*, as well as *V. amurensis*. His most recent work is dedicated to incorporating powdery mildew resistance into newly bred varieties. Korbuly has been evaluating seedlings since 1980 with the hope of improving resistance to fungal diseases and frost using *V. amurensis* for winegrape production. Kocsis has been important with respect to rootstock breeding. Since the late 1990s, he has been breeding to increase lime and drought tolerance while preserving phylloxera resistance in rootstocks. Italian breeding efforts, led by the University of Milan, have also been developing rootstocks centered around drought tolerance and lime (Bavaresco et al. 2015). Additionally, the Research and Innovation Centre at Fondazione Edmund Mach (FEM) S. Michele all' Adige (Bavaresco et al. 2015; Emanuelli et al. 2013) and the University of Udine are developing new winegrape cultivars resistant to downy and powdery mildew (Bavaresco et al. 2015; Coleman et al. 2009; Di Gaspero et al. 2012; Venuti et al. 2013).

One of the binding features of rootstock trials is that rootstock performance is deeply affected by the site they are tested on. Resistance to phylloxera was the primary focus of these breeding programs, and detailed classification of rootstock effects on scions across sites was not thoroughly investigated, which left a substantial void in our understanding of rootstock selection.

From the late 1980s to the early 2000s, James Wolpert at UC Davis conducted an extensive rootstock evaluation program to understand the role rootstock selection has on scion performance. The program tested a genetically diverse group of eighteen rootstocks across six vineyard sites of varying soil composition and analyzed fruit yield and pruning weight data. The factors varied at each trail site reflecting differing management decisions appropriate for the particular site and scion cultivar. The primary objective was to determine if trends in yield and pruning weights existed across diverse environments. The secondary objective was to develop a guide that would allow rootstock recommendations for specific soil types and vineyard sites.

14.2 Site Selection and Trial Details

The selected commercial vineyard sites were all located in California and were organized as a completely randomized design with respect to rootstock. Vineyard management practices, such as degree of pruning, irrigation amounts and timing, fertilization regimens and canopy management, were executed in accordance with vineyard collaborator practices. Shoots from lateral buds as well as positions that had double primary shoots were thinned. Cluster thinning was not performed at any of the sites. Yield at the time of harvest and pruning weight data was taken from the rootstocks planted across the six sites. There were between five and eight vines, depending on the vineyard site availability, for treatment replicates. The data for yields and pruning weights were collected on a per-vine basis from the center three to six vines in each set of treatment replicates and an average value was used for analysis.

Table 14.1 summarizes the trial information with scion cultivars, nearest California Irrigation Management Information System (CIMIS) weather station, location, cumulative degree-days and yearly average rainfall for the years of data collection. Table 14.1 also includes the site summary for each trial location including site name, scion, number of rootstocks present, soil type and depth and any abiotic or biotic stress present. Table 14.2 lists the rootstocks present at a given site and the corresponding parentage of each rootstock. Paul Anamosa (1998) verified the soil profile descriptions at each vineyard site for accuracy with the United States Soil Survey Descriptions (Soil Survey Division Staff 1993).

Table 14.1 Detailed site descriptions including name, scion cultivar, location, years' data was collected, cumulative degree-days, average rainfall, row × vine spacing, soil type, soil depth and site-specific stress factors

Trial site	Scion cultivar	Row × vine spacing (meter)	Soil type	Soil depth (cm)	Abiotic or biotic stress	Geoposition coordinates	Years' studied	Average degree-days	Average rainfall
Sacramento Delta	Chardonnay	3.0 × 2.7	Egbert clay	121.92		N 38° 19.788′ W 121° 30.504′	1993–1997	2082	419
Sacramento Delta	Cabernet Sauvignon	3 × 3	Tinnin loamy sand	167.64		N 38° 20.178′ W 121° 31.362′	1992–1998	2071	490
Amador Montevina	Zinfandel	3 × 1.8	Sierra course sandy loam	182.88		N 38° 30.576′ W 120° 48.210′	1994, 1996–2000	1894	1397
Mendocino La Ribera	Cabernet Sauvignon	3.7 × 2.4	Russian loam gravelly substratum	81.28	*Xiphinema index* (Dagger), *Meloidogyne incognita* (Root-knot), *Mesocriconema xenoplax* (Ring)	N 39° 04.566′ W 123° 09.677′	1993–1998, 2001	1889	1071
Sonoma Chalk Hill	Merlot	3.7 × 1.8	Clear Lake/Haire clay	76.2–114.3	Potassium deficient	N 38° 29.286′ W 122° 47.516′	1997–2001[a] 1996–2001[a]	1532[a] 1548[b]	811[a] 779[b]
Napa Rutherford Bench	Cabernet Sauvignon	3 × 1.8	Cortina very gravely loam	152.4	History of Phylloxera Biotype B	N 38° 27.406′ W 122° 22.888′	1994–1996	1794	1250

[a]Years the yield data was collected differing from pruning weight collection years
[b]Years the pruning weight data was collected differing from yield collection years

Table 14.2 Rootstocks present at individual trial sites, including scion cultivar present

Rootstock	Parentage	Sacramento Delta Chardonnay	Sacramento Delta Cabernet Sauvignon	Amador Montevina Zinfandel	Mendocino La Ribera Cabernet Sauvignon	Sonoma Chalk Hill Merlot	Napa Rutherford Cabernet Sauvignon
Teleki 5C	*V. berlandieri* × *V. riparia*	X	X	X	X	X	
Kober 5BB	*V. berlandieri* × *V. riparia*	X	X	X		X	
420A MGT	*V. berlandieri* × *V. riparia*	X	X	X	X	X	X
(SO4)	*V. berlandieri* × *V. riparia*						X
Richter 110 (110R)	*V. berlandieri* × *V. rupestris*	X	X	X	X	X	X
Paulsen 1103 (1103P)	*V. berlandieri* × *V. rupestris*	X	X	X		X	X
Ruggeri 140 (140Ru)	*V. berlandieri* × *V. rupestris*					X	
101-14 MGT	*V. riparia* × *V. rupestris*	X	X	X	X	X	X
Couderc 3309 (3309C)	*V. riparia* × *V. rupestris*	X	X	X	X	X	X
Riparia Gloire	*V. riparia*						X
St. George	*V. rupestris*	X	X	X			

(continued)

Table 14.2 (continued)

Rootstock	Parentage	Sacramento Delta Chardonnay	Sacramento Delta Cabernet Sauvignon	Amador Montevina Zinfandel	Mendocino La Ribera Cabernet Sauvignon	Sonoma Chalk Hill Merlot	Napa Rutherford Cabernet Sauvignon
Malègue 44-53	*V. riparia* × *V. vulpina* × *V. rupestris)*	X	X	X		X	
Couderc 1616 (1616C)	*V. solonis* × *V. riparia*	X	X	X			
O39-16	*V. vinifera* × *M. rotundifolia*	X	X	X	X		X
O43-43	*V. vinifera* × *M. rotundifolia*				X		
Harmony	1613 (*V. solonis* × Othello) × Dog Ridge OP Seedling	X	X	X	X		X
Freedom	1613 (*V. solonis* × Othello) × Dog Ridge OP Seedling	X	X	X	X		X
Ramsey	*V. champinii*	X	X	X			

A principal component analysis (PCA) was run on the pruning weights, and yields for the various years' data were collected for each of the sites. The figures were designed so that the X-axis represents pruning or yield weight and the Y-axis represents the change in yields or pruning weights over time. Statistical Analysis Software (SAS) version 9.1.3 (SAS Institute Inc., Cary, NC, USA) was used to perform the analyses, which found that rootstocks tended to separate into three distinctive groupings with regard to pruning and fruit weights. These groupings were subsequently designated as high, medium and low (Table 14.3).

14.2.1 Findings and Interpretation

This study was conducted across multiple vineyard sites throughout California over the course of nine years. However, it is important to note that in this rootstock survey, not all rootstocks were present at each site and the years in which the data were collected were not always consistent across all sites. Data should be considered on a site-by-site basis as it is difficult to make across site comparisons due variation in pests, disease or soil factor variation at each of the vineyard site locations. Data are presented for each site

Table 14.3 Rootstock performance, classified as high, medium or low, as a function of site for yield and pruning weights across all years of the study

Trial site	Scion cultivar	Factor	Low	Medium	High
Sonoma Chalk Hill	Merlot	Pruning Weight	44-53, 420A	101-14, 110R, 5C, 140Ru, 5BB, 3309	1103P
		Yield	44-53	420A, 101-14	1103P, 140Ru, 5BB, 3309, 110R, 5C
Mendocino La Ribera	Cabernet Sauvignon	Pruning Weight	AXR1, 101-14, 420A, 3309C	5C, Harmony, Freedom	O39-16, 110R
		Yield	101-14, AXR1	420A, 3309C, 5C, Harmony, Freedom	110R, O39-16
Napa Rutherford	Cabernet Sauvignon	Pruning Weight	101-14, SO4, 420A, 3309C	Harmony, Riparia, 110R, 1103P, Freedom	O39-16
		Yield	101-14, SO4	420A, 3309C, Harmony, Riparia, 110R, 1103P, Freedom	O39-16
Sacramento Delta	Chardonnay	Pruning Weight	44-53, 420A, O39-16	5BB, 5C, 3309C, 110R, St. George, 1616C, Ramsey, 101-14, Harmony	Freedom, 1103P
		Yield	44-53	O39-16, 420A, St. George, 5BB, 1616C, 5C, Harmony	1103P, Ramsey, Freedom, 3309C, 101-14, 110R
Sacramento Delta	Cabernet Sauvignon	Pruning Weight	44-53, 3309C, O39-16	St. George, 420A, 101-14	1616C, Freedom, 5BB, Harmony, 1103P, Ramsey, 5C, 110R
		Yield	St. George, 1103P, 44-53	Ramsey, 3309C, Freedom, 5C, 101-14, 110R	Harmony, O39-16, 1616C, 420A, 5BB
Amador Montevina	Zinfandel	Pruning Weight	101-14, 420A, 44-53, 3309C, Harmony, 110R	Ramsey, O39-16, 5C, St. George	1616C, Freedom, 1103P, 5BB
		Yield	5BB, 420A, 110R, 1103P	Harmony, 5C, Freedom, Ramsey, St. George, 44-53, 1616	3309C, 101-14, O39-16

individually; meaning rootstocks within a site were ranked against others at the same site location for crop yields and pruning weights. Trends in groupings that transcend a specific site are discussed, which allows the performance of a range of rootstocks on a given vineyard site with similar location, soil type, or pest and disease pressures to be compared.

Rootstocks were found to have an impact on yield and pruning weights and, within each site, were classified as low, moderate or high producers based on harvest weight and pruning weight (Table 14.3). Although statistical comparisons among sites could not be made, trends were apparent. 43-53 was consistently lower in yields and pruning weights regardless of site conditions. St. George was consistently lower in yields and pruning weights in the trial regardless of site, which is unusual given its previously documented performance (Christensen et al. 2003). 420A, 5BB and 5C usually were clustered tightly, but 420A tended to be the lowest of the three for yields and pruning weights. 3309C had relatively high yields and lower pruning weights when compared to the other rootstocks.

Rootstock performance is impacted by the site environment and management practices (Rogiers and Clarke 2013), available soil water (Ozden et al. 2010) and soil fertility and structure (Lambert et al. 2008; Wolf and Pool 1988). Scion performance is impacted by temperature, and light (Bergqvist et al. 2001), management practices such as leaf and lateral shoot removal (Koblet et al. 1994; Bledsoe et al. 1988), pruning practices (Lider et al. 1973), irrigation (McCarthy et al. 1997; Ozden et al. 2010) and fertilization (Keller et al. 2001; Dalbo et al. 2011; Neilsen et al. 2010). Despite influences from various management techniques and site environmental profiles, rootstock behavior and suitability are also driven, to a certain extent, by parentage. Particular genetic backgrounds are better adapted to dealing with specific soil moistures and texture types than others.

Until relatively recently, there was a general assumption that little to no rootstock scion interaction existed and instead that scion genotypes performed much the same way whether or not they were grafted to rootstock or growing on own roots (Christensen 1984). However, it is now clear rootstock–scion interactions exist and that they can have large impacts (Virgona et al. 2003; Vrsic et al. 2015; Dodson Peterson and Walker 2017), particularly in regard to mineral nutrition (Koblet and Keller 1996; Lambert et al. 2008), which varies site to site as well as yield (Li et al. 2019), vine vigor (Li et al. 2019) and physiochemical (weight, size, pH, soluble solids, titratable acidity) quality attributes of the grape berries (Rodrigues da Silva et al. 2018). For example, grafting to a rootstock that is associated with poor magnesium uptake can result in deficiency symptoms. The rootstock 44-53 has a higher affinity for potassium (K) than magnesium (Mg), which can be compounded if the soil is rich in K, limiting the ability of 44-53 to take up Mg from the soil (Brancadoro et al. 1994). In contrast to 44-53's preference for K, the rootstock 1103 is known to have a higher affinity for Mg (Scienza et al. 1986).

Another example of rootstock variation is in response to soil lime content. Calcareous soils can have a large impact on the ability of some rootstocks to take up iron. The ability to deal with this type of environmental challenge seems to be based on genetic background. *Vitis berlandieri*-based rootstocks (140 Ru, 110R, 420A) tend to be more lime tolerant, while *V. riparia*-based rootstocks (101-14) are generally more sensitive to lime (calcium carbonate) (Bavaresco et al. 1993). Additionally, rootstocks with a *V. berlandieri* background are associated with lower petiole potassium content at bloom (Wolpert et al. 2005; Lambert et al. 2008), which can result in developmental issues of the reproductive and vegetative organs. There are also examples in which rootstock or scion cultivar selection impacted vine performance in response to salinity (Bybordi 2012), soil texture (Morano and Kliewer 1994), soil moisture (Paranychianakis et al. 2004), irrigation amount (Williams 2010; Nelson et al. 2016) and various disease pressures (Goodman et al. 1993; Harris 1984). There are also documented effects of scion selection effects on rootstock performance (Virgona et al. 2003).

Although many of the sites in this study did exhibit clear rootstock trends based on genetic parentage, it was difficult to determine which site and environment factors were influencing rootstock–scion behavior and to what degree; a determination that was confounded by the differences in soils, management practices, environmental conditions, pest and disease pressures and scion cultivars (Table 14.1). The noted differences in rootstock behavior and the resulting variation in scion yield and pruning weights emphasize the impact that soils and nutrients have on vineyard development and resource allocation (Lambert et al. 2008).

Most of the rootstocks in commercial use derive from crosses among *V. riparia*, *V. rupestris* and *V. berlandieri*. The first two were utilized because they are resistant to phylloxera and easy to propagate, and *V. berlandieri* was included because of its lime tolerance. Crosses are generally made between these three native North American species to produce the commercial rootstocks used to combat grapevine pests, diseases and soil-related challenges. *Vitis riparia* *V. rupestris*-derived rootstocks are generally best suited for fertile soils, without excess lime issues, and however, the nematode resistance varies greatly. They typically induce moderate-to-low vigor in the scions and are generally good candidates for higher density plantings, but will not do well on dry-farmed vineyard sites. Common examples include 3309C (nematode susceptible) and 101-14 (moderate nematode resistance). *Vitis berlandieri* × *V. riparia*-derived rootstocks are generally more phylloxera resistant, have moderate nematode resistance and are lime tolerant. Rootstocks in this grouping have been found to have higher fine root hydraulic conductivity, which can be traced, in part, to higher aquaporin expression and activity (Gambetta et al. 2012). Generally, this hybrid category is considered to induce moderate vigor rootstocks (420A is the exception with low vigor) and generally have moderate to shallow rooting architecture. Examples include Teleki 5C, 5BB, SO4 and 420A. Furthermore, although no difference in yields was found, Blank et al. (2018) recently found that SO4 produced almost double the amount of pruning mass compared to that of lower vigor stocks such as Riparia Gloire and Schwarzmann when grafted to Pinot noir (Blank et al. 2018). *Vitis berlandieri* × *V. rupestris*-derived rootstocks are known for their drought (Yildirim et al. 2018) and lime tolerance, deeper rooting architecture (drought avoidance), minimal nematode resistance and moderate to good phylloxera resistance. This grouping is considered to be more difficult to root and graft with the exception of 1103P which is easy to propagate. Well-known examples include 110R, 1103P and 140Ru, all of which are considered to induce high vegetative vigor when planted on deep fertile soils. Most rootstocks in this hybrid cross are considered to be better suited to deficit irrigation regimes (Sabir and Sahin 2018).

Other species have been used in rootstock breeding including *V. champinii*, *V. aestivalis*, *M. rotundifolia*, *V. labrusca* and *V. candicans*. Rootstocks from these less utilized species are also not fully understood when it comes to performance in commercial vineyards. *Vitis champinii*-derived rootstocks have high vigor, are drought tolerant due to the plunging root system and have broad nematode resistance. These characteristics make these rootstocks useful in soils with low fertility and high populations of root-knot nematodes. Freedom and Harmony were produced by crossing 1613C (*V. solonis* ((*V. riparia* × *V. labrusca*) × *V. vinifera*) OP seedling × Dog Ridge (possibly *V. candicans* × *V. berlandieri*) OP Seedling for the purpose of providing a nematode-resistant rootstock for low fertility soils. They have low phylloxera resistance due to the *V. vinifera* parentage of the 1613C parent and are also sensitive to virus infections. The majority of these characteristics are anecdotal, based on unreplicated observations in vineyards.

The *V. vinifera* × *M. rotundifolia* siblings 039-16 and O43-43 are the only sources of tolerance to fanleaf degeneration. Unfortunately, O43-43 is susceptible to phylloxera and O39-16 is susceptible to root-knot nematodes. Despite being difficult to propagate, they are both considered to induce high vigor. O43-43 is no longer commercially available.

14.2.2 Sacramento County, Delta: Chardonnay

The Sacramento County's Delta Chardonnay rootstock site was on a clay soil in the Egbert clay loam series and was found to be below the potassium-to-CEC ratio predicted value of 2.5 referenced in the literature for this type of soil (Champagnol 1984; Etourneaud and Loue 1986), and thus, the available soil potassium and soil fertility of this site were lower than what would typically be found in an Egbert clay loam soil. In a soil of this nature (depths of 1.22 m, no documented pest pressure), the *V. berlandieri* × *V. rupestris*, *V. riparia* × *V. rupestris* and the *V. champinii* rootstocks produced the highest fruit yields (Fig. 14.1a). Despite the lower than expected potassium level, 420A (*V. berlandieri* × *V. riparia*), thought to be susceptible to potassium deficiency (Pongrácz 1983), maintained moderate yield output and was clustered with the other *V. berlandieri* × *V. riparia*-derived rootstocks. The low rainfall at this site put *V. rupestris* (deep anchoring root system) and *V. berlandieri* (deep rooting)-based rootstocks at an advantage over *V. riparia*-based (shallow rooting) rootstocks (Guillon 1905; Pongrácz 1983).

Pruning weights were clustered by parentage as well, but the groupings often overlapped (Fig. 14.1b). The *V. champinii*-based rootstocks were clustered in the high and the higher end of the moderate spectrum (Fig. 14.1b). The *V. berlandieri* × *V. riparia*-based rootstocks also were clustered, but overlapped the low and lower end of the moderate spectrum (Fig. 14.1b). The *V. riparia* × *V. rupestris* rootstocks, 3309C and 101-14, were clustered closely within the moderate grouping and varied less from year to year compared to the vines on St. George or 1616C (Fig. 14.1b). It is interesting to note that the *V. riparia* × *V. rupestris* rootstocks produced high yields in comparison to the rest of the rootstocks, but only moderate pruning weights. This allocation of resources directed to reproductive versus vegetative growth may result in more light penetration into the canopy, promoting more fruitful buds (May et al. 1976). This would lead one to infer that the less dense canopies of the *V. riparia*-based rootstocks would produce higher yields than those rootstocks that induce more vigorous canopies and more shaded buds. However, the higher vigor rootstocks typically have higher yields despite having denser canopies and more shaded buds. One possible explanation for this might be the

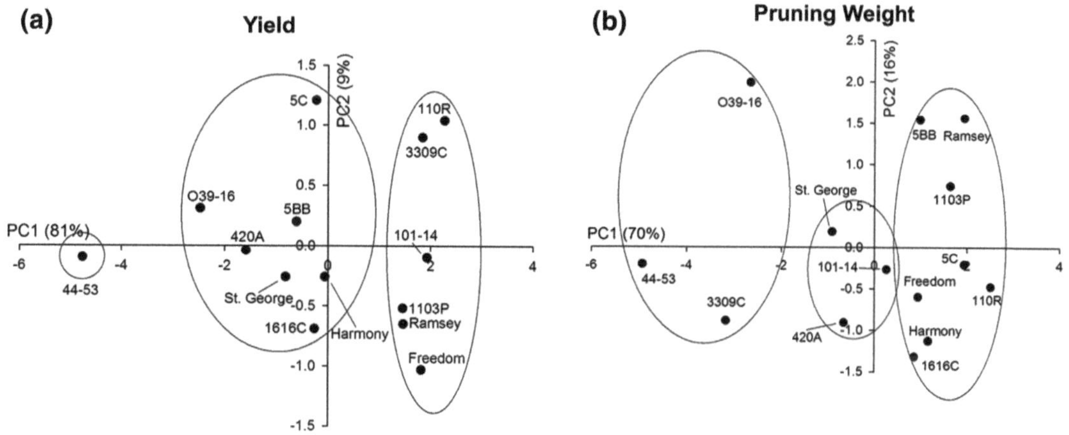

Fig. 14.1 Sacramento County, Delta Chardonnay principal component analysis for **a** yield and **b** pruning weights of the different rootstocks. The first principal component (X-axis) correlated in all cases to either pruning weight or yield and accounted for the majority of the variation. The second principal component (Y-axis) correlated to year and accounted for relatively little variation

variation in spatial rooting patterns and distribution (deep versus shallow) (Morlet and Jacquet 1993), as well as the actual rooting density of a given rootstock (Swanepoel and Southey 1989). Another possible explanation might be due to the difference in affinity for certain nutrients among rootstocks (Bavaresco et al. 1991; Grant and Matthews 1996; Lambert et al. 2008; Romero et al. 2018) which impacts overwintering mineral nutrient stores, perhaps giving vines grafted to more densely rooted rootstocks a greater pool of resources to allocate to both vegetative and reproductive growth.

14.2.3 Sacramento County, Delta: Cabernet Sauvignon

The Sacramento County's Delta Cabernet Sauvignon rootstock site was on a Tinnin loamy sand soil that increased in sand content with depth. This site was also slightly potassium deficient with a relatively low CEC. Unlike in the heavier clay soils of the Delta Chardonnay site, 1103P had one of the lowest yields compared to the rest of the rootstocks (Fig. 14.2a), but had one of the highest pruning weight productions (Fig. 14.2a) and these observations agreed with Williams' (2010) findings in which 1103P also had the highest pruning weights regardless of irrigation treatment imposed. However, the results do conflict with Keller et al. (2012), who found that 1103P reduced pruning weights in a trial on Shano silt loam soil comparing multiple rootstocks. Although higher vegetative vigor is typically associated with lower light penetration to the buds (May et al. 1976), it does not always result in reduced fruitfulness (Sanchez and Dokoozlian 2005). Clearly, the interaction between rootstock selection, accessible resources, canopy density and bud fruitfulness is not fully understood and warrant further exploration.

O39-16 (*V. vinifera* × *M. rotundifolia*) behaved similar to 1103P with yield and pruning weights at the opposite ends of the classification scale (Fig. 14.2a, b). O39-16 had one of the highest yields and the second to lowest pruning weight, but did have more variation from year to year in pruning weight compared to the rest of the rootstocks examined at this site.

14.2.4 Amador County, Montevina: Zinfandel

The Amador County's Montevina Zinfandel site was a Sierra coarse sandy loam. The sandy texture of this site resulted in a low water holding

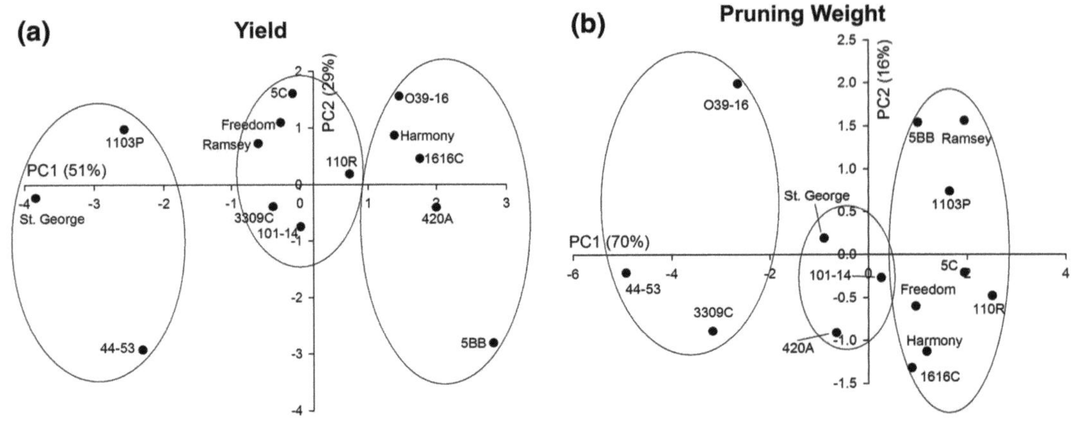

Fig. 14.2 Sacramento County, Delta Cabernet Sauvignon principal component analysis for **a** yield and **b** pruning weights of the different rootstocks. The first principal component (X-axis) correlated in all cases to either pruning weight or yield and accounted for the majority of the variation. The second principal component (Y-axis) correlated to year and accounted for relatively little variation

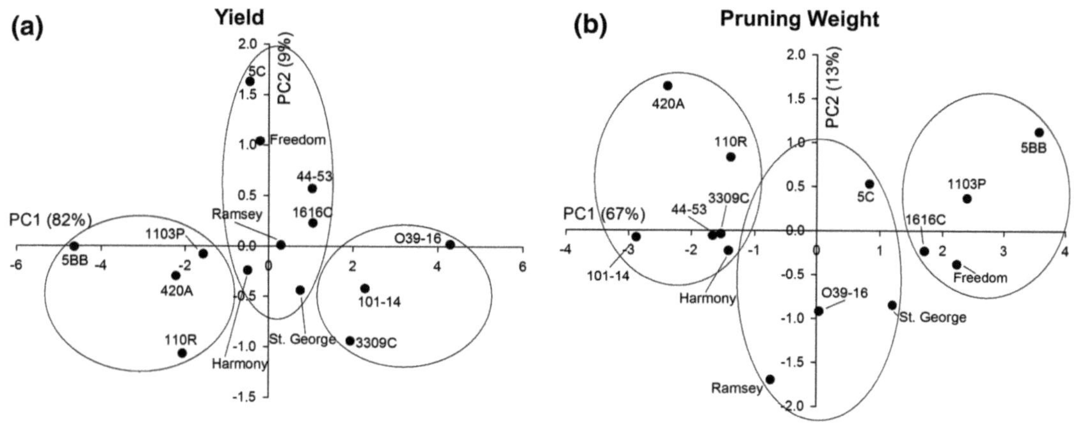

Fig. 14.3 Amador County, Montevina Zinfandel principal component analysis for **a** yield and **b** pruning weights of the different rootstocks. The first principal component (X-axis) correlated in all cases to either pruning weight or yield and accounted for the majority of the variation. The second principal component (Y-axis) correlated to year and accounted for relatively little variation

capacity, and it was found to be high in potassium due to the fertilization practices at this site (Lambert et al. 2008). There was also a high calcium-to-magnesium ratio, which resulted in a lower degree of exchangeable magnesium (Lambert et al. 2008).

Yields at this site were clustered closely by genetic background. High yields and low pruning weights were correlated with *V. riparia* × *V. rupestris* rootstocks, 101-14 and 3309C (Fig. 14.3a, b). 101-14 produced lower pruning weights than 3309C, which was similar to the findings of Keller et al. (2012) who examined pruning weights on three different sites. Typically, lower yields would be expected with lower vegetative growth; however, more resources were diverted to reproductive growth rather than vegetative growth. Although the Montevina Zinfandel was not under deficit irrigation conditions, the *V. berlandieri* × *V. riparia* rootstocks 5C, 5BB and 420A had very different pruning weights from one another (Fig. 14.3b): 420A, had the lowest pruning weights; 5C had moderate pruning weights; and 5BB had high pruning weights. This ranking was consistent with that of Christensen et al. (2003). Although closely clustered as having moderate yields, *V. champinii*-derived rootstocks, Freedom, Harmony and Ramsey separated into different categories for pruning weights (Fig. 14.3a, b)—Harmony is considered to be less vigorous than Freedom (Christensen et al. 2003).

14.2.5 Mendocino County, La Ribera: Cabernet Sauvignon

The Mendocino County's La Ribera Cabernet Sauvignon site is a Russian loam with gravelly substratum. This site has a history of nematode infestation including dagger, ring and root-knot nematodes. Despite being considered susceptible to most nematodes, 110R had consistently high yields and pruning weights on this site (Fig. 14.4a, b). O39-16 is resistant to *X. index* and ring nematode, but moderately susceptible to root-knot nematode, but it also had high yields and pruning weights (Fig. 14.4a, b). Nematode-resistant rootstocks, Harmony and Freedom, produced moderate yields and pruning weights compared to the other rootstocks at this site (Fig. 14.4a, b). 420A and 101-14 are considered to have moderate to low nematode resistance, respectively (Christensen et al. 2003). At this site, they performed similar to 3309C, a nematode-susceptible rootstock (Fig. 14.4a, b). These results suggest that either the nematode population is sporadically distributed at this site, no longer an issue, or that 110R and

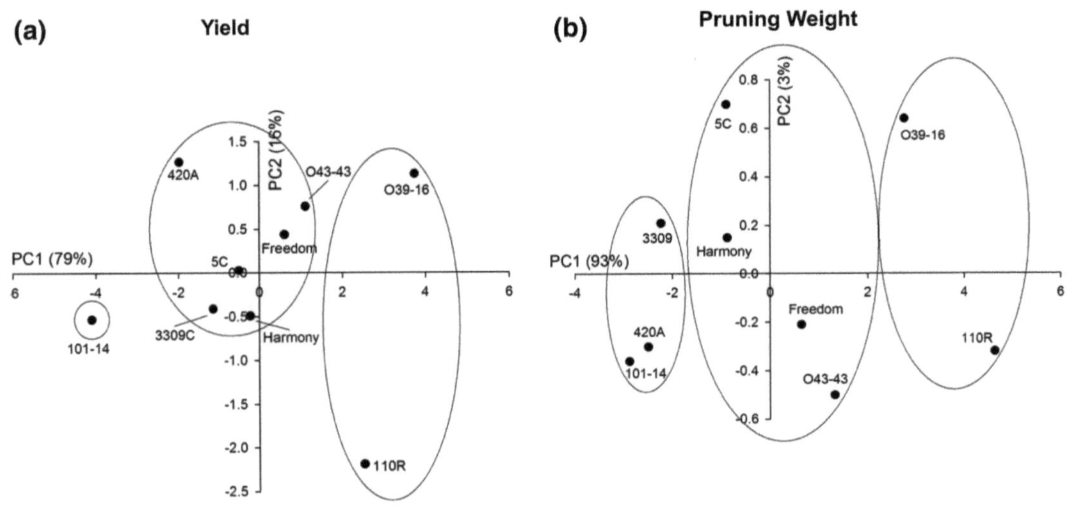

Fig. 14.4 Mendocino County, La Ribera Cabernet Sauvignon principal component analysis for **a** yield and **b** pruning weights of the different rootstocks. The first principal component (X-axis) correlated in all cases to either pruning weight or yield and accounted for the majority of the variation. The second principal component (Y-axis) correlated to year and accounted for relatively little variation

O39-16 are vigorous enough that the damage by the nematode population is not having a profound effect yet.

14.2.6 Sonoma County, Chalk Hill: Merlot

The Sonoma County's Chalk Hill Merlot site is considered to be a Clear Lake/Haire clay soil and was found to be potassium deficient (Lambert et al. 2008). Rootstocks 420A, 110R, 5BB, 5C and 1103P are sensitive to soil potassium deficiencies (Christensen et al. 2003). 420A is particularly sensitive to potassium deficiency, and its low yields and pruning weights were consistent with low potassium levels (Fig. 14.5a, b). Despite the deficiency, *V. berlandieri* × *V. rupestris*-based rootstocks (110R and 1103P) performed well on this heavy clay site with high yields and moderate to high pruning weights (Fig. 14.5a, b). 5C and 5BB (*V. berlandieri* × *V. riparia*) also appeared to be unaffected by the potassium deficiency. 101-14 was less vigorous and produced less yield than 3309C (Fig. 14.5a, b). 44-53 is known to have a high affinity for potassium and boron uptake and poor magnesium uptake. Despite the ability to scavenge potassium in a low potassium environment, 44-53 had the lowest pruning and fruit weights at this site (Fig. 14.5a, b).

14.2.7 Napa County, Rutherford: Cabernet Sauvignon

The Napa County's Rutherford Cabernet Sauvignon site was on Cortina very gravelly loam and had a known history of biotype B phylloxera infestation (Anamosa 1998). Harmony and Freedom are considered to have only low to moderate resistance to phylloxera (Christensen et al. 2003). Despite phylloxera's presence, both were classified as having moderate yields (Fig. 14.6a and pruning weights (Fig. 14.6b). It is possible that without phylloxera, both would have had high yields and pruning weights, as both are considered to be highly productive and vigor inducing (Christensen et al. 2003). O39-16 was the only rootstock with high yield and high pruning weights. Rootstocks considered to be drought susceptible clustered together in the low pruning weight category: 420A, 101-14, 3309C and SO4 (Fig. 14.6b).

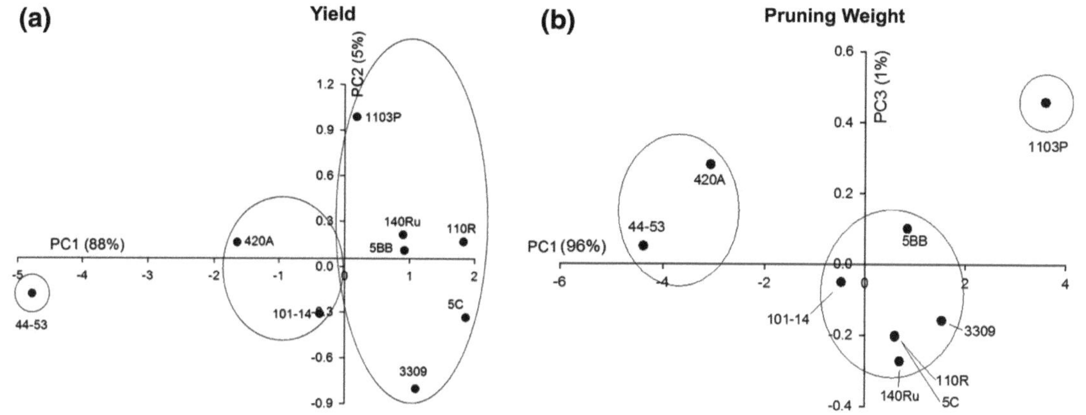

Fig. 14.5 Sonoma County, Chalk Hill Merlot principal component analysis for **a** yield and **b** pruning weights of the different rootstocks. The first principal component (X-axis) correlated in all cases to either pruning weight or yield and accounted for the majority of the variation. The second principal component (Y-axis) correlated to year and accounted for relatively little variation

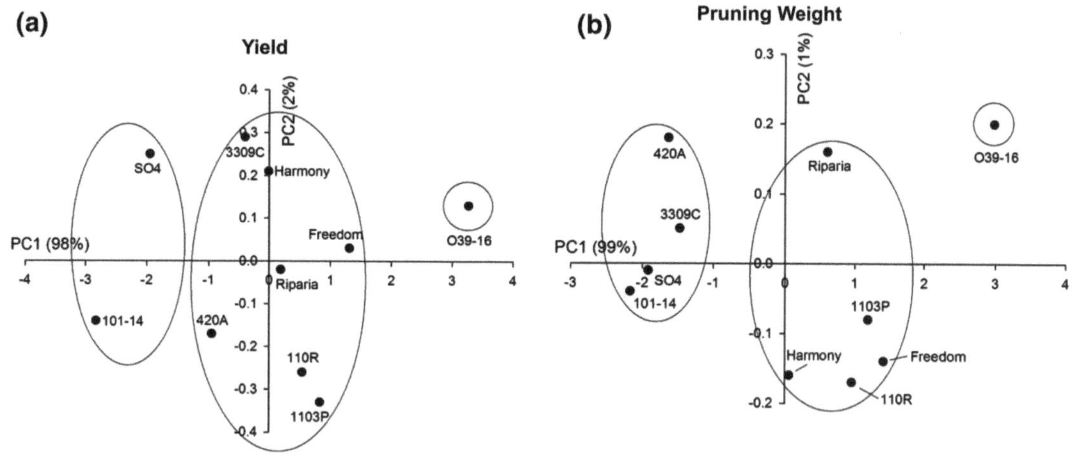

Fig. 14.6 Napa County, Rutherford Cabernet Sauvignon principal component analysis for **a** yield and **b** pruning weights of the different rootstocks. The first principal component (X-axis) correlated in all cases to either pruning weight or yield and accounted for the majority of the variation. The second principal component (Y-axis) correlated to year and accounted for relatively little variation

14.3 Conclusions

Differences among the rootstocks in the various trial sites emphasize the role soil texture, water and nutrient availability, pest and disease pressure, trellising and scion cultivar have on the productivity of grapevines, specifically with regard to fruit yields and pruning weights. These rootstock trials serve to guide to continued rootstock breeding efforts focusing on site-specific production and site adaptation goals.

Despite genetically driven differences in rootstock behavior, site soil conditions are paramount in selecting an appropriate rootstock. On the heavily textured clay soils of the Delta Chardonnay, the highest pruning weights and fruit yields were associated with the rootstocks with strong affinity for deep fertile clay soils such as 1103P and 110R. On lower fertility gravelly soils such as the Mendocino and Napa sites, the more poorly adapted *V. riparia*-driven rootstocks, 101-14, 3309C, 420A and SO4, had lower yields and pruning weights. On the sandy soil sites of the Delta Cabernet Sauvignon, Montevina and Lodi sites, rootstocks such as 110R and 5BB typically had the higher pruning weights and yields.

References

Anamosa PR (1998) Characterization of soil beneath the field evaluation of winegrape rootstocks research trials. Master's Thesis. University of California, Davis

Bavaresco L, Fregoni M, Fraschini P (1991) Investigations on iron uptake and reduction by excised roots of different grapevine rootstocks and V. *vinifera* cultivar. Plant Soil 130:109–113

Bavaresco L, Fraschini P, Perino A (1993) Effect of the rootstock on the occurrence of lime-induced chlorosis of potted *Vitis vinifera* L. cv. 'Pinot blanc'. Plant Soil 157:305–311

Bavaresco L, Gardiman M, Brancadoro L, Espen L, Failla O, Scienza A, Vezzulli S, Zulini L, Velasco R, Stefanini M, Di Gaspero G, Testolin R (2015) Grapevine breeding in Italy. In: Reynolds A (ed) Grapevine breeding programs for the wine industry. Woodhead Publishing, Cambridge

Bergqvist J, Dokoozlian N, Ebisuda N (2001) Sunlight exposure and temperature effects on berry growth and composition of Cabernet Sauvignon and Grenache in the central San Joaquin Valley of California. Am J Enol Vitic 52:1–7

Blank M, Tittmann S, Ghozlen NB, Stoll M (2018) Grapevine rootstocks result in differences in leaf composition (*Vitis vinifera* L. cv. Pinot Noir) detected through non-invasive fluorescence sensor technology. Aust J Grape Wine Res 24:327–334

Bledsoe AM, Kliewer WM, Marois JJ (1988) Effects of timing and severity of leaf removal on yield and fruit composition of Sauvignon blanc grapevines. Am J Enol Vitic 39:49–54

Brancadoro L, Valenti L, Reina A, Scienza A (1994) Potassium content of grapevine during the vegetative period: the role of rootstock. J Plant Nutr 17:2165–2175

Bybordi A (2012) Study effect of salinity on some physiologic and morphologic properties of two grape cultivars. Life Sci J 9:1092–1101

Campbell C (2005) The Botanist and the Vintner: how wine was saved for the world. Algonquin Books, Chapel Hill, New York

Champagnol F (1984) Elements of the physiology of the vine and of general viticulture. Saint Gely du Fesc, France, Champagnol F

Christensen LP (1984) Nutrient level comparisons of leaf petioles and blades in twenty-six grape cultivars over three years (1979 through 1981). Am J Enol Vitic 36:124–133

Christensen LP, Dokoozlian NK, Walker MA, Wolpert JA (2003) Rootstock selection. In: Reynolds A (ed) Winegrape varieties of California. ANR Pub 3419. Oakland, CA

Coleman C, Copetti D, Cipriani G, Hoffmann S, Kozman P, Kovács L, Morgante M, Testolin R, Di Gaspero G (2009) The powdery mildew resistance gene REN1 co-segregates with an NBS-LRR gene cluster in two Central Asian grapevines. BMC Genet 10:89

Csepregi P, Zilai J (1955) Our vine varieties (Ampelography). Agricultural Publishing House, Budapest

Dalbo MA, Schuck E, Basso C (2011) Influence of rootstock on nutrient content in grape petioles. Rev Bras Frutic 33:941–947

Di Gaspero G, Copetti D, Coleman C, Castellarian SD, Eibach R, Kozman P, Lacombe T, Gambetta G, Zvyagin A, Cindrić P, Kovács L, Morgante M, Testolin R (2012) Selective sweep at the Rpv3 locus during grapevine breeding for downy mildew resistance. Theor Appl Genet 124:277–286

Dodson Peterson JC, Walker MA (2017) Grapevine rootstock influence on scion development and initiation of senescence. Catalyst 1:48–54

Doyle JT (1894) Report of the Board of State Viticultural Commissioners for 1893-94, p 208

Emanuelli F, Lorenzi S, Grzeskowiak L, Catalano V, Stefanini M, Troggio M, Myles S, Martinez-Zapater JM, Zyprian E, Moreira FM, Grando MS (2013) Genetic diversity and population structure assessed by SSR and SNP markers in a large germplasm collection of grape. BMC Plant Biol 13:39

Etourneaud F, Loue A (1986) Petiolar diagnosis of grapevine in relation to the interpretation of soil analysis for potassium and magnesium. In: International symposium on vine physiology. Bordeaux, France, pp 240–246

Gambetta GA, Manuck CM, Drucker ST, Shaghasi T, Fort K, Matthews MA, Walker MA, McElrone AJ (2012) The relationship between root hydraulics and scion vigour across *Vitis* rootstocks: what role do root aquaporins play? J Exp Bot 63:6445–6455

Granett J, Goheen AC, Lider LA (1987a) Evaluation of grape rootstocks for resistance to type A and type B grape phylloxera. Am J Enol Vitic 38:298–300

Granett J, Goheen AC, Lider LA (1987b) Grape phylloxera in California. Calif Agric 41:10–12

Grant RS, Matthews MA (1996) The influence of phosphorous availability and rootstock on root system characteristics, phosphorus uptake, phosphorus partitioning, and growth efficiency. Am J Enol Vitic 47:403–409

Goodman RN, Grimm R, Frank M (1993) The influence of grape rootstocks on the crown gall infection process and on tumor development. Am J Enol Vitic 44:22–26

Guillon JM (1905) General study of the vine: historic vineyards and vintages, anatomy and physiology, soil and climate. Masson, Paris

Hajdu E (2015) Grapevine breeding in Hungary. In: Reynolds A (ed) Grapevine breeding programs for the wine industry. Woodhead Publishing, Cambridge

Harris AR (1984) Resistance of some *Vitis* rootstocks to *Xiphinema index*. J Nematol 15:405–409

Keller M, Kummer M, Carmo Vasconcelos M (2001) Reproductive growth of grapevines in response to nitrogen supply and rootstock. Aust J Grape Wine Res 7:12–18

Keller M, Mills LJ, Harbertson JF (2012) Rootstock effects on deficit-irrigated winegrapes in a dry climate: vigor, yield formation, and fruit ripening. Am J Enol Vitic 63:29–39

Koblet W, Candolfi-Vasconcelos CM, Zweifel W, Howell GS (1994) Influence of leaf removal, rootstock, and training system on yield and fruit composition of Pinot noir grapevines. Am J Enol Vitic 45:181–187

Koblet W, Keller M (1996) Effects of training system, canopy management practices, crop load and rootstock on grapevine photosynthesis. Acta Hortic 427: 133–140

Lambert JJ, Anderson MM, Wolpert JA (2008) Vineyard nutrient needs vary with rootstocks and soils. Calif Agric 62:202–207

Li M, Guo Z, Jia N, Yuan J, Han B, Yin Y, Sun Y, Liu C, Zhao S (2019) Evaluation of eight rootstocks on the growth and berry quality of 'Marselan' grapevines. Sci Hortic 248:58–61

Lider LA (1958) Phylloxera-resistant grape rootstocks for the coastal valleys of California. Hilgardia 27: 287–318

Lider LA, Kasimatis AN, Kliewer WM (1973) Effect of pruning severity and rootstock on growth and yield of two grafted, cane-pruned wine grape cultivars. J Am Soc Hortic Sci 98:8–12

May P, Clingeleffer PR, Brien CJ (1976) Sultana (*Vitis vinifera* L.) canes and their exposure to light. Vitis 14:278–288

McCarthy MG, Cirami RM, Furkaliev DJ (1997) Rootstock response of Shiraz (*Vitis vinifera*) grapevines to dry and drip-irrigated conditions. Aust J Grape Wine Res 3:95–98

Morano L, Kliewer WM (1994) Root distribution of three grapevine rootstocks grafted to Cabernet Sauvignon grown on a very gravelly clay loam soil in Oakville, California. Am J Enol Vitic 45:345–347

Morlet R, Jacquet A (1993) The soil effects on the grapevine root system in several vineyards of the Loire valley (France). Vitis 32:35–42

Neilsen GH, Neilsen D, Bowen P, Bogdanoff C, Usher K (2010) Effect of timing, rate, and form of N fertilization on nutrition, vigor, yield, and berry yeast-assimilable N of grape. Am J Enol Vitic 61:327–336

Nelson CC, Kennedy JA, Zhang Y, Kurtural SK (2016) Applied water and rootstock affect productivity and anthocyanin composition of Zinfandel in Central California. Am J Enol Vitic 67:18–28

Ozden M, Vardin H, Simsek M, Karaasian M (2010) Effects of rootstocks and irrigation levels on grape quality of *Vitis vinifera* L. cv. Shiraz. Afr J Biotechnol 9:3801–3807

Paranychianakis NV, Aggelides S, Angelakis AN (2004) Influence of rootstock, irrigation level and recycled water on growth and yield of Sultanina grapevines. Agric Water Manag 69:13–27

Pongrácz DP (1983) Rootstocks for grape-vines. David Philip, Cape Town

Ramming DW (2010) Greenhouse screening of grape rootstock populations to determine inheritance of resistance to phylloxera. Am J Enol Vitic 61:234–239

Reynolds AG (2015) Grapevine breeding in France—a historical perspective. In: Reynolds A (ed) Grapevine breeding programs for the wine industry. Woodhead Publishing, Cambridge

Rodrigues da Silva MJ, Maia Paiva AP, Pimentel A Jr, Pereira Contreras Sánchez CA, Callili D, Moura MF, Leonel S, Tecchio MA (2018) Yield performance of new juice grape varieties grafted onto different rootstocks under tropical conditions. Sci Hortic 241:194–200

Rogiers SY, Clarke SJ (2013) Vegetative growth and cluster development in Shiraz grapevines subjected to partial root-zone cooling. AoB Plants. https://doi.org/10.1093/aobpla/plt036

Romero P, Batía P, Navarro JM (2018) Selecting rootstocks to improve vine performance and vineyard sustainability in deficit irrigated Monastrell grapes under semiarid conditions. Agric Water Manag 209:73–93

Ruehl E, Schmid J, Eibach R, Töpfer R (2015) Grapevine breeding programmes in Germany. In: Reynolds A (ed) Grapevine breeding programs for the wine industry. Woodhead Publishing, Cambridge

Sabir A, Sahin Z (2018) The response of soilless grown 'Michele Palieri' (*Vitis vinifera* L.) grapevine cultivar to deficit irrigation under the effects of different rootstocks. Erwerbs-Obstau 60:21–27

Sanchez LA, Dokoozlian NK (2005) Bud microclimate and fruitfulness in *Vitis vinifera* L. Am J Enol Vitic 56:319–329

Scienza A, Failla O, Romano F (1986) Investigations on the variety-specific uptake of minerals by grapevines. Vitis 25:160–168

Soil Survey Division Staff (1993) Soil Survey Manual. USDA Handbook No. 18, Issued October 1993. US GPO, Washington, DC

Swanepoel JJ, Southey JM (1989) The influence of rootstock on the rooting pattern of the grapevine. S Afr J Enol Vitic 10:23–28

Venuti S, Copetti D, Foria S, Falginella L, Hoffmann S, Bellin D, Cindric P, Kozma P, Scalabrin S, Morgante M, Testolin R, Di Gaspero G (2013) Historical introgression of the downy mildew resistance gene Rpv12 from the Asian species *Vitis amurensis* into grapevine varieties. PLoS ONE 8:e61228

Virgona JM, Smith JP, Holzapfel BP (2003) Scions influence apparent transpiration efficiency of *Vitis vinifera* (cv. Shiraz) rather than rootstocks. Aust J Grape Wine Res 9:183–185

Vrsic S, Pulko B, Kocsis L (2015) Factor influencing grafting success and compatibility of grape rootstocks. Sci Hortic 181:168–173

Walker MA, Wolpert JA, Weber E (1994) Viticultural characteristics of VR hybrid rootstocks in a vineyard site infected with grapevine fanleaf virus. Vitis 33:19–23

Williams LE (2010) Interaction of rootstock and applied water amounts at various fractions of estimated evapotranspiration (ETc) on productivity of Cabernet Sauvignon. Aust J Grape Wine Res 16:434–444

Wolf TK, Pool RM (1988) Effects of rootstock and nitrogen fertilization on the growth and yield of Chardonnay grapevines in New York. Am J Enol Vitic 39:29–37

Wolpert JA, Smart DR, Anderson M (2005) Lower petiole potassium concentration at bloom in rootstocks with Vitis berlandieri genetic backgrounds. Am J Enol Vitic 56:163–169

Yildirim K, Yagci A, Sucu S, Tunc S (2018) Responses of grapevine rootstocks to drought through altered root system architecture and root transcriptomic regulations. Plant Phys Biochem 127:256–268

Scion Breeding for Resistance to Biotic Stresses

15

Ian Dry, Summaira Riaz, Marc Fuchs, Mark Sosnowski and Mark Thomas

Abstract

The majority of grapevine cultivars used for wine, table grape and dried-fruit production are derived from the Eurasian grape species *Vitis vinifera* because of its superior aroma and flavour characteristics. However, this species has little or no genetic resistance against the major pests and pathogens that attack above-ground parts of the grapevine including the trunk, canopy and bunches. As a result, grape production is highly dependent on the frequent use of fungicides and pesticides, which has significant implications for the economic and environmental sustainability of grape production. This chapter will summarize our current knowledge of the different resistance loci/genes that have been identified in wild grapevine species that could potentially be used to develop new grapevine cultivars with enhanced genetic resistance by marker-assisted selection.

15.1 Introduction

The Eurasian grape species, *Vitis vinifera*, which is the predominant species used for wine, table grape and dried grape production all over the world is susceptible to numerous pests and pathogens including fungi, oomycetes, bacteria, viruses, phytoplasma, insects and arachnids. All parts of the grapevine plant are subject to attack by these organisms including the roots, trunk, arms, cordons, canes, shoots, leaves, rachis and berries. One reason for the susceptibility of *V. vinifera* cultivars to many of the major pests and pathogens is that these organisms are not indigenous to Eurasia, and as such, there has been no selection pressure to evolve resistance.

At the moment, most of these pests and pathogens are controlled by the frequent application of fungicides and pesticides. Depending on the region and the season, grape growers may be applying anywhere between 10 and 25

I. Dry (✉) · M. Thomas
CSIRO Agriculture and Food, WIC West Building, Urrbrae, SA 5064, Australia
e-mail: ian.dry@csiro.au

M. Thomas
e-mail: mark.r.thomas@csiro.au

S. Riaz
Department of Viticulture and Enology, University of California Davis, One Shields Ave, Davis, CA 95616, USA
e-mail: snriaz@ucdavis.edu

M. Fuchs
Section of Plant Pathology and Plant-Microbe Biology, School of Integrative Plant Science, New York State Agricultural Experiment Station, Cornell University, Geneva, NY 14456, USA
e-mail: marc.fuchs@cornell.edu

M. Sosnowski
South Australian Research and Development Institute and the University of Adelaide, Waite Campus, Urrbrae, SA 5064, Australia
e-mail: mark.sosnowski@sa.gov.au

D. Cantu and M. A. Walker (eds.), *The Grape Genome*, Compendium of Plant Genomes, https://doi.org/10.1007/978-3-030-18601-2_15

applications per season (Butault et al. 2010). Not only does this translate into increased production costs for growers, but there are also potential negative impacts of these chemicals on the health of vineyard soils (Komarek et al. 2010; Brunetto et al. 2016), beneficial organisms (Gadino et al. 2011), vineyard workers (Le Moal et al. 2014) and even surrounding populations (Raanan et al. 2017). In a recent study, Esteve-Turrillas et al (2016) analysed 250 commercial wines from the major wine-growing regions of the world for the presence of five of the most commonly used new-generation organic fungicides and found 44% of wines contained at least one of the fungicides at concentrations > 10 µg per L and more than 100 µg per L in 8.4% of the wines tested. This clearly has potential implications for consumer health and international trade.

A more economically and environmental sustainable method to reduce the susceptibility of grapevines to attack by these pests and pathogens would be to breed new cultivars with enhanced genetic resistance. Indeed, European grape breeders started introgressing resistance (*R*) loci from wild North American *Vitis* spp. into *V. vinifera* in the late 1880s, in response to the accidental introduction of powdery mildew and downy mildew from North America. While this resulted in the generation of many *Vitis* interspecific hybrids with improved resistance to powdery and downy mildew, the reduced quality of wine made from these resistant hybrids has significantly limited their adoption for wine production (Pedneault and Provost 2016). This is because the time and cost involved in mounting a grapevine breeding program meant that breeders, in the main, did not undertake sufficient backcrossing to remove deleterious wine-quality traits while preserving the resistance loci.

With the publication of the PN40024 grape genome in 2007, the use of more efficient breeding techniques, such as marker-assisted selection (MAS), to introgress resistance loci from wild grape species from North America and East Asia is now possible. Chapter 14 has covered the breeding of new rootstocks with resistance to belowground pathogens; this chapter will focus specifically on breeding strategies to improve the resistance of the grapevine scion.

15.2 Breeding for Resistance to Mildews

Plant pathogens can be divided into biotrophs, hemi-biotrophs and necrotrophs, according to their lifestyles (Glazebrook 2005). Biotrophs derive nutrients from living host tissues, whereas necrotrophs derive nutrients from dead or dying cells. Some pathogens can be clearly assigned as biotrophs or necrotrophs. However, many others behave as both biotrophs and necrotrophs, depending on the conditions in which they find themselves or the stages of their life cycles. Such pathogens are called hemi-biotrophs.

Microbes become pathogens when they evolve the capacity to breach the first line of plant defence called pathogen-associated molecular pattern-triggered immunity (PTI). They do this by secreting small proteins into plant cells, called effectors, which suppress PTI and facilitate infection (Dodds and Rathjen 2010). Overtime, certain plant species in which PTI had been compromised, evolved resistance genes (*R-genes*) that encode proteins that specifically recognize these effectors, leading to effector-triggered immunity (ETI). Effector-triggered immunity is commonly associated with programmed cell death (PCD) (observed as a hypersensitive response), which kills the invaded cell and thereby prevents biotrophic pathogens from obtaining the nutrition required for further growth and development.

The world grape industry is based predominately on cultivars of the Eurasian grape species, *V. vinifera*, which were bred in Europe some 200–600 years ago (Robinson et al. 2012). However, powdery mildew (*Erysiphe necator syn. Uncinula necator*) and downy mildew (*Plasmopora viticola*) were only introduced into Europe from North America in the mid-nineteenth century (Gessler 2011; Gadoury et al. 2012). As a result, the important *V. vinifera* cultivars have little or no genetic resistance to these pathogens and grape growers rely on the frequent use of agrochemicals to minimize the potentially devastating impact of these pathogens on grape yield and quality. While the use of older contact fungicides based on sulphur and copper remain effective, the development of *E. necator* and *P. viticola* isolates with resistance to systemic fungicide chemistries introduced since the 1960s presents a serious management problem in some viticultural regions (Colcol and Baudoin 2016). In contrast, many wild grapevine species endemic to North America and China display significant levels of resistance to these pathogens (Wan et al. 2007; Cadle-Davidson 2008; Cadle-Davidson et al. 2011), which offers the potential to generate new mildew-resistant grape cultivars.

15.2.1 Powdery Mildew Resistance Loci

To date, 12 loci have been identified from a range of different grape species native to North America, China and Central Asia, to confer resistance to *E. necator* (Table 15.1).

Run1 (Resistance to *Uncinula necator* 1) was the first locus identified from the wild North American grapevine species *M. rotundifolia* (syn. *V. rotundifolia*) cv. Thomas that could confer strong resistance to powdery mildew following introgression into *V. vinifera* (Bouquet 1986; Pauquet et al. 2001). The gene responsible for powdery mildew resistance at the *Run1* locus was cloned and functionally characterized by Feechan et al. (2013) and shown to encode a Toll/interleukin-1 receptor (TIR)—nucleotide-binding site (NB)—leucine-rich repeat domain (LRR) domain protein which represents the most important class of *R* proteins in plants (Gururani et al. 2012). These NB-LRR proteins specifically recognize pathogen effector molecules secreted during infection and initiate effector-triggered immunity, which is highly effective against biotrophic pathogens such as powdery mildew. Interestingly, the genomes of perennial woody plants appear to possess a larger number of NB-LRR resistance genes than annual herbaceous plants which most probably reflects the more diverse range of pathogens that perennial plants have to deal with over their lifespan (Tobias and Guest 2014). The gene, designated *MrRUN1*, confers complete resistance against isolates from Australia, North America and France by rapidly inducing PCD in penetrated epidermal cells (Feechan et al. 2013). However, a powdery mildew isolate (Musc4) collected from the south-eastern region of North America (Brewer and Milgroom 2010) to which *M. rotundifolia* is native, was found to be capable of breaking *MrRUN1* resistance (Feechan et al. 2013) indicating that the effector recognized by the MrRUN1 protein has either been mutated or completely lost from the Musc4 isolate.

Two other powdery mildew *R* loci have also been mapped to different chromosomes in other *M. rotundifolia* cultivars. Allelic variants of the *Run2* locus, *Run2.1* and *Run2.2* on chr18, have been identified in the *M. rotundifolia* cultivars ‘Magnolia’ and ‘Trayshed’, respectively (Riaz et al. 2011). Like *Run1*, powdery mildew resistance mediated by *Run2.1* and *Run2.2* appears to be mediated via programmed cell death (PCD) (Feechan et al. 2015). However, whereas *Run2.1* was able to mount a resistance response against as the Musc4 isolate, *Run2.2* was completely susceptible. The resistance conferred by *Run2.1* against the Musc4 isolate makes it a good candidate for pyramiding with *Run1*. *Ren5* (Resistance to *Erysiphe necator*) was mapped to chr14 in *M. rotundifolia* cv. ‘Regale’ (Blanc et al. 2012) and appears to exert its action after the formation of the first appressorium and by stopping further mycelium development.

Other North American Vitis species have also been shown to be potential sources of powdery mildew resistance, but the level of resistance appears to be weaker than that conferred by powdery mildew *R* genes from *M. rotundifolia.* The *Ren2* locus, from *V. cinerea,* provides partial resistance to powdery mildew including the Musc4 isolate (Feechan et al. 2015). The *Ren3* locus was originally reported by Welter et al. (2007) to confer partial resistance to powdery mildew and was mapped to chr15. This locus was originally identified as coming from the interspecific hybrid ‘Regent’ which has a

Table 15.1 Resistance loci in grapevine species that confer resistance to scion pathogens

Locus	Pathogen species	Chr	Resistance type	Source of resistance	Origin of *R* locus	References
Powdery mildew						
Run1	*Erysiphe necator*	12	Major	*M. rotundifolia*	North America	Pauquet et al. (2001)
Run2		18	Major	*M. rotundifolia*	North America	Riaz et al. (2011)
Ren1		13	Major	*V. vinifera subsp. sylvestris*	Central Asia	Hoffmann et al. (2008)
Ren2		14	Partial	*V. cinerea*	North America	Dalbo et al. (2001)
Ren3		15	Partial	unknown	North America	Zendler et al. (2017)
Ren4		18	Major	*V. romanetii*	China	Ramming et al. (2011)
Ren5		14	Major	*M. rotundifolia*	North America	Blanc et al. (2012)
Ren6		9	Major	*V. piasezkii*	China	Pap et al. (2016)
Ren7		19	Partial	*V. piasezkii*	China	Pap et al. (2016)
Ren8		18	Minor	unknown	North America	Zyprian et al. (2016)
Ren9		15	Partial	unknown	North America	Zendler et al. (2017)
Ren10		2	Minor	unknown	North America	Teh et al. (2017)
Downy mildew						
Rpv1	*Plasmopara viticola*	12	Partial	*M. rotundifolia*	North America	Merdinoglu et al. (2003)
Rpv2		18	Major	*M. rotundifolia*	North America	Merdinoglu, 2018, pers. comm.
Rpv3		18	Partial	Multiple Vitis species (see text)	North America	Welter et al. (2007), Bellin et al. (2009), Di Gaspero et al. (2012)
Rpv4		4	Minor	unknown	North America	Welter et al. (2007)
Rpv5		9	Minor	*V. riparia*	North America	Marguerit et al. (2009)
Rpv6		12	Minor	*V. riparia*	North America	Marguerit et al. (2009)
Rpv7		7	Minor	unknown	North America	Bellin et al. (2009)
Rpv8		14	Major	*V. amurensis*	China	Blasi et al. (2011)

(continued)

Table 15.1 (continued)

Locus	Pathogen species	Chr	Resistance type	Source of resistance	Origin of *R* locus	References
Rpv9		7	Minor	*V. riparia*	North America	Moreira et al. (2010)
Rpv10		9	Partial	*V. amurensis*	China	Schwander et al. (2012)
Rpv11		5	Minor	unknown	North America	Fischer et al. (2004)
Rpv12		14	Major	*V. amurensis*	China	Venuti et al. (2013)
Rpv13		12	Minor	*V.riparia*	North America	Moreira et al. (2011)
Rpv14		5	Minor	*V. cinerea*	North America	Ochssner et al. (2016)
Rpv15		18	Major	*V. piasezkii*	China	Pap et al. (unpublished)
Rpv16		9	Minor	*V. piasezkii*	China	Pap et al. (unpublished)
Rpv17		8	Minor	unknown	North America	Divilov et al. (2018)
Rpv18		11	Minor	unknown	North America	Divilov et al. (2018)
Rpv19		14	Minor	*V. rupestris*	North America	Divilov et al. (2018)
Rpv20		6	Minor	unknown	North America	Divilov et al. (2018)
Rpv21		7	Minor	unknown	North America	Divilov et al. (2018)
Rpv22		15	Partial	*V. amurensis*	China	Song et al. (2018)
Rpv23		2	Minor	*V. amurensis*	China	Song et al. (2018)
Rpv24		18	Minor	*V. amurensis*	China	Song et al. (2018)
Rpv25		15	Partial	*V. amurensis*	China	Lin et al. (2019)
Rpv26		15	Partial	*V. amurensis*	China	Lin et al. (2019)
Rpv27		18	Partial	*V. aestivalis* cv. *'Norton'*	North America	Sapkota et al. (2019)
Botrytis bunch rot						
Unnamed QTL	*Botrytis cinerea*	2	Major	*V. aestivalis* cv. *'Norton'*	North America	Hwang et al. (2018)
Non-Botrytis bunch rots						
Rgb1	*Guignardia bidwellii*	14	Major	*V. cinerea*	North America	Rex et al. (2014)
Rgb2		16	Minor	*V. cinerea* or *V. riparia*	North America	Rex et al. (2014)

(continued)

Table 15.1 (continued)

Locus	Pathogen species	Chr	Resistance type	Source of resistance	Origin of *R* locus	References
Bacterial diseases						
Pdr1	*Xylella fastidiosa*	14	Major	*V. arizonica*	North America	Riaz et al. (2006)
Rcg1	*Agrobacterium spp.*	15	Major	*V. amurensis*	China	Kuczmog et al. (2012)
Trunk disease						
Rda1	*Diaporthe ampelina*	15	Major	*V. cinerea*	North America	Barba et al. (2018)
Rda2		7	Minor	Multiple Vitis species	North America	Barba et al. (2018)

complex pedigree involving *V. vinifera* and North American *Vitis* species. Thus, the original source of *Ren3* is still unknown. Further fine mapping of the *Ren3* locus indicated that the original locus may, in fact, be composed of two adjacent *R* loci, *Ren3* and *Ren9* (Zendler et al. 2017). The mechanism of resistance conferred by the *Ren3* and *Ren9* loci is unknown, but involves a post-invasion response as demonstrated by the fact that *E. necator* is still able to develop a dense mycelial network on the leaf surface, but only infrequently forms conidia. Two other minor powdery mildew *R* loci, *Ren8* and *Ren10,* from North American Vitis species of unknown origin have also been reported (Zyprian et al. 2016; Teh et al. 2017).

Wild Chinese *Vitis* species also represent an important source of major dominant *R* loci against powdery mildew with resistance levels similar to that observed for *Run1*. *Ren4* has been successfully introgressed into *V. vinifera* from the *V. romanetii* and shown to segregate as a single dominant locus (Ramming et al. 2011; Mahanil et al. 2012). *Ren4* resistance was initially reported to be associated with high levels of penetration resistance and did not appear to be dependent on the induction of PCD (Ramming et al. 2011). However, more recent studies indicate that *Ren4*-mediated resistance occurs post-penetration and may involve two different mechanisms; penetrated epidermal cells either undergo PCD or the haustoria becomes encased in callose (Fig. 15.1), thereby effectively blocking nutrient uptake (Dry IB, unpublished). We have also confirmed that *Ren4* resistance is not broken by the Musc4 isolate (Fig. 15.1).

Another wild Chinese grapevine species, *V. piasezkii*, has also been shown to contain at least two powdery mildew *R* loci, designated *Ren6* and *Ren7*, on chr 9 and 19, respectively, which mediate a PCD-based resistance response (Pap et al. 2016). The *Ren6* resistance response was found to be even stronger than that mediated by the *Run1* locus, when compared in the same *Vitis* background with 92–95% of epidermal cells displaying effective PCD, i.e. no development of secondary hyphae, after 2 dpi. In contrast, the resistance response of *Ren7* genotypes was much slower than *Ren6* resulting in a high percentage of penetrated epidermal cells in which either no PCD is observed or the PCD induction can be considered ineffective because the fungus is still able to produce a secondary hyphae (Pap et al. 2016).

Finally, it is now clear that certain accessions of *V. vinifera* from Central Asia also contain a major *R* locus that, while less effective than *Run1*-mediated resistance, still significantly restricts powdery mildew growth and sporulation. The *Ren1* locus has been mapped to chr13 in two *V. vinifera* cultivars, 'Kishmish vatkana' and 'Dzhandzhal kara', originating from Uzbekistan (Hoffmann et al. 2008; Coleman et al. 2009). The speed of PCD induction in

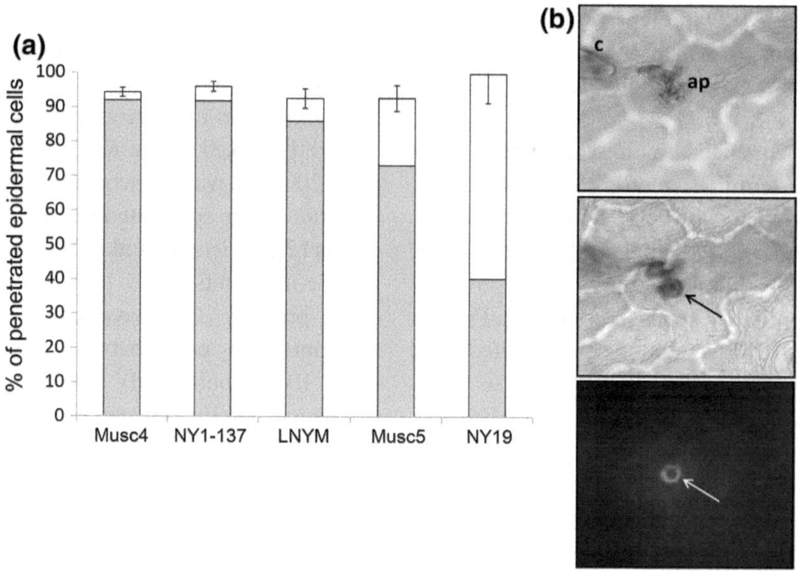

Fig. 15.1 *Ren4*-mediated resistance involves the induction of both programmed cell death and callose encasement of the haustorial complex. **a** Grapevine powdery mildew spores of a various North American isolates (Feechan et al. 2015) were inoculated onto detached leaves of a BC2-*Ren4* genotype and samples after 2 days. Visualization and scoring for PCD and callose deposition were performed as described by Feechan et al. (2011). Results are expressed as the percentage of successful infections that resulted in either PCD induction (grey bar) or callose deposition (white bar). Note that isolates Musc4 and Musc5 are virulent and NY1-137 partially virulent on *Run1* genotypes, whereas the *Ren4* genotype is highly resistant to all isolates tested. **b** Photographs of infection by NY19 isolate. Top panel shows germinated conidium (c) and appressoria (ap) on the surface of the leaf. Middle panel is focussed below the appressorium to show the globular papillum (arrow). Bottom panel is the same view as middle panel but viewed using a blue light filter set to visualize callose deposition around the haustorial complex

penetrated epidermal cells appears to be slower than that observed in *Run1* genotypes, and as a result, more powdery mildew hyphal growth and sporulation is observed on *Ren1* plants than on *Run1* plants (Hoffmann et al. 2008). The syntenous region in the *V. vinifera* PN40024 reference genome contains a cluster of genes that encode putative coiled-coil (CC)-NB-LRR proteins (Coleman et al. 2009). However, no data has yet been published to indicate what candidate *R* genes are present in this region in 'Kishmish vatkana' or 'Dzhandzhal kara'. Riaz and co-workers (2013) subsequently identified an additional six *V. vinifera* and two *V. vinifera* subsp. *sylvestris* accessions from Central Asia that also contained a *Ren1*-like locus. Based on genetic marker analysis, they concluded that the *Ren1*-like resistance in *V. vinifera* subsp. *sylvestris* was most likely the progenitor of the resistance in the Central Asian *V. vinifera* accessions.

15.2.2 Breeding for Reduced Susceptibility to Powdery Mildew?

Successful penetration of a plant host by an adapted powdery mildew species has been shown to be dependent on the presence of a functional allele of the *Mildew resistance locus O* (*MLO*) in a number of crop species (Kusch and Panstruga 2017). This therefore represents an example of a pathogen susceptibility gene. MLO proteins belong to large gene families, which are unique to plants and encode seven-transmembrane domain proteins of unknown biochemical activity localized in the plasma membrane (Acevedo-Garcia et al. 2014). Significantly, only specific *MLO* genes within the family are capable of acting as powdery mildew susceptibility genes and these appear to encode proteins with conserved motifs within the cytoplasmic C-terminal domain of the MLO protein (Panstruga 2005). The mechanism by which MLO proteins act as powdery mildew susceptibility factors is yet to be resolved.

Based on sequence homology, the presence of conserved C-terminal motifs and expression kinetics following powdery mildew infection, Feechan et al. (2008) identified three members of the *VvMLO* gene family that may act as powdery mildew susceptibility genes in grapevine. Pessina et al. (2016) subsequently demonstrated that RNAi-mediated silencing of one of these genes *VvMLO-7* (designated *VvMLO17* in (Feechan et al. 2008)) significantly increased resistance to powdery mildew in the grapevine.

MLO-based resistance is recessive and non-race specific. As such, it is likely to be much more durable in the field than the resistance conferred by to the dominant race-specific resistance conferred by the powdery mildew *R-genes* described above. However, being a recessive trait poses significant challenges for strategies based on conventional breeding techniques in comparison to targeted gene-editing approaches, which are currently still considered as transgenic in some countries. Indeed, gene editing has already been shown to be effective in generating powdery mildew-resistant bread wheat through the simultaneous editing of three MLO homoalleles (Wang et al. 2014).

Thus, for the foreseeable future, the only way to generate a non-transgenic powdery mildew-resistant *MLO* grapevine mutant is to employ techniques such as EcoTILLING (Mejlhede et al. 2006) to search *V. vinifera* germplasm collections for point mutations and/or small insertions/deletions in *VvMLO7,* where the powdery mildew resistance phenotype is masked by the presence of the wild-type *MLO* allele.

15.2.3 Downy Mildew Resistance Loci

As many as 27 *R* loci have been reported from wild grapevine species that are capable of conferring some level of increased resistance to *P. viticola* when introgressed into *V. vinifera* (Table 15.1).

The first downy mildew *R* locus to be identified from a wild grape species was *Rpv1* (Resistance to *Plasmopora viticola* 1) from *M. rotundifolia* cv. *'Trayshed'* and was found to be tightly linked to the *Run1* locus (Merdinoglu et al. 2003). It was subsequently shown that *Rpv1* and *Run1* were co-located within the same region on chr12 (Anderson et al. 2011). Indeed, to date, no recombinants have been identified in over 4000 progeny that have been analysed for a recombination event between *Run1* and *Rpv1* (Dry IB, unpublished). Therefore, these two resistance specificities can effectively be considered as being part of the same genetic locus, the *Run1/Rpv1* locus. Subsequent sequencing of the *Run1/Rpv1* locus showed it to contain seven genes that encode TIR-NB-LRR proteins, one of which confers resistance to powdery mildew (*MrRUN1*) and one which confers resistance to downy mildew (*MrRPV1*) (Feechan et al. 2013). While the mechanism of resistance mediated by both *Run1* and *Rpv1* in *V. vinifera* appears to be based on induction of PCD following penetration, the level of resistance conferred by these two loci is different. *Run1* resistance is found to be qualitative in most *V. vinifera* backgrounds with little or no hyphal development or sporulation, whereas *Rpv1* resistance can be considered as quantitative, typically reducing downy mildew sporulation by 70–80%. Even so, *Rpv1* still confers strong resistance to downy mildew under field conditions (Fig. 15.2). Interestingly, Feechan et al. (2013) demonstrated that the level of downy mildew resistance in *MrRPV1* transgenic vines was significantly higher than that observed in the *Rpv1* backcross 5 breeding line BC5:3294-R23 which may be the result of the much higher levels of *MrRPV1* transcription in transgenic vines relative to the BC5 line.

A second major downy mildew *R* locus, designated *Rpv2,* has been introgressed into *V. vinifera* from *M. rotundifolia cv.* Trayshed (D. Merdinoglu, 2018, personal communication). In contrast to *Rpv1*, *Rpv2* confers total resistance to downy mildew with no sporulation and the appearance of small localized necrotic lesions. The *Rpv2* locus has been mapped to a region on chr18, which contains a cluster of five genes

Fig. 15.2 *Rpv1* confers strong resistance against downy mildew in the field. Comparison of impact of heavy downy mildew infection on the performance of the susceptible *V. vinifera* cultivar 'Chasan' and the Bouquet breeding line BC4:3082-1-42 containing the *Rpv1* downy mildew resistance locus grown in an unsprayed vineyard at INRA Pech Rouge, France (Photograph courtesy of Alain Bouquet, deceased)

encoding putative TIR-NB-LRR proteins within the syntenous region on chr18 of the PN40024 *V. vinifera* reference genome.

Numerous minor downy mildew *R* loci have been identified from a range of North American *Vitis* species (Fischer et al. 2004; Welter et al. 2007; Bellin et al. 2009; Marguerit et al. 2009; Moreira et al. 2010; Ochssner et al. 2016; Divilov et al. 2018) (Table 15.1). While many of these minor *R* loci could be useful, when used in combination with other major downy mildew *R* loci to enhance the durability of resistance in the field, it is not clear from the data available as to whether they would provide significant field resistance if deployed on their own. For this reason, this chapter will not consider these minor downy mildew *R* loci in any further detail.

One exception is *Rpv3*, which confers partial resistance to downy mildew characterized by the induction of PCD and the development of sparse sporangiophores around the site of attempted infection (Welter et al. 2007; Bellin et al. 2009; Di Gaspero et al. 2012; Zyprian et al. 2016). Since the late 1800s, the *Rpv3* locus has been introgressed into numerous hybrid wine grape cultivars from the complex interspecific hybrid 'Villard blanc' (Bellin et al. 2009). This has led to the generation of seven different *Rpv3* haplotypes originating from at least four different North American donor species: *V. labrusca*, *V. lincecumii*, *V. riparia* and *V. rupestris* (Di Gaspero et al. 2012). The widespread use of these downy mildew-resistant hybrid cultivars in Eastern Europe, since the early twentieth century, may also be responsible for the appearance of downy mildew isolates that are avirulent on *Rpv3* genotypes (Peressotti et al. 2010). However, it is also worth noting that downy mildew isolates virulent on the *Rpv3*-containing cultivar 'Regent' were observed in Bordeaux within only 5 years after planting (Delmotte et al. 2014) raising questions about the durability of the *Rpv3* locus for breeding purposes. The same isolates that were found to be virulent on 'Regent' were still avirulent on the *Rpv1* genotype Mtp3082-1-42 demonstrating that *Rpv1* and *Rpv3* have different pathogen specificities (Delmotte et al. 2014).

A recent report has also highlighted the influence of the genetic background on the intensity of *Rpv3*-dependent downy mildew resistance. Foria et al. (2018) analysed the level of downy mildew resistance in the field of 76 grape cultivars into which the *Rpv3* locus had been introgressed. Their results demonstrated that while all cultivars exhibited a PCD-mediated resistance response, some genotypes exhibited high resistance under all conditions, whereas others performed well under low disease pressure, but suffered substantial damage with higher disease pressure.

A clue to what other grapevine genes might be important in modulating the effectiveness of the *Rpv3* locus comes from analysis of the segregation of downy mildew resistance in a population derived from a cross between the downy mildew-resistant cultivar 'Merzling' (a complex hybrid with both *V. rupestris* and *V. lincecumii* in its background) and susceptible *V. vinifera* cv. Teroldego (Vezzulli et al. 2018). QTL mapping showed that downy mildew resistance in this population was not only associated strongly with inheritance of the *Rpv3-3* locus, but was also associated with a number of other QTLs linked to stilbenoid production. This led to the conclusion that an important component of *Rpv3-3*-mediated downy mildew resistance may involve the action of stilbene phtyoalexins which have previously been shown to have toxic effects on downy mildew growth (Pezet et al. 2004a, b; Alonso-Villaverde et al. 2011).

As with powdery mildew *R* loci, a number of wild Chinese *Vitis* species have also been identified as potential source of major *R* loci against *P. viticola*. Some accessions of *V. amurensis* display a high level of resistance to *P. viticola* (Staudt and Kassemeyer 1995; Wan et al. 2007; Cadle-Davidson 2008). Using a population of 232 progeny from a selfing of *V. amurensis,* Blasi et al. (2011) reported the mapping of a major downy mildew *R* locus on the upper arm of chr14 which they designated *Rpv8*. The *Rpv8* locus mediates a strong induction of PCD resulting in a low level of sporulation. Subsequently, Venuti et al. (2013) using a different, and much larger segregating population of 2532 individuals, also mapped a major downy mildew *R* locus in *V. amurensis*, which they designated as *Rpv12*, to the same approximate location on chr14. *Rpv12* conferred the ability to establish a HR within 24–48 h post-inoculation and significantly restricted sporulation of *P. viticola*. A direct comparison of the resistance phenotype with genotypes containing *Rpv12* and *Rpv3* indicated that the restriction of *P. viticola* sporulation by *Rpv12* was more significant than *Rpv3*. Further investigation will be required to confirm whether the resistance conferred by *Rpv8* and *Rpv12* is mediated by the same R-gene or paralogous genes. Analysis of the syntenous region in the PN40024 reference genome identified a cluster of 13 CC-NB-LRR genes which form part of a more complex structure of 46 clustered NB-LRRs in the upper arm of chr14 (Venuti et al. 2013).

Another downy mildew R locus, *Rpv10*, has also been mapped to chr9 of *V. amurensis* (Schwander et al. 2012). *Rpv10* confers partial resistance to downy mildew, equal to or slightly better than that observed for *Rpv3*. No information is available as to the mechanism of *Rpv10*-mediated resistance, although Schwander et al. (2012) noted that the syntenous region on chr9 in the PN40024 reference genome contains a large CC-NBS-LRR gene cluster.

More recently, five more *downy mildew R* loci have been identified in two specific *V. amurensis* cultivars. Three QTLs (*Rpv22*, *Rpv23* and *Rpv24*) were identified in 'ShuangHong' while two QTLs (*Rpv25* and *Rpv26*) were mapped in 'Shuangyou'. Both *Rpv25* and *Rpv26* map to chr15, and it is still not certain if they represent the same locus or two different loci. It is interesting to note that *Rpv22* also maps to chr15 and that both Rpv22 and Rpv25/26 confer partial resistance. Given that both 'ShuangHong' and 'Shuangyou' are derived from the same parent (*V. amurensis* cv. Shuangqing) (Huang et al. 1988; Song et al. 1998), there is a possibility that *Rpv22*, *Rpv25* and *Rpv26* are actually the same locus.

Finally, another Chinese species *V. piasezkii* is also reported to have two downy mildew *R* loci (D. Pap, 2018, personal communication). This includes a major *R* locus, designated *Rpv15*, on chr18 and a minor downy mildew R locus designated *Rpv16*, which maps to chr9. Preliminary results indicate that *Rpv15* confers strong PCD-mediated resistance similar to that observed for *Rpv12* (Dry IB, unpublished).

15.2.4 Potential Sources of New Mildew *R* Loci

In addition to the wild grapevine species listed in Table 15.1, there are a number of other *Vitis* species that have been reported to show good resistance to powdery mildew and downy mildew that warrant further investigation as potential sources of new major *R* loci to be used for future grapevine breeding programs.

In terms of powdery mildew resistance, this includes the three North American species *V. doaniana*, *V. palmata* and *V. shuttleworthii* and the three Chinese species *V. davidii*, *V. davidii var.* 'cyanocarpa', *V. pseudoreticulata* var. 'Baihe-35-1' and *V. quinquangularis* (Staudt 1997; Wan et al. 2007; Cadle-Davidson et al. 2011). However, a more recent survey of powdery mildew resistance in Chinese *Vitis* species, carried out by Gao et al. (2016), indicated that the level of powdery mildew resistance observed for *V. davidii*, *V. davidii cv.* cyanocarpa and *V. quinquangularis* was lower than had been previously reported (Wan et al. 2007). These differences may be the result of differences in powdery mildew isolates used in each study or the different assay systems used, i.e. inoculated detached leaves versus natural field infections. A number of these same species also show good resistance to downy mildew including *V. davidii var.* 'cyanocarpa', *V. pseudoreticulata*, *V. quinquangularis* and *V. shuttleworthii as well as V. romanetii and V. yeshanensis* (Staudt and Kassemeyer 1995; Wan et al. 2007; Cadle-Davidson 2008).

One important point to note about the results of these surveys of powdery mildew and downy mildew resistance of wild *Vitis* species is that not all accessions of a particular species show the resistance phenotype (Staudt and Kassemeyer 1995; Wan et al. 2007; Cadle-Davidson 2008). It is therefore critical to confirm the resistance phenotype of any new germplasm with local powdery mildew and/or downy mildew isolates before using it as a parent to generate new resistant genotypes.

15.3 Breeding for Resistance to Bunch Rots

The most common and economically significant bunch rot for grape production is caused by *Botrytis cinerea*. Any bunch rot caused by organisms other than *B. cinerea* is classified as a non-Botrytis bunch rot. They can be caused by a range of fungi, yeasts and some bacteria, including acetic acid bacteria. The majority of the organisms involved are fungi that spread through the formation of fungal spores and can be carried in the wind or by rain splash. Many are opportunistic pathogens that infect berries through wounds (e.g. berry splitting after rain events). Bunch rots reduce grape yields and have negative effects on grape and wine quality. Some bunch rots infect berries directly including alternaria rot (*Alternaria* spp.), bitter rot (*Greeneria uvicola*), black rot (*Guignardia bidwellii*), botryosphaeria rot (*Botryosphaeria* spp.), cladosporium rot (*Cladosporum* spp.) and ripe rot (*Colletotrichum* spp.). Other bunch rots are secondary invaders that enter the berry through wounds or following infection by a primary invader and include aspergillus rot (*Aspergillus* spp.), penicillium rot (*Penicillium* spp.), rhizopus rot (*Rhizopus* spp.), sour rot (various fungi, yeasts and bacteria) and white rot (*Coniella diplodiella*). For further information about these different non-botrytis rots, the reader is directed to the comprehensive review of Wilcox et al. (2015).

In general, the pathogens responsible for bunch rots fall into the category of necrotrophic pathogens, and in contrast to the success of plants in evolving major *R-genes* to resist or reduce infection by biotrophic pathogens by ETI, such a defence response will clearly not be successful against necrotrophic pathogens that colonize dead or dying tissue. It is not surprising therefore that no major *R-genes* have been found in any crop species that confer strong resistance against necrotrophic pathogens. Instead, plants generally rely on the contribution of many minor defence genes to try and restrict the development of necrotrophic pathogens.

15.3.1 Resistance to Botrytis Bunch Rot

Botrytis bunch rot can result in a reduction in wine quality by causing oxidation, off-flavours and other biochemical changes (Ribéreau-Gayon et al. 1980). Economic loss for grape production worldwide is estimated to be at least 2 billion $US per annum (Elmer and Michailides 2004). The level of *B. cinerea* infection observed in the vineyard can be considered the result of at least two major factors. The first is the expression of any genetic resistance within the developing grape berry. Analysis of genetic inheritance of Botrytis resistance in tomato, Arabidopsis and gerbera has demonstrated that it is quantitative and genetically complex, requiring the contribution of multiple loci to reduce disease severity (Finkers et al. 2007; Rowe and Kliebenstein 2008; Fu et al. 2017). This can only be accurately assessed in grapevine by the inoculation of individual grape berries.

The second major factor determining the susceptibility of different grape cultivars to Botrytis bunch rot is bunch architecture. Cultivars with tight (compact) bunches develop severe rot, whereas those with loose (open) bunches are less susceptible (Vail and Marois 1991; Smithyman et al. 1998; Vail et al. 1998; Zabadal and Dittmer 1998). This heightened susceptibility in tight bunches is most likely due to the fact that the inner surface of the cluster is exposed to high water vapour concentrations and, possibly, extended periods of surface wetness (Vail and Marois 1991). This can lead to an increase in micro-cracking of the berry cuticular membrane (Becker and Knoche 2012) which is thought to play a critical role as a barrier to *B. cinerea* infection. A significant correlation has also been demonstrated between cuticular fractures on the surface of sweet cherries and incidence of *B. cinerea* infections (Borve et al. 2000). The reader is referred to an excellent review summarizing the current knowledge around the genetic and environmental factors influencing grapevine bunch compactness by Tello and Ibanez (2018).

15.3.1.1 Genetic Basis of Resistance of Grape Berries to *Botrytis cinerea* Infection

Gabler et al. (2003) investigated morphological, anatomical and chemical characteristics of 42 genetically diverse grape cultivars and selections with various levels of resistance to *B. cinerea* to determine which features were associated with resistance. Little or no resistance exists in berries of *V. vinifera* cultivars, whereas North American grape species or hybrids such as *M. rotundifolia*, *V. labrusca* or *V. labrusca* × *V. vinifera* hybrids were found to be highly resistant. Similar results were obtained by Naegele (2018). Highly resistant cultivars were characterized by a number of properties including (a) a low number of surface pores on berries, (b) a thick cuticle and a high wax content and (c) the number and thickness of epidermal and hypodermal cell layers. The importance of the berry skin features with regard to the mechanical protection against *B. cinerea* was further supported by the observation of a positive correlation between the electrical impedance of cuticle and epicuticular waxes and resistance of berries to *B. cinerea* (Herzog et al. 2015).

Another North American grape which is highly resistant to Botrytis bunch rot is 'Norton' which is thought to be hybrid between *V. vinifera* and *V. aestivalis* (Ambers 2013). Mature Norton berries inoculated with *B. cinerea* spores showed only a low level of disease incidence (7.5%) and disease severity (3.7%) after 10 days compared to *V. vinifera* cv. Cabernet Sauvignon berries which were highly infected exhibiting an average disease incidence and severity of greater than 90% (Sapkota et al. 2015). Subsequent QTL analysis of a Norton × *V. vinifera* Cabernet Sauvignon mapping population indicates the presence of a major *R* locus on chr2 for Botrytis bunch rot (Chin-Feng et al. 2018).

A survey of wild Chinese *Vitis* species for resistance to *B. cinerea* has also been undertaken but using grapevine leaves instead of berries (Wan et al. 2015). Little or no resistance was observed with *V. vinifera* cultivars, whereas eighteen of the thirty Chinese *Vitis* species were resistant to the fungus with the highest levels of resistance observed with selected accessions of *V. amurensis*, *V. adstricta* and *V. yenshanensis*. However, how leaf resistance relates to berry resistance in these genotypes requires further examination, especially as the two *V. labrusca* × *V. vinifera* hybrids tested were found to be highly susceptible to *B. cinerea*, which would appear to be at odds with the results of other *B. cinerea* assays using individual berries (Gabler et al. 2003; Naegele 2018).

15.3.1.2 Genetics of Bunch Architecture

The breeding of new grape cultivars with more open bunches is likely to result in a significant reduction in the incidence of Botrytis bunch rot. Clones of the cultivars Chardonnay (Vail et al. 1998) and Albarino (Alonso-Villaverde et al. 2008) with the least compact clusters were found to have the lowest levels of Botrytis bunch rot in the field. In a pruning trial with Seyval blanc, treatments that led to reduced fruit set, and consequently more open bunches, were shown to significantly reduce Botrytis bunch rot across two seasons (Smithyman et al. 1998).

To identify genes that regulate bunch architecture, it is first necessary to identify the key structural characteristics that determine whether a bunch is compact or loose. Shavrukov et al. (2004) identified inflorescence length (in particular rachis internode length) as the major trait responsible for the difference in bunch architecture between two compact (Chardonnay and Riesling) and two loose (Exotic and Sultana) cultivars. In a more recent study, Tello et al. (2015) analysed the genetic variability of bunch compactness of 125 table and wine grape cultivars across three consecutive seasons and showed that the main components determining bunch compactness were length of the rachis, the number of berries per bunch and, to a lesser extent, berry size.

To date, three studies have been published on the genetic analysis of bunch architecture. Correa et al. (2014) analysed a segregating population derived from a cross of 'Ruby Seedless' × 'Sultanina' and identified 19 QTLs across chr5, 8, 9, 14, 17 and 18. Using an association analysis with 114 cultivars, Tello et al. (2016) identified a number of SNPs associated with rachis internode length and bunch compactness, including four that were recurrently associated with the rachis internode length across the three seasons evaluated. However, it is not known how these SNPs relate to the QTLs identified by Correa et al. (2014).

A third study has been undertaken across two seasons on 150 F1 progeny derived from a cross between GF.GA-47-42 ('Bacchus' × 'Seyval blanc') which has loose clusters, crossed with 'Villard blanc'(Richter et al. 2017). More than 20 QTLs related to key determinants of bunch architecture including rachis length, peduncle length and pedicel length were reproducibly found over two seasons and they are dispersed throughout the genome. No information was provided as to how these QTLs link to the results of the two previous studies.

In summary, the results of these published studies indicate that the genetic control of bunch architecture is likely to be highly complex with many genes contributing minor effects. At this point in time, there appear to be no obvious candidates for use in MAS of new cultivars with more open clusters.

15.3.2 Resistance to Non-Botrytis Bunch Rots

To date, genetic resistance within wild grapevine species for non-botrytis bunch rots has only been reported for black rot and ripe rot.

Black rot (*G. bidwellii*) is a hemibiotrophic fungus native to North America. The fungus infects all green parts of the plant, and complete crop loss can occur in warm, humid climates. All *V. vinifera* cultivars are highly susceptible, but resistance has been observed in North American *Vitis* species. Barrett (1953) tested several wild North American species for black rot resistance including *V. cinerea, V. rupestris* and *V. lincecumii* and found *V. cinerea* to have the highest level of resistance, with nearly every accession tested free of black rot both on foliage and fruit. Dalbó et al. (2000) subsequently mapped three QTLs linked to black rot resistance in a population derived from a cross between 'Horizon' (whose pedigree includes *V. vinifera*, *V. labrusca*, *V. aestivalis* and *V. rupestris*) × 'Illinois 547–1' (*V. rupestris* × *V. cinerea* B9). The hybrid cultivar 'Börner' (*V. riparia* Gm183 × *V. cinerea* Arnold) was also shown to display a high level of resistance to black rot (Rex et al. 2014). A major QTL was detected based on the results of phenotyping a mapping population generated from a cross of the susceptible breeding line V3125 (Schiava grossa × Riesling) with Börner. The QTL designated *Rgb1* (Resistance to *Guignardia bidwellii* 1) is located on chr14 and explained up to 21.8% of the phenotypic variation. A second minor QTL, designated *Rgb2*, was mapped to chr16 and explained 8.5% of the phenotypic variation. Rex et al. (2014) concluded that the *Rgb1* locus derived from *V. cinerea* Arnold is most likely allelic to the QTL mapped by Dalbó et al. (2000) from *V. cinerea* B9. Recent analysis of *V. amurensis* hybrids also indicates that wild Chinese species are a potential source of strong resistance to black rot (Roznki et al. 2017).

Ripe rot is associated with vineyards that experience warm and wet conditions close to harvest and is more frequently found in open canopies where the fruit is exposed. There are two species of the fungus responsible for ripe rot, *Colletotrichum acutatum* and *Colletotrichum gloeosporioides*, but *C. acutatum* is the predominant species found in vineyards. There is limited published information available about the resistance of grape species or cultivars to ripe rot. However, the available information tends to suggest that the resistance varies widely within species rather than between species. For example, a survey of *M. rotundifolia* cultivars showed that the incidence of ripe rot on the bronze-fruited cultivars ('Carlos', 'Fry', 'Magnolia', 'Scuppernong') ranged from 6.7 to 33.5%, while symptoms of ripe rot were never observed on the black-fruited cultivars ('Noble', 'Tarheel', 'Pride') (Daykin and Milholland 1984). Similarly, a survey of table grape cultivars from South Korea showed that four were resistant to ripe rot; two were derived from *V. vinifera;* and two were interspecific hybrids with North American species (Jang et al. 2011). Further work needs to be undertaken to examine the inheritance of this ripe rot resistance.

15.4 Other Fungal Diseases

Grapevines are also susceptible to a number of other fungal pathogens which cause symptoms other than bunch rot. Anthracnose, or black spot, caused by the fungus *Elsinoe ampelina* is a very damaging disease in viticultural regions with a warm, humid climate and results in lesions which destroy leaves, shoots and fruit. It has been reported that *V. vinifera* cultivars are highly susceptible, whereas accessions of North American species including *V. aestivalis*, *V. shuttleworthii*, *V. labrusca*, *V. rupestris* and *M. rotundifolia* are completely resistant (Mortensen 1981). Wang et al. (1998) surveyed multiple accessions of thirteen different wild Chinese species including *V. amurensis*, *V. davidii*, *V. piasezkii*, *V. pseudoreticulata*, *V. quinquangularis* and *V. romanetii* and showed them to also be resistant to *E. ampelina.* Genetic analysis of inheritance of resistance to anthracnose from crosses involving *V. labrucsa*, *V. rupestris* and *V. riparia* confirmed that anthracnose resistance is controlled by a single dominant locus (Kim et al. 2008). Although no information is currently available as to the location of this locus, a RAPD marker was identified that predicted the presence/absence of the anthracnose resistance across *V. labrucsa*-, *V. rupestris*- and *V. riparia*-derived genotypes.

Grapevine leaf rust, caused by the fungus *Phakopsora euvitis,* occurs mainly in warm temperate and subtropical grape growing regions. It infects leaves causing chlorotic spots in the infected area and necrosis in older infections, but can also infect the fruit, stems and rachis. Severe disease may cause defoliation, reducing the vigour and yield of infected vines (Hennessy et al. 2007). A number of North American grape species have been assessed as being highly resistant (asymptomatic) to grape rust including cultivars of *M. rotundifolia, V. labrusca, V. berlandieri, V. candicans, V. champini* and *V. palmata* (Clayton and Ridings 1970; Patil et al. 1998). However, a number of these species were assessed as being susceptible in a later study (Hennessy et al. 2007) highlighting the need for further phenotypic analysis before considering using these genotypes for resistance breeding.

15.5 Breeding for Resistance to Bacterial Diseases

15.5.1 Pierce's Disease

Pierce's disease (PD) is a serious impediment to viticulture in North America (Hopkins and Purcell 2002; Kyrkou et al. 2018). It is caused by the xylem-limited bacterium, *Xylella fastidiosa*, which is classified as a single species with multiple subspecies and strains that cause disease in over 100 monocotyledonous and dicotyledonous plants (Hopkins and Purcell 2002; Newman et al. 2003). It is transmitted to host plants by specialized insect vectors; in the case of grapevines, it is the glassy-winged sharpshooter. Symptoms are expressed as xylem vessels become blocked by bacterial aggregation and the formation of gums and tyloses, leading to desiccation. Infected grapevines show distinct symptoms (Fig. 15.3) including marginal and inter-vein leaf scorch, leaf blades that drop leaving attached petioles ('matchsticks'), irregular shoot maturation referred to as 'green islands', shrivelled fruit in late summer and eventual plant death within one to five years of infection (Hopkins 1989; Krivanek and Walker 2005; Fritschi et al. 2007). Until recently, the pathogen was primarily found in North America, but in 2013 the first European outbreak was recorded in olive trees in Italy (Saponari et al. 2013).

Pierce's disease was first reported in the mid-nineteenth century when an outbreak destroyed thousands of acres of vineyards in Anaheim California (Pierce 1892). It was later reported across the southern USA (Stoner 1953; Crall and Stover 1957; Hewitt 1958; Perry et al. 1974) and Mexico (Raju et al. 1979). Grape species from the south-eastern USA, such as *V. aestivalis* and *V. shuttleworthii*, are resistant to PD and early breeding efforts utilized them to develop resistant cultivars. However, these cultivars had limited acceptance because their fruit characters were less favourable than pure *V. vinifera* cultivars (Loomis 1958; Mortensen 1968; Mortensen et al. 1977; Mortensen 1988; Halbrooks and Mortensen 1989).

A PD resistance breeding program was initiated at the UC Davis in 1990 s based on breeding populations developed from crosses of *V. rupestris* × *M. rotundifolia* originally made by Dr. Harold P. Olmo. However, genetic mapping of these populations indicated that the majority of the seedlings were not true to type and that PD resistance in this population actually originated from *V. arizonica* (Riaz et al. 2007). The first PD *R* locus, *PdR1*, was identified on chr14 of b43-17, a hybrid of *V. arizonica* × *V. candicans*, collected by Olmo near Monterrey, Mexico (Krivanek et al. 2006; Riaz et al. 2006, 2007, 2008). The *PdR1* locus is the foundation of the PD-resistant wine grape breeding program at the UC-Davis, in which MAS has been used to facilitate the introgression of PD resistance into elite *V. vinifera* selections using a two-year seed-to-seed cycle (Riaz et al. 2009). Certified virus-free plant material of five superior resistant lines was released to the nurseries in 2017, and public release is scheduled in 2020.

One objective of the breeding program is to expand the genetic base of PD resistance and develop lines that incorporate resistance from more than one genetic background. To meet this objective, a large portion of the germplasm collected from Mexico and south-western USA was evaluated and many accessions with strong resistance to PD were identified based on the greenhouse screening method optimized by

Fig. 15.3 Typical symptoms of Pierce's disease. **a** Cane from a grapevine infected with Pierce's disease (left) compared to a cane from a healthy vine (right) showing loss of leaves and irregular shoot maturation with 'green islands'. **b** Close up of 'green islands' on cane from infected vine. **c** Leaf from infected vine showing marginal and inter-vein leaf scorch

Krivanek et al. (2005). A recent study by Riaz et al. (2018a) utilized a limited mapping strategy that combines greenhouse phenotyping of small breeding populations with genotyping data from SSR markers linked to *PdR1*. The study identified nine accessions with a major QTL within the *PdR1* genomic region and three accessions whose PD resistance was not associated with *PdR1*-linked markers. Comparative sequence analysis is currently being used to determine whether the resistant accessions possess different alleles of the same candidate resistance gene and/or different genes, making them good candidates for future sequence analysis studies aimed at understanding the evolution of PD *R-genes*. Interestingly, the physical map of the *PdR1* locus has revealed a large cluster of putative LRR receptor kinase genes. This is significant because resistance to bacterial blight, a vascular disease of rice caused by the bacterium *Xanthomonas oryzae* pv. *oryzae*, is conferred by the *XA21* gene from the wild rice species *Oryza longistaminata* which encodes a LRR receptor kinase (Park et al. 2010). Transformation studies with two of these candidate LRR-RK genes are currently underway to determine if they confer resistance to PD in susceptible grapevines (Riaz et al. unpublished).

15.5.2 Crown Gall

Crown gall of grapevine is caused mainly by *Agrobacterium vitis* and occasionally by *Agrobacterium tumefaciens* and occurs in most parts of the world where grapes are grown (Kuczmog et al. 2012). Infected plants may remain symptomless until they are injured by freezing, pruning, grafting or other mechanical treatments used in maintaining the vineyard. As the gall forms, vascular bundle tissues become highly disorganized and lose their ability to transfer water and photosynthetic products. Large galls girdle the stem and result in significant grape decline and may even lead to plant death.

Cultivars of *V. vinifera* are highly susceptible to *Agrobacterium* infections and crown gall formation, but certain wild *Vitis* species, including *V. labrusca* and *V. amurensis,* have been shown to be resistant (De Cleene and De Ley 1976). Introgression of crown gall resistance from *V. amurensis* into *V. vinifera* demonstrated it to be inherited as a single dominant locus that provided resistance against both *A. vitis* and *A. tumefaciens* (Szegedi and Kozma 1984). The *R* locus, designated *Rcg1* (Resistance to crown gall 1), was found to be tightly linked to the SSR markers VVIV67, VVS16 and UDV015 on chr15. At present, there is no information available about the mechanism of *Rcg1*-mediated resistance.

15.6 Breeding for Resistance to Trunk Diseases

Grapevine trunk diseases (GTDs), which include Eutypa, Botryosphaeria and Phomopsis dieback, as well as esca, Petrie and black foot diseases, are caused by a wide and complex range of wood invading fungal species (Gramaje et al. 2018). GTDs threaten the sustainability of viticulture, with an estimated worldwide economic impact of US$1.5 billion per year based on the annual replacement of 1% of vines (Hofstetter et al. 2012).

There have been reports of varying susceptibility of *V. vinifera* cultivars to GTDs. Field surveys have reported varying levels of foliar symptoms of Eutypa dieback on a wide range of cultivars (Carter 1991; Highet and Wicks 1998; Loschiavo et al. 2007). The colonization by *Eutypa lata*, the primary causal agent of Eutypa dieback, of wood of different cultivars also varied significantly (Sosnowski et al. 2007). For species that cause Botryosphaeria dieback, lesion length varied in canes of different cultivars (Billones-Baaijens et al. 2014; Guan et al. 2016). Sosnowski et al. (2016) reported a large variation in GTD dieback symptoms on mature vines in a *V. vinifera* germplasm repository and subsequently found significant differences between cultivars in the rate of pathogen colonization of grapevine canes by *E. lata* and *Diplodia seriata* (Botryosphaeria dieback). Esca symptoms in the vineyard have been reported with varying incidence between cultivars (Marchi 2001; Fussler et al. 2008; Murolo and Romanazzi 2014). Furthermore, inoculations with *Phaeoacremonium minimum* and *Phaeomoniella chlamydospora* (casual agents of esca and Petrie disease, respectively) indicated variable susceptibility of grapevine cultivars (Feliciano et al. 2004; Landi et al. 2012). Greenhouse screening of cultivated and wild *Vitis* spp. for the length of wood discolouration by the causal agents of Botryosphaeria dieback, esca, Eutypa dieback and Phomopsis dieback revealed significant variation for all diseases (Travadon et al. 2013).

Interestingly, in contrast to observations for many other fungal pathogens of grapevine, Travadon et al. (2013) found that some of the North American grape species examined actually showed higher susceptibility to *E. lata* than most of the *V. vinifera* cultivars tested. This led them to speculate that because the centre of origin for *E. lata* is thought to be Europe (Travadon et al. 2012) that co-evolution of the pathogen and *V. vinifera* may have enriched cultivars of this species for increased resistance to *E. lata* relative to the North American *Vitis* species.

In summary, no single *V. vinifera* cultivar or *Vitis* spp. have been reported to be completely resistant to any of the grapevine trunk diseases, but variation in the expression of disease symptoms suggests differences in tolerance. Current research is exploring clonal variation within *V. vinifera* cultivars, and preliminary results are promising, with the likelihood of identifying low susceptibility germplasm for future plantings (Berlanas et al. 2017; Sosnowski 2018, unpublished data).

There is very little known about the mechanisms of resistance to GTDs. Enhanced resistance to *E. lata* toxin was reported in transgenic 'Richter 110' grapevines that were constitutively expressing a eutypine detoxyfing gene (Vr-ERE) (Legrand et al. 2003). It was believed that eutypine was responsible for foliar symptom expression; however, the most damaging symptom of trunk disease is the death of wood tissue. Relatively high lignin levels have been associated with wood and cane tissue of grapevine cultivars having less susceptibility to *E. lata* infection (Rolshausen et al. 2008; Hamblin 2015). Furthermore, cultivar susceptibility was correlated to xylem vessel diameter for esca pathogens (Pouzoulet et al. 2014) and *E. lata* (Hamblin 2015). Bertsch et al. (2013) reviewed the biochemical defence mechanisms that have been reported for GTDs, but these are yet to be specifically correlated to cultivar susceptibility.

Barba et al. (2018) recently identified two major *R* loci that reduce the severity and incidence of Phomopsis cane and cluster symptoms (caused by *Diaporthe ampelina* which also causes Phomopsis dieback). These two loci, designated *Rda1* (Resistance to *Diaporthe ampelina*) and *Rda2*, originate from *V. cinerea* B9 and 'Horizon', respectively. Grapevines with either the *Rda1* or *Rda2* locus showed either no symptoms or only small, discrete lesions compared to susceptible vines. These dominant *R* loci offer the best potential yet for the breeding of new cultivars with resistance to GTD.

To date, no single *R* locus has been identified that is capable of controlling all GTDs. This is not surprising as it would be unlikely that a single gene would be effective against all pathogens within the GTD complex. Continued efforts to identify sources of tolerance or resistance to GTD pathogens are required. The genomes of the GTD fungi *Botryosphaeria dothidea* (Joint Genomics Institute (JGI), http://1000.fungalgenomes.org), *D. seriata* (Morales-Cruz et al. 2015), *E. lata* (Blanco-Ulate et al. 2013c), *Neofusicoccum parvum* (Blanco-Ulate et al. 2013a), *P. minimum* (Blanco-Ulate et al. 2013b) and *P. chlamydospora* (Antonielli et al. 2014) have now been sequenced in their entirety. This substantially improves our ability to locate, compare and manipulate the genes associated with the mechanisms of pathogenesis and virulence in these pathogens (Morales-Cruz et al. 2015, 2018) and, ultimately, the resistance of grapevines.

15.7 Transgenic Approaches for Virus Resistance

Viruses, viroids and phytoplasmas threaten grape production and vineyard profitability by reducing vigour, yield, fruit quality and the productive lifespan of vineyards. While viruses can cause serious economic losses worldwide, phytoplasmas are only problematic in certain grape-producing regions, and viroids only have a limited detrimental effect (Maliogka et al. 2014; Wilcox et al. 2015; Dermastia et al. 2017; Mannini and Digiaro 2017; Martelli 2017).

To date, breeding efforts for resistance in grapevine have focused primarily on viruses. Conventional breeding is currently not an option because *Vitis* species with virus resistance are yet to be identified (Oliver and Fuchs 2011; Maliogka et al. 2014). Therefore, genetic engineering techniques have been employed in an attempt to generate new grapevine cultivars with resistance to viruses (Laimer et al. 2009; Maliogka et al. 2014; Saporta et al. 2016; Fuchs and Lemaire 2017).

Early on the most commonly used approach to achieve virus resistance consisted of expressing virus-derived gene constructs in susceptible *Vitis* cultivars, as an application of pathogen-derived resistance (Sanford and Johnston 1985). The coat protein (CP) gene was the most routinely used viral gene to engineer resistance (Laimer et al. 2009; Maliogka et al. 2014; Saporta et al. 2016) based on a successful application of the same approach in other perennial crops such as papaya (Gonsalves et al. 2008), plum (Scorza et al. 2016) and citrus (Soler et al. 2011). The movement protein (MP) gene was another viral gene used to confer virus resistance in *Vitis* cultivars. Several transgenic cultivars expressing *CP* or *MP* were produced including the wine grape cultivars 'Chardonnay' (Mauro et al. 1995; Dal Bosco et al. 2018), 'Nebbiolo', 'Lumassina' and 'Blaufränkish' (Gambino et al. 2005, 2010), and the table grape cultivars 'Thompson Seedless' (Scorza et al. 1996), 'Superior Seedless' (Martinelli et al. 2000) and 'Russalka' (Gölles et al. 2000; Maghuly et al. 2006). These cultivars were engineered for resistance to Arabis mosaic virus, grapevine fanleaf virus (GFLV), tomato ringspot virus, grapevine leafroll-associated virus 3, grapevine virus A and grapevine virus B (Laimer et al. 2009; Saporta et al. 2016; Dal Bosco et al. 2018). The insertion and expression of *CP* or *MP* genes, as well as the level of methylation of transgenes and their regulatory sequence elements, were extensively characterized in transgenic cultivars. However, information on virus resistance is extremely limited although transgenic 'Nebbiolo' and 'Blaufränkish' were not immune to GFLV infection following challenge inoculation by grafting (Gambino et al. 2010).

More recently, RNA interference (RNAi), a potent defence mechanism of plants against viruses (Duan et al. 2012), was applied to achieve resistance to GFLV in the table grape cultivar 'Arich Dressé' using inverted repeat *MP* constructs (Jardak-Jamoussi et al. 2009). Another RNAi method using modified microRNA (miRNAs) precursor genes was developed to express artificial miRNAs (amiRNAs) targeting the *CP* of GFLV in somatic embryos of 'Chardonnay' (Jelly et al. 2012). Yet resistance has to be reported for RNAi and amiRNAs transgenic *Vitis* cultivars.

It is anticipated that future efforts to develop virus resistance in *Vitis* cultivars will rely extensively on RNAi approaches and on genome editing techniques, as documented for 'Chardonnay' (Ren et al. 2016) and 'Thompson Seedless' (Wang et al. 2018), to target genes required for essential steps of the virus infectious cycle. Nonetheless, until resistant cultivars are available, the adoption of prophylactic measures will remain indispensable to mitigate the impact of viruses in vineyards (Maliogka et al. 2014).

Flavescence dorée (FD) is a severe epidemic disease of grapevine in Europe caused by FD-phytoplasma (FDp): a small wall-less bacteria transmitted by the leafhopper vector *Scaphoideus titanus* (Eveillard et al. 2016). During summer, infected grapevines show leaf yellowing or reddening, depending on the cultivar, downward leaf curling, drying of inflorescence and bunches, lack of cane lignification, presence of black spots on the new canes and premature leaf fall (Caudwell 1990). No evidence for resistance to FD was observed in thirteen wild North American and Chinese *Vitis* species examined, but differences in susceptibility were observed between *V. vinifera* cultivars. (Eveillard et al. 2016). Cabernet Sauvignon was found to be highly susceptible, with a high proportion of symptomatic branches and high FDp titres. Merlot, on the other, had much lower FPp titres in symptomatic branches, and this characteristic appears to have been inherited from its maternal parent, Magdeleine Noire des Charentes.

15.8 Insects/Arachnids

The most destructive and economically significant insect pests of grapevine are the root pests, root-knot nematode and phylloxera, that were discussed in Chap. 14. However, the grapevine scion is also subject to attack by a range of other insect pests including leafhoppers, grape berry moths, mealybugs, thrips and mites (member of the arachnid family) (Creasy and Creasy 2009; Wilcox et al. 2015; Thiery et al. 2018). Currently, there are no published reports detailing any variation in resistance of any grapevine species or cultivars to these insect/arachnid pests. However, a novel transgenic approach has been developed which may offer the prospect of reducing the susceptibility of grapevines to grape berry moths.

The European grapevine moth (*Lobesia botrana*), the American berry moth (*Polychrosis viteana*) and the Australian light brown apple moth (*Epiphyas postvittana*) all produce larvae that feed on grape flowers and fruits, causing direct damage as the larvae penetrate the berry and hollow out the grapes. Damage is further compounded by the invasion of damaged berries with secondary infections such as Botrytis bunch rot. Current control systems are based primarily on the use of insecticides or on mating disruption. Previous studies have shown that the European grapevine moth host uses a ratio-specific blend of three ubiquitous plant volatiles to find the grapevine host. The odour signal that attracts mated females to grape consists of the terpenoids (E)-beta-caryophyllene, (E)-beta-farnesene and (E)-4,8-dimethyl-1,3,7-nonatriene (Tasin et al. 2006). Furthermore, when the specific ratio of these compounds is disrupted there is a significant decrease in attractiveness. Building on this observation, Salvagnin and co-workers (Salvagnin et al. 2018) created stable grapevine transgenic lines which produced an altered ratio of these three terpenoids and demonstrated that these transgenic lines were less attractive to the European grapevine moth. They proposed that a strategy based on volatile ratio modification could form the basis for the development of new environmentally friendly approaches for berry moth control in grapevines.

15.9 Pyramiding Resistance Genes for Increased Durability

One of the most effective ways to increase the durability of *R-genes* in the field is to combine or 'stack' *R-genes* from different wild species, within the same plant (Mundt 2018). This is because plant resistance proteins, such as *MrRUN1* and *MrRPV1*, are activated by the recognition of effectors that are secreted into the plant cell by the invading pathogen (Dodds and Rathjen 2010). Activation of the host resistance protein initiates ETI which prevents further infection by biotrophic pathogens. Thus, if a mutation occurs in an effector that is normally recognised by the plant resistance protein, such that recognition can no longer take place, a defence response will not be initiated upon infection and the pathogen will be able to colonize the plant.

It is generally assumed that *R-genes* from different sources or regions have evolved to recognize different pathogen effector proteins, and that by combining these *R-genes* in the one genotype, the likelihood of a single pathogen isolate simultaneously mutating both effectors to overcome both resistance loci, at the same time, is significantly reduced. McDonald and Linde (2002) have highlighted the fact that the risk posed by different pathogens for breaking resistance in the field depends more on the characteristics of the pathogen, than on the *R-gene*. For example, pathogens that reproduce exclusively via asexual spores and have limited potential for spread because they are soil-borne, have the lowest risk category. In contrast, pathogens such as powdery mildew that have a mixed reproduction system and produce large amounts of asexual spores that are disseminated over long distances by wind, are in the highest risk category.

At present, the only grapevine pathogens for which sufficient *R* loci have been identified to enable a pyramiding strategy to be employed are powdery and downy mildew (Table 15.1). However, limited information is available regarding the race-specificity of these different mildew *R* loci. Feechan et al. (2015) demonstrated that the powdery mildew resistance conferred by *Run2.1* and *Ren2* is not overcome by the *Run1*-breaking Musc4. Similarly, preliminary studies with *Ren4* and *Ren6* also suggest that the resistance conferred by these two loci is not compromised by the Musc4 isolate (Dry IB, unpublished) making these *R* loci good candidates for pyramiding with *Run1*. Much more work is required to determine the race-specificity of the mildew *R* loci that have been identified.

Assuming that the *R-genes* to be employed in a pyramiding strategy do produce proteins that recognize different pathogen effectors, how many *R-genes* need to be combined within the same cultivar to ensure long-term durability in the field? This is especially important for a perennial crop, like grapevine, that is expected to be in the field for 20+ years. Stam and McDonald (2018) estimated that for a cereal powdery mildew population, the probability of two mutations occurring simultaneously within the same pathogen isolate that would enable it to overcome two *R-genes* in the plant host at the same time corresponds to ten double mutants per hectare per day. However, if three *R-genes* were present, they estimated that only one triple mutant capable of overcoming this triple resistance would be produced each day in 10,000 infected hectares. Finally, they hypothesized that a pyramid of four *R-genes* would be expected to be virtually impregnable.

Many wine grape and table grape breeding programs around the world in France (Merdinoglu et al. 2018), Germany (Eibach et al. 2007),

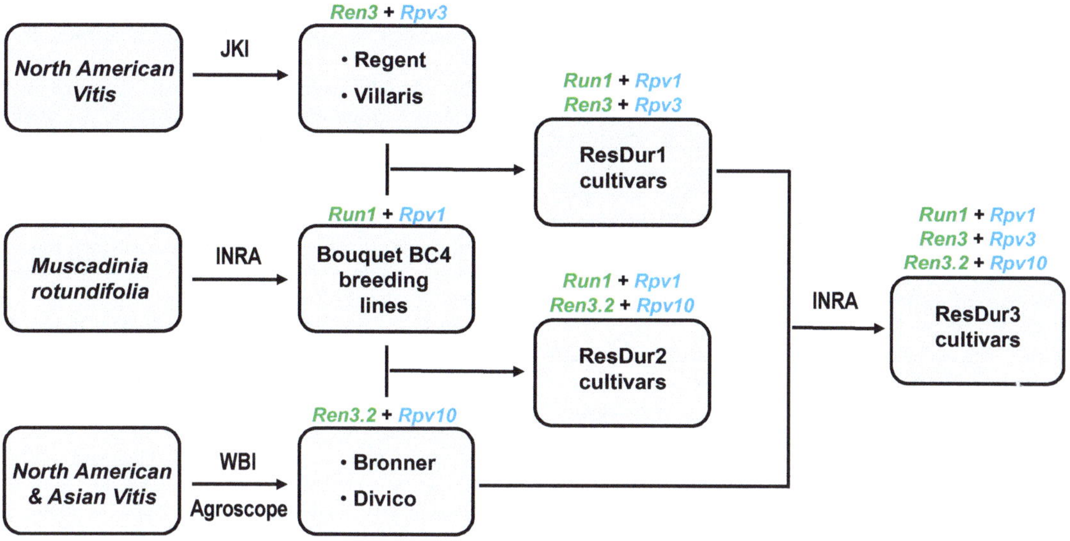

Fig. 15.4 Strategy of the INRA-ResDur breeding program. Powdery mildew resistance loci are shown in green and downy mildew resistance loci in blue. The mildew-resistant genotypes, with single powdery and downy mildew resistance loci, used in the strategy include selected Bouquet BC4 breeding lines (bred by INRA), 'Regent' and 'Villaris' (bred by the Julius Kuhn-Institut, Germany), 'Bronner' (bred by the Staatliches Weinbauinstitut, Germany) and 'Divico' (bred by Agroscope, Switzerland)

Italy (Cipriani et al. 2018), USA (Riaz et al. 2018b), Australia (Dry and Thomas 2015), Chile (Agurto et al. 2017) and Brazil (Sánchez-Mora et al. 2017) are currently developing new mildew-resistant cultivars containing at least two powdery mildew and/or downy mildew *R* loci. The most advanced of these programs is the INRA-ResDur breeding program (Merdinoglu et al. 2018) (Fig. 15.4). Four new mildew-resistant ResDur1 cultivars 'Artaban', 'Floreal', 'Vidoc' and 'Volti' containing two PM R loci (*Run1* + *Ren3*) and two DM *R* loci (*Rpv1* + *Rpv3*) have already been released. These new dual PM and DM *R* loci genotypes were created by combining the *Run1*/*Rpv1* locus from selected Bouquet backcross 4 resistant breeding selections with the *Ren3* and *Rpv3* loci from the inter-specific resistant hybrids 'Regent' and 'Villaris'. Future crosses will see the introduction of a third powdery mildew *R* locus (*Ren3.2*) and a third downy mildew *R* locus (*Rpv10*). However, the race-specificities of the different *R* loci used in this pyramiding strategy are yet to be determined.

References

Acevedo-Garcia J, Kusch S, Panstruga R (2014) Magical mystery tour: MLO proteins in plant immunity and beyond. New Phytol 204:273–281

Agurto M, Schlechter RO, Armijo G, Solano E, Serrano C, Contreras RA, Zuniga GE, Arce-Johnson P (2017) *RUN1* and *REN1* pyramiding in grapevine (*Vitis vinifera* cv. Crimson seedless) displays an improved defense response leading to enhanced resistance to powdery mildew (*Erysiphe necator*). Front Plant Sci 8:758

Alonso-Villaverde V, Boso S, Luis Santiago J, Gago P, Martínez M-C (2008) Relationship between susceptibility to Botrytis bunch rot and grape cluster morphology in the *Vitis vinifera* L. cultivar Albariño. Int J Fruit Sci 8:251–265

Alonso-Villaverde V, Voinesco F, Viret O, Spring JL, Gindro K (2011) The effectiveness of stilbenes in resistant Vitaceae: ultrastructural and biochemical events during Plasmopara viticola infection process. Plant Physiol Biochem 49:265–274

Ambers CP (2013) A historical hypothesis on the origin of the Norton grape. J Wine Res 24:85–95

Anderson C, Choisne N, Adam-Blondon A-F, Dry IB (2011) Positional cloning of disease resistance genes in grapevine. In: Kole C (ed) Genetics, genomics, and breeding of grapes. CRC Press, Versailles, pp 186–210

Antonielli L, Compant S, Strauss J, Sessitsch A, Berger H (2014) Draft genome sequence of *Phaeomoniella chlamydospora* strain RR-HG1, a grapevine trunk disease (esca)-related member of the Ascomycota. Genome Announc 2:e00098-00014

Barba P, Lillis J, Luce RS, Travadon R, Osier M, Baumgartner K, Wilcox WF, Reisch BI, Cadle-Davidson L (2018) Two dominant loci determine resistance to Phomopsis cane lesions in F1 families of hybrid grapevines. Theor Appl Genet 131:1173–1189

Barrett HC (1953) A survey of black rot resistance of the foliage of wild grape species. Proc Am Soc Hortic Sci 62:319–322

Becker T, Knoche M (2012) Water induces microcracks in the grape berry cuticle. Vitis 51:141–142

Bellin D, Peressotti E, Merdinoglu D, Wiedemann-Merdinoglu S, Adam-Blondon AF, Cipriani G, Morgante M, Testolin R, Di Gaspero G (2009) Resistance to *Plasmopara viticola* in grapevine 'Bianca' is controlled by a major dominant gene causing localised necrosis at the infection site. Theor Appl Genet 120:163–176

Berlanas C, Songy A, Clément C, Fontaine F, Gramaje D (2017) Variation amongst 'Tempranillo' clones in susceptibility to neofusicoccum parvum. Phytopathol Meditter 56:545

Bertsch C, Ramírez-Suero M, Magnin-Robert M, Larignon P, Chong J, Abou-Mansour E, Spagnolo A, Clément C, Fontaine F (2013) Grapevine trunk diseases: complex and still poorly understood. Plant Pathol 62:243–265

Billones-Baaijens R, Jones EE, Ridgway HJ, Jaspers MV (2014) Susceptiblity of common rootstock and scion varieties of grapevines to Botryosphaeriaceae species. Australas Plant Pathol 43:25–31

Blanc S, Wiedemann-Merdinoglu S, Dumas V, Mestre P, Merdinoglu D (2012) A reference genetic map of *Muscadinia rotundifolia* and identification of *Ren5*, a new major locus for resistance to grapevine powdery mildew. Theor Appl Genet 125:1663–1675

Blanco-Ulate B, Rolshausen PE, Cantu D (2013a) Draft genome sequence of *Neofusicoccum parvum* isolate UCR-NP2, a fungal vascular pathogen associated with grapevine cankers. Genome Announc 1:e00339-00313

Blanco-Ulate B, Rolshausen PE, Cantu D (2013b) Draft genome sequence of the ascomycete *Phaeoacremonium aleophilum* strain UCR-PA7, a causal agent of the esca disease complex in grapevines. Genome Announc 1:e00390-00313

Blanco-Ulate B, Rolshausen PE, Cantu D (2013c) Draft genome sequence of the grapevine dieback fungus *Eutypa lata* UCR-EL1. Genome Announc 1: e00228-00213

Blasi P, Blanc S, Wiedemann-Merdinoglu S, Prado E, Ruhl EH, Mestre P, Merdinoglu D (2011) Construction of a reference linkage map of *Vitis amurensis* and genetic mapping of *Rpv8*, a locus conferring resistance to grapevine downy mildew. Theor Appl Genet 123:43–53

Borve J, Sekse L, Stensvand A (2000) Cuticular fractures promote postharvest fruit rot in sweet cherries. Plant Dis 84:1180–1184

Bouquet A (1986) Introduction dans l'espe`ce *Vitis vinifera* L. d'uncaracte`re de re´sistance a` l'oidium (*Uncinula necator* Schw. Burr.) issu de l'espe`ce *Muscadinia rotundifolia* (Michx.) Small. Vignevini 12:141–146

Brewer MT, Milgroom MG (2010) Phylogeography and population structure of the grape powdery mildew fungus, *Erysiphe necator*, from diverse Vitis species. BMC Evolut Biol 10:268

Brunetto G, de Melo GWB, Terzano R, Del Buono D, Astolfi S, Tomasi N, Pii Y, Mimmo T, Cesco S (2016) Copper accumulation in vineyard soils: rhizosphere processes and agronomic practices to limit its toxicity. Chemosphere 162:293–307

Butault J-P, Dedryver C-A, Gary C, Guichard L, Jacquet F, Meynard JM, Nicot P, Pitrat M, Reau R, Sauphanor B, Savini I, Volay T (2010) Synthèse du rapport d'étude Écophyto R&D: quelles voies pour réduire l'usage des pesticides? INRA Editions 978-2-7380-1272-2

Cadle-Davidson L (2008) Variation within and between Vitis spp. for foliar resistance to the downy mildew pathogen *Plasmopara viticola*. Plant Dis 92:1577–1584

Cadle-Davidson L, Chicoine DR, Consolie NH (2011) Variation within and among *Vitis* spp. for foliar resistance to the powdery mildew pathogen *Erysiphe necator*. Plant Dis 95:202–211

Carter MV (1991) The status of *Eutypa lata* as a pathogen. C.A.B., Wallingford

Caudwell A (1990) Epidemiology and characterization of Flavescence doree (Fd) and other grapevine yellows. Agronomie 10:655–663

Cipriani G, Foria S, Monte C, Testolin R, Di Gaspero G (2018) Pyramidizing resistance genes in grape: a breeding program for the selection of 'elite' cultivars. Paper presented at the XIIth international grapevine breeding and genetics conference, Bordeaux, France

Clayton CN, Ridings WH (1970) Grape Rust, *Physopella ampelopsidis*, on *Vitis rotundifolia* in North Carolina. Phytopathology 60:1022

Colcol JF, Baudoin AB (2016) Sensitivity of *Erysiphe necator* and *Plasmopara viticola* in Virginia to QoI fungicides, Boscalid, Quinoxyfen, Thiophanate Methyl, and Mefenoxam. Plant Dis 100:337–344

Coleman C, Copetti D, Cipriani G, Hoffmann S, Kozma P, Kovacs L, Morgante M, Testolin R, Di Gaspero G (2009) The powdery mildew resistance gene *REN1* co-segregates with an NBS-LRR gene cluster in two Central Asian grapevines. BMC Genet 10:89

Correa J, Mamani M, Munoz-Espinoza C, Laborie D, Munoz C, Pinto M, Hinrichsen P (2014) Heritability and identification of QTLs and underlying candidate genes associated with the architecture of the grapevine cluster (*Vitis vinifera* L.). Theor Appl Genet 127:1143–1162

Crall JM, Stover LH (1957) The significance of Pierce's disease in the decline of bunch grapes in Florida. Phytopathology 47:518

Creasy G, Creasy LL (2009) Chapter 9—Grapevine pests, diseases and disorders. In: Grapes crop production science in horticulture. CABI Publishing, UK, pp 229–296

Dal Bosco D, Sinski I, Ritschel PS, Camargo UA, Fajardo TVM, Harakava R, Quecini V (2018) Expression of disease resistance in genetically modified grapevines correlates with the contents of viral sequences in the T-DNA and global genome methylation. Transgenic Res 27:379–396

Dalbó MA, Weeden NF, Reisch BI (2000) QTL analysis of disease resistance in interspecific hybrid grapes. Acta Hortic 528:215–219

Daykin ME, Milholland RD (1984) Histopathology of Ripe Rot caused by *Colletotrichum gloeosporioides* on Muscadine grape. Phytopathology 74:1339–1341

De Cleene M, De Ley J (1976) The host range of crown gall. Bot Rev 442:389–466

Delmotte F, Mestre P, Schneider C, Kassemeyer HH, Kozma P, Richart-Cervera S, Rouxel M, Deliere L (2014) Rapid and multiregional adaptation to host partial resistance in a plant pathogenic oomycete: evidence from European populations of *Plasmopara viticola*, the causal agent of grapevine downy mildew. Infect Genet Evol 27:500–508

Dermastia M, Bertaccini A, Constable F, Mehle N (2017) Grapevine yellow diseases and their phytoplasma agents: biology and Detection. Springer briefs in agriculture. Springer, Cham, pp 1–95

Di Gaspero G, Copetti D, Coleman C, Castellarin SD, Eibach R, Kozma P, Lacombe T, Gambetta G, Zvyagin A, Cindric P, Kovacs L, Morgante M, Testolin R (2012) Selective sweep at the *Rpv3* locus during grapevine breeding for downy mildew resistance. Theor Appl Genet 124:277–286

Divilov K, Barba P, Cadle-Davidson L, Reisch BI (2018) Single and multiple phenotype QTL analyses of downy mildew resistance in interspecific grapevines. Theor Appl Genet 131:1133–1143

Dodds PN, Rathjen JP (2010) Plant immunity: towards an integrated view of plant-pathogen interactions. Nat Rev Genet 11:539–548

Dry IB, Thomas MR (2015) Disease resistance: fast-tracking grape breeding for disease resistance. Wine Vitic J 30:52–55

Duan C-G, Wang C-H, Guo H-S (2012) Application of RNA silencing to plant disease resistance. Silence 3:5

Eibach R, Zyprian E, Welter L, Töpfer R (2007) The use of molecular markers for pyramiding resistance genes in grapevine breeding. Vitis 46:120–124

Elmer PAG, Michailides TJ (2004) Epidemiology of *Botrytis cinerea* in orchard and vine crops. In: Elad Y, Williamson B, Tudzynski P, Delen N (eds) Botrytis: biology, pathology and control. Kluwer Academic Publishers, Dordrecht, pp 243–272

Esteve-Turrillas FA, Agullo C, Abad-Somovilla A, Mercader JV, Abad-Fuentes A (2016) Fungicide multiresidue monitoring in international wines by immunoassays. Food Chem 196:1279–1286

Eveillard S, Jollard C, Labroussaa F, Khalil D, Perrin M, Desque D, Salar P, Razan F, Hevin C, Bordenave L, Foissac X, Masson JE, Malembic-Maher S (2016) Contrasting susceptibilities to Flavescence doree in *Vitis vinifera*, rootstocks and wild Vitis species. Front Plant Sci 7:1762

Feechan A, Jermakow AM, Torregrosa L, Panstruga R, Dry IB (2008) Identification of grapevine *MLO* gene candidates involved in susceptibility to powdery mildew. Funct Plant Biol 35:1255–1266

Feechan A, Kabbara S, Dry IB (2011) Mechanisms of powdery mildew resistance in the Vitaceae family. Mol Plant Path 12:263–274

Feechan A, Anderson C, Torregrosa L, Jermakow A, Mestre P, Wiedemann-Merdinoglu S, Merdinoglu D, Walker AR, Cadle-Davidson L, Reisch B, Aubourg S, Bentahar N, Shrestha B, Bouquet A, Adam-Blondon AF, Thomas MR, Dry IB (2013) Genetic dissection of a TIR-NB-LRR locus from the wild North American grapevine species *Muscadinia rotundifolia* identifies paralogous genes conferring resistance to major fungal and oomycete pathogens in cultivated grapevine. Plant J 76:661–674

Feechan A, Kocsis M, Riaz S, Zhang W, Gadoury DM, Walker MA, Dry IB, Reisch B, Cadle-Davidson L (2015) Strategies for *RUN1* deployment using *RUN2* and *REN2* to manage grapevine powdery mildew informed by studies of race specificity. Phytopathology 105:1104–1113

Feliciano AJ, Eskalen A, Gubler WD (2004) Differential susceptibility of three grapevine cultivars to *Phaeoacremonium aleophilum* and *Phaeomoniella chlamydospora* in California. Phytopathol Mediterr 43:66–69

Finkers R, van den Berg P, van Berloo R, ten Have A, van Heusden AW, van Kan JAL, Lindhout P (2007) Three QTLs for *Botrytis cinere*a resistance in tomato. Theor Appl Genet 114:585–593

Fischer BM, Salakhutdinov I, Akkurt M, Eibach R, Edwards KJ, Topfer R, Zyprian EM (2004) Quantitative trait locus analysis of fungal disease resistance factors on a molecular map of grapevine. Theor Appl Genet 108:501–515

Foria S, Magris G, Morgante M, Di Gaspero G (2018) The genetic background modulates the intensity of *Rpv3*-dependent downy mildew resistance in grapevine. Plant Breed 137:220–228

Fritschi FB, Lin H, Walker MA (2007) *Xylella fastidiosa* population dynamics in grapevine genotypes differing in susceptibility to Pierce's disease. Am J Enol Vitic 58:326–332

Fu YQ, van Silfhout A, Shahin A, Egberts R, Beers M, van der Velde A, van Houten A, van Tuyl JM, Visser RGF, Arens P (2017) Genetic mapping and QTL analysis of Botrytis resistance in *Gerbera hybrida*. Mol Breed 37:13

Fuchs M, Lemaire O (2017) Novel approaches for virus disease management. In: Meng B, Martelli GP,

Golino DA, Fuchs MF (eds) Grapevine viruses: molecular biology, diagnostics and management. Springer, Berlin, pp 599–621

Fussler L, Kobes N, Bertrand F, Mauray M, Grosman J, Savary S (2008) A characterization of grapevine trunk diseases in France from data generated by the National Grapevine Wood Diseases Survey. Phytopathology 98:571–579

Gabler FM, Smilanick JL, Mansour M, Ramming DW, Mackey BE (2003) Correlations of morphological, anatomical, and chemical features of grape berries with resistance to *Botrytis cinerea*. Phytopathology 93:1263–1273

Gadino AN, Walton VM, Dreves AJ (2011) Impact of vineyard pesticides on a beneficial arthropod, *Typhlodromus pyri* (Acari: Phytoseiidae), in laboratory bioassays. J Econ Entomol 104:970–977

Gadoury DM, Cadle-Davidson L, Wilcox WF, Dry IB, Seem RC, Milgroom MG (2012) Grapevine powdery mildew (*Erysiphe necator*): a fascinating system for the study of the biology, ecology and epidemiology of an obligate biotroph. Mol Plant Pathol 13:1–16

Gambino G, Gribaudo I, Leopold S, Schartl A, Laimer M (2005) Molecular characterization of grapevine plants transformed with GFLV resistance genes: I. Plant Cell Rep 24:655–662

Gambino G, Perrone I, Carra A, Chitarra W, Boccacci P, Marinoni DT, Barberis M, Maghuly F, Laimer M, Gribaudo I (2010) Transgene silencing in grapevines transformed with GFLV resistance genes: analysis of variable expression of transgene, siRNAs production and cytosine methylation. Transgenic Res 19:17–27

Gao YR, Han YT, Zhao FL, Li YJ, Cheng Y, Ding Q, Wang YJ, Wen YQ (2016) Identification and utilization of a new *Erysiphe necator* isolate NAFU1 to quickly evaluate powdery mildew resistance in wild Chinese grapevine species using detached leaves. Plant Physiol Biochem 98:12–24

Gessler C (2011) *Plasmopora viticola*: a review of knowledgr on downy mildew of grapevine and effective disease management. Phytopathol Mediterr 50:3–44

Glazebrook J (2005) Contrasting mechanisms of defense against biotrophic and necrotrophic pathogens. Annu Rev Phytopathol 43:205–227

Gölles R, Moser R, Puhringer H, Katinger H, Da Camara Laimer, Machado M, Minafra A, Savino V, Saldarelli P, Da Camara Machado A (2000) Transgenic grapevines expressing coat protein gene sequences of grapevine fanleaf virus, arabisa mosaic virus, grapevine virus A and grapevine virus B. Acta Hortic 528:305–311

Gonsalves D, Ferreira S, Suzuki J, Tripathi S (2008) Papaya. Compendium of transgenic crop plants. Blackwell Publishing, Oxford

Gramaje D, Urbez-Torres JR, Sosnowski MR (2018) Managing grapevine trunk diseases with respect to etiology and epidemiology: current strategies and future prospects. Plant Dis 102:12–39

Guan X, Essakhi S, Laloue H, Nick P, Bertsch C, Chong J (2016) Mining new resources for grape resistance against Botryosphaeriaceae: a focus on *Vitis vinifera* subsp. sylvestris. Plant Pathol 65:273–284

Gururani MA, Venkatesh J, Upadhyaya CP, Nookaraju A, Pandey SK, Park SW (2012) Plant disease resistance genes: current status and future directions. Physiol Mol Plant Pathol 78:51–65

Halbrooks MC, Mortensen JA (1989) Origin and significance of Florida hybrid bunch grapes and rootstocks. HortScience 24:546–550

Hamblin J (2015) Factors affecting grapevine susceptibility to Eutypa dieback. University of Adelaide, Adelaide

Hennessy CR, Daly AM, Hearnden MN (2007) Assessment of grapevine cultivars for resistance to *Phakopsora euvitis*. Aust Plant Pathol 36:313–317

Herzog K, Wind R, Topfer R (2015) Impedance of the grape berry cuticle as a novel phenotypic trait to estimate resistance to *Botrytis cinerea*. Sensors (Basel) 15:12498–12512

Hewitt WB (1958) The probable home of Pierce's disease virus. Am J Enol Vitic 9:94–98

Highet A, Wicks T (1998) The incidence of eutypa dieback in South Australian vineyards. Aust N Z Grapegrower Winemaker 414:135–136

Hoffmann S, Di Gaspero G, Kovacs L, Howard S, Kiss E, Galbacs Z, Testolin R, Kozma P (2008) Resistance to *Erysiphe necator* in the grapevine 'Kishmish vatkana' is controlled by a single locus through restriction of hyphal growth. Theor Appl Genet 116:427–438

Hofstetter V, Buyck B, Croll D, Viret O, Couloux A, Gindro K (2012) What if esca disease of grapevine were not a fungal disease? Fungal Divers 54:51–67

Hopkins DL (1989) *Xylella fastidiosa*—Xylem-limited bacterial pathogen of plants. Annu Rev Phytopathol 27:271–290

Hopkins DL, Purcell AH (2002) *Xylella fastidiosa*: cause of Pierce's disease of grapevine and other emergent diseases. Plant Dis 86:1056–1066

Huang P, Xiu J, Zhang H, Guo X, Li C, Guo Z, Qian W (1988) A preliminary report on hermaphrodites *Vitis amurensis* cv ShuangYou. Acta Agric Univ Jilinensis 10:31–33

Hwang C-F, Sapkota S, Chen L-L, Yang S, Cadle-Davidson L (2018) QTL mapping of downy mildew and botrytis bunch rot resistance in a *Vitis aestivalis*-derived 'Norton'-based population. Paper presented at the XIIth international grapevine breeding and genetics conference, Bordeaux, France

Jang MH, Moon YS, Noh JH, Kim SH, Hong SK, Yun HK (2011) In vitro evaluation system for varietal resistance against Ripe rot caused by *Colletotrichum acutatum* in grapevines. Hortic Environ Biotechnol 52:52–57

Jardak-Jamoussi R, Winterhagen P, Bouamama B, Dubois C, Mliki A, Wetzel T, Ghorbel A, Reustle GM (2009) Development and evaluation of a GFLV inverted repeat construct for genetic transformation of grapevine. Plant Cell Tissue Organ 97:187–196

Jelly NS, Schellenbaum P, Walter B, Maillot P (2012) Transient expression of artificial microRNAs targeting *Grapevine fanleaf virus* and evidence for RNA silencing in grapevine somatic embryos. Transgenic Res 21:1319–1327

Kim GH, Yun HK, Choi CS, Park JH, Jung YJ, Park KS, Dane F, Kang KK (2008) Identification of AFLP and RAPD markers linked to anthracnose resistance in grapes and their conversion to SCAR markers. Plant Breed 127:418–423

Komarek M, Cadkova E, Chrastny V, Bordas F, Bollinger JC (2010) Contamination of vineyard soils with fungicides: a review of environmental and toxicological aspects. Environ Int 36:138–151

Krivanek AF, Walker MA (2005) Vitis resistance to Pierce's disease is characterized by differential *Xylella fastidiosa* populations in stems and leaves. Phytopathology 95:44–52

Krivanek AF, Stevenson JF, Walker MA (2005) Development and comparison of symptom indices for quantifying grapevine resistance to Pierce's disease. Phytopathology 95:36–43

Krivanek A, Riaz S, Walker MA (2006) Identification and molecular mapping of *PdR1*, a primary resistance gene to Pierce's disease in Vitis. Theor Appl Genet 112:1125–1131

Kuczmog A, Galambos A, Horvath S, Matai A, Kozma P, Szegedi E, Putnoky P (2012) Mapping of crown gall resistance locus *Rcg1* in grapevine. Theor Appl Genet 125:1565–1574

Kusch S, Panstruga R (2017) mlo-based resistance: an apparently universal "weapon" to defeat powdery mildew disease. Mol Plant Microbe Interact 30:179–189

Kyrkou I, Pusa T, Ellegaard-Jensen L, Sagot MF, Hansen LH (2018) Pierce's disease of grapevines: a review of control strategies and an outline of an epidemiological model. Front Microbiol 9:2141

Laimer M, Lemaire O, Herrbach E, Goldschmidt V, Minafra A, Bianco P, Wetzel T (2009) Resistance to viruses, phytoplasmas and their vectors in the grapevine in Europe: a review. J Plant Pathol 91:7–23

Landi L, Murolo S, Romanazzi G (2012) Colonization of *Vitis* spp. wood by sGFP-transformed *Phaeomoniella chlamydospora*, a tracheomycotic fungus involved in esca disease. Phytopathology 102:290–297

Le Moal J, Rolland M, Goria S, Wagner V, De Crouy-Chanel P, Rigou A, De Mouzon J, Royere D (2014) Semen quality trends in French regions are consistent with a global change in environmental exposure. Reproduction 147:567–574

Legrand V, Dalmayrac S, Latche A, Pech JC, Bouzayen M, Fallot J, Torregrosa L, Bouquet A, Roustan JP (2003) Constitutive expression of Vr-ERE gene in transformed grapevines confers enhanced resistance to eutypine, a toxin from *Eutypa lata*. Plant Sci 164:809–814

Lin H, Leng H, Guo Y, Kondo S, Zhao Y, Shi G, Guo X (2019) QTLs and candidate genes for downy mildew resistance conferred by interspecific grape (*V. vinifera* L. × *V. amurensis* Rupr.) crossing. Sci Hortic 244:200–207

Loomis NH (1958) Performance of *Vitis* species in the south as an indication of their relative resistance to Pierce's disease. Plant Dis Rep 42:833–836

Loschiavo A, Sosnowski M, Wicks T (2007) Incidence of eutypa dieback in the Adelaide hills. Aust NZ Grapegrower Winemaker 519:26–29

Maghuly F, Leopold S, Machado AD, Fernandez EB, Khan MA, Gambino G, Gribaudo I, Schartl A, Laimer M (2006) Molecular characterization of grapevine plants transformed with GFLV resistance genes: II. Plant Cell Rep 25:546–553

Mahanil S, Ramming D, Cadle-Davidson M, Owens C, Garris A, Myles S, Cadle-Davidson L (2012) Development of marker sets useful in the early selection of *Ren4* powdery mildew resistance and seedlessness for table and raisin grape breeding. Theor Appl Genet 124:23–33

Maliogka V, Martelli GP, Fuchs M, Katis N (2014) Control of viruses infecting grapevine. Adv Virus Res 91:175–227

Mannini F, Digiaro M (2017) The effects of viruses and viral diseases on grapes and wine. In: Meng B, Martelli GP, Golino DA, Fuchs M (eds) Grapevine viruses: molecular biology, diagnostics and management. Springer, Berlin

Marchi G (2001) Susceptibility to esca of various grapevine (*Vitis vinifera*) cultivars grafted on different rootstocks in a vineyard in the province of Siena (Italy). Phytopathol Mediterr 40:27–36

Marguerit E, Boury C, Manicki A, Donnart M, Butterlin G, Nemorin A, Wiedemann-Merdinoglu S, Merdinoglu D, Ollat N, Decroocq S (2009) Genetic dissection of sex determinism, inflorescence morphology and downy mildew resistance in grapevine. Theor Appl Genet 118:1261–1278

Martelli GP (2017) An overview of grapevine viruses, viroids and the diseases they cause. In: Meng B, Martelli GP, Golino DA, Fuchs MF (eds) Grapevine viruses: molecular biology, diagnostics and management. Springer, Berlin, pp 31–46

Martinelli L, Buzkhan N, Minafra A, Saldarelli P, Costa D, Poletti V, Festi S, Perl A, Martelli GP (2000) Genetic transformation of tobacco and grapevines for resistance to viruses related to the rugose wood disease complex. Acta Hortic 528:321–327

Mauro MC, Toutain S, Walter B, Pinck L, Otten L, Coutos-Thevenot P, Deloire A, Barbier P (1995) High efficiency regeneration of grapevine plants transformed with the GFLV coat protein gene. Plant Sci 112:97–106

McDonald BA, Linde C (2002) Pathogen population genetics, evolutionary potential and durable resistance. Annu Rev Phytopathol 40:349–379

Mejlhede N, Kyjovska Z, Backes G, Burhenne K, Rasmussen SK, Jahoor A (2006) EcoTILLING for the identification of allelic variation in the powdery mildew resistance genes *mlo* and *Mla* of barley. Plant Breed 125:461–467

Merdinoglu D, Wiedemann-Merdinoglu S, Coste P, Dumas V, Haetty S, Butterlin G, Greif C (2003) Genetic analysis of downy mildew resistance derived from Muscadinia rotundifolia. Acta Hortic 603:451–456

Merdinoglu D, Schneider C, Prado E, Wiedemann-Merdinoglu S, Mestre P (2018) Breeding for durable resistance to downy and powdery mildew in grapevine. OENO One 52:189–195

Morales-Cruz A, Amrine KCH, Blanco-Ulate B, Lawrence DP, Travadon R, Rolshausen PE, Baumgartner K, Cantu D (2015) Distinctive expansion of gene families associated with plant cell wall degradation, secondary metabolism, and nutrient uptake in the genomes of grapevine trunk pathogens. BMC Genom 16:469

Morales-Cruz A, Allenbeck G, Figueroa-Balderas R, Ashworth VE, Lawrence DP, Travadon R, Smith RJ, Baumgartner K, Rolshausen PE, Cantu D (2018) Closed-reference metatranscriptomics enables in planta profiling of putative virulence activities in the grapevine trunk disease complex. Mol Plant Pathol 19:490–503

Moreira FM, Madini A, Marino R, Zulini L, Stefanini M, Velasco R, Kozma P, Grando MS (2010) Genetic linkage maps of two interspecific grape crosses (*Vitis* spp.) used to localize quantitative trait loci for downy mildew resistance. Tree Genet Genomes 7:153–167

Mortensen JA (1968) The inheritance of resistance to Pierce's disease in Vitis. J Am Soc Hortic Sci 92:331–337

Mortensen JA (1981) Sources and inheritance of resistance to Anthracnose in Vitis. J Hered 72:423–426

Mortensen JA (1988) Blanc Du Bois grape. HortScience 23:418–419

Mortensen JA, Stover LH, Balerdi CF (1977) Sources of resistance to Pierce's disease in Vitis. J Am Soc Hortic Sci 102:695–697

Mundt CC (2018) Pyramiding for resistance durability: theory and practice. Phytopathology 108:792–802

Murolo S, Romanazzi G (2014) Effects of grapevine cultivar, rootstock and clone on esca disease. Australas Plant Pathol 43:215–221

Naegele RP (2018) Evaluation of host resistance to Botrytis bunch rot in Vitis spp. and its correlation with botrytis leaf spot. HortScience 53:204–207

Newman KL, Almeida RPP, Purcell AH, Lindow SE (2003) Use of a green fluorescent strain for analysis of *Xyella fastidiosa* colonization of *Vitis vinifera*. Appl Environ Microb 69:7319–7327

Ochssner I, Hausmann L, Topfer R (2016) Rpv14, a new genetic source for Plasmopara viticola resistance conferred by *Vitis cinerea*. Vitis 55:79–81

Oliver JE, Fuchs M (2011) Tolerance and resistance to viruses and their vectors in Vitis sp.: a virologist's perspective of the literature. Am J Enol Vitic 62:438–451

Panstruga R (2005) Discovery of novel conserved peptide domains by ortholog comparison within plant multi-protein families. Plant Mol Biol 59:485–500

Pap D, Riaz S, Dry IB, Jermakow A, Tenscher AC, Cantu D, Olah R, Walker MA (2016) Identification of two novel powdery mildew resistance loci, *Ren6* and *Ren7*, from the wild Chinese grape species *Vitis piasezkii*. BMC Plant Biol 16:170

Park CJ, Han SW, Chen XW, Ronald PC (2010) Elucidation of XA21-mediated innate immunity. Cell Microbiol 12:1017–1025

Patil S, Honrao B, Karmkar S (1998) Reaction of some grape germplasm against the rust diseases. J Maharashtra Agric Univ 23:138–140

Pauquet J, Bouquet A, This P, Adam-Blondon AF (2001) Establishment of a local map of AFLP markers around the powdery mildew resistance gene *Run1* in grapevine and assessment of their usefulness for marker assisted selection. Theor Appl Genet 103:1201–1210

Pedneault K, Provost C (2016) Fungus resistant grape varieties as a suitable alternative for organic wine production: benefits, limits, and challenges. Sci Hortic 208:57–77

Peressotti E, Wiedemann-Merdinoglu S, Delmotte F, Bellin D, Di Gaspero G, Testolin R, Merdinoglu D, Mestre P (2010) Breakdown of resistance to grapevine downy mildew upon limited deployment of a resistant variety. BMC Plant Biol 10:147

Perry RL, Mollenhauer HH, Bowen HH (1974) Electron photomicroscopy verification of Pierce's disease on grape plants from Texas. Plant Dis Rep 58:780–782

Pessina S, Lenzi L, Perazzolli M, Campa M, Dalla Costa L, Urso S, Vale G, Salamini F, Velasco R, Malnoy M (2016) Knockdown of *MLO* genes reduces susceptibility to powdery mildew in grapevine. Hortic Res 3:16016

Pezet R, Gindro K, Viret O, Richter H (2004a) Effects of resveratrol, viniferins and pterostilbene on *Plasmopara viticola* zoospore mobility and disease development. Vitis 43:145–148

Pezet R, Gindro K, Viret O, Spring JL (2004b) Glycosylation and oxidative dimerization of resveratrol are respectively associated to sensitivity and resistance of grapevine cultivars to downy mildew. Physiol Mol Plant Pathol 65:297–303

Pierce N (1892) The California vine disease: a preliminary report of investigations. US Dept of Agriculture, Bulletin No. 2, Washington, MD

Pouzoulet J, Pivovaroff AL, Santiago LS, Rolshausen PE (2014) Can vessel dimension explain tolerance toward fungal vascular wilt diseases in woody plants? Lessons from Dutch elm disease and esca disease in grapevine. Front Plant Sci 5:253

Raanan R, Gunier RB, Balmes JR, Beltran AJ, Harley KG, Bradman A, Eskenazi B (2017) Elemental sulfur use and associations with pediatric lung function and respiratory symptoms in an agricultural community (California, USA). Environ Health Perspect 125:087007

Raju BC, Goheen AC, Teliz D, Nyland G (1979) Occurrence of Pierce's disease of grapevines in Mexico. Phytopathology 69:919

Ramming DW, Gabler F, Smilanick J, Cadle-Davidson M, Barba P, Mahanil S, Cadle-Davidson L (2011) A single dominant locus, ren4, confers rapid non-race-specific resistance to grapevine powdery mildew. Phytopathology 101:502–508

Ren C, Liu XJ, Zhang Z, Wang Y, Duan W, Li SH, Liang ZC (2016) CRISPR/Cas9-mediated efficient targeted mutagenesis in Chardonnay (*Vitis vinifera* L.). Sci Rep 6:32289

Rex F, Fechter I, Hausmann L, Topfer R (2014) QTL mapping of black rot (*Guignardia bidwellii*) resistance in the grapevine rootstock 'Borner' (*V. riparia* Gm183 × *V. cinerea* Arnold). Theor Appl Genet 127:1667–1677

Riaz S, Krivanek AF, Xu K, Walker MA (2006) Refined mapping of the Pierce's disease resistance locus, *PdR1*, and *Sex* on an extended genetic map of *Vitis rupestris* × *V arizonica*. Theor Appl Genet 113:1317–1329

Riaz S, Vezzulli S, Harbertson ES, Walker MA (2007) Use of molecular markers to correct grape breeding errors and determine the identity of novel sources of resistance to *Xiphinema index* and Pierce's disease. Am J Enol Vitic 58:494–498

Riaz S, Tenscher AC, Graziani R, Walker MA (2008) Using marker-assisted selection to breed for Pierce's disease resistance in grape. Am J Enol Vitic 59:341a–341a

Riaz S, Tenscher AC, Graziani R, Walker MA (2009) Breeding winegrapes with resistance to Pierce's disease. Am J Enol Vitic 60:388a–389a

Riaz S, Tenscher AC, Ramming DW, Walker MA (2011) Using a limited mapping strategy to identify major QTLs for resistance to grapevine powdery mildew (*Erysiphe necator*) and their use in marker-assisted breeding. Theor Appl Genet 122:1059–1073

Riaz S, Boursiquot JM, Dangl GS, Lacombe T, Laucou V, Tenscher AC, Walker MA (2013) Identification of mildew resistance in wild and cultivated Central Asian grape germplasm. BMC Plant Biol 13:149

Riaz S, Huerta-Acosta K, Tenscher AC, Walker MA (2018a) Genetic characterization of Vitis germplasm collected from the southwestern US and Mexico to expedite Pierce's disease-resistance breeding. Theor Appl Genet 131:1589–1602

Riaz S, Pap D, Tenscher A, Walker A (2018b) Durable powdery mildew resistance in grapevines: myth or reality. Paper presented at the XIIth International Grapevine Breeding and Genetics Conference, Bordeaux, France

Ribéreau-Gayon J, Ribéreau-Gayon P, Seguin G (1980) *Botrytis cinerea* in enology. In: Coley-Smith JR, Verhoff K, Jarvis WR (eds) The biology of Botrytis. Academic Press, London, pp 251–274

Richter R, Rossmann S, Töpfer R, Theres K, Zyprian E (2017) Genetic analysis of loose cluster architecture in grapevine. BIO Web Conf 9:01016

Robinson J, Harding J, Vouillamoz J (2012) Wine grapes: a complete guide to 1,368 vine varieties, including their origins and flavours. Penguin Books Ltd, London

Rolshausen PE, Greve LC, Labavitch JM, Mahoney NE, Molyneux RJ, Gubler WD (2008) Pathogenesis of *Eutypa lata* in grapevine: Identification of virulence factors and biochemical characterization of cordon dieback. Phytopathology 98:222–229

Rowe HC, Kliebenstein DJ (2008) Complex genetics control natural variation in *Arabidopsis thaliana* resistance to *Botrytis cinerea*. Genetics 180:2237–2250

Roznki D, Hoffmann S, Kozma P (2017) Screening a large set of grape accessions for resistance against black rot (*Guignardia bidwellii*/(Ell.)). Mitteilungen Klosterneuburg, Rebe und Wein, Obstbau und Früchteverwertung 67:149–157

Salvagnin U, Malnoy M, Thoming G, Tasin M, Carlin S, Martens S, Vrhovsek U, Angeli S, Anfora G (2018) Adjusting the scent ratio: using genetically modified *Vitis vinifera* plants to manipulate European grapevine moth behaviour. Plant Biotechnol J 16:264–271

Sánchez-Mora FD, Saifert L, Zanghelini J, Assumpção WT, Guginski-Piva CA, Giacometti R, Novak EI, Klabunde GH, Eibach R, Dal Vesco L, Nodari RO, Welter LJ (2017) Behavior of grape breeding lines with distinct resistance alleles to downy mildew (*Plasmopara viticola*). Crop Breed Appl Biotechnol 17:141–149

Sanford JC, Johnston SA (1985) The concept of parasite-derived resistance—deriving resistance genes from the parasites own genome. J Theor Biol 113:395–405

Sapkota S, Chen L-L, Schreiner K, Ge H, Hwang C-F (2015) A phenotypic study of Botrytis bunch rot resistance in *Vitis aestivalis*-derived 'Norton' grape. Trop Plant Pathol 40:279–282

Saponari M, Boscia D, Nigro F, Martelli GP (2013) Identification of DNA sequences related to *Xylella fastidiosa* in Oleander, Almond and Olive trees exhibiting leaf scorch symptoms in Apulia (Southern Italy). J Plant Pathol 95:668

Saporta R, San Pedro T, Gisbert C (2016) Attempts at grapevine (*Vitis vinifera* L.) breeding through genetic transformation: the main limiting factors. Vitis 55:173–186

Schwander F, Eibach R, Fechter I, Hausmann L, Zyprian E, Topfer R (2012) *Rpv10*: a new locus from the Asian Vitis gene pool for pyramiding downy mildew resistance loci in grapevine. Theor Appl Genet 124:163–176

Scorza R, Cordts JM, Gray DJ, Gonsalves D, Emershad RL, Ramming DW (1996) Producing transgenic 'Thompson Seedless' grape (*Vitis vinifera* L) plants. J Am Soc Hortic Sci 121:616–619

Scorza R, Ravelonandro M, Callahan A, Zagrai I, Polak J, Malinowski T, Cambra M, Levy L, Damsteegt V, Krska B, Cordts J, Gonsalves D, Dardick C (2016) 'HoneySweet' (C5), the first genetically engineered plum pox virus-resistant Plum (*Prunus domestica* L.) cultivar. HortScience 51:601–603

Shavrukov YN, Dry IB, Thomas MR (2004) Inflorescence and bunch architecture development in *Vitis vinifera* L. Aust J Grape Wine Res 10:116–124

Smithyman RP, Howell GS, Miller DP (1998) The use of competition for carbohydrates among vegetative and reproductive sinks to reduce fruit set and Botrytis bunch rot in Seyval blanc grapevines. Am J Enol Vitic 49:163–170

Soler N, Fagoaga C, Chiibi S, López C, Moreno P, Navarro L, Flores R, Peña L (2011) RNAi-mediated protection against *Citrus tristeza virus* in transgenic *Citrus* plants non coding RNAs in Plants, RNA technologies. Springer, Berlin

Song R, Lu W, Wang J, Shen Y, Shi G, Li W (1998) A new grapevine variety of *Vitis amurensis*. China Fruits 4:5–7

Song S, Fu P, Lu J (2018) Downy mildew resistant QTLs in *Vitis amurensis* 'Shuang Hong' grapevine. Paper presented at the XIIth International Grapevine Breeding and Genetics Conference, Bordeaux, France

Sosnowski MR, Lardner R, Wicks TJ, Scott ES (2007) The influence of grapevine cultivar and isolate of *Eutypa lata* on wood and foliar symptoms. Plant Dis 91:924–931

Sosnowski M, Ayres M, McCarthy M, Wicks T, Scott E (2016) Investigating the potential for resistance to grapevine trunk diseases. Wine Vitic J 31:41–45

Stam R, McDonald BA (2018) When resistance gene pyramids are not durable—the role of pathogen diversity. Mol Plant Pathol 19:521–524

Staudt G (1997) Evaluation of resistance to grapevine powdery mildew (*Uncinula necator* [Schw.] Burr., anamorph *Oidium tuckeri* Berk.) in accessions of Vitis species. Vitis 36:151–154

Staudt G, Kassemeyer HH (1995) Evaluation of downy mildew resistance in various accessions of wild Vitis species. Vitis 34:225–228

Stoner WN (1953) Leafhopper transmission of a degeneration of grape in Florida and its relation to Pierce's disease. Phytopathology 43:611–615

Szegedi E, Kozma P (1984) Studies on the inheritance of resistance to Crown Gall disease of grapevine. Vitis 23:121–126

Tasin M, Backman AC, Bengtsson M, Ioriatti C, Witzgall P (2006) Essential host plant cues in the grapevine moth. Naturwissenschaften 93:141–144

Teh SL, Fresnedo-Ramirez J, Clark MD, Gadoury DM, Sun Q, Cadle-Davidson L, Luby JJ (2017) Genetic dissection of powdery mildew resistance in interspecific half-sib grapevine families using SNP-based maps. Mol Breed 37:1

Tello J, Ibanez J (2018) What do we know about grapevine bunch compactness? A state-of-the-art review. Aust J Grape Wine Res 24:6–23

Tello J, Aguirrezabal R, Hernaiz S, Larreina B, Montemayor MI, Vaquero E, Ibanez J (2015) Multicultivar and multivariate study of the natural variation for grapevine bunch compactness. Aust J Grape Wine Res 21:277–289

Tello J, Torres-Perez R, Grimplet J, Ibanez J (2016) Association analysis of grapevine bunch traits using a comprehensive approach. Theor Appl Genet 129:227–242

Thiery D, Louapre P, Muneret L, Rusch A, Sentenac G, Vogelweith F, Iltis C, Moreau J (2018) Biological protection against grape berry moths. A review. Agron Sustain Dev 38:15

Tobias PA, Guest DI (2014) Tree immunity: growing old without antibodies. Trends Plant Sci 19:367–370

Travadon R, Baumgartner K, Rolshausen PE, Gubler WD, Sosnowski MR, Lecomte P, Halleen F, Peros JP (2012) Genetic structure of the fungal grapevine pathogen *Eutypa lata* from four continents. Plant Pathol 61:85–95

Travadon R, Rolshausen PE, Gubler WD, Cadle-Davidson L, Baumgartner K (2013) Susceptibility of cultivated and wild Vitis spp. to wood infection by fungal trunk pathogens. Plant Dis 97:1529–1536

Vail ME, Marois JJ (1991) Grape cluster architecture and the susceptibility of berries to *Botrytis cinerea*. Phytopathology 81:188–191

Vail ME, Wolpert JA, Gubler WD, Rademacher MR (1998) Effect of cluster tightness on Botrytis bunch rot in six Chardonnay clones. Plant Dis 82:107–109

Venuti S, Copetti D, Foria S, Falginella L, Hoffmann S, Bellin D, Cindric P, Kozma P, Scalabrin S, Morgante M, Testolin R, Di Gaspero G (2013) Historical introgression of the downy mildew resistance gene *Rpv12* from the Asian species *Vitis amurensis* into grapevine varieties. PLoS ONE 8:e61228

Vezzulli S, Malacarne G, Masuero D, Vecchione A, Haile ZM, Banchi E, Velasco R, Stefanini M, Vhrovsek U, Zulini L, Franceschi P, Moser C (2018) The *Rpv3-3* locus and stilbenoid induction mediate downy mildew resistance in a grapevine inter-specific population. In: XIIth international grapevine breeding and genetics conference, Bordeaux, France, 2018

Wan Y, Schwaniniger H, He P, Wang Y (2007) Comparison of resistance to powdery mildew and downy mildew in Chinese wild grapes. Vitis 46:132–136

Wan R, Hou X, Wang X, Qu J, Singer SD, Wang Y, Wang X (2015) Resistance evaluation of Chinese wild Vitis genotypes against *Botrytis cinerea* and different responses of resistant and susceptible hosts to the infection. Front Plant Sci 6:854

Wang Y, Liu Y, He P, Lamikanra O, Lu J (1998) Resistance of Chinese Vitis species to *Elsinoe ampelina* (de Bary) Shear. HortScience 33:123–126

Wang YP, Cheng X, Shan QW, Zhang Y, Liu JX, Gao CX, Qiu JL (2014) Simultaneous editing of three homoeoalleles in hexaploid bread wheat confers heritable resistance to powdery mildew. Nat Biotechnol 32:947–951

Wang XH, Tu MX, Wang DJ, Liu JW, Li YJ, Li Z, Wang YJ, Wang XP (2018) CRISPR/Cas9-mediated efficient targeted mutagenesis in grape in the first generation. Plant Biotechnol J 16:844–855

Welter LJ, Gokturk-Baydar N, Akkurt M, Maul E, Eibach R, Topfer R, Zyprian EM (2007) Genetic mapping and localization of quantitative trait loci affecting fungal disease resistance and leaf morphology in grapevine (*Vitis vinifera* L). Mol Breed 20:359–374

Wilcox WF, Gubler WD, Uyemoto JK (2015) Diseases caused by biotic factors. In: Compendium of grape diseases, disorders, and pests, 2nd edn. APS Press, St Paul, MN, pp 17–146

Zabadal TJ, Dittmer TW (1998) Vine management systems affect yield, fruit quality, cluster compactness, and fruit rot of 'Chardonnay' grape. HortScience 33:806–809

Zendler D, Schneider P, Töpfer R, Zyprian E (2017) Fine mapping of *Ren3* reveals two loci mediating hypersensitive response against *Erysiphe necator* in grapevine. Euphytica 213:68

Zyprian E, Ochssner I, Schwander F, Simon S, Hausmann L, Bonow-Rex M, Moreno-Sanz P, Grando MS, Wiedemann-Merdinoglu S, Merdinoglu D, Eibach R, Topfer R (2016) Quantitative trait loci affecting pathogen resistance and ripening of grapevines. Mol Genet Genom 291:1573–1594

Grape Biotechnology: Past, Present, and Future

16

Humberto Prieto, María Miccono, Carlos Aguirre, Evelyn Sánchez and Álvaro Castro

Abstract

Genetic improvement of grapevine relies on conventional breeding and genetic engineering, but the latter often seems far from having a significant impact. A small but important difference with previous breeding efforts is that, today, genome studies and technology advances in grapevine genetic engineering have become available in such a way that new varieties can be developed that are compatible with market challenges. Since the completion of the first reference grapevine genome sequence, relevant information has been gathered that allows for the identification of novel genes, analysis of structural gene variants, and discovery of SNPs. Also, regulatory regions for coding sequences, analyses of small RNA populations, and modulation processes coupled to DNA modification (i.e., methylations) have started to be elucidated, thereby enabling the New Breeding Techniques (NBTs), also referred to as precision breeding. RNA interference (RNAi) and RNA-guided editing of genomes are among the most promising new techniques for RNA-based systems that affect gene expression. Also, both RNAi and RNA-guided editing of DNA are expanding technical platforms by which DNA methylation can also be proposed, thus adding possibilities for epigenetic regulation. Here, we will present and discuss advances in gene transfer procedures from a NBTs' perspective. We will use a chronological arrangement of gene transfer experimentation carried out over the last 10 years as a complementary view to recent excellent works already available. Also, our experience in the use of the editing systems will be introduced.

H. Prieto (✉) · M. Miccono · C. Aguirre
Biotechnology Laboratory, La Platina Station, Instituto de Investigaciones Agropecuarias, Santa Rosa 11610, La Pintana, Santiago de Chile, Chile
e-mail: hprieto@inia.cl

M. Miccono
e-mail: maria.miccono@inia.cl

C. Aguirre
e-mail: caguirre.d@gmail.com

E. Sánchez
Doctoral Program in Integrative Genomics, Campus Huechuraba, Universidad Mayor, Camino La Pirámide 5750, Huechuraba, Santiago de Chile, Chile
e-mail: evelyn.sanchez.s@mayor.cl

Á. Castro
Life Sciences Innovation Center, University of California-Davis Chile, Avenida Andrés Bello 2299, Piso 11, Providencia, Chile
e-mail: alvcastro@ucdavis.edu

D. Cantu and M. A. Walker (eds.), *The Grape Genome*,
Compendium of Plant Genomes, https://doi.org/10.1007/978-3-030-18601-2_16

16.1 Cell Tissue Culture Techniques for Grape Gene Transfer

16.1.1 Somatic Embryogenesis

While both organogenesis and embryogenesis approaches can be induced for adventitious regeneration in grapevines, somatic embryogenesis (SE) is currently the focus of efforts, aiming at its application in germplasm propagation and storage, sanitation, and gene transfer. The SE process is a remarkable developmental switch in which the induction of embryo development from differentiated plant cells integrates stress, hormonal, and developmental pathways (Fehér 2015).

Grapevine SE is not a routine procedure that can be easily and efficiently reproduced in different cultivars (Martinelli and Gribaudo 2001; Araya et al. 2008). Baseline experimental procedures were established several years ago, using as source explant anthers (Rajasekaran and Mullins 1983; Araya et al. 2008), unfertilized ovules (Mullins and Srinivasan 1976), ovaries (Martinelli and Gribaudo 2001), leaves (Martinelli et al. 1993), petioles (Martinelli et al. 1993), and tendrils (Salunkhe et al. 1997). SE feasibility depends on parameters such as selection and treatment of the original explant tissue and the procedures for the generation, selection, and maintenance of cell lines; this includes considering the physicochemical culture conditions of the growth and the differentiation/production phases of the full process, as well as analyses regarding factors such as induction media and genotype, source, and developmental stage of the explants (Saporta et al. 2016; Vidal et al. 2009). 'Chardonnay' and 'Thompson Seedless' have been identified as by far the best responding genotypes for SE establishment leading to whole-plant production, mostly based on the treatment of floral explants (anthers and ovaries) with the combination of the plant growth regulators 2,4-dichlorophenoxyacetic acid (2,4-D) plus the cytokinin 6-Benzylaminopurine (BAP). These systems represent starting points, and important derivative procedures have allowed SE establishment in other *V. vinifera* cultivars (San Pedro et al. 2017; Carra et al. 2016; Araya et al. 2008) and *Vitis* hybrids—including common rootstocks (Oláh et al. 2009) and some *Vitis* species described until very recently as recalcitrant (Li et al. 2014).

Routine SE for grapevine germplasm conservation, propagation, and gene transfer has also been improved using bioreactors (Tapia et al. 2009). After the initiation step using solid media, 'Thompson Seedless' pro-embryogenic masses have been conducted into their propagation and development using liquid versions of SE cultures, assisted by use of agitated (i.e., permanently oxygenated) containers. Expansion of this platform to other genotypes has been successfully assayed for other varieties and rootstocks (Fig. 16.1). Similarly, SE yields have been improved using recurrent cycles of secondary embryogenesis induced over torpedo and mid-cotyledonary cells in 'Thompson Seedless' (Zhou et al. 2014).

Although these improvements in SE procedures continue to develop, gene transfer experiments have added new factors to regular embryo development, which need to be considered. The condition for optimal *Agrobacterium tumefaciens*-grapevine somatic embryo interaction is well described as genotype-dependent (Saporta et al. 2016). In general, the quality of the embryogenic culture plays key roles in successful transformation, and cells at the stages between pro-embryonic masses, embryogenic cells, and globular somatic embryos can be considered the most suitable explants for *Agrobacterium* infection. Also, transformation efficiency is determined by several other factors, including co-cultivation conditions (Vidal et al. 2009), *Agrobacterium* strain (Saporta et al. 2016; Torregrosa et al. 2002), and selection regime (Saporta et al. 2016; Wang et al. 2005).

16.1.2 Protoplasts

The multicellular status of somatic embryos represents a limiting step for genetic engineering techniques, such as gene transfer and DNA-editing techniques. Thus, the use of protoplasts

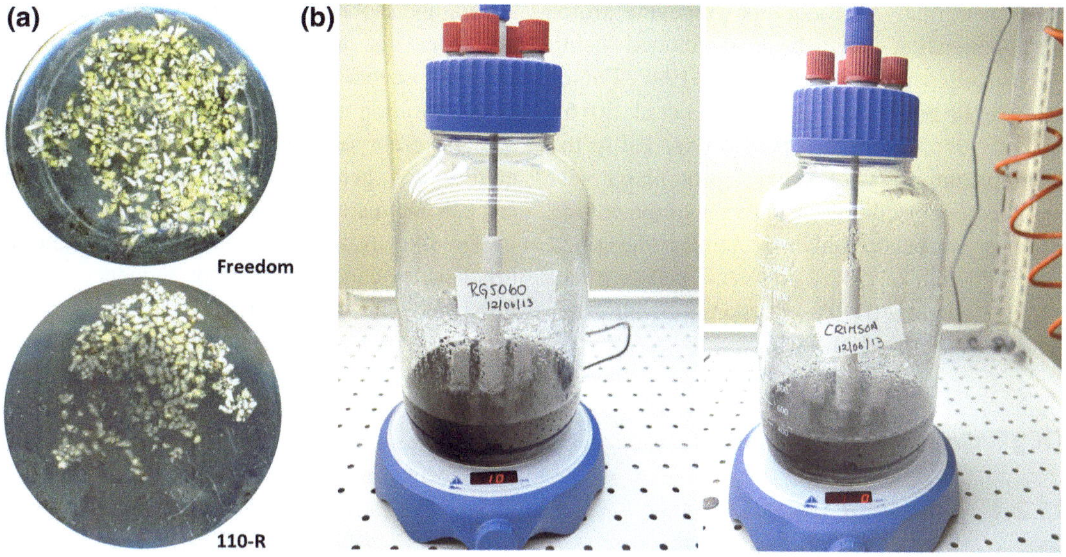

Fig. 16.1 High-throughput somatic embryogenesis systems in different *Vitis* genetic backgrounds. Somatic embryogenesis (SE) is today the most convenient procedure for gene transfer experimentation. Although the basis of SE was developed during the 1970s and 1980s, these protocols have led to improved approaches by which SE can be optimized and scaled up. Routine gene transfer experimentation can be achieved using liquid media (**a**) and automatic permanent immersion systems (bioreactors, **b**). These systems improve the nutrients transference coefficients into cells, thereby accelerating the rates for embryo development. The current goal is to spread SE platforms to diverse *Vitis* genotypes. Systems for SE based on liquid media in 'Red Globe,' 'Crimson Seedless,' Freedom (1613 (solonis × Othello) × Dogridge) and 110 Richter (Berlandieri × rupestris) are shown. Details can be obtained from Tapia *et al.* (2009)

has regained attention in recent years. Currently, plant protoplasts represent an important tool for studying many aspects of plant biology, including plant defense mechanisms, protein activity, and genetic engineering. The first successful procedures obtaining fully regenerative grapevine protoplasts were described as early as in the second half of the 1990s. The first system used 'Seyval Blanc' embryogenic calli obtained from leaf disks cultured in solid Nitsch's medium supplemented with naphthoxyacetic acid (NOA) and thidiazuron (TDZ) and subsequent sub-cultivation in hormone-free medium (Reustle et al. 1995); an improvement on those approaches that relied on early stage 'Koshusanjaku' pro-embryogenic masses made friable by several subsequent sub-cultures in modified Nitsch's medium (lacking vitamins, inositol, and glycine) supplemented with 2,4-D, sucrose, and activated charcoal (Zhu et al. 1997). Both approaches used gellan gum as a gelling agent during the embryogenesis steps, and cellulase, celiulysin, and macerozyme R-10 as enzymatic disruption treatments of these cells led to viable protoplasts which after regeneration led to whole plants with no morphology change. Grapevine protoplast generation, biochemistry, recalcitrance, and ultrastructure have been exhaustively reviewed by Papadakis et al. (2001, 2009).

The grapevine protoplast technology was soon applied to gene transfer experiments. Electroporation of the CAT gene into 'Cabernet Sauvignon' protoplast was described during the same period (Kovalenko and Schuman 1997). In a period when pathogen-derived resistance received much attention, and the insertion of viral coat protein (and other viral sequences) genes was intensively used to generate resistant individuals, Grapevine fanleaf virus (GFLV) viral RNA was delivered into leaf-derived 41B rootstocks protoplasts (Valat et al. 2000) as part of a rapid screening assay for the identification of virus resistant transgenic 41B (*Vitis vinifera* × *Vitis berlandieri*) individuals (Valat et al. 2006).

Since then, many protocols for grapevine protoplast generation have been developed using several cell sources (Nunan et al. 1997; Fontes et al. 2010; Wang et al. 2015; Zhao et al. 2016). Recently, these same procedures have led to the development of 'Chardonnay' protoplasts susceptible to gene editing by delivery of the editing reagents, a work in which the authors described the production of edited grapevine cells without stable exogenous DNA incorporation into their genome (Malnoy et al. 2016).

16.2 Genetic Transformation

16.2.1 Transient Gene Transfer Systems

Advances in the understanding (annotation) of grapevine genomes have enabled deeper studies of the proposed coding sequences of key candidate genes and the functional role of their regulatory sequences. For this reason, experimental approaches regarding both the overexpression of such coding sequences and RNAi assays targeting specific sequences and gene regions, ideally at their specific tissues or organs, have begun to be evaluated. Transient expression assays have become relevant as screening systems prior to experimentation linked to precise breeding as they avoid the technical difficulty and lengthy periods associated with whole-plant generation. A complete summary of transient expression assays in *V. vinifera* is given by Jelly et al. (2014).

Ten years ago, transient gene expression in grapevine was limited to particle bombardment of cell suspensions. Thus, factors influencing leaf agroinfiltration procedures started to be studied and optimized (Santos-Rosa et al. 2008). Today, the most immediate approach in transient expression assays is based on the development of agroinfiltration techniques. Both in vitro plantlets and young plants are the most regular sources used in these assays. Attached (Zottini et al. 2008) or detached leaves (Santos-Rosa et al. 2008) and roots (Terrier et al. 2009) have been also used from in vitro plantlets (Chialva et al. 2018), and isolated leaves from potted individuals (Merz et al. 2015). In addition to leaf explants, several other works describe the use of agroinfiltration procedures in cell suspensions, somatic embryos, and protoplasts. The explant quality and genotype have been found to be critical for this type of experimentation.

The increasing amount of information and the urgency for candidate gene or sequence evaluation in vivo has led transient gene transfer toward a more systemic approach. A recent example was the functional elucidation of the role of *VviAGL11* in seed formation. Overexpression and RNAi constructs for this gene allowed the analysis of seed formation in berries by direct injection of clusters' peduncles at their base end in the seeded 'Prosecco,' 'Alvarinho,' 'Chardonnay,' 'Italia,' 'Moscato Giallo,' 'Pinot Noir,' 'Ruby,' and 'Trebbiano' plants and in the apirenic 'Clara,' 'Linda,' and 'Thompson Seedless' plants (Malabarba et al. 2018). More recently, a whole-plant agroinfiltration protocol was reported in which promoter regions of members of the *VviSTS* gene family were evaluated and the effect of different elicitors pursued (Chialva et al. 2018). The whole-plant agroinfiltration procedure was applicable in 'Thompson Seedless,' 'Chardonnay,' 'Pinot Meunier,' and 'Carménere' plants, and in the rootstocks Harmony (1613 (solonis Othello) × Dogridge) and Salt Creek (Ramsey; *Vitis* parentage *champinii*) (Fig. 16.2).

As we indicated in the previous section, the edited grapevine protoplasts allowed the corroboration of editing tools with no foreign DNA integration and increased the need for screening/reporter systems coupled to precise breeding. As we will discuss later in this chapter, CRISPR/Cas9 technology requires the development of multiple accessory techniques in order to be applied to produce stable edited individuals.

16.2.2 Stable Gene Transfer

Regardless of the public acceptance of genetically modified crops, stably transformed grapevine plants and their field trials have represented a limiting factor for an effective application of

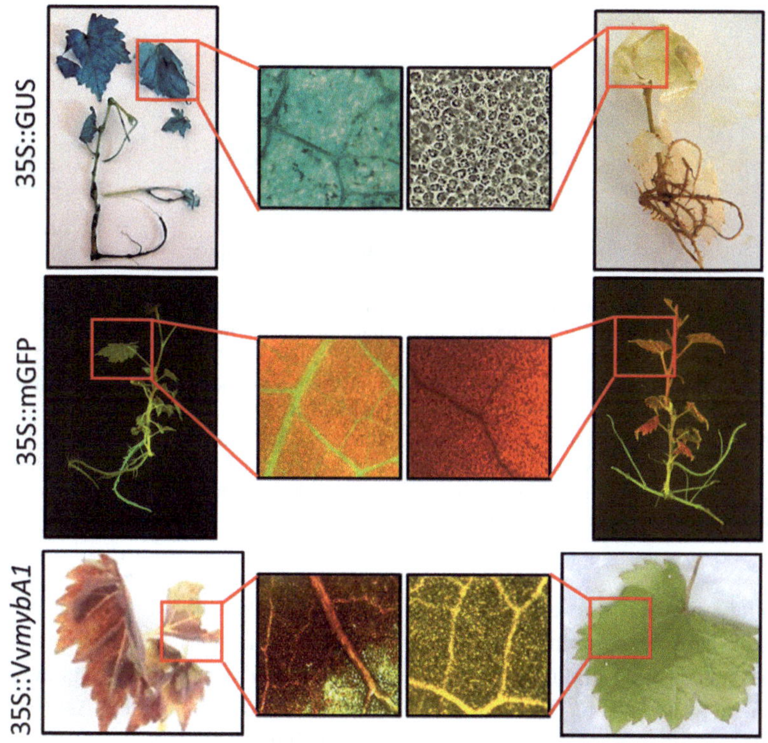

Fig. 16.2 A whole-plant agroinfiltration technique for transient expression. Using vacuum infiltrations, whole in vitro plants can transiently express different DNA sequences. The most relevant advantage of this approach is that it produces lower mechanical wounding on the infiltrated tissues, thus allowing for plants maintenance under culture conditions and further experimentation, if required. For this reason, this technique is suitable for overexpression of CDS, artificial miRNA expression, or evaluation of promoter responses upon a battery of challenges. Expression of reporter genes, such as *GUS*, *GFP*, and *VvimybA1*, is shown as representative examples of the level of effectiveness on cells. Inset boxes correspond to images captured by optical microscopy (10×) for *GUS* and *VvimybA1* expression and under epifluorescence microscopy for *GFP* expression (10×). WT, wild-type individuals; 35S, Cauliflower virus 35S RNA promoter. Details can be obtained in Chialva et al. (2018)

genetic engineering in breeding. While some research was scaled up to field tests during the 2000s (Gray et al. 2011), there are currently only a few active field trials for genetically modified grapevines, including some studies addressing Pierce's disease and *Xylella fastidiosa*, virus diseases, such as *Grapevine leafroll-associated virus* 3 (APHIS 2018), and tolerance to fungal diseases, such as powdery mildew (Rubio et al. 2015). Long-term maintenance of a transgenic collection in field plots not only allowed the evaluation of durability of disease tolerance, but also led to the identification of other interesting phenotypes, such as individuals showing reproducible recess delay (Fig. 16.3).

The first attempts to overcome the drawbacks in transgenic development came from the elimination of selection marker genes. The use of the *Cre/lox P* recombination system as an element in transformation vectors to remove the undesired vector and reporter gene sequences has been formally described in model species (Ow 2007). In grapevines, using the cultivar 'Brachetto,' the estrogen receptor-based fusion transactivator XVE (XVE System; Zuo et al. 2001) showed that 17-β-estradiol supply was useful in *npt*II gene removal in plantlets derived from three morphogenic in vitro systems. This system was very effective in the roots and partially effective in leaves (Martinelli et al. 2009; Dalla Costa et al.

Fig. 16.3 Phenotypical variations on grapevine transgenic events. Genetic transformation in grapevine has generated several reports for field trials of individuals during recent years. However, policy and public perception issues have affected these assays, and most of them have already declined. One of these trials is located at La Platina Station in INIA-Chile, which is still maintained as a collection. While, the main goal of this research was the generation of individuals with improved tolerance to fungal diseases (Rubio et al. 2015), the long-term analysis has also shown interesting new traits as presented in this set of pictures. 'Thompson seedless' lines transformed with chitinases and glucanases derived from *Trichoderma* spp. were established on the field since 2004 against fungal pathogens. After 14 years of field evaluations, a single event (115) shows a delay in recess relative to *wild-type* and other transgenic plants placed in the block

2010). In a similar approach, the expression of the yeast flippase (*Flp*) recombinase (Lyznik et al. 2003) was successfully induced in 'Brachetto' (Dalla Costa et al. 2016). By using the heat-shock *Gmhsp17.5-E* promoter (Czarnecka et al. 1989), *Flp* expression was induced in the transgenic plant subjected to temperature increases (40–42 °C), which resulted in the efficient excision of the *npt*II marker gene flanked by flippase recognition target (FRT) sites.

Removal of non-grape sequences (marker and selection genes) in gene transfer-derived individuals can be considered as a technical effort focused on producing individuals transformed with only-grape sequences, i.e., intragenics. The approach toward cisgenics (Nielsen 2003; Schouten et al. 2006), using genetic elements derived from the grape genome, has also begun to be considered for the plasmids used in gene transfer. Selected from the group of pathogenesis-related proteins produced by grapevines under fungal challenge, a thaumatin-like protein *VviTL*1 has been used in the generation of 'Thompson Seedless' plants with delayed powdery mildew infection under greenhouse conditions, tolerance to the fungus *Guignardia bidwellii* (grape black rot) at field level, and decreased incidence of sour-rot in berries under storage at room temperature (Dhekney et al. 2011).

The availability of genome sequences opens the possibility to advance toward improved concepts in grapevine cisgenics by defining plant-derived transfer DNAs (Rommens et al. 2005). These elements can be assembled into gene transfer plasmids adequate for plant (in general) and grapevine (in particular) genetic transformation experiments. Conserved and variable regions in the *Agrobacterium tumefaciens* Ti plasmid accessions have been reviewed and mimicked by synthetic blocks that can be studied for their effect in the DNA transfer process (Conner et al. 2007; Rommens et al. 2005; Holme 2013). For instance, the 25-nt *Agrobacterium* T-DNA right borders function in the initiation of DNA transfer and have a highly conserved 13-bp 5′-ATATATCCTG-[C/T]-CA motif preceded by a more degenerate 12-bp consensus 5′-[A/C/G][A/T]-[A/T]-[G/T]-AC-[A/C/T]-N-[C/G/T]-[A/C/G]-[A/C/G]-N (Rommens et al. 2005). Similarly, several conserved blocks can be deduced from both T-DNAs and Ti plasmids. These 'codes' have allowed the design of synthetic elements, for example, possible right borders, and their evaluation of their integration capability. According to that information, plant DNA sequences can be searched in datasets and proposed as plant-derived transfer DNA modules equivalent to those used by *Agrobacterium*. Figure 16.4 shows the proof of concept for an

'All Grape' vector developed in our laboratory, expressing a GFP cassette in *Nicotiana benthamiana* explants, thus combining the high functionality as transfer DNAs of the selected elements found in the grapevine genome.

In the development of the 'All Grape' vector, we relied on several key elements found in the T-DNA and surrounding segments, which—based on the available information—can be mimicked by the grapevine genome. Nucleotide sequences belonging to right and left T-DNA borders were compared among *Agrobacterium* strains and used to create search patterns based on regular expressions to search over various databases that include grapevine genome information (e.g., NCBI, Gramene, and Genoscope). These alignments were carried out following special BLAST parameters, as indicated by Rommens et al. (2005). From this, we built a table of candidate regions in the grapevine genome that could accomplish this function (Table 16.1).

In addition to these positions, we recorded down- and upstream blocks in these sequences to obtain additional 'codes' found in the T-DNA. Some of these blocks include the right T-border (RB) sequence and their upstream sequences (such as overdrive sequences that promote DNA transference such as AC-rich (ACR) domain) and RB downstream sequences (10-nt motifs (decamers; DR) and CCCG blocks (Tzfira et al. 2004; Tzfira and Citovsky 2008; Toro et al. 1988; van Haaren et al. 1987; Culianez-Macia and Hepburn 1988; Lee and Gelvin 2008), Left T-border (LB) sequences and their LB upstream sequences (AT-Rich domain and UL motif), and LB downstream sequences (C-cluster) (Rommens et al. 2005; Rommens 2004; Tzfira et al. 2004; An et al. 2013; Gelvin 2000); a summary with the sequence requirements is shown in Table 16.2.

Although cisgenics has been classified among the New Breeding Techniques, its closeness to transgenics has stalled the progress of this technology. The advantages associated with vectors based on plant DNA most probably the use of RNAi and gene-editing approaches will seem likely to gain traction in near future.

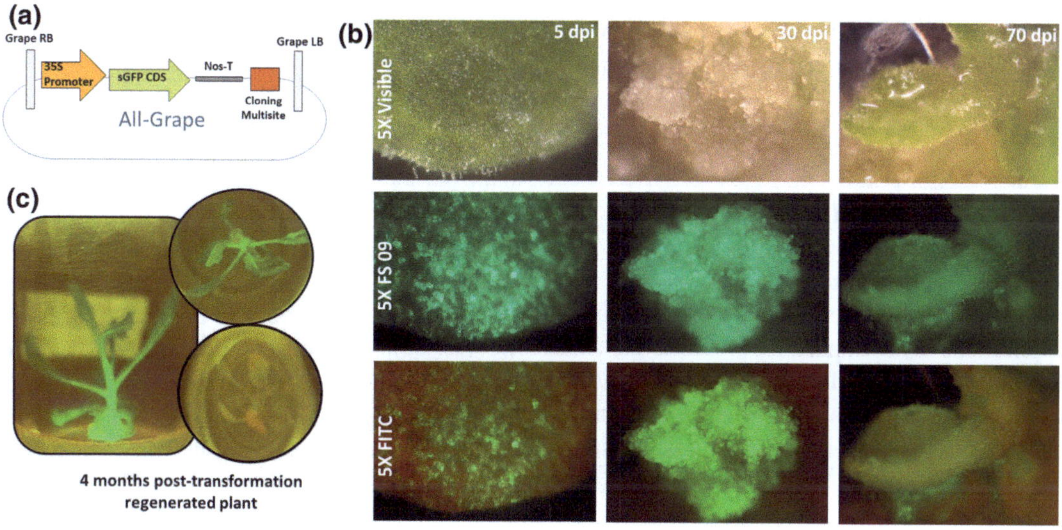

Fig. 16.4 Functional evaluation of grapevine-derived transfer DNAs in *Nicotiana benthaminana*. Using the genome draft information, bioinformatic analyses allowed us to identify several grapevine genome modules that can mimic the T-DNA function in gene transfer experiments. As a proof of concept, a green fluorescent protein (GFP) expression cassette was cloned into the All Grape vector (**a**) and used in *N. benthamiana* gene transfer experiments. A transformed explant time course (**b**) showed GFP expression from the early stages of the organogenesis process up to 70 days post-infection and subsequently, in the fully regenerated plants (**c** four months after infection)

Table 16.1 Putative grapevine transfer DNA-mimicking T-DNA borders

Genome region*	Identity
Chr18: 19096056-19096979	VIT_18s0072g00080
Chr15: 18461532-18461555	Intergenic
Chr04: 22043123-22043145	Intergenic
Chr13: 7204421-7204442	Intergenic
Chr13: 4497591-4497610	Intergenic
Chr16: 3096123-3096140	VIT_16s00039g02720
Chr18: 615525-615544	VIT_18s0122g00850
Chr07: 18380158-18380177	VIT_07s0031g02240
Chr04: 19365963-19365987	Intergenic
Chr18: 18140575-18140593	VIT_12s0059g00150
Chr01: 10177649-10177665	VIT_01s0026g01170
Chr03: 96578-96598	Intergenic
Chr10: 4999352-4999371	VIT_10s0003g02880
Chr06: 11633925-11633943	Intergenic

*Locations according to Phytozome v12.1

Table 16.2 Summary of nucleotide motifs involved in T-DNA function

Motif name	Characteristic
ACR	C or T, separated by an A-rich tri-nucleotide segment. Pyrimidine hexa-nucleotide close to the border
DR	[A/C/T]-[A/C]-[A/C/T]-[A/G/T]-[A/T]-T-[A/C]-G-[G/T]-[G/T]
AT-Rich	[A/G]TTTACA[A/C/T][A/C/T][A/C/T] [C/G]AATATATCCTGCC[A/G]
UL	A[C/T]T[C/G]A[A/T]T[G/T][C/T][G/T][C/G]A[C/T][C/T][A/T]
C-cluster	$CCN_{1-11}CCN_{1-11}CCN_{1-11}CC$

16.3 New Breeding Techniques

We will focus this section on only two technologies that are considered major current applications with impact, both based on small RNAs: RNAi and CRISPR. Both technologies have several elements and concepts in common and are derived from regular cell processes.

16.3.1 RNA Interference (RNAi)

Biotechnological uses of silencing mechanisms began with the work of Fire and colleagues (Fire et al. 1998), where the authors demonstrated that a double-stranded RNA can be used to interfere with the expression of an endogenous gene by producing RNAi. Later, one of the most used systems inducing RNAi was designed by the generation of hairpin RNAs, composed of a sense RNA (with respect to the target gene) and its corresponding antisense, separated by a spacer region (Wang and Waterhouse XXX; Wesley et al. 2001). These double-stranded RNAs (dsRNAs) induced vectors and then trigger the biosynthesis of small interfering RNA (siRNA).

16.3.2 Small Interfering RNAs (siRNAs)

RNAi is triggered by the formation of endogenous or exogenous dsRNA precursors via the activity of one or more cellular RNA-dependent RNA polymerase (Incarbone and Dunoyer 2013). These precursors are processed by the Dicer or Dicer-like ribonuclease III (DCL) enzymes to form siRNA duplexes, which are protected from exonucleolytic degradation (Li et al. 2005) and loaded into Argonaute effectors through the selection of one strand as a guide for target identification and the removal of the passenger strand (Vazquez et al. 2006). These events facilitate the formation of the RNA Interference Silencing Complex (RISC) for cleavage of the target RNA. Of the total pool of siRNA species generated, molecules between 21- and 24-nt long play roles in gene silencing (Montes et al., 2014).

The accumulation of different RNAs species and its double-strain structures (aberrant RNAs) due to several cellular process involved in siRNA biosynthesis can be triggered, from a technology point-of-view, using virus-based vectors expressing high amounts of specific RNAs or using double-stranded RNA hairpin-inducing plasmids, in which 200–600-bp segments of specific genes are arranged in sense and antisense orientation in an expression vector.

In grapevines, siRNAs generation has been strongly supported by transient expression assays. Based on the single-stranded RNA genome of the *Grapevine leafroll-associated virus*-2 (GLRaV-2), an assembled DNA version of a GLRaV-2-based vector was built harboring an expression cassette for foreign sequences (Kurth et al. 2012). First assayed for the GFP gene expression, this expression cassette was then modified and fused downstream of this reporter with a sense fragment of the 'Syrah' phytoene desaturase (*PDS*) gene or antisense of the 'Syrah' subunit I of magnesium-protoporphyrin IX chelatase (*ChlI*) gene. In both cases, constructs led to leaf bleaching due to the loss of chlorophyll in the agroinfiltrated grapevine leaves. Additional examples for RNAi in grapevines have relied on hairpin-inducing vectors. Agroinfiltration procedures were used in 'Carbenet Franc' leaves for the expression of a hairpin-inducing construct to decrease mRNA levels of the defense-related gene *VviPGIP*1, encoding a polygalacturonase-inhibiting protein (PGIP) (Bertazzon et al. 2012). More recently, a silencing construct based on sequences of the *VviAGL*11 gene, which is involved in seed formation in grapevines, was used in infiltration experiments of peduncles of the seeded 'Italia' and 'Ruby' and conducted for the downregulation of *VviAGL*11 mRNAs, also resulting in seedless and seed rudiments in clusters in the infiltrated individuals (Malabarba et al. 2018).

Stable transformation assays using hairpin-inducing constructs have also been carried out. The cloned somatic embryos of the reference accession PN40024 were used to develop a silencing construct directed against the GFP reporter gene (Romon et al. 2013). In these experiments, the authors also analyzed the effect of low temperatures on the silencing capability in the plants and found that the process is unaffected by temperature treatments of these stably transformed grapevine plants up to 4 °C of incubation, in contrast to *Arabidopsis* transgenic lines subjected to similar situations, whose GFP fluorescence recovered. Stable transformed plants were generated with a dsRNA hairpin that induced silencing constructs for different gene isoforms of the susceptibility gene Mildew Locus O (MLO) gene family (Pessina et al. 2016). Specific silencing constructs for the *VviMLO*6, *VviMLO*7, *VviMLO*11, and *VviMLO*13 gene versions were built and used in gene transfer experiments of 'Brachetto' somatic embryos, and successfully regenerated plants were evaluated for fungal tolerance. Using these data, the authors propose the *VviMLO*6 and *VviMLO*7 genes as responsible in the fungus–plant interaction and suggest them as a target for gene editing to generate powdery mildew-resistant grapevines.

16.3.3 MicroRNAs (miRNAs)

Directly linked to the genomic knowledge of the species and the use of next-generation sequencing (NGS) systems, the regulatory role of miRNAs has become a key topic of genetic engineering.

MicroRNAs are the processed version of nuclear genes (miRNA genes) and are responsible for gene regulation in several key pathways in organisms. During their processing, the primary miRNA gene transcripts (pri-miRNAs) form a partially double-stranded stem-loop structure (pre-miRNA) that is processed by DCL1 proteins to release mature and functional miRNAs (Bartel 2004). Mature miRNAs are recruited to the RISC, where they become single-stranded to execute different functions.

Today, we know that these molecules play a crucial role in the genetic programming and fine tuning of plant biology (Bartel 2004; Kurihara and Watanabe 2004; Brodersen and Voinnet 2006), which is of major relevance in grapevine, a species with a life cycle spanning a two-season period with multiple developmental stages. One of the first maps for these molecules came from miRVine (Belli-Kullan et al. 2015), a dedicated miRNA database for the species, obtained from NGS of small RNAs obtained from 'Corvina' and from the 'Pinot Noir' derived reference grape genome sequence accession PN0024, leading the announcement of over 285 miRNAs.

In comparison to siRNA species, miRNA pathways could operate through less transitivity (secondary siRNA biogenesis), thus the practical application of gene silencing via the design and use of artificial miRNAs (amiRNAs) could represent a powerful alternative in RNAi in terms of specificity (Montes et al. 2014; Castro et al. 2016). However, this involves the technical inconvenience of requiring a backbone sequence that ensures the generation of the correct miRNA. Several examples of silencing using amiRNAs exist in species such as *Arabidopsis thaliana*, *Oryza sativa*, *Medicago truncatula*, and *Chlamydomonas reinhardtii* (Schwab et al. 2006; Warthmann et al. 2008; Álvarez et al. 2006; Molnar et al. 2009; Devers et al. 2013). These studies relied on the use of the simple stem-loop structure as backbones (pre-miRNAs) for the processed, final miRNA, such as the *A. thaliana* pre-miRNA319a (ath-miR319a). The pre-miRNA319a has been incorporated into the plasmid pRS300, becoming an amiRNA assembly-expression tool (Schwab et al. 2006). In a recent work, we described the use of the simplest member of the grapevine miR319 family in terms of size and structure, the vvi-miRNA319e, as a suitable artificial miRNA template for genetic engineering (Castro et al. 2016). The simplicity observed in this molecule and its precursor makes it an easy-to-handle silencing tool, which can also be proposed as a part of vectors harboring multiple expression cassettes for simultaneous gene silencing and replacement (Fig. 16.5) experiments. We have also designed a Web tool by which the primers for both synthesis and stem-loop detection can be deduced (available as 'Plant amiRNA designer' at www.fruit-tree-genomics.com, tab 'Biotools').

16.3.4 Clustered Regularly Interspaced Short Palindromic Repeats (CRISPR)/*Cas9* in Grapevines

In the last 10 years, programmable DNA-binding proteins (effectors), such as zinc finger and transcription activator-like effectors, have emerged as an alternative to conventional editing, mainly based on random mutagenesis techniques. An advantage of these tools is their recognition capability of specific target DNA sequences based on the joining of tailored (customized) arrangements of one (TALE) or three (ZF) nucleotides, thereby bringing to these places a nuclease (generally C-terminal domain of *Fok*I) that disrupt DNA adjacent to the recognition zones. Both ZF- and TALE-nucleases, ZFN and TALEN, respectively, require two effectors (left and right) to define a nuclease (*Fok*I) cutting site, located between the left and right effectors (Gaj et al. 2013).

While many research teams have started to use ZFNs and TALENs, in June 2012, Doudna and Carpentier (Jinek et al. 2012) adapted RNA-guided machinery to direct the nuclease *Cas9* to cleave DNA. The Clustered Regularly Interspaced Short Palindromic Repeat (CRISPR)/*Cas9* system has been a revolutionary molecular tool since its discovery as an adaptive line of defense against viral infection in *Archae* (Mojica et al. 2000). CRISPR/*Cas9* is currently one of the most relevant

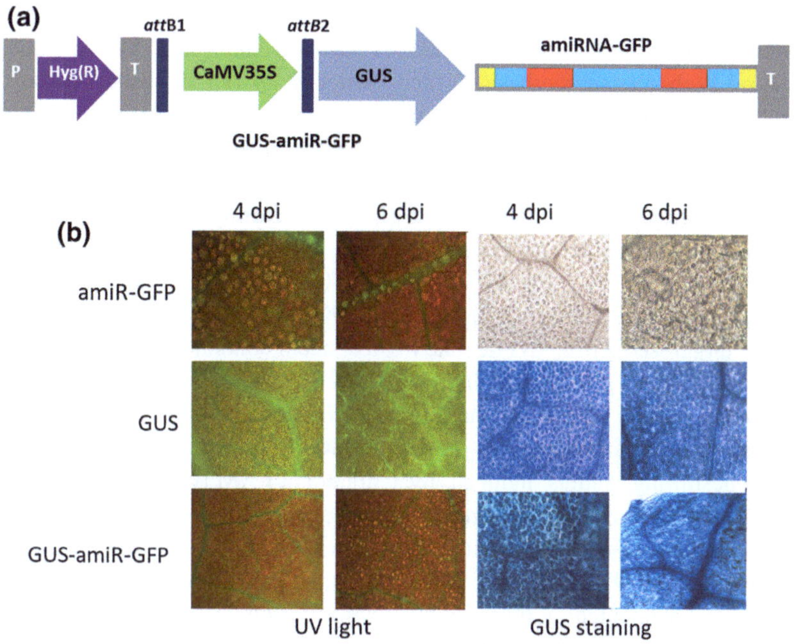

Fig. 16.5 Gene replacement using an artificial grapevine microRNA (amiRNA). An amiRNA based on the VvimiRNA319e backbone (Castro et al. 2016) was synthesized targeting the green fluorescent protein (GFP) gene and 3′-fused to a GUS CDS directed by the 35S CaMV promoter, generating the gene replacement construct GUS-amiR-GFP (**a**). This construct was used in transient expression assays of leaves from 'Thompson Seedless' plants stably expressing GFP (**b**). Leaves were evaluated at 4 and 6 days post-infiltration (dpi) using epifluorescence microscopy (UV light) and GUS staining. In (**b**) amiR-GFP, leaves agroinfiltrated with an amiR319e construct targeting the GFP gene; GUS, leaves agroinfiltrated with a GUS expressing construct; GUS-amiR-GFP, leaves with the GUS-amiR319-GFP construct

gene-editing techniques that allow for the direct generation of sequence modifications in the genome. First described as an adaptive immune response in *Streptococcus pyogenes*, the main elements of this system have been adopted as biotechnology tools for targeted mutagenesis. This involves making guide RNAs (gRNAs) that target customized sequences in the genome to direct the *Cas*9 nuclease activity to generate double-stranded breaks adjacent to the gRNA joining location.

Currently, CRISPR/Cas9 has been used successfully in humans (Baumann 2017), insects (Taning et al. 2017), fungi (Nødvig et al. 2015), and plants (Demirci et al. 2017). Delivery of CRISPR/*Cas*9 components into the plant cell has been achieved by their stable integration into the genome using gene transfer techniques, which has mostly relied on the use of binary, Ti-derived plasmids convenient for *Agrobacterium*-mediated transformation.

The term gRNA refers a short synthetic RNA composed of a scaffold sequence necessary for *Cas*-binding and a user-defined 20 nucleotide spacer that determines the genomic target to be modified. The target sequence recognized by the spacer will be a protospacer sequence, located contiguous to an adjacent motif recognized by *Cas*9 required for DNA cleavage, which is an NGG nucleotide arrangement called protospacer adjacent motif (PAM). For practical uses in the next paragraphs, we will use gRNA to indicate the user-defined spacer sequence only.

16.3.4.1 Theoretical Gene Editing

The challenge of setting CRISPR/*Cas*9 technology involves a previous knowledge of genome information. Several pipelines based on the use of gRNA + PAM target site datasets have been described and used for on-target and off-target predicting activity of gRNAs inside a genome

(Pulido-Quetglas et al. 2017). In the case of grapevine, genome tools for CRISPR design are available in several Web sites on the basis of the 'PN40024' reference grape genome (Brazelton et al. 2015). However, the use of the reference genome can generate problems if the target genome differs in homology, entailing loss of PAM sites or the occurrence of non-predicted off-targets. Also, these tools have been focused on a single-gRNA approach, leading to discrete mutations in the target sequence. While early studies utilizing individual gRNAs to induce mutations in protein coding regions frequently resulted in complete loss-of-function gene mutations by frameshifting, indels induced by single gRNA targeting non-coding regions are less likely to produce loss-of-function mutations (Shalem et al. 2014; Hart et al. 2015). With these considerations, we implemented a more dedicated tool to process the genome information of several woody fruit crops, including grapevines. Our systems allow the generation of 'gRNA pairs' that lead to gene deletions flanking a target region (Ho et al. 2015; Aparicio-Prat et al. 2015).

One of our first aims was to establish gene editing in the laboratory and to develop essential tools for correct candidate gene visualization and analyses, which means an *ex-ante* visualization of on-target and off-target activities of the designed gRNAs. The analysis of an on-target activity is closely related to the gRNA design. There are several online tools that allow the design of single gRNAs for different plant genomes; these tools are based on thermodynamic parameters and different predictive models that lead to defining a ranking of the best candidate molecules. Among these, the empirical logistic regression model by Doench et al. (2014) is noteworthy; this model was initially trained with experimental assays for 1841 gRNAs for human and mouse genes and considers the nucleotide sequence features of the [protospacer + NGG] region plus four and three nucleotides upstream and downstream, respectively.

The computing of PAMs for *Cas*9 over a target genome allows for a first-dimensional analysis of the putative cut sites for the nuclease upon gRNA leadership. Wang et al. (2016) described the occurrence of more than 35 million PAMs in the grapevine genome. In the construction of our dedicated system for *Vitis*, we found a total of 36,505,702 potential CRISPR sites [protospacer + PAM] in the reference genome (Genoscope 12X, Jaillon et al. 2007). They were uniformly distributed among the 20 chromosomes, ranging from 1,397,855 (chr17) to 3,040,562 (chrUn). Of all protospacer sequences, 22,994,707 (63%) were unique and 37,047 protospacers contained an 'N' (an unknown residue) and, for this reason, were not useful during gRNAs design. Four types of PAM were scored (AGG, TGG, CGG, and GGG). The most abundant PAM type across the genome was TGG (32.8%), followed by AGG, GGG, and CGG, with the latter accounting for just 7.4% of the total. Interestingly, expanding this study to other woody fruits, we have found similar results in the analysis of the apple genome (GDDH13 Version 1.1).

Subsequently, gRNA target sequences (i.e., protospacers) added to this dataset of PAM sites will allow the generation of every possibility in the genome for the gRNA/*Cas*9 complex. This collection of combinations constitutes the potential target sites within a genome, which can also be considered as potential off-targets of a specific gRNA if mismatches are allowed within their native sequences. The generation of off-target databases (previously calculated for different genomes) represents a useful approach for fast genome scoring of unwanted editing activity using CRISPR/*Cas*9. The main property of this approach is to represent a relatively 'light processing' workflow for computers and online work.

The system was based on CRISPR-Analyzer (Shen et al. 2014) and CRISPETa (Pulido-Quetglas et al. 2017). CRISPETa is a suite of command-line Python scripts that find all possible gRNAs given a target genome region. This tool also ranks each gRNA according to its predicted on-target activity using an empirical logistic regression model (Doench et al. 2014). Alternatively, CRISPR-Analyzer, a collection of command-line C++ scripts, enabled us to find and index all the possible [protospacer + NGG sequences] along the *V. vinifera* reference genome. Next, we generated the possible 'CRISPR sites' for this *Vitis* reference. This database was additionally processed to compute the possible off-targets sites with 0–4 mismatches for each recorded site and stored them in a MySQL database, which is used to quickly find the number of off-targets upon gRNA query. The pipeline for the *V. vinifera* genome is presented as a Web application and integrates this information using the JBrowse Genome Browser (Skinner et al. 2009) and SequenceServer (Priyam et al. 2015). As in CRISPETa, the user can also establish advanced parameters, such as the maximum allowed number of off-targets with 0, 1, 2, 3, and 4 mismatches and the minimal individual and paired scores for the gRNAs. The results for the off-targets are individualized according to the chromosome, sequence coordinates, mismatch number and position, and location (exonic, intronic, or intergenic). Our grapevine 'CRISPR search tool,' as well as similar tools for other woody fruit crops, is available at Genome Browser at https://www.fruit-tree-genomics.com (tab 'Biotools').

16.3.4.2 Experimental Gene Editing

Woody fruit species, such as grapevine, present their own difficulties when subjected to gene transfer experiments; these include low efficiency in regeneration and transformation, chimerism, recalcitrance, and a long-time regeneration process. These conditions are also predictable for CRISPR/Cas9 experimentation. Once a set of candidate gRNAs are derived from the grapevine CRISPR search tool or from a different available tool, these molecules must be evaluated for their efficiency at directing *Cas*9 to the target site in the grapevine genome. While some works have proposed the use of RNAi for fast screening of the target candidate sequences (Pessina et al. 2016), we think that a recall for transient expression systems could also be an extremely useful tool. In our experience, the use of transient expression systems is the most suitable, convenient, and fast procedure to carry out gRNA evaluation processes. From the above-referred strategies, the fastest experimentally effective assays have been the use of agroinfiltration procedures, using either whole plants (Chialva et al. 2018) or leaf agroinfiltrations (Zottini et al. 2008; Miccono et al. 2018) (Fig. 16.6).

Several reports have described grapevine genome editing by generating transgenic individuals expressing the editing reagents gRNAs and *Cas*9 (Ren et al. 2016; Nakajima et al. 2017; Wang et al. 2017). These works demonstrated the feasibility of gene editing in the species and encouraged additional efforts into editing individuals without foreign DNA insertion into the genome, as for instance, the delivery of the already assembled ribonucleoprotein-editing reagents (Malnoy et al. 2016).

Using a different approach, we have conducted several trials that finally allowed for the establishment of a DNA-replicons strategy (Baltes et al. 2014), based on the *Bean yellow dwarf virus* (BeYDV) genome structure in the absence of proteins required for its multiplication (i.e., disarmed virus). This allows a high copy number in the cell without the insertion of the replicon into the plant genome.

By assembling the LSL *cis*-elements (Baltes et al. 2014) from the geminivirus genome, we designed and built a universal plasmid capable of replicational release of a BeYDV-derived replicon from a regular T-DNA. These elements allow the simultaneous expression of *Cas*9 nuclease and up to four different gRNAs. The vector, called pGMV-U (Addgene plasmid #112797), was used in the proof-of-concept study for the editing of the *VviSWEET4* gene, a sugar transporter up-regulated under several biotic stresses in grapevine, as well as in fungi and bacteria (Chong et al. 2014). Using 'Thompson Seedless,' SE, and pGMV-U, a conventional SE *Agrobacterium*-mediated gene transfer experimentation has allowed for the generation and identification of the first edited non-transgenic individuals for the *VviSWEET4* gene (Miccono et al. 2018), which are currently under functional evaluation.

Fig. 16.6 Fast screening system for the evaluation of guide RNAs. A preliminary system for *in vivo* evaluation of designed gRNAs was applied using transient expression assays. In our experience, the generation of edited grapevine individuals requires long-term experimentation based on somatic embryogenesis gene transfer experiments. Before proceeding with this experimentation, functional analyses of the designed gRNAs are recommended. Vectors carrying the designed gRNAs are infiltrated using a needleless syringe in leaves of acclimatized plants (**a**). After infiltrations, tissue samples from the exposed areas are used for genomic DNA extraction and analysis. A paired gRNAs-editing strategy in grapevines (Miccono et al. 2018) enables to cleave a considerable fragment of a target gene (**b**); in this case, gene editing can be monitored by conventional PCR of the extracted DNA (**c**). In the (**c**), the three bands correspond to edited genes; amplicons of 121 bp (indicated by the white arrows) confirmed the targeted deletion (wild-type amplicon size: 1542 bp)

16.4 Future Prospects

The advance in gene transfer technology has been considerable. We are witnessing an era in which the fundamentals of tissue culture, genomics, and the promising genetic engineering seem to be advancing faster than ever. A relevant aspect of these facts is to realize that the extent of the advances has been based on a return to the works and techniques developed more than 30 years ago. Tissue culture techniques have regained interest following the development of the new technologies involving precise breeding, such as trans-grafting, protoplast techniques, and SE (in the case of grapevines). It has become evident that a fusion between genome knowledge and genetic engineering is finally challenging our laboratory skills to translate those hypotheses arising from bioinformatics.

There is little doubt that the gene-editing era has driven these different routes into complementary workflows in which efforts in genetic transformation could contribute to grapevine breeding. The refinement of gene-editing technology is also accelerating future developments in gene transfer technologies. Recently, a high-throughput assembly module, available as a Web-based toolkit, has been released for custom design of vectors for gene editing in plants (Čermák et al. 2017). In the toolkit, TALEN or CRISPR/*Cas*9 reagents for creating targeted DNA sequence modifications can be assembled, enabling the use of CRISPR/*Cas*9 technology assisted by up to 12 gRNAs at a time, based on the use of a polycistronic mRNA. Additional improvements to the CRISPR/Cas9 vectors considered in the kit were the fusion of a single C-terminal nuclear localization signal for the nuclease, thereby improving *Cas*9 efficiency. On the other hand, new approaches in CRISPR effector nucleases by which improved novel *Cas*-like proteins are leading to new approaches in gene editing (Zetsche et al. 2015) and related techniques, such as Base Editing (Komor et al. 2016) and even RNA editing by *Cas*13 (Cox et al. 2017).

Finally, while the commercial success of new individuals derived from precise breeding will be subject to public concerns and criticism, as happened in the transgenics era, the scope of new techniques from a technical point-of-view is clear and will certainly represent a boost in the genome knowledge for *Vitis*.

References

Álvarez JP, Pekker I, Goldshmidt A, Blum E, Amsellem Z, Eshed Y (2006) Endogenous and synthetic microRNAs stimulate simultaneous efficient and localized regulation of multiple targets in diverse species. Plant Cell 18:1134–1151

Araya S, Prieto H, Hinrichsen P (2008) An efficient buds culture method for the regeneration via somatic embryogenesis of table grapes 'Red Globe' and 'Flame Seedless'. Vitis 47:251–252

An C, Orbović V, Mou Z (2013) An efficient intragenic vector for generating intragenic and cisgenic plants in *Citrus*. Am J Plant Sci 4:2131–2137

Aparicio-Prat E, Arnan C, Sala I, Bosch N, Guigo R, Johnson R (2015) DECKO: single-oligo, dual-CRISPR deletion of genomic elements including long non-coding RNAs. BMC Genom 16:846

Baltes NJ, Gil-Humanes J, Čermák T, Atkins PA, Voytas DF (2014) DNA replicons for plant genome engineering. Plant Cell 26:151–163. https://doi.org/10.1105/tpc.113.119792

Bartel DP (2004) MicroRNAs: genomics, biogenesis, mechanism, and function. Cell 116:281–297

Baumann K (2017) Genome editing: CRISPR–Cas becoming more human. Nat Rev Mol Cell Biol 18:591

Belli-Kullan J, Lopes Paim Pinto D, Bertolini E, Fasoli M, Battista Tornielli G, Pezzotti M, Meyers BC, Farina L, Zenoni S, Pè ME, Mica E (2015) miRVine: a microRNA expression atlas of grapevine based on small RNA sequencing. BMC Genom 16:393. https://doi.org/10.1186/s12864-015-1610-5

Bertazzon N, Castiglioni C, Angelini E, Raiola A, Gardiman M, Borgo M, Ferrari S (2012) Transient silencing of the grapevine gene VvPGIP1 by agroinfiltration with a construct for RNA interference. Plant Cell Rep 31:133–143. https://doi.org/10.1007/s00299-011-1147-2

Brazelton VA, Zarecor S, Wright DA, Wang Y, Liu J, Chen K, Yang B, Lawrence-Dill CJ (2015) A quick guide to CRISPR sgRNA design tools. GM Crops Food 6:266–276

Brodersen P, Voinnet O (2006) The diversity of RNA silencing pathways in plants. Trends Genet 22:268–280

Carra A, Sajeva M, Abbate L, Siragusa M, Pathirana R, Carimi F (2016) Factors affecting somatic embryogenesis in eight Italian grapevine cultivars and the genetic stability of embryo-derived regenerants as assessed by molecular markers. Sci Hortic 204:123–127

Castro Á, Quiroz D, Sánchez E, Miccono M, Aguirre C, Ramírez A, Montes C, Prieto H (2016) Synthesis of an artificial *Vitis vinifera* miRNA 319e using overlapping long primers and its application for gene silencing. J Biotechnol 233:200–210. https://doi.org/10.1016/j.jbiotec.2016.06.028

Čermák T, Belanto JJ, Curtin SJ, Starker CG, Gil-Humanes J, Mathre JW, Cegan R, Greenstein RL, Kono TJY, Koneĉná E, Voytas DF (2017) A multipurpose toolkit to enable advanced genome engineering in plants. Plant Cell 29:1196–1217

Chialva C, Muñoz C, Miccono M, Eichler E, Calderón L, Prieto H, Lijavetzky D (2018) Differential expression patterns within the grapevine stilbene synthase gene family revealed through their regulatory regions. Plant Mol Biol Rep 36:225–238. https://doi.org/10.1007/s11105-018-1073-3

Chong J, Piron M-C, Meyer S, Merdinoglu D, Bertsch C, Mestre P (2014) The SWEET family of sugar transporters in grapevine: VvSWEET4 is involved in the interaction with *Botrytis cinerea*. J Exp Bot 65:6589–6601

Conner A, Barrell P, Baldwin S, Lokerse A, Cooper P, Erasmuson A, Nap JP, Jacobs J (2007) Intragenic vectors for gene transfer without foreign DNA. Euphytica 154:341–353

Cox DBT, Gootenberg JS, Abudayyeh OO, Franklin B, Kellner MJ, Joung J, Zhang F (2017) RNA editing with CRISPR-Cas13. Science 358:1019–1027. https://doi.org/10.1126/science.aaq0180

Culianez-Macia FA, Hepburn A (1988) Right-border sequences enable the left border of an *Agrobacterium tumefaciens* nopaline Ti-plasmid to produce single-stranded DNA. Plant Mol Biol 11:389–399

Czarnecka E, Key JL, Gurley WB (1989) Regulatory domains of the Gmhsp17.5-E heat shock promoter of soybean. Mol Cell Biol 9:3457–3463

Dalla Costa L, Mandolini M, Poletti V, Martinelli L (2010) Comparing 17-β-estradiol supply strategies for applying the XVE-Cre/loxP system in grape gene transfer (*Vitis vinifera* L.). Vitis 49:201–208

Dalla Costa L, Piazza S, Campa M, Flachowsky H, Hanke M-V, Malnoy M (2016) Efficient heat-shock removal of the selectable marker gene in genetically modified grapevine. Plant Cell Tissue Organ Cult 124:471–481

Demirci Y, Zhang B, Unver T (2017) CRISPR/Cas9: an RNA-guided highly precise synthetic tool for plant genome editing. J Cell Physiol 233:1844–1859

Dhekney SA, Li ZT, Gray DJ (2011) Grapevines engineered to express cisgenic *Vitis vinifera* thaumatin-like protein exhibit fungal disease resistance. Vitro Cell Dev Biol Plant 47:458–466

Devers EA, Teply J, Reinert A, Gaude N, Krajinski F (2013) An endogenous artificial microRNA system for unraveling the function of root endosymbioses related genes in *Medicago truncatula*. BMC Plant Biol 13:82. https://doi.org/10.1186/1471-2229-13-82

Doench JG, Hartenian E, Graham DB, Tothova Z, Hegde M, Smith I, Sullender M, Ebert BL, Xavier RJ, Root DE (2014) Rational design of highly active sgRNAs for CRISPR-Cas9-mediated gene inactivation. Nat Biotechnol 32:1262–1267

Fehér A (2015) Somatic embryogenesis—stress-induced remodeling of plant cell fate. Biochim Biophys Acta 1849:385–402

Fire A, Xu S, Montgomery M, Kostas S, Driver S, Mello C (1998) Potent and specific genetic interference by double-stranded RNA in *Caenorhabditis elegans*. Nature 391:806–811

Fontes N, Gerós H, Papadakis AK, Delrot S, Roubelakis-Angelakis KA (2010) Isolation and use of protoplasts from grapevine tissues. In: Delrot S, Medrano H, Or E, Bavaresco L, Grando S (eds) Methodologies and results in grapevine research. Springer, Dordrecht. https://doi.org/10.1007/978-90-481-9283-0_18

Gaj T, Gersbach CA, Barbas CF III (2013) ZFN, TALEN, and CRISPR/Cas-based methods for genome engineering. Trends Biotechnol 31:397–405. https://doi.org/10.1016/j.tibtech.2013.04.004

Gelvin S (2000) *Agrobacterium* and plant genes involved in T-DNA transfer and integration. Annu Rev Plant Physiol Plant Mol Biol 51:223–256

Gray DJ, Dhekney SA, Li ZT, Cordts JM (2011) Genetic engineering of grapevine and progress toward commercial deployment. In: Mou B, Scorza R (eds) Transgenic horticultural crops, challenges and opportunities. CRC Press, Boca Raton, pp 317–331

Hart T, Chandrashekhar M, Aregger M, Steinhart Z, Brown KR, MacLeod G, Mis M, Zimmermann M, Fradet-Turcotte A, Sun S, Mero P, Dirks P, Sidhu S, Roth FP, Rissland OS, Durocher D, Angers S, Moffat J (2015) High-resolution CRISPR screens reveal fitness genes and genotype-specific cancer liabilities. Cell 163:1515–1526. https://doi.org/10.1016/j.cell.2015.11.015

Ho TT, Zhou N, Huang J, Koirala P, Xu M, Fung R, Wu F, Mo YY (2015) Targeting non-coding RNAs with the CRISPR/Cas9 system in human cell lines. Nucleic Acids Res 43(3):e17

Holme IWT (2013) Intragenesis and cisgenesis as alternatives to transgenic crop development. Plant Biotechnol J 11:395–407

Incarbone M, Dunoyer B (2013) RNA silencing and its suppression: novel insights from in plant analyses. Trends Plant Sci 18:382–392

Jelly NS, Valat L, Walter B, Maillot P (2014) Transient expression assays in grapevine: a step towards genetic improvement. Plant Biotechnol J 12:1231–1245. https://doi.org/10.1111/pbi.12294

Jaillon O, Aury JM, Noel B, Policriti A, Clepet C, Casagrande A, Choisne N, Aubourg S, Vitulo N, Jubin C, Vezzi A, Legeai F, Hugueney P, Dasilva C, Horner D, Mica E, Jublot D, Poulain J, Bruyère C, Billault A, Segurens B, Gouyvenoux M, Ugarte E, Cattonaro F, Anthouard V, Vico V, Del Fabbro C, Alaux M, Di Gaspero G, Dumas V, Felice N, Paillard S, Juman I, Moroldo M, Scalabrin S, Canaguier A, Le Clainche I, Malacrida G, Durand E, Pesole G, Laucou V, Chatelet P, Merdinoglu D, Delledonne M, Pezzotti M, Lecharny A, Scarpelli C, Artiguenave F, Pè ME, Valle G, Morgante M, Caboche M, Adam-Blondon AF, Weissenbach J, Quétier F, Wincker P (2007) The grapevine genome sequence suggests ancestral hexaploidization in major angiosperm phyla. Nature 449:463–467. https://doi.org/10.1038/nature06148

Jinek M, Chylinski K, Fonfara I, Hauer M, Doudna JA, Charpentier E (2012) A programmable dual-RNA-guided DNA endonuclease in adaptive bacterial immunity. Science 337:816–821. https://doi.org/10.1126/science.1225829

Komor AC, Kim YB, Packer MS, Zuris JA, Liu DR (2016) Programmable editing of a target base in genomic DNA without double-stranded DNA cleavage. Nature 533:420–424

Kovalenko PG, Schuman NV (1997) Biotechnological advances of electroporation of grapevine and sugar beet cells. Bioelectrochem Bioenerg 43:165–168

Kurihara Y, Watanabe Y (2004) *Arabidopsis* micro-RNA biogenesis through Dicer-like 1 protein functions. Proc Natl Acad Sci USA 101:12753–12758

Kurth EG, Peremyslov VV, Prokhnevsky AI, Kasschau KD, Miller M, Dolja VV (2012) Virus-derived gene expression and RNA interference vector for grapevine. J Virol 86:6002–6009

Lee LY, Gelvin SB (2008) T-DNA binary vectors and systems. Plant Physiol 146:325–332. https://doi.org/10.1104/pp.107.113001

Li ZT, Kim KH, Dhekney SA, Jasinski JR, Creech MR, Gray DJ (2014) An optimized procedure for plant recovery from somatic embryos significantly facilitates the genetic improvement of *Vitis*. Hortic Res 1:14027–14033. https://doi.org/10.1038/hortres.2014.27

Li J, Yang Z, Yu B, Liu J, Chen X (2005) Methylation protects miRNAs and siRNAs from a 3'-end uridylation activity in *Arabidopsis*. Curr Biol 15:1501–1507

Lyznik LA, Gordon-Kamm WJ, Tao Y (2003) Site-specific recombination for genetic engineering in plants. Plant Cell Rep 21:925–932. https://doi.org/10.1007/s00299-003-0616-7

Malabarba J, Buffon V, Mariath JEA, Maraschin FS, Margis-Pinheiro M, Pasquali G, Revers LF (2018) Manipulation of VviAGL11 expression changes the seed content in grapevine (*Vitis vinifera* L.). Plant Sci 269:126–135

Malnoy M, Viola R, Jung M-H, Koo O-J, Kim S, Kim J-S, Velasco R, Kanchiswamy CN (2016) DNA-free genetically edited grapevine and apple protoplast using CRISPR/Cas9 ribonucleoproteins. Front Plant Sci 7:1904. https://doi.org/10.3389/fpls.2016.01904

Martinelli L, Bragagna P, Poletti V, Scienza A (1993) Somatic embryogenesis from leaf- and petiole-derived callus of *Vitis rupestris*. Plant Cell Rep 12:207–210

Martinelli L, Gribaudo I (2001) Somatic embryogenesis in grapevine. In: Roubelakis-Angelakis KA (ed) Molecular biology and biotechnology of the grapevine. Kluwer Academic Publishers, Dordrecht, pp 393–410

Martinelli L, Dalla Costa L, Vaccari I, Poletti V, Gribaudo I, Gambino I, Guzzo F, Saldarelli P, Minafra A, Turturo C (2009) Application of a site-specific DNA excision strategy for marker gene removal during gene transfer in *Vitis* spp. Acta Hortic 827:399–403

Merz PR, Moser T, Höll J, Kortekamp A, Buchholz G, Zyprian E, Bogs J (2015) The transcription factor VvWRKY33 is involved in the regulation of grapevine (*Vitis vinifera*) defense against the oomycete pathogen *Plasmopara viticola*. Physiol Plant 153: 365–380

Miccono MA, Madrid G, Aguirre C, Olivares F, Olmedo B, Mora R, Sánchez E, Quiroz D, Prieto H (2018) DNA replicon-mediated genome editing in grapevine using CRISPR/Cas9 and a paired gRNA strategy. In: Plant biology conference, Montreal, Québec, Canada, July 14–18

Mojica FJ, Díez-Villaseñor C, Soria E, Juez G (2000) Biological significance of a family of regularly spaced repeats in the genomes of Archaea, Bacteria and mitochondria. Mol Microbiol 36:244–246

Molnar A, Bassett A, Thuenemann E, Schwach F, Karkare S, Ossowski S, Weigel D, Baulcombe D (2009) Highly specific gene silencing by artificial microRNAs in the unicellular alga *Chlamydomonas reinhardtii*. Plant J 58:165–174. https://doi.org/10.1111/j.1365-313X.2008.03767.x

Montes C, Castro Á, Barba P, Rubio J, Sánchez E, Carvajal D, Aguirre C, Tapia E, Dell Orto P, Decroocq V, Prieto H (2014) Differential RNAi responses of *Nicotiana benthamiana* individuals transformed with a hairpin-inducing construct during Plum pox virus challenge. Virus Genes 49:325–338. https://doi.org/10.1007/s11262-014-1093-5

Mullins M, Srinivasan C (1976) Somatic embryos and plantlets from an ancient clone of grapevine (cv. Cabernet-Sauvignon) by apomixes in vitro. J Exp Bot 27:1022–1030

Nakajima I, Ban Y, Azuma A, Onoue N, Moriguchi T, Yamamoto T, Toki S, Endo M (2017) CRISPR/Cas9-mediated targeted mutagenesis in grape. PLoS ONE 12(5):e0177966. https://doi.org/10.1371/journal.pone.0177966

Nielsen K (2003) Transgenic organisms: time for conceptual diversification? Nat Biotechnol 21: 227–228

Nødvig CS, Nielsen JB, Kogle ME, Mortensen UH (2015) A CRISPR-Cas9 system for genetic engineering of filamentous fungi. PLoS ONE 10(7):e0133085

Nunan KJ, Sims IM, Bacic A, Robinson SP, Fincher GB (1997) Isolation and characterization of cell walls from the mesocarp of mature grape berries (*Vitis vinifera*). Planta 203(1):93–100. https://doi.org/10.1007/s004250050169

Oláh R, Zok A, Pedryc A, Howard S, Kovács LA (2009) Somatic embryogenesis in a broad spectrum of grape genotypes. Sci Hortic 120:134–137

Ow DW (2007) GM maize from site-specific recombination technology, what next? Curr Opin Biotechnol 18:115–120

Papadakis AK, Siminis CI, Roubelakis-Angelakis KA (2001) Reduced activity of antioxidant machinery is correlated with suppression of totipotency in plant protoplasts. Plant Physiol 126:434–444

Papadakis A, Fontes N, Gerós H, Roubelakis-Angelakis K (2009) Progress in grapevine protoplast technology. In: Roubelakis-Angelakis KA (ed) Grapevine molecular physiology & biotechnology. Springer, Dordrecht

Pessina S, Lenzi L, Perazzolli M, Campa M, Dalla Costa L, Urso S, Valè G, Salamini F, Velasco R, Malnoy M (2016) Knockdown of MLO genes reduces susceptibility to powdery mildew in grapevine. Hortic Res 3:16016–16024. https://doi.org/10.1038/hortres.2016.16

Priyam A, Woodcroft BJ, Rai V, Munagala A, Moghul I, Ter F, Gibbins MA, Moon H, Leonard G, Rumpf W, Wurm Y (2015). Sequenceserver: a modern graphical user interface for custom BLAST databases. bioRxiv 033142. https://doi.org/10.1101/033142

Pulido-Quetglas C, Aparicio-Prat E, Arnan C, Polidori T, Hermoso T, Palumbo E, Ponomarenko E, Guigo R, Johnson R (2017) Scalable design of paired CRISPR guide RNAs for genomic deletion. PLoS Comput Biol 13(3):e1005341. https://doi.org/10.1371/journal.pcbi.1005341

Rajasekaran K, Mullins M (1983) Influence of genotype and sex-expression on formation of plantlets by cultured anthers of grapevines. Agronomie 3:233–238

Ren C, Liu X, Zhang Z, Wang Y, Duan W, Liang Z (2016) CRISPR/Cas9-mediated efficient targeted mutagenesis in Chardonnay (*Vitis vinifera* L.). Sci Rep 6:32289. https://doi.org/10.1038/srep32289

Reustle G, Harst M, Alleweldt G (1995) Plant regeneration of grapevine (*Vitis* sp.) protoplasts isolated from embryogenic tissue. Plant Cell Rep 15:238–241

Rommens CM, Humara JM, Ye J, Yan H, Richael C, Zhang L, Perry R, Swords K (2004) Crop improvement through modification of the plant's own genome. Plant Physiol 135:421–431

Rommens CM, Bougri O, Yan H, Humara JM, Owen J, Swords K, Ye J (2005) Plant-derived transfer DNAs. Plant Physiol 139:1338–1349

Romon M, Soustre-Gacougnolle I, Schmitt C, Perrin M, Chevalier E, Mutterer J, Himber C, Zervudacki J, Burdloff Y, Montavon T, Zimmermann A, Elmayan T, Vaucheret H, Dunoyer P, Masson JE (2013) RNA silencing is resistant to low-temperature in grapevine. PLoS ONE 8(12):e82652. https://doi.org/10.1371/journal.pone.0082652

Rubio J, Montes C, Castro Á, Álvarez C, Olmedo B, Munoz M, Tapia E, Reyes F, Ortega M, Sánchez E, Miccono M, Dalla Costa L, Martinelli L, Malnoy M, Prieto H (2015) Genetically engineered thompson seedless grapevine plants designed for fungal tolerance: selection and characterization of the best performing individuals in a field trial. Transgenic Res 24:43–60

Salunkhe C, Rao P, Mhatre M (1997) Induction of somatic embryogenesis and plantlets in tendrils of *Vitis vinifera* L. Plant Cell Rep 17:65–67

Santos-Rosa M, Poutaraud A, Merdinoglu D, Mestre P (2008) Development of a transient expression system in grapevine via agro-infiltration. Plant Cell Rep 27:1053–1063

Shalem O, Sanjana NE, Hartenian E, Shi X, Scott DA, Mikkelsen TS, Heckl D, Ebert BL, Root DE, Doench JG, Zhang F (2014) Genome-scale CRISPR-Cas9 knockout screening in human cells. Science 343:84–87. https://doi.org/10.1126/science.1247005

Shen B, Zhang W, Zhang J, Zhou J, Wang J, Chen L, Wang L, Hodgkins A, Iyer V, Huang X, Skarnes WC (2014) Efficient genome modification by CRISPR-Cas9 nickase with minimal off-target effects. Nat Methods 11:399–402. https://doi.org/10.1038/nmeth.2857

San Pedro T, Gammoudi N, Peiró R, Olmos A, Gisbert C (2017) Somatic embryogenesis from seeds in a broad range of *Vitis vinifera* L. varieties: rescue of true-to-type virus-free plants. Plant Biol 17:226. https://doi.org/10.1186/s12870-017-1159-3

Saporta R, San Pedro T, Gisbert C (2016) Attempts at grapevine (*Vitis vinifera* L.) breeding through genetic transformation: the main limiting factors. Vitis 55:173–186. https://doi.org/10.5073/vitis.2016.55.173-186

Schouten HS, Krens FA, Jacobsen E (2006) Cisgenic plants are similar to traditionally bred plants: international regulations for genetically modified organisms should be altered to exempt cisgenesis. EMBO Rep 7:750–753

Schwab R, Ossowski S, Riester M, Warthmann N, Weigel D (2006) Highly specific gene silencing by artificial microRNAs in *Arabidopsis*. Plant Cell 18:1121–1133

Skinner ME, Uzilov AV, Stein LD, Mungall CJ, Holmes IH (2009) JBrowse: a next-generation genome browser. Genome Res 19:1630–1638. https://doi.org/10.1101/gr.094607.109

Taning CNT, Van Eynde B, Ma NYS, Smagghe G (2017) CRISPR/Cas9 in insects: applications, best practices and biosafety concerns. J Insect Physiol 18:245–257

Tapia E, Sequeida A, Castro Á, Montes C, Zamora P, López R, Acevedo F, Prieto H (2009) Development of grapevine somatic embryogenesis using an air-lift bioreactor as an efficient tool in the generation of transgenic plants. J Biotechnol 139:95–101. https://doi.org/10.1016/j.jbiotec.2008.09.009

Terrier N, Torregrosa L, Ageorges A, Vialet S, Verrie C, Cheynier V, Romieu C (2009) Ectopic expression of VvMybPA2 promotes proanthocyanidin biosynthesis in grapevine and suggests additional targets in the pathway. Plant Physiol 149:1028–1041

Toro N, Datta A, Yanofsky M, Nester E (1988) Role of the overdrive sequence in T-DNA border cleavage in *Agrobacterium*. Proc Natl Acad Sci USA 85:8558–8562

Torregrosa L, Iocco P, Thomas MR (2002) Influence of *Agrobacterium* strain, culture medium, and cultivar on the transformation efficiency of *Vitis vinifera* L. Am J Enol Vitic 53:183–190

Tzfira T, Citovsky V (2008) *Agrobacterium*: from biology to biotechnology, Springer edn. Springer, New York

Tzfira T, Li J, Lacroix B, Citovsky V (2004) Agrobacterium T-DNA integration: molecules and models. Trends Genet 20:375–383

Valat L, Toutain S, Courtois N, Gaire F, Decout E, Pinck L, Mauro M, Burrus M (2000) GFLV replication in electroporated grapevine protoplasts. Plant Sci 155:203–212

Valat L, Fuchs M, Burrus M (2006) Transgenic grapevine rootstock clones expressing the coat protein or movement protein genes of grapevine fanleaf virus: characterization and reaction to virus infection upon protoplast electroporation. Plant Sci 170:739–747. https://doi.org/10.1016/j.plantsci.2005.11.005

van Haaren MJ, Sedee NJ, Schilperoort RA, Hooykaas PJ (1987) Overdrive is a T-region transfer enhancer which stimulates T-strand production in *Agrobacterium tumefaciens*. Nucl Acids Res 15:8983–8997

Vazquez F, Legrand S, Windels D (2006) The biosynthetic pathways and biological scopes of plant small RNAs. Trends Plant Sci 15:337–345

Vidal JR, Rama J, Taboada L, Martín C, Ibáñez M, Segura A, González-Benito ME (2009) Improved somatic embryogenesis of grapevine (*Vitis vinifera*) with focus on induction parameters and efficient plant regeneration. Plant Cell Tissue Organ Cult 96:85–94

Wang MB, Abbott DC, Waterhouse PM (2000) A single copy of a virus-derived transgene encoding hairpin RNA gives immunity to barley yellow dwarf virus. Mol Plant Pathol 1:347–356

Wang Q, Li P, Hanania U, Sahar N, Mawassi M, Gafny R, Selac I, Tannea E, Perlb A (2005) Improvement of *Agrobacterium*-mediated transformation efficiency and transgenic plant regeneration of *Vitis vinifera* L. by optimizing selection regimes and utilizing cryopreserved cell suspensions. Plant Sci 168:565–571

Wang HL, Wang W, Zhan JC, Huang WD, Xu HY (2015) An efficient PEG-mediated transient gene expression system in grape protoplasts and its application in subcellular localization studies of flavonoids biosynthesis enzymes. Sci Hortic 191:82–89. https://doi.org/10.1016/j.scienta.2015.04.039

Wang X, Tu M, Wang D, Liu J, Li Y, Li Z, Wang X (2017) CRISPR/Cas9-mediated efficient targeted mutagenesis in grape in the first generation. Plant Biotechnol J 16:844–855. https://doi.org/10.1111/pbi.12832

Wang Y, Liu X, Ren C, Zhong G-Y, Yang L, Li S, Liang Z (2016) Identification of genomic sites for CRISPR/Cas9-based genome editing in the *Vitis vinifera* genome. BMC Plant Biol 16:96. https://doi.org/10.1186/s12870-016-0787-3

Warthmann N, Chen H, Ossowski S, Weigel D, Herve P (2008) Highly specific gene silencing by artificial miRNAs in rice. PLoS ONE 3:e1829

Wesley SV, Helliwell CA, Smith NA, Wang MB, Rouse DT, Liu Q, Gooding PS, Singh SP, Abbott D, Stoutjesdijk PA (2001) Construct design for efficient, effective and highthroughput gene silencing in plants. Plant J 27:581–590

Zetsche B, Gootenberg JS, Abudayyeh OO, Slaymaker IM, Makarova KS, Essletzbichler P, Volz SE, Joung J, van der Oost J, Regev A, Koonin EV, Zhang F (2015) Cpf1 is a single RNA-guided endonuclease of a class 2 CRISPR-Cas system. Cell 163:759–771. https://doi.org/10.1016/j.cell.2015.09.038

Zhao F-L, Li Y-J, Hu Y, Gao Y-R, Zang X-W, Ding Q, Wang Y-J, Wen Y-Q (2016) A highly efficient grapevine mesophyll protoplast system for transient gene expression and the study of disease resistance proteins. Plant Cell Tissue Organ Cult 125:43. https://doi.org/10.1007/s11240-015-0928-7

Zhou Q, Dai L, Cheng S, He J, Wang D, Zhang J, Wang Y (2014) A circulatory system useful both for long-term somatic embryogenesis and genetic transformation in *Vitis vinifera* L. cv. Thompson Seedless. Plant Cell Tissue Organ Cult 118:157. https://doi.org/10.1007/s11240-014-0471-y

Zhu Y-M, Hoshino Y, Nakano M, Takahashi E, Miia M (1997) Highly efficient system of plant regeneration from protoplasts of grapevine (*Vitis vinifera* L.) through somatic embryogenesis by using embryogenic callus culture and activated charcoal. Plant Sci 123:151–157. https://doi.org/10.1016/S0168-9452(96)04557-8

Zottini M, Barizza E, Costa A, Formentin E, Ruberti C, Carimi F, Lo Schiavo F (2008) Agroinfiltration of grapevine leaves for fast transient assays of gene expression and for long-term production of stable transformed cells. Plant Cell Rep 27:845–853. https://doi.org/10.1007/s00299-008-0510-4

Zuo J, Niu QW, Geir Møller S, Chua NH (2001) Chemical-regulated, site-specific DNA excision in transgenic plants. Nat Biotechnol 19:157–161

MIX
Papier aus verantwortungsvollen Quellen
Paper from responsible sources
FSC® C105338

If you have any concerns about our products,
you can contact us on
ProductSafety@springernature.com

In case Publisher is established outside the EU,
the EU authorized representative is:
Springer Nature Customer Service Center GmbH
Europaplatz 3, 69115 Heidelberg, Germany

Printed by Libri Plureos GmbH
in Hamburg, Germany